The Dictionary of CELL AND MOLECULAR BIOLOGY

Third Edition

J. M. Lackie and J. A. T. Dow

Yamanouchi Research Institute, Littlemore Park, Oxford OX4 4SX, UK

and

University of Glasgow, IBLS, Division of Molecular Genetics, Glasgow G11 6NU, UK

Contributing Authors

S. E. Blackshaw
C. T. Brett
A. S. G. Curtis
J. G. Edwards
A. J. Lawrence
G. R. Moores

ACADEMIC PRESS

Harcourt Publishers
San Diego London New York
Boston Sydney Tokyo Toronto

This book is printed on acid-free paper.

First published 1989

Academic Press
24–28 Oval Road, London NW1 7DX, UK
http://www.hubk.co.uk/ap/

Academic Press
A Harcourt Science and Technology
525 B Street, Suite 1900, San Diego, California 92101-4495, USA
http://www.apnet.com

ISBN 0-12-432565-3

A catalogue for this book is available from the British Library

Library of Congress Card Number: 99–63830

Cover illustration

Co-staining of filamentous actin (red) and vinculin (green) in HEK 293 cells. Vinculin which is associated with focal contacts localizes to the ends of the actin cables. Acquired digitally with a Zeiss Axiovert 135 fluorescence microscope, a Hamamatsu CCD camera and OpenLab 2.0 (Improvision, Coventry, England).

Typeset by M Rules, London, UK
Printed and bound in Great Britain by The Bath Press, Avon, UK

99 00 01 02 03 04 BP 9 8 7 6 5 4 3 2 1

Preface to the Third Edition

Although the title has changed, we as Editors have talked of this as the Third Edition of *The Dictionary of Cell Biology* – and the change in title simply reflects the changes that have happened to Cell Biology in the last decade. In the Preface to the First Edition, we commented that the boundaries of 'cell biology' were difficult to define and this has certainly not changed – if anything the territory has grown! But, the molecular entries have continued to increase in number as more detail of cell structure and the complexity of signalling pathways is known. The new title reflects the content more accurately – though we have also added a number of entries that are neither cellular nor molecular.

Now that the Human Genome Project is nearing completion and we have the complete genome sequence of yeast and the nematode *Caenorhabditis elegans*, the big problem becomes that of assigning function to genes. Inevitably this will draw heavily on cell biology and many molecular biologists will find themselves entering new territory: and cell biologists they meet will need to talk the language of molecular biology. Cell biology encompasses an extremely wide range of experimental systems and has a very diverse vocabulary: this Dictionary should help to provide some guidance.

In preparing this volume, we have been much influenced by the usage of the Internet version of the Second Edition of *The Dictionary of Cell Biology*. Putting the Dictionary on the Net was an experiment – and one that has proved fascinating. For the first time, perhaps, it has been possible to monitor how people used the 'book' and what they searched for. Approximately one-third of a million visits have been made to the site and it has been cross-referenced from various other web pages. We have maintained a log of abortive searches and also gave an e-mail address for feedback. Though we had put a tear-out sheet for comment in the First and Second paper editions, we had almost no response, however, by e-mail it was different!

Many abortive searches were a result of inability to spell or perhaps to type accurately – we can do little about that! But many searches were for things that we felt really should have been in –sometimes we had omitted to write an entry, sometimes the search was for something new, sometimes the search was for things that are part of the wider vocabulary of those trained in the older disciplines. Thus there were quite a lot of searches for fixatives well known to the histologist, for Latin names of species where the common name is well known, for syndromes that have emerged or been diagnosed (we suspect). We have tried to put in some entries to help on these aspects, tried to put in new things from the literature but, in the case of diseases, we have tended to put entries only where there is a known cell or molecular basis for the disease. This edition has more than 7000 entries and we have been more comprehensive with cross-referencing of synonyms and from the text making this Dictionary almost double the size of the First Edition! The entries from the First and Second editions have been scrutinised and modified where necessary, particularly when there has been comment from readers, and many of the new entries are for words sought by on-line users. A variety of sources have been used and a brief list is appended. Again, Dr Ann Lackie helped with the final preparation and cross-checking.

We have always tried to provide short, clear definitions that are helpful to people with the widest range of backgrounds, and the huge volume of feedback we have received from the Internet edition suggests that these efforts are well received. Not only career cell biologists, but school teachers, high school students and journalists have all sent glowing testimonials. Our aim is to continue to develop this resource, both on-line and, given the speed at which the discipline is evolving, to produce another in due course. Meanwhile we hope that people will find this edition useful and find it much easier, and quicker to search adjacent entries. We also hope that people will not hesitate to send new entries, suggest amendments and help in the evolution of this unusually interactive resource.

The on-line Dictionary is to be found on *http://www.mblab.gla.ac.uk/dictionary*

John Lackie
Julian Dow
1999

A note regarding entries

The main entry word is followed by synonyms in brackets. Words in ***bold italic*** in the definition are cross-references to other entries which might contribute usefully to the entry being consulted, although other words within the definition may well have entries.

Generally speaking entry words which have a Greek letter or numerical prefix have been alphabetized ignoring the prefix.

John Lackie

Preface to the Second Edition

In the preface to the first edition we commented that 'the subject was far from static', and we have not been disappointed. The last few years have seen rapid progress and an increasing reliance upon the powerful tools of molecular biology. In this second edition we have extended the coverage, particularly in molecular biology and neurobiology, though the sense of barely keeping up with the emergence of new names for proteins is ever more pressing. 'Cytokine' has overtaken 'Interleukin' as the numbers of cytokines have almost doubled and chemokines have arrived. The table of CD numbers has grown impressively and the G-proteins have proliferated; far more genes have been named and new proteins are legion. A measure of the rate of change is that we have had to include more than 1000 new entries; by the next revision we will have to start deleting entries - and anybody who scans the old literature (pre-1990!) will find names that have disappeared without trace.

As before, the choice of entries partly reflects our own interests, but we have tried to be broadminded with our inclusions. This time Dr Susanna Blackshaw has written entries rather than acting as an external adviser, in order to increase the neurobiological content, and both editors have become much more involved with molecular biology. In the first edition we included a tear-off page for users to send us their neologisms and revisions but, disappointingly, had little response. We continue to hope that interested users of this book will provide suggestions for future improvements.

Several people have helped with the revision of the manuscript itself and in particular we wish to thank Mrs Lynn Sanders and Mrs Joanne Noble for help with typing. Many colleagues who have not actually been directly involved have nevertheless been plagued with questions, and we are grateful for their patience. We hope that this new edition will be as useful as the first and that it will date no faster. On past experience, however, our rapidly growing subject will doubtless keep this project alive for years to come. We have taken the unusual step of making a searchable version of the Dictionary available on the Internet. Although undoubtedly less convenient than the paper version, this will carry incremental changes and updates – including those provided electronically by readers – until the third edition is published. Details are provided on the revision form at the back of the book: we hope you find this service useful.

John Lackie
Julian Dow
July 1994

Preface to the First Edition

The stimulus to write this dictionary came originally from our teaching of a two-year Cell Biology Honours course to undergraduates in the University of Glasgow. All too often students did not seem to know the meanings of terms we felt were commonplace in cell biology, or were unable, for example, to find out what compounds in general use were supposed to do. But before long it became obvious that although we all considered ourselves to be cell biologists, individually we were similarly ignorant in areas only slightly removed from our own – though collectively the knowledge was there. It was also clear that many of the things we considered relevant were not easy to find, and that an extensive reference library was needed. In that we have found the exercise of preparing the Dictionary informative ourselves, we feel that it may serve a useful purpose.

An obvious problem was to decide upon the boundaries of the subject. We have not solved this problem: modern biology is a continuum and any attempt to subdivide it is bound to fail. 'Cell Biology' implies different things to zoologists, to biochemists, and indeed to each of the other species of biologists. There is no sensible way to set limits, nor would we wish to see our subject crammed into a well-defined niche. Inevitably, therefore the contents are somewhat idiosyncratic, reflecting our current teaching, reading, prejudices, and fancies.

It may be of some interest to explain how we set about preparing the Dictionary. The list of entry words was compiled largely from the index pages of several textbooks, and by scanning the subject indexes of cell-biological journals. To this were added entries for words we cross-referenced. The task of writing the basic entries was then divided amongst us roughly according to interests and expertise. We all wrote subsets of entries which were then compiled and alphabetised before being edited by one of us. Marked copies were then sent out to a panel of colleagues who scrutinised entries in their own fields. All entries were looked at by one or more of this panel, and then the annotated entries were re-edited, corrections made on disc, and the files copy-edited for consistency of style. A very substantial amount of the handling of the compiled text and the preparation of the final discs was done by Dr A M Lackie who also acted as copy-editor.

Glasgow is a major centre for Life Sciences, and we are fortunate in having many colleagues to whom we could turn for help. We are very grateful to them for the work which they put in and for the speed with which they checked the entries that we sent. Although we have tried hard to avoid errors and ambiguities, and to include everything that will be useful, we apologise at this stage for the mistakes and omissions, and emphasise that blame lies with the authors and not with our panel (though they have saved us from many embarassments).

Since there is no doubt Cell Biology is developing rapidly as a field, it is inevitable that usages will change, that new terms will become commonplace, that new proteins will be christened on gels, and that the dictionary will soon have omissions. Were the subject static this dictionary would not be worth compiling – and we cannot anticipate new words.

Because the text is on disc, it will be relatively easy to update: please let us have your comments, suggestions for entries (preferably with a definition), and (perhaps) your neologisms. A sheet is included at the back of the dictionary for this purpose.

John Lackie

Tables

Numerical (Pre-A)

14-3-3 proteins Family of adapter proteins able to interact with a range of signalling molecules including c-Raf, Bcr, PI-3-kinase, polyoma middle T-antigen. Bind to phosphorylated serine residues in cdc25C and block its further activity, may bind Bad (death inducer) thereby blocking heteromeric interaction with Bcl-XL, and in plants bind and inhibit activity of phosphorylated nitrate reductase. Basic mode of action may be to block specific protein–protein interactions.

2,4-D See ***dichlorophenoxyacetic acid***.

43 kD postsynaptic protein See ***postsynaptic protein***.

464.1 (744) Human ***macrophage inflammatory protein 1*** β (MIP1-β).

5-hydroxytryptamine, 5-HT See ***hydroxytryptamine***.

744 See ***464.1***.

9E3 (pCEF-4) An avian cytokine produced by chicken cells infected with ***Rous sarcoma virus***.

α Entry prefix is given as 'alpha'; alternatively look for main portion of word.

A23187 A monocarboxylic acid extracted from *Streptomyces chartreusensis* that acts as a mobile-carrier calcium ***ionophore***. See Table I3.

A band That portion of the ***sarcomere*** in which the thick myosin filaments are located. It is anisotropic in polarized light.

A-cells (α-cells) Cell of the ***endocrine*** pancreas (islets of Langerhans) that form approximately 20% of the population; their opaque spherical granules may contain ***glucagon***. See ***B-cells, D-cells***.

A chain Shorter of the 2 polypeptide chains of insulin (21 residues compared to 30 in the B chain). Many other heterodimeric proteins have their smaller chain designated the A chain, so the term cannot be used without qualification. Other A chains: ***abrin, activin***, C1q (see ***complement***), ***diphtheria toxin, inhibin, laminin***, mistletoe lectin, ***PDGF, relaxin, ricin, tPA***.

A channel Type of potassium-selective ion channel that is activated by depolarization but only after a preceding hyperpolarization, ie. they are inactivated at rest. Important for repetitive firing of cells at low frequencies – see ***Shaker*** mutant.

A-DNA Right-handed double-helical ***DNA*** with approximately 11 residues per turn. Planes of base pairs in the helix are tilted 20° away from perpendicular to the axis of the helix. Formed from ***B-DNA*** by dehydration.

α **hairpin** See ***hairpin***.

A kinase cAMP-regulated protein kinase, sometimes abbreviated PKA to contrast with ***PKC***.

A kinase anchoring protein See ***AKAP79***.

A layer ***S layer*** in *Aeromonas* sp.

'A' motif See ***ATP-binding site***.

A site Site on the ***ribosome*** to which aminoacyl tRNA attaches during the process of peptide synthesis. See also ***P site***.

A-type particles Retrovirus-like particles found in cells. Non-infectious. The mouse genome contains around 1000 copies of homologous sequences.

α-1-antitrypsin See ***antitrypsin***.

α-1-iduronidase See ***iduronidase***.

α-2-antiplasmin See ***antiplasmin***.

α-actinin See ***alpha-actinin***.

α-cell See ***alpha-cell***.

α-fetoprotein See ***alpha-fetoprotein***.

α-helix See ***alpha-helix***.

α-L-iduronic acid See ***iduronic acid***.

A4 protein See ***amyloidogenic glycoprotein***.

A9 cells Established line of heteroploid mouse fibroblasts that are deficient in *HGPRT*.

AA-tRNA See ***aminoacyl tRNA***.

Ab Common abbreviation for antibody. See ***immunoglobulins***.

AB toxin Multi-subunit toxin in which there are two major components, an active (A) portion and a portion that is involved in binding (B) to the target cell. The A portion can be effective in the absence of the B subunit(s) if introduced directly into the cytoplasm. In the well-known examples, the A subunit has ADP-ribosylating activity. See ***cholera toxin, diphtheria toxin, pertussis toxin, ADP ribosylation***.

abaecins Proline-rich basic antibacterial peptides (4 kD) found in the ***haemolymph*** of the honeybee. See ***apidaecins***.

ABC (1) Antigen-binding cell or antigen binding capacity. (2) Avidin–biotin peroxidase complex. Used in visualizing antigen. Primary (antigen-specific) antibody is bound first, a second biotinylated anti-immunoglobulin antibody is then bound to the first antibody, then ABC complex that has excess biotin-binding capacity is bound to the biotin on the second antibody and finally the peroxidase used to catalyse a colorimetric reaction generating brown staining. The method gives substantial signal enhancement.

ABC proteins Membrane proteins involved in active transport or regulation of ion channel function and having an ATP-binding cassette. Most examples are prokaryotic, but important eukaryotic examples are the P-glycoprotein (multidrug resistance transporter), the cystic fibrosis transmembrane conductance regulator (***CFTR***), and ***SUR***.

ABC-excinuclease See ***ABC-exonuclease***.

ABC-exonuclease (ABC-excinuclease) Enzyme complex, product of *uvrA, uvrB* and *uvrC* genes from *E. coli* that mediates incision and excision steps of DNA excision-repair. Enzyme has the ability to recognize distortion in DNA structure caused by, eg. ultraviolet irradiation.

Abelson leukaemia virus A replication-defective virus originating from the Moloney murine leukaemia virus by acquisition of *c-abl*. The virus induces B-cell lymphoid leukaemias within a few weeks. The *v-abl* product has tyrosine kinase activity.

abenzyme See ***catalytic antibody***.

aberration Departure from normal; in microscopy two common forms of optical aberration cause problems, ***spherical aberration*** in which there is distortion of the image of the magnified object and ***chromatic aberration*** that leads to coloured fringes – a consequence of the unequal refraction of light of different wavelengths.

abetalipoproteinaemia Autosomal recessive defect in which there is total absence of apoprotein B (a component of LDL, VLDL and chylomicrons). Characteristic feature is presence of ***acanthocytes***; later in life neurological disorders and retinitis pigmentosa develop and death is usually a consequence of cardiomyopathy.

abiogenesis Spontaneous generation of life from non-living material.

abl An oncogene, identified in a mouse leukaemia, encoding a tyrosine ***protein kinase***. See also ***ABLV*** and Table O1.

ABLV The Abelson murine ***leukaemia*** virus, a mammalian ***retrovirus***. Its transforming gene, *abl*, encodes a protein with ***tyrosine kinase*** activity closely related to *src*.

ABM paper Aminobenzyloxy methylcellulose paper: paper to which single-stranded nucleic acid can be covalently coupled.

ABO blood group system Probably the best known of the blood group systems, involves a single gene locus that codes for a fucosyl transferase. If the H-gene is expressed then fucose is added to the terminal galactose of the precursor oligosaccharide on the red cell surface and the A- or B-gene products, also glycosyl-transferases, can then add N-acetyl galactosamine or galactose to produce the A- or B-antigens respectively. Antibodies to the ABO-antigens occur naturally and make this an important set of antigens for blood transfusion. Transfusion of mismatched blood with surface red cell antigens that elicit a response leads to a transfusion reaction. The natural antibodies are usually IgM. See ***Rhesus, Kell, Duffy*** and ***MN***.

abortive infection Viral infection of a cell in which the virus fails to replicate fully, or produces defective progeny. Since part of the viral replicative cycle occurs, its effect on the host can still be cytopathogenic.

abortive transformation Temporary transformation of a cell by a virus that fails to integrate into the host DNA.

ABP See ***actin-binding proteins***.

ABP-50 Actin-binding protein (50 kD) from *Dictyostelium* that crosslinks actin filaments into tight bundles. Identical to ***elongation factor*** EF-1a. Calcium insensitive; localized near cell periphery and in protrusions from moving cells.

ABP-67 Homologue of ***fimbrin***. In yeast encoded by SAC6 gene, mutations in which lead to disruption of the actin ***cytoskeleton***.

ABP-120 Actin-binding protein (857 amino acids; 92 kD) from *Dictyostelium*. A small rod-shaped molecule (35–40nm long), dimeric, capable of crosslinking filaments. Has strong sequence similarities with ***ABP-280***.

ABP-280 Actin-binding protein (2647 amino acids; 280 kD) from *Dictyostelium*, but very similar to filamin from other sources. A long rod-shaped molecule (80nm long), dimeric, with the two monomers associated end to end making a very long crosslinker of microfilaments. Has actin-binding domain similar to that in ***ABP-120, spectrin, filamin, α-actinin*** and ***fimbrin***; also has binding site for platelet ***von Willebrand factor***.

abrin Toxic ***lectin*** from seeds of *Abrus precatorius* that has a binding site for galactose and related residues in carbohydrate but, because it is monovalent, is not an ***agglutinin*** for ***erythrocytes***.

abscess A cavity within a tissue occupied by pus (chiefly composed of degenerating inflammatory cells), generally caused by bacteria that resist killing by phagocytes.

abscisic acid A growth-inhibiting plant hormone found in vascular plants. Originally believed to be important in abscission (leaf fall), now known to be involved in a number of growth and developmental processes in plants, including, in some circumstances, growth promotion. Primary plant hormone that mediates responses to stress. Downstream signalling through ***cyclic ADP-ribose*** as a second messenger.

absorption coefficient Any of four different coefficients that indicate the ability of a substance to absorb electromagnetic radiation. Absorbance is defined as the logarithm of the ratio of incident and transmitted intensity and thus it is necessary to know the base of the logarithm used. Scattering and reflectance are generally ignored when dealing with solutions.

absorption spectrum Spectrum of wavelengths of electromagnetic radiation (usually visible and UV light) absorbed by substance. Absorption is determined by existence of atoms that can be excited from their ground state to an excited state by absorption of energy carried by a photon at that particular wavelength.

Acanthamoeba Soil amoebae 20–30μm in diameter that can be grown under ***axenic*** conditions and have been extensively used in biochemical studies of cell motility. They have been isolated from cultures of monkey kidney cells and are pathogenic when injected into mice or monkeys.

acanthocyte Cell with projecting spikes; most commonly applied to erythrocytes where the condition may be caused naturally by ***abetalipoproteinaemia*** or experimentally by manipulating the lipid composition of the plasma membrane.

acanthocytosis Condition in which red cells of the blood show spiky deformation; symptomatic of ***abetalipoproteinaemia***.

acanthosome (1) Spinous membranous ***organelle*** found in skin ***fibroblasts*** from ***nude mice*** as a result of chronic ultraviolet irradiation. (2) Sometimes used as a synonym for ***coated vesicle*** (should be avoided).

Acanthus Genus of spiny-leaved Mediterranean plants

acapnia Medical condition in which there is a low concentration of carbon dioxide in the blood.

ACC (1-aminocyclopropane-1-carboxylic acid) Immediate precursor of the plant hormone ethylene in most vascular plants. Synthesized from S-adenosyl methionine by ACC synthase.

ACC synthase (ACC methylthioadenosine lyase) Enzyme (65 kD; EC 4.1.1.14) that catalyses conversion of S-adenosylmethionine to ACC; first step in production of the plant hormone ethylene.

accelerin Activated blood factor V which acts on prothrombin to generate thrombin during blood coagulation.

accessory cells Cells that interact, usually by physical contact, with T-lymphocytes and which are necessary for induction of an immune response. Include antigen-presenting cells, antigen-processing cells, etc. They are usually MHC Class II positive (see ***histocompatibility antigens***). Monocytes, macrophages, dendritic cells, ***Langerhans cells*** and B-lymphocytes may all act as accessory cells.

accessory pigments In ***photosynthesis***, pigments that collect light at different wavelengths and transfer the energy to the primary system.

ACE See ***angiotensin converting enzyme***.

acellular Not made of cells; commonest use is in reference to slime moulds such as *Physarum* that are multinucleate syncytia.

acellular slime moulds Protozoa of the order ***Eumycetozoida*** (also termed true slime moulds). Have a multinucleate plasmodial phase in the life cycle.

acentric Descriptive of pieces of ***chromosome*** that lack a ***centromere***.

Acetabularia Giant single-celled ***alga*** of the order Dasycycladaceae. The plant is 3–5cm long when mature and consists of *rhizoids* at the base of a stalk, at the other end of which is a cap that has a shape characteristic of each species. The giant cell has a single nucleus, located at the tip of one of its rhizoids, which can be removed easily by cutting off that rhizoid. Nuclei can also be transplanted from one cell to another.

Acetobacter Genus of aerobic bacilli that will use ethanol as a substrate to produce acetic acid – thus will convert wine to vinegar.

acetyl CoA Acetylated form of coenzyme A that is a carrier for ***acyl*** groups, particularly in the ***tricarboxylic acid cycle***.

acetyl salicylate See ***aspirin***.

acetylation Addition, either chemically or enzymically, of acetyl groups.

acetylcholine (ACh) Acetyl ester of choline. Perhaps the best characterized ***neurotransmitter***, particularly at neuromuscular junctions. ACh can be either excitatory or inhibitory, and its receptors are classified as ***nicotinic*** or ***muscarinic***, according to their pharmacology. In ***chemical synapses*** ACh is rapidly broken down by ***acetylcholine esterases***, thereby ensuring the transience of the signal.

acetylcholine esterase An enzyme, found in the ***synaptic clefts*** of cholinergic synapses, that cleaves the ***neurotransmitter*** acetylcholine into its constituents, acetate and choline, thus limiting the size and duration of the ***postsynaptic potential***. Many nerve gases and insecticides are potent acetylcholine esterase inhibitors, and thus prolong the time course of postsynaptic potentials.

acetylcholine receptor See ***nicotinic acetylcholine receptor, muscarinic acetylcholine receptor***.

ACh See ***acetylcholine***.

achondroplasia Failure of endochondral ossification responsible for a form of dwarfism; caused by an ***autosomal dominant mutation***. Relatively high incidence (1:20 000 live births), mostly (90%) new mutations. Also known as chondrodystrophia fetalis.

Achyla Genus of aquatic fungi with a branched coenocytic mycelium.

acid-citrate-dextrose Citric acid/sodium citrate buffered glucose solution used as an anticoagulant for blood (citrate complexes calcium).

acid hydrolases ***Hydrolytic enzymes*** that have a low pH optimum. The name usually refers to the ***phosphatases, glycosidases, nucleases*** and ***lipases*** found in the ***lysosome***. They are secreted during ***phagocytosis***, but are considered to operate as intracellular digestive enzymes.

acid phosphatase Enzyme (EC 3.1.3.2) with acidic pH optimum that catalyses cleavage of inorganic phosphate from a variety of substrates. Found particularly in ***lysosomes*** and ***secretory vesicles***. Can be localized histochemically using various forms of the ***Gomori procedure***.

acid protease Proteolytic enzyme with an acid pH optimum, characteristically found in lysosomes. See ***proteases***.

acid-secreting cells Large specialized cells of the epithelial lining of the stomach (parietal or oxyntic cells) that secrete 0.1N HCl, by means of H^+ ***antiport*** ATPases on the luminal cell surface.

acidic FGF See ***fibroblast growth factor***.

acidophilia Having an affinity for acidic dyes, particularly eosin; may be applied either to tissues or to bacteria.

acidophils One class of cells found in the pars distalis of the adenohypophysis.

acidosome Non-lysosomal vesicle in which receptor–ligand complexes dissociate because of the acid pH.

acinar cells Epithelial ***secretory cells*** arranged as a ball of cells around the lumen of a gland (as in the pancreas).

acinus (plural, acini) Small sac or cavity surrounded by ***secretory cells***.

acoelomate Animal without a coelom. The Acoelomate phyla include sponges, coelenterates and lower worms such as nematodes and platyhelminths.

aconitase Enzyme of the ***tricarboxylic acid cycle*** that catalyses isomerization of citrate/isocitrate. Isoforms are found both in mitochondrial matrix and cytoplasm.

Acquired immune deficiency syndrome See ***AIDS***.

acquired immunity Classically, the reaction of an organism to a new antigenic challenge and the retention of a memory of this, as opposed to innate immunity. In modern terms, the ***clonal selection*** and expansion of a population of immune cells in response to a specific antigenic stimulus and the persistence of this clone.

acrasiales See ***Acrasidae***.

Acrasidae Order of ***Protozoa*** also known as the cellular slime moulds. They normally exist as free-living phagocytic soil amoebae (vegetative cells), but when bacterial prey become scarce, they aggregate to form a pseudoplasmodium (cf. true ***plasmodium*** of Eumycetozoida) that is capable of directed motion. The grex, or slug, migrates until stimulated by environmental conditions to form a fruiting body or sorocarp. The slug cells differentiate into elongated stalk cells and spores, where the cells are surrounded by a cellulose capsule. The spores are released from the sporangium at the tip of the stalk and, in favourable conditions, an ***amoeba*** emerges from the capsule, feeds, divides and so establishes a new population. They can be cultured in the laboratory and are widely used in studies of cell–cell adhesion, cellular ***differentiation***, ***chemotaxis*** and ***pattern formation***. The commonest species studied are *Dictyostelium discoideum*, *D. minutum*, and *Polysphondylium violaceum*.

acrasin Name originally given to the ***chemotactic*** factor produced by cellular slime moulds (***Acrasidae***): now known to be ***cAMP*** for *Dictyostelium discoideum*.

acridine orange A fluorescent vital dye that intercalates into nucleic acids. The nuclei of stained cells fluoresce green; cytoplasmic RNA fluoresces orange. Acridine orange also stains acid mucopolysaccharides and is widely used as a pH-sensitive dye in studies of acid secretion. May be carcinogenic.

acridines Heterocyclic compounds with a pyridine nucleus. Usually fluorescent and reactive with double stranded DNA as intercalating agents at very low concentrations. Hence dsDNA can be detected on gels by fluorescence after acridine staining. Mutagenic (causing frame-shift mutations), cytostatic (and hence antimicrobial). They also affect RNA synthesis and have been used for cell marking.

acrocentric See ***metacentric***.

acromegaly Enlargement of the extremities of the body as a result of the overproduction of growth hormone (***somatotropin***), eg. by a pituitary tumour.

acrosin Serine protease stored in the ***acrosome*** of a sperm as an inactive precursor.

acrosomal process A long process actively protruded from the acrosomal region of the spermatozoon following contact with the egg and which assists penetration of the gelatinous capsule.

acrosome Vesicle at the extreme anterior end of the spermatozoan, derived from the ***lysosome***.

Act-2 Human ***macrophage inflammatory protein 1β***.

ACT1, ACT2 Actin genes from yeast; ACT1 is the essential (conventional) actin, 89% homologous in sequence with mouse cytoplasmic actin; ACT2 encodes a 44 kD protein 47% identical to yeast actin that is required for vegetative growth. Divergence from conventional actin by ACT2 is in regions associated with actin polymerization, DNAase I and myosin binding.

ActA Major surface protein (90 kD) of

Listeria monocytogenes that acts as the nucleating site for actin polymerization at one pole of the bacterial cell; assembly of the bundle of microfilaments pushes the bacterium through the cell – though the appearance is like a comet with a tail. A functionally similar protein, ***IcsA***, is found in *Shigella*.

ACTH See ***adrenocorticotrophin***.

actin A protein of 42 kD, very abundant in eukaryotic cells (8–14% total cell protein) and one of the major components of the ***actomyosin*** motor and the cortical microfilament meshwork. First isolated from ***striated muscle*** and often referred to as one of the muscle proteins. ***G-actin*** is the globular monomeric form of actin, 6.7 × 4.0nm: it polymerizes to form filamentous F-actin.

actin-binding proteins A diverse group of proteins that bind to ***actin*** and that may stabilize ***F-actin*** filaments, nucleate filament formation, crosslink filaments, lead to bundle formation, etc. See Table A1.

actin meshwork ***Microfilaments*** inserted proximally into the plasma membrane and crosslinked by ***actin-binding proteins*** to form a mechanically resistive network that may support protrusions such as ***pseudopods*** (sometimes referred to as the ***cortical meshwork***).

actin-RPV Vertebrate actin-related protein; 42 kD with 54% sequence homology with muscle actin and 69% similarity with cytoplasmic actin – more similar to conventional actin than to ***act-2***; generally the core structure of the protein is conserved and differences are in surface loops. Forms part of the ***dynactin*** complex, an activator of ***dynein***-driven vesicle movement.

Actinia equina Common beadlet anemone; a ***coelenterate***. See ***equinatoxins***.

actinic keratoses See ***keratoses***.

actinic keratosis Thickened area of skin as a result of excessive exposure to sunlight, particularly common in those with very fair skin.

actinogelin Protein (115 kD) from ***Erlich ascites cells*** that gelates and bundles ***microfilaments***.

Actinomycetales Order of ***Gram-positive bacteria***, widespread in soil, compost and aquatic habitats. Most are saprophytic, but there are a few pathogens; some produce important antibiotics. Important genera include *Actinomyces*, Corynebacterium, Frankia, Mycobacterium, Streptomyces.

actinomycin C A mixture of ***antibiotics***, actinomycins C1, C2 and ***actinomycin D*** elaborated by a species of ***Streptomyces***.

actinomycin D ***Antibiotic*** from *Streptomyces* spp. that by binding to DNA blocks the movement of ***RNA polymerase*** and prevents RNA synthesis in both pro- and eukaryotes.

Actinophrys sol Species of ***Heliozoa*** often used in studies on microtubule stability: the axopodia are supported by a bundle of crosslinked microtubules arranged in a complex double-spiral pattern when viewed in cross-section.

Actinosphaerium Species of ***Heliozoa*** that is multinucleate. Remarkable for its long radial protruding axopodia that contain complex double-spiral arrangements of many microtubules. It catches prey by protrusion and retraction of the axopodia.

actinotrichia (singular, *actinotrichium*) Aligned ***collagen*** fibres (ca 2μm diameter) that provide a guidance cue for ***mesenchyme*** cells in the developing fin of teleost fish.

actinotrichium (singular) See ***actinotrichia***.

action potential An electrical pulse that passes along the membranes of ***excitable cells***, such as ***neurons***, ***muscle cells***, fertilized eggs and certain plant cells. The precise shape of action potentials varies, but action potentials always involve a large ***depolarization*** of the cell membrane, from its normal ***resting potential*** of –50 to –90mV. In a neuron, action potentials can reach +30mV, and last

Table A1. Actin-binding proteins

Protein	M_r(kD)	Source/comment
Monomer sequestering (bind G-actin) or depolymerizing		
19 kD Protein	19	Pig brain
Actobindin	9.7	*Acanthamoeba castellani*
Actolinkin	20	Echinoderm eggs
Actophorin	13–19	*Toxoplasma gondii*
ADF	19	Various
Coactosin	17	*Dictyostelium*
Cofilin	19	Various. Depolymerizes
Depactin	18	Starfish oocytes
Destrin	19	Various
DNAase I	31	Pancreas
Drebin	140	Neuronal cells
Profilin	12--15	Various
Twinfilin	37	Yeast
Vitamin D-binding protein	57	Plasma
End-blocking and nucleating		
ActA	90	*Listeria*. Promotes polymerization
β-actinin	37, 35	Kidney and striated muscle. End-blocker
Acumentin	65	Mammalian leucocytes. End-blocker
Adseverin	74	Adrenal medulla. Calcium-sensitive
Aginactin	70	*Dictyostelium*. Barbed-end cap
Capping protein	31, 28	*Acanthamoeba*
Fragmin/severin	40–45	*Physarum, Dictyostelium*, sea urchin eggs
Gelsolin	90	Mammalian cells; same as brevin and ADF Calcium-sensitive
Hisactophilin	13.5	*Dictyostelium*. Promotes polymerization
IcsA		*Shigella flexneri*. Promotes assembly
Ponticulin	17	*Dictyostelium*
Radixin	82	Adherens junction and cleavage furrow
Tensin/insertin	30	Focal adhesions. Bind barbed end
Villin	95	Amphibian eggs, avian and mammalian epithelium
Crosslinking		
Isotropic gelation		
Actin-binding protein	2 × 270	Macrophages, platelets, *Xenopus* eggs
ABP-120	120	*Dictyostelium*
ABP-280	280	*Dictyostelium*
Cortactin	80/85	Various. Has SH3 domain
Filamin	2 × 250	Smooth muscle
Spectrin	2 × 240, 2 × 220	Erythrocytes
Fodrin	2 × 260, 2 × 240	Brain
Transgelin	21	Highly conserved
TW 260/240	2 × 260, 2 × 240	Intestinal epithelium
Anisotropic bundling		
ABP-50	50	*Dictyostelium*
ABP-67	67	Yeast
α-actinin	2 × 95	Various
Actinogelin	2 × 115	Ehrlich ascites tumour cells
Coronin	55	*Dictyostelium*
Dematin	52	Erythrocytes
Fascin	53–57	Pig brain, echinoderm gametes
Fimbrin	68	Intestinal epithelium
Plastin/p65	65	Various mammalian cells
Villin	95	Intestinal epithelium

Table A1. (Continued)

Protein	M_r(kD)	Source/comment
Miscellaneous		
Caldesmon	88	Non-muscle cells
Calpactins	35, 36	Various
Connectin	70	Binds both laminin and actin
Gelactins	23–38	4 types; from *Acanthamoeba*
Hsp90	90	Stressed cells
MAP2	300	Brain, microtubule-associated
MARCKS	80	Various, substrate for PKC
Nebulin	600–900	N-line of sarcomere
Nuclear actin BP	2 × 34	Various
Scruin	102	*Limulus* acrosomal process
Tau	50–68	Microtubule-associated

Specific entries should be seen for further details.

1ms. In muscles, action potentials can be much slower, lasting up to 1s.

action spectrum The relationship between the frequency (wavelength) of a form of radiation and its effectiveness in inducing a specific chemical or biological effect.

activated macrophage A ***macrophage*** (***mononuclear phagocyte***) that has been stimulated by ***lymphokines*** and which has greatly enhanced cytotoxic and bactericidal potential

activation (of egg) Normally brought about by contact between spermatozoon and egg membrane. Activation is the first stage in development and occurs independently of nuclear fusion. The first observable change is usually the cortical reaction that may involve elevation of the fertilization membrane; the net result is a block to further fusion and thus to polyspermy. In addition to the morphological changes, there are rapid changes in metabolic rate and an increase in protein synthesis from maternal mRNA.

activation energy The energy required to bring a system from the ground state to the level at which a reaction will proceed.

active immunity Immunity resulting from the normal response to antigen. Only really used to contrast with ***passive immunity***, in which antibodies or sensitised lymphocytes are transferred from the reactive animal to the passive recipient.

active site The region of a protein that binds to substrate molecule(s) and facilitates a specific chemical conversion. Produced by juxtaposition of amino acid residues as a consequence of the protein's ***tertiary structure***.

active transport Often defined as transport up an electrochemical gradient. More precisely defined as unidirectional or vectorial transport produced within a membrane-bound protein complex by coupling an energy-yielding process to a tranport process. In primary active transport systems the transport step is normally coupled to ***ATP*** hydrolysis within a single protein 'complex'. In secondary active transport the movement of one species is coupled to the movement of another species down an electrochemical gradient established by primary active transport.

active zone Site of transmitter release on presynaptic terminal at chemical synapses. At the neuromuscular junction, active zones are located directly across the synaptic cleft from clusters of ***acetylcholine receptors***. Evidence from ***conotoxin*** binding studies suggests that presynaptic ***calcium channels*** are exclusively localized at active zones.

activin Dimeric growth factors of the TGF family with effects on a range of cell types in addition to its original role (FSH-releasing) in gonadal sites. Composed of two of the β subunits of ***inhibin*** (which is an αβ heterodimer); since there are two isoforms, A and B, there are three forms of activin, AA, BB and AB. Receptor has serine/threonine kinase activity in its cytoplasmic domain.

activin-response factor (ARF) Induced in early *Xenopus* blastomeres by ***activin***, Vg-1 and ***TGF***-β, binds to activin-response element in the ***mix***-2 homeobox gene. The ARF complex contains XMAD2, a *Xenopus* homologue of the *Drosophila Mad* product, and ***FAST***-1, a winged-helix transcription factor.

actobindin Protein (9.7 kD) from *Acanthamoeba castellani*. Potent inhibitor of actin polymerization under certain conditions, possibly by binding to actin oligomers rendering them non-nucleating. 88 amino acids with a nearly identical repeat sequence of 33 amino acids.

actolinkin Monomeric protein (20 kD) from echinoderm eggs. Seems to link actin filaments to inner surface of plasma membrane by their barbed ends.

actomere Site of actin filament nucleation in sperm of some echinoderms in which the acrosomal process is protruded by rapid assembly of a parallel microfilament bundle.

actomyosin Generally: a motor system that is thought to be based on ***actin*** and ***myosin***. The essence of the motor system is that myosin makes transient contact with the actin filaments and undergoes a conformational change before releasing contact. The hydrolysis of ATP is coupled to movement, through the requirement for ATP to restore the configuration of myosin prior to repeating the cycle. More specifically: a viscous solution formed when actin and myosin solutions are mixed at high salt concentrations. The viscosity diminishes if ATP is supplied and rises as the ATP is hydrolyzed. Extruded threads of actomyosin will contract in response to ATP.

actophorin An actin depolymerizing factor (ADF; 13–15 kD) from protozoa (*Toxoplasma gondii, Acanthamoeba castellanii*) that will sever actin filaments and sequester ***G-actin***. Binds ADP–G-actin with higher affinity than ATP-actin and binding is very sensitive to the divalent cation present on the actin. Has high sequence homology with vertebrate ***cofilin*** and ***destrin***, echinoderm ***depactin*** and some plant ADFs but lacks the nuclear localization sequence found in the vertebrate ADFs.

acumentin Protein (65 kD) originally thought to cap pointed end of microfilaments, isolated from vertebrate ***macrophages***.

acute (1) Sharp or pointed. (2) Of diseases; coming rapidly to a crisis – not persistent (***chronic***).

acute inflammation Response of vertebrate body to insult or infection; characterized by redness (rubor), heat (calor), swelling (tumour), pain (dolour), and sometimes loss of function. Changes occur in local blood flow, and ***leucocytes*** (particularly ***neutrophils***) adhere to the walls of postcapillary venules (margination) and then move through the ***endothelium*** (diapedesis) towards the damaged tissue. Although acute inflammation is usually short term, there are situations in which acute-type inflammation persists.

acute lymphoblastic leukaemia (ALL) See ***leukaemia***.

acute myeloblastic leukaemia (AML) See ***leukaemia***.

acute-phase protein Proteins found in the serum of organisms showing ***acute inflammation***. In particular ***C-reactive protein*** and ***serum amyloid*** A protein.

acute-phase reaction Response to acute inflammation involving the increased synthesis of various plasma proteins (***acute phase proteins***).

acutely transforming virus *Retrovirus* that rapidly ***transforms*** cells, by virtue of possessing one or more ***oncogenes***. Archetype: ***Rous sarcoma virus***.

ACV (1) ***acyclovir***. (2) α-aminoadipylcysteinyl-valine, precursor for isopenicillin synthesis.

ACV synthase (α-aminoadipylcysteinyl-valine synthase) Enzyme responsible for an early step in cephalosporin synthesis. ACV is acted upon by isopenicillin N synthase to produce isopenicillin N.

acyclovir (ACV; 2-(hydroxyethoxy) methyl guanine) Antiviral agent that is an analogue of ***guanosine*** and inhibits ***DNA replication*** of viruses. Particularly successful against herpes simplex infections.

acylation Introduction of an acyl (RCO-) group into a molecule: for example the formation of an ester between glycerol and fatty acid to form mono-, di-, or triacylglycerol, or the formation of an aminoacyl-tRNA during protein synthesis. Acyl-enzyme intermediates are transiently formed during covalent catalysis.

adaptation A change in sensory or excitable cells upon repeated stimulation, which reduces their sensitivity to continued stimulation. Those cells that show rapid adaptation are known as phasic; those which adapt slowly are known as tonic. Can also be used in a more general sense for any system that changes responsiveness with time, eg. by downregulation of receptors (***tachyphylaxis***) or through internal modulation of the signalling system, as in ***bacterial chemotaxis***.

adaptins Proteins of 100–110 kD found as part of the adaptor complex.

adaptor A protein complex associated with coated vesicles. Adaptors have been shown to promote the ***in vitro*** assembly of *clathrin* cages and to bind to the cytoplasmic domains of specific membrane proteins. It has been proposed that adaptors link selected membrane proteins to clathrin, causing them to be packaged into a ***coated vesicle***. Two adaptors have been identified (HA-1 and HA-2); they are hetero-tetramers composed of two proteins of 100–110 kD, termed adaptins, and two smaller polypeptides of around 50 and 20 kD. The HA-1 subunits are γ-adaptin, β'-adaptin with polypeptides of 50 kD and 17 kD. HA-1 is associated with coated pits formed from the Golgi complex. The HA-2 subunits are α-adaptin, β-adaptin with polypeptides of 47 kD and 20 kD. HA-2 is associated with coated pits formed from the plasma membrane. (β '-and β-adaptin are closely related.)

Addison's disease Chronic insufficiency of the adrenal cortex as a result of tuberculosis or, specific ***autoimmune*** destruction of the ***ACTH***-secreting cells.

addressins Cell surface molecules, particularly on endothelial cells believed to be involved in controlling the location of migrating cells, especially in lymphocyte homing. Probably act as cell adhesion molecules or their receptors. All have lectin, EGF, and complement-regulatory domains extracellularly in addition to the membrane spanning and cytoplasmic domains. Examples are ELAM-1, GMP-140 (PADGEM), gp90mel; also known as ***selectins***.

adducin ***Calmodulin***-binding protein associated with the membrane skeleton of erythrocytes. A substrate for ***protein kinase C***, it binds to ***spectrin–actin*** complexes (but only weakly to either alone) and promotes the assembly of spectrin onto spectrin–actin complexes unless micromolar calcium is present. Has subunits of 102 and 97 kD and is distinguishable from band 4.1.

adductor muscle Large muscle of bivalve molluscs that is responsible for holding the two halves of the shell closed. Its unusual feature is its ability to maintain high tension with low energy expenditure by using a 'catch' mechanism, and the high content of ***paramyosin***.

adenine (6-aminopurine), one of the bases found in ***nucleic acids*** and ***nucleotides***. In DNA, it pairs with ***thymine***.

adeno- Prefix indicating association with, or similarity to, glandular tissue.

adenocarcinoma Malignant neoplasia of a glandular epithelium, or ***carcinoma*** showing gland-like organization of cells.

adenohypophysis Anterior lobe of the ***pituitary*** gland; responsible for secreting a number of hormones and containing a comparable number of cell types.

adenoma ***Benign tumour*** of glandular epithelium.

adenomatous polyposis coli See ***polyposis coli***.

adenosine (9-β-D-ribofuranosyladenine) The ***nucleoside*** formed by linking adenine to ***ribose***.

adenosine 5′ triphosphate See *ATP*.

adenosine diphosphate See *ADP*.

adenosine monophosphate See *AMP*.

adenosine receptors Four adenosine receptors have been identified, A1, A2A, A2B, and A3. All are seven membrane-spanning G-protein coupled. There is considerable difference in properties of receptors from different species. A2A receptors regulate voltage-sensitive ***calcium channels***.

adenosine triphosphate See *ATP*.

adenosquamous Benign tumour of epithelial origin (***adenoma***) in which cells have flattened morphology as opposed to cuboidal or columnar.

Adenoviridae (adenovirus) Large group of viruses first isolated from cultures of adenoids. The ***capsid*** is an icosahedron of 240 hexons and 12 pentons and is in the form of a base and a fibre with a terminal swelling; the genome consists of a single, linear molecule of double-stranded DNA. They cause various respiratory infections in humans. Some of the avian, bovine, human and simian adenoviruses cause tumours in newborn rodents, generally hamsters. They can be classified into highly, weakly and non-***oncogenic viruses*** from their ability to induce tumours ***in vivo*** though all of these groups will transform cultured cells. The viruses are named after their host species and subdivided into many serological types, eg. human Adenovirus type 3.

adenylate cyclase Enzyme that produces cAMP (***cyclic AMP***) from ***ATP*** and acts as a signal-transducer coupling hormone binding to change of cytoplasmic cAMP levels. The name strictly refers to the catalytic moiety, but it is usually applied to the complex system that includes the hormone receptor and the GTP-binding modulator protein (see ***GTP-binding proteins***).

ADF (1) Actin depolymerizing factor (19 kD); regulates actin polymerization in developing chick skeletal muscle. Has the ability to depolymerize microfilaments and bind G-actin, but does not cap filaments. ***Destrin*** is apparently identical, ***cofilin*** similar in function but the product of a different gene. Has a nuclear localization domain and interaction with actin is regulated by phosphoinositides. (2) Adult T-cell leukaemia-derived factor. Inducer of interleukin-2 receptor α (IL-2Rα). Homologue of thioredoxin. Autocrine growth factor produced by HTLV-1 or EBV-transformed cells.

ADF-H domain (ADF-homology domain) An actin-binding module found in an extensive family of proteins with three phylogenetically distinct classes, ADF/cofilins, twinfilins and drebin/ABP-1s.

ADH See ***antidiuretic hormone***.

adhalin See ***sarcoglycan***.

adherens junction Specialized cell–cell junction into which are inserted ***microfilaments*** (in which case also known as ***zonula adherens***), or ***intermediate filaments*** (macula adherens or spot ***desmosomes***).

adhesins General term for molecules involved in adhesion, but its use is restricted in Microbiology where it refers to bacterial surface components.

adhesion plaque Another term for a ***focal adhesion***, a discrete area of close contact between a cell and a non-cellular substratum, with cytoplasmic insertion of ***microfilaments*** and considerable electron-density adjacent to the contact area. On the cytoplasmic face are local concentrations of various proteins such as ***vinculin*** and ***talin***.

adhesion site (1) In ***Gram-negative bacteria***, a region where the outer membrane and the plasmalemma appear to fuse. May be important in export of proteins or viral entry. (2) Used rather generally of any region of a cell specialized for adhesion.

Adiantum capillus-veneris A species of fern, the source of a number of drugs.

adipocyte Mesenchymal cell in fat tissue that has large lipid-filled vesicles. There may be distinct types in white and brown fat. 3T3-L1 cells are often used as a model system; they can be induced to differentiate with dexamethasone/insulin/IBMX treatment.

adipofibroblast ***Adipocytes*** from subcutaneous fat lose fat globules and develop a fibroblastic appearance when grown in culture. Unlike skin fibroblasts they will take up fat from serum taken from obese donors, and probably retain a distinct differentiated state.

adipose tissue Fibrous connective tissue with large numbers of fat-storing cells, ***adipocytes***.

adipsin A serine protease with complement factor D activity. Synthesized by ***adipocytes***. Altered levels are characteristic of some genetic and acquired obesity syndromes.

adjuvant Additional components added to a system to affect action of its main component, typically to increase ***immune response*** to an ***antigen***. See ***Freund's adjuvant***.

adocia sulphate-2 (AS-2) Inhibitor of ***kinesin***, isolated from sponge (*Haliclona* spp.). Binds to motor domain of kinesin, mimicking tubulin.

adoptive immunity Immunity acquired as a result of the transfer of lymphocytes from another animal.

ADP Adenosine diphosphate. Unless otherwise specified is the nucleotide 5′ADP, ***adenosine*** bearing a diphosphate (pyrophosphate) group in ribose-O-phosphate ester linkage at position 5′ of the ribose moiety. Adenosine 2′5′ and 3′5′ diphosphates also exist, the former as part of ***NADP*** and the latter in ***coenzyme A***.

ADP-ribosylation A form of ***post-translational*** modification of protein structure involving the transfer to protein of the ADP-ribosyl moiety of ***NAD***. Believed to play a part in normal cellular regulation as well as in the mode of action of several bacterial toxins.

ADP-ribosylation factor (ARF) Ubiquitous ***GTP-binding protein***, approximately 20 kD, N-myristoylated, stimulates ***cholera toxin*** ADP-ribosylation. Mediates binding of non-clathrin coated vesicles and AP1 (adaptor-protein 1) of clathrin-coated vesicles to Golgi membranes. At least six isoforms have been identified.

adrenal Endocrine gland adjacent to the kidney. Distinct regions (cortex and medulla) produce different ranges of hormones.

adrenaline (epinephrine) A hormone secreted (with ***noradrenaline***) by the medulla of the ***adrenal*** gland, and by ***neurons*** of the ***sympathetic nervous system***, in response to stress. The effects are those of the classic 'fight-or-flight' response, including increased heart function, elevation in blood sugar levels, cutaneous vasoconstriction making the skin pale, and raising of hairs on the neck.

adrenergic neuron A neuron is adrenergic if it secretes ***adrenaline*** at its terminals. Many neurons of the ***sympathetic nervous system*** are adrenergic.

adrenergic receptors Receptors for ***noradrenaline*** and ***adrenaline***. All are seven membrane spanning G-protein coupled receptors linked variously either to adenylate cyclase or phosphoinositide second messenger pathways. Three subgroups are usually recognized; the β-adrenergic receptors linked to Gs, the α1 linked to Gi, and the α2 linked to Gq. The β-adrenegic receptor gene is unusual in having no introns.

adrenocorticotrophin (ACTH) A peptide hormone produced by the pituitary gland in response to stress (mediated by corticotrophin-releasing factor, a 41-residue peptide, from the hypothalamus). Stimulates the release of adrenal cortical hormones, mostly ***glucocorticoids***. Derived from a larger precursor, ***pro-opimelanocortin***, by the action of an endopeptidase, which also releases ***β-lipotropin***. See also Table H2.

adrenoleucodystrophy (X-linked Schilder's disease) Demyelinating disease with childhood onset.

adrenomedullin One of the ***calcitonin family peptides***.

adriamycin An antibiotic that ***intercalates*** into RNA and DNA. Related to daunomycin.

adseverin Actin-regulating protein (74 kD) isolated from adrenal medulla. Has severing, nucleating and capping activities similar to those of ***gelsolin***, but does not cross-react immunologically. Has phospholipid binding domain and its properties are regulated by phosphatidyl inositides and by calcium. May be identical to scinderin.

adsorption coefficient A constant, under defined conditions, that relates the binding of a molecule to a matrix as a function of the weight of matrix, for example in a column.

adult respiratory distress syndrome (acute respiratory distress syndrome; ARDS) See ***septic shock***.

adventitia Outer coat of the wall of ***vein*** or ***artery***, composed of loose ***connective tissue*** that is vascularized. Generally used to mean outer covering of an organ.

Aedes Genus of mosquitos, several of which transmit diseases of man. *A. aegypti* is the vector of yellow fever.

Aequorea victoria Hydrozoan jellyfish (a ***coelenterate***) from which ***green fluorescent protein*** (GFP) was isolated. ***Aequorin*** can be isolated from *A. victoria* and *A. forskaolea*.

aequorin Protein (30 kD) extracted from jellyfish (*Aequorea aequorea*) that emits light in proportion to the concentration of calcium ions. Used to measure calcium concentrations, but has to be microinjected into cells. Contains ***EF-hand*** motif. See also ***bioluminescence***.

aerenchyma Form of ***parenchyma*** with large air spaces that gives buoyancy to aquatic plants.

aerobes Organisms that rely on oxygen.

aerobic respiration Controlled process by which carbohydrate is oxidized to carbon dioxide and water, using atmospheric oxygen, to yield energy.

aerolysin Channel-forming bacterial exotoxin produced by *Aeromonas hydrophila* as a 50 kD protoxin that later has a 43-residue C-terminal peptide cleaved off to generate the active toxin. Binds to specific receptor on target cells (probably glycophorin on human erythrocytes, but may be other proteins in other cells) and polymerizes to form a heptameric complex that inserts into the plasma membrane and has a pore of approximately 1.5nm diameter with some properties similar to ***porin*** channels.

Aeromonas Genus of ***Gram-negative bacteria*** some species of which are pathogenic. *A. salmonicida* causes furunculosis in fish.

aerotaxis A ***taxis*** in response to oxygen (air).

AF2 (1) Activation function domain 2 of steroid receptors. (2) Antiflammin-2, a synthetic peptide inhibitor of PLA2.

afferent Leading towards; afferent nerves lead towards the central nervous system, afferent lymphatics towards the lymph node. The opposite of efferent.

affinity An expression of the strength of interaction between two entities, eg. between receptor and ligand or between enzyme and substrate. The affinity is usually characterized by the equilibrium constant (***association constant*** or ***dissociation constant***) for the binding, this being the concentration at which half the receptors are occupied.

affinity chromatography ***Chromatography*** in which the immobile phase (bed material) has a specific biological affinity for the substance to be separated or isolated, such as the affinity of an antibody for its antigen, or an enzyme for a substrate analogue.

affinity labelling Labelling of the active site of an enzyme or the binding site of a receptor by means of a reactive substance that forms a covalent linkage once having bound. Linkage is often triggered by a change in conditions, for example in photoaffinity labelling as a result of illumination by light of an appropriate wavelength.

aFGF See ***fibroblast growth factor***.

aflatoxins A group of highly toxic substances produced by the fungus *Aspergillus flavus*, and other species of *Aspergillus*, in stored grain or mouldy peanuts. They cause enlargement and death of liver cells if ingested, and may be carcinogenic.

Agalenopsis aperta American funnel web spider. See ***agatoxins***.

agammaglobulinaemia Sex-linked genetic defect that leads to the complete absence of immunoglobulins (IgG, IgM, and IgA) in the plasma as a result of the failure of pre-B cells to differentiate. Failure to produce a humoral antibody response leads to high incidence of opportunistic infections but cell-mediated immunity is unimpaired.

agar A ***polysaccharide*** complex extracted from seaweed (Rhodophyceae) and used as an inert support for the growth of cells, particularly bacteria and some cancer cell lines (eg. sloppy agar).

agarose A galactan polymer purified from ***agar*** that forms a rigid gel with high free water content. Primarily used as an electrophoretic support for separation of macromolecules. Stabilized derivatives are used as 'macroporous' supports in ***affinity chromatography***. See ***Sepharose***.

agatoxins Toxins from the American funnel web spider, ***Agalenopsis aperta***. The μ-toxins are 36–38 residue peptides that act on insect but not vertebrate voltage-sensitive sodium channels. The ω-agatoxins are more diverse (5–9 kD) and act on various calcium channels, mostly in neuronal cells, blocking release of neurotransmitters. Affect vertebrate and invertebrate channels.

agglutination The formation of adhesions by particles or cells to build up multicomponent aggregates, otherwise termed agglutinates or flocs. Distinguished from ***aggregation*** by the fact that agglutination phenomena are usually very rapid. Usually caused by addition of extrinsic agents such as ***antibodies***, ***lectins*** or other bi- or polyvalent reagents. See ***aggregation***.

agglutinins Agents causing ***agglutination***, eg. antibodies, ***lectins***, ***polylysine***.

agglutinogen The antigen (in the case of antibody) or ligand (in the case of lectin) with which an ***agglutinin*** reacts.

aggrecan The major structural proteoglycan of cartilage. It is a very large and complex macromolecule, comprising a core protein of 210 kD to which are linked around 100 chondroitin sulphate chains and several keratan sulphate chains as well as O- and N-linked oligosaccharide chains. It binds to a link protein (around 40 kD) and to hyaluronic acid, forming large aggregates, hence its name.

aggregation The process of forming ***adhesions*** between particles such as cells. Aggregation is usually distinguished from ***agglutination*** by the slow nature of the process in that not every encounter between the cells is effective in forming an adhesion.

aginactin Agonist-regulated actin-filament barbed-end capping protein (70 kD) that inhibits microfilament polymerization, isolated from ***Dictyostelium***. Interestingly, it is regulated by cAMP, a chemotactic factor for *Dictyostelium*.

agitoxins Scorpion toxins (peptides) that inhibit potassium channels. Closely related to ***kaliotoxin***.

agmatinase Enzyme that degrades ***agmatine*** to ***putrescine***.

agmatine (1-amino-4-guanidobutane) A metabolite of arginine via arginine decarboxylase, is metabolized to ***putrescine*** by agmatinase. Suppresses polyamine biosynthesis and polyamine uptake by cells by inducing ***antizyme***. Binds to ***imidazoline receptors*** and α-2-adrenoreceptors but with different affinities and is thought to be the endogenous ligand for imidazoline (I1) receptors. On the basis of its distribution in the brain, it is proposed to be a neurotransmitter involved in behavioural and visceral control. An increasing range of biological activities are being described.

agonist (1) In neurobiology, of a ***neuron*** or ***muscle***; one that aids the action of another. If the two effects oppose each other, then they are known as antagonistic. (2) In pharmacology, a compound that acts on the same receptor, and with a similar effect, to the natural ligand. (3) In ethology, 'agonistic behaviour' means aggressive behaviour towards a conspecific animal.

agorins Major structural proteins of the membrane matrix, constituting approximately 15% of total plasma membrane proteins of P815 mastocytoma cells. They form large detergent-insoluble structures when the membranes are extracted with Triton X-100 and EGTA. Agorin I, 20 kD; Agorin II, 40 kD.

agouti Central American rodent that has given its name to a grey flecked coat coloration in mice caused by alternate light and dark bands on individual hairs. The gene codes for a 131 residue secreted protein that regulates phaeomelanin synthesis in melanocytes but associated with the locus are genes important in embryonic development. Mice with dominant mutation at the Ay locus develop diabetes and obesity. The agouti gene product binds to melanocortin receptor 1 but does not antagonize α-MSH and has similar antiproliferative effects on melanoma cells in culture. The dark agouti (DA) rat has a high susceptibility to developing arthritis.

AGP (arabinoglycan-protein) A class of extracellular ***proteoglycan***, found in many higher-plant tissues, and secreted by many suspension-cultured plant cells. Contains 90–98% ***arabinogalactan*** and 2–10% protein. Related to arabinogalactan II of the cell wall.

agranular vesicles Synaptic vesicles that do not have a granular appearance in EM; 40–50nm in diameter, with membrane only 4–5nm thick. Characteristic of peripheral cholinergic ***synapses***; see also ***neurotransmitter***. Some are located very close to ***presynaptic membrane***.

agranulocytosis Severe acute deficiency of ***granulocytes*** in blood.

agretope Portion of antigen that interacts with an MHC molecule.

agrin Secreted protein (200 kD) isolated from the synapse-rich electric organ of *Torpedo californica* that induces the formation of synaptic specializations on ***myotubes*** in culture. Present in muscle cells before innervation, and concentrated at the ***neuromuscular junction*** once AChR clustering occurs. The release of agrin from motor axon terminals is thought to trigger the formation of the postsynaptic apparatus at developing and regenerating neuromuscular junctions. Has several EGF repeats and a protease inhibitor-like domain.

Agrobacterium tumefaciens A Gram-negative, rod-shaped, flagellated bacterium responsible for ***crown gall*** tumour in plants. Following infection the T1 ***plasmid*** from the bacterium becomes integrated into the host plant's DNA and the presence of the bacterium is no longer necessary for the continued growth of the tumour.

Ah receptor Cytoplasmic receptor for aryl hydrocarbons: once the ligand is bound it translocates to the nucleus and binds to the ***xenobiotic response element***. Contains basic ***helix-turn-helix*** motif.

AIDS (acquired immune deficiency syndrome) Disease caused by infection with HIV (also called LAV or HTLV-3 in the early literature) virus, resulting in a deficiency of ***T-helper cells*** with resulting ***immunosuppression*** phenomena further resulting in susceptibility to other infectious diseases and to certain types of tumour, particularly ***Kaposi's sarcoma***.

air-lift fermenter A fermenter in which circulation of the culture medium and aeration is achieved by injection of air into some lower part of the fermenter. Usually not suitable for animal cell production. Related to gas-lift systems where an inert gas is used to achieve circulation in anaerobic conditions.

AKAP79 (A kinase-anchoring protein) Scaffold protein from mammalian cells to which ***protein kinase A*** (PKA), ***calcineurin*** and ***protein kinase C*** (PKC) all bind. PKC apparently binds at a different site so that PKC and calcineurin can, for example, be bound simultaneously. Through anchorage to AKAP79, kinases and phosphatases are co-localizedat postsynaptic densities in neurons. AKAP79 resembles yeast STE5.

Akt (PKB) Product of the normal gene homologue of v-*akt*, the transforming oncogene of AKT8 virus. A serine/threonine kinase (58 kD) with SH2 and PH domains, activated by PI3 kinase downstream of insulin and other growth factor receptors. AKT will phosphorylate ***GSK*** 3 and is involved in stimulation of Ras and control of cell survival. Three members of the Akt/PKB family have been identified, Akt/PKBα, AKT2/PKBβ, AKT3/PKBγ. Only AKT2 has been shown to be involved in human malignancy.

AKV A replication-competent murine leukaemia virus occurring endogenously in some mouse strains.

ALA synthase Enzyme responsible for the synthesis of 5-aminolevulinic acid. Defects in the enzyme cause microcytic anaemia because activity is essential for haem formation.

alamethicin A polyene pore-forming ionophore that forms relatively non-specific anion- or cation-transporting pores in plasma membranes or artificial lipid membranes. These pores may be potential gradient sensitive.

alanine Normally refers to L-α-alanine, the aliphatic ***amino acid*** found in proteins. See Table A2. The isomer β-alanine is a component of the vitamin ***pantothenic acid*** and thus also of ***coenzyme A***.

alarmone A small signal molecule in bacteria that induces an alteration of metabolism as a response to stress. Many metabolic responses may be altered by a single alarmone.

albinism Condition in which no ***melanin*** (or other pigment) is present.

albino Organism deficient in melanin biosynthesis. Hair and skin are unpigmented and the retinal pigmented epithelium is transparent, making the eyes appear red.

albolabrin See ***disintegrin***.

albumin The term normally refers to serum albumins, the major protein components of the serum of vertebrates. They have a single polypeptide chain, with multidomain structure containing multiple binding sites for many lipophilic metabolites, notably fatty acids and bile pigments. In the embryo their functions are fulfilled by ***alpha-fetoproteins***. See ***bovine serum albumin***. The

Table A2. Amino acids (L-amino acids specified by the biological code for proteins)

Name	Abbreviation	Single letter	Side chain	pKa[a]	M_r (D)	Hydropathy index[b] (Kyte & Doolittle)	Codons
Alanine	ala	A	$—CH_3$		89.1	1.8	GC(X)
Arginine	arg	R	$—CH_2\,CH_2\,CH_2\,NH\,(CN^+H_2)\,NH_2$	12	174.2	– 4.5	CG(X) AGA AGG
Aspartic acid	asp	D	$—CH_2\,COO^-$	4.4	133.1	– 3.5	GAU GAC
Asparagine	asn	N	$—CH_2\,CONH_2$		132.2	– 3.5	AAU AAC
Cysteine	cys	C	$—CH_2\,SH$	8.5	121.2	2.5	UGU UGC
Glutamic acid	glu	E	$—CH_2\,CH_2\,COO^-$	4.4	147.2	– 0.4	GG(X)
Glutamine	gln	Q	$—CH_2\,CH_2\,CONH_2$		146.2	– 3.5	CAA CAG
Glycine	gly	G	—H		75.1	– 3.5	GG(X)
Histidine	his	H	$—CH_2$–(imidazolium ring: +HN, NH)	6.5	155.2	– 3.2	CAU CAC
Iso-leucine	ile	I	$—CH\,(CH_3)\,CH_2\,CH_3$		131.2	4.5	AUU AUC AUA
Leucine	leu	L	$—CH_2\,CH\,(CH_3)_2$		131.2	3.8	CU(X) UUA UUG
Lysine	lys	K	$—CH_2\,CH_2\,CH_2\,CH_2\,NH_3^+$	10	146.2	– 3.9	AAA AAG
Methionine	met	M	$—CH_2\,CH_2\,SCH_3$		149.2	1.9	AUG
Phenylalanine	phe	F	$—CH_2$–(benzene ring)		165.2	2.8	UUU UUC
Proline	pro	P	$NH_2^+—CH—COO^-$ (ring: CH_2, CH_2, CH_2)		115.1	– 1.6	CC(X)
Serine	ser	S	$—CH_2\,OH$		105.1	– 0.8	UC(X)
Threonine	thr	T	$—CH\,(OH)\,CH_3$		119.1	– 0.7	AC(X)
Tryptophan	trp	W	$—CH_2$–(indole ring, N–H)		204.2	– 0.9	UGG (UGA mitochondria)
Tyrosine	tyr	Y	$—CH_2$–(benzene ring)–OH	10	181.2	– 1.3	UAU UAC
Valine	val	V	$—CH\,(CH_3)_2$		117.2	4.2	GU(X)

[a]The value for side chain ionization when the amino acid residue is present in a polypeptide.
[b]A measure of the tendency for the residue to be buried within the interior of a folded protein.

viability of analbuminaemic mutants (those deficient in albumin) raises serious questions about the biological role of albumin.

Alcaligenes Widespread genus of ***Gram-negative*** aerobic bacilli found in the digestive tract of many vertebrates and on skin. They occasionally cause opportunistic infections.

alcian blue Water-soluble copper phthalocyanin stain used to demonstrate acid ***mucopolysaccharides***. By varying the ionic strength some differentiation of various types is possible.

aldose reductase Enzyme that mediates conversion of glucose to sorbitol and the rate-limiting enzyme in the polyol pathway. Altered activity of aldose reductase is thought to play a part in the alterations in the vasculature seen as a complication of diabetes.

aldosterone A steroid hormone produced by the ***adrenal*** cortex, which controls salt and water balance in the kidney.

aleurone grain (body) Membrane-bounded ***storage granule*** within plant cells that usually contains protein. May be an ***aleuroplast*** or just a specialized ***vacuole***.

aleuroplast A semi-autonomous organelle (***plastid***) within a plant cell that stores protein.

Alexandrium **spp.** Genus of ***dinoflagellates*** that produce toxins associated with shellfish poisoning.

algae A non-taxonomic term used to group several phyla of the lower plants, including the ***Rhodophyta*** (red algae), ***Chlorophyta*** (green algae), ***Phaeophyta*** (brown algae), and ***Chrysophyta*** (diatoms). Many algae are unicellular or consist of simple undifferentiated colonies, but red and brown algae are complex multicellular organisms, familiar to most people as seaweeds. Blue-green algae are a totally separate group of ***prokaryotes***, more correctly known as *Cyanophyta*, or ***Cyanobacteria***.

alginate Salts of alginic acids, occurring in the cell walls of some algae. Commercially important in food processing, swabs, some filters, fire-retardants, etc. Calcium alginates form gels. Alginic acid is a linear polymer of mannuronic and glucuronic acids.

algorithm A process or set of rules by which a calculation or process can be carried out – usually referring to calculations that will be done by a computer.

aliphatic Carbon compound in which the carbon chain is open (non-cyclic).

aliphatic amino acids The naturally occurring amino acids with aliphatic side chains are glycine, alanine, valine, leucine and isoleucine.

aliquot Small portion. It is common practice to subdivide a precious solution of reagent into aliquots that are used when needed without handling the total sample.

alkaline phosphatase Enzyme (EC 3.1.3.1) catalysing cleavage of inorganic phosphate non-specifically from a wide variety of phosphate esters, and having a high (> 8) pH optimum. Found in bacteria, fungi and animals but not in higher plants.

alkaloid A nitrogenous base. Usually refers to biologically active (toxic) molecules, produced as allelochemicals by plants to deter grazing. Examples: ***ouabain***, ***digitalis***

alkaptonuria Congenital absence of homogentisic acid oxidase, an enzyme that breaks down tyrosine and phenylalanine. Accumulation of homogentisic acid in homozygotes causes brown pigmentation of skin and eyes and damage to joints; urine blackens on standing.

alkylating agent A reagent that places an alkyl group, eg. propyl in place of a nucleophilic group in a molecule. Alkylating reagents include a number of cytotoxic drugs, some of which react fairly specifically with N7 of the purine ring and lead to depurination of DNA, eg. the agent ethyl ethanesulphonic acid, and thus to mutagenesis.

ALL See ***acute lymphoblastic leukaemia***.

all-or-nothing Of an action potential, meaning that action potentials once triggered are of a stereotyped size and shape, irrespective of the size of stimulus that triggered them.

allantois Outgrowth from the ventral side of the hindgut in embryos of reptiles, birds and mammals. Serves the embryo as a store for nitrogenous waste and in chick embryos fuses with the chorion to form the ***chorioallantoic membrane*** (CAM).

allatostatin Peptide hormones produced by the corpora allata of insects that reversibly inhibit the production of ***juvenile hormone***. Similar peptides are found in other phyla. Allatostatin-4, smallest of the family, is DRLYSFGL-amide. Allatostatins may also be produced in other insect tissues, particularly midgut.

alleles Different forms or variants of a ***gene*** found at the same place, or ***locus***, on a ***chromosome***. Assumed to arise by ***mutation***.

allelic exclusion The process whereby one or more loci on one of the ***chromosome*** sets in a ***diploid*** cell is inactivated (or destroyed) so that the locus or loci is (are) not expressed in that cell or a clone founded by it. For example in mammals one of the X chromosome pairs of females is inactivated early in development (see ***Lyon hypothesis***) so that individual cells express only one allelic form of the product of that locus. Since the choice of chromosome to be inactivated is random, different cells express one or other of the X chromosome products resulting in mosaicism. The process is also known to occur in ***immunoglobulin*** genes so that a clone expresses only one of the two possible allelic forms of immunoglobulin.

allelochemical Substances effecting allelopathic reactions. See ***allelopathy***.

allelomorph One of several alternative forms of a gene: commonly shortened to ***allele***.

allelopathy The deleterious interaction between two organisms or cell types that are allogeneic to each other (the term is often applied loosely to interactions between ***xenogeneic*** organisms). Allelopathy is seen between different species of plant, between various individual sponges, and between sponges and gorgonians.

allelotype Occurence of an allele in a population or an individual with a particular allele. The allelotype of a tumour, the expression of particular microsatellite markers or isoenzymes, can indicate whether it is of polyclonal or monoclonal origin and the extent to which there is development of aneuploidy.

Allen video enhanced contrast (AVEC microscopy) A method for enhancing microscopic images pioneered by R. D. Allen. The digitized image has the background (an out-of-focus image of the same microscopic field with comparable unevenness of illumination, etc.) subtracted, and the contrast expanded to utilize the potential contrast range. Interestingly, it is possible to produce images of objects that are below the theoretical limit of resolution, eg. ***microtubules***.

allergic encephalitis See ***experimental allergic encephalomyelitis***.

allergy In an animal, a ***hypersensitivity*** response to some ***antigen*** that has previously elicited an immune response in the individual, producing a large and immediate ***immune response***. Allergies, eg. to bee venom, are occasionally fatal in humans.

Allium Genus that includes onions (*A. cepa*), leeks (*A. porrum*) and garlic (*A. sativum*).

alloantibody (alloserum) Antibody raised in one member of a species that recognizes genetic determinants in other individuals of the same species. Common in multiparous women and multiply transfused individuals who tend to have alloantibodies to MHC or blood group antigens.

alloantigen Individuals of a species differ in alleles (are ***allogeneic***) and the antigenic differences will cause an immune response to allografts. The antigens concerned are often of the ***histocompatibility complex*** and are referred to as alloantigens.

allochthonous Anything found at a site remote from that of its origin.

allogeneic Two or more individuals (or strains) are stated to be allogeneic to one another when the genes at one or more loci are not identical in sequence in each organism. Allogenicity is usually specified with reference to the locus or loci involved.

allograft Grafts between two or more individuals allogeneic at one or more loci (usually with reference to ***histocompatibility*** loci). As opposed to ***autograft*** and ***xenograft***.

allomone Compound produced by one organism that affects, detrimentally, the behaviour of a member of another species. If the benefit is to the recipient the substance is referred to as a ***kairomone***, if both organisms benefit then it is a ***synomone***.

allopolyploidy ***Polyploid*** condition in which the contributing genomes are dissimilar. When the genomes are doubled, fertility is restored and the organism is an amphidiploid. Common in plants but not in animals.

allopurinol (4-hydroxypyrazolo-(3,4-d) pyrimidine) A ***xanthine oxidase*** inhibitor used in the treatment of gout.

allosomes One or more chromosomes that can be distinguished from ***autosomes*** by their morphology and behaviour. Synonyms: accessory chromosomes, heterochromosomes, sex chromosomes.

allosteric Of a binding site in a protein, usually an enzyme. The catalytic function of an enzyme may be modified by interaction with small molecules, not only at the ***active site***, but also at a spatially distinct (allosteric) site of different specificity. Of a protein, a protein possessing such a site. An allosteric effector is a molecule bound at such a site that increases or decreases the activity of the enzyme.

allotetraploidy Example of ***allopolyploidy*** in which the hybrid diploid genome (formed from two chromosome sets) doubles in chromosome number.

allotope (allotypic determinant) The structural region of an ***antigen*** that distinguishes it from another ***allotype*** of that antigen.

allotype Products of one or more ***alleles*** that can be detected as inherited variants of a particular molecule. Usually the usage is restricted to those ***immunoglobulins*** that can be separately detected antigenically. See also ***idiotype***. In humans light chain allotypes are known as Km (Inv) allotypes and heavy chain allotypes as Gm allotypes.

allotypic determinant See ***allotope***.

alloxan Used to produce ***diabetes mellitus*** in experimental animals. Destroys ***B-cells*** in the pancreas by a mechanism involving ***superoxide*** production.

allozyme Variant of an enzyme coded by a different allele. See ***isoenzyme***.

alopecia Baldness. Can take various forms: alopecia areata in which hair loss is patchy, alopecia universalis in which hair loss is complete.

alpha hairpin See ***hairpin***.

alpha-2-macroglobulin (α-2-macroglobulin) Large (725 kD) plasma antiprotease with very broad spectrum of inhibitory activity against all classes of proteases. Apparently works by trapping the protease within a cage that closes when the protease-sensitive bait sequence is cleaved. The protease is still active against small substrates that can diffuse into the cage, and the conformational change that closes the trap alters the properties of the α-2-macroglobulin molecule so that it is rapidly removed from circulation. ***Plasminogen activator*** is one of the few proteases against which α-2-macroglobulin is ineffective.

alpha-actinin (α-actinin) A protein of 100 kD normally found as a dimer and which may link actin filaments end-to-end with opposite polarity. Originally described in the ***Z disc***, now known to occur in ***stress fibres*** and at ***focal adhesions***. Non-muscle isoform contains ***EF-hand*** motif.

alpha-amylase (α-amylase) An endo-amylase enzyme that rapidly breaks down starch to dextrins.

alpha-cell (α-cell) See ***A-cells*** of endocrine pancreas.

alpha-fetoprotein (α-fetoprotein) Proteins from the serum of vertebrate embryos, which probably fulfil the function of ***albumin*** in the mature organism. Found in both glycosylated and non-glycosylated forms. Presence in the fluid of the ***amniotic sac*** is diagnostic of spina bifida in the human foetus.

alpha-helix (α-helix) A particular helical folding of the polypeptide backbone in protein molecules (both fibrous and globular), in which the carbonyl oxygens are all hydrogen-bonded to amide nitrogen atoms three residues along the chain. The translation of amino acid residues along the long axis is 0.15nm, and the rotation per residue, 100°, so that there are 3.6 residues per turn.

alpha-neurotoxins (α-neurotoxins) Post-synaptic neurotoxins, many varieties of which are found in snake venoms. Two subclasses, short (four disulphides, 60–62 residues) and long (five disulphides and 66–74 residues). Examples include ***α-bungarotoxin***, ***α-cobratoxin***, ***erabutotoxins***.

alpha-sarcin (α-sarcin) Anti-tumour factor (ribotoxin; 17 kD) from *Aspergillus giganteus*. Is generally cytotoxic and has high specificity as an RNAase, binding to the 28S rRNA. One of a family that includes mitogillin, restrictocin and AspFl.

alpha-synuclein Protein that accumulates in the brain in Parkinson's disease particularly in ***Lewy bodies***. Mutations in the gene are associated with some familial forms of ***Parkinsonism***.

Alphavirus Genus of the Togaviridae family of RNA viruses. Sindbis and Semliki Forest viruses are the best-known examples.

Alport's syndrome Commonest of the hereditary nephropathies. Associated with nerve deafness and variable ocular disorders. Seems to be due to a defect in basement membrane.

Alsever's solution A solution used for preserving red blood cells.

altered self-hypothesis The hypothesis that the ***T-cell*** receptor in MHC-mediated phenomena recognizes a ***syngeneic*** MHC Class I or Class II molecule after modification by a virus or certain chemicals. See ***MHC restriction***.

alternative oxidase pathway Pathway of mitochondrial electron transport in higher plants, particularly in fruits and seeds, which does not involve cytochrome oxidase and thus is resistant to cyanide.

alternative pathway See ***complement*** activation.

alternative splicing Eukaryotic genes are composed of ***exon***s and ***intron***s, the latter being removed by ***RNA splicing*** before transcribed ***mRNA*** leaves the nucleus. Commonly, a single ***gene*** can encode several different mRNA transcripts, caused by cell- or tissue-specific combination of different exons. This is known as alternative splicing.

Alu (1) Type II ***restriction endonuclease***, isolated from *Arthrobacter luteus*. The recognition sequence is 5′-AG/CT-3′. (2) Alu sequences are highly repetitive sequences found in large numbers (100 000–500 000) in the human genome, and which are cleaved more than once within each sequence by the Alu endonuclease. The Alu sequences look like DNA copies of mRNA because they have a 3′ ***poly-A tail*** and flanking repeats.

alveolar cell Cell of the air sac of the lung.

alveolar macrophage ***Macrophage*** found in lung and which can be obtained by lung lavage; responsible for clearance of inhaled particles and lung surfactant. Metabolism slightly different from peritoneal macrophages (more oxidative metabolism); often have ***multivesicular bodies*** that may represent residual undigested lung surfactant.

Alzheimer's disease A presenile dementia characterized cellularly by the appearance of unusual helical protein filaments in nerve cells (neurofibrillary tangles), and by degeneration in cortical regions of brain, especially frontal and temporal lobes. See also ***senile plaques***.

amacrine cell A class of ***neuron*** of the middle layer of the ***retina***, with processes parallel to the plane of the retina. They are thought to be involved in image processing.

Amanita phalloides Poisonous mushroom, the death cap. Contains ***amanitin*** and ***phalloidin***.

amanitin (α-, β-, γ-amanitin) Group of cyclic peptide toxins. The most toxic components of *Amanita phalloides* (death cap). Specific inhibitors of ***RNA polymerase*** II in eukaryotes thus inhibiting protein synthesis by blocking the production of ***mRNA***.

amantadine Used as an antiviral agent (especially against influenza virus). Produces some symptomatic relief in ***Parkinsonism***.

amber codon One of the three ***termination codons***. Its sequence is UAG. See also ***ochre codon, opal codon***.

Ambystoma mexicanum Mexican axolotl (amphibian). A salamander that shows ***neoteny***. The adult may retain the larval form, but can reproduce. The neotenous, aquatic axolotl will metamorphose into the terrestrial form if injected with thyroid or pituitary gland extract.

ameloblasts Columnar epithelial cells that secrete the enamel layer of teeth in mammals. Their apical surfaces are tapering (Tomes processes) and are embedded within the enamel matrix.

amelogenins Extracellular matrix proteins (20 and 25 kD) of developing dental enamel; regulate form and size of hydroxyapatite crystallites during mineralization. Hydrophobic and proline-rich, produced by ***ameloblasts***.

Ames test One of a number of procedures used to test substances for likely ability to cause cancer that combines the use of animal tissue to generate active metabolites of the substance with a test for mutagenicity in bacteria.

amethopterin See ***aminopterin***.

amidation site A C-terminus consensus sequence, required for C-terminus amidation of peptides. Consensus is glycine, followed by 2 basic amino acids (arg or lys).

amiloride Drug that blocks sodium/proton ***antiport***; used clinically as a potassium-sparing diuretic.

amino acid permease A widely distributed group of large integral membrane proteins, required for the entry of amino acids into cells.

amino acid receptors Ligand-gated ***ion channel***s with specific receptors for amino acid ***neurotransmitter***s. An extended protein superfamily that also includes subunits of the ***nicotinic acetylcholine receptor***.

amino acid transmitters Amino acids released as neurotransmitter substances from nerve terminals and acting on postsynaptic receptors, eg. γ-aminobutyric acid (***GABA***) and glycine that are fast inhibitory transmitters in the mammalian central nervous system. Glutamate and aspartate mediate fast excitatory transmission. ***Strychnine*** (for glycine) and ***bicuculline*** (for GABA) are blocking agents for amino acid action.

amino acids Organic acids carrying amino groups. The L-forms of about 20 common amino acids are the components from which proteins are made. See Table A2, also Table C5 for the ***codon*** assignment.

amino sugar Monosaccharide in which an OH group is replaced with an amino group; often acetylated. Common examples are D-galactosamine, D-glucosamine, neuraminic acid, muramic acid. Amino sugars are important constituents of bacterial cell walls, some antibiotics, blood group substances, milk oligosaccharides and chitin.

aminotransferases (transaminase) A family of enzymes (EC 2.6.1.x) that transfer an amino group from an amino acid to an α-keto acid, as in the transfer from glutamic acid to oxaloacetic acid, to form aspartic acid and α-ketoglutarate. One reactant is often glutamic acid, and the reactions employ pyridoxal phosphate as coenzyme.

aminoacyl tRNA Complex of an ***amino acid*** to its ***tRNA***, formed by the action of aminoacyl tRNA synthetase. Requires ATP, which forms the linkage between the two molecules.

aminoacyl tRNA synthetases Enzymes that attach an amino acid to its specific tRNA. An intermediate step is the formation of an activated amino acid complex with AMP; the AMP is released following attachment to the tRNA.

aminoglycoside antibiotics Group of antibiotics active against many aerobic ***Gram-negative*** and some ***Gram-positive*** bacteria. Composed of two or more amino sugars attached by a glycosidic linkage to a hexose nucleus; polycationic and highly polar compounds. Inhibit bacterial protein synthesis by binding to a site on the 30S ribosomal subunit thereby altering codon–anticodon recognition. Common examples are streptomycin, gentamycin, amikacin, kanamycin, tobramycin, netilmicin, neomycin, framycetin. See also Table A4.

aminopeptidase Enzymes that remove the N-terminal amino acid from a protein or peptide.

aminophylline An inhibitor of cAMP ***phosphodiesterase***.

aminopterin A ***folic acid*** analogue and inhibitor of ***dihydrofolate reductase***. A potent cytotoxic agent used in the treatment of acute ***leukaemia***.

amitosis An unusual form of nuclear division, in which the nucleus simply constricts, rather like a cell, without chromosome condensation or spindle formation. Partitioning of daughter chromosomes is haphazard. Observed in some Protozoa.

AML See ***acute*** myeloblastic ***leukaemia***.

ammodytin L See ***ammodytoxins***.

ammodytoxins (atx A, B and C; ammodytin L) Group II secretory phospholipases A2 (122 residues) found in the venom of *Vipera ammodytes*. Act specifically at peripheral nerve endings in the neuromuscular junction. Binding site may be a subunit of potassium channels.

amniocentesis Sampling of the fluid in the ***amniotic sac***. In humans this is carried out between the 12th and 16th weeks of pregnancy, by inserting a needle through the abdominal wall into the uterus. By ***karyotyping*** the cells and determining the proteins present, it is possible to determine the sex of the foetus and whether it is suffering from certain congenital diseases such as ***Down's syndrome*** or spina bifida.

amnion Terrestrial vertebrates have embryos that develop in fluid-filled sacs formed by the outgrowth of the extra-embryonic ***ectoderm*** and ***mesoderm*** as projecting folds. These folds fuse to form two epithelia separated by mesoderm and ***coelom***. The inner layer is the amnion and encloses the amniotic sac in which the embryo is suspended. The outer layer is the ***chorion***.

amniotic sac Sac, enclosing the embryo of amniote vertebrates, that provides a fluid environment to prevent dehydration during development of land-based animals. See ***amnion***.

amoeba Genus of protozoa, but also an imprecise name given to several types of free-living unicellular phagocytic organism. Giant forms (eg. *Amoeba proteus*) may be up to 2mm long, and crawl over surfaces by protruding ***pseudopods*** (***amoeboid movement***). Amoebae exhibit great plasticity of form and conspicuous ***cytoplasmic streaming***.

amoebocytes Phagocytic cells found circulating in the body cavity of coelomates (particularly annelids and molluscs), or crawling through the interstitial tissues of sponges. A fairly non-committal classification.

amoeboid movement Crawling movement of a cell brought about by the protrusion of ***pseudopods*** at the front of the cell (one or more may be seen in monopodial or polypodial amoebae, respectively). The pseudopods form distal anchorages with the surface.

AMP (adenosine monophosphate) Unless otherwise specified, 5′AMP, the nucleotide bearing a phosphate in ribose-O-phosphate ester linkage at position 5 of the ribose moiety. Both 2′ and 3′ derivatives also exist. See also ***cyclic AMP*** (adenosine 3′5′-cyclic monophosphate).

AMP-PNP (5′-adenylyl imidodiphosphonate) Non-hydrolysable analogue of ATP used in isolation of some motor proteins.

AMPA (α-amino-3-hydroxy-5-methyl-4-isoxazoleproprionate) Synthetic agonist for metabotropic glutamate receptors.

AMPA receptor ***Glutamate***-operated ***ion channel***. See ***excitatory amino acid*** receptor channels.

amphetamine Drug of abuse that acts by increasing extraneuronal ***dopamine*** in midbrain. Thought to displace dopamine in ***synaptic vesicles***, leading to increased synaptic levels.

amphibolic Description of a pathway that functions not only in ***catabolism***, but also to provide precursors for ***anabolic*** pathways.

amphimixis Sexual reproduction resulting in an individual having two parents. Invariably the case in most animals, with the exception of a few hermaphrodite organisms, but not uncommon in plants where a single individual may produce both male and female gametes (be monoecious) and be self-fertile.

amphipathic Of a molecule, having both ***hydrophobic*** and ***hydrophilic*** regions. Can apply equally to small molecules, such as phospholipids, and macromolecules such as proteins.

amphiphilic Having affinity for two different environments, eg. a molecule with hydrophilic (polar) and lipophilic (non-polar) regions. Detergents are classic examples. Antonym of ***amphipathic***.

amphiphysin Protein of the nerve terminal that associates with synaptic vesicles, probably through AP2 and ***synaptotagmin***.

amphiregulin A heparin-binding ***growth factor*** containing an ***EGF-like domain***. See ***HB-EGF***.

amphitrophic Of organisms that can grow either photosynthetically or chemotrophically.

ampholyte Substance with ***amphoteric*** properties. Most commonly encountered as descriptive of the substances used in setting up electrofocusing columns or gels.

amphoteric Having both acidic and basic characteristics. This is true of proteins since they have both acidic and basic side groups (the charges of which balance at the isoelectric point).

amphotericin B (Fungizone) Polyene antibiotic from *Streptomyces* spp. Used as a fungicide, it is cytolytic by causing the formation of pores (5–10 molecules of amphotericin in association with cholesterol) that allow passage of small molecules through the plasma membrane and thus to cytolysis. Only acts on membranes containing sterols (preferentially ergosterol, hence selectivity for fungi). See also ***filipin***.

ampicillin ***Penicillin*** derivative with broad spectrum activity; ampicillin resistance is often used as a marker for ***plasmid*** transfer in genetic engineering (eg. pBR322 is ampicillin resistant).

amplicon The DNA product of a ***polymerase chain reaction***.

amygdala Almond-shaped body in the lateral ventricle of the brain.

amylin Natural hormone produced by pancreatic ***beta-cells*** that moderates the glucose-lowering effects of insulin. One of the ***calcitonin family peptides***.

amylobarbitone See ***amytal***.

amyloid ***Glycoprotein*** deposited extracellularly in tissues in ***amyloidosis***. The glycoprotein may either derive from ***light chain*** of ***immunoglobulin*** (AIO (amyloid of immune origin): 5–18 kD glycoprotein; product of a single clone of ***plasma cells***, the N-terminal part of lambda or kappa ***L chain***) or, in what used to be referred to as AUO, amyloid of unknown origin, from serum amyloid A (SAA), one of the acute phase proteins that increases many-fold in inflammation. The polypeptides are organized as a ***beta-pleated sheet*** making the material rather inert and insoluble. Minor protein components are also found. Should be distinguished from β-amyloid deposited in the brain and that is derived from ***amyloid precursor protein*** (see ***amyloidogenic glycoprotein***).

amyloid precursor protein (APP) Individuals with ***Alzheimer's disease*** are characterized by extensive accumulation of amyloid in the brain, referred to as ***senile plaques***. These consist of a core of amyloid fibrils surrounded by dystrophic neurites. The principal component of the amyloid fibrils is B/A4, a peptide derived from the larger APP. The specific role of amyloid protein is unclear but it is thought that amyloid deposits may cause neurons to degenerate. Amyloid deposits also occur in brains of older ***Down's syndrome*** patients.

amyloidogenic glycoprotein (A4 protein) An integral membrane ***glycoprotein*** of the brain, and related to the *Drosophila vnd*-gene product. A precursor of β-amyloid, which accumulates in ***Alhzeimer's disease*** and ***Down's syndrome***. See ***amyloid precursor protein***.

amyloidosis Deposition of ***amyloid***. A common complication of several diseases (leprosy, tuberculosis); often associated with perturbation of the immune system, although there may be immunosuppression or enhancement.

amylopectin Component of ***starch*** in which glucose chain is α-1,4 linked (α-1,6 at branch points).

amyloplast A plant ***plastid*** involved in the synthesis and storage of starch. Found in many cell types, but particularly in storage tissues. Characteristically has starch grains in the plastid ***stroma***.

amylose A linear polysaccharide formed from α-D-glucopyranosyl units in α-1,4 linkage. Found both in starch (starch amylose) and glycogen (glycogen amylose).

amyotrophic lateral sclerosis (Lou Gehrig's disease; motor neuron disease) Progressive degenerative disease of motor neurons in the brain stem and spinal cord that leads to weakening of the voluntary muscles.

amytal (amylobarbitone) A barbiturate that inhibits respiration.

Anabaena A genus of ***Cyanobacteria*** that forms filamentous colonies with specialized cells (***heterocysts***), capable of nitrogen fixation. Ecologically important in wet tropical soils and forms symbiotic associations with the fern *Azolla*.

anabolic Of a process, route or reaction. ***Metabolic*** pathways are classically divided into anabolic and ***catabolic*** types. The former are synthetic processes, frequently requiring expenditure of phosphorylating ability of ***ATP***, and reductive steps; the latter, degradative processes, often oxidative, with attendant regeneration of ATP.

anabolism Synthesis; opposite of ***catabolism***.

anaemia (USA, anemia) Reduced level of ***haemoglobin*** in blood for any of a variety of reasons including abnormalities of mature red cells (***sickle cell anaemia***, ***spherocytosis***), iron deficiency, haemolysis of erythrocytes, reduced ***erythropoiesis***, haemorrhage (to name the most common).

anaerobic The absence of air (specifically of free oxygen). Used to describe a biological habitat or an organism that has very low tolerance for oxygen.

anaerobic respiration Metabolic processes in which organic compounds are broken down to release energy in the absence of oxygen. Requires inorganic oxidizing agents or accumulation of reduced coenzymes.

analgesia A state of insensitivity to pain, even though the subject is fully conscious.

analogous Of genes or gene products, performing a similar role in different organisms. cf. ***homologous***.

anamnestic response Archaic term now replaced by such terms as ***secondary immune response***, ***immunological memory***.

anandamide (arachidonyl ethanolamide) Endogenous agonist for ***cannabinoid receptors***.

anaphase The stage of ***mitosis*** or ***meiosis*** beginning with the separation of sister ***chromatids*** (or homologous ***chromosomes***) followed by their movement towards the poles of the ***spindle***.

anaphylatoxin Originally used of an antigen that reacted with an ***IgE*** antibody thus precipitating reactions of ***anaphylaxis***. Now restricted to defining a property of ***complement*** fragments C3a and C5a, both of which bind to the surfaces of ***mast cells*** and ***basophils*** and cause the release of inflammatory mediators.

anaphylaxis As opposed to ***prophylaxis***. A system or treatment that leads to damaging effects on the organism. Now reserved for those inflammatory reactions resulting from combination of a soluble antigen with ***IgE*** bound to a ***mast cell*** that leads to degranulation of the mast cell and release of ***histamine*** and histamine-like substances, causing localized or global immune reponses. See ***hypersensitivity***.

anaplasia Lack of differentiation, characteristic of some tumour cells.

anaplerotic Reactions that replenish ***tricarboxylic cycle*** intermediates and allow respiration to continue; eg. carboxylation of ***phosphoenolpyruvate*** in plants.

Anas platyrhynchos Mallard – from which the domestic duck is derived by breeding (traditional genetic engineering).

anastomosis Joining of two or more cell processes or multicellular tubules to form a branching system. Anastomosis of blood vessels allows alternative routes for blood flow.

ANCA (anti-neutrophil cytoplasmic antibodies) ANCA-positivity is seen in patients with a variety of inflammatory disorders including IBD, Wegener's granulomatosis and hepatobiliary disorders. Two forms are recognized: peripheral ANCA (p-ANCA) where the antigen seems to reside at the periphery of the nucleus; and cytoplasmic ANCA (c-ANCA) where the antigen is distributed throughout the cytoplasm of the neutrophil.

anchorage Attachment, not necessarily adhesive in character; because the mechanism is not assumed the term ought to be more widely used.

anchorage dependence The necessity for attachment (and spreading) in order that a cell will grow and divide in culture. Loss of anchorage dependence seems to be associated with greater independence from external growth control and is probably one of the best correlates of ***tumorigenic*** events ***in vivo***. Anchorage independence is usually detected by ***cloning*** cells in soft agarose; only anchorage-independent cells will grow and divide (as they will in suspension).

anchored PCR (anchored polymerase chain reaction) Variety of ***polymerase chain reaction*** in which only enough information is known to make a single primer. A known sequence is thus added to the end of the DNA, perhaps by enzymic addition of a polynucleotide stretch or by ligation of a known piece of DNA. The PCR can then be performed with the gene-specific primer and the anchor primer.

Androctonus mauretanicus mauretanicus Moroccan scorpion. See ***kaliotoxin***.

androgen General term for any male sex hormone in vertebrates.

anemia See ***anaemia***.

anemone toxins Polypeptide toxins (around 5 kD) from sea anemones (anthozoan ***coelenterates***), most of which act on voltage-gated sodium channels. Some, however, block voltage-regulated potassium channels.

anergy Failure of lymphocytes that have been primed to respond to second exposure to the antigen. Consequence is a depression or lack of normal immunological function.

aneuploid Having a chromosome complement that is not an exact multiple of the haploid number. Chromosomes may be present in multiple copies (eg. trisomy) or one of a homologous pair may be missing in a diploid cell.

aneurysm Balloon-like swelling in the wall of an ***artery***.

ANF See ***atrial natriuretic peptide***.

Angel dust See ***phencyclidine***.

Angelman syndrome Syndrome in which there is severe mental retardation and ataxic movement associated with absence of maternal 15q11q13, and the absence of the β3 subunit of ***GABA receptor***-A.

angiogenesis The process of ***vascularization*** of a tissue involving the development of new capillary blood vessels.

angiogenin Polypeptide (14 kD) that induces the proliferation of endothelial cells; one of the components of ***tumour angiogenesis factor***. It has sequence homology with pancreatic ribonuclease, and has ribonucleolytic activity, although the biological relevance of this is unclear.

angiokeratoma See ***Fabry disease***.

angioma A knot of distended blood vessels atypically and irregularly arranged. Most are not tumours but ***haematomas***.

angioplasty (PTCA) Surgical distension of an occluded blood vessel. Percutaneous transluminal coronary angioplasty (PTCA) is commonly used as a method for restoring patency to occluded coronary arteries (the cause of angina); a catheter is passed from a vessel in the arm through to the coronary vessels and a balloon at the end of the catheter is then inflated to dilate the vessel.

angiopoietin Angiopoetin-1 (498 residues) is the ligand for ***Tie2***; angiopoietin-2 (496 residues) is a natural antagonist. Angiopoietin-1, but not Ang-2, is chemotactic for endothelial cells: neither have effects on proliferation.

Angiospermae (angiosperm) A large class of flowering plants that bear seeds in a closed fruit.

angiostatin Inhibitor of angiogenesis.

angiotensin A peptide ***hormone***. Angiotensinogen (renin substrate) is a 60 kD polypeptide released from the liver and cleaved in the circulation by ***renin*** to form the biologically inactive decapeptide angiotensin I. This is in turn cleaved to form active angiotensin II by angiotensin converting enzyme (ACE). Angiotensin II causes contraction of vascular smooth muscle, and thus raises blood pressure, and stimulates ***aldosterone*** release from the adrenal glands. Angiotensin is finally broken down by angiotensinases.

angiotensin converting enzyme (ACE) See ***angiotensin***.

angiotensinase See ***angiotensin***.

angiotensinogen See ***angiotensin***.

Ångström unit Small unit of measurement (10^{-10}m) named after Swedish physicist and astronomer. Much used as a unit in early electron microscopy though since it is not in the approved mks system should probably be avoided (but nanometres, which are 10 times larger, are sometimes less convenient).

animal pole In most animal ***oocytes*** the nucleus is not centrally placed and its position can be used to define two poles. That nearest to the nucleus is the animal pole, and the other is the ***vegetal pole***, with the animal–vegetal axis between the poles passing through the nucleus. During ***meiosis*** of the ***oocyte*** the ***polar bodies*** are expelled at the animal pole. In many eggs there is also a graded distribution of substances along this axis, with pigment granules often concentrated in the animal half, and yolk (where present) largely in the vegetal half.

animalized cells The 8–16 cell early blastula of sea urchins has ***animal*** and ***vegetal poles***; by manipulating the environmental conditions it is possible to shift more cells from vegetal to animal in their characteristics.

anion exchanger Family of integral membrane proteins that perform the exchange of chloride and bicarbonate across the plasma membrane. Best known is ***band III*** of the red blood cell.

anionic detergents Detergents in which the hydrophilic function is fulfilled by an anionic grouping. ***Fatty acids*** are the best-known natural products in this class, but it is doubtful if they have a specific detergent function in any biological system. The important synthetic species are aliphatic sulphate esters, eg. sodium dodecyl sulphate (***SDS*** or SLS).

anisogamy Mode of sexual reproduction in which the two gametes are of different sizes.

anisotropic Not the same in all directions.

ANK repeat Amino acid motif found in diverse proteins including ***ankyrin*** (hence the name), the ***Notch*** product, transcriptional regulators, cell cycle regulatory proteins and a toxin produced by the black widow spider. The motif is about 33 amino acids long and is generally found as a tandem array of 2–7 repeats, though ankyrins contain 24 repeats. Their role is not established, but they may be involved in protein–protein binding.

ankylosing spondylitis Poly***arthritis*** involving spine, which may become more-or-less rigid. Interestingly the disease seems to be associated with HLA-B27: those with this ***histocompatibility antigen*** are 300 times more likely to get the disease; 90% of sufferers have HLA-B27

ankylosis Fusion of bones across a joint. Complication of chronic ***inflammation***. See ***ankylosing spondylitis***.

ankyrin Globular protein (200 kD) that links ***spectrin*** and an integral membrane protein (***Band III***) in the erythrocyte plasma membrane. Isoforms exist in other cell types.

ankyrin repeat See ***ANK repeat***.

anlage Region of the embryo from which a specific organ develops.

annealing (1) Toughening upon slow cooling. (2) Used in the context of DNA renaturation after temperature dissociation of the two strands. Rate of annealing is a function of complementarity. (3) Fusion of microtubules or microfilaments end to end.

Annelida Phylum of segmented (metameric) coelomate worms. Common earthworm (*Lumbricus terrestris*) is a familiar example of the phylum.

annexins Group of calcium-binding proteins that interact with acidic membrane

Table A3. Annexins

Name	Synonyms	Repeats
Annexin 1	Lipocortin 1, calpactin 2, p35, chromobindin 9	4
Annexin 2	Lipocortin 2, calpactin 1, protein I, p36, chromobindin 8	4
Annexin 3	Lipocortin 3, PAP-III	4
Annexin 4	Lipocortin 4, endonexin I, protein II, chromobindin 4	4
Annexin 5	Lipocortin 5, endonexin 2, VAC-a, anchorin CII, PAP-I	4
Annexin 6	Lipocortin 6, protein III, chromobindin 20, p68, p70	8
Annexin 7	Synexin	4
Annexin 8	Vascular anticoagulant b, VAC-b	4

phospholipids in membranes. They contain 4 or 8 repeats of a 61-amino acid domain that folds into five α-helices. Also known by several other names (eg. lipocortins, endonexins), reflecting the history of their discovery in different contexts. See Table A3, ***lipocortin***, ***endonexin*** I and II, ***calpactin***, ***p70***, and ***calelectrin***.

annulate lamellae Organelles described in the oocytes of several animal species. Associated with the nuclear envelope; may be associated with tubulin synthesis from mRNA accumulated in these organelles.

annulus (*adj*., annulate) Ring-like structure.

anoikis ***Apoptosis*** in normal epithelial and endothelial cells. Process is important in the regulation of cell number in skin.

anomers The α-and β-forms of hexoses. Interconversion (***mutarotation***) is anomerization and is promoted by mutarotases (aldose epimerases).

Anopheles Genus of mosquitos (order ***Diptera***) that carry the ***Plasmodium*** parasites which causes malaria.

anosmia Condition of being unable to smell. Can be transiently induced by osmic acid.

anoxia Total lack of oxygen, cf. ***hypoxia***.

ANP See ***atrial natriuretic peptide***.

ANP receptor Family of three receptors for ***atrial natriuretic peptide***. ANP-A and ANP-B have intracellular ***guanylate cyclase*** and ***protein kinase***-like domains. ANP-C shares the extracellular ligand-binding and transmembrane domains, but lacks the functional intracellular domains, and is not thought to be involved in ***signal transduction***.

antagonist Compound that inhibits the effect of a hormone or drug; the opposite of ***agonist***.

antennal complex Light-harvesting complexes (LHC) of protein and pigment in or on photosynthetic membranes of bacteria are organized into arrays, called antennae. They transfer photon energy to reaction centres.

antennapedia *(antp)* ***Homeotic gene*** of ***Drosophila***, controlling thoracic/head fate determination. In addition to the ***homeobox***, there is also a 6-amino acid *antennapedia*-specific consensus, shared by a range of homeotic genes in human, mouse, chicken, *Xenopus*, newt, zebrafish and *Caenorhabditis*.

anterograde transport Movement of material from the cell body of a ***neuron*** into axons and dendrites (retrograde axoplasmic transport also occurs).

anthocyanins Red plant pigments not directly involved in ***photosynthesis***. Can mask the green of ***chlorophyll***, and give the plant a red-purple colour.

Anthopleura Genus of sea anemone (an anthozoan ***coelenterate***). See ***anthopleurins***.

anthopleurins Peptide toxins (Anthopleurin A, B and C, 49, 50, 47 residues) from sea anemone, *Anthopleura*. Affect sodium channel of nerve and muscle, increase the duration of the action potential.

anthrax Highly contagious disease of man and domestic animals caused by *Bacillus anthracis*. Onset is rapid and disease often fatal. A variety of ***anthrax toxins*** are known.

anthrax toxins (1) (Anthrax oedema factor) Multi-subunit toxin produced by *Bacillus anthracis*. Active subunit is a calmodulin-dependent adenylyl cyclase. (2) Anthrax lethal toxin (LeTx) Toxin responsible for the shock-like effects of infection with *Bacillus anthracis*. Two subunits, protective antigen (PA, 83 kD) and lethal factor (LF, 90 kD). PA binds to the target cell, is proteolytically cleaved to a 63 kD subunit that then oligomerizes to form a ring that allows LF access to the cytoplasm. LF acts on macrophages to induce a massive oxidative burst and also the release of IL-1β and TNFα followed by lysis of the cell.

anti-idiotype antibody An antibody directed against the antigen-specific part of the sequence of an antibody or T-cell receptor. In principle an anti-idiotype antibody should inhibit a specific immune response.

anti-oncogene See ***tumour suppressor*** gene.

antibiotic Substance produced by one microorganism that selectively inhibits the growth of another. Many wholly synthetic antibiotics have been produced; see also Table A4.

antibiotic resistance gene Gene that encodes an enzyme that degrades or excretes an antibiotic, so conferring resistance. Frequently found in cloning vectors like plasmids, and sometimes in natural populations of bacteria. Example: bacterial ampicillin resistance is conferred by expression of the β-lactamase gene.

antibody General term for an ***immunoglobulin***.

antibody-dependent cell-mediated cytotoxicity (ADCC) Killing of target cells by lymphocytes or other leucocytes that carry antibody specific for the target cell, attached to their ***Fc receptors***. The cell involved in the killing may be a passive carrier of the antibody.

antibody-induced lysis See ***complement*** lysis, also see ***natural killer cells***. The term is imprecise and should not be used since there is confusion as to which mechanism is involved, ie. natural killing or complement-lysis.

antibody-producing cell A ***lymphocyte*** of the B series synthesizing and releasing ***immunoglobulin***. Equivalent to plasmacyte and ***plasma cell***.

anticoagulant Substance that inhibits the clotting of blood. The most commonly used are EDTA and citrate (both of which work by chelating calcium) and ***heparin*** (that interferes with ***thrombin***, probably by potentiating ***antithrombins***). Other compounds such as warfarin and dicoumarol act as anticoagulants ***in vivo*** by interfering with clotting factors.

anticodon Nucleotide triplet on ***transfer RNA*** that is complementary to the ***codon*** of the ***messenger RNA***.

antidiuretic hormone (ADH) See ***vasopressin***.

antidromic Running in the opposite direction along an axis. Thus an antidromic signal in a neuron would travel from the presynaptic region back towards the cell body, the opposite of normal.

antigen A substance inducing and reacting in an ***immune response***. Normally antigens have molecular weights greater than about 1 kD. The ***antigenic determinant*** group is termed an ***epitope*** and the association of this with a carrier molecule (which may be part of the same molecule) makes it active as an antigen. Thus dinitrophenol-modified human serum albumin is antigenic to humans, dinitrophenol being the hapten. Usually antigens are foreign to the animal in which they produce immune reactions.

antigen presentation See ***antigen-presenting cell***.

Table A4. Mode of action of various antibiotics

Antibiotic	Source organism	Mode of action
Penicillins		
Ampicillin	Semi-synthetic	Inhibitors of cell wall synthesis.
Benzylpenicillin	*Penicillium notatum*	Penicillins kill growing bacteria probably by binding to active site of the enzyme that crosslinks the peptidoglycan wall. Most effective against Gramo-positive bacteria. Resistant strains of bacteria produce penicillinases (β-lactamase) that cleave the 4-membered β-lactam ring.
Phenoxymethyl-penicillin		
Cephalosporins		
Cephaloridine	*Cephalosporium acremonium* (fungi)	Inhibitors of cell wall synthesis. Similar mode of action to penicillins
Cephalothin	*Cephalosporium* species	
Aminoglycosides		
Gentamicin	*Micromonospora purpurea*	Inhibitors of bacterial protein synthesis.
Kanamycin	*Streptomyces kanamyceticus*	Bind to the 30S subunit of the 70S (bacterial) ribosome, though at a number of different sites. They prevent the transition from an initiating complex to - a chain-elongating ribosome
Neomycin	*Streptomyces fradiae*	
Streptomycin	*Streptomyces griseus*	
Macrolides		
Erythromycin	*Streptomyces erythraeus*	Inhibit bacterial protein synthesis by binding to 50S subunit preventing the translocation step
Oleandomycin	*Streptomyces antibioticus*	
Cycloheximide	*Streptomyces noursei*	Inhibits eukaryote, but not prokaryote, protein synthesis by preventing the peptidyl transferase reaction
Peptides		
Bacitracin	*Bacillus subtilis*	Inhibitor of cell wall synthesis
Gramicidin	*Bacillus brevis*	Ionophore. Forms 'pore' in cell membrane, causing loss of K^+
Polymyxin	*Bacillus polymyxa*	Acts on cell membrane causing leakage of small molecules
Polyenes		
Amphotericin B	*Streptomyces nodosus*	Form complex with cholesterol in the plasma membrane. These complexes form a ring in each half bilayer, producing a 'pore'. In the case of amphotericin B, eight complexes give a pore of diameter 8Å, large enough to leak glucose. Only act on membranes containing cholesterol, so have no effect on bacteria, but kill eukaryotes including the fungi
Nystatin	*Streptomyces noursei*	
Tetracyclines		
Chlortetracycline	*Streptomyces aureofaciens*	Inhibit bacterial protein synthesis by preventing aminoacyl tRNA binding to the A site of the 30S ribosomal subunit
Tetracycline	*Streptomyces aureofaciens* (mutant)	
Others		
Actinomycin D	*Streptomyces parvullus*	Inhibits RNA synthesis by binding to DNA blocking movement of RNA polymerase
Chloramphenicol	*Streptomyces venezuelae*	Inhibits prokaryote, but not eukaryote, protein synthesis by preventing the peptidyl transferase reaction
Puromycin	*Streptomyces albo-niger*	Inhibits protein synthesis. Analogue of 3′ end of aminoacyl tRNA. Is added to growing end of peptide chain but as it has no aminoacyl group causes premature chain termination
Rifamycin	*Streptomyces mediterranei*	Inhibits bacterial RNA synthesis. Binds to RNA polymerase and prevents initiation of transcription. Effective against acid-fast as well as Gram positive bacteria
Valinomycin	*Streptomyces fulvissimus*	Ionophore. Causes leakage of K^+

antigen-presenting cell A cell that carries on its surface antigen bound to MHC Class I or Class II molecules, and presents the antigen in this 'context' to T-cells. Includes macrophages, endothelium, dendritic cells and ***Langerhans cells*** of the skin. See also ***MHC restriction, histocompatibility*** antigens.

antigen processing Modification of an antigen by ***accessory cells***. This usually involves endocytosis of the antigen and either minimal cleavage or unfolding. The processed antigen is then presented in modified form by the accessory cell.

antigen shift Abrupt change in surface antigens expressed by a species or variety of organisms. Usually seen in microorganisms where the change may allow escape from immune recognition. Antigenic drift is a more gradual change. See ***antigenic variation***.

antigenic determinant (epitope) That part of an antigenic molecule against which a particular immune response is directed, eg. a tetra- to pentapeptide sequence in a protein; a tri- to pentaglycoside sequence in a polysaccharide. See also ***hapten***. In the animal most ***antigens*** will present several or even many antigenic determinants simultaneously.

antigenic variation The phenomenon of changes in surface ***antigens*** in parasitic populations of ***Trypanosoma*** and ***Plasmodium*** (and some other parasitic protozoa) in order to escape immunological defence mechanisms. At least 100 different surface proteins have been found to appear and disappear during antigenic variation in a clone of trypanosomes. Each antigen is encoded in a separate gene. Antigenic variation is also known to occur in free-living protozoa and certain bacteria.

antilymphocyte serum ***Immunoglobulins*** raised ***xenogeneically*** against lymphocyte populations. Referring particularly to ***antisera*** recognizing one or more ***antigenic determinants*** on ***T-cell*** populations. Of use in experimental ***immunosuppression***.

antimitotic drugs Drugs that block mitosis; the term is often used of those which cause metaphase arrest such as ***colchicine*** and the ***vinca alkaloids***. Many anti-tumour drugs are antimitotic, blocking proliferation rather than being cytotoxic.

antimycin Inhibitor of QH2 cytochrome C-reductase.

antineoplaston A naturally occurring cytodifferentiating agent that has been tested for anti-tumour activity and used to induce differentiation of astrocytes in rat models of neurodegenerative disease.

antioxidant Any substance that inhibits oxidation, usually because it is preferentially oxidized itself. Common examples are vitamin E (***α-tocopherol***) and vitamin C. Important for trapping free radicals generated during the ***metabolic burst*** and possibly for inhibiting ageing.

antiparallel Having the opposite ***polarity*** (eg. the two strands of a DNA molecule).

antiplasmin (α-2-antiplasmin) Plasma protein (65 kD) that inhibits plasmin (and factors XIa, XIIa, plasma kallikrein, thrombin and trypsin) and therefore acts to regulate fibrinolysis.

antiplectic Pattern of ***metachronal*** coordination of the beating of ***cilia***, in which the waves pass in the opposite direction to that of the active stroke.

antipodal cells Three cells of the embryo sac in angiosperms, found at the end of the embryo away from the point of entry of the pollen tube.

antiport (exchanger) A membrane protein that transports two different ions or molecules in opposite directions across a lipid bilayer. Energy may be required, as in the ***sodium*** pump; or it may not, as in ***Na^+/H^+ antiport***.

antiproteases (antiproteinases) Substances that inhibit proteolytic enzymes.

antiproteinases See ***antiproteases***.

Antirrhinum majus A flowering plant (the common snapdragon), widely used as a model system for plant molecular genetics.

antisense In general the complementary strand of a coding sequence of DNA or of mRNA. Antisense RNA hybridizes with and inactivates mRNA.

antisepsis Processes, procedures or chemical treatments that kill or inhibit microorganisms in contrast to ***asepsis*** where microorganisms are excluded.

antiserum (plural, antisera) Serum containing ***immunoglobulins*** against specified ***antigens***.

antitermination During transcription, failure of an ***RNA polymerase*** to recognize a termination signal: can be of significance in regulation of gene expression.

antithrombins Plasma ***glycoproteins*** of the α-2-globulin class that inhibit the proteolytic activity of ***thrombin*** and serve to regulate the process of blood clotting. See also Table F1.

antitoxin An ***antibody*** reacting with a toxin, eg. anti-***cholera toxin*** antibody.

antitrypsin (α-1-antitrypsin) Better named α-1-antiprotease (α-1-protease inhibitor). A (54 kD) major protein of blood ***plasma*** (3mg/ml in human), part of the α-globulin fraction, and able to inhibit a wide spectrum of ***serine proteases***.

antizyme Repressor of ornithine decarboxylase. Antizyme (29 kD) is a polyamine-inducible protein involved in feedback regulation of cellular polyamine levels. The N terminus of antizyme is not required for the interaction with ODC but is necessary to induce its degradation. Antizyme can be induced by IL-1. The elaborate regulation of ODC activity in mammals still lacks a defined developmental role but an antizyme-like gene in *Drosophila, gutfeeling* (*guf*), is required for proper development of the embryonic peripheral nervous system.

antp See ***antennapedia***.

anucleate Having no nucleus.

anucleolate Literally, having no nucleoli. An anucleolate mutant of *Xenopus* (viable when ***heterozygous***) is used in nuclear transplantation experiments because nuclei are of identifiable origin.

Anura Class of amphibians; the frogs and toads.

anxiolytic Drug that reduces anxiety, eg. benzodiazepines and barbiturates.

AP-1 (1) A ***transcription factor***, formed from a heterodimer of the products of the ***protooncogenes fos*** and ***jun***. Binds the palindromic DNA sequence TGACTCA. See also Table O1. (2) Adaptor protein found in the *trans*-Golgi network that links membrane proteins to clathrin (see ***AP-2***).

AP-2 (1) *Cis*-acting transcription activator. (2) One of the multimeric adaptor proteins (APs; ca 270 kD) found in clathrin-associated complexes. AP-2 is found at the plasma membrane and may bind preferentially to the cytoplasmic tail of the EGF receptor. Also associates with the EGF-R tyrosine kinase substrate eps15. See ***adaptins***.

AP-3, AP-4, AP-5 (amino-3-phosphonopropanoate,2-amino-4-phosphonobutanoate, 2-amino-5-phosphonopentanoic acid) Selective antagonists for ***NMDA receptors***.

APA (amino pimelic acid) Low-affinity rapidly dissociating competitive antagonist of ***NMDA receptors***.

APAF-1 (apoptosis protease activating factor 1) Protein that binds to cytochrome c that has been released from mitochondria and links with ***caspase***-9 which then activates caspase-3, initiating a cascadeof events that end in apoptotic death of the cell.

apamin A small (2027 D) basic peptide present in the venom of the honeybee (*Apis mellifera*). Blocks calcium-activated potassium channels and has an inhibitory action in the central nervous sytem.

APC (1) ***Antigen presenting cell***. (2) adenomatous ***polyposis coli***.

aphidicolin (1) Inhibitor of eukaryotic DNA polymerases. (2) tetracyclic diterpenoid from *Cephalosporium.*

apical dominance Growth-inhibiting effect exerted by actively growing apical bud of higher-plant shoots, preventing the growth of buds further down the shoot. Thought to be mediated by the basipetal movement of ***auxin*** from the apical bud.

apical meristem (primary meristem) The meristem at the tips of stems and roots. Composed of undifferentiated cells, many of which divide to add to the plant body but the central mass (the quiescent centre) remains inert and only becomes active if the meristem is damaged. Also known as eumeristem.

apical plasma membrane The term used for the cell membrane on the apical (inner or upper) surface of transporting epithelial cells. This region of the cell membrane is separated, in vertebrates, from the basolateral membrane by a ring of ***tight junctions*** that prevents free mixing of membrane proteins from these two domains.

apidaecins Proline-rich basic antibacterial peptides (2 kD) found in the immune ***haemolymph*** of the honeybee.

Apis mellifera Common honeybee, a Hymenopteran.

aplasia Defective development of an organ or tissue so that it is totally or partially absent from the body.

aplastic anaemia ***Anaemia*** due to loss of most or all of the ***haematopoietic*** bone marrow. Usually all haematopoietic cells are equally diminished in number.

Aplysia (sea hare) Opisthobranch mollusc with reduced shell; favourite source of ganglia for neurophysiological study.

apoA, etc. (apoA, apoB, apoC, apoE) Plasma apolipoproteins. ApoE is the specific ligand for uptake of lipoprotein by the LDL receptor and different alleles of the *ApoE* gene are associated with variations in plasma cholesterol levels.

apocrine Form of secretion in which the apical portion of the cell is shed, as in the secretion of fat by cells of the mammary gland. The fat droplet is surrounded by apical plasma membrane, and this has been used experimentally as a source of plasma membrane.

apoenzyme An enzyme without its ***cofactor***. See ***apoprotein***.

apolipoprotein The protein component of serum lipoproteins. Small proteins containing multiple copies of the ***kringle*** domain.

apomixis Type of degenerate sexual reproduction in some fungi: meiosis and gamete formation do not occur even though an ***ascus*** containing identical diploid spores is formed.

apoplast Since the ***protoplasts*** of cells in a plant are connected through ***plasmodesmata***, plants may be described as having two major compartments: the apoplast, which is external to the plasma membrane and includes cell walls, xylem vessels, etc., through which water and solutes passes freely; and the ***symplast***, the total cytoplasmic compartment.

apoprotein When a protein can exist as a complex between polypeptide and a second moiety of non-polypeptide nature, the term apoprotein is sometimes used to refer to the molecule divested of the latter. For example, ***ferritin*** lacking its ferric hydroxide core may be referred to as apoferritin.

apoptosis The most common form of physiological (as opposed to pathological) cell death. Apoptosis is an active process requiring metabolic activity by the dying cell; often characterized by shrinkage of the cell, cleavage of the DNA into fragments that give a so-called 'laddering pattern' on gels and by condensation and margination of chromatin. Often called ***programmed cell death***, though this is not strictly accurate. Cells that die by apoptosis do not usually elicit the inflammatory responses that are associated with necrosis, though the reasons are not clear. See also: ***ced mutant***, ***bcl-2***.

apoptosis protease-activating factor 1 See *APAF-1*.

APP See ***amyloid precursor protein***.

App(NH) p ((β, γ-imido) ATP) Non-hydrolysable analogue of ATP.

applagin See ***disintegrin***.

apple domain A consensus sequence, composed of 90 amino acids including six cysteines, that forms a characteristic, vaguely apple-shaped pattern via disulphide bridges. Shared by plasma kallikrein and coagulation factor XI, both ***serine proteases***.

aprotinin Basic polypeptide that inhibits several serine proteases (including trypsin, chymotrypsin, kallikrein, pepsin).

aptamers Double-stranded DNA or single-stranded RNA molecules that bind to specific molecular targets.

APUD cells Acronym for amine-precursor uptake and decarboxylation cells: ***paracrine*** cells of which ***argentaffin cells*** are an example. Usage neither helpful nor memorable.

apurinic sites Sites in DNA from which purines have been lost by cleavage of the deoxy-ribose N-glycosidic linkage.

apyrase Enzyme (EC 3.6.1.5) that catalyses breakdown of ATP to AMP; usually extracted from plants, but aortic and placental forms have also been described.

aquaporin (AQP; CHIP28) Integral membrane protein (28 kD) with six transmembrane domains that greatly increases water permeability. Found especially in kidney, red blood cells. AQP1 forms a homotetramer of four independent channels. *Arabidopsis* has at least 20 distinct AQP genes. Members of the ***major intrinsic protein*** family.

ara* C** In bacteria, the arabinose ***ara operon regulatory protein. One of a large group of bacterial ***transcription*** factors with the ***helix-turn-helix*** motif.

ara operon Operons involved in arabinose metabolism, especially the *araBAD* operon of *E. coli*. Two other *ara* operons are known in *E. coli*.

Arabidopsis thaliana The common wall cress. A small plant, adopted as a model system for plant molecular biology, because of its small genome (7×10^7 bp), and short generation time (5–8 weeks).

arabinogalactans Plant cell-wall polysaccharides containing predominantly arabinose and galactose. Two main types are recognized: arabinogalactan 1, found in the pectin portion of angiosperms and containing α-(1-4)-linked galactan and α-arabinose side chains; arabinogalactan II, a highly branched polymer containing β (1-3)- and β (1-6)-linked galactose and peripheral α arabinose residues. Arabinogalactan II is found in large amounts on some gymnosperms, especially larches, and is related to ***AGP***.

arabinoglycan-protein See ***AGP***.

arabinose A ***pentose*** monosaccharide that occurs in both D- and L-configurations. D-arabinose is the 2-epimer of ***D-ribose***, ie. differs from D-ribose by having the opposite configuration at carbon 2. D-arabinose occurs *inter alia* in the polysaccharide arabinogalactan, a neutral ***pectin*** of the cell wall of plants, and in the metabolites cytosine arabinoside and adenine arabinoside.

arabinoxylan Polysaccharide with a backbone of ***xylose*** (β-1,4 linked) with side chains of ***arabinose*** (α-1,3 linked): constituent of ***hemicellulose*** of angiosperm cell wall.

arachidonic acid (5,8,11,14 eicosatetraenoic acid) An essential dietary component for mammals. The free acid is the precursor for biosynthesis of the signalling molecules ***prostaglandins***, ***thromboxanes***, hydroxyeicosatetraenoic acid derivatives including ***leukotrienes*** and is thus of great biological significance. Within cells the acid is found in the esterified form as a major acyl component of membrane ***phospholipids*** (especially ***phosphatidyl inositol***) and its release from phospholipids is thought to be the limiting step in the formation of its active metabolites.

arboviruses Diverse group of single-stranded RNA viruses that have an envelope surrounding the ***capsid***. Arthropod borne, hence the name, and multiply in both invertebrate and vertebrate host, causing eg. yellow fever and encephalitis. The group is very heterogeneous and three major families are recognized: ***Togaviridae***, ***Bunyaviridae***, and ***Arenaviridae***.

Archaea Alternative name suggested for the ***Archaebacteria*** to emphasize the difference of this subkingdom from the Eubacteria.

Archaebacteria One of two major subdivisions of the prokaryotes. There are three main orders: extreme ***halophiles***, ***methanobacteria***, and sulphur-dependent extreme- ***thermophiles***. Archaebacteria differ from ***Eubacteria*** in ribosomal structure, the possession (in some cases) of ***introns***, and in a number of other features including membrane composition.

archaeocyte An amoeboid cell type of sponges (Porifera).

archegonium Female sex organ of liverworts, mosses, ferns and most gymnosperms.

ARD-1 Bifunctional protein (64 kD) that has 18 kD GTP-binding ADP-ribosylation factor (ARF) domain and 46 kD GTPase activating (GAP) domain.

ARE ***Activin***-response element to which ***ARF*** binds.

Arenaviridae Family of ssRNA viruses including Lassa virus, lymphocytic choriomeningitis virus, and the Tacaribe group of viruses; not all require arthropods for transmission, despite their inclusion in the ***arbovirus*** group.

areolar connective tissue Loose connective tissue of the sort found around many organs in vertebrates. Does not have marked anisotropy or particularly pronounced content of any one matrix protein.

AREs (AU-rich elements) Cytoplasmic mRNA stability is mediated by proteins that bind to AU-rich elements (AREs) in the 3′ untranslated region of transcripts. This has been shown for mRNA encoding oncoproteins, cytokines and transcription factors.

ARF See ***ADP-ribosylation factor***.

arg ***Oncogene***, related to ***abl***, that encodes a ***tyrosine kinase***.

Arg See ***arginine***.

argentaffin cells So-called because they will form cytoplasmic deposits of metallic silver from silver salts. Their characteristic histochemical behaviour arises from 5-HT (***serotonin***), which they secrete. Found chiefly in the epithelium of the gastrointestinal tract (though possibly of neural crest origin) their function is rather obscure, although there is a widely distributed family of such ***paracrine*** (local endocrine) cells (***APUD cells***).

argentation chromatography Modified form of standard thin layer chromatography in which the solid phase includes silver salts. Used for lipid analysis.

arginine (Arg; R) An essential amino acid (174 D); a major component of proteins and contains the guanido group that has a pKa of greater than 12, so that it carries a permanent positive charge at physiological pH. See Table A2.

argyrophil cells Neuroendocrine cells that take up silver ions from a staining solution but require the addition of a reducing agent to precipitate metallic silver (unlike ***argentaffin cells*** which do not). Carcinoids of the foregut tend to be argyrophilic whereas those of the lower intestine tend to be argentaffinic.

ARIA Polypeptide purified from chick brain that promotes the accumulation of ***acetylcholine receptors*** in chick ***myotubes***.

armadillo (β-catenin; beta-catenin; arm) ***Drosophila*** gene encoding ***β-catenin***, a component of ***adherens junctions***. Links junctional complex to the cytoskeleton.

armadillo repeat Protein motif comprising 42 amino acids, originally described in the *Drosophila* ***armadillo*** protein. Usually found in multiple repeats that form a superhelix of helices with a positively charged groove. Mediates interactions with proteins such as ***cadherins***, Tcf-family transcription factors, and the tumour suppressor gene ***adenomatous polyposis coli***.

ARNO (ARF nucleotide-binding site opener) Guanine nucleotide exchange factor for ***ARFs*** that will stimulate nucleotide exchange on both ARF1 and ARF6. Closely related to ***cytohesin*** and GRP-1. Exchange-factor activity resides in the sec7-like domain.

aromatase Microsomal enzyme complex that converts testosterone to oestradiol.

ARP-1 (apolipoprotein regulatory protein-1) Nuclear receptor that binds to response element with two core motifs, 5′-RG(G/T) TCA, as do various retinoic acid receptors such as ***COUP-TF*** I, and ***PPAR***.

arrestin Family of inhibitory proteins that bind to tyrosine-phosphorylated receptors, thereby blocking their interaction with G-proteins and effectively terminating the signalling. Arrestin (S antigen; 48 kD, from retinal rods) competes with ***transducin*** for light-activated ***rhodopsin***, thus inhibiting the response to light (adaptation). Immune responses to arrestin lead to autoimmune uveitis. Similarly, β-arrestin binds to phosphorylated ***β-adrenergic receptors***, inhibiting their ability to activate the ***G-protein*** Gs.

Arrhenius plot A plot of the logarithm of reaction rate against the reciprocal of absolute temperature. For a single-stage reaction this gives a straight line from which the activation energy and the frequency factor can be determined. Often applied to data from complex biological systems when the form observed is frequently a series of linear portions with sudden changes of slope. Great caution must be observed in interpreting such slopes in terms of activation energies for single processes.

arrhythmia Lack of normal ordered rhythm, particularly in the case of the heart where arrhythmia can be a prelude to cardiac arrest.

arrowheads Fanciful description given to the pattern of ***myosin*** molecules attached to a filament of ***F-actin***. Easier to see if tannic acid is added to the fixative. The arrowheads indicate the polarity of the filament; the barbed (attachment) end is the site of major subunit addition.

ARS (autonomously replicating sequence) A DNA sequence originally isolated from *Saccharomyces cerevisiae* that when linked to a non-replicating sequence can confer on the latter the ability to be replicated in a yeast cell. Transformations effected with the use of ARS occur at relatively high frequency but are unstable. Homologous recombination of the DNA of interest with the host cell chromosomes is not required for expression when ARS routes are used.

Artemia salina Brine shrimp, a crustacean of the order Anostraca.

arteriole Finest branch of an artery before capillary bed.

arteriosclerosis Imprecise term for various disorders of arteries, particularly hardening due to fibrosis or calcium deposition; often used as a synonym for ***atherosclerosis***.

artery Blood vessel carrying blood away from the heart; walls have smooth muscle and are innervated by the ***sympathetic nervous system***.

arthritis General term for inflammation of one or more joints. Many diseases may cause arthritis, although in most cases the cause of the inflammation is not understood. This is particularly true of ***rheumatoid arthritis***, though knowledge of other forms is not much better.

Arthrobacter Genus of obligate aerobic bacteria of irregular shape, found extensively in soil.

arthropathy Any disease affecting a joint. Care should be taken not to confuse arthrosclerosis (stiffness of joints) with atherosclerosis.

arthropod The largest phylum of the animal kingdom, containing several million species. Arthropods are characterized by a rigid external skeleton, paired and jointed legs, and a haemocoel. The phylum Arthropoda includes the major classes Insecta, Crustacea, Myriapodia and Arachnida.

Arthus reaction A localized ***inflammation*** due to injection of ***antigen*** into an animal that has a high level of circulating ***antibody*** against that antigen. A haemorrhagic reaction with oedema occurs due to the destruction of small blood vessels by thrombi. It may occur, as in 'farmer's lung' , as a reaction to natural exposure to antigen.

articulins Membrane-associated protein complex of *Euglena;* two isoforms of 80 and 86 kD, completely unlike spectrin, though functionally analogous. Have a core domain of 12-residue repeats, rich in valine and proline. May attach directly to membrane proteins.

Artiodactyla Order of herbivorous even-toed mammals that includes antelope, pig, cow, giraffe and hippopotamus.

aryl sulphatase (EC 3.1.6.1) Aryl sulphatases A, B and C comprise a group of enzymes originally assayed by their ability to hydrolyze O-sulphate esters of aromatic substrates. Aryl sulphatase A, substrate cerebroside 3-sulphate, is deficient in metachromatic leukodystrophy. Aryl sulphatase B, substrate acetylhexosamine 4-sulphate in glycosaminoglycans, is deficient in Maroteaux–Lamy syndrome. Aryl sulphatase C hydrolyzes oestrogen sulphates. All 3 are deficient in multiple sulphatase deficiency.

AS-2 See ***adocia sulphate-2***.

asbestosis ***Fibrosis*** of the lung as a result of the chronic inhalation of asbestos fibres. The needle-like asbestos fibres are phagocytosed by ***alveolar*** macrophages but burst the phagosome (***phagocytic vesicle***) and kill the macrophage and the cycle is repeated. ***Mesothelioma***, a rare tumour of the mesothelial lining of the pleura, is associated with intense chronic exposure to asbestos dust, particularly that of crocidolite asbestos.

Ascaris Genus of nematodes (Aschelminthes). *Ascaris suum* is the common roundworm of pigs; *Ascaris lumbricoides* causes ascariasis in humans; though the worms are restricted to the gut they divert a substantial proportion of food intake and heavy infestation can cause growth retardation in children or in extreme cases may cause intestinal blockage. One of the WHO's six major diseases, and used in developmental studies.

Ascaris lumbricoides A parasitic gut-dwelling nematode worm, of major medical significance.

Aschheim–Zondek test Old pregnancy-testing method that involved injecting specimen of urine into mice: a positive sample will cause swelling of the ovaries.

Aschoff bodies Small ***granulomas*** composed of ***macrophages, lymphocytes*** and multinucleate cells grouped around eosinophilic ***hyaline*** material derived from collagen. Characteristic of the ***myocarditis*** of rheumatic fever.

Ascidiacea Class of simple or compound tunicates that have a motile larva but sedentary adult form that filter-feeds. Sea squirts are the commonly known examples.

ascites Accumulation of fluid in the peritoneal cavity causing swelling; causes include infections, portal hypertension and various tumours.

ascites tumour Tumour that grows in the peritoneal cavity as a suspension of cells. Obviously such cells have lost ***anchorage dependence***, and they can easily be isolated and passaged. ***Hybridomas*** are sometimes grown as ascites tumours, and the ascites fluid can then be used as the crude 'antiserum'.

Ascomycotina Ascomycete fungi that produce spores, usually eight, in a structure known as an ascus. Includes yeasts and *Neurospora*.

ascorbic acid (vitamin C) A requisite in the diet of humans and guinea pigs. May act as a reducing agent in enzymic reactions, particularly those catalysed by ***hydroxylases***. See also Table V1.

ascospore Diploid spore formed by ascomycete fungi, contained within an ***ascus***.

ascus Elongated spore case containing four or eight haploid sexual ascospores of ascomycete fungi (which include most yeasts).

asepsis State in which harmful microorganisms are absent. Aseptic technique aims to avoid contamination of sterile systems.

asexual Reproducing without a sexual process and thus without formation of gametes or reassortment of genetic characters.

ASGP Membrane-associated mucin present on rat mammary carcinoma cells. ASGP-1 and ASGP-2 are generated from a single precursor; ASGP-2 acts as a membrane anchor for ASGP-1. Though ASGP is thought to have similar functions, it has no sequence homology with ***episialin***.

asialoglycoprotein The carbohydrate moiety of many vertebrate glycoproteins bears terminal residues of ***sialic acid***. If such residues are removed, eg. by treatment with a ***neuraminidase***, the resulting proteins are known as asialoglycoproteins. In the case of certain plasma proteins, the asialo-derivatives are specifically bound by a receptor on the surface of liver parenchymal cells (the ***scavenger receptor***).

ASK (*Arabidopsis* SHAGGY-related protein kinase) Gene family, *Arabidopsis* SHAGGY-related protein kinases, have homology to mammalian ***GSK***-3 and *Drosophila* SHAGGY. There are at least 10 ASK genes in the haploid *Arabidopsis* genome.

Askenazy cells Abnormal thyroid epithelial cells found in autoimmune ***thyroiditis***. The cubical cells line small ***acini*** and have ***eosinophilic*** granular cytoplasm and often bizarre nuclear morphology. Also known as Hurthle cells, oxyphil cells or oncocytes.

Asn See ***asparagine***.

Asp See ***aspartate***.

asparaginase Enzyme (EC 3.5.1.1) that hydrolyzes L-asparagine to L-aspartate and ammonia that is used as an antitumour agent especially against lymphosarcoma and lymphatic leukaemia.

asparagine (β-asparagine; Asn; N) The β-amide of aspartic acid (132 D); the L-form is one of the 20 amino acids directly coded in proteins. Coded independently of aspartic acid. See Table A2.

aspartame Trademark for Asp-Phe Methyl Ester, an artificial sweetener.

aspartate (aspartic acid; Asp; D) L-aspartate is one of the 20 amino acids directly coded in proteins (133D); the free amino acid is a neurotransmitter. See Table A2.

aspartic acid See ***aspartate***.

aspartokinase Enzyme that phosphorylates L-aspartate to produce aspartyl phosphate.

aspergillins Family of toxins (17 kD) produced by *Aspergillus*. All are ribonucleases and disrupt protein biosynthesis. Includes ***alpha-sarcin***, mitogillin, restrictocin and Asp fl.

aspergillosis Lung disease caused by fungi of the genus *Aspergillus*.

Aspergillus A genus of common ascomycete fungi found in soil. Industrially important in production of organic acids, and a popular fungus for genetic study (esp. *A. niger*).

aspirin (acetyl salicylate) An analgesic, antipyretic and antinflammatory drug. It is a potent ***cyclooxygenase*** inhibitor and blocks the formation of ***prostaglandins*** from ***arachidonic acid***.

association constant (K_a; K_{ass}) Reciprocal of ***dissociation constant***. A measure of the extent of a reversible association between two molecular species at equilibrium.

astacin Astacin, a zinc-endopeptidase from crayfish (*Astacus*), is the prototype for the astacin family of metallo-endopeptidases. Family includes ***BMP***-1, Meprin A, ***stromelysin*** 1, and ***thermolysin***.

aster Star-shaped cluster of microtubules radiating from the polar ***microtubule organizing centre*** at the start of mitosis.

asthma Inflammatory disease of the airways involving marked eosinophil infiltration and remodelling of the airways. Attacks can be triggered by allergic responses, physical exertion, inhaled chemicals or stress and involve wheezing, breathlessness and coughing.

astroblast An embryonic ***astrocyte***.

astrocyte A ***glial cell*** found in vertebrate brain, named for its characteristic star-like shape. Astrocytes lend both mechanical and metabolic support for neurons, regulating the environment in which they function. See ***oligodendrocytes***.

astrocytoma A neuro-ectodermal tumour (***glioma***) arising from ***astrocytes***. Probably the commonest glioma, it has a tendency to become ***anaplastic***.

astroglia See ***astrocytes***.

astrogliosis Hypertrophy of the ***astroglia***, usually in response to injury.

Astropectinidae Family of echinoderms that includes many starfish species with long spines.

astrotactin Neuronal surface glycoprotein (100–105 kD; three ***EGF-like*** repeat domains, two ***fibronectin*** III repeats), that functions in murine cerebellar granule cell migration ***in vitro***, acting as the ligand for neuron–glial cell binding. Message has been detected in neuronal precursors in the cerebellum, hippocampus, cerebrum and olfactory bulb in the ***brain***. See ***weaver*** mutant.

astroviruses Spherical viruses with 5- or 6-pointed star-shaped surface pattern. May be associated with enteritis in various vertebrates.

ataxia telangiectasia Louis Bar syndrome; a hereditary ***autosomal*** recessive disease in humans characterized by a high frequency of spontaneous chromosomal aberrations, neurological deterioration and susceptibility to various cancers. Ataxia: imbalance of muscle control; telangectasia: dilated capillary vessels. In part an immune deficiency disease and in part one of DNA repair; it is believed to be due to hypersensitivity to background ionizing radiation.

ATCase (aspartate transcarbamylase) Enzyme (EC 2.1.3.2) that catalyses the first step in pyrimidine biosynthesis, condensation of aspartate and carbamyl phosphate. Positively allosterically regulated by ATP and negatively by CTP; classic example of an allosterically regulated enzyme. Bacterial ATCases exist in three forms: class A (ca 450–500 kD), class B (ca 300 kD) and class C (ca 100 kD).

ATCC The American Type Culture Collection, repository of many eukaryotic cell lines (which may be purchased). Comparable collections of microorganisms, protozoa etc. are kept.

atheroma Degeneration of the walls of the arteries because of the deposition of fatty plaques in the ***intima*** of the vessel wall, and scarring and obstruction of the lumen.

atherosclerosis Condition caused by the deposition of lipid in the wall of arteries in (atheromatous plaques). Migration of smooth muscle cells from media to intima, smooth muscle cell proliferation, the formation of ***foam cells*** and extensive deposition of extracellular matrix all contribute to the formation of the lesions that may ultimately occlude the vessel or, following loss of the endothelium, trigger the formation of thrombi. Should be distinguished from ***arteriosclerosis*** which is a more general term usually applied to arterial hardening through other causes. Atherosclerosis is a major medical problem in most of the developed world.

ATM Protein product of the gene mutated in ***ataxia telangiectasia*** (AT), a member of the phosphatidylinositol-3-kinase family. ATM constitutively binds to the SH3 domain of the tyrosine kinase c-***Abl*** in normal but not AT cells and ATM seems to activate DNA damage-induced activation of c-Abl (which is deficient in AT cells).

atomic force microscopy (AFM) A form of ***scanning probe microscopy***, in which a microscopic probe is mechanically tracked over a surface of interest in a series of x-y scans, and the force encountered at each coordinate measured with piezoelectric sensors. This provides information about the chemical nature of a surface at the atomic level.

atopy Allergic (***hypersensitive***) response at a site remote from the stimulus (eg. food-induced dermatitis).

ATP (adenosine 5′ triphosphate) Synthesis in cells from ADP is driven by energy-yielding processes. Enzymic transfer of the terminal phosphate or pyrophosphate to a wide variety of substrates provides a means of transferring chemical free energy from metabolic to catabolic processes.

ATPase An enzyme capable of releasing the terminal (γ) phosphate from ATP, yielding ***ADP*** and inorganic phosphate. The description could mislead, because in most cases the enzymic activity is not a straightforward hydrolysis, but is part of a coupled system for achieving an energy-requiring process, such as ion-pumping or the generation of motility.

ATP-binding site ('A' motif) A consensus domain found in a number of ATP- or GTP-binding proteins, for example ***ATP synthase, myosin heavy chain, helicases,*** thymidine kinase, ***G-protein*** α-subunits, GTP-binding ***elongation factors, Ras*** family. Consensus is: (A or G)-XXXXGK-(S or T); this is thought to form a flexible loop (the P-loop) between α-helical and ***beta-pleated sheet*** domains.

ATP synthase A proton-translocating ***ATPase***, found in the inner membrane of ***mitochondria, chloroplasts*** and the plasmalemma of ***bacteria***. It can be known as the F1/Fo or CF1/CFo ATPase, or as the class of F-type ATPases. In all these cases, the enzyme is driven in reverse by the large ***proton motive force*** generated by the ***electron transport chain***, and thus synthesizes, rather than uses, ***ATP***. See also ***chemiosmosis, V-type ATPase, P-type ATPase***.

atria (plural) See ***atrium***.

atrial natriuretic factor Obsolete name for ***atrial natriuretic peptide***.

atrial natriuretic peptide (ANP) A polypeptide hormone found mainly in the atrium of many species of vertebrates. It is released in response to atrial stretching, and thus to elevated blood pressure. ANP acts to reduce blood pressure through stimulating the rapid excretion of sodium and water in the kidneys (reducing blood volume), by relaxing vascular smooth muscle (causing vasodilation), and through actions on the brain and adrenal glands.

atrium (plural, atria) A cavity in the body, especially either of the two upper chambers of the heart in higher vertebrates.

atrophy Wasting away of tissue.

atropine An alkaloid, isolated from deadly nightshade, *Atropa belladonna*, that inhibits muscarinic acetylcholine receptors. Applied to the eye it causes dilation of the pupil that is said to enhance the beauty of a woman, hence belladonna as the specific name of the plant from which the ancients extracted the drug.

attachment constriction See ***centromere***.

attachment plaques Specialized structures at the ends of a chromosome by which it is attached to the nuclear envelope at ***leptotene*** stage of mitosis.

attacins Antibacterial proteins (20–22 kD) produced by insect haemocytes following bacterial challenge. May be basic or acidic.

attennuation Viruses that have been ***passaged*** extensively may become attenuated (non-virulent), and can be used as a vaccine.

atx A, B & C See ***ammodytoxins***.

AUG The ***codon*** in ***messenger RNA*** that specifies initiation of a polypeptide chain, or within a chain, incorporation of a ***methionine*** residue.

Aurelia aurita Common jellyfish: transparent disc with four blue/purple horseshoe-shaped gonads clearly visible. Phylum Cnidaria; class Scyphozoa.

aurosome Gold-containing secondary lysosome found in patients treated with gold complexes.

aurovertin Inhibitor of the ***respiratory chain*** that binds to ATPase.

Australia antigen An envelope antigen now known as HBsAg of ***hepatitis B*** virus. Appearance of the antigen in serum is associated with a phase of high infectivity.

autacoids Local hormones such as ***histamine, serotonin, angiotensin, eicosanoids***.

autoantibody Antibody that reacts with an antigen that is a normal component of the body. Obviously this can lead to some problems, and autoimmunity has been proposed as a causative factor in a number of diseases such as rheumatoid arthritis. See also ***systemic lupus erythematosus, Hashimoto's thyroiditis, myasthenia gravis***.

autocatalytic A compound that catalyses its own chemical transformation. More commonly a reaction that is catalysed by one of its products or an enzyme-catalysed reaction in which one of the products functions as an enzyme activator.

autochthonous Found in the place where it was originally formed, indigenous.

autocrine Secretion of a substance, such as a ***growth factor***, that stimulates the secretory cell itself. One route to independence of ***growth control*** is by autocrine growth factor production.

autofluorescence Property of a compound or material that will fluoresce in its own right, without the addition of an exogenous fluorophore. A common problem in fluorescence microscopy and in assays where the read-out is fluorescence.

autogamy Self-fertilization, common in plants and also in some ciliate protozoa where gametic nuclei from a single micronucleus subsequently fuse to form the zygote nucleus.

autogenous Generated without external influence or input.

autograft Graft taken from one part of the body and placed in another site on the same individual.

autoimmune Adjective describing a situation in which the immune system responds to normal components of the body. Several diseases are thought to have an autoimmune component, but it is often not clear whether this is causative.

autologous Derived from an organism's tissues or DNA. cf ***heterologous, homologous***.

autolysis Spontaneous ***lysis*** (rupture) of cells or organelles produced by the release of internal hydrolyic enzymes. Normally associated with the release of lysosomal enzymes.

autonomic nervous system ***Neurons*** that are not under conscious control, comprising two antagonistic components, the ***sympathetic*** and ***parasympathetic nervous systems***. Together, they control the heart, viscera, smooth muscle, etc.

autonomously replicating sequence See ***ARS***.

autophagy Removal of cytoplasmic components, particularly membrane bounded organelles, by digesting them within ***secondary lysosomes*** (autophagic vacuoles). Particularly common in embryonic development and senescence.

autophosphorylation Addition of a phosphate to a protein kinase (possibly affecting its activity) by virtue of its own enzymic activity.

autoradiography Technique in which a specimen containing radioactive atoms is overlaid with a photographic emulsion, which is subsequently developed, revealing the localization of radioactivity as a pattern of silver grains. Resolution is determined by the path length of the radiation, and so the low-energy β-emitting isotope, tritium, is usually used.

autoregulation Regulation of a gene encoding a ***transcription factor*** by its own gene product: a feedback process.

autosomal dominant mutation Mutation of a gene located on an ***autosome*** that has a ***phenotype*** even when one normal gene copy is present. Often attributed to a ***gain-of-function***.

recessive mutation Mutation carried on an ***autosome*** that is deleterious only in ***homozygotes***.

autosomes Chromosomes other than the sex chromosomes.

autotroph (lithotroph) Organisms that synthesize all their organic molecules from inorganic materials (carbon dioxide, salts, etc.). May be photo-autotrophs or chemo-autotrophs, depending upon the source of the energy. Also known as lithotrophic organisms.

auxesis Growth by increase in cell size rather than by increasing cell numbers.

auxilin A novel ***adaptin*** found associated with the coated vesicles isolated from brain cells.

auxins A group of ***plant growth substances*** (often called phytohormones or plant hormones), the most common example being ***indole acetic acid*** (IAA), responsible for raising the pH around cells, making the cell wall less rigid and allowing elongation.

auxotroph Mutant that differs from the wild-type in requiring a nutritional supplement for growth. A deficiency mutant.

auxotyping Method for strain-typing *Neisseria* by checking their requirements for specific nutrients in defined media.

AVEC microscopy See ***Allen video-enhanced contrast***.

Avena sativa Cultivated oat.

avermectin B1 (abamectin) Metabolite of *Streptomyces avermitilis* used as an acaricide, insecticide and anthelminthic.

avian erythroblastosis virus See ***avian leukaemia virus***.

avian leukaemia virus Group of C-type RNA tumour viruses (Oncovirinae) that cause various leukaemias and other tumours in birds. The acute leukaemia viruses, which are replication-defective and require helper viruses, include avian erythroblastosis (AEV), myeloblastosis (AMV) and myelocytomatosis viruses. AEV carries two transforming genes, v-erbA and v-erbB; the cellular homologue of the latter is the structural gene for the ***epidermal growth factor*** receptor. AMV carries v-myb and causes a myeloid leukaemia; avian myelocytomatosis virus carries v-myc. The avian lymphatic leukaemia viruses (ALV) are also ***Retroviridae*** but are replication-competent and induce neoplasia only after several months; they often occur in conjunction with replication-defective leukaemia viruses.

avian myeloblastosis virus (AMV) Retrovirus of the subfamily Oncornaviridae. Causes myelocytomatosis, osteopetrosis, lymphoid leukosis and nephroblastoma. May be a mixture of viruses.

avidin ***Biotin***-binding protein (68 kD) from egg-white. Binding is so strong as to be effectively irreversible: a diet of raw egg-white leads to biotin deficiency.

avidity Strength of binding, usually of a small molecule with multiple binding sites by a larger; particularly the binding of a complex antigen by an antibody.

(*Affinity* refers to simple receptor–ligand systems.)

axenic A situation in which only one species is present. Thus an axenic culture is uncontaminated by organisms of other species, an axenic organism does not have commensal organisms in the gut, etc. Some organisms have obligate symbionts and cannot be grown axenically.

axil Member of the ***axin*** family. Interacts with ***GSK***-3 and β-catenin. By enhancing phosphorylation and thus the subsequent degradation of β-catenin, inhibits axis formation in *Xenopus* embryos.

axin A negative regulator of the ***Wnt*** signalling pathway. Binds to APC (adenomatous ***polyposis coli*** protein) and to β-***catenin*** and regulates the stability of the catenin. Interacts directly with glycogen synthase kinase-3 (***GSK***-3) and promotes GSK-3 phosphorylation of β-acatenin, which is then degraded.

axokinin Axonemal protein (56 kD) that, when phosphorylated by a cAMP-dependent protein kinase, reactivates the axoneme.

axolemma ***Plasma membrane*** of an axon.

axon Long process, usually single, of a ***neuron***, that carries efferent (outgoing) ***action potentials*** from the cell body towards target cells. See ***dendrite***.

axon hillock Tapering region between a ***neuron***'s cell body and its axon. This region is responsible for summating the graded inputs from the ***dendrites***, and producing ***action potentials*** if the threshold is exceeded.

axonal guidance General term for mechanisms that ensure correct projections by nerve cells in developing and regenerating nervous systems. Implies accurate navigation by ***growth cones***, the highly motile tips of growing neuronal processes. See ***growth cone collapse***.

axoneme The central ***microtubule*** complex of eukaryotic ***cilia*** and flagella with the characteristic '9 + 2' arrangement of tubules when seen in cross-section.

axonin Chick homologue of TAG-1. See ***tax-1***.

axonogenesis The growth and differentiation of axonal processes by developing neurons. See ***axon***.

axoplasm The ***cytoplasm*** of a ***neuron***.

axopod (plural, axopodia) Thin processes (a few μm in diameter but up to 500μm long), supported by complex arrays of ***microtubules*** that radiate from the bodies of *Heliozoa*.

axopodia (plural) See ***axopod***.

axostyles Ribbon-like bundles of ***microtubules*** found in certain parasitic protozoa that may generate bending waves by ***dynein***-mediated sliding of microtubules.

azacytidine (5-azacytidine; β-ribofuranosyl 5-azacytidine) The ribonucleoside of **5-*azacytosine***.

azacytosine (5-azacytosine) An analogue of the ***pyrimidine*** base cytosine, in which carbon 5 is replaced by a nitrogen. In DNA, unlike cytosine, it cannot be methylated.

azaserine An analogue of glutamine that competitively inhibits various pathways in which glutamine is metabolized, hence an antibiotic and anti-tumour agent.

azide Usually the sodium salt NaN_3, an inhibitor of electron transport that blocks electron flow from cytochrome oxidase to oxygen. Frequently used to prevent growth of microorganisms, eg. in refrigerated antisera or chromatography columns.

Azotobacter Genus of free-living rod-shaped bacilli capable of fixing atmospheric nitrogen.

azurin (1) Blue copper-containing protein from *Pseudomonas aeruginosa*. (2) Histochemical dye.

azurophil granules ***Primary lysosomal*** granules found in ***neutrophil granulocytes***;

contain a wide range of hydrolytic enzymes. Sometimes referred to as primary granules to distinguish them from the secondary or ***specific granules***.

B

β Entry prefix is given as 'beta'; alternatively look for main portion of word.

B-cells (of pancreas) Cells within discrete endocrine islands (***islets of Langerhans***) embedded in the major exocrine tissue of vertebrate pancreas. The B- or β-cells (originally distinguished by differential staining from A, C and D), are responsible for synthesis and secretion into the blood of the hormone insulin. See also ***B-lymphocytes***.

B-chromosome Small acentric chromosome; part of the normal genome of some races and species of plants.

(β, γ-imido) ATP See *App(NH) p*.

b-c1 complex A part of the ***mitochondrial electron transport chain*** that accepts electrons from ***ubiquinone***, and passes them on to ***cytochrome c***. The b.c1 complex consists of two cytochromes.

b-COP See ***beta-COP***.

B-DNA The structural form of ***DNA*** originally described by Crick & Watson. It is the form normally found in hydrated DNA and is strictly an average, approximate stucture for a family of B forms. In B-DNA, the double helix is a right-handed helix with about 10 residues per turn and has a major and a minor groove. The planes of the base pairs are perpendicular to the helix axis.

B-lymphocyte See ***lymphocyte***.

Babes–Ernst granules Metachromatic intracellular deposits of polyphosphate found in *Corynebacterium diphtheriae* when the bacteria are grown on suboptimal media. Stain reddish with methylene blue or toluidine blue.

Babesia Genus of protozoa that are found as parasites within red blood cells of mammals and are transmitted by ticks.

baby hamster kidney cells See ***BHK cells***.

BAC library (bacterial artificial chromosome library) Library constructed in a vector with an ***origin of replication*** that allows its propagation in bacteria as an extra chromosome. Advantageous in constructing ***genomic libraries*** with relatively large DNA fragments (100–300 kb). See also ***bacterial artificial chromosome***.

Bac7 Proline- and arginine-rich antimicrobial peptide (7 kD) isolated from bovine neutrophils. Bac-5 is similar. The upstream region of proBac-5 and -7 both have sequence homology with similar regions of other neutrophil antimicrobial peptides (CAP18 from rabbit neutrophils and bovine indolicidin). The pro-region also has similarity to porcine ***cathelin***. Member of the ***protegrin*** family of peptides.

Bacille Calmette-Guerin Attenuated ***mycobacteria*** derived from *Mycobacterium tuberculosis*, used in tuberculosis vaccination. Extracts of the bacterium have remarkable powers in stimulation of lymphocytes and leucocytes and are used in ***adjuvants***.

bacillus Cylindrical (rod-shaped) bacterium. Bacilli are usually 0.5–1.0μm long, 0.3–1μm wide.

Bacillus thuringiensis Soil-living bacterium that produces a ***delta-endotoxin*** that is deadly to insects. Many strains exist, each with great specificity as to target Orders of insects. In general, the mode of action involves solubilization at the high pH within the target insect's gut, followed by proteolytic cleavage; the activated peptides form pores in the gut cell apical plasma membranes, causing lysis of the cells.

bacitracin Branched cyclic peptides produced by strains of *Bacillus licheniformis*. Interfere with murein (peptidoglycan) synthesis in ***Gram-positive*** bacteria.

bactenecin Highly cationic polypeptides found in lysosomal granules of bovine neutrophil granulocytes. They are thought to be involved in bacterial killing and occur in a third class of granules, the large granules, not found in the neutrophils of most species.

bacteraemia (USA, bacteremia) The presence of living bacteria in the circulating blood: usually implies the presence of small numbers of bacteria that are transiently present without causing clinical effects, in contrast to ***septicaemia***.

bacteria One of the two major classes of prokaryotic organism (the other being the ***Cyanobacteria***). Bacteria are small (linear dimensions of around 1μm), non-compartmentalized, with circular DNA, and ribosomes of 70S. Protein synthesis differs from that of eukaryotes, and many antibacterial antibiotics interfere with protein synthesis, but do not affect the infected host. Recently bacteria have been subdivided into ***Eubacteria*** and ***Archaebacteria***, although some would consider the Archaebacteria to be a third kingdom, distinct from both Eubacteria and Eukaryotes. The Eubacteria can be further subdivided on the basis of their staining using ***Gram stain***. Since the difference between ***Gram-positive*** and ***Gram-negative*** depends upon a fundamental difference in cell wall structure it is therefore more soundly based than classification on gross morphology alone (into cocci, bacilli, etc.).

bacterial artificial chromosome (BAC) Method of construction of ***genomic library***, in which the vector contains sites necessary for the DNA to be handled and replicated as a bacterial chromosome. Like ***YACs***, this allows clones to contain very large pieces of DNA (around 200 kb), so aiding rapid, low resolution ***physical mapping***.

bacterial chemotaxis The response of bacteria to gradients of attractants or repellents. In a gradient of attractant the probability of deviating from a smooth forward path is reduced if the bacterium is moving upgradient. Since the opposite is true if moving downgradient, the effect is to bias displacement towards the source of attractant. Strictly should perhaps be considered a ***klinokinesis*** with adaptation.

bacterial flagella Thin filaments composed of ***flagellin*** subunits that are rotated by the basal motor assembly and act as propellors. If rotating anticlockwise (as viewed from the flagellar tip) the bacterium moves in a straight path, if clockwise the bacterium 'tumbles'.

bacteriochlorophyll Varieties of ***chlorophyll*** (bacteriochlorophylls a, b, c, d, e and g) found in ***photosynthetic bacteria*** and differing from plant chlorophyll in the substituents around the tetrapyrrole nucleus of the molecule, and in the absorption spectra.

bacteriocide A substance that kills bacteria.

bacteriocins Exotoxins, often ***plasmid*** coded, produced by bacteria and which kill other bacteria (not eukaryotic cells). ***Colicins*** are produced by about 40% of *E. coli* strains: colicin E2 is a DNAase, colicin E3 an RNAase.

bacteriophaeophytin-b One of the components of the bacterial photosynthetic ***reaction centre***. (See also ***ubiquinone***.)

bacteriophages (phages) Viruses that infect bacteria. The bacteriophages that attack *Escherichia coli* are termed coliphages, examples of these are lambda phage and the T-even phages, T2, T4 and T6. Basically, phages consist of a protein coat or ***capsid*** enclosing the genetic material, DNA or RNA, which is injected into the bacterium upon infection. In the case of virulent phages, all synthesis of host DNA, RNA and proteins ceases and the phage genome is used to direct the synthesis of phage nucleic acids and proteins using the host's transcriptional and translational apparatus. These phage components then self-assemble to form new phage particles. The synthesis of a phage lysozyme leads to rupture of the bacterial cell wall releasing, typically, 100–200 phage progeny. The temperate phages, such as lambda, may also show this lytic cycle when they infect a cell, but more frequently they induce ***lysogeny***. The study of bacteriophages has been important for our understanding of gene structure and regulation. Lambda has been extensively used as a ***vector*** in recombinant DNA studies.

bacteriorhodopsin A light-driven proton-pumping protein (248 residues, 26 kD), similar to ***rhodopsin***, found in 'purple patches' in the cytoplasmic membrane of the bacterium *Halobacterium halobium*. It is composed of seven transmembrane helices, and contains the light-absorbing ***chromophore, retinal***. Light absorption maxima: 568nm (light-adapted); 558nm (dark-adapted). Each photon results in the movement of two protons from cytoplasmic to extracellular sides of the membrane. The resulting proton gradient is used (amongst other things) to drive synthesis of ATP by ***chemiosmosis***.

bacteriostatic Adjective applied to substances that inhibit the growth of bacteria without necessarily killing them.

bacteroid Small, often irregularly rod shaped bacterium, eg. those found in root nodules of nitrogen-fixing plants.

Baculovirus Viruses specialized as pathogens of lepidopteran larvae. Widely used as eukaryotic ***expression vectors*** for proteins requiring post-translational modifications such as ***glycosylation***, proteolytic ***cleavage*** and fatty acylation.

bafilomycin Microbial toxin that is a specific inhibitor of the ***V-type ATPase***.

bag cell neurons Cluster of electrically coupled neurons in the abdominal ganglion of *Aplysia* that are homogeneous, easily dissected out and release peptides that stimulate egg laying.

BALB/c Inbred strain of white (albino) mice. Used as a source for one of the various 3T3 cell lines.

Balbiani ring The largest ***puffs*** seen on the ***polytene chromosomes*** of Diptera are called Balbiani rings after the 19-century microscopist who first described polytene chromosomes.

balloon cell Non-specific description of any cell with abundant clear cytoplasm. May arise through a variety of causes and includes some carcinoid cells, hepatocytes following some forms of toxic insult, neurons or other cells in storage diseases and cells in some melanomas.

Bam* H I** (BamHI) Common ***restriction enzyme (from *Bacillus amyloliquefaciens* H) that cuts the sequence G|GATCC. See Table R1.

band cells Immature ***neutrophils*** released from the bone marrow reserve in response to acute demand.

band III A 90 kD protein of the human erythrocyte membrane, identified as the major anion transport/exchange protein. Analogous proteins exist in other erythrocytes. A dimeric transmembrane glycoprotein, with binding sites for many cytolasmic proteins, including ***ankyrin***, on its cytoplasmic domain.

banding patterns Chromosomes stained with certain dyes, commonly quinacrine (Q banding) or Giemsa (G banding), show a pattern of transverse bands of light and heavy staining that is characteristic for the individual chromosome. The basis of the differential staining, which is the same in most tissues, is not understood: each band represents 5–10% of the length, about 10^7 base pairs, although this is not true for ***polytene chromosomes*** in *Drosophila* that show more than 4000 bands.

bandshift assay (gel shift assay) An assay for proteins, such as ***transcription factors***, that bind specific DNA sequences. A labelled oligonucleotide corresponding to the recognition sequence is incubated with an appropriate nuclear protein extract, and run on a non-denaturing acrylamide gel. Oligonucleotides that have been bound by proteins are retarded relative to those that are unbound.

BAPTA (1,2-bis(o-aminophenoxy) ethane tetraacetate) Calcium chelator with low affinity for magnesium. Absorption maximum shifts when calcium is bound so it can be used as an indicator of intracellular calcium concentration (though it will chelate calcium and therefore alter the situation). See ***MAPTAM***.

barnase Bacterial ribonuclease.

baroreceptor (baroceptor) In an organism, a receptor that is sensitive to pressure.

Barr body Small dark-staining inactivated X chromosome seen in female (XX) cells. According to the ***Lyon hypothesis***, random inactivation occurs.

basal body Structure found at the base of eukaryotic ***cilia*** and ***flagella*** consisting of a continuation of the nine outer sets of axonemal ***microtubules*** but with the addition of a C-tubule to form a triplet (like the ***centriole***). May be self-replicating and serves as a nucleating centre for axonemal assembly. Anchored in the cytoplasm by ***rootlets***. Synonymous with ***kinetosome***.

basal cell carcinoma (BCC; rodent ulcer) Common ***carcinoma*** derived from the basal cells of the epidermis. Often a consequence of exposure to sunlight and much more common in those with fair skin; rarely metastasizes.

basal cells General term for relatively undifferentiated cells in an epithelial sheet that give rise to more specialized cells (act as ***stem cells***). In the ***stratified squamous epithelium*** of mammalian skin the basal cells of the epidermis (stratum basale) give rise by an unequal division to another basal cell and to cells that progress through the spinous, granular and horny layers, becoming progressively more keratinized, the outermost being shed as ***squames***. In olfactory mucosa the basal cells give rise to olfactory and sustentacular cells. In the epithelium of epididymis their function is unclear, but they probably serve as stem cells.

basal ganglia Three large subcortical nuclei of the vertebrate brain: the putamen, the caudate nucleus and the globus pallidus. They participate in the control of movement along with the ***cerebellum***, the corticospinal system and other descending motor systems. Lesions of the basal ganglia occur in a variety of motor disorders including ***Parkinsonism*** and ***Huntington's chorea***.

basal lamina See ***basement membrane***.

base analogues Purine and pyrimidine bases that can replace normal bases used in DNA synthesis and hence can be included in DNA, eg. 5-bromouracil (replacing thymine) or 2-aminopurine (replacing adenine). May be used for inducing mutations, including point mutations.

base pairing The specific hydrogen-bonding between ***purines*** and ***pyrimidines*** in double-stranded nucleic acids. In DNA the pairs are ***adenine*** and ***thymine***, and ***guanine*** and ***cytosine***, while in RNA they are adenine and ***uracil***, and guanine and cytosine. Base-pairing leads to the formation of a DNA double helix from two complementary single strands.

basement membrane Extracellular matrix characteristically found under epithelial cells. There are two distinct layers: the basal lamina, immediately adjacent to the cells, is a product of the epithelial cells themselves and contains collagen type IV; the reticular lamina is produced by fibroblasts of the underlying ***connective tissue*** and contains fibrillar collagen.

baseplate A hypothetical cell adhesion molecule possibly involved in sponge cell adhesion, existence unproven.

Basic Blue 9 See ***Methylene Blue***.

basic leucine zipper (bZIP) Family of proteins having a basic region and a ***leucine zipper***. The basic region is the DNA-binding domain and the leucine zipper is involved in protein-protein interactions to form homo- or heterodimers. Includes ***AP-1***, ATF and ***CREB*** transcription factors.

Basidiomycetes Group of fungi that includes rusts, smuts and edible fungi. Produce basidiospores.

basidiospore Spores of Basidiomycete fungi. These spores are usually uninucleate and haploid.

basidium Club-shaped organ involved in sexual reproduction in basidiomycete fungi (mushrooms, toadstools etc.). Bears four haploid basidiospores at its tip.

basilar membrane A thin layer of tissue covered with mesothelial cells that separates the cochlea from the scala tympani in the ear.

basket cells Cerebellar neurons with many small dendritic branches that enclose the cell bodies of adjacent ***Purkinje cells*** in a basket-like array.

basolateral plasma membrane The plasma membrane of epithelial cells that is adjacent to the ***basal lamina*** or the adjoining cells of the sheet. Differs both in protein and phospholipid composition from the ***apical plasma membrane*** from which it is isolated by ***tight junctions***.

basophil Mammalian ***granulocyte*** with large heterochromatic basophilic granules that contain ***histamine*** bound to a protein and heparin-like mucopolysaccharide matrix. They are not phagocytic. Very similar to ***mast cells*** though it is not clear whether they have common lineage.

basophilia (1) Having an affinity for basic dyes. (2) Condition in which there is an excess of ***basophils*** in the blood.

batrachotoxin (BTX) ***Neurotoxin*** from the Columbian poison frog *Phyllobates*. A steroidal alkaloid that affects sodium channels; batrachotoxin R is more effective than related batrachotoxin A.

batroxostatin See ***disintegrin***.

bax Protein related to ***bcl-2*** that promotes apoptosis in cultured cells. Mice deficient in bax have selective hyperplasias. Bax seems to act as a tumour suppressor and is induced by ***p53***, though is not solely responsible for p53-mediated apoptosis.

Bayer's junctions See ***Bayer's patches***.

Bayer's patches (Bayer's junctions) Sites of adhesion between the outer and cytoplasmic membranes of ***Gram-negative bacteria***.

Bayesian statistics Statistical theory, based on Bayes' decision rule, that outlines a framework for producing decisions based on relative payoffs of different outcomes. Used in genetic counselling.

BBB See ***blood–brain barrier***.

BCG See ***Bacille Calmette-Guerin***.

bcl-2 ***Protooncogene***, activated by chromosome translocation in human B-cell lymphomas (hence 'bcl'). Encodes a plasma membrane protein. The gene product inhibits programmed cell death (***apoptosis***) and is homologous with the worm gene *ced-9*; see **ced *mutant***.

bcl-3 ***Oncogene*** associated with some cases of B-cell chronic lymphocytic leukaemias. The protein product contains seven 'ankyrin-repeats' very similar to those found in I κ B. Seems to interfere with binding of 50 kD subunit of NF κ B to DNA.

bcr (breakpoint cluster region) Region on chromosome 22 involved in the ***Philadelphia chromosome*** translocation.

BDGF See ***brain-derived growth factor***.

BDNF See ***brain-derived neurotrophic factor***.

beaded filaments (beaded-chain filaments) Intermediate filaments found in the lens fibre cells of the eye, composed of ***filensin*** and ***phakinin***.

Becker muscular dystrophy Benign X-linked muscular dystrophy with later onset and lower severity than Duchenne dystrophy.

Beckwith–Wiedemann syndrome Rare developmental disorder with a complex pattern of inheritance suggesting a defect in maternal ***imprinting***. Characteristics are all growth abnormalities: enlarged tongue, gigantism, enlarged adrenal glands, enlarged visceral organs, advanced ageing and predisposition to childhood tumours. Possibly due to a defect in the ***cyclin-dependent kinase inhibitor***, $p57^{KIP2}$, though in some cases there are two copies of the IGF-2 gene

Becquerel (Bq) The Systeme Internationale (SI, MKS) unit of radioactivity, named after the discoverer of radioactivity, and equal to 1 disintegration per second. Use is fairly recent, superseding the Curie (Ci). 1Ci = 37 GBq.

beige mouse A mouse strain typified by beige hair and ***lymphadenopathy***, reticulum cell neoplasms, and giant lysosomal granules in ***leucocytes***. May be the murine equivalent of ***Chediak–Higashi syndrome*** in humans.

belt desmosome Another name for the zonula adherens or ***adherens junction***.

Bence–Jones protein Dimers of ***immunoglobulin*** light chains, normally produced by ***myelomas***. Bence–Jones proteins are sufficiently small to be excreted by the kidney.

benign tumour A clone of ***neoplastic*** cells that does not invade locally or cause ***metastasis***, having lost ***growth control*** but not positional control. Usually surrounded by a fibrous capsule of compressed tissue.

benzodiazepine Drugs widely used in medical practice as CNS depressants, eg. diazepam (the tranquillizer Valium). Enhance the inhibitory action of ***GABA*** by modulating $GABA_A$ receptors.

benzopyrene Polycyclic aromatic compound. Potent mutagen and carcinogen.

Berk–Sharp technique (S1 mapping) A technique of genetic mapping in which ***mRNA*** is hybridized with ***single-stranded DNA*** and the non-hybridized DNA is then digested with S1 ***nuclease***; the residual DNA that hybridized with the messenger is then characterized by ***restriction mapping***.

Bernard–Soulier syndrome Genetic deficiency in platelet membrane glycoprotein Ib (CD42); platelets aggregate normally (cf. ***Glanzmann's thrombasthenia***) but do not stick to collagen of subendothelial basement membrane.

Best's carmine Stain that can be used to demonstrate the presence of glycogen, which stains deep red.

beta-actinin (β-actinin) See ***capZ***.

beta-alpha-beta motif (β-α-β motif) Protein ***motif*** comprising a beta strand-loop-helix-loop-strand arrangment, with the strands lying parallel.

beta-amylase (β-amylase) A terminal amylase that cleaves starch to maltose units from the end of starch chains.

beta arch (β arch) Protein ***motif*** comprising two adjacent antiparallel beta strands joined by a coil that are part of different sheets, usually forming a ***beta sandwich***.

beta barrel (β barrel) Protein motif in which a series of (typically ***amphipathic***) ***beta sheets*** is arranged around a central pore. Example: ***voltage-gated ion channel***.

beta-blocker (β-blocker) Inhibitor of β-adrenergic receptors; causes decrease in heart rate and blood pressure. Those that inhibit β1 receptors have effects primarily on the heart.

beta bulge (β bulge) Protein ***motif*** comprising a disruption of a ***beta sheet***, usually by the insertion of a single residue.

beta-cells (pancreas) (β-cells) See ***B cells of pancreas***.

beta-COP (β-COP) Major component (110 kD) of coat of non-clathrin coated vesicles derived from Golgi. Has homology with β-adaptin.

beta-emitter (β-emitter) A radionuclide whose decay is accompanied by the emission of β particles, most commonly negatively charged electrons. Many isotopes used in biology, such as ^{3}H, ^{14}C, ^{35}S, and ^{32}P are pure β-emitters.

beta-galactosidase (β-galactosidase) Enzyme (EC 3.2.1.23) encoded by the *LacZ* gene, that is widely used as a ***reporter gene***, as a variety of coloured or fluorescent compounds can be produced from appropriate substrates (typically ***Xgal***, which produces a blue colour). *LacZ* is incorporated in many plasmid ***vectors*** to allow blue–white colour ***selection***.

beta-glucosidase (β-glucosidase) EC 3.2.1.21. Enzyme catalysing the release of glucose by hydrolysis of the glycosidic link in various β-D-glucosides, R-β -D-glucose, where the group R may be alkyl, aryl, mono- or oligosaccharide. Favoured source: almonds, from which enzyme is known as emulsin.

beta-glucuronidase (β-glucuronidase) Enzyme (EC 3.2.1.31) that catalyses hydrolysis of a β-D-glucuronoside to D-glucuronate and the compound to which it was attached. Often used as a marker enzyme for lysosomes.

beta hairpin (β hairpin) Protein ***motif*** describing one possible arrangement of strands in a ***beta sheet***. Strands are antiparallel and hydrogen-bonded, lying adjacent in the sheet. See ***hairpin***.

beta-helix (β-helix; solenoid) Protein ***motif*** comprising a large right-handed coil (or superhelix), containing either 2 or 3 ***beta sheets***.

beta-lactamase (β-lactamase) Specifically, the plasmid-coded enzyme secreted by many bacteria that inactivates penicillins by opening the lactam ring.

beta-lactams Class of antibiotics that includes penicillins, ampicillin, cloxacillin, piperacillin, cephalosporins such as cephalothin, cephamycins such as cefoxitin and monolactams such as clavulanic acid (component of Augmentin).

beta-oxidation (β-oxidation) The process whereby fatty acids are degraded in steps, losing two carbons as (acetyl)-CoA. Involves CoA ester formation, desaturation, hydroxylation and oxidation before each cleavage.

beta-pleated sheet (β-pleated sheet) Beta secondary structure in proteins consists of two almost fully extended polypeptide chains lying side by side, linked by interchain hydrogen bonds between peptide C=O and N–H groups. When multiple chains are involved, an extended sheet, the β-pleated sheet, is formed, which can consist of parallel or antiparallel sheets (where the chains run in the same or opposite directions) or mixed sheets.

beta prism (β prism) Protein ***motif*** comprising three antiparallel ***beta sheets*** arranged in a triangular, prism shape. In the orthogonal prism, strands are orthogonal to the prism access; in the aligned prism, the strands and prism axis are parallel.

beta propellor (β propellor) Protein ***motif*** comprising 4–8 antiparallel ***beta sheets*** arranged like the blades of a propellor.

beta sandwich (β sandwich) Protein ***motif*** comprising two ***beta sheets*** that pack together face to face, in a layered arrangement.

beta sheet (β sheet) See ***beta-pleated sheet***.

beta strand (β strand) Region of polypeptide chain that forms part of a ***beta sheet***.

beta trefoil (β trefoil) Protein ***motif*** consisting of three ***beta hairpins*** forming a triangular shape.

beta turn (β turn) Protein ***motif*** which consists of an abrupt 180° reversal in direction of a polypeptide chain. The turn is defined as being complete within four residues.

beta-2-microglobulin (β-2-microglobulin) Immunoglobulin-like polypeptide (12 kD, homologous with the constant region of Ig) that is found on the surfaces of most cells, associated non-covalently with Class I ***histocompatibility antigens***.

betaine A derivative of glycine characterized by high water solubility. Can function as an osmotic agent in plant tissues. See ***biogenic amines***.

bFGF See ***fibroblast growth factor***.

BFU-E See ***burst forming unit***-erythrocytic.

BHK cells (baby hamster kidney cells) A quasi-diploid established line of Syrian hamster cells, descended from a clone (Clone 13) isolated by Stoker & McPherson from an unusually rapidly growing primary culture of newborn hamster kidney tissue. Usually described as fibroblastic, although smooth muscle-like in that they express the muscle intermediate filament protein ***desmin***. Widely used as a viral host, in studies of oncogenic transformation and of cell physiology.

Biacore Proprietary name for an instrument that uses ***surface plasmon resonance*** to detect the binding of a substance to the surface of a flow chamber. Using this machine it is possible to measure the on- and off-rates for the binding of a molecule to a defined surface, eg. the binding of an antibody to the antigen-coated surface of the flow cell or of ligand to an immobilized receptor.

bicoid An ***egg-polarity gene*** in *Drosophila*, concentrated at the anterior pole of the egg, and required for subsequent anterior structures. A ***maternal-effect gene***.

bicuculline From *Dicentra cucullaria* and herbs of the genus *Corydalis*. Specific blocking agent for the action of the amino acid transmitter γ-aminobutyric acid (*GABA*). See ***amino acid transmitter***; ***amino acid receptor*** superfamily, ***GABA receptor***.

big brain A ***neurogenic gene*** of *Drosophila*, believed to encode a product involved in cell–cell communication, perhaps via ***gap junctions***. Member of the ***major intrinsic protein*** family.

biglycan A small proteoglycan, 150–240 kD, of the extracellular matrix. The core protein has a mass of around 42 kD and is very similar to the core protein of ***decorin*** and ***fibromodulin***. All three have highly conserved sequences containing 10 internal homologous repeats of around 25 amino acids with leucine-rich motifs. Biglycan has two glycosaminoglycan chains, either chondroitin sulphate or dermatan sulphate and N-linked oligosaccharides.

bilharzia (schistosomiasis) Disease caused by the blood fluke *Schistosoma* spp., a digenean Platyhelminth.

biliproteins See ***phycobilins***.

bilirubin Red-brown pigment found in bile, formed by breakdown of haemoglobin.

biliverdin Green bile pigment formed by haemoglobin breakdown; can be converted into ***bilirubin*** by reduction.

bindin Molecule of around 30 kD normally sequestered in the ***acrosome*** of a sea-urchin spermatozoon, and which through its specific ***lectin***-like binding to the ***vitelline membrane*** of the egg confers species-specificity in fertilization.

bioassay An assay for the activity or potency of a substance that involves testing its activity on living material.

bioautography The use of cells to detect by their attachment or other reaction the presence of a particular substance, eg. an adhesion protein on an electrophoretic gel.

bioavailability Relative amount of a drug (or other substance) that will reach the systemic circulation when administered by a route other than direct intravenous injection.

bioblasts When Altmann first observed mitochondria he considered them to be intracellular parasites and christened them bioblasts.

bioflavonoids Group of coloured phenolic pigments originally considered vitamins (vitamins P, C2) but not shown to have any nutritional role. Responsible for the red/purple colours of many higher plants.

biogenic amines Amines found in both animals and plants that are frequently involved in signalling. There are several groups: ethanolamine derivatives include ***choline***, ***acetylcholine*** and muscarine; catecholamines include ***adrenaline***, ***noradrenaline*** and ***dopamine***; polyamines include ***spermine***; indolylalkylamines include tryptamine and ***serotonin***; betaines include ***carnitine***; polymethyline diamines include cadaverine and ***putrescine***.

bioinformatics The discipline of using computers to collate and form datasets of interest to biologists. Usually used to refer to databases of DNA and protein sequences, and of mutations, disease and gene functions, in the context of genome projects.

bioluminescence Light produced by a living organism. The best known system is firefly luciferase (an ATPase), which is used routinely as a sensitive ATP assay system. Many other organisms, particularly deep-sea organisms, produce light and even leucocytes emit a small amount of light when their oxidative metabolism is stimulated. Does not really differ from ***chemiluminescence***, except that the light-emitting molecule occurs naturally and is not a synthetic compound like luminol or lucigenin.

biosynthesis Synthesis by a living system (as opposed to chemical synthesis)

biotin (vitamin H) A prosthetic group for carboxylase enzymes. Important in fatty acid biosynthesis and catabolism and has found widespread use as a covalent label for macromolecules which may then be detected by high-affinity binding of labelled ***avidin*** or ***streptavidin***. Essential ***growth factor*** for many cells.

BiP Molecular chaperone (78 kD) found in endoplasmic reticulum and related to hsp70 family of heat-shock proteins. Originally described as immunoglobulin heavy chain binding protein.

bipolar cells A class of retinal ***interneurons***, named after their morphology, that receive input from the photoreceptors and send it to the ***ganglion cells***. Bipolar cells are ***non-spiking neurons***; their response to light is evenly graded, and shows ***lateral inhibition***.

bipolar filaments Filaments that have opposite polarity at the two ends: classic example is the ***thick filament*** of striated muscle.

Birbeck granules Characteristic inclusion bodies seen by electron microscopy in histiocytes (Langerhans cells) of patients with histiocytosis X, a group of diseases with uncertain pathogenesis.

birefringence Optical property of a material in which the refractive index is different for light polarized in one plane compared to the orthogonal plane. See **birefringent**.

birefringent Any material that has different refractive index according to the plane of polarization of the light. The effect is to rotate the plane of the refracted light so that, using crossed Nicholl prisms (polarizers set at right angles to give complete extinction), the birefringent material appears bright. The birefringence can arise through anisotropy of structure (form birefringence) or through orientation of molecules either because of mechanical stretching (stress birefringence) or because of alignment in flow (flow birefringence). A classic example often used to demonstrate the effect is a hair which shows form birefringence because of the orientation of the keratin.

bistatin A ***aisintegrin*** found in the venom of the puff adder *Bitis arietans*.

Biston betularia Peppered moth; famous for the shift to the melanized form as industrial pollution turned trees black and gave the melanotic form a selective advantage, and for reversion to the lighter form following the Clean Air Act.

bithorax complex A group of ***homeotic*** mutations of *Drosophila* that map to the bithorax region on chromosome III. The mutations all cause the third thoracic segment to develop like the second thoracic segment to varying extents. The genes of the bithorax complex are thought to determine the differentiation of the posterior thoracic segments and the abdominal segments.

Bittner agent Earlier name, now superseded, for the mouse ***mammary tumour virus***.

bivalent Used of two homologous chromosomes when they are in synapsis during ***meiosis***.

black membrane An artificial (phospho)-lipid membrane formed by 'painting' a solution of phospholipid in organic solvent over a hole in a hydrophobic support immersed in water. Drainage of the solvent from the film produces diffraction colours until the thickness falls below the wavelength of light–it then appears to be black. The structure is an extended bimolecular leaflet.

black widow spider venom Potent ***neurotoxin*** that induces catastrophic release of ***acetylcholine*** from ***presynaptic*** terminals of cholinergic ***chemical synapses***.

blast cells Cells of a proliferative compartment in a cell lineage.

blast transformation The morphological and biochemical changes in B- and T-lymphocytes on exposure to ***antigen*** or to a ***mitogen***. The cells appear to move from G0 to G1 stage of the cell cycle. They usually enlarge and proceed to S phase and mitosis later. The process probably involves receptor crosslinking on the plasma membrane.

blastema A group of cells in an organism that will develop into a new individual by asexual reproduction, or into an organized structure during regeneration.

blastocoel (USA, blastocele) The cavity formed within the mass of cells of the ***blastula*** of many animals during the later stages of cleavage.

blastocyst In mammalian development, cleavage produces a thin-walled hollow sphere, whose wall is the ***trophoblast***, with the embryo proper being represented by a mass of cells at one side. The blastocyst is formed before implantation and is equivalent to the ***blastula.***

blastoderm In many eggs with a large amount of yolk, cell division (cleavage) is restricted to a superficial layer of the fertilized egg (meroblastic cleavage). This layer is termed the blastoderm. In birds it is a flat disc of cells at one pole of the egg and in insects an outer layer of cells surrounding the yolk mass.

blastomere One of the cells produced as the result of cell division (cleavage) in the fertilized egg.

blastopore During ***gastrulation*** cells on the surface of the embryo move into the interior to form the ***mesoderm*** and ***endoderm***. The opening formed by this invagination of cells is the blastopore. It is an opening from the archenteron, the primitive gut, to the exterior. In some animals this opening becomes the anus, whilst in others it closes up and the anus opens at the same spot or nearby. In some animals, eg. the chick, invagination occurs without a true blastopore and the site at which the cells move in, the (***primitive streak***), may be termed a virtual blastopore.

blastula Stage of embryonic development of animals near the end of cleavage but before ***gastrulation***. In animals where cleavage (cell division) involves the whole egg, the blastula usually consists of a hollow ball of cells.

Blattella germanica German cockroach.

bleb Protrusion from the surface of a cell, usually approximately hemispherical; may be filled with fluid or supported by a meshwork of microfilaments.

bleomycin Any of a group of glycopeptide antibiotics from *Streptomyces verticillus.* Blocks cell division in G2: used to synchronize the division of cells in culture and as an antiproliferative agent in oncology.

Blepharisma Genus of ciliate protozoans of the order Heterotricha.

blepharoplast Alternative name for a ***basal body***. An organelle derived from the ***centriole*** and giving rise to the ***flagella***. Found chiefly in Protozoa and Algae.

blocking antibody An antibody used in a reaction to prevent some other reaction taking place, for example one antibody competing with another for a cell surface receptor. See also ***desensitization***.

blood group antigens The set of cell surface antigens found chiefly, but not solely, on blood cells. More than fifteen different blood group systems are recognized in humans. There may be naturally occurring antibodies without immunization, especially in the case of the ***ABO*** system, and matching blood groups is important for safe transfusion. In most cases the antigenic determinant resides in the carbohydrate chains of membrane glycoproteins or glycolipids. See also ***Rhesus, Duffy, Kell, Lewis*** and ***MN*** blood groups.

blood vessels All the vessels lined with ***endothelium*** through which blood circulates.

blood–brain barrier (BBB) The blood vessels of the brain (and the retina) are much more impermeable to large molecules (like antibodies) than blood vessels elsewhere in the body. This has important implications for the ability of the organism to mount an immune response in these tissues, although the basis for the difference in endothelial permeability is not well understood.

Bloom's syndrome Rare human autosomal recessive defect associated with genomic instability causing short stature, immunodeficiency and increased risk of all types of cancer. Caused by mutation of BLM locus on chromosome 15q. BLM protein has homology to ***helicases***.

blotting General term for the transfer of protein, RNA or DNA molecules from a relatively thick acrylamide or agarose ***gel*** to a paper-like membrane (usually nylon or nitrocellulose) by capillarity or an electric field, preserving the spatial arrangment. Once on the membrane, the molecules are immobilized, typically by baking or by ultraviolet irradiation, and can then be detected at high sensitivity by ***hybridization*** (in the case of DNA and RNA), or antibody labelling (in the case of protein). RNA blots are called ***Northern*** blots; DNA blots, ***Southern***; protein blots, ***Western***. See also ***dot*** and ***slot blots***.

blue naevus A non-malignant accumulation of highly pigmented ***melanocytes*** deep in the ***dermis***.

blue-green algae Group of prokaryotes that should now be referred to as ***Cyanobacteria***. See ***Cyanophyta***.

blue-white colour selection Method for identifying bacterial clones containing plasmids with inserts. Many modern ***vectors*** have their ***polycloning site*** within a part of the *LacZ* gene encoding ***beta-galactosidase***, which provides ***α-complementation*** in an appropriate mutant *E. coli* strain. This means that a re-ligated (empty) vector will produce blue colonies when grown on plates containing ***IPTG*** and ***Xgal***, but colonies with a substantial insert in their plasmid's polycloning site are unable to produce functional β-galactosidase, and so produce white colonies.

Bluescript (pBluescript) Proprietary plasmid, sold by Stratagene. Very widely used.

Bluetongue virus Reovirus that causes serious disease of sheep and milder disease in cattle and pigs. Transmitted by biting flies.

blunt end End of double-stranded DNA that has been cut at the same site on both strands by a ***restriction enzyme*** that does not produce ***sticky ends***.

Blym An oncogene, identified in lymphoma of ***bursa of Fabricius***. See also Table O1.

BM-40 See ***osteonectin***.

Bmax Amount of drug required to saturate a population of receptors and a measure of the number of receptors present in the sample. Usually derived from ***Scatchard plot*** of binding data. Analogous to Vmax in enzyme kinetics.

BMP (bone morphogenetic protein) Multifunctional cytokines, members of the TGF-β superfamily. Activity of BMPs are regulated by BMP-binding proteins ***noggin*** and chordin. Receptors are serine-threonine kinase receptors (types I and II) that link with ***smads***. *Drosophila* decapentaplegic (Dpp) is a homologue of mammalian BMPs. BMP2 is involved in regulating bone formation, BMP4 acts during development as a regulator of mesodermal induction and is overexpressed in fibrodysplasia ossificans. Follistatin inhibits BMP function in early *Xenopus* embryos.

BNLF-1 An ***Oncogene*** from ***Epstein–Barr virus***. Encodes a plasma membrane protein. See also Table O1.

BNP See ***brain natriuretic peptide***.

Bohr effect Decrease in oxygen affinity of ***haemoglobin*** when pH decreases or concentration of carbon dioxide increases.

bombesin Tetradecapeptide ***neurohormone*** with both ***paracrine*** and ***autocrine*** effects first isolated from skin of fire-bellied toad (*Bombina bombina*); mammalian equivalent is ***gastrin-releasing peptide*** (GRP).

Bombesin cross-reacts with GRP receptors. Both are ***mitogenic*** for Swiss 3T3 fibroblasts at nanomolar levels. Neuropeptides of this type are found in many tissues and at high levels in pulmonary (small cell carcinoma) and thyroid tumours.

Bombyx mori Commercial silkmoth.

bone marrow Tissue found in the centre of most bones; site of ***haematopoiesis***. The most radiation-sensitive tissue of the body.

bone morphogenetic protein See ***BMPs***.

Bordetella pertussis A small, aerobic, ***Gram negative*** bacillus, causative organism of whooping cough. Produces a variety of toxins including a dermonecrotizing toxin, an adenyl cyclase, an ***endotoxin***, and ***pertussis toxin***, as well as surface components such as fimbrial haemagglutinin.

Borna disease Virally induced T-cell dependent immunopathological disorder of central nervous system. There are suggestions that Borna disease virus (a broadly distributed unclassified arthropod-borne virus that infects domestic animals and humans) may be associated with some psychiatric disorders.

Borrelia burgdorferi Spirochaete, responsible for ***Lyme disease***. Can be isolated from midgut of ticks (*Ixodes*).

Bos taurus Domestic cow.

bottle cells The first cells to migrate inwards at the ***blastopore*** during amphibian ***gastrulation***. The 'neck' of the bottle is at the outer surface of the embryo.

botulinolysin ***Cholesterol-binding toxin*** from *Clostridium botulinum*.

botulinum toxin Neurotoxin (50 kD; 7 distinct serotypes) produced by certain strains of *Clostridium botulinum*. The bacterium produces the toxin as a complex with a haemagglutinin that prevents toxin inactivation in the gut. Proteolysis in the body results in cleavage into two fragments A and B; B binds to gangliosides and may stimulate the endocytosis of fragment A. See ***synaptobrevin, tetanus toxin***.

botulinus toxin C2 Binary toxin with binding subunit (100 kD) and enzymatic subunit (50 kD) that ADP-ribosylates monomeric G-actin and blocks the formation of microfilaments. Produced by C and D strains of *Clostridium botulinum*.

botulinus toxin C3 (exoenzyme C3) Toxin (24 kD) produced by C and D strains of *Clostridium botulinum*. An ADP-ribosyl transferase that inactivates ***Rho***. Needs to be injected into cells and is a laboratory tool rather than a true toxin.

Bouin's solution Picric acid-based fixative that also contains formaldehyde and acetic acid. It has the advantage that specimens can be stored indefinitely and generally preserves nuclear morphology quite well.

bovine serum albumin BSA ***Albumin*** derived from bovine serum. Frequently used in cell culture, or as a carrier protein in biochemistry.

Bowman–Birk protease inhibitors Family of ***serine protease*** inhibitors found in seeds of leguminous plants and cereals.

box Casual term for a DNA sequence that is a characteristic feature of regions that bind regulatory proteins, eg. ***homeobox, TATA box*** and ***CAAT*** box.

Boyden chamber Simple chamber used to test for chemotaxis, especially of leucocytes. Consists of two compartments separated by a millipore filter (3-8μm pore size); chemotactic factor is placed in one compartment and the gradient develops across the thickness of the filter (ca 150μm). Cell movement into the filter is measured after an incubation period less than the time taken for the gradient to decay. See also ***checkerboard assay***.

Brachydanio rerio (*Danio rerio*) See ***zebrafish***.

brachyury Mouse gene encoding a transcription factor, one of the ***T-box genes***. Product of the gene is important in tissue specification, morphogenesis and organogenesis. Mouse mutant has a short tail.

Bradford method Method for estimating the protein content by using the change in absorption of ***Coomassie blue*** dye when it binds to proteins.

bradycardia Condition in which the heart beats unusually slowly. Opposite of tachycardia.

bradykinin Vasoactive nonapeptide (RPPGFSPFR) formed by action of proteases on kininogens. Very similar to ***kallidin*** (which has the same sequence but with an additional N-terminal lysine). Bradykinin is a very potent vasodilator and increases permeability of postcapillary venules; it acts on endothelial cells to activate phospholipase A2. It is also spasmogenic for some smooth muscle and will cause pain.

brain natriuretic peptide (BNP) Brain peptide that induces diuresis; related to ***atrial natriuretic peptide***. See also ***natriuretic peptides.***

brain regions The central nervous system of mammals is complex and the terminology often confusing. In development the brain is generated from the most anterior portion of the neural tube and there are three main regions, fore-, mid- and hindbrain. The lumen of the embryonic nervous system persists in the adult as the cerebral ventricles, filled with cerebrospinal fluid, which are connected to the central canal of the spinal cord. The forebrain develops to produce the cerebral hemispheres and basal ganglia and the diencephalon which forms the thalamus and hypothalamus. The cerebrum consists of two hemispheres, connected by the corpus callosum, the outer part being greatly expanded in humans with the increased surface being thrown into fold (ridges are gyri, valleys are sulci). The outer layer (cerebral cortex) is responsible for so-called higher-order functions such as memory, consciousness and abstract thought, the deeper layers (basal ganglia) include the caudate nucleus and putamen (collectively the striatum), amygdaloid nucleus and hippocampus. The hypothalamus controls endocrine function (hunger, thirst, emotion, behaviour, sleep), the thalamus coordinates sensory input and pain perception. The midbrain is relatively small and develops to form corpora quadrigemina and the cerebral peduncle. The hindbrain develops into two regions, the more anterior being the metencephalon, the region nearest the spinal cord being the myelencephalon. The metencephalon contains the cerebellum, responsible for sensory input and coordination of voluntary muscles, and the pons. The myelencephalon contains the medulla oblongata, which regulates blood pressure, heart rate and other basic involuntary functions, and dorsally the choroid plexus.

brain-derived growth factor (BDGF) See ***brain-derived neurotrophic factor.***

brain-derived neurotrophic factor (BDNF; BDGF) Small basic protein purified from pig brain; a member of the family of ***neurotrophins*** that also includes ***nerve growth factor*** and ***neurotrophin-3***. In contrast to NGF, BDNF is predominantly (though not exclusively) localized in the central nervous system. It supports the survival of primary ***sensory neurons*** originating from the ***neural crest*** and ectodermal ***placodes*** that are not responsive to NGF.

Branchiostoma Genus of lancelets-includes *Amphioxus*.

Brassica napus Oilseed rape (canola in USA). Source of edible oil (see ***erucic acid***).

brca (breast cancer-related gene) Two genes (BRCA-1 and BRCA-2) associated with familial breast carcinoma have now been identified.

breakpoint cluster region See ***bcr***.

brefeldin A A macrocyclic lactone synthesized from palmitic acid by several fungi including *Penicillium brefeldianum*. It was initially described as an antiviral antibiotic, but it was later found to inhibit protein secretion at an early stage, probably blocking secretion in a pre-Golgi compartment. Its exact site of action is still a matter of debate, one suggestion being that it inhibits the binding of regulatory coat proteins to organellar membranes. Nevertheless, it has proved a valuable tool for studying membrane traffic and the control of organelle structure.

Brevibacterium Genus of ***Gram-positive*** aerobic coryneform bacteria.

bride of sevenless (boss) In *Drosophila* eye development, the ligand for the ***sevenless*** tyrosine kinase receptor. Boss is expressed by the central R8 cell. It is unusual as a ligand for a tyrosine receptor kinase in that it is on the surface of another cell and has, in addition to a large extracellular domain, seven transmembrane segments and a C-terminal cytoplasmic tail.

bright field Type of ***light microscopy*** in which the sample is directly illuminated by transmitted light, and the object imaged in terms of differences in transmittance (brightness) or colour. See also Table L2.

bright-field microscopy Optical ***microscopy***, in which absorption to a great extent and diffraction to a minor extent give rise to the image, as opposed to ***phase contrast*** or ***interference*** methods of microscopy.

Brilliant Blue C See ***Brilliant Cresyl Blue***.

Brilliant Blue R See ***Coomassie Brilliant Blue***.

Brilliant Cresyl Blue (Brilliant Blue C) Dye used in staining of bone marrow smears.

bromelain Thiol protease (EC 3.4.22.4) from pineapple.

bromo-deoxyuridine See ***BUdR***.

bromophenol blue Dye used as pH indicator: changes from yellow to blue in the range 3.0–4.6

bromophenol red Dye used as pH indicator: changes from yellow to red in range 5.2–6.8

Bromoviruses Plant viruses with a genome of three linear, positive sense ssRNA molecules. Named originally after brome grass.

brown fat cells Brown fat is specialized for heat production and the ***adipocytes*** have many mitochondria in which an inner-membrane protein can act as an uncoupler of ***oxidative phosphorylation*** allowing rapid thermogenesis.

Brownian motion Random motion of small objects as a result of intermolecular collisions. First described by the 19th-century microscopist, Brown.

Brownian ratchet Mechanism proposed to explain protein translocation across membranes and force generation by polmerizing actin filaments. Relies upon asymmetry of *cis* and *trans* sides of the membrane or biased thermal motion as a result of polymerization. Still an hypothesis.

Brucella Genus of ***Gram-negative*** aerobic bacteria which occur as intracellular parasites or pathogens in man and other animals. *Brucella abortus* is responsible for spontaneous abortion in cattle and causes undulent fever (brucellosis), a persistent recurrent acute fever, in humans.

brush border The densely packed ***microvilli*** on the apical surface of, eg. intestinal epithelial cells.

Bruton's disease Sex-linked recessive ***agammaglobulinaemia*** caused by a deficiency in ***B-lymphocyte*** function. See ***btk***.

Bruton's tyrosine kinase See ***btk***.

Brx ***Dbl*** family member that modulates oestrogen receptor activity. May thus integrate cytoplasmic signalling mediated by rho (for which it is a ***GEF***) and nuclear receptors.

Bryophyta Plant phylum that includes mosses and liverworts.

bryostatin General name for a group of compounds isolated from bryozoans; activate protein kinase C (***PKC***), though after longer-term exposure cells down-regulate their PKC.

BSE (Bovine spongiform encepalopathy) Transmissible encephalopathy that affected large numbers of cattle in the UK during the 1990s and is widely believed to have arisen through consumption, by cattle, of feedstuff containing sheep tissues from animals with scrapie. A link with 'new variant CJD' is strongly suspected.

btk (Bruton's tyrosine kinase) ***Tyrosine kinase*** of ***tec*** family, defective in Bruton's agammaglobulinaemia. Mutations in btk lead to B-cell immunodeficiencies XLA in humans, Xid in mice. Overexpression of btk enhances calcium influx following B-cell antigen receptor crosslinking. ***Sab*** selectively binds the SH3 domain of btk. Btk interacts with membrane through ***PH domain***, and ***SHIP***, by reducing PIP3 levels, regulates this association.

BTX See ***batrachotoxin***.

budding A type of cell division in fungi and in protozoa in which one of the daughter cells develops as a smaller protrusion from the other. Usually the position of the budding cell is defined by polarity in the mother cell. In some protozoa the budded daughter may lie within the cytoplasm of the other daughter.

BUdR (bromo-deoxyuridine; deoxynucleoside of 5-bromo-uracil) Analogue of thymidine that induces point mutations because of its tendency to tautomerization: in the enol form it pairs with G instead of A. It is used as a mutagen, and also as a marker for DNA synthesis (the incorporation of BUdR can be recognized because the staining pattern differs: an even more sensitive method uses a monoclonal antibody staining procedure.)

buffer A system that acts to minimize the change in concentration of a specific chemical species in solution against addition or depletion of this species. pH buffers: weak acids or weak bases in aqueous solution. The working range is given by pKa ±1. Metal ion buffers: a metal ion chelator, eg. ***EDTA***, partially saturated by the metal ion acs as a buffer for the metal ion.

buffy coat Thin yellow-white layer of leucocytes on top of the mass of red cells when whole blood is centrifuged.

bufotenine (3-(2-(dimethylamino) ethyl)-1H-indol-5-ol; mappine) An indole alkaloid with hallucinogenic effects, isolated from *Piptadenia* spp. (Mimosidae); first isolated from skin glands of toad (*Bufo* sp.).

bullous pemphigoid Form of pemphigoid (which also affects mucous membranes), in which blisters (bulli) form on the skin. Patients have circulating antibody (usually IgG) to ***basement membrane*** of ***stratified epithelium***, although the antibody titre does not correlate with the severity of the disease.

bungarotoxins Toxins found in the venom of *Bungarus multicinctus*. α-bungarotoxin: polypeptide toxin (74 residues). A powerful antagonist of ***acetylcholine***, it causes a virtually irreversible block of the vertebrate neuromuscular junction by binding (as a monomer) to each of the α-subunits of the postsynaptic nicotinic acetycholine receptors (nAChR). Has been much used in identifying, quantifying and localizing these receptors on muscle cells. Will also bind some neuronal nAChR; β-bungarotoxin: a two-chain phospholipase A2 neurotoxin that acts at the presynaptic site of motor nerve terminals and blocks transmitter release. Subunit A (120 residues) is structurally homologous to other PLA2s, the B subunit (60 residues) has homology with Kunitz-type serine protease inhibitors and ***dendrotoxins***. Binds to subtype of voltage-sensitive potassium channels. κ-bungarotoxin: (bungarotoxin 3.1; toxin F, neuronal bungarotoxin) is a polypeptide (66 residues) from the venom of *Bungarus multicinctus*. Has considerable homology with α-bungarotoxin. Functional toxin a homodimer. Potent antagonist for a subset of neuronal nAChR but is much less active against muscle receptors.

Bungarus multicinctus Formosan snake (banded krait). See ***bungarotoxins***.

Bunyaviridae ssRNA enveloped viruses infecting vertebrates and arthropods. Genome consists of negative sense RNA molecules. Virion spherical or oval, 90–100µm diameter. Some genera contain organisms causing serious disease, eg. viral haemorrhagic fever.

Burkitt's lymphoma Malignant tumour of ***lymphoblasts*** derived from B-lymphocytes. Most commonly affects children in tropical Africa: both ***Epstein–Barr virus*** and immunosuppression due to malarial infection are involved.

burr cells Triangular helmet-shaped cells found in blood, usually indicative of disorders of small blood vessels.

bursa of Fabricius A ***lymphoid tissue*** found at the junction of the cloaca and the gut of birds giving rise to the so-called ***B-lymphocyte*** series.

burst-forming unit (BFU-E) A bone marrow ***stem cell*** lineage detected in culture by its mitotic response to ***erythropoietin*** and subsequent erythrocytic differentiation in about 12 mitotic cycles into erythrocytes.

butyric acid ($CH_3.CH_2.CH_2.COOH$) Acid from which butyrate ion is derived. Smells of rancid butter, hence the name.

butyrophilin Integral membrane glycoprotein of the immunoglobulin superfamily (59 kD) of mammary secretory epithelium. Secreted in association with milk-fat globule membrane. The extracellular domain of butyrophilin has features that suggest it may have a receptor function.

bystander help Lymphokine-mediated non-specific help by T-lymphocytes, stimulated by one antigen, to lymphocytes stimulated by other antigens.

bZip See ***basic leucine zipper***.

C

c- Prefix used to denote the normal cellular form of, eg., a gene such as *src* that is also found as a viral gene.

C1-C9 Proteins of the mammalian ***complement*** system. See also under individual numbered components.

C1 First component of ***complement***; actually three subcomponents, C1q, C1r and C1s, which form a complex in the presence of calcium ions. C1q, the recognition subunit, has an unusual structure of collagen-like triple helices forming a stalk for its Ig-binding globular heads. Upon binding to immune complexes the C1 complex becomes an active protease that cleaves and activates C4 and C2.

C2 Second component of ***complement***. A β-2-globulin.

C3 Third component of ***complement***, present in plasma at around 0.5–1mg.ml^{-1}. Both classical and alternate pathways converge at C3, which is cleaved to yield C3a, an ***anaphylatoxin***, and C3b, which acts as an opsonin and is bound by ***CR1***; C3b in turn can be proteolytically cleaved to iC3b (ligand for ***CR3***) and C3dg by C3b-inactivator. C3b complexed with factor B (to form C3bBb) will cleave C3 to give more C3b, although the C3bBb complex is unstable unless bound to ***properdin*** and a carbohydrate-rich surface. The C3b–C4b2a complex and C3bBb are both C5 convertases (cleave ***C5***). Cobra venom factor is homologous with C3b but the complex of cobra venom factor, properdin and factor Bb is insensitive to C3b-inactivator.

C4 Fourth component of ***complement***, although the third to be activated in the classical pathway. Becomes activated by cleavage (by C1) to C4b, which complexes with C2a to act as a C3 convertase, generating C3a and C3b. The C4b2a3b complex acts on C5 to continue the cascade.

C5 Fifth component of ***complement***, which is cleaved by C5-convertase to form C5a, a 74-residue anaphylotoxin and potent chemotactic factor for leucocytes, and C5b. C5a rapidly loses a terminal arginine to form C5a desarg, which retains chemotactic but not anaphylotoxin activity. C5b combines with C6, C7, C8 and C9 to form a membranolytic complex.

C6, C7 Sixth and seventh components of the ***complement*** cascade. Contain EGF-like motifs. See ***C5*** and ***C9***.

C8 Eighth component of ***complement***: 3 peptide chains, α, β and γ.

C9 Ninth component of ***complement***. Complexed with C5b, 6, 7, 8 it forms a potent membranolytic complex (sometimes referred to as the membrane attack complex, MAC). Membranes that have bound the complex have toroidal 'pores'; a single pore may be enough to cause lysis.

C2-kinin A kinin-like fragment generated from ***complement*** C2; causes vasodilation and increased vascular permeability. Distinct from ***bradykinin***.

C3 plants Plants that fix CO_2 in photosynthesis by the ***Calvin-Benson cycle***. The enzyme responsible for CO_2 fixation is ***RuDP carboxylase***, whose products are compounds containing three carbon atoms. C3 plants are typical of temperate climates. ***Photorespiration*** in these plants is high.

C3G Guanine nucleotide exchange factor (***GEF***) that activates ***Rap1***. C3G is involved in signalling from ***Crk to JNK***.

C4 plants Plants found principally in hot climates whose initial fixation of CO_2 in photosynthesis is by the ***HSK pathway***. The enzyme responsible is ***PEP carboxylase***, whose products contain four carbon atoms. Subsequently the CO_2 is released and refixed by the ***Calvin–Benson cycle***. The presence of the ***HSK pathway*** permits efficient photosynthesis at high light intensities and low CO_2 concentrations. Most species of this type have little or no ***photorespiration***.

C banding (centromeric banding) Method of defining chromosome structure by staining with ***Giemsa*** and looking at the ***banding pattern*** in the heterochromatin of the ***centromeric*** regions. Giemsa banding (G banding) of the whole chromosome gives higher resolution. Q banding is done with quinacrine.

C-EBP (CCAAT-enhancer binding protein) Group of transcription factors (α–δ) particularly implicated in adipocyte differentiation.

c-myc tag ***Epitope tag*** (EQKLISEEDL) derived from the c-myc protein.

C polysaccharide (C substance) Polysaccharide released by pneumococci which contains galactosamine-6-phosphate and phosphoryl choline. ***C-reactive protein*** is so called because it will precipitate this polysaccharide through an interaction with the phosphoryl choline.

C-proteins Striated muscle thick filament-associated proteins (140–150 kD) that show up in the C zone of the A band as 43nm transverse stripes. Structurally related to various other myosin-binding proteins (twitchin, ***titin***, ***myosin light chain kinase***, skelemin, 86 kD protein, projectin, ***M-protein***).

C-reactive protein A protein of the ***pentraxin*** family found in serum in various disease conditions particularly during the acute phase of immune response. C reactive protein is synthesized by ***hepatocytes*** and its production may be triggered by ***prostaglandin*** E1 or parogen. It consists of five polypeptide subunits forming a molecule of total molecular weight 105 kD. It binds to polysaccharides present in a wide range of bacterial, fungal and other cell walls or cell surfaces and to ***lecithin*** and to phosphoryl-or choline-containing molecules. It is related in structure to serum ***amyloid***. See also ***acute phase proteins*** and ***C polysaccharide***.

C-region The parts of the heavy or light chains of ***immunoglobulin*** molecules that are of constant sequence, in contrast to variable or V regions. The constancy of sequence is relative because there are several constant region genes and alleles thereof (see ***allotypes***), but within one animal homozygous at the light and heavy chain constant region genes all immunoglobulin molecules of any one class have constant sequences in their C regions. The constant region sequences for the various different types of immunoglobulin, eg. IgG, IgA, will vary.

c-strand An abbreviation for the term 'complementary strand' used of nucleic acids.

C-subfibre The third partial microtubule associated with the A- and B-tubules of the outer axonemal doublets in the basal body (and in the centriole) to form a triplet structure.

C substance See ***C polysaccharide***.

C-type lectins One of two classes of ***lectin*** produced by animal cells, the other being the ***S-type***. The C-type lectins require disulphide-linked cysteines and Ca^{2+} ions in order to bind to a specific carbohydrate (cf. S-type lectins). The carbohydrate recognition domain of C-type lectins consists of about 130 amino acids which contains 18 invariant residues in a highly conserved pattern. These invariant residues include cysteines which probably form disulphide bonds. So far, all identified C-type lectins are extracellular proteins and include both integral membrane proteins, such as the ***asialoglycoprotein*** receptor, and soluble proteins.

C-type virus Originally C-type particles identified in mouse tumour tissue and later shown to be oncogenic RNA viruses (Oncovirinae) that bud from the plasma membrane of the host cell starting as a characteristic electron-dense crescent. Include feline leukaemia virus, murine leukaemia and sarcoma viruses.

C value paradox Comparison of the amount of DNA present in the haploid genome of different organisms (the C value) reveals two problems: the value can differ widely between two closely related species, and there seems to be far more DNA in higher organisms than could possibly be required to code for the modest increase in complexity.

CAAT box Nucleotide sequence in many eukaryotic promoters usually about 75bp upstream of the start of transcription. Binds ***NF-1***.

cachectin Protein produced by macrophages that is responsible for the wasting (cachexia) associated with some tumours. Now known to be identical to ***tumour necrosis factor*** (TNF). Has three 17 kD subunits, all derived from a single highly conserved gene.

Caco cells (CAC0) Cell line derived from a primary colonic carcinoma of a 72-year-old male Caucasian. Epithelial morphology.

CADASIL (cerebral autosomal dominant arteriopathy with subcortical infarcts and leukoencephalopathy) Hereditary adult-onset condition causing stroke and dementia, mapped to Chr 19 and thought to be a defect in ***Notch***-3, though how a defect in this signalling pathway leads to the pathological effect remains unclear.

cadaverine (1,5-pentanediamine) Substance formed by microbial action in decaying meat and fish by decarboxylation of lysine. The smell can be imagined. Like many of the other diamines (eg. ***putrescine***) has effects on cell proliferation and differentiation.

cadherins Integral membrane proteins involved in calcium dependent cell adhesion. There are three types, named after their distributions: N-cadherin (neural); E-cadherin (epithelial) (equivalent to uvomorulin and L-CAM); and P-cadherin (placental). Formed of a 600 amino acid extracellular domain, containing four repeats believed to contain the Ca^{2+} binding sites, a transmembrane domain, and a 150 amino acid intracellular domain.

Caenorhabditis elegans Nematode much used in lineage studies since the number of nuclei is determined, and the nervous system is relatively simple. The organism can be maintained axenically and there are mutants in behaviour, in muscle proteins and in other features. Sperm are amoeboid and move by an unknown mechanism which does not seem to depend upon actin or tubulin.

caerulin Amphibian peptide hormone related to ***gastrin*** and ***cholecystokinin***.

caesium chloride (USA, cesium chloride) Salt that yields aqueous solutions of high density. When equilibrium has been established between sedimentation and diffusion during ultracentrifugation, a linear density gradient is established in which macromolecules such as DNA band at a position corresponding to their own buoyant density.

caffeine A ***xanthine*** derivative that elevates cAMP levels in cells by inhibiting phosphodiesterases.

caged-ATP A derivative of ATP that is not biologically active until a photosensitive bond has been cleaved.

Cairns mechanism A mechanism for the replication of a double-stranded circular DNA molecule. Replication is initiated at a fixed point and proceeds either uni- or bidirectionally.

calbindin Vitamin-D-induced calcium-binding protein (28 kD) found in primate striate cortex and other neuronal tissues. Contains an ***EF-hand*** motif.

calcein A calcium-chelating agent that fluoresces brightly in the presence of bound calcium. The acetomethoxy derivative can be transported into live cells and the reagent is useful as a viability test and for short-term marking of cells.

calcicludine Polypeptide toxin (60 residues) from *Dendroaspis angusticeps*. Blocks most high-threshold calcium channels (L-, N- or P-type). Structurally homologous to Kunitz-type serine protease inhibitors and ***dendrotoxins***.

calciferol (vitamin D) See Table V1.

calcimedins Annexins; see Table A3.

calcineurin Calmodulin-stimulated protein-phosphatase (EC 3.1.3.16), the major calmodulin-binding protein in brain. Enzymic activity is inhibited by binding of ***immunophilin***-ligand complex (immunophilin alone does not bind) and therefore may play a part in the mechanism of action of ***cyclosporin A*** and FK506. Thought also to be involved in the control of sperm motility.

calcinosis See ***CREST***.

calciosome Now discredited. A membrane compartment proposed to contain the intracellular calcium store released in response to hormonal activity and thought to be distinct from the endoplasmic reticulum.

calciseptine Polypeptide toxin (60 residues) from *Dendroaspis polylepis*. Specific blocker of some L-type calcium channels that will cause relaxation of smooth muscle and inhibition of cardiac muscle but has no effect on skeletal muscle.

calcitonin A polypeptide hormone produced by C-cells of the thyroid that causes a reduction of calcium ions in the blood.

calcitonin family peptides (ADM; CGRP1; CGRP2; amylin; calcitonin) Family of small (32–51 residue) highly homologous peptides that act through seven-transmembrane G-protein coupled receptors. Adrenomedullin (ADM; 51 residues) is a potent vasodilator and has receptorson astrocytes; amylin (37 residues) is thought to regulate gastric emptying and carbohydrate metabolism; calcitonin (32 residues) is involved in control of bone metabolism; calcitonin gene-related peptides 1 and 2 (37 residues) regulate neuromuscular junctions, antigen presentation, vascular tone and sensory neurotransmission. Receptors are themselves regulated by ***RAMPs***.

calcitonin gene-related peptide (CGRP) Neuropeptide of 37 amino acids with structural homology to salmon calcitonin. Colocalizes with ***substance P*** in neurons. Intracerebral administration of CGRP leads to a rise in noradrenergic sympathetic outflow, a rise in blood pressure and a fall in gastric secretion. A family of related peptides exist (***calcitonin family peptides***). See ***RAMPs***.

calcitriol (1α, 25-dihydroxyvitamin D_3) The form of vitamin D_3 that is biologically active in intestinal transport and calcium resorption by bone.

calcium ATPase Usually used of the calcium-pumping ATPase present in high concentration as an integral membrane protein of the sarcoplasmic reticulum of muscle. This pump lowers the cytoplasmic calcium level and causes contraction to stop. Normal function of the pump seems to require a local phospholipid environment from which cholesterol is excluded.

calcium-binding proteins There are two main groups of calcium-binding proteins, those that are similar to ***calmodulin***, and are called ***EF-hand*** proteins, and those that bind calcium and phospholipid (eg. ***lipocortin***) and which have been grouped under the generic name of ***annexins***. Many other proteins will bind calcium, although the binding site usually has considerable homology with the calcium binding domains of calmodulin.

calcium channel Membrane channel that is specific for calcium. Probably the best characterized is the voltage-gated channel of the sarcoplasmic reticulum which is ryanodine-sensitive. See ***voltage-sensitive calcium channels.***

calcium current Inflow of calcium ions through specific ***calcium channels***. Critically important in release of transmitter substance from presynaptic terminals.

calcium-dependent regulator protein (CDRP) Early name for ***calmodulin***.

calcium pump A ***transport protein*** responsible for moving calcium out of the cytoplasm. See ***calcium ATPase***.

calcivirus Genus of ***Picornaviridae***.

calcyclin Prolactin receptor associated protein, one of a family of small (around 10 kD) calcium-binding proteins containing the ***EF-hand*** motif, originally isolated from Erlich ascites tumour cells, but human and rat forms now identified. Regulated through the cell cycle. Binds to annexin II (p36) and to glyceraldehyde-3-phosphate dehydrogenase.

calcyphosin (thyroid protein p24) Calcium-binding protein that contains an ***EF-hand*** motif.

caldesmon Protein originally isolated from smooth muscle (h-caldesmon; 120–150 kD on gels but 88.7 kD from sequence) also found in non-muscle cells (l-caldesmon; 70-80 kD on gels but 58.8 kD from sequence). Normally dimeric, binds to ***F-actin*** blocking the myosin binding site. Calcium–calmodulin binding to caldesmon causes its release from actin, though phosphorylation of caldesmon may also affect the link with actin. Caldesmon can block the effect of ***gelsolin*** on ***F-actin*** and will dissociate actin-gelsolin complexes and actin–profilin complexes.

calelectrins Membrane-associated proteins (70 kD and 32 kD) of the annexin family. Originally from *Torpedo*, but subsequently found in bovine liver. May regulate exocytosis.

calgranulins Calcium binding myeloid-associated proteins (8 kD and 14 kD; also known as p8,14 and as MRP-8, MRP-14) expressed at high levels in neutrophils and monocytes but lost during differentiation to macrophages. Related to migration inhibitory factor (MIF). Associated with ***keratinocyte*** cytoskeleton. Part of the ***S100*** family.

calitoxin Small toxic peptide (46 residues) from sea anemone *Calliactis parasitica* that acts on neuronal sodium channels.

Calliactis parasitica Sea anemone (an anthozoan ***coelenterate***) that lives commensally with hermit crabs. See ***calitoxin***.

callose A plant cell-wall polysaccharide (a β-(1-3)-***glucan***) found in phloem ***sieve plates***, wounded tissue, pollen tubes, cotton fibres and certain other specialized cells.

callus (1) *Bot*. Undifferentiated plant tissue produced at wound edge: callus tissue can be grown *in vitro* and induced to differentiate by varying the ratio of the hormones ***auxin*** and ***cytokinin*** in the medium. (2) *Path*. Mass of new bony trabeculae and cartilaginous tissue formed by ***osteoblasts*** early in the healing of a bone fracture.

calmidazolium (Compound R24571) Inhibitor of calmodulin-regulated enzymes; also blocks sodium channel and voltage-gated calcium channel.

calmodulin Ubiquitous and highly conserved calcium-binding protein (17 kD) with four ***EF-hand*** binding sites for calcium (three in yeast). Ancestor of ***troponin*** C, ***leiotonin*** C, and ***parvalbumin***.

calmodulin-dependent kinase I See ***CaMKI***.

calnexin Calcium-binding lectin-like protein (67 kD, 592 residues) of endoplasmic reticulum that couples glycosylation of newly synthesized proteins with their folding. Calnexin and ***calreticulin*** act together as chaperones for newly synthesized proteins and prevent ubiquitinylation and proteosomal degradation. Can be phosphorylated by casein kinase II.

calpactins Calcium-binding proteins from cytoplasm. Calpactin II is identical to ***lipocortin***, and is one of the major targets for phosphorylation by pp60src. See ***annexin***

calpain Calcium-activated cytoplasmic proteases containing the ***EF-hand*** motif. calpain I is activated by micromolar calcium, calpain II by millimolar calcium. Calpain has two subunits, the larger (80 kD) has four domains: one homologous with ***papain***: one with ***calmodulin***, the smaller (30 kD) has one domain homologous with calmodulin. First isolated from erythrocytes, but now described from other cells.

calpastatin Cytoplasmic inhibitor of calcium-activated protease ***calpain***.

calphobindins ***Annexins*** V and VI (35 kD) found in placenta (see Table A3). Have substantial sequence homology with ***lipocortin*** and may function like ***calelectrin***.

calphostin C One of a group of compounds isolated from *Cladosporium cladosporioides* that will inhibit ***protein kinase C*** with some specificity, though inhibits other classes of kinases if present in high concentration.

calponin Calcium- and calmodulin binding troponin T-like protein (34 kD) isolated from chicken gizzard and bovine aortic smooth muscle. Interacts with ***F-actin*** and tropomyosin in a calcium-sensitive manner and acts as a regulator of smooth muscle contraction (inhibits when not phosphorylated). Distinct from ***caldesmon*** and ***myosin light chain kinase***, but has some antigenic cross-reactivity with cardiac troponin-T.

calregulin See ***calreticulin***.

calreticulin (calregulin) ***Calcium-binding protein*** of the ***endoplasmic reticulum***. Acts as a chaperone for newly synthesized proteins, possibly in conjunction with ***calnexin***. May be more selective in the proteins with which it associates than calnexin.

calretinin Neuronal protein (29 kD) of the ***calmodulin*** family isolated from chick retina. Has 58% sequence homology with calbindin, the intestinal cell isoform. Contains an ***EF-hand*** motif.

calsequestrin Protein (44 kD) found in the cisternae of sarcoplasmic reticulum: sequesters calcium.

calspectin Non-erythroid ***spectrin***.

calspermin High-affinity calcium/calmodulin-binding protein found in post-meiotic male germ cells. Represents the C-terminal 169 amino acids of ***protein kinase IV*** and is produced from a promoter located in an intron of the protein kinase IV gene. Lacks the kinase domain.

caltractin Calcium-binding (***EF-hand***) protein (20 kD) from *Chlamydomonas reinhardtii*. Major component of the contractile striated rootlet system that links basal bodies to the nucleus in *Chlamydomonas*. Part of the calmodulin/troponin C family, it has sequence homology with 20 kD calcium-binding proteins (***centrins***) found in other basal body-associated structures and with the *cdc31* gene product associated with spindle pole body duplication in the yeast *Saccharomyces cerevisiae*.

caltrin Inhibitor of calcium ion transport found in bovine seminal plasma (47 amino acids, MW 5411, on gels 10 kD app.) and that resembles seminal antibacterial protein (confusingly called plasmin, though not related to the protease).

calvarium One of the bones that makes up the vault of the skull (in humans these are the frontal, two parietals, occipital and two temporals). Calvaria are often used in organ culture to investigate bone catabolism or synthesis.

Calvin–Benson cycle (Calvin cycle) Metabolic pathway responsible for photosynthetic CO_2 fixation in plants and bacteria. The enzyme that fixes CO_2 is ***RuDP carboxylase***. The cycle is the only photosynthetic pathway in ***C3 plants*** and the secondary pathway in ***C4 plants***. The enzymes of the pathway are present in the stroma of the chloroplast.

calyculin A Toxin from marine sponge, *Discodermia calyx*; potent tumour promoter and an inhibitor of protein phosphatases of types 1 and 2a.

CAM See ***crassulacean acid metabolism*** or ***cell adhesion molecule***.

CAM See ***crassulacean acid metabolism***.

CaM kinase II (calcium/calmodulin-dependent kinase II; CaMKII) A multi-substrate calcium-sensitive kinase composed of four homologous subunits(α, β, γ, δ) all encoded by different genes. The heteromultimeric holoenzyme (500–600 kD) has 10 or 12 subunits with the ratio of α,β,γ and δ reflecting that present in the cell. Ca^{2+}/calmodulin-dependent protein kinase II (CaMKII) has been implicated in various neuronal functions, including ***synaptic plasticity***. It is highly concentrated in the postsynaptic region and undergoes autophosphorylation at several sites in a manner that depends on the frequency and duration of Ca^{2+} spikes. Constitutively active CaMKII produces dendritic exocytosis in the absence of calcium stimulus. CaMKII activation is the primary event leading to inactivation of both CSF (cytostatic factor) and ***MPF*** (maturation-promoting factor) in mammalian eggs.

cambium (1) Bot. ***Meristematic*** plant tissue, commonly present as a thin layer which forms new cells on both sides. Located either in vascular tissue (vascular cambium), forming xylem on one side and phloem on the other, or in cork (cork cambium or phellogen). (2) *Anat*. Inner region of the ***periosteum*** from which ***osteoblasts*** differentiate.

camera lucida Attachment for a microscope that permits both a view of the object and, simultaneously, of the viewer's hand and drawing implement, thus facilitating accurate drawing of the object of interest.

CaMKI (calmodulin-dependent kinase I) Calcium-regulated kinase (37–42 kD) that is known to phosphorylate ***synapsins***, ***CREB*** and ***CFTR***. Widely distributed. See ***CaMKII***.

cAMP See ***cyclic AMP***.

cAMP- and cGMP-dependent protein kinase phosphorylation site Both cAMP- and cGMP-dependent protein kinases phosphorylate exposed serine or threonine residues near at least two consecutive N-terminal basic residues, with a consensus pattern: ***RK(2)-x-ST***.

camptothecin Cytotoxic plant alkaloid originally isolated from *Camptotheca acuminata*. Anti-cancer drug, inhibits DNA ***topoisomerase*** I.

Campylobacter Genus of ***Gram-negative*** microaerophilic motile bacteria with a single flagellum at one or both poles. Found in reproductive and intestinal tracts of mammals. Common cause of food poisoning and can also cause opportunistic infections, particularly in immunocompromised patients.

canaliculi In bone, channels that run through the calcified matrix between lacunae containing ***osteocytes***. In liver, small channels between ***hepatocytes*** through which bile flows to the bile duct and thence to the intestinal lumen.

cancellous bone Adult bone consisting of mineralized regularly ordered parallel collagen fibres more loosely organized than the lamellar bone of the shaft of adult long bones. Found in the end of long bones; also known as trabecular bone.

cancer Originally descriptive of breast carcinoma, now a general term for diseases caused by any type of malignant tumour.

cancer susceptibility gene See ***tumour suppressor*** gene.

Candida albicans A dimorphic fungus that is an opportunistic pathogen of humans (causing candidiasis).

candidiasis Infection by ***Candida albicans***, common on mucous membranes ('thrush'), but in immunosuppressed patients can opportunistically infect many tissues.

Canis familiaris Dog.

cannabinoid Group of compounds, all derivatives of 2-(2-isopropyl-5-methylphenyl)-5-pentylresorcinol, found in cannabis. Most important members of the group are cannabidiol, cannabidol and various tetrahydrocannabinols (THCs). Bind to the ***cannabinoid receptors*** and mimic actions of endogenous agonists ***anandamide*** and palmitoyl ethanolamine.

cannabinoid receptors (CB1, CB2) Seven membrane-spanning G-protein coupled receptors for cannabinoids (and endogenous agonists such as anandamide). CB1 receptors are mostly found in brain and may mediate the psychotropic activities, the CB2 receptors are more peripheral and found extensively in the immune system.

canonical Classical, archetypal or prototypic. For example, the canonical polyadenylation sequence is AATAAA.

CAP (catabolite gene activator protein) Protein from *E. coli* that regulates the expression of genes for the use of alternative carbon sources if glucose is not available. For example, CAP and the ***lactose repressor*** protein act together to enable lactose to be utilized. If glucose is present CAP will not bind to DNA; in the absence of glucose cAMP levels rise, CAP binds cAMP, undergoes a conformational change, and is then capable of binding to DNA and promoting transcription of derepressed genes.

cap-binding protein Protein (24 kD) with affinity for cap structure at 5′-end of mRNA that probably assists, together with other initiation factors, in binding the mRNA to the 40S ribosomal subunit. Translation of mRNA ***in vitro*** is faster if it has a cap-binding protein.

CAP proteins CAP1 = ***FADD***; CAP2 = hyperphosphorylated FADD; CAP3 = unknown; CAP4 = pro-FLICE; CAP5, CAP6 = cleaved prodomains of ***FLICE***.

CAP-18 Lipopolysaccharide-binding protein (18 kD) isolated from rabbit neutrophils. May mediate interaction of antimicrobial ***protegrins*** with surfaces of ***Gram-negative*** bacteria.

capacitance flicker Brief closings of an ***ion channel*** during its open phases, observed during ***patch clamp***; or rapid transition of an ion channel between open and closed states such that the individual channel openings cannot be distinguished properly due to the limited bandwidth of the patch clamp amplifier.

capacitation A process occurring in mammalian sperm after exposure to secretions in the female genital tract. Surface changes take place probably involved with the ***acrosome*** which are necessary before the sperm can fertilize an egg.

capillary The small blood vessels that link arterioles with venules. Lumen may be formed within a single endothelial cell and have a diameter smaller than that of an erythrocyte, which must deform to pass through. Blood flow through capillaries can be regulated by precapillary sphincters, and each capillary probably only carries blood for part of the time.

capnine Sulphonolipid isolated from the envelope of the *Cytophaga/Flexibacter* group of ***Gram-negative*** bacteria. The acetylated form of capnine seems to be necessary for gliding motility.

capping (1) Movement of crosslinked cell surface material to the posterior region of a moving cell, or to the perinuclear region. (2) The intracellular accumulation of intermediate filament protein in the pericentriolar region following microtubule disruption by colchicine. (3) The blocking of further addition of subunits by binding of a cap protein to the free end of a linear polymer such as actin. See also ***cap-binding protein***.

capsaicin (8-methyl-N-vanillyl-6-nonenamide) Molecule in chilli peppers that makes them hot and will stimulate release of neurogenic peptides (***substance P***, neurokinins) from sensory neurons. Acts on the ***vanilloid receptor-1*** (VR1). Can be used to desensitize nociceptors to which it binds, and which may eventually be killed.

capsazepine Competitive ***capsaicin*** antagonist.

Capsicum Genus that includes red peppers, pimentoes and green peppers. See ***capsaicin***.

capsid A protein coat that covers the nucleoprotein core or nucleic acid of a virion. Commonly shows icosahedral symmetry and may itself be enclosed in an envelope (as in the ***Togaviridae***). The capsid is built up of subunits (some integer multiple of 60, the number required to give strict icosahedral symmetry) that self-assemble in a pattern typical of a particular virus. The subunits are often packed, in smaller capsids, into 5- or 6-membered rings (pentamers or hexamers) that constitute the morphological unit (capsomere). The packing of subunits is not perfectly symmetrical in most cases and some units may have strained interactions and are said to have quasi-equivalence of bonding to adjacent units.

capsomeres See ***capsid***.

capsule *Bact.* Thick gel-like material attached to the wall of ***Gram-positive*** or ***Gram-negative*** bacteria, giving colonies a 'smooth' appearance. May contribute to pathogenicity by inhibiting phagocytosis. Mostly composed of very hydrophilic acidic polysaccharide, but considerable diversity exists. *Path.* Cellular response in invertebrate animals to a foreign body too large to be phagocytosed. A multicellular aggregate of ***haemocytes*** or ***coelomocytes*** isolates the foreign object. In some insects the capsule is apparently acellular and composed of ***melanin***.

capZ Microfilament capping protein (32–36 kD) found in *Dictyostelium* and *Acanthamoeba* and which binds to the barbed ends of thin filaments in the ***Z disc*** of striated muscle. Widely distributed in vertebrate cells, though in non-muscle cells is predominantly in the nucleus. Identical to β-actinin. Some isoforms may bind to the pointed ends of microfilaments.

Carassius auratus The goldfish, one of the carp family.

carbachol (carbamoyl choline) Parasympathomimetic drug formed by substituting the acetyl of acetylcholine with a carbamoyl group; acts on both ***muscarinic*** and ***nicotinic acetylcholine receptors*** and is not hydrolyzed by acetylcholine esterase.

carbamoyl Acyl group -CO-NH2.

carbamoyl choline See ***carbachol***.

carbamoylcholine See ***carbachol***.

carbamyl (carbamoyl) Obsolete: use carbamoyl.

carbohydrates Very abundant compounds with the general formula $C_n(H_2O)_n$. The smallest are monosaccharides like glucose; polysaccharides (eg. starch, cellulose, glycogen) can be large and indeterminate in length.

carboxyglutamate (γ-carboxyglutamate) An amino acid found in some proteins, particularly those that bind calcium. Formed by post-translational carboxylation of glutamate.

carboxypeptidase Enzymes (particularly of pancreas) that remove the C-terminal amino acid from a protein or peptide. Carboxypeptidase A (EC 3.4.17.1) will remove any amino acid; carboxypeptidase B (EC 3.4.17.2) is specific for terminal lysine or arginine.

carboxysome Inclusion body (polyhedral body; 90–500nm diameter) found in some Cyanobacteria and autotrophic bacteria; contains ***ribulose bisphosphate carboxylase*** (RUBISCO) and is involved in carbon dioxide fixation.

carcinoembryonic antigen (CEA) Antigen found in blood of patients suffering from cancer of colon and some other diseases, that is otherwise normally found in foetal gut tissue.

carcinogen An agent capable of initiating development of malignant tumours. May be a chemical, a form of electromagnetic radiation, or an inert solid body.

carcinogenesis The generation of cancer from normal cells, correctly the formation of a ***carcinoma*** from epithelial cells, but often used synonymously with ***transformation, tumorigenesis***.

carcinoid Intestinal tumour arising from specialized cells with paracrine functions (APUD cells), also known as argentaffinoma. The primary tumour is commonly in the appendix, where it is clinically benign; hepatic secondaries may release large amounts of vasoactive amines such as serotonin to the systemic circulation.

carcinoma Malignant neoplasia of an epithelial cell: by far the commonest type of tumour. Those arising from glandular tissue are often called ***adenocarcinomas***. Carcinoma cells tend to be irregular with increased basophilic staining of the cytoplasm, have an increased nuclear: cytoplasmic ratio and polymorphic nuclei.

carcinosarcoma A mixed tumour with features of both carcinoma and sarcoma.

cardiac cell Strictly speaking, any cell of or derived from the cardium of the heart, but often used loosely to describe heart cells.

cardiac glycoside Specific blockers of the ***Na^+/K^+ATPase*** especially of heart muscle, eg. ***strophanthin***.

cardiac jelly Gelatinous extracellular material that lies between endocardium and myocardium in the embryo.

cardiac muscle See ***muscle***.

cardiolipin A diphosphatidyl glycerol that is found in the membrane of *Treponema pallidum* and is the antigen detected by the Wasserman test for syphilis.

cardiotrophin-1 (CT-1) Cytokine (201 amino acids) belonging to the ***IL-6 cytokine*** family. Binds to hepatocyte cell lines and induces synthesis of various ***acute-phase proteins***, is a potent cardiac survival factor and supports long-term survival of spinal motoneurons.

cardiovirus Genus of viruses belonging to the family picornaviridae, isolated mostly from rodents, which cause encephalitis and myocarditis.

carditis Inflammation of the heart, including pericarditis, myocarditis and endocarditis, according to whether the enveloping outer membrane, the muscle or the inner lining is affected.

carnitine (β-hydroxy-β-trimethyl-aminobutyric acid) Compound that transports long-chain fatty acids across the inner mitochondrial membrane in the form of acyl-carnitine. Sometimes referred to as vitamin B_t or vitamin B_7. See Table V1.

carnosine (β-Ala-His) Dipeptide found at millimolar concentration in vertebrate muscle.

Carnoy Fixative containing ethanol, chloroform and acetic acid. Better for nuclear structure than for cytoplasm.

carotenes Hydrocarbon carotenoids usually with nine conjugated double bonds. Beta-carotene is the precursor of vitamin A, each molecule giving rise to two vitamin A molecules.

carotenoids Accessory lipophilic photosynthetic pigments in plants and bacteria, including ***carotenes*** and ***xanthophylls***; red, orange or yellow, with broad absorption peaks at 450–480nm. Act as secondary pigments of the ***light-harvesting system***, passing energy to ***chlorophyll*** and as protective agents, preventing photoxidation of chlorophyll. Found in chloroplasts and also in plastids in some non-photosynthetic tissues, eg. carrot root.

carotid body cell Cells derived from the neural crest, involved in sensing pH and oxygen tension of the blood.

carrageenan (carrageenin) Sulphated cell wall polysaccharide found in certain red algae. Contains repeating sulphated disaccharides of galactose and (sometimes) anhydrogalactose. It is used commercially as an emulsifier and thickener in foods, and is also used to induce an inflammatory lesion when injected into experimental animals (probably activates ***complement***).

carrageenin See ***carrageenan***.

cartilage ***Connective tissue*** dominated by ***extracellular matrix*** containing ***collagen*** type II and large amounts of ***proteoglycan***, particularly ***chondroitin sulphate***. Cartilage is more flexible and compressible than bone and often serves as an early skeletal framework, becoming mineralized as the animal ages. Cartilage is produced by ***chondrocytes*** that come to lie in small lacunae surrounded by the matrix they have secreted.

Cas (p130Cas) Protein encoded by *Crkas* gene (*Crk*-associated protein). Adaptor molecule with SH3 domain, multiple YXXP motifs and proline-rich region. Involved in induction of cell migration and apparently contributes to tumour invasiveness. Highly homologous to p105HEF.

caseation See ***caseous necrosis***.

casein Group of proteins isolated from milk. α- and β-caseins are amphipathic polypeptides of around 200 amino acids with substantial hydrophobic C-terminal domains that associate to give micellar polymers in divalent cation-rich medium. κ-casein is a glycoprotein rather different from α- and β-casein.

casein kinase (CKII) Casein kinase II is thought to regulate a broad range of transcription factors in which it binds the ***basic leucine zipper*** (bZIP) DNA-binding domains. CKII is present in brain and has been associated with process of ***long-term potentiation*** by phosphorylating proteins important for neuronal plasticity.

casein kinase II phosphorylation site Casein kinase II phosphorylates exposed Ser or sometimes Thr residues, provided that an acidic residue is present three residues from the phosphate acceptor site. Consensus pattern: (S/T)-x-x-(D/E).

caseous necrosis (caseation) The development of a necrotic centre (with a cheesy appearance) in a tuberculous lesion.

Casparian band Region of plant cell wall specialized to act as a seal to prevent back-leakage of secreted material (analogous to ***tight junction*** between epithelial cells). Found particularly where root parenchymal cells secrete solutes into xylem vessels.

caspases Family of proteases involved in processing of ***IL-1*** β and in ***apoptosis***. See Table C1.

cassette A pre-existing structure into which an insert can be moved. Fashionably used to refer to certain vectors.

cassette mechanism Term used for genes such as the a- and α-genes that determine ***mating-type*** in yeast; either one or the other is active. In this ***gene conversion*** process, a double-stranded ***nuclease*** makes a cut at a specific point in the MAT locus, the old gene is replaced with a copy of a silent gene from one or other flanking region, and the new copy becomes active. As the process involves replacing one ready-made construct with another in an active 'slot', it is called a cassette mechanism.

castanospermine Alkaloid inhibitor of α-glucosidase I, of which the effect is to leave N-linked oligosaccharides in their 'high-mannose', unmodified state.

Castleman's disease Disease characterized by lymph node swelling, hypergammaglobulinaemia, increased levels of ***acute-phase proteins*** and increased numbers of platelets. Probably caused by excess ***IL-6*** production.

CAT See ***chloramphenicol acetyltransferase***.

catabolin Protein, later shown to be interleukin-1 ***IL-1***, that stimulates the breakdown of connective tissue extracellular matrix.

catabolism The sum of all degradative processes, the opposite of ***anabolism***.

Table C1. Caspases

Name	Synonyms	Substrate
Caspase 1	ICE	Pro-IL-1β
Caspase 2	Ich-1_L	
Caspase 3	CPP32, Yama, apopain	PARP, PKCδ, actin, Gas2, PAK2, procaspases 6, 9, U1-SnRNP
Caspase 4	Tx/Ich-2, ICE_{rel}-II	pPro-ICE
Caspase 5	ICE_{rel}-III, Ty	
Caspase 6	Mch2	Lamins A, C, B1
Caspase 7	Mch3, CMH-1, ICE-LAP3	PARP
Caspase 8	Mch5, MACH, FLICE	Procaspases 3, 4, 7, 9
Caspase 9	Mch6, ICE-LAP6	
Caspase 10	Mch4	Procaspases 3, 7

Based on Villa P, Kaufman SH and Earnshaw WC *Trends in Biochemical Sciences* 1997, 22: 388.

catabolite Product of catabolism, the breakdown of complex molecules into simpler ones.

catabolite gene activator protein See ***CAP***.

catabolite repression Inducible enzyme systems in some microorganisms (such as the ***lac operon***) that are repressed when a more favoured carbon source, such as glucose, is available. Repression in *E. coli* is partially relieved if cAMP is bound to the cAMP-catabolite activator protein (cAMP receptor protein; CRP) that binds to DNA upstream of the repressed operon concerned. Catabolite repression (of the respiratory system) is seen in yeast in high glucose concentrations, though the mechanism is different.

catalase Tetrameric haem enzyme (EC 1.11.1.6; 245kD) that breaks down hydrogen peroxide.

catalytic antibody (abenzyme) Antibody raised against a transition-state analogue (eg. a phosphate analogue of a carboxylic acid ester transition state) that can then catalyse the analogous chemical reaction, though not as effectively as a true enzyme.

cataract Opacity of the lens of the eye.

catch muscle See ***adductor muscle***.

catecholamine A type of ***biogenic amine*** derived from tyramine, characterized as alkylamino derivatives of *o*-dihydroxybenzene. Catecholamines include ***adrenaline, noradrenaline*** and ***dopamine***, with roles as ***hormones*** and ***neurotransmitters***.

catenate Two or more circular DNA molecules where one or more circles run through the enclosed space of another like links in a chain.

catenins Proteins associated with the cytoplasmic domain of ***uvomorulin*** and presumably involved in linking to the cytoskeleton (α-catenin, 102 kD; β-catenin, 88 kD; γ-catenin, 80 kD).

cathelin Protein (11 kD) isolated from porcine neutrophils that was originally described as being a cysteine protease inhibitor though subsequent reports have suggested that this is due to contamination with ***PLCPI***. Cathelin-like sequences are found upstream of several antimicrobial peptides (protegrins) and are expressed in the pro-peptides.

cathepsins Intracellular proteolytic enzymes of animal tissues, such as cathepsin B (EC 3.4.22.1), a lysosomal thiol proteinase; C, dipeptidyl peptidase (EC 3.4.14.1); D (EC 3.4.23.5) which has pepsin-like specificity; G (EC 3.4.23.5), similar to chymotrypsin; H, which possesses aminopeptidase activity; N, which attacks N-terminal peptides of collagen, and so on.

cationic proteins Proteins of azurophil granules of neutrophils, rich in arginine. A chymotrypsin-like protease found in azurophil granules is also very cationic as is cathepsin G and neutrophil elastase. Eosinophil cationic protein (21 kD) is particularly important because it damages ***schistosomula*** *in vitro*.

cationized ferritin Ferritin, treated with dimethyl propanediamine, and used to show, in the electron microscope, the distribution of negative charge on the surface of a cell. The amount of cationic ferritin binding is very approximately related to the surface charge.

caudate nucleus The most frontal of the basal ganglia in the brain. Damage to caudate neurons is characteristic of ***Huntington's chorea*** and other motor disorders.

Caulobacter Genus of ***Gram-negative*** aerobic bacteria that have a stalk or holdfast. Found in soil and fresh water.

caveola (plural, caveolae) Small invagination of the plasma membrane characteristic of many mammalian cells and associated with endocytosis. The membrane of caveolae contain integral membrane proteins, ***caveolins*** (21-24 kD) that interact with heterotrimeric G-proteins. Caveolar membranes are enriched in cholesterol and sphingolipids and may be the efflux route for newly synthesized lipids. ***Clathrin*** is not associated with caveolae.

caveolin Family of integral membrane proteins including VIP21-caveolin, M-caveolin (from muscle), some of which are tissue specific, and which are associated with ***caveolae***.

Cavia porcellas Guinea pig.

cbl Oncogene of a murine retrovirus that induces lymphomas and leukaemias. Protein has N-terminal transforming region with phosphotyrosine-binding (PTB) domain and C-terminal region with RING finger motif, large proline-rich region and a leucine zipper. Cbl serves as a substrate for receptor and non-receptor tyrosine kinases and binds to Grb2, Crk and p85 of PI-3-kinase. May act as a negative regulator of tyrosine kinase signalling. Signals to JNK through C3G. See also Table O1.

CBP See ***CREB-binding protein***.

CC10 See ***uteroglobin***.

CCAAT box Consensus sequence for RNA polymerase, found at about –80 bases relative to the transcription start site. Less well conserved than the ***TATA box***.

CCAAT box-binding transcription factor See ***CTF***.

CCCP (m-chloro-carbonylcyanide-phenylhydrazine) An ***uncoupling agent*** that dissipates proton gradients across membranes.

CCK (cholecystokinin) Polypeptide hormone (33 residues) secreted by the duodenum. Stimulates secretion of digestive enzymes by the pancreas and contraction of the gall bladder. The C-terminal octapeptide is found in some dorsal root ganglion neurons where it presumably acts as a peptide neurotransmitter.

CD See ***circular dichroism***.

CD antigens See table of 'cluster of differentiation' antigens (Table C2).

CD-21 See ***CR2***.

CD11b/CD18 See ***CR3***.

CD43 See ***leukosialin***.

cdc genes Cell division cycle genes, of which many have now been defined, especially in yeasts. See ***cyclin***. The cyclin-dependent kinases are also known as cdc2 kinases.

CDEP Human protein containing ***ezrin***-like domain of band 4.1 superfamily and also ***DH domain***. Found in differentiated chondrocytes and various foetal and adult tissues.

cdk See ***cyclin-dependent kinases***.

cDNA (complementary DNA) Viral ***reverse transcriptase*** can be used to synthesize DNA that is complementary to RNA (eg. an isolated mRNA). The cDNA can be used, eg. as a probe to locate the gene or can be cloned in the double-stranded form.

CDP See ***cytidine 5′ diphosphate***.

CDRP See ***calmodulin***.

CEA See ***carcinoembryonic antigen***.

cecropin One of a group of inducible antibacterial proteins purified and characterized from *Hyalophora cecropia* (silkmoth) pupae, and now found in several other species of endopterygote insects. Small basic proteins that cause ***lysis*** of both ***Gram-positive*** and ***Gram-negative bacteria***.

ced mutant Genes identified in ***Caenorhabditis elegans*** after studies of developmental mutations in which cells did not die when expected. *ced* ('cell death') genes are thus thought to be involved in the pathways that control ***apoptosis***.

celiac See ***coeliac***.

celiac disease (USA) See ***coeliac disease***.

cell An autonomous self-replicating unit (in principle) that may constitute an organism (in the case of unicellular organisms) or be a subunit of multicellular organisms in which individual cells may be more or less specialized (differentiated) for

Table C2. CD antigens

Cluster designation	Main cellular expression of antigen	Other names	Antigen (kDa) based on sequence	Comment
CD1	Cortical thymocytes (strong), Langerhans cells, B-cell subset, dendritic cells	T6	35	Associated with β2 microglobulin. Similar to MHC Class I, probably have role in presentation
CD2	T-cells, thymocytes, NK cells	E rosette receptor, leucocyte function antigen 2 (LFA-2)	37	Interacts with CD58. Ig superfamily
CD3	Thymocytes, mature T-cells	T-cell receptor complex	16, 20, 25-28	Complex of several proteins
CD4	T-helper/inducer cells, monocytes, macrophages	Receptor for MHC class II and HIV antigens	48	Ig superfamily member. Binds MHC Class II and is an important accessory molecule
CD5	Thymocytes, T-cells, B-cell subset	Leu-1/Ly-1	52	Coprecipitates with TCR
CD6	Thymocytes, T-cells, B-cell subset	T12	69	Extracellular domains similar to CD5. Ligand CD166
CD7	Majority of T-cells	gp40	23	Unknown function
CD8	T-cytoxic/suppressor cells	MHC Class 1 coreceptor (with TCR)	$\alpha = 23$, $\beta = 21$	Coreceptor in antigen recognition
CD9	Pre-B cells, monocytes, platelets	MRP-1	25	Major component of platelet surface, TM4 superfamily. Role in adhesion
CD10	Lymphoid progenitor cells, granulocytes	CALLA (common acute lymphoblastic leukaemia antigen); neutral endopeptidase (NEP)	100	Zinc-binding metalloproteinase/enkephalinase
CD11a	Leucocytes	Leucocyte function antigen (LFA-1) integrin α_L subunit	126	Associates with CD18. Binds ICAM-1,2,3
CD11b	Granulocytes, monocytes, NK cells	Integrin α_M subunit, Mac-1; CR3, C3biR	126	Associates with CD18, upregulated in inflammation
CD11c	Granulocytes, monocytes, NK cells, B-cell subset, T-cell subset	Integrin α_X subunit, p150/95	126	Associates with CD18, upregulated in inflammation
CD11d	PBLs, splenic macrophages, foam cells	Integrin α_D subunit	125	Binds ICAM-3 but not ICAM-1 or VCAM

Table C2. (Continued)

Cluster designation	Main cellular expression of antigen	Other names	Antigen (kDa) based on sequence	Comment
CDw12	Granulocytes, monocytes	—	(90–120)	A phosphoprotein, function unknown
CD13	Myeloid cells and various tissue cells	Aminopetidase N	109	May inactivate small signalling peptides
CD14	Monocytes, some granulocytes and macrophages	—	35	GPI-linked, LPS receptor
CD15	Granulocytes, monocytes	—	Carbohydrate	Present on many proteins
CD15s	Neutrophils	sialyl-Lewis x (sLe-x)	(Pentasaccharide)	Ligand for CD62 structures
CD16a	Macrophage, NK cells (neutrophils)	Transmembrane form, FcgRIIIA/FcgRIIIB	27	Low-affinity receptor for aggregated IgG, structural similarities with CD64, CDw32
CD16b	Granulocyte form only	GPI-linked form, FcγRIIIb	21	
CDw17	Granulocytes, monocytes, platelets	Lactosylceramide	Not determined	
CD18	Leucocytes; platelets negative	Integrin β_2 subunit	82	Associates with CD11a,b,c,d
CD19	Pan B-cell except plasma cells	B4	59	Involved in regulation of B-cell proliferation
CD20	Pan B-cell	B1	33	May be B-cell calcium channel
CD21	Mature B-cells, follicular dendritic cells	C3d/EBV-receptor (CR2)	117	Forms complexes with CD35 and CD19
CD22	Mature B-cells, hairy cell leukaemia cells	BL-CAM	α 71/β 93	Heterodimeric (α and β subunits). IgSF. Reduces B-cell activation threshold. Interacts with Shp-1
CD23	Activated B-cells, activated macrophages, eosinophils, platelets	IgE Fc low affinity receptor (FcεRII)	36	Has C-type lectin domain. Involved in regulation of IgE production
CD24	B-cells, granulocytes, normal epithelium	Heat-stable antigen (HSA)	31	GPI-linked sialoglycoprotein
CD25	Activated T-cells, B-cells and macrophages	IL-2 receptor α, Tac	55	Associates with CD122 to form high-affinity receptor
CD26	Activated T-cells and B-cells, macrophages	Dipeptidyl peptidase IV	88	T-cell co-stimulation, adhesion, proteolytic function

Table C2. (Continued)

Cluster designation	Main cellular expression of antigen	Other names	Antigen (kDa) based on sequence	Comment
CD27	Thymocyte subset, mature T-cells, EBV transformed B-cells	—	27	Member of the NGFR superfamily
CD28	T-cell subset, activated B-cells	Tp44	23	Binds CD80 (B7) and CD86. Important co-stimulator of T cell activation
CD29	Ubiquitous (but not on erythrocytes)	Integrin β1 subunit	85-89	Associates with CD49
CD30	Activated T- and B-cells, Reed–Sternberg cells	Ki-1 antigen	62	NGFR Superfamily. Binds CD153 and co-stimulates T-cell proliferation
CD31	Platelets, monocytes, macrophages, granulocytes, B-cells, some T-cells, endothelial cells	PECAM-1 (platelet endothelial cell adhesion molecule 1)	80	Homotypic adhesion and heterotypic interaction with integrin αvβ3
CD32	Monocytes, granulocytes, B-cells, eosinophils	Fcγ receptor II	27–31	Low affinity receptor for aggregated IgG
CD33	Myeloid progenitor cells, monocytes	—	38	May mediate cell–cell adhesion
CD34	Haematopietic precursor cells, capillary endothelial cells	Sgp90	37	Sialylated forms are ligand for L- and E-selectin
CD35	Granulocytes (basophils negative), monocytes, B-cells, erythrocytes, some NK cells, follicular dendritic cells	Complement Receptor 1 (CR1), C3b Receptor	220	Binds complement C3b and C4b, enhances FcR-mediated phagocytosis
CD36	Monocytes, macrophages, platelets, B-cells (weakly)	Platelet GPIV (IIIb)	53	Binds collagen, LDL, *P. falciparum* infected rbcs. Interacts with src kinases fyn, lyn and yes
CD37	Mature B-cells; T-cells and myeloid (weakly)	gp 52-40	32	TM4 superfamily (four1 membrane-spanning domains)
CD38	Plasma cells, thymocytes, activated T-cells	T10	34	Cyclic ADP-ribose hydrolase/ADP-ribosyl cyclase
CD39	Activated B- and NK cells, EBV-transformed B cells	gp80	58	Mediates B-cell homotypic adhesion. Has ecto-apyrase activity

Table C2. (Continued)

Cluster designation	Main cellular expression of antigen	Other names	Antigen (kDa) based on sequence	Comment
CD40	B-cells, monocytes (weakly), some epithelia	gp50	27	NGFR superfamily, binds CD154 and essential for secondary immune response
CD41	Platelets, megakaryocytes	GPIIb, integrin αIIb	110	Associates with CD61 (β3 Integrin). Important for adhesion, defective in Glanzmann's thrombasthenia
CD42a	Platelets, megakaryocytes	GPIX	17	Defective in Bernard–Soulier syndrome
CD42b	Platelets, megakaryocytes	GPIB	$\alpha = 67$, $\beta = 19$	Complexed with CD42a, binds von Willebrand factor
CD43	Leucocytes but not resting B-cells	Leukosialin, sialophorin	39	Anti-adhesion molecule
CD44	Leucocytes, erythrocytes, platelets (weakly), brain cells	Pgp-1, H-CAM, extracellular matrix receptor III	37/variable	Hermes antigen; multiple splice variants known
CD45	Pan leucocyte	Leucocyte common antigen (L-CA); T200	127, 146	Protein tyrosine phosphatase, essential for signalling through TCR. Variants CD45-RO, RA, RB, RC
CD46	Haematopoietic and non-haematopoietic cells (not erythrocytes)	MCP (membrane cofactor protein)	37–40	Regulator of complement activation; binds C3b and C4b. Receptor for measles virus and *Strep. pyogenes*
CD47	All cell types	Rh group associated; integrin-associated protein (IAP)	35	Associates with β3 integrins, binds thrombospondin. Has role in leucocyte–endothelial interactions
CD48	Many leucocytes, not granulocytes, platelets or rbcs	Blast-1	22	Sequence similarities with CD2, CD58, CD152
CD49a	Activated T-cells, monocytes	Very late antigen-1 (VLA-1); β1 integrins	128	Associates with CD29. Binds laminin and collagen
CD49b	B-cells, monocytes, platelets	VLA-2	126	Mediates Mg^{2+}-dependent adhesion of platelets to collagen

Table C2. (Continued)

Cluster designation	Main cellular expression of antigen	Other names	Antigen (kDa) based on sequence	Comment
CD49c	B-cells	VLA-3	113	Binds various ligands
CD49d	Thymocytes, B-cells	VLA-4	111	Binds VCAM-1 and CS-1 domain of fibronectin
CD49e	Memory T-cells, monocytes	VLA-5	110	Fibronectin receptor, enhances Fcγ-mediated phagocytosis
CD49f	Memory T-cells, thymocytes, monocytes	VLA-6	117/120	Laminin receptor on platelets, monocytes and T-cells
CD50	Leucocytes (platelets and erythrocytes negative)	ICAM-3	56	Ligand for LFA-1. May also have signalling function
CD51	Platelets, endothelial cells, various other cells	VNRα (vitronectin receptor)	121	α_V integrin chain (associates with CD61)
CD52	Leucocytes (not plasma cells, platelets or rbcs)	CAMPATH-1	1.2	Unknown. GPI-anchored
CD53	Pan leucocyte	—	24	TM4 superfamily; possibly has signalling function
CD54	Endothelial cells, many activated cell types	ICAM-1	55	Binds CD11a/b/c /CD18 (β2 integrins)
CD55	Many haematopoietic cells and cells in contact with serum.	DAF (decay accelerating factor)	35	GPI-anchored and transmembrane forms. Defective in paroxysmal nocturnal haemoglobinuria. Interacts with CD97
CD56	NK cells	NKH1, isoform of N-CAM	140	Function unclear
CD57	NK cells, T-cells, B-cell subsets	HNK-1	Oligosaccharide	Binds L- and P-selectin. Function obscure
CD58	Many haematopoietic and non-haematopoietic cells	LFA-3	25	Binds to CD2
CD59	Many haematopoietic and non-haematopoietic cells	p18, gP18, MAC inhibitor, MACIF	8961 D	Protective against complement lysis. GPI linked

Table C2. (Continued)

Cluster designation	Main cellular expression of antigen	Other names	Antigen (kDa) based on sequence	Comment
CD60	Platelets, T-cell subset	UM4D4	Carbohydrate	
CD61	Platelets, megakaryocytes, monocytes, macrophages, endothelium	Integrin β3, GPIIIa, vitronectin receptor	84	Associates with CD41 or CD51
CD62E	Endothelium	E-selectin, ELAM-1, LECAM-2	64	Binds sialyl-Lewis X (CD15s)
CD62L	B- and T-cells, monocytes, NK cells	L-selectin, LECAM-1, LAM-1	37	Mel-14 antigen
CD62P	Platelets, activated endothelial cells,	P-selectin, GMP-140, PADGEM	86	Mediates neutrophil rolling
CD63	Activated platelets, monocytes, macrophages	Platelet activation antigen, granulophysin	26	TM4 superfamily
CD64	Monocytes, macrophages	High affinity Fcγ receptor 1, FCγR1	A = 30 B = 41	Will bind monomeric IgG
CD65	Granulocytes, monocytes	Fucoganglioside	Not confirmed	
CD66a	Neutrophil lineage cells	Biliary glycoprotein (BGP-1)	54	
CD66b	Granulocytes	CEA gene member 6 (CGM6)	31	Previously CD67
CD66c	Neutrophils, colon carcinoma	Non-specific cross-reacting antigen (NCA)	31	
CD66d	Neutrophils	CEA gene member 1 (CGM1)	23	
CD66e	Adult colon, epithelia, colon carcinoma	Carcinoembryonic antigen (CEA)	180–200	
CD68	Monocytes, macrophages, granulocytes, large lymphocytes	Macrosialin (in mouse)	35	Same family as CD107a and b (Lamp1 and 2)
CD69	Activated T- and B-cells, activated macrophages, NK cells	gp34/28, AIM (activation inducer molecule)	22	May function in signalling
CD70	Activated T- and B-cells, Reed–Sternberg cells, macrophages (weakly)	Ki-24, CD27L	21	TNF superfamily. Co-stimulator of T and B cell activation
CD71	Activated T- and B-cells, macrophages, proliferating cells	T9, transferrin receptor	85	Found as homodimer

Table C2. (Continued)

Cluster designation	Main cellular expression of antigen	Other names	Antigen (kDa) based on sequence	Comment
CD72	Pan B-cell	LyB in mouse	40	Expressed as homodimer. Ligand for CD5
CD73	B- and T-cell subsets	ecto-5′-nucleotidase, lymphocyte vascular adhesion protein 2	61	GPI-linked
CD74	B-cells, macrophages, monocytes	gp41/35/33, Ii	41, 35, 33	MHC Class II associated invariant chain
CDw75	Mature B-cells, T-cell subset (weak expression)	—	Not confirmed	Now known not to be ligand for CD22
CDw76	Mature B-cells, T-cell subset	—	(gel) 86	Epitope is apparently carbohydrate
CD77	Activated B-cells, follicular centre	BLA, Gb3, b^k	Not confirmed	Burkitt's lymphoma associated antigen
CDw78	B-cells	Ba antigen	Not confirmed	
CD79a	B-cell specific	Igα	21	Forms heterodimer with CD79b associated with membrane Ig. B cell receptor
CD79b	B-cell specific	B29, BCR, Igβ	23	Interacts with CD79a
CD80	B-cell subset *in vivo*, most activated B-cells *in vitro*	B7, B7-1	30	CD28 ligand
CD81	B-cells (broad expression including lymphocytes), T-cells, neutrophils	TAPA-1 (target of antiproliferative antibody)	26	Has a role in early T-cell development
CD82	Broad expression on leucocytes (weak), platelets, not erythrocytes	R2	30	Can transduce signals, will suppress metastasis
CD83	Marker for circulating dendritic cells, activated B- and T-cells, germinal centre cells	HB15	23	Function unknown
CDw84	Platelets and monocytes (strong), B-cells		73	Function unknown
CD85	Circulating B-cells (weak), monocytes (strong)		(gel) 83	Function unknown

Table C2. (Continued)

Cluster designation	Main cellular expression of antigen	Other names	Antigen (kDa) based on sequence	Comment
CD86	Circulating monocytes, germinal centre cells, activated B-cells	B7-2	35	Ig superfamily. Structurally related to CD80. Has role in co-stimulation by interaction with CD28
CD87	Granulocytes, monocytes, macrophage, activated T-cells	UPa-R (urokinase plasminogen activator receptor)	31	May be important in extravasation
CD88	Polymorphonuclear leucocytes, mast cells, macrophage, smooth muscle	C5a receptor	39	Seven-membrane spanning receptor
CD89	Neutrophils, monocytes, macrophage, T- and B-cell subpopulation	FcαR, IgA receptor	19–30	
CD90	CD34$^+$, subset on bone marrow, cord blood, fetal liver	Human Thy-1	13	Associated with fyn. GPI anchored
CD91	Monocytes and some non haemopoietic cell lines	α2 macroglobin receptor, LDL receptor-related protein	502	May have a clearance function
CDw92	Neutrophils, monocytes, endothelial cells, platelets		(gel) 70	Function unknown
CD93	Neutrophils, monocytes, enothelial cells		(gel) 120	Function unknown
CD94	NK-cells, α/β,γ/δ, T-cell subsets	KP43	20	Role in recognition of MHC Class I
CD95	Variety of cell lines including myeloid and T-lymphoblastoid	APO-1, FAS	36	NGFR superfamily. Antibody to Fas kills by apoptosis
CD96	Activated T-cells	TACTILE (T-cell activation increased late expression)	61	Function unknown
CD97	Activated T- and B-cells, granulocytes, monocytes	GR1, BL—KDD/F12	79	Function unknown but has seven transmembrane domains
CD98	T-cells and B-cells (weak), monocytes (strong), most human cell lines	4F2	57	Found as heterodimer with glycosylated heavy chain and non-glycosylated light chain

Table C2. (Continued)

Cluster designation	Main cellular expression of antigen	Other names	Antigen (kDa) based on sequence	Comment
CD99	Peripheral blood lymphocytes, thymocytes	MIC2, E2	17	
CD99R	T-cells, NK cells, myeloid cells, some leukaemias	MIC2, E2	17	
CD100	Broad expression on haemopoietic cells		94	A semaphorin
CD101	Granulocytes, macrophage, activated T-cells	V7, p126	115	Role in TCR/CD3 co-stimulation of T-cells
CD102	Resting lymphocytes, monocytes, vascular endothelial cells (strongest)	ICAM-2	28	Ligand for CD11a / CD18. Constitutively expressed
CD103	Intraepithelial lymphocytes	α_E integrin, α_6, HML-1	127	Binds E-cadherin
CD104	Epithelia, Schwann cells, some tumour cells	β4 integrin chain, β4	192–197	Interacts with CD49f
CD105	Endothelial cells, bone marrow cell subset, *in vitro* activated macrophage	Endoglin, receptor for TGFβ1 and β3	68	
CD106	Endothelial cells	VCAM-1, INCAM-110	79	Ligand for VLA4 (CD49d)
CD107a	Activated platelets	LAMP 1 (lysosomal associated membrane protein)	42	Ligand for galaptin
CD107b	Activated platelets	LAMP 2	42	Ligand for galaptin
CDw108	Activated T-cells in spleen, some stromal cells	GR2	(Ag) 80	Function unknown. GPI linked
CD109	Activated T-cells, platelets, endothelial cells	Platelet activation factor, GR56	(Ag) 170/150	Function unknown. GPI linked
CD110/113	Nothing yet assigned	—	—	
CD114	Granulocytes, monocytes	G-CSF receptor	(Ag) 95, 139	

Table C2. (Continued)

Cluster designation	Main cellular expression of antigen	Other names	Antigen (kDa) based on sequence	Comment
CD115	Monocytes, macrophage, placenta	M-CSFR (Macrophage colony-stimulating factor receptor)	(Ag) 150	Encoded by c-fms protooncogene, a receptor tyrosine kinase
CD116	Monocytes, neutrophils, esoinophils, fibroblasts, endothelial cells	GM-CSF R α chain (granulocyte, macrophage colony-stimulating factor receptor)	(Ag) 75–85	Shares β subunit with IL-3 and IL-5 receptors
CD117	Bone marrow progenitor cells	Stem cell factor receptor, (SCF-R), c-KIT	107	Receptor tyrosine kinase
CD118	Broad cellular expression	IFNα, β receptor	(Ag) 90	
CD119	Macrophage, monocyte, B-cells, epithelial cells	IFNγ R (Interferon γ receptor)	(Ag) 90	Interacts with JAK
CD120a	Most cell types, higher levels on epithelial cell lines	TNFαR (Tumour necrosis factor α receptor) type I	(Ag) 55	NGFR superfamily
CD120b	Most cell types, higher levels on myeloid cell lines	TNFα-R type II	(Ag) 75	NGFR superfamily
CDw121a	T-cells, thymocytes, fibroblasts, endothelial cells	IL-1R (Interleukin-1 receptor) type 1	80	
CD121b	B-cells, macrophages, monocytes	IL-1R, type II	68	
CD122	NK-cells, resting T-cells subpopulation, some B-cell lines	IL-2Rβ	75	Associates with CD25
CDw123	Bone marrow stem cells, granulocytes, monocytes, megakaryocytes	IL-3Rα	70	
CD124	Mature B- and T-cells, haemopoietic precursor cells	IL-4R, IL-13 receptor	140	
CDw125	Eosinophils and basophils	IL-5Rα chain	60	Associates with CDw131
CD126	Activated B-cells and plasma cells (strong), most leucocytes (weak)	IL-6Rα chain	80 (α subunit)	gp130 (CDw130) is the β subunit

Table C2. (Continued)

Cluster designation	Main cellular expression of antigen	Other names	Antigen (kDa) based on sequence	Comment
CD127	Bone marrow lymphoid precursors, Pro-B-cells, mature T-cells, monocytes	IL-7Rα chain	(Ag) 65–75	
CDw128	Neutrophils, basophils, T-cell subset	IL-8R	(Ag) 58–67	Seven-membrane spanning G-protein coupled receptor
CD129	Not assigned	—	—	
CD130	Activated B-cells and plasma cells (strong), most leucocytes (weak), endothelial cells	gp130, Common β subunit for IL-6, IL-11, LIF, OSM, LNTF, CT-1	130	Associates with CD126
CDw131	Myeloid cells	Common β chain for IL-3, IL-5, GM-CSF-R	(Ag) 120	
CD132	Broad	Common γ chain for IL-2, IL-4, IL-7, IL-9, IL-15 receptor	(Ag) 64	
CD133	Not assigned	—	—	
CD134	Activated T subset	OX40	26	TNF-R superfamily
CD135	Progenitor cell subset	STK-1, Flt3, Flk2	110	Involved in growth and differentiation of primitive haematopoietic cells
CDw136	Macrophages	Macrophage stimulating factor receptor	(Ag) 180	
CDw137	T-cell subset	4-1BB	26	Coprecipitates with lck
CD138	Plasma cells, epithelial cells	syndecan-1	30	ECM receptor, possibly a coreceptor for FGF
CD139	Germinal centre B-cells		(Ag) 209, 228	
CD140a	Undetectable on most cells	PDGF-Rα	(Ag) 180	
CD140b	Endothelial cell subset, stromal cells	PDGF-Rβ	(Ag) 180	
CD141	Endothelial cells	Thrombomodulin	(Ag) 100	

Table C2. (Continued)

Cluster designation	Main cellular expression of antigen	Other names	Antigen (kDa) based on sequence	Comment
CD142	Activated monocytes and endothelial cells	Tissue factor	(Ag) 45	
CD143	Endothelial cell subsets	Angiotensin-converting enzyme (ACE)	(Ag) 170	
CD144	Endothelial cells	VE-cadherin	(Ag) 135	
CDw145	Endothelial cells, some stromal cells		(Ag) 25, 90, 110	
CD146	Endothelium, activated T-cells	MUC18	(Ag) 118	
CD147	Endothelium, monocytes, T-cell subset, platelets, erythrocytes	neurothelin, basigin, EMMPRIN	28	Extracellular matrix metalloproteinase inducer
CD148	Haematopoietic cells	HPTP-η/DEP-1	142	PTPase. May have role in growth regulation of epithelial cells
CDw149	Broad			
CDw150	T- and B-cells, thymocytes	IPO-3/SLAM	34	Structural features similar to CD2, CD48, CD58
CD151	Platelets, endothelium	PETA-3	28	Member of TM4 superfamily
CD152	Activated T-cells	CTLA-4	20	Negative regulator of T-cell activation. Similar to CD28, binds CD80 and CD86
CD153	Activated T-cells	CD30 ligand	26	Member of TNF superfamily
CD154	Activated T-cells, mast cells, basophils	CD40 ligand	29	Member of TNF superfamily
CD155	Monocytes, macrophages, thymocytes	Poliovirus receptor	(Ag) 80–90	
CD156	Monocytes, macrophages, granulocytes	ADAM-8	(Ag) 60	Cell surface metallo-proteinase
CD157	Monocytes, neutrophils, endothelium	BST-1	(Ag) 42-45	
CD158a	NK cells, T-cell subset	p58.1, P50.1	(Ag) 58, 50	

Table C2. (Continued)

Cluster designation	Main cellular expression of antigen	Other names	Antigen (kDa) based on sequence	Comment
CD158b	NK cells, T-cell subset	p58.2, P50.2	(Ag) 58, 50	
CD159,160	Not assigned	—		
CD161	NK cells, T-cell subset	NKR-P1 (NK receptor P1 family)	25	Has C-type lectin domain
CD162	T-cells, monocytes, granulocytes, B-cell subset	PSGL-1 (P selectin glycoprotein ligand 1)	41	Mediates neutrophil and T-cell tethering and rolling
CD163	Monocytes	KiM4	116	Unknown function
CD164	T-cells, monocytes, granulocytes, B-cell subset	MGC-2	(Ag) 80	
CD165	Thymocytes, thymic epithelium	AD2/gp37	(Ag) 37	
CD166	Activated lymphocytes, endothelium, fibroblasts	ALCAM, CD6 ligand	62	Ig superfamily. May have role in axonal guidance

Where the molecular weight of the polypeptide portion is known, that is the value quoted. In other cases it is the apparent molecular weight from behaviour on reducing gels.

particular functions. All living organisms are composed of one or more cells. Implicit in this definition is that viruses are not living organisms, and since they cannot exist independently, this seems reasonable.

cell adhesion See ***adhesins, cadherins, cell adhesion molecules*** (CAMs), ***contact sites A, DLVO theory, integrins, sorting out, uvomorulin*** and various specialized junctions (***adherens junctions, desmosomes, focal adhesions, gap junction*** and ***zonula occludens***).

cell adhesion molecule (CAM) Although this could mean any molecule involved in cellular adhesive phenomena, it has acquired a more restricted sense, namely a molecule on the surface of animal tissue cells, antibodies (or Fab fragments) against which specifically inhibit some form of intercellular adhesion. Examples are LCAM (liver cell adhesion molecule) and NCAM (neural cell adhesion molecule), both named from tissues in which first detected, although their occurrence is not in fact restricted to these.

cell behaviour General term for activities of whole cells such as movement, adhesion and proliferation, by analogy with animal behaviour.

cell body Used in reference to ***neurons***; the main part of the cell around the nucleus excluding long processes such as ***axons*** and ***dendrites***.

cell centre ***Microtubule organizing centre*** (MTOC) of the cell, the pericentriolar region.

cell culture General term referring to the maintenance of cell strains or lines in the laboratory. See Table C3.

cell cycle The sequence of events between mitotic divisions. The cycle is conventionally divided into G0, G1 (G standing for gap), S (synthesis phase during which the DNA is replicated), G2 and M (mitosis). Cells that will not divide again are considered to be in G0, and the transition from G0 to G1 is thought to commit the cell to completing the cycle and dividing.

cell death Cells die (non-accidentally) either when they have completed a fixed number of division cycles (around 60; the ***Hayflick limit***) or at some earlier stage when programmed to do so, as in digit separation in vertebrate limb morphogenesis. Whether this is due to an accumulation of errors or a programmed limit is unclear; some transformed cells have undoubtedly escaped the limit. See ***apoptosis***.

cell division The separation of one cell into two daughter cells, involving both nuclear division (***mitosis***) and subsequent cytoplasmic division (***cytokinesis***).

cell electrophoresis Method for estimating the surface charge of a cell by looking at its rate of movement in an electrical field; almost all eukaryotic cells have a net negative surface charge. Measurement is complicated by the streaming potential at the wall of the chamber itself, and by the fact that the cell is surrounded by a layer of fluid (see ***double layer***). The electrical potential measured (the zeta potential) is actually some distance away from the plasma membrane. One of the more useful modifications is to systematically vary the pH of the suspension fluid to determine the pK of the charged groups responsible (mostly carboxyl groups of sialic acid).

cell fate Of an embryonic parent (progenitor) cell or cell type, the range and distribution of differentiated tissues formed by its daughter cells. For example, cells of the ***neural crest*** differentiate to form (among other things) cells of the peripheral nervous system.

cell fractionation Strictly this should mean the separation of homogeneous sets from a heterogeneous population of cells (by a method such as ***flow cytometry***), but the term is more frequently used to mean subcellular fractionation, ie. the separation of different parts of the cell by differential centrifugation, to give nuclear, mitochondrial, microsomal and soluble fractions.

cell-free system Any system in which a normal cellular reaction is reconstituted in the absence of cells, for example, ***in vitro*** translation systems that will synthesize protein from mRNA using a lysate of rabbit reticulocytes or wheatgerm.

cell fusion Fusion of two previously separate cells occurs naturally in fertilization and in the formation of vertebrate skeletal muscle, but can be induced artificially

Table C3. Common cell lines
Although there are a great many cell lines available through the cell culture repositories and from trade suppliers, there are a few 'classic' lines that will be met fairly frequently. Many of these well-known lines are listed below, but the table is not comprehensive.

Name	Tissue of origin	Cell type	Comment
Human			
MRC5	Embryonic lung	Fib	Diploid, susceptible to virus infection
WI38	Embryonic lung	Fib	Diploid, finite division potential
HeLa	Cervical carcinoma	Epi	Established line
HEp2	Laryngeal carcinoma	Epi	
Raji	Burkitt lymphoma	Lym	Grows in suspension. EB virus undetectable
Daudi	Burkitt lymphoma	Lym	
HL60	—	Myl	Will differentiate to granulocytes or macrophages
J111	Monocytic leukaemia	Myl	
U937	Monocytic leukaemia	Myl	Will differentiate to macrophages
Monkey			
BSC-1	Kidney	Fib	Derived from African green monkey, often used in virus studies
Vero	Kidney	Fib	
Hamster			
BHK21	Baby kidney	Fib	Syrian hamster. Usually C13 (clone 13)
CHO	Ovary	Epi	Chinese hamster ovary
Don	Lung	Fib	Chinese hamster
Potoroo			
PtK1	Female kidney	Epi	Small number of large chromosomes
PtK2	Male kidney	Epi	Cells stay flat during mitosis
Dog			
MDCK	Kidney	Epi	Madin–Darby canine kidney
Mouse			
A9	From L929	Fib	HGPRT negative
L1210	Ascites fluid	Lym	Grows in suspension; DBA/2 mouse
MOPC31C	Plasmacytoma	Lym	Grows in suspension; secretes IgG
L929	Connective tissue	Fib	Clone of L-cell
3T3	Whole embryo	Fib	Swiss or Balb/c types; very density-dependent
SV40-3T3	From 3T3	Fib	Transformed by SV40 virus
P388D1	—	Lym	Grows in suspension
EAT	Ascites tumour	Fib	
S180	Sarcoma	Fib	Invasive; maintained *in vivo*
MCIM	Sarcoma	Fib	Methylcholanthrene-induced
B16	Melanoma	Mel	High and low metastatic variants (F1 and F10)
Rat			
GH1	Pituitary tumour	Fib	Secrete growth hormone
WRC-256	Carcinoma	—	Walker carcinoma; many variants
PC12	Adrenal	Neur	Phaeochromocytoma; can be induced to produce neurites

Epi = epithelial; Fib = fibroblastic; Lym = lymphocytic; Mel = melanin containing; Myl = myeloid; Neur = neural.

by the use of ***Sendai virus*** or fusogens such as polyethylene glycol. Fusion may be restricted to cytoplasm or nuclei may fuse as well. A cell formed by the fusion of dissimilar cells is often referred to as a heterokaryon.

cell growth Usually used to mean increase in the size of a population of cells though strictly should be reserved for an increase in cytoplasmic volume of an individual cell.

cell junctions Specialized junctions between cells. See ***adherens junctions, desmosomes, tight junctions, gap junctions***.

cell line A cell line is a permanently established cell culture that will proliferate indefinitely given appropriate fresh medium and space. Lines differ from cell strains in that they have escaped the ***Hayflick limit*** and become immortalized. Some species, particularly rodents, give rise to lines relatively easily, whereas other species do not. No cell lines have been produced from avian tissues, and the establishment of cell lines from human tissue is difficult. Many cell biologists would consider that a cell line is by definition already abnormal and that it is on the way towards becoming the culture equivalent of a neoplastic cell.

cell lineage The lineage of a cell relates to its derivation from the undifferentiated tissues of the embryo. Committed embryonic progenitors give rise to a range of differentiated cells: in principle it should be possible to trace the ancestry (lineage) of any adult cell.

cell locomotion Movement of a cell from one place to another.

cell-mediated immunity Immune response that involves effector T-lymphocytes and not the production of humoral antibody. Responsible for ***allograft*** rejection, delayed ***hypersensitivity*** and in defence against viral infection and intracellular protozoan parasites.

cell membrane Rather imprecise term usually intended to mean ***plasma membrane***.

cell migration Implies movement of a population of cells from one place to another, as in the movement of neural crest cells during morphogenesis.

cell movement A more general term than locomotion, that can include shape-change, cytoplasmic streaming, etc.

cell plate Region in which the new cell wall forms after the division of a plant cell. In the plane of the equator of the spindle a disc-like structure, the ***phragmoplast***, forms, into which are inserted pole-derived microtubules. Golgi-derived vesicles containing pectin come together and fuse at the plate which develops from the centre outwards and eventually fuses with the plasma membrane thereby separating the daughter cells.

cell polarity (1) In epithelial cells the differentiation of apical and basal specializations. In many epithelia the apical and basolateral regions of plasma membrane differ in lipid and protein composition, and are isolated from one another by tight junctions. The apical membrane may, for example, be the only region where secretory vesicles fuse, or have a particular ionic pumping system. (2) A motile cell must have some internal polarity in order to move in one direction at a time: a region in which protrusion will occur (the front) must be defined. Locomotory polarity may be associated with the pericentriolar ***microtubu***le ***organizing centre***, and can be perturbed by drugs that interfere with microtubule dynamics.

cell proliferation Increase in cell number by division.

cell recognition Interaction between cells that is possibly dependent upon specific adhesion. Since the mechanism is not entirely clear in most cases, the term should be used with caution.

cell renewal Replacement of cells, eg. those in the skin, by the proliferative activity of basal stem cells.

cell sap Effectively equivalent to the term 'cytosol'.

cell signalling Release by one cell of substances that transmit information to other cells.

cell sorting The process or processes whereby mixed populations of cells, eg. in a reaggregate, separate out into two or more populations that usually occupy different parts of the same aggregate or separate into different aggregates. Cell sorting probably takes place in the development of certain organs. See ***differential adhesion, flow cytometry***.

cell strain Cells adapted to culture, but with finite division potential. See ***cell line***.

cell synchronization A process of obtaining (either by selection, or imposition of a reversible blockade) a population of growing cells that are to a greater or lesser extent in phase with each other in the cycle of growth and division.

cell wall Extracellular material serving a structural role. In plants the primary wall is pectin-rich, the secondary wall mostly composed of ***cellulose***. In bacteria, cell wall structure is complex: the walls of ***Gram-positive*** and ***Gram-negative*** bacteria are distinctly different. Removal of the wall leaves a ***protoplast*** or ***spheroplast***.

cell-surface marker Any molecule characteristic of the plasma membrane of a cell or in some cases of a specific cell type. 5′-***nucleotidase*** and Na^+/K^+ATPase are often used as plasma membrane markers.

cellobiose Reducing ***disaccharide*** composed of two D-glucose moieties β-1,4 linked. The disaccharide subunit of cellulose, though not found as a free compound ***in vivo***.

cellubrevin Protein involved in regulating vesicle fusion. Has 60% sequence identity with ***synaptobrevin*** (VAMP-2) and is a target for tetanus toxin.

cellular engineering The use of techniques for constructing replacement or additional or experimental parts of cells and tissues for both fundamental investigation and as prosthetic devices. Often involves the interfacing of cells and nonliving structures.

cellular immunity Immune response that involves enhanced activity by phagocytic cells and does not imply lymphocyte involvement. Since the term is easily confused with ***cell-mediated immunity***, its use in this sense should be avoided.

cellular retinoic acid-binding protein (CRABP) A cytoplasmic ***fatty acid-binding protein*** that acts as an initial receptor for the putative ***morphogen, retinoic acid***.

cellular slime mould See *Acrasidae*.

cellulases Enzymes that break down ***cellulose***, and are involved in cell wall breakdown in higher plants, especially during abscission. Produced in large amounts by certain fungi and bacteria. Degradation of cellulose ***microfibrils*** requires the concerted action of several cellulases.

cellulitis Inflammation of the subcutaneous connective tissues (dermis), mostly affecting face or limbs. *Streptococcus pyogenes* is commonly the causative agent. Also known as erysipelas.

cellulose A straight-chain polysaccharide composed of β (1-4)-linked glucose subunits. A major component of plant cell walls where it is found as microfibrils laid down in orthogonal layers.

cellusome A cellulose-binding, cellulase-containing, cell surface organelle in certain prokaryotes.

cenocyte See ***coenocyte***.

CENP antigens Proteins of the ***kinetochore*** (CENP-A 27 kD, CENP-B 80 kD, CENP-C 140 kD and CENP-D 50 kD) that react strongly with antibodies from ***CREST*** sera.

centractin An actin homologue (50% homology with muscle actin, 70% if conservative substitutions are taken into account) associated with the vertebrate ***centrosome***. Highly conserved between species.

central lymphoid tissue See ***lymphoid tissue***.

central nervous system (CNS) In vertebrates, the brain and spinal cord. In invertebrates, the CNS is composed of the segmental ganglia of the ventral nerve cord together with the fused ganglia or brain at the anterior end.

centrifugation The process of separating fractions of systems in a centrifuge. The most basic separation is to sediment a pellet at the bottom of the tube, leaving a supernatant at a given centrifugal force. In this case sedimentation is determined by size and density of the particles in the system amongst other factors. Density may be used as a basis for sedimentation in ***density gradient*** centrifugation. At very high *g* values molecules may be separated, ie. ***ultracentrifugation***. In continuous centrifugation the supernatant is removed continuously as it is formed.

centrins Acidic phosphoproteins (20 kD), homologous to ***caltractin,*** found in striated flagella roots of various flagellated algae, centrosomal region of some mammalian cells and basal bodies of human sperm.

centriolar region See ***pericentriolar region*** or ***centrosome***.

centriole Organelle of animal cells that is made up of two orthogonally arranged cylinders each with nine microtubule triplets composing the wall. Almost identical to ***basal body*** of cilium. The pericentriolar material, but not the centriole itself, is the major ***microtubule organizing centre*** of the cell. Centrioles divide prior to mitosis and the daughter centrioles and their associated pericentriolar material come to lie at the poles of the spindle.

centromere The region in eukaryote chromosomes where daughter chromatids are joined together. The kinetochore, to which the spindle chromosomes are attached, lies adjacent to the centromere. The centromeric DNA codes for the kinetochore.

centromeric banding See ***C banding***.

centrophilin A microtubule-binding protein identified by the production of monoclonal antibodies raised against isolated centromeres. In mitotic cells centrophilin is not restricted to the centromeres, but is a major antigen of the spindle polar bodies.

centrosome The ***microtubule*** organizing centre that, in animal cells, surrounds the ***centriole,*** and which will divide to organize the two poles of the ***mitotic spindle***. By directing the assembly of a cell's skeleton, this organelle controls division, motility and shape.

centrosphere Alternative (rare) name for ***centrosome***.

Centruroides margaritatus Scorpion. See ***margaratoxin***.

Centruroides noxius Scorpion. See ***noxiustoxin***.

Cepaea Genus of land snails. Two species, *C. hortensis* and *C. nemoralis* have been much studied as convenient examples of polymorphism in colour and banding pattern.

cephalosporin Tetracyclic triterpene antibiotics isolated from culture filtrates of the fungus *Cephalosporium* sp. Effective against ***Gram-positive bacteria***.

ceramide An N-acyl ***sphingosine,*** the lipid moiety of ***glycosphingolipids***.

cercidosome Specialized organelle of trypanosomes, site of terminal oxidative metabolism.

cerebellum Part of the vertebrate hindbrain, concerned primarily with somatic motor function, the control of muscle tone and the maintenance of balance. Important model for cell migration in developing mammalian brain owing to well-studied migratory pathway of the ***granule cell*** and to the existence of the neurological mutant mouse ***weaver*** in which granule cell migration fails.

cerebroside ***Glycolipid*** found in brain (11% of dry matter). ***Sphingosine*** core with fatty amide or hydroxy fatty amide and a single monosaccharide on the alcohol group (either glucose or galactose).

cereolysin Cytolytic (haemolytic) toxin released by *Bacillus cereus*. Inactivated by oxygen, reactivated by thiol reduction (hence thiol-activated cytolysin). Binds to cholesterol in the plasma membrane and rearrangement of the toxin–cholesterol complexes in the membrane leads to altered permeability.

ceruloplasmin A blue copper-containing ***dehydrogenase*** protein (135 kD) found in serum (200–500μg/ml). Apparently involved in copper detoxification and storage, and possibly also in mopping up excess oxygen radicals or ***superoxide*** anions.

CFAG See ***cystic fibrosis antigen***.

CFTR See ***cystic fibrosis transmembrane conductance regulator***.

CFU-E ***Colony forming unit*** for cell lines of ***erythrocytes***.

CFU-S See ***colony-forming unit***.

CG island See ***CpG island***.

CGD See ***chronic granulomatous disease***.

cGMP See ***cyclic GMP***.

CGRP See ***calcitonin gene-related peptide***.

chaeotropic (chaotropic) An agent that causes chaos, usually in the sense of disrupting or denaturing macromolecules. For example, iodide is often used in protein chemistry to break up and randomize ***micelles***. In molecular biology guanidium isothiocyanate is used to provide a denaturing environment in which RNA can be extracted intact without exposure to ***RNAases***.

chaetoglobosins Chaetoglobosin J is a fungal metabolite related to cytochalasins that will inhibit elongation at the barbed end of an actin microfilament. Chaetoglobosin A is produced by *Chaetomium globosum*. Chaetoglobosin K is a plant growth inhibitor and toxin from the fungus *Diplodia macrospora*.

Chagas disease South American trypanosomiasis caused by *Trypanosoma cruzi* and transmitted by blood-sucking reduviid bugs such as ***Rhodnius***.

chalone Cell-released tissue-specific inhibitor of cell proliferation thought to be responsible for regulating the size of a population of cells. Contentious.

Chang liver cells Derived from non-malignant human tissue. Extensively used in virology and biochemistry. Cells are epithelial in morphology and grow to high density.

channel gating See ***gating current***.

channel protein A protein that facilitates the diffusion of molecules/ions across lipid membranes by forming a hydrophilic pore. Most frequently multimeric with the pore formed by subunit interactions.

channel-forming ionophore An ***ionophore*** that makes an amphipathic pore with hydrophobic exterior and hydrophilic interior. Most known types are cation selective.

Chaos chaos Giant multinucleate freshwater amoeba (up to 5mm long) much used for studies on the mechanism of cell locomotion.

chaperones Cytoplasmic proteins of both prokaryotes and eukaryotes (and organelles such as mitochondria) that bind to nascent or unfolded polypeptides and ensure correct folding or transport. Chaperone proteins do not covalently bind to their targets and do not form part of the finished product. Heat-shock proteins are an important subset of chaperones. Three major families are recognized, the ***chaperonins*** (groEL and hsp60), the hsp70 family and the hsp90 family. Outside these major families are other proteins with similar functions including ***nucleoplasmin***, secB, and T-cell receptor-associated protein.

chaperonins Subset of ***chaperone*** proteins found in prokaryotes, mitochondria and plastids. A major example is prokaryotic GroEL (the eukaryotic equivalent of which is hsp60).

CHAPS (3-((3-cholamidopropyl) dimethylammonio)-1-propane sulphonate) Zwitterionic detergent used for membrane solubilization.

Chara See ***Characean algae***.

Characean algae (Charophyceae) Class of filamentous green algae exemplified by the genus *Chara*, in which the mitotic spindle is not surrounded by a nuclear envelope. Probably the closest relatives, among the algae, to higher plants. The giant internodal cells (up to 5cm long) exhibit dramatic ***cyclosis*** and have been much used for studies on ion transport and cytoplasmic streaming.

Charcot–Marie–Tooth disease Hereditary degenerative motor and sensory neuropathy affecting peripheral nerves and inherited as an autosomal dominant.

Charophyceae See ***Characean algae***.

chartins Microtubule-associated proteins (***MAPs***) of 64, 67 and 80 kD, distinct from ***tau protein.*** Isolated from neuroblastoma cells. They are regulated by ***nerve growth factor*** (NGF) and may influence microtubule distribution.

charybdotoxin Peptide (37 residues) isolated from *Leiurus quinquestriatus hebraeus* (scorpion) venom that is a selective blocker of high conductance Ca^{2+}-activated K^+-channel.

checkerboard assay Variant of the Boyden chamber assay for leucocyte chemotaxis introduced by Zigmond. By testing different concentrations of putative chemotactic factor in non-gradient conditions, it is possible to calculate the enhancement of movement expected due simply to ***chemokinesis*** and to compare this with the distances moved in positive and negative gradients. Good experimental design thus allows chemotaxis to be distinguished from chemokinesis.

checkpoint Any stage in the ***cell cycle*** at which the cycle can be halted and entry into the next phase postponed. Two major checkpoints are at the G1/S and G2/M boundaries. These are the points at which ***cdc*** proteins act.

Chediak–Higashi syndrome Autosomal recessive disorder characterized by the presence of giant lysosomal vesicles in phagocytes and in consequence poor bactericidal function. Some perturbation of microtubule dynamics seems to be involved. Reported from humans, albino Hereford cattle, mink, beige mice and killer whales.

chelation Binding of a metal ion by a larger molecule such as EDTA or protein (iron in haem is held as a chelate). The binding is strong but reversible and chelating agents can be used to buffer the free concentration of the ion in question.

chemical potential The work required (in J mol^{-1}) to bring a molecule from a standard state (usually infinitely separated in a vacuum) to a specified concentration. More usually employed as chemical potential difference, the work required to bring one mole of a substance from a solution at one concentration to another at a different concentration, Δ m = RT. ln (c_2/c_l). This definition is useful in studies of ***active transport***; note that, for charged molecules, the electrical potential difference must also be considered (see ***electrochemical potential***).

chemical synapse A nerve–nerve or nerve–muscle junction where the signal is transmitted by release from one membrane of a chemical transmitter that binds to a receptor in the second membrane. Importantly, signals only pass in one direction.

chemiluminescence Light emitted as a reaction proceeds. Becoming used increasingly to assay ATP (using firefly ***luciferase***) and the production of toxic oxygen species by activated phagocytes (using ***luminol*** or ***lucigenin*** as bystander substrates that release light when oxidized). See also ***bioluminescence***.

chemiosmosis A theoretical mechanism (proposed by Mitchell) to explain energy transduction in the mitochondrion. As a general mechanism it is the coupling of one enzyme-catalysed reaction to another using the transmembrane flow of an intermediate species, eg. cytochrome oxidase pumps protons across the mitochondrial inner membrane and ATP synthesis is 'driven' by re-entry of protons through the ATP-synthesizing protein complex. The alternative model is production of a chemical intermediate species, but no compound capable of coupling these reactions has ever been identified.

chemiosmotic hypothesis See ***chemiosmosis***.

chemoattractant A substance that elicits accumulation of cells.

chemoattraction Non-committal description of cellular response to a diffusible chemical – not necessarily by a tactic response. Term preferable to 'chemotaxis' when the mechanism is unknown.

chemoautotroph (chemotroph) Chemotrophic ***autotroph***. Organism in which energy is obtained from endogenous light-independent reactions involving inorganic molecules.

chemodynesis Induction of cytoplasmic streaming in plant cells by chemicals rather than by light (photodynesis).

chemokine receptors (CCR) Chemokine receptors are G-protein linked ***serpentine receptors*** that, in addition to binding ***chemokines***, are used as coreceptors for the binding of immunodeficiency viruses (***HIV, SIV, FIV***) to leucocytes. CXCR4 is a coreceptor for T-tropic viruses, CCR5 for macrophage-tropic (M-tropic) viruses. Individuals deficient in particular CCRs seem to be resistant to HIV-1 infection. See Tables C4a and C4b (chemokines and chemokine receptors).

chemokines Small secreted proteins that stimulate ***chemotaxis*** of leucocytes. Chemokines can be subdivided into classes on the basis of conserved cysteine residues: the α-chemokines (IL-8, NAP-2, Gro-α, Gro-γ, ENA-78 and GCP-2) have conserved C-X-C motif and are mainly chemotactic for neutrophils; the β-chemokines (MCP-1-5, MIP-1α, MIP-1β, eotaxin, RANTES) have adjacent cysteines (C-C) and attract monocytes, eosinophils or basophils; the γ-chemokines have only one cysteine pair and are chemotactic for lymphocytes (lymphotactin); the δ-chemokines are structurally rather different being membrane-anchored, have a C-X-X-X-C motif and are restricted (so far) to brain (***neurotactin***). Human genes for the α-chemokines are on Chr 4 and 10, for β-chemokines on Chr 17, for lymphotactin on Chr 1 and for neurotactin on Chr16. The receptors are ***G-protein*** coupled.

chemokinesis A response by a motile cell to a soluble chemical that involves an increase or decrease in speed (positive or negative ***orthokinesis***) or of frequency of movement, or a change in the frequency or magnitude of turning behaviour (***klinokinesis***).

chemolithotroph Alternative name for a ***chemoautotroph***.

chemoreceptor A cell or group of cells specialized for responding to chemical substances in the environment.

chemorepellant Opposite of chemoattractant.

chemostat Apparatus for maintaining a bacterial population in the exponential phase of growth by regulating the input of a rate-limiting nutrient and the removal of medium and cells.

chemosynthesis Synthesis of organic compounds by an organism using energy derived from oxidation of inorganic molecules rather than light (see ***chemotrophy*** and ***photosynthesis***).

chemotactic See ***chemotaxis***.

chemotaxis A response of motile cells or organisms in which the direction of movement is affected by the gradient of a diffusible substance. Differs from chemokinesis in that the gradient alters probability of motion in one direction only, rather than rate or frequency of random motion.

chemotherapy Treatment of a disease with drugs that are designed to kill the causative organism or, in the case of tumours, the abnormal cells.

chemotroph See ***chemoautotroph***.

chemotrophy Systems of metabolism in which energy is derived from endogenous

Table C4a. Chemokines

Chemokine	Synonyms	Attracts	Produced by:
CXC family	**α-chemokines**		
IL-8	NAP-1, MONAP, MDNCF, NAF, LAI, GCP	Neutrophils	Many cells, including monocytes, lymphocytes, fibroblasts, endothelial cells, mesangial cells
γIP-10	CRG-2, C7	Monocytes	Keratinocytes, monocytes, T-cells, endothelial cells and fibroblasts
PBP	CTAP-III, NAP-2, β-TG, low affinity platelet factor 4	Fibroblasts (βTG), neutrophils (NAP-2)	Platelets
PF-4	Oncostatin A, platelet factor 4	Neutrophils	Aggregated platelets, activated T-cells
MIP-2α, β	GROβ, GROγ, respectively, NAP-3, KC	Neutrophils	Activated monocytes, fibroblasts, epithelial and endothelial cells.
PBSF	SDF-1α	?	Fibroblasts, bone marrow stromal cells.
CC family	**β-chemokines**		
MCPs	MCAF, JE, LDCF, GDCF, HC14, MARC	Monocytes, basophils	Monocytes, T-cells, fibroblasts, endothelial cells, smooth muscle, some tumours. Upregulated by IFN-γ
RANTES	sisδ	Monocytes, memory T-cells, eosinophils	T-cells, macrophages
Eotaxin		Eosinophils	
I-309	TCA-3	Monocytes	T-cells, mast cells
MIP1α	pLD78, pAT464, GOS19	Eosinophils	T-cells, B-cells, Langerhans cells, neutrophils, macrophages
TARC			Activated T-cells
MDC	Monocyte-derived chemokine		Monocytes
MIP1β	ACT-2, pAT744, hH400, hSISα, G26, HC21, MAD-5, HIMAP	Memory T-cells	T-cells, B-cells, macrophages
C family	**γ chemokines**		
Lymphotactin		Lymphocytes	
CX3C family	**δ chemokines**		
Fractalkine		Monocytes, lymphocytes	
Neurotactin		Neutrophils	Brain

Based on Callard R and Gearing A (1994) *Cytokine Facts Book*. Academic Press.

chemical reactions rather than from food or light energy, eg. in deep-sea hot-spring organisms.

chemotropism Growth or possibly bending of an organism in response to an external chemical gradient. Sometimes used in error when the terms ***chemotaxis*** or ***chemokinesis*** should have been used.

CHF See ***chick heart fibroblasts***.

chi-squared (χ-squared) Common statistical test to determine whether the observed values of a variable are significantly different from those expected on the basis of a null hypothesis.

chiasma (plural, chiasmata) Junction points between non-sister ***chromatids*** at the first ***diplotene*** of ***meiosis***, the consequence of a ***crossing-over*** event between maternal and paternally derived ***chromatids***. A

Table C4b. Chemokine receptors

Receptor	Ligands	Cell types
CXCR1	IL-8, CGP-2	Neutrophils
CXCR2	IL-8, NAP-2; GROα, ENA-78	Neutrophils
CXCR3	IP10, MIG	Activated Th1
CXCR4	SDF-1	Naïve T; B-cells
CXCR5	BCA-1	B-cells
CCR1	RANTES, MIP-1α, MCP-3	Activated T; monocytes; eosinophils; dendritic cells
CCR2	MCP-1-5	Monocytes; macrophages; activated T
CCR3	Eotaxin, MCP-3, MCP-4, RANTES	Activated Th2; eosinophils; basophils
CCR4	TARC, MDC, MIP-1α, RANTES	Activated Th2; basophils; platelets
CCR5	MIP-1β, RANTES, MIP-1α	Activated T; monocytes; macrophages; dendritic cells
CCR6	MIP-3α	Dendritic cells; T-cells
CCR7	MIP-3β, SLC	B- and T-cells
CCR8	I309	Monocytes; macrophages
CCR9	CC chemokines	Non-haematopoietic
Duffy Ag	IL-8; GROα, RANTES, MCP-1	Erythrocytes
CX3CR	Fractalkine	NK cells, $CD8^+$ T-cells

Based on TNC Wells, CA Power & AEI Proudfoot (1998) *TiPS* 19: 376.

chiasma also serves a mechanical function and is essential for normal equatorial alignment at meiotic ***metaphase*** I in many species. Frequency of chiasmata is very variable between species.

chick heart fibroblasts The cells that emigrate from an explant of embryonic chick heart maintained in culture. Often used as archetypal normal cell.

chicken ovalbumin upstream promoter transcription factors I and II See ***COUP-TFs***.

chimera Organism composed of two genetically distinct types of cells. Can be formed by the fusion of two early blastula stage embryos or by the reconstitution of the bone marrow in an irradiated recipient, or by somatic segregation.

Chinese hamster ovary cells (CHO cells) Cell line that is often used for growing viruses or for transfection; see also Table C3.

Chironomus Genus of flies (midges). Larvae live in fresh water and have been much studied because of the giant ***polytene chromosomes*** in the salivary glands; ***haemolymph*** contains haemoglobin in solution.

chitin Polymer (β-1,4 linked) of N-acetyl-D-glucosamine, extensively cross-linked and the major structural component of arthropod exoskeletons and fungal cell walls. Widely distributed in plants and fungi.

chitinase Enzyme (EC 3.2.1.14) that catalyses the hydrolysis of 1,4-β linkages of ***chitin***.

chitosan A polymer of 1-4-β-D-glucosamine found in the cell wall of some fungi.

chitosome Membrane-bound vesicular organelle (40–70nm diameter) found in many fungi. Contains chitin sythetase that produces chitin microfibrils that are released and incorporated into the cell wall.

Chlamydia Genus of minute prokaryotes that replicate in cytoplasmic vacuoles within susceptible eukaryotic cells. Genome about one-third that of *E. coli*. *C. trachomatis* causes trachoma in man; *C. psittaci* causes economically important diseases of poultry.

Chlamydomonas A genus of unicellular green algae, usually flagellated. Easily grown in the laboratory and have often been used in studies on flagellar function: a range of paralysed flagellar (pf) mutants has been isolated and studied extensively.

chloragosome Cytoplasmic granule of unknown function found in the coelomocytes of annelids.

chloramphenicol An antibiotic from *Streptomycetes venezuelae* that inhibits protein synthesis in prokaryotes and in mitochondria and chloroplasts by acting on the 50S ribosomal subunit. It is relatively toxic but finds wide application in medicine.

chloramphenicol acetyltransferase (CAT) Enzyme that inactivates the antibiotic ***chloramphenicol*** by acetylation. Widely used as a ***reporter gene***.

Chlorella Genus of green unicellular algae extensively used in studies of photosynthesis.

chlorenchyma Form of ***parenchyma*** tissue active in photosynthesis, in which the cells contain many ***chloroplasts***; found especially in leaf ***mesophyll***.

chloride channel ***Ion channels*** selective for chloride ions. Various types including ***ligand-gated*** Cl-channels at synapses (the ***GABA***-and ***glycine***-activated channels), as well as ***voltage-gated*** Cl-channels found in a variety of plant and animal cells. See also ***CFTR, MDR***.

chloride current Flow of chloride ions through chloride-selective ***ion channels***.

chlorophyll The photosynthetic pigments of higher plants, but closely related to bacteriochlorophylls. Magnesium complexes of tetrapyrolles.

Chlorophyta (Green algae) Division of algae containing photosynthetic pigments similar to those in higher plants and having a green colour. Includes unicellular forms, filaments and leaf-like thalluses (eg. *Ulva*). Some members form ***coenobia***, and the ***Characean algae*** have branched filaments.

chloroplast Photosynthetic organelle of higher plants. Lens-shaped and rather variable in size but approximately 5μm long. Surrounded by a double membrane and contains circular DNA (though not enough to code for all proteins in the chloroplast). Like the mitochondrion, it is semi-autonomous. It resembles a cyanobacterium from which, on the endosymbiont hypothesis, it might be derived. The photosynthetic pigment, chlorophyll, is associated with the membrane of vesicles (thylakoids) that are stacked to form grana.

chloroquine Antimalarial drug that has the interesting property of increasing the pH within the ***lysosome*** when added to intact cells in culture.

chlorosis Yellowing or bleaching of plant tissues due to the loss of chlorophyll or failure of chlorophyll synthesis. Symptomatic of many plant diseases, also of deficiencies of light or certain nutrients.

chlorosome Elongated membranous vesicles attached to the plasma membrane of green photosynthetic bacteria; contain the light-harvesting antenna complexes of bacteria in the sub-order Chlorobiineae. Pigments include bacteriochlorophylls and carotenoids.

chlorpromazine Neuroleptic aliphatic phenothiazine, thought to act primarily as dopamine antagonist, but also antagonist at α-adrenergic, H1 histamine, muscarinic and serotonin receptors. Used clinically as an antiemetic. Has been shown to alter fibroblast behaviour.

CHO cells See ***Chinese hamster ovary cells***, and Table C3.

choanocytes Cells that line the radial canals of sponges. Have long flagella that are responsible for generating the feeding current.

cholate In practice, the sodium salt of cholic acid, that has strong detergent properties and can replace membrane lipids to generate soluble complexes of membrane proteins.

cholecalcin See ***calbindin***.

cholecystitis Inflammatory condition of the wall of the gall bladder caused by *Salmonella typhi*.

cholecystokinin See ***CCK***.

cholera toxin A multimeric protein toxin from *Cholera vibrio*. The toxic A subunit (27 kD) activates adenyl cyclase irreversibly by ***ADP-ribosylation*** of a Gs protein. The B subunit (57 kD) has five identical monomers, binds to GM1 ganglioside and facilitates passage of the A subunit across the cell membrane.

cholesterol The major ***sterol*** of higher animals. An important component of cell membranes, especially of the plasma (outer) membrane, most notably the ***myelin*** sheath. Transported in the esterified form via plasma lipoproteins.

cholesterol-binding toxins Family of 50–60 kD pore-forming toxins from various genera of bacteria including *Streptococcus, Listeria,* Bacillus and *Clostridium*. Apparently bind to cholesterol and oligomerize to form a pore: as a result cause cell lysis and are lethal. See ***Streptolysin O***. Other examples include Pneumolysin from *S. pneumoniae*, Cereolysin O from *Bacillus cereus*, Thuringolysin O from *B. thuringiensis*, Tetanolysin from *Clostridium* tetani, Botulinolysin from *C. botulinum*, Perfringolysin O from *C. perfringens*, Listeriolysin O from *L. monocytogenes*.

choline Esterified in the head group of phospholipids (phosphatidyl choline and sphingomyelin) and in the neurotransmitter acetylcholine. Otherwise a biological source of methyl groups.

cholinergic neurons Neurons in which actylcholine is the neurotransmitter.

chondro- Prefix: cartilage related/associated.

chondroblast Embryonic cartilage-producing cell.

chondrocyte Differentiated cell responsible for secretion of extracellular matrix of ***cartilage***.

chondroitin sulphates Major components of the extracellular matrix and connective tissue of animals. They are repeating polymers of glucuronic acid and sulphated N-acetyl glucosamine residues that are highly hydrophilic and anionic. Found in association with proteins.

chondronectin A 180 kD protein isolated from chick serum that specifically favours attachment of ***chondrocytes*** to type II ***collagen*** if present with the appropriate cartilage ***proteoglycan***.

chorioallantoic membrane (1) Protective membrane around the eggs of insects and fishes. (2) Extra-embryonic membrane surrounding the embryo of amniote vertebrates. The outer epithelial layer of the chorion is derived from the trophoblast, by the apposition of the ***allantois*** to the inner face of the ***chorion***. The chorioallantoic membrane is highly vascularized, and is used experimentally as a site upon which to place pieces of tissue in order to test their invasive capacity.

choriocarcinoma Malignant tumour of trophoblast.

chorion (1) Protective membrane around the eggs of insects and fishes. (2) Extra-embryonic membrane surrounding the embryo of amniote vertebrates. The outer epithelial layer of the chorion is derived from the trophoblast.

chorionic gonadotrophin A glycoprotein hormone synthesized in the placenta that controls the size of the gonads and the synthesis of sex hormones.

choroid Middle layer of the vertebrate eye, between ***retina*** and sclera. Well vascularized and also pigmented to throw light back onto the retina (the tapetum is an iridescent layer in the choroid of some eyes). Not to be confused with the choroid plexus, a highly vascularized region of the roof of the ventricles of the vertebrate brain that secretes cerebrospinal fluid.

choroid plexus Mass of highly branched blood vessels in margin of cerebral ventricles that produces cerebrospinal fluid.

Christmas disease Congenital deficiency of blood-clotting factor IX (first described in the Christmas issue of the *British Medical Journal*, 1952). Inherited in similar sex-linked way to classical haemophilia.

chromaffin cell See ***granins***.

chromaffin tissue Tissue in medulla of adrenal gland containing two populations of cells, one producing adrenaline, the other noradrenaline. The ***catecholamine*** is associated with carrier proteins (chromogranins) in membrane vesicles (chromaffin granules).

chromatic aberration When using white light through a lens system, it is inevitable that different wavelengths (colours) are brought to a focus at slightly different points. As a consequence, there are chromatic aberrations in the image; good microscope objectives are therefore corrected for this at two wavelengths (achromats) or at three wavelengths (apochromats), as well as for ***spherical aberration***.

chromatid Single chromosome containing only one DNA duplex. Two daughter chromatids become visible at mitotic metaphase, though they are present throughout G2.

chromatin Stainable material of interphase nucleus consisting of nucleic acid and associated histone protein packed into ***nucleosomes***. Euchromatin is loosely packed and accessible to RNA polymerases, whereas heterochromatin is highly condensed and probably transcriptionally inactive.

chromatin body Barr body; condensed X chromosome in female mammalian cell.

chromatography Techniques for separating molecules based on differential absorption and elution. Term for separation methods involving flow of a fluid carrier over a non-mobile absorbing phase.

chromatophores (1) Pigment-containing cells of the dermis, particularly in teleosts and amphibians. By controlling the intracellular distribution of pigment granules the animal can blend with the background. ***Melanocytes*** and ***melanophores*** are melanin-containing chromatophores. (2) Term occasionally used for chloroplasts in the chromophyte algae.

chromobindin See ***annexin***.

chromocentre Condensed heterochromatic region of a chromosome that stains particularly strongly although in the polytene chromosomes of *Drosophila* the chromocentre is of under-replicated heterochromatin and stains lightly.

chromogranins See ***chromaffin tissue*** and ***granins***.

chromomere Granular region of condensed ***chromatin***. Used of chromosomes at leptotene and zygotene stages of meiosis, of the condensed regions at the base of loops on lampbrush chromosomes, and of condensed bands in polytene chromosomes of Diptera.

chromophore The part of a visibly coloured molecule responsible for light absorption over a range of wavelengths thus giving rise to the colour. By extension the term may be applied to UV- or IR-absorbing parts of molecules. Do not confuse with ***chromatophores***.

chromoplast Plant chromatophore filled with red/orange or yellow carotenoid pigment. Responsible for the colour of carrots and of many petals.

chromosome The DNA of eukaryotes is subdivided into chromosomes, presumably for convenience of handling, each of which has a long length of DNA associated with various proteins. The chromosomes become more tightly packed at mitosis and become aligned on the ***metaphase plate***. Each chromosome has a characteristic length and banding pattern. See ***C banding, G banding.***

chromosome condensation The tight packing of DNA into ***chromosomes*** in ***metaphase***, in preparation for ***cell division***.

chromosome painting See ***fluorescence in situ hybridization***.

chromosome segregation The orderly separation of one copy of each chromosome into each daughter cell at ***mitosis***.

chromosome synapsis The close apposition of homologous chromosomes before cell division, or permanently in giant ***polytene chromosomes***.

chromosome translocation The fusion of part of one chromosome onto part of another. Largely sporadic and random, there are some translocations at 'hotspots' that occur often enough to be clinically significant. See ***Philadelphia chromosome***.

chromosome walking A procedure to find and sequence a gene whose approximate position in a chromosome is known by classical genetic linkage studies. Starting with the known sequence of a gene shown by classical genetics to be near to the novel gene, new clones are picked from a ***genomic library*** by ***hybridization*** with a short probe generated from the appropriate end of the known sequence. The new clones are then sequenced, new probes generated, and the process repeated until the gene of interest is reached.

chronic Persistent, long-lasting (as opposed to ***acute***). Chronic ***inflammation*** is generally a response to a persistant antigenic stimulus.

chronic granulomatous disease (CGD) Disease, usually fatal in childhood, in which the production of hydrogen peroxide by ***phagocytes*** does not occur because of a lesion in an NADP-dependent oxidase. Catalase-negative bacteria are not killed and there is no luminol-enhanced ***chemiluminescence*** when the cells are tested. The absence of the oxygen-dependent killing mechanism is not itself fatal but seriously compromises the primary defence system. At least three separate lesions can cause the syndrome, the commonest being a defect in plasma membrane cytochrome.

chronic lymphocytic or myelogenous leukaemia See ***leukaemia***.

Chrysophyceae: golden algae See ***Chrysophyta***.

Chrysophyta (Chrysophyceae: golden algae) Division or class of algae, coloured golden-brown due to high levels of the ***xanthophyll***, fucoxanthin. Mostly single-celled or colonial. Also called Chrysomonadida by protozoologists.

CHUK (conserved helix-loop-helix ubiquitous kinase) Kinase responsible for phosphorylation of ***IκB*** thus triggering degradation of IκB and allowing ***NFκB*** to move to the nucleus.

Churg–Strauss syndrome ***ANCA***-associated vasculitis in which there is no complement consumption and no deposition of immune complexes. Affects predominantly small vessels.

chylomicron Colloidal fat globule found in blood or lymph; used to transport fat from the intestine to the liver or to adipose tissue. Has a very low density, a low protein and high triacylglyceride content.

chymosin (EC 3.4.23.4) Protease from abomasum (fourth stomach) of calf that has properties similar to pepsin. Will cleave casein to paracasein and is used in cheese making.

chymostatin Low molecular weight peptide–fatty acid compound of microbial origin that inhibits ***chymotrypsins*** and ***papain***.

chymotrypsin Serine proteases from pancreas. Preferentially hydrolyze Phe, Tyr, or Trp peptide and ester bonds.

cicatrization Contraction of fibrous tissue, formed at a wound site by ***fibroblasts***, reducing the size of the wound but causing tissue distortion and disfigurement. Once thought to be due to contraction of collagen but now known to be due to cellular activity.

ciclosporine See ***cyclosporin A***.

CIG (cold insoluble globulin) Obsolete synonym for ***fibronectin***.

cilia (plural) See ***cilium***.

ciliary body Tissue that includes the group of muscles that act on the eye lens to produce accommodation and the arterial circle of the iris. The inner ciliary epithelium is continuous with the ***pigmented retinal epithelium***; the outer ciliary epithelium secretes the aqueous humour.

ciliary ganglion ***Neural crest***-derived ganglion acting as relay between parasympathetic neurons of the oculomotor nucleus in the midbrain and the muscles regulating the diameter of the pupil of the eye.

ciliary neurotrophic factor (CNTF) ***Neurotrophin*** originally characterized as a survival factor for chick ciliary neurons *in vitro*. Subsequently shown to promote the survival of a variety of other neuronal cell types, and to promote the differentiation of bipotential ***O-2A progenitor*** cells to ***type-2 astrocytes in vitro***. Developmental expression and regional distribution studies show that, unlike NGF, CNTF is not a target-derived neurotrophic factor. Now considered to be one of the ***IL-6 cytokine family*** since it acts through a receptor containing gp130.

cilium (plural, cilia) Motile appendage of eukaryotic cells that contains an ***axoneme***, a bundle of microtubules arranged in a characteristic fashion with nine outer doublets and a central pair ('9 + 2' arrangement). Active sliding of doublets relative to one another generates curvature, and the asymmetric stroke of the cilium drives fluid in one direction (or the cell in the other direction).

CINC See ***melanoma growth-stimulatory activity.***

cingulin Rod-shaped dimeric protein (108 kD subunit) found in cytoplasmic domain of vertebrate tight junctions.

circadian rhythm Regular cycle of behaviour with a period of approximately 24 hours. In most animals the endogenous periodicity, which may be of longer or shorter duration, is entrained to 24 hours by environmental cues. See ***periodic*** and ***timeless***.

circular dichroism (CD) Differential absorption of right-hand and left-hand circularly polarized light resulting from molecular asymmetry involving a chromophore group. CD is used to study the conformation of proteins in solution.

circular DNA DNA arranged as a closed circle. This brings serious topological problems for replication that are solved with ***DNA topoisomerase***. Characteristic of prokaryotes but also found in mitochondria, chloroplasts and some viral genomes.

cirrhosis Irreversible condition affecting the whole liver involving loss of parenchymal cells, inflammation, disruption of the normal tissue architecture, and eventually hepatic failure.

cirri (singular, cirrus) Large motor organelles of hypotrich ciliates: formed from fused ***cilia***.

cirrus (singular) See ***cirri***.

CIS (cytokine-inducible immediate early gene) Gene activated by cytokine signals, the product of which may inhibit the signalling pathway. One of the ***SOCS*** family.

***cis*-activation** Activation of a gene by an activator located on the same chromosome, ie. not by a diffusible product.

***cis*-dominance** When a gene or promoter affects only gene activity in the DNA duplex molecule in which it is placed, the effect is referred to as *cis*, as opposed to *trans* effects when a gene or promoter on one DNA molecule can affect genes on another DNA molecule. *cis*-dominance is seen only when the appropriate pair or set of genes are all *cis* to each other.

***cis-trans* test** The complementation test with two or more interacting genes placed in *cis* and in *trans* relationships to each other. A double mutant genome is used in the *cis* test made from the two single mutant genomes used in the *trans* test by recombination. If the wild-type phenotype is restored by both *cis* and *trans* arrangements it is concluded that the two mutations are in different genes and hence that the phenotype is determined by more than one gene. If the *trans* test is negative and the *cis* positive this means that the two mutations are in the same gene. If both tests are negative then at least one of the mutations must be dominant. Thus the double test provides a means of fine mapping of genes.

cisplatin (*cis*-diamineplatinum dichloride) Cytotoxic drug used in tumour chemotherapy. Binds to DNA and forms platinum–nitrogen bonds with adjacent guanines.

cisterna (plural, cisternae) Membrane-bounded sacs of smooth and rough ***endoplasmic reticulum*** and ***Golgi apparatus***.

cistron A genetic element defined by means of the *cis-trans* ***complementation*** test for functional allelism; broadly equivalent to the sequence of DNA that codes for one polypeptide chain, including adjacent control regions.

citric acid cycle Also known as ***tricarboxylic acid cycle*** or Krebs cycle.

citrulline (2-amino-5-ureiodovaleric acid) An α-amino acid not found in proteins. L-citrulline is an intermediate in the urea cycle.

CJD See ***Creutzfeldt–Jacob disease***.

clamp connection In many ***basidiomycete*** fungi a short lateral branch of a binucleate cell develops. This is the developing clamp connection. One of the nuclei migrates into it. Both nuclei then undergo simultaneous mitosis so that one end of the cell contains two daughter nuclei from each of the parental nuclei. The nucleus in the branch and the two nuclei are separated off from the centre of the cell by ***septa***. A single nucleus remains in the central region. The clamp connection then extends towards and fuses with the central section so that a binucleate cell is reformed.

class switching Phenomenon that occurs during the maturation of an immune response in which, for example, B-cells cease making IgM and begin making IgG that has the same antigen specificity. Switching between other immunoglobulin classes can occur.

clathrin Protein composed of three heavy chains (180 kD) and three light chains (34 and 36 kD), that forms the basketwork of 'triskelions' around a ***coated vesicle***. There are two genes for light chains, each of which can generate two distinct transcripts by tissue-specific alternative splicing.

clathrin adaptor proteins (HA1 and HA2 adaptors) Family of proteins that bind to clathrin and promote its assembly into vesicle coats. Different ***adaptor*** proteins are associated with coated vesicles of Golgi or plasma membrane origin.

cleavage The early divisions of the fertilized egg to form blastomeres. The cleavage pattern is radial in some phyla, spiral in others.

Cleland's reagent Dithiothreitol.

cli See ***clusterin***.

climacteric A particular stage of fruit ripening, characterized by a surge of respiratory activity, and usually coinciding with full ripeness and flavour in the fruit. Its appearance is hastened by ethylene at low concentrations.

CLIP-170 (cytoplasmic linker protein-170) Phosphorylation-regulated microtubule-binding protein that accumulates towards the plus ends of cytoplasmic microtubules; isolated from HeLa cells.

CLL Chronic lymphocytic ***leukaemia***.

CLN2 A yeast cyclin. See ***FAR1***.

cloche Zebrafish mutation that affects differentiation of endothelial and haematopoietic cells and probably acts upstream of ***flk-1***.

clock gene A gene with a level of expression that varies cyclically, and which might therefore be involved in the generation of biological rhythyms. Examples: ***period***, ***timeless***.

clonal deletion One of the two main hypotheses advanced to explain the absence of autoimmune responses: now generally accepted. Clonal deletion is the programmed death of inappropriately stimulated, autoreactive, clones of T-cells.

clonal selection The process whereby one or more clones, ie. cells expressing a particular gene sequence, are selected by naturally occurring processes from a mixed population. Generally the clonal selection is for general expansion by mitosis, particularly with reference to ***B-lymphocytes*** where selection with subsequent expansion of clones occurs as a result of antigenic stimulation only of those lymphocytes bearing the appropriate receptors.

clone A propagating population of organisms, either single cell or multicellular, derived from a single progenitor cell. Such organisms should be genetically identical, though mutation events may abrogate this.

clonidine Centrally acting antihypertensive agent that works by inhibiting activity of the sympathoexcitatory neurons that regulate arterial pressure. Considered to be a mixed agonist acting on both α-2-adrenoreceptors and ***imidazoline receptors*** (I1-R). Moxonidine and rilmenidine are similar but more specific for I1 receptors.

cloning The process whereby clones are established. The term covers various manipulations for isolating and establishing clones: in simple systems single cells may be isolated without precise knowledge of their genotype; in other systems (see ***gene cloning***) the technique requires partial or complete selection of chosen genotypes; in plants the term refers to natural or artificial vegetative propagation.

cloning vector A plasmid ***vector*** that can be used to transfer DNA from one cell type to another. Cloning vectors are usually designed to have convenient restriction sites that can be cut to generate sticky ends to which the DNA that is to be cloned can be ligated easily.

Clonorchis sinensis Chinese liver fluke. Can infect man if inadequately cooked fish is eaten and can cause biliary obstruction as a result of liver infestation.

Clostridium Genus of ***Gram-positive*** anaerobic spore-forming bacilli commonly found in soil. Many species produce exotoxins of great potency, the best known being *C. botulinum*, and *C. tetani*. Among the toxins produced by *C. perfringens* are ***perfringolysin*** (θ-toxin), an α-toxin (phospholipase C), β, ∊ and ι-toxins (act on vascular endothelium to cause increased vascular permeability), δ-toxin (a haemolysin), and κ-toxin (a collagenase).

Clostridium difficile* toxins** *C. difficile* enterotoxin A (308 kD) is secreted, enters eukaryotic cell by ***receptor-mediated endocytosis and once in cytoplasm glucosylates small G-proteins of π family thereby inactivating them and leading to loss of actin filament bundles. Cytotoxin B (270 kD) is similar.

cluster of differentiation (CD) antigens See Table C2.

clusterin Vertebrate glycoprotein of uncertain function. Secreted as a 400 amino acid peptide, then cleaved to form two 200 amino acid peptides that are linked by a disulphide bridge. Also known as complement-associated protein SP-40, complement cytolysis inhibitor (CLI), apolipoprotein J, sulphated glycoprotein 2 (SGP-2), dimeric acid glycoprotein (DAG), glycoprotein III (GpIII).

clustering Of ***acetylcholine receptors***: aggregation of the receptors in developing ***myotubes*** in the vicinity of the presynaptic terminal, induced by nerve contact. See ***agrin***.

CMC (cell-mediated cytotoxicity) The term is applied to T-lymphocytes that react to antigen by mitogenesis and develop into clones of specific T-effector cells.

CMD1 Chick homologue of ***myoD***.

CML (1) Cell-mediated lympholysis: the process of target cell lysis by CMCs. (2) Chronic myeloid ***leukaemia***.

CMV See ***cytomegalovirus***.

CNBr See ***cyanogen bromide***.

CNFs See ***cytotoxic necrotizing factors***.

Cnidaria (Coelenterata) Diverse phylum of diploblastic animals that includes Classes

Hydrozoa (freshwater polyps, small jellyfish), Scyphozoa (large jellyfish), and Anthozoa (sea anemones and stony corals). Characteristically the ectoderm has specialized stinging cells containing ***nematocysts***.

cnidoblast Developing form of ***cnidocyte***.

cnidocyst See ***nematocyst***.

cnidocyte Ectodermal cell of Cnidaria (coelenterates) specialized for defence or capturing prey. Each cell has a ***nematocyst*** that can be replaced once discharged.

CNS See ***central nervous system***.

CNTF See ***ciliary neurotrophic factor***.

co-carcinogens Substances that, though not carcinogenic in their own right, potentiate the activity of a carcinogen. Strictly speaking they differ from ***tumour promoters*** in requiring to be present concurrently with the carcinogen.

co-culture Growth of distinct cell types in a combined culture. In order to get some cells to grow at low (clonal) density it is sometimes helpful to grow them together with a ***feeder layer*** of ***macrophages*** or irradiated cells. The mixing of different cell types in culture is otherwise normally avoided, although it is possible that this could prove an informative approach to modelling interactions ***in vivo***.

co-transport In membrane transport describes tight coupling of the transport of one species (generally Na^+) to another (eg. a sugar or amino acid). The transport of Na_+ from high to low concentration can provide the energy for transport of the second species up a concentration gradient. See secondary ***active transport***.

coacervate Colloidal aggregation containing a mixture of organic compounds. One theory of the evolution of life is that the formation of coacervates in the primaeval soup was a step towards the development of cells.

coagulation factor Group of plasma proteins, many of which contain ***EGF***-like domains. See Table F1.

coagulation factor XI A plasma serine ***protease*** with an ***apple*** domain. See Table F1.

coated pit First stage in the formation of a ***coated vesicle***.

coated vesicle Vesicle formed as an invagination of the plasma membrane (a ***coated pit***), and which is surrounded by a basket of ***clathrin***. Associated with receptor-mediated pinocytosis and receptor recycling.

cobalamin Vitamin B12. See Table V1.

cobra venom factor See ***C3***.

cobratoxin (α-cobratoxin) Polypeptide toxin (71 residues) from *Naja kaouthia*. One of the ***alpha-neurotoxins*** (curaremimetics), it binds to nicotinic ***acetylcholine receptors*** with high affinity.

cocaine Drug of abuse and psychostimulant that acts to increase extraneuronal ***dopamine*** in midbrain by binding to the dopamine uptake transporter and hence inhibiting dopamine reuptake at the plasma membrane.

cocci Bacteria with a spherical shape.

cochlear hair cell The sound-sensing cell of the inner ear. The cells have modified ciliary structures (hairs), that enable them to produce an electrical (neural) response to mechanical motion caused by the effect of sound waves on the cochlea. Frequency is detected by the position of the cell in the cochlea and amplitude by the magnitude of the disturbance.

codominant Genes in which both alleles of a pair are fully expressed in the heterozygote as, for example, AB blood group in which both A and B antigens are present.

codon The coding unit of DNA that specifies the function of the corresponding messenger RNA. A triplet of bases recognized by anticodons on transfer RNA and hence specifying an amino acid to be incorporated into a protein sequence. The code is degenerate, ie. each amino acid has more than one codon. The stop-codon determines the end of a polypeptide. See Table C5.

Table C5. The codon assignments of the genetic code

First position (5′ end)	Second position U	C	A	G	Third position (3′ end)
U	Phe, F	Ser, S	Tyr, Y	Cys, C	U
	Phe, F	Ser, S	Tyr, Y	Cys, C	C
	Leu, L	Ser, S	Stop (ochre)	Stop: (Trp)[a]	A
	Leu, L	Ser, S	Stop (amber)	Trp, W	G
C	Leu, L	Pro, P	His, H	Arg, R	U
	Leu, L	Pro, P	His, H	Arg, R	C
	Leu, L	Pro, P	Gln, Q	Arg, R	A
	Leu, L	Pro, P	Gln, Q	Arg, R	G
A	lle, I	Thr, T	Asn, N	Ser, S	U
	lle, I	Thr, T	Asn, N	Ser, S	C
	lle 1: (Met)[a]	Thr, T	Lys, K	Arg, R: (stop)[a]	A
	Met, M (start)	Thr, T	Lys, K	Arg, R: (stop)[a]	G
G	Val, V	Ala, A	Asp, D	Gly, G	U
	Val, V	Ala, A	Asp, D	Gly, G	C
	Val, V	Ala, A	Glu, E	Gly, G	A
	Val, V: (Met)[b]	Ala, A	Glu, E	Gly, G	G

[a]Unusual codons used in human mitochondria.
[b]Normally codes for valine but can code for methionine to initiate translation from an mRNA chain.

coelenterate Animal of the Phylum ***Cnidaria***. Mostly marine, diploblastic and with radial symmetry. Sea anemones and *Hydra* are well-known examples.

coeliac disease (USA, celiac disease) Gluten enteropathy: atrophy of ***villi*** in small intestine leads to impaired absorption of nutrients. Caused by sensitivity to ***gluten*** (protein of wheat and rye). Sufferers have serum antibodies to gluten and show delayed hypersensitivity to gluten; the risk factor is 10 times greater in HLA-B8 positive individuals.

coelom Body cavity characteristic of most multicellullular animals (all ***coelomates***). Arises within the embryonic mesoderm, that is thereby subdivided into ***somatic mesoderm*** and ***splanchnic mesoderm***, and is lined by the mesodermally derived peritoneum. May be secondarily lost and it is unclear whether it evolved once or several times.

coelomocyte Cell found in the ***coelom***, eg. immune cells of sea urchins.

coenobia (plural) See ***coenobium***.

coenobium (plural, coenobia) Colony of cells formed by certain green algae, in which little or no specialization of the cells occurs. The cells are often embedded in a mucilaginous matrix. Examples: *Volvox, Pandorina.*

coenocyte Organism that is not subdivided into cells but has many nuclei within a mass of cytoplasm (a syncytium), as for example some fungi and algae, and the acellular slime mould *Physarum*.

coenzyme Either: low molecular weight intermediate that transfers groups between reactions (eg. NAD), or: catalytically active low molecular weight component of an enzyme (eg. haem). Coenzyme and apoenzyme together constitute the holoenzyme.

coenzyme A A derivative of adenosine triphosphate and pantothenic acid that can carry acyl groups (usually acetyl) as thioesters. Involved in many metabolic pathways, eg. citric acid cycle and in fatty acid oxidation.

coenzyme M (2-mercaptoethanesulphonic acid.) This substance is involved in the formation of methane from carbon dioxide by methanogenic bacteria.

coenzyme Q See ***ubiquinone***.

cofactor Inorganic complement of an enzyme reaction, usually a metal ions. See ***coenzyme***.

cofilin Actin-severing protein (19 kD), related to ***destrin***. Binds to the side of filaments and is pH sensitive. Shares with tropomyosin a 13 amino acid ***F-actin*** binding domain. Very similar to ***ADF*** (actin depolymerizing factor).

coil Protein secondary structure ***motif*** that does not qualify as ***alpha-helix, beta sheet*** or ***beta turn***.

coiled body A ubiquitous nuclear organelle containing the mRNA splicing machinery (U1, U2, U4, U5, U6 and U7 snRNAs), the U3 and U8 snRNPs that are involved in pre-rRNA processing, together with nucleolar proteins such as ***fibrillarin*** and ***coilin***. Ultrastructurally they appear to consist of a tangle of coiled threads and are spherical, between 0.5 and 1μm in diameter.

coilin (p80-coilin) Protein, M(r) 80 kD, (62.6 kD calculated from 576-residue sequence) found in ***coiled body***. A relatively short portion of the N-terminus seems to target the protein to the organelle.

Col-V A plasmid of *E. coli* that codes for ***colicin*** V, that confers resistance to complement-mediated killing, for a siderophore to scavenge iron, and for F-like pili that permit conjugation.

colcemid Methylated derivative of ***colchicine***.

colchicine Alkaloid (400 D) isolated from the Autumn crocus (*Colchicum autumnale*) that blocks microtubule assembly by binding to the ***tubulin*** heterodimer (but not to tubulin). As a result of interfering with microtubule, reassembly will block mitosis at ***metaphase***.

Colchicum Genus of crocuses. *C. autumnale*, the autumn crocus, is the source of ***colchicine***.

cold agglutinins ***Antibodies*** that agglutinate particles with greater activity below 32°C. They are IgM antibodies specifically reactive with blood groups I and i in humans, and agglutinate red blood cells on cooling, causing Raynaud's phenomenon *in vivo*. See ***agglutination***.

cold insoluble globulin (CIG) Name, now obsolete, originally given to fibronectin prepared from ***cryoprecipitate***.

coleoptericin Inducible antibacterial peptide found in the haemolymph of a tenebrionid beetle following the injection of heat-killed bacteria. Peptide A (glycine-rich, 74 residues) is active against ***Gram-negative bacteria***; peptides B and C are isoforms of a 43 residue cysteine-rich peptide that has sequence homology with ***defensins*** and is active against ***Gram-positive*** bacteria. See ***diptericins, cecropins, apidaecins, abaecins***.

coleoptile Closed hollow cylinder or sheath of leaf-like tissue surrounding and protecting the plumule (shoot axis and young leaves) in grass seedlings.

coleorhiza Closed hollow cylinder or sheath of leaf-like tissue surrounding and protecting the radicle (young root) in grass seedlings.

colicins Bacterial exotoxins (***bacteriocins***) that affect other bacteria. Colicins E2 and E3 are ***AB toxins*** with DNAase and RNAase activity respectively. Most other colicins are channel-forming transmembrane peptides. Coded on plasmids which can be transferred at conjugation.

coliform ***Gram-negative*** rod-shaped bacillus. (1) May be used loosely of any rod-shaped bacterium. (2) Any Gram negative enteric bacillus. (3) More specifically, bacteria of the genera *Klebsiella* or *Escherichia*.

colipase Protein ***cofactor*** for ***lipase***.

collagen Major structural protein (285 kD) of extracellular matrix. An unusual protein both in amino acid composition (very rich in glycine (30%), proline, ***hydroxyproline***, lysine, and ***hydroxylysine***; no tyrosine or tryptophan), structure (a triple helical arrangement of 95 kD polypeptides giving a ***tropocollagen*** molecule, dimensions 300 x 0.5nm), and resistance to proteases. Most types are fibril-forming with characteristic quarter-stagger overlap between molecules producing an excellent tension-resisting fibrillar structure. Type IV, characteristic of ***basal lamina*** does not form fibrils. Many different types of collagen are now recognized. Some are glycosylated (glucose-galactose dimer on the hydroxylysine), and nearly all types can be crosslinked through lysine side chains. See ***dermatosparaxis, Ehlers–Danlos syndrome, scurvy***.

collagenase Proteolytic enzyme capable of breaking native collagen. Once the initial cleavage is made, less specific proteases will complete the degradation. Collagenases from mammalian cells are metalloenzymes and are collagen-type specific. May be released in latent (proenzyme) form into tissues and require activation by other proteases before they will degrade fibrillar matrix. Bacterial collagenases are used in tissue disruption for cell harvesting.

collapsin Glycoprotein (100 kD) from chick brain that may act as a repulsive cue in development and inhibit regeneration of mature neurons. Causes the collapse of the nerve growth cone at picomolar concentrations. Has a domain with sequence homology to fasciclin IV and Ig-like domains.

collectin Family of collagenous ***lectins*** believed to play an important part in first-line defence by binding to viruses and by opsonizing yeasts and bacteria. Contain a collagen-like region and a C-type lectin domain. Pulmonary surfactant proteins A and D (SP-A, SP-D), CL-43, serum mannan-binding protein (MBP) and conglutinin are all members of the family. Complement C1q is structurally related to the family.

collenchyma Plant tissue in which the ***primary cell walls*** are thickened, especially at the cell corners. Acts as a supporting tissue in growing shoots, leaves and petioles. Often arranged in cortical 'ribs', as seen prominently in celery and rhubarb petioles. ***Lignin*** and ***secondary walls*** are absent; the cells are living and able to grow.

collenocytes Stellate cells with long thin processes that ramify through the inhalent canal system of sponges.

colligative properties Properties that depend upon the numbers of molecules present in solution rather than their chemical characteristics.

collimating lens Lens that produces a non-divergent beam of light or other electromagnetic radiation. Simpler collimators involve slits. Essential in obtaining good illumination in microscopy and for many measuring instruments.

colonization factors The pili on enteropathogenic forms of *E. coli* facilitate adhesion of the bacteria to receptors (probably GM1 gangliosides) on gut epithelial cells and are often referred to as colonization or adherence factors. Colonization factor antigens may be plasmid coded, are essential for pathogenicity and are strain-specific, for example K88 (diarrhoea in piglets), CFAI and CFAII on strains causing similar disease in man.

colony-forming unit (CFU-S) Irradiated mice can have their immune systems reconstituted by the injection of bone marrow cells from a non-irradiated animal. The injected cells form colonies in the spleen (hence-S), each colony representing the progeny of a pluripotent stem cell. Operationally, therefore, the number of colony-forming units is a measure of the number of stem cells.

colony-stimulating factor ***Cytokines*** involved in the maturation of various leucocyte, macrophage, monocyte lines.

colony-stimulating factor-1 See ***CSF-1***.

colostrum The first milk secreted by an animal coming into lactation. May be especially rich in maternal lymphocytes and Ig and thus transfer immunity passively.

comb plates Large flat organelles formed by the fusion of many cilia. Vertical rows of comb-plates form the motile appendages of Ctenophores.

combinatorial chemistry Method by which large numbers of compounds can be made, usually utilizing solid-phase synthesis. In the simplest form, carrier beads would be treated separately so as to couple subunits A, B and C, mixed, redivided, and then subunits A, B and C added in the three reaction mixtures. Thus bead +AA, AB, AC, BA, BB, BC, etc. would have been synthesized, though the products are mixed and deconvolution of an active mixture will be necessary to identify the active molecule. Using a relatively small number of reactions, enormous diversity can be generated rapidly. Increasingly the term is used loosely for any procedure that generates highly diverse sets of compounds; the more recent tendency is to prefer high-speed parallel synthesis in which each reaction chamber contains only one compound.

combined immunodeficiency (severe combined immunodeficiency syndrome) Congenital immunodeficiency with thymic agenesis, lymphocyte depletion and hypogammaglobulinaemia: both cellular and humoral immune systems are affected, and life expectancy is low unless marrow transplantation is successful.

combining site Any region of a molecule that binds or reacts with a given compound, especially the region of immunoglobulin that combines with the determinant of an appropriate antigen.

communicating junction Another name for a ***gap junction***.

compaction Process that occurs during the morula stage of embryogenesis in which blastomeres increase their cell–cell contact area and develop gap junctions. In mice compaction occurs at the 8-cell stage and after this the developmental fate of each cell becomes restricted.

companion cell Relatively small plant cell, with little or no vacuole, found adjacent to a phloem ***sieve tube*** and originating with the latter from a common mother cell. Thought to be involved in translocation of sugars in and out of the sieve tube.

compartments In the insect wing, for example, there are two compartments, anterior and posterior, each containing several clones, but clones do not cross the boundary. It seems from studies with ***homeotic mutants*** that cells in different compartments are expressing different sets of genes. The evidence for such developmental compartments in vertebrates is sparse at present.

competent cells (1) Bacterial cells with enhanced ability to take up exogenous DNA and thus to be transformed. Competence can arise naturally in some bacteria (*Pneumococcus, Bacillus* and *Haemophilus* spp); a similar state can be induced in *E. coli* by treatment with calcium chloride. Once competence has been induced the cells can be stored at low temperature in cryoprotectant and used when needed. (2) Cells capable of responding to an inducer in embryonic development.

competitive inhibition Inhibitor that occupies the active site of an enzyme or the binding site of a receptor and prevents the normal substrate or ligand from binding. At sufficiently high concentration of the normal ligand inhibition is lost: the Km is altered by the competitive inhibitor, but the Vmax remains the same.

complement A heat-labile system of enzymes in plasma associated with response to injury. Activation of the complement cascade occurs through two convergent pathways. In the classical pathway the formation of antibody–antigen complexes leads to binding of C1, the release of active esterase that activates C4 and C2 that in turn bind to the surface. The C42 complex splits C3 to produce C3b, an opsonin, and C3a (anaphylatoxin). C423b acts on C5 to release C5a (anaphylatoxin and chemotactic factor) leaving C5b which combines with C6789 to form a cytolytic ***membrane attack complex***. In the alternate pathway C3 cleavage occurs without the involvement of C142, and can be activated by IgA, endotoxin, or polysaccharide-rich surfaces (eg. yeast cell wall, zymosan). Factor B combines with C3b to form a C3 convertase that is stabilized by factor P, generating a positive feedback loop. The alternate pathway is presumably the ancestral one upon which the sophistication of antibody recognition has been superimposed in the classical pathway. The enzymatic cascade amplifies the response, leads to the activation and recruitment of leucocytes, increases phagocytosis and induces killing directly. It is subject to various complex feedback controls that terminate the response.

complement cytolysis inhibitor See ***clusterin***.

complement fixation Binding of ***complement*** as a result of its interaction with immune complexes (the classical pathway) or particular surfaces (alternative pathway).

complementary base pairs The crucial property of DNA is that the two strands are complementary: guanine and cytosine are complementary and pair up through their hydrogen bonds, as are adenine and thymine that only form two hydrogen bonds (adenine and uracil in RNA).

complementary DNA See ***cDNA***.

complementation The ability of a mutant chromosome to restore normal function to a cell that has a mutation in the homologous chromosome when a hybrid or heterokaryon is formed, the explanation being that the mutations are in different cistrons and between the two a complete set of normal information is present.

complexins Nerve terminal ***syntaxin***-binding proteins.

Con A See ***concanavalin A***.

Con A binding sites See ***Con A receptors***.

Con A receptors A common misuse of the term receptor. Con A binds to the mannose residues of many different glycoproteins and glycolipids and the binding is therefore not to a specific site. It could be argued that the receptor is the Con A and cells have Con A ligands on their surfaces; certainly this would be less confusing.

conalbumin (ovotransferrin) Non-haem iron-binding protein found in chicken plasma and egg white.

conantokins Class of small peptides (17-21 residues) from *Conus* spp. that inhibit NMDA class of glutamate receptors.

concanamycin Specific inhibitor of vacuolar H^+-ATPase.

concanavalin A (Con A) A ***lectin*** isolated from the jack bean, *Canavalia ensiformis*. See Table L1.

concatamer Two or more identical linear molecular units covalently linked in tandem. Especially used of nucleic acid molecules and of units in artificial polymers.

condensing vacuole Vacuole formed from the *cis* face of the ***Golgi apparatus*** by the fusion of smaller vacuoles. Within the condensing vacuole the contents are concentrated and may become semicrystalline (***zymogen granules*** or ***secretory vesicles***).

conditional mutation A mutation that is only expressed under certain environmental conditions, eg. temperature-sensitive mutants.

conditioned medium Cell culture medium that has already been partially used by cells. Although depleted of some components, it is enriched with cell-derived material, probably including small amounts of growth factors; such cell-conditioned medium will support the growth of cells at much lower density and, mixed with some fresh medium, is therefore useful in ***cloning***.

cone cell See ***retinal cone***.

confluent culture A cell culture in which all the cells are in contact and thus the entire surface of the culture vessel is covered. It is also often used with the implication that the cells have also reached their maximum density, though confluence does not necessarily mean that division will cease or that the population will not increase in size.

confocal microscopy A system of (usually) ***epifluorescence*** light microscopy in which a fine laser beam of light is scanned over the object through the objective lens. The technique is particularly good at rejecting light from outside the plane of focus, and so produces higher effective resolution than is normally achieved.

conformational change Alteration in the shape – usually the tertiary structure of a protein – as a result of alteration in the environment (pH, temperature, ionic strength) or the binding of a ligand (to a receptor) or binding of substrate (to an enzyme).

congenic Organisms that differ in ***genotype*** at (ideally) one specified locus. Strictly speaking these are conisogenics. Thus one homozygous strain can be spoken of as being congenic to another.

conglutinin Protein present in serum that causes ***agglutination*** of antibody–antigen–complement complexes; binds C3b.

Congo red Naphthalene dye that is pH sensitive (blue-violet at pH 3, red at pH 5). Used as vital stain, also in staining for amyloid.

conidium Asexual spore of fungus, borne at the tip of a specialized ***hypha*** (conidiophore).

conjugation Union between two gametes or between two cells leading to the transfer of genetic material. In eukaryotes the classic examples are in *Paramecium* and *Spirogyra*. Conjugation between bacteria involves an F^+ bacterium (with F-pili) attaching to an F^-; transfer of the F-plasmid then occurs through the sex pilus. In Hfr mutants the F-plasmid is integrated into the chromosome and so chromosomal material is transferred as well. Conjugation occurs in many ***Gram-negative bacteria*** (*Escherichia, Shigella, Salmonella, Pseudomonas* and *Streptomyces*).

Conn's syndrome Uncontrolled secretion of ***aldosterone*** usually by an adrenal ***adenoma***.

connectin Cell surface protein (70 kD) from mouse ***fibrosarcoma*** cells that binds ***laminin*** and ***actin***.

connective tissue Rather general term for mesodermally derived tissue that may be more or less specialized. Cartilage and bone are specialized connective tissue, as is blood, but the term is probably better reserved for the less specialized tissue that is rich in extracellular matrix (collagen, proteoglycan, etc.) and which surrounds other more highly ordered tissues and organs.

connective tissue diseases A group of diseases including rheumatoid arthritis, systemic lupus erythematosus, rheumatic fever, scleroderma and others, that are sometimes referred to as rheumatic diseases. They probably do not affect solely connective tissues but the diseases are linked in various ways and have interesting immunological features which suggest that they may be autoimmune in origin.

connective tissue-activating peptide III (CTAP III) ***Cytokine***, produced from ***platelet basic protein***, that acts as a ***growth factor***.

connexin Generic term for proteins isolated from gap junctions. It has been proposed that connexins are the major structural proteins of the connexon. However, this is still a matter of debate and connexins vary from tissue to tissue.

connexon The functional unit of gap junctions. An assembly of six membrane-spanning proteins ***connexins*** having a water-filled gap in the centre. Two connexons in juxtaposed membranes link to form a continuous pore through both membranes.

conotoxins Toxins from cone shells (*Conus* spp). The α-conotoxins (small peptides, 13–18 residues) are competitive inhibitors of nicotinic acetylcholine receptors. The μ-conotoxins are small (22 residue) peptides that bind voltage-sensitive sodium channels in muscle, causing paralysis. The ω-conotoxins are similar in size and inhibit voltage-gated calcium channels, thereby blocking synaptic transmission.

consensus sequence Of a series of related DNA, RNA or protein sequences, the sequence that reflects the most common choice of base or amino acid at each position. Areas of particularly good agreement often represent conserved functional domains. The generation of consensus sequences has been subjected to intensive mathematical analysis.

conservative substitution In a gene product, a substitution of one amino acid with another with generally similar properties (size, hydrophobicity, etc.), such that the overall functioning is likely not to be seriously affected.

constitutive Constantly present, whether there is demand or not. Thus some enzymes are constitutively produced, whereas others are inducible.

constriction ring The equatorial ring of ***microfilaments*** that diminishes in diameter probably both by contraction and disassembly as ***cytokinesis*** proceeds.

contact following Behaviour shown by individual ***slime mould*** cells when they join a stream moving towards the aggregating centre. ***Contact sites A*** at front and rear of cell may be involved in *Dictyostelium*.

contact guidance Directed locomotory response of cells to an anisotropy of the environment, eg. the tendency of fibroblasts to align along ridges or parallel to the alignment of collagen fibres in a stretched gel.

contact inhibition of growth/division The inhibition of cell division when cells are in contact, eg. when they become ***confluent*** in ***cell culture***, or when they form differentiated tissues *in vivo*. Loss of contact inhibition of growth/division is a necessary step in ***carcinogenesis***.

contact inhibition of locomotion/movement Reaction in which the direction of motion of a cell is altered following collision with another cell. In heterologous contacts both cell may respond (mutual inhibition), or only one (non-reciprocal). Type I contact inhibition involves paralysis of the locomotory machinery; type II is a consequence of adhesive preference for the substratum rather than the dorsal surface of the other cell.

contact inhibition of phagocytosis Phenomenon described in sheets of kidney epithelial cells that, when confluent, lose their weak phagocytic activity, probably because of a failure of adhesion of particles to the dorsal surface in the absence of ruffles.

contact sensitivity Allergic response to contact with irritant, usually a ***hypersensitivity***.

contact sites A Developmentally regulated adhesion sites that appear on the ends of aggregation-competent *Dictyostelium* discoideum (see ***Acrasidae***) at the stage when the starved cells begin to come together to form the ***grex***. Originally detected by the use of Fab fragments of polyclonal antibodies, raised against aggregation-competent cells and adsorbed against vegetative cells, to block adhesion in EDTA-containing medium. (Cell–cell adhesion mediated by contact sites A, unlike that mediated by contact sites B, is not divalent cation-sensitive). The fact that a mutant deficient in csA behaves perfectly normally in culture is puzzling.

contact sites B See ***contact sites A***.

contact-induced spreading The response in which contact between two ***epithelial cells*** leads to a stabilized contact and the increased spreading of the cells so that the area covered is greater than that covered by the two cells in isolation.

contactin A 130 kD ***glycoprotein*** attached to the ***cytoskeleton*** via its cytoplasmic domain; concentrated in areas of interneuronal contact. Its sequence contains both immunoglobulin-like domains and ***fibronectin*** type III repeats. Its close homology with NCAM suggests that it is a ***CAM***. Like ***L1, F11, neurofascin*** and ***TAG-1*** in vertebrate nervous systems and ***fasciclin*** II in insects, thought to be associated with the process of selective ***fasciculation***. Sometimes ***GPI-anchored***.

contactinhibin Plasma membrane ***glycoprotein*** of 60–70 kD isolated from human diploid fibroblasts, which when immobilized on silica beads has been reported to reversibly inhibit the growth of cultured cells.

contig DNA sequence assembled from overlapping shorter sequences to form one large contigous sequence.

contractile ring See ***constriction ring***.

contractile vacuole A specialized vacuole of eukaryotic cells, especially Protozoa, that fills with water from the cytoplasm and then discharges this externally by the opening of a permanent narrow neck or a transitory pore. Function is probably osmoregulatory.

contrapsin Trypsin inhibitor (***serpin***) from rat.

control element Generic term for a region of DNA, such as a ***promoter*** or ***enhancer*** adjacent to (or within) a gene that allows the regulation of gene expression by the binding of ***transcription factors***.

control region General name for genomic DNA that, though binding of ***transcription factors*** to its promoters, enhancers and repressors, modulates the expression level of nearby genes.

Conus Genus of gastropod molluscs, cone snails. See ***conotoxins*** and ***conantokins***.

Kenacid blue (Brilliant Blue R; Coomassie Blue; Coomassie Brilliant Blue) Trademark name for a dye that binds nonspecifically to proteins, used in ***Bradford method*** for protein estimation and for detecting proteins on gels.

cooperativity Phenomenon displayed by enzymes or receptors that have multiple binding sites. Binding of one ligand alters the affinity of the other site(s). Both positive and negative cooperativity are known; positive cooperativity gives rise to a sigmoidal binding curve. Cooperativity is often invoked to account for non-linearity of binding data, although it is by no means the only possible cause.

coordination complex Complex held together by coordinate (dipolar) bonds, covalent bonds in which the two shared electrons derive from only one of the two participants.

Coprinus Genus of fungi with gills that autodigest once spores have been discharged giving rise to a black inky fluid.

copy number The number of molecules of a particular type on or in a cell or part of a cell. Usually applied to specific genes, or to plasmids within a bacterium.

cord blood Blood taken postpartum from the umbilical cord.

cord factor Glycolipid (trehalose-6,6′-dimycolate) found in the cell walls of Mycobacteria (causing them to grow in serpentine cords) and important in virulence, being toxic and inducing granulomatous reactions identical to those induced by the whole organism.

cornea Transparent tissue at the front of the eye. The cornea has a thin outer squamous epithelial covering and an endothelial layer next to the aqueous humour, but is largely composed of avascular collagen laid down in orthogonal arrays with a few fibroblasts. Transparency of the cornea depends on the regularity of spacing in the collagen fibrils.

cornified epithelium Epithelium in which the cells have accumulated keratin and died. The outer layers of vertebrate skin, hair, nails, horn and hoof are all composed of cornified cells.

coronal section A cross-section of the brain taken effectively where the edge of a crown would touch.

Coronaviridae Family of single-stranded RNA viruses responsible for respiratory diseases. The outer envelope of the virus has club-shaped projections that radiate outwards and give a characteristic corona appearance to negatively stained virions.

coronin Actin-binding protein (55 kD) of ***Dictyostelium discoideum***. Associated with crown-shaped cell surface projections in growth phase cells. Accumulates at front of cells responding to a chemotactic gradient of cAMP. Amino-terminal domain has similarity to β–subunits of heterotrimeric G-proteins; C-terminal has high α-helical content.

corpus callosum Band of white matter at the base of the longitudinal fissure dividing the two cerebral hemispheres of the brain.

corpus luteum (plural, corpora lutea) Glandular body formed from the Graafian follicle in the ovary following release of the ovum. Secretes ***progesterone***.

corralling The proposed confinement of membrane proteins within a diffusion barrier, thereby limiting long-range translational diffusion rates without affecting short range properties (eg. rotation rates).

cortactin A p80/85 protein first identified as a substrate for src kinase. An ***F-actin*** binding protein that redistributes to membrane ruffles as a result of growth factor-induced ***Rac*** 1 activation. Has proline-rich and SH3 domains. Overexpression of cortactin increases cell motility and invasiveness.

cortex (1) *Bot*. Outer part of stem or root, between the vascular system and the epidermis; composed of ***parenchyma***. (2) Region of cytoplasm adjacent to the plasma membrane. (3) *Histol*. Outer part of organ.

cortical granule Specialized secretory vesicles lying just below the plasma membrane of the egg, which fuse and release their contents immediately after fertilization (activation) to prevent polyspermy.

cortical layer See ***cortical meshwork***.

cortical meshwork Subplasmalemmal layer of tangled microfilaments anchored to the plasma membrane by their barbed ends. This meshwork contributes to the mechanical properties of the cell surface and probably restricts the access of cytoplasmic vesicles to the plasma membrane.

corticostatin Name given to some ***defensins*** because they inhibit ***corticotropin***-induced ***corticosteroid*** production.

corticosteroids ***Steroid*** hormones produced in the ***adrenal*** cortex. Formed in response to ***adrenocorticotrophin*** (ACTH). Regulate both carbohydrate metabolism and salt/water balance. Glucocorticoids (eg. cortisol, cortisone) predominantly affect the former and minerocorticoids (eg. aldosterone) the latter.

corticotrophin (corticotropin) See ***adrenocorticotrophin***.

corticotrophin-releasing factor See ***adrenocorticotrophin***.

corticotropin See ***corticotrophin***.

cortisol The major adrenal glucocorticoid; stimulates conversion of proteins to carbohydrates, raises blood sugar levels and promotes glycogen storage in the liver.

cortisone (11-dehydroxy-cortisol) Natural ***glucocorticoid*** formed by 11 β-hydroxysteroid dehydrogenase action on hydrocortisone; inactive until converted into ***hydrocortisone*** in the liver.

Corynebacteria Genus of ***Gram-positive*** non-motile rod-like bacteria, often with a club-shaped appearance. Most are facultative anaerobes with some similarities to ***mycobacteria*** and ***nocardia***. *C. diphtheriae* is the causative agent of diphtheria and produces a potent exotoxin, ***diphtheria toxin***.

COS cells Simian fibroblasts (CV-1 cells) transformed by SV40 that is deficient in the origin of replication region. Express ***large T-antigen*** constitutively and if transfected with a vector containing a normal SV40 origin have all the other early viral genes necessary to generate multiple copies of the vector and thus to give very high levels of expression.

cosegregation Of two genotypes, meaning that they tend to be inherited together, implying close linkage.

cosmid A type of ***bacteriophage*** lambda vector. Often used for construction of ***genomic libraries***, because of their ability to carry relatively long pieces of DNA insert, compared with ***plasmids***.

costa Rod-shaped intracellular organelle lying below the undulating membrane of

Trichomonas. Generates active bending associated with local loss of ***birefringence*** at the bending zone, probably as a result of conformational change in the longitudinal lamellae. Major protein approximately 90 kD.

costamere Regular periodic submembranous arrays of ***vinculin*** in muscle cells; link sarcomeres to the membrane and are associated with links to extracellular matrix.

Cot curve Physicochemical technique for measuring the complexity (or size) of DNA. The DNA is heated to make it single stranded, then allowed to cool. The renaturation of the DNA is followed spectroscopically: larger DNA molecules take longer to reanneal.

cotranslational transport Process whereby a protein is moved across a membrane as it is being synthesized. This process occurs during the translation of the message at membrane-associated ***ribosomes*** in ***rough endoplasmic reticulum*** during the synthesis of secreted proteins in eukaryotic cells.

Coturnix coturnix japonica Japanese quail. Used extensively in developmental biology because quail nuclei can easily be distinguished from those of the chicken and this facilitates grafting experiments for fate mapping.

cotyledon Modified leaf ('seed leaf'), found as part of the embryo in seeds, involved in either storage or absorption of food reserves. Dicotyledonous seeds contain two, monocotyledonous seeds only one May appear above ground and show photosynthetic activity in the seedling.

Coulter counter Particle counter used for bacteria or eukaryotic cells; works by detecting change in electrical conductance as fluid containing cells is drawn through a small aperture. (The cell, a non-conducting particle, alters the effective cross-section of the conductive channel.)

coumarin (O-hydroxycinnamic acid) Pleasant-smelling compound found in many plants and released on wilting (probably a major component of the smell of fresh hay). Has anticoagulant activity by competing with vitamin K. Coumarin derivatives have anti-inflammatory and antimetastatic properties and inhibit xanthine oxidase and the production of 5-HETE by neutrophils and macrophages. Various derivatives have these activities including esculentin, esculin (6,7-dihydroxycoumarin 6-O-D-glucoside), ***fraxin***, umbelliferone (7-hydroxy-coumarin) and scopoletin (6-methoxy-7-hydroxy-coumarin).

counterstain Rather non-specific stain used in conjunction with another histochemical reagent of greater specificity to provide contrast and reveal more of the general structure of the tissue. Light Green is used as a counterstain in the Mallory procedure, for example.

COUP-TFs (chicken ovalbumin upstream promoter transcription factors I and II) Nuclear orphan receptors. See ***ARP-1***.

coupling The linking of two independent processes by a common intermediate, eg. the coupling of electron transport to oxidative phosphorylation or the ATP–ADP conversion to transport processes.

coupling factors Proteins responsible for coupling transmembrane potentials to ATP synthesis in ***chloroplasts*** and ***mitochondria***. Include ATP-synthesizing enzymes (F1 in mitochondrion), that can also act as ATPases.

Cowden disease Germ-line mutations in ***PTEN*** are responsible for Cowden disease, a rare autosomal dominant multiple-***hamartoma*** syndrome.

Coxsackie viruses Species of enteroviruses of the ***Picornaviridae*** first isolated in Coxsackie, New York. Coxsackie A produces diffuse myositis; Coxsackie B produces focal areas of degeneration in brain and skeletal muscle. Similar to polioviruses in chemical and physical properties.

CpG island (CG island) Region of genomic DNA rich in the dinucleotide C-G. Methylation of the C in the dinucleotide is maintained through cell divisions, and profoundly affects the degree of transcription of the nearby genes, and is important in developmental regulation of gene expression. There are around 30 000 CpG islands in a typical mammalian genome, and these tend to be undermethylated and upstream of ***house keeping genes***.

CPS (carbamoyl phosphate synthetase) Enzyme responsible for production of carbamoyl phosphate, key substrate for pyrimidine biosynthesis (see ***ATCase***). In *Saccharomyces cerevisiae* the multifunctional protein Ura2 carries out both CPSase and ATCase activities.

CR1 Complement receptor 1 (CD35). Binds particles coated with C3b. Present on neutrophils, mononuclear phagocytes, B-lymphocytes and Langerhans' cells, and involved in the opsonic phagocytosis of bacteria and uptake of immune complexes. Also present on follicular dendritic cells and glomerular podocytes.

CR2 (CD-21) Receptor for complement fragment C3d. Present only on B-lymphocytes, follicular dendritic cells and some B- and T-cell lines and is the site to which the ***Epstein–Barr virus*** binds.

CR3 (CD11b/CD18; MAC-1) Receptor for complement fragment C3bi (iC3b), present on neutrophils and mononuclear phagocytes. A β_2 integrin.

CR4 (CD) Receptor for C3dg, the ***complement*** fragment that remains when C3b is cleaved to C3bi. Thought to be present on monocytes, macrophages and neutrophils, but there is some disagreement at present.

CRABP See ***cellular retinoic acid-binding protein***.

crassulacean acid metabolism (CAM) Physiological adaptation of certain succulent plants, in which CO_2 can be fixed (non-photosynthetically) at night into malic and other acids. During the day the CO_2 is regenerated and then fixed photosynthetically into the ***Calvin–Benson cycle***. This adaptation permits the stomata to remain closed during the day, conserving water.

cre Gene of *E. coli* ***bacteriophage*** P1 that mediates ***site-specific recombination*** at loxP sites. Now used in vertebrate transgenics: see ***lox-Cre system***.

creatine kinase (creatine phosphokinase) (EC 2.7.3.2.) Dimeric enzyme (82 kD) that catalyses the formation of ATP from ADP and creatine phosphate in muscle.

creatine phosphokinase See ***creatine kinase***.

CREB Cyclic AMP response element binding factor. ***Basic leucine zipper*** (bZip) transcription factor involved in activating genes through cAMP; binds to CRE element TGANNTCA. Phosphorylation by cAMP-dependent protein kinase (PKA) at serine-119 is required for interaction with DNA and phosphorylation at serine-133 allows CREB to interact with CBP (***CREB-binding protein***) leading to interaction with RNA polymerase II.

CREB-binding protein (CBP) Transcriptional co-activator (265kD) of ***CREB*** and of c-*Myb*. Only binds the phosphorylated form of CREB.

crenation Distortion of the erythrocyte membrane giving a spiky, echinocyte, morphology. Results from ATP depletion or an excess of lipid species in the external lipid layer of the membrane.

CREST (calcinosis; Reynaud's phenomenon; oesophageal dysmotility; sclerodactyly; telangielactasia) A complex syndrome characterized by the presence of autoantibodies toward proteins of the ***centromere***, largely the CENT A, B, C and D antigens.

Creutzfeldt–Jakob disease (CJD) Rare fatal presenile dementia of humans, similar to ***kuru*** and other slow viruses. Method of transmission unknown. Will induce neurological disorder in goats 3–4 years after inoculation with CJD brain extract. Classified pathologically as a subacute ***spongiform encephalopathy*** of man. A new variant, vCJD, has recently been recognized and associated with ***bovine spongiform encephalopathy***. See ***prions***.

cri-du-chat **syndrome** Syndrome produced by loss of part of the short arm of chromosome 5 in humans. Results in severe congenital malformation and affected infants produce a curious mewling sound said to resemble the cry of a cat.

Cricetulus griseus Chinese hamster. See ***CHO cells*** and ***Mesocricetus auratus***.

crinophagy Digestion of the contents of secretory granules following their fusion with lysosomes.

critical point drying A method for preparing specimens for the scanning electron microscope that avoids the problems of shrinkage caused by normal drying procedures. Water in the specimen is replaced by an intermediate fluid, eg. liquid carbon dioxide, avoiding setting up a liquid–gas interface, and then the second fluid is allowed to vaporize by raising the temperature above the critical point, the temperature at which the liquid state no longer occurs.

crk An ***oncogene***, identified in a chicken ***sarcoma***, encoding an activator of tyrosine ***protein kinase***. Protein product is a member of the family of adaptor-type signalling molecules that have SH2 and SH3 domains (see ***Grb2***) and may recruit cytoplasmic proteins to associate with the cytoplasmic domain of receptor tyrosine kinases. See also Table O1.

cro-protein Protein synthesized by bacteriophage lambda in the lytic state. The cro-protein blocks the synthesis of the lambda repressor (that is produced in the lysogenic stage, and inhibits cro-protein synthesis). Production of the cro-protein in turn controls a set of genes associated with rapid virus multiplication.

Crohn's disease Inflammatory bowel disease that seems to have both genetic and environmental causes; not well understood.

crossing over Recombination as a result of DNA exchange between homologous chromatids in meiosis, giving rise to ***chiasmata***.

crossover Protein ***motif*** that describes the connection between strands in a parallel ***beta sheet***. In principle, can be extended to the region between adjacent parallel ***alpha-helices***.

croton oil Oil from the seeds of the tropical plant *Croton tiglium* (Euphorbiaceae), causes severe skin irritation and contains a potent ***tumour promoter*** (co-carcinogen), ***phorbol ester***.

crotoxins Toxins from *Crotalus* spp. (rattlesnake) venoms. Heterodimeric phospholipase A2s, similar to secreted PLA2. Bind to specific proteins on presynaptic membranes and alter transmitter release.

crown gall Gall, or tumour, found in many dicotyledonous plants, caused by the bacterium ***Agrobacterium tumefaciens***.

CRP See ***C-reactive protein***.

CRP55 See ***calreticulin***.

crumbs *Drosophila* gene involved in epithelial development. Gene product contains 26 repeats of the ***EGF-like domain***.

CRY-1 (cryptochrome-1) Blue-light receptor from *Arabidopsis*. A flavoprotein coded by *Hy4*, a gene that is part of a small family that also encodes CRY-2, another blue-light photoreceptor. Mutants lacking both CRY-1 and CRY-2 are deficient in phototropism.

cryofixation Fixation processes for microscopy carried out at low temperature to improve the quality of fixation. Often very low temperatures and fast cooling are used to prevent formation of ice crystals. Cooling rates of 10 000° per minute may be used and liquid nitrogen or even liquid helium temperatures used. Especially for preparing specimens for scanning electron microscopy. Frequently no chemical treatment is used.

cryoglobulin Abnormal plasma globulin (IgG or IgM) that precipitates when serum is cooled.

cryoprecipitate The precipitate that forms when plasma is frozen and then thawed; particularly rich in ***fibronectin*** and blood-clotting factor VIII.

cryoprotectant Substance that is used to protect from the effects of freezing, largely by preventing large ice-crystals from forming. The two commonly used for freezing cells are ***DMSO*** or glycerol.

crypt Deep pit that protrudes down into the connective tissue surrounding the small intestine. The epithelium at the base of the crypt is the site of stem cell proliferation and the differentiated cells move upwards and are shed 3–5 days later at the tips of the villi.

cryptobiont An organism that lives hidden away or with all signs of life disguised as in dormancy.

cryptochrome See ***CR1***.

Cryptomonas Genus of flagellate protozoa with two slightly unequal flagella and a large chromatophore in some species.

Cryptosporidium Coccidian parasite found in the gut of various vertebrates that causes severe diarrhoea in man.

crystallins Major proteins of the vertebrate lens. Range from high-MW oligomeric species to low MW monomeric species. Immunological cross-reactivity suggests that the sequences of crystallin subunits are relatively highly conserved in evolution.

CSAT Monoclonal antibody defining integral membrane protein of chick fibroblasts. Originally thought to recognize a trimeric complex, now thought to recognize two different β1 integrins (with different α chains).

CSIF (cytokine synthesis-inhibiting factor) Usually in reference to the gene for this activity which is present in normal cells and in HIV. May play a role in immunosuppression.

Csk Protein tyrosine kinase that phosphorylates a tyrosine residue in ***src family*** kinases, thereby allowing an inhibitory interaction with src kinase SH2 domain. (It is loss of the tyrosine residue phosphorylated by csk that makes v-src unregulated.)

CSP See ***cysteine string protein***.

CTAP III See ***connective tissue-activating peptide III***.

CTD (carboxy-terminal domain) Protein domain unique to Pol-II that contains multiple repeats of the YSPTSPS sequence. The CTD plays an important part in organizing the various protein factors that make up the RNA-processing 'factory' that regulates the processing of the 3′ end of mRNAs made by Pol-II.

Ctenophora Phylum of biradially symmetrical triploblastic coelomates. Lack nematocysts and cilia though have comb plates (costae) arranged in eight rows. The comb jelly or sea gooseberry is the best known example.

CTF (CCAAT box-binding transcription factor; TGGCA-binding proteins) Large family of vertebrate nuclear protein ***transcription factors***, around 400–600 amino acids, that bind to the palindrome TGGCAnnnTGCCA in a range of cellular promoters. Includes nuclear factor-1 ***NF-1***.

CTL See ***cytotoxic T-cells***.

CTLA-4 (CD152) Type I transmembrane protein (20 kD) of the immunoglobulin superfamily. Found on activated T-cells and binds to CD80 on B-cells. Resembles CD28 but acts as a negative regulator of T-cell activation. Cytoplasmic domain interacts with SH2 domain of Shp2 (protein tyrosine phosphatase) and possibly with PI3-kinase. CTLA-4-deficient mice develop a severe lymphoproliferative disorder.

Culex pipiens Most widely distributed species of mosquito. Salivary glands have giant ***polytene chromosomes***.

cullin (CDC53) Gene family that is involved in cell cycle control and when mutated may contribute to tumour progression. Cullin (cdc53) is part of the SCF ubiquitin protein ligase complex.

culture To grow ***in vitro***.

curare Curare alkaloids are the active ingredients of arrow poisons used by South American Indians; they have muscle-relaxant properties because they block motor ***endplate*** transmission, acting as competitive antagonists for acetylcholine.

CURL The compartment for uncoupling of receptors and ligands: internalized receptor–ligand complexes are stripped of the ligand and recycled.

Cushing's syndrome A type of hypertensive disease in humans due probably to the oversecretion of ***cortisol*** due in turn to excessive secretion of ***adrenocorticotrophin*** (ACTH). Adrenal tumours are the usual primary cause.

cutin Waxy hydrophobic substance deposited on the surface of plants. Composed of complex long-chain fatty esters and other fatty acid derivatives. Impregnates the outer wall of epidermal cells and also forms a separate layer, the cuticle, on the outer surface of the ***epidermis***.

Cyanobacteria (Cyanophyta) Modern term for the blue-green algae, prokaryotic cells that use chlorophyll on intracytoplasmic membranes for photosynthesis. The blue-green colour is due to the presence of ***phycobilins***. Found as single cells, colonies or simple filaments. In *Anabaena*, in which the cells are arranged as a filament, heterocysts capable of nitrogen fixation occur at regular intervals. According to the ***endosymbiont hypothesis*** Cyanobacteria are the progenitors of ***chloroplasts***.

cyanocobalamin (vitamin B12) Usual form of vitamin B12. See table V1.

cyanogen bromide (CNBr) Agent that cleaves peptide bonds at methionine residues. The peptide fragments so generated can then, for example, be tested to locate particular activities.

Cyanophyta See ***Cyanobacteria***.

cyanosis Blueish appearance of skin due to insufficient oxygenation of blood in capillaries. May be natural (response to cold) or pathological (cyanide poisoning, among other things).

cyclic ADP-ribose (cADPR; adenosine 5′-cyclic diphosphoribose) Second messenger synthesized by the multifunctional transmembrane ectoenzyme CD38 in various systems particularly platelets, microsomes and sea urchin eggs. Endogenous regulator of intracellular calcium. May act by regulating ***ryanodine*** receptor though other mechanisms are suggested.

cyclic AMP (cAMP) 3′5′-cyclic ester of AMP. The first second messenger hormone signalling system to be characterized. Generated from ATP by the action of adenyl cyclase that is coupled to hormone receptors by ***G-proteins*** (***GTP-binding proteins***). cAMP activates a specific (cAMP-dependent) protein kinase and is inactivated by phosphodiesterase action giving 5′AMP. Also functions as an extracellular morphogen for some slime moulds.

cyclic GMP (cGMP) 3′5′-cyclic ester of GMP. A second messenger generated by guanylyl cyclase. See ***ANP***, ***nitric oxide***.

cyclic inositol phosphates 1,2-cyclic derivatives of inositol phosphatide that are invariably formed during enzymic hydrolysis of phosphatidyl inositol species. Have been proposed as second messengers in hormone-activated pathways.

cyclic nucleotide phosphodiesterases Often casually referred to simply as phosphodiesterases. Multiple isoenzymes are known. PDE-I is calcium/ calmodulin-regulated and important in CNS and vasorelaxation, PDE-II is cGMP-stimulated and hydrolyzes cAMP. PDE-III regulates vascular and airway dilation, platelet aggregation, cytokine production and lipolysis; PDE-IV (inhibited by rolipram) is important in control of airway smooth muscle and inflammatory mediator release but also has a role in CNS and in regulation of gastric acid secretion. PDE-V and VI are cGMP-specific. PDE-V is involved in platelet aggregation; PDE-VI is regulated by interaction with ***transducin*** in photoreceptors. PDE-VII is abundant in skeletal muscle and present in heart and kidney.

cyclic phosphorylation Any process in which a phosphatide ester forms a cylic diester by linkage to a neighbouring hydroxyl group.

cyclic photophosphorylation Process by which light energy absorbed by ***photosystem I*** in the chloroplast can be used to generate ATP without concomitant reduction of ***NADP***$^{+}$ or other electron acceptors. Energized electrons are passed from PS-I to ferredoxin, and thence along a chain of electron carriers and back to the reaction centre of PS-I, generating ATP *en route*.

cyclin Proteins (A and B forms known) whose levels in a cell varies markedly during the ***cell cycle***, rising steadily until ***mitosis***, then falling abruptly to zero. As cyclins reach a threshold level, they are thought to drive cells into ***G2*** phase and thus towards mitosis. Cyclins combine with p34 kinase (cdc2) to form maturation-promoting factor (MPF). See also ***M-phase promoting factor***, ***cyclin-dependent kinase*** (cdk).

cyclin-dependent kinase (cdk) Family of kinases including cdc28, cdc2 and $p34^{cdc2}$ that are only active when they form a complex with cyclins. The complex is maturation-promoting factor (MPF) and its activity is necessary for cells to leave G2 and enter mitosis. Catalytic domain resembles that of cAMP-dependent kinase (PKA)

cyclin-dependent kinase-activating kinase (CAK) Kinase that activates cdks by phosphorylation. CAK phosphorylates a threonine residue of several cdks, and a tyrosine on cdc2 (phosphatase cdc25 reverses this).

cyclin-dependent kinase inhibitors (CKIs) Two classes of CKIs are known in mammals, the $p21^{CIP1/Waf}1$ class that includes $p27^{KIP1}$ and $p57^{KIP2}$ and which inhibit all G1/S ***cyclin-dependent kinases*** (cdks), and the $p16^{INK4}$ class that bind and inhibit only Cdk4 and Cdk6. The $p21^{CIP1}$ inhibitor is transcriptionally regulated by p53 tumour suppressor, is important in G1 DNA-damage checkpoint, and its expression is associated with terminally differentiating tissues. Deletion of $p21^{CIP1}$ is non-lethal in mice; deletion of $p27^{KIP1}$ leads to relatively normal mice but with some proliferation disorders; deletion of $p57^{KIP2}$ causes fairly major developmental abnormalities similar to ***Beckwith–Wiedemann syndrome***. See also ***ICK1***, ***Waf-1***.

cyclooxygenase Enzyme complex present in most tissues that produces various prostaglandins and thromboxanes from arachidonic acid; inhibited by aspirin-like drugs, probably accounting for their anti-inflammatory effects.

cyclodextrins Cyclic polymers of 6, 7 or 8 α-1,4-linked D-glucose residues. The toroidal structure allows them to act as hydrophilic carriers of hydrophobic molecules.

cycloheximide Antibiotic (MW 281) isolated from *Streptomyces griseus*. Blocks eukaryotic (but not prokaryotic) protein synthesis by preventing initiation and elongation on 80S ribosomes. Commonly used experimentally.

cyclolysin Protein from ***Bordetella pertussis*** that is both an ***adenylate cyclase*** and a ***haemolysin***.

cyclophilin Enzyme with ***PPIase*** activity; binds the immunosuppressive drug ***cyclosporin A***. See ***immunophilin***.

cyclophosphamide An alkylating agent and important immunosuppressant. Acts by alkylating SH and NH2 groups, especially the N7 of guanine.

cyclosis Cyclical streaming of the cytoplasm of plant cells, conspicuous in giant internodal cells of algae such as *Chara*, in pollen tubes and in stamen hairs of *Tradescantia*. Term also used to denote cyclical movement of food vacuoles from mouth to ***cytoproct*** in ciliate protozoa.

cyclosporin A (ciclosporine) Cyclic undecapeptide isolated from *Tolypocladium inflatum*, that has potent immunosuppressant activity on both humoral and cellular systems. The use of cyclosporin has made transplant surgery much easier, although the long-term consequences of suppressing immune function are not yet clear. Used widely as an an anti-rejection drug in transplant surgery. See also ***cyclophilin***.

CYP See ***cytochrome P450***.

cypris Larval stage of Cirrepedia (barnacles) following nauplius stage.

Cys See ***cysteine***.

cyst (1) A resting stage of many prokaryotes and eukaryotes in which a cell or several cells are surrounded with a protective wall of extracellular materials. (2) A pathological fluid-filled sac bounded by a cellular wall, often of epithelial origin, found on occasion in all species of multicellular animal. May result from a wide range of insults or be of embryological origin.

cystatins A group of natural cysteine-protease inhibitors (approximately 13 kD) widely distributed both intra- and extracellularly. See ***stefin***.

cysteine (Cys: C) The only amino acid to contain a thiol (SH) group. In intracellular enzymes the unique reactivity of this group is frequently exploited at the catalytic site. In extracellular proteins found only as half-cystine in disulphide bridges or fatty acylated.

cysteine proteinase (thiol proteinase) Any protease of the subclass EC 3.4.22. Have a cysteine residue in the active site that can be irreversibly inhibited by sulphydryl reagents. Includes ***cathepsins*** and ***papain***.

cysteine string protein (CSP) Peripheral membrane protein, containing more than 10 palmitoylated cysteines and a DNA-J homology domain. See also ***palmitoylation***.

cystic fibrosis Generalized abnormality of exocrine gland secretion that affects pancreas (blockage of the ducts leads to cyst formation and to a shortage of digestive enzymes), bowel, biliary tree, sweat glands and lungs. The production of abnormal mucus in the lung predisposes to respiratory infection, a major problem in children with the disorder. A fairly common (1 in 2000 live births in Caucasians) ***autosomal recessive*** disease.

cystic fibrosis antigen (CFAG; MIF-related protein 8) Now known to be MRP-8. See ***calgranulins***.

cystic fibrosis transmembrane conductance regulator (CFTR) Gene believed to be defective in cystic fibrosis. Gene encodes a chloride channel, homologous to a family of proteins that actively transport small solutes in an ATP-dependent manner (***ABC proteins***).

cystine The amino acid formed by linking two cysteine residues with a disulphide linkage between the two sulphydryl (SH) groups. The analogous compound present within proteins is termed two half cysteines.

cytidine Nucleoside consisting of D-ribose and the pyrimidine base cytosine.

cytidine 5′ diphosphate (CDP) CDP (derived from cytidine 5′ triphosphate) is important in phosphatide biosynthesis; activated choline is CDP-choline.

cytidylic acid Ribonucleotide of ***cytosine***.

cytocalbins ***Calmodulin***-binding proteins associated with the cytoskeleton.

cytochalasins A group of fungal metabolites that inhibit the addition of G-actin to a nucleation site and therefore perturb labile microfilament arrays. Cytochalasin B inhibits at around 1μg/ml but at about 5μg/ml begins to inhibit glucose transport. Cytochalasin D affects only the microfilament system and is therefore preferable.

cytochemistry Branch of histochemistry associated with the localization of cellular components by specific staining methods, as for example the localization of acid phosphatases by the Gomori method. Immunocytochemistry involves the use of labelled antibodies as part of the staining procedure.

cytochrome m See ***cytochrome P450***.

cytochrome oxidase Terminal enzyme of the electron transport chain that accepts electrons from (ie. oxidizes) cytochrome C and transfers electrons to molecular oxygen.

cytochrome P450 (cytochrome m; CYP) Large group of mixed-function oxidases of the cytochrome b type, involved, among other things, in steroid hydroxylation reactions in the adrenal cortex. In liver they are found in the microsomal fraction and can be induced for the detoxification of foreign substances. Found in most animal cells and organelles, in plants, and in microorganisms.

cytochromes Enzymes of the electron transport chain that are pigmented by virtue of their ***haem*** prosthetic groups. Very highly conserved in evolution.

cytohesin-1 (B2-1) Guanine nucleotide exchange factor for human ADP-ribosylation factor (***ARF***) GTPases. Abundant in cells of the immune system where it mediates PI3-kinase activation of β 2 integrin (particularly LFA-1) through interaction with the cytoplasmic domain. Closely related to ***ARNO*** and ***GRP-1***; all three have a central Sec7-like domain and C-terminal ***pleckstrin*** homology (PH) domain. The PH domain and C-terminal polybasic sequence are important for membrane association and function.

cytokeratins Generic name for the intermediate filament proteins of epithelial cells.

cytokine synthesis inhibiting factor See ***CSIF***.

cytokines Small proteins (in the range of 5–20 kD) released by cells which affect the behaviour of other cells. Not really different from hormones, but the term tends to be used as a convenient generic shorthand for ***interleukins***, ***lymphokines*** and several related signalling molecules such as ***TNF*** and ***interferons***. Generally growth factors would not be classified as cytokines, though TGF is an exception. Rather an imprecise term, though in very common usage. ***Chemokines*** are a subset of cytokines; see Table C4a.

cytokinesis Process in which the cytoplasm of a cell is divided after nuclear division (mitosis) is complete.

cytokinins Class of ***plant growth substances*** (plant hormones) active in promoting cell division. Also involved in cell growth and differentiation and in other physiological processes. Examples: ***kinetin***, ***zeatin***, benzyl adenine.

cytolipin K See ***globoside***.

cytology The study of cells. Implies the use of light or electron microscopic methods for the study of morphology.

cytolysis Cell ***lysis***.

cytolysosome Membrane-bounded region of cytoplasm that is subsequently digested.

Cytomegalovirus Probably the most widespread of the Herpetoviridae group. Infected cells enlarge and have a characteristic inclusion body (composed of virus particles) in the nucleus. Causes disease only *in utero* (leading to abortion or stillbirth or to various congenital defects), although can be opportunistic in an immunocompromised host.

cytoplasm Substance contained within the plasma membrane excluding, in eukaryotes, the nucleus.

cytoplasmic bridge (plasmodesmata) Thin strand of cytoplasm linking cells as in higher plants, *Volvox*, between ***nurse cells*** and developing eggs, and between developing sperm cells. Unlike gap junctions, allows the transfer of large macromolecules.

cytoplasmic inheritance Inheritance of parental characters through a non-chromosomal means; thus mitochondrial DNA is cytoplasmically inherited since the information is not segregated at mitosis. In a broader sense the organization of a cell may be inherited through the continuity of structures from one generation to the next. It has often been speculated that the information for some structures may not be encoded in the genomic DNA, particularly in Protozoa that have complex patterns of surface organelles. See ***maternal inheritance***.

cytoplasmic linker protein-170 See ***CLIP-170***.

cytoplasmic streaming Bulk flow of the cytoplasm of cells. Most conspicuous in large cells such as amoebae and the internodal cells of *Chara* where the rate of movement may be as high as 100μm/s. See ***cyclosis***.

cytoplast Fragment of cell with nucleus removed (in ***karyoplast***); usually achieved by cytochalasin B treatment followed by mild centrifugation on a step gradient.

cytoproct Cell anus: region at posterior of a ciliate where exhausted food vacuoles are expelled.

cytosine Pyrimidine base found in DNA and RNA. Pairs with guanine. Glycosylated base is ***cytidine***.

cytosine arabinoside (cytarabine) Cytotoxic drug used in oncology (particularly *AML*) and against viral infections. Blocks DNA synthesis.

cytoskeleton General term for the internal components of animal cells which give them structural strength and motility: plant cells and bacteria use an extracellular ***cell wall*** instead. The major components of cytoskeleton are the ***microfilaments*** (of ***actin***), ***microtubules*** (of ***tubulin***) and ***intermediate filament*** systems in cells.

cytosol That part of the cytoplasm that remains when organelles and internal membrane systems are removed.

cytosome A specialized region of various protozoans in which phagocytosis is likely to occur. Often there is a clear concentration of microtubules or/and microfilaments in the region of the cytostome. In ciliates there may be a specialized arrangement of cilia around the cytostome.

cytotactin See ***tenascin***.

cytotoxic lymphocyte maturation factor See ***NK stimulatory factor***.

cytotoxic necrotising factors (CNFs) Toxins (110 kD, monomeric) produced by some strains of *E. coli*. Induce ruffling and stress fibre formation in fibroblasts and block cytokinesis by acting on p21 Rho.

cytotoxic T-cells (CTL) Subset of T-lymphocytes (mostly $CD8^+$) responsible for lysing target cells and for killing virus-infected cells (in the context of Class I ***histocompatibility*** antigens).

cytotrophic Descriptive of any substance that promotes the growth or survival of cells. Not commonly used except in the tissue-specific case of ***neurotrophic*** factors.

cytotropic Having affinity for cells: not to be confused with ***cytotrophic***.

cytotropism Movement of cells towards or away from other cells.

D

δ Either look under delta or main portion of word.

D-cells (δ-cells; delta-cells) Cells of the pancreas; about 5% of the cells present in primate pancreas with small argentaffin-positive granules. Their function is unclear, but they may release ***somatostatin***.

D-gene segment (diversity gene segment) Part of the gene for the immunoglobulin ***heavy chain***, it codes for part of the ***hypervariable region*** of the VH domain and is located between the VH and JH segments. There are probably about 20 different D segments.

D loop (displacement loop) Structure formed when an additional strand of DNA is taken up by a duplex so that one strand is displaced and sticks out like a D-shaped loop. Tends to happen in negatively supercoiled DNA, particularly in mitochondrial DNA as an intermediate during recombination.

D-mannose See ***mannose***.

D²O See ***heavy water***.

DAG See ***diacylglycerol***.

Dane particle 42nm particle, the complete infective virion of ***hepatitis B***.

Danio rerio See ***Brachydanio rerio***.

dansyl chloride (1-dimethyl-amino-naphthalene-5-sulphonyl chloride) A strongly fluorescent compound that will react with the terminal amino group of a protein. After acid hydrolysis of all the other peptide bonds, the terminal amino acid is identifiable as the dansylated residue.

DAP Diabetes related peptide. See ***islet amyloid peptide***.

DAP-12 Disulphide-linked homodimeric protein (12 kD) that interacts with ***KIR*** 2DS2. Resembles γ chain of Fc ε RI and ζ chain of T-cell receptor. Cytoplasmic tail has an ***ITAM*** that will bind ***zap-70*** and ***syk*** and thus activate the NK cell.

Daphnia magnus Cladoceran crustacean, the water flea.

DAPI stain (4,6-diamidino-2-phenylindole) Fluorochrome that binds to DNA and is used biochemically for detection of DNA and to stain the nucleus in fluorescence microscopy.

dapsone Drug related to the sulphonamides (diaminodiphenyl sulphone) that is used to treat leprosy (causative agent is *Mycobacterium leprae*). May act by inhibiting folate synthesis.

dark current (of retina) Current caused by constant influx of sodium ions into the ***rod outer segment*** of retinal photoreceptors, and which is blocked by light (leading to hyperpolarization). The plasma membrane sodium channel is controlled through a cascade of amplification reactions initiated by photon capture by ***rhodopsin*** in the disc membrane.

dark field microscopy A system of microscopy in which particles are illuminated at a very low angle from the side so that the background appears dark and the objects are seen by diffracted and reflected patches of light against a dark background.

dark reaction The reactions in phostosynthesis that occur after NADPH and ATP production, and that take place in the stroma of the chloroplast. By means of the reaction, CO_2 is incorporated into carbohydrate.

Datura stramonium Jimson weed or thornapple. Source of ***scopolamine***.

Daudi B-lymphoblastoid cell line from Burkitt's lymphoma in 16-year-old male

Negro. Have surface complement receptors and IgG and are ***EBV*** marker-positive.

dauer larva Semidormant stage of larval development in nematodes (eg. *Caenorhabditis elegans*), triggered by a pheromone: essentially a survival strategy.

dbl Human ***oncogene*** originally identified by transfection of NIH-3T3 cells with DNA from human diffuse B-cell lymphoma. A guanine nucleotide exchange factor (***GEF***) for rho-family members. Protein contains a domain of around 250 amino acids, the Dbl-homology (DH) domain. See also Table O1.

Dbl family (Dbl, Dbs, Brx, Lfc, Lsc, Ect2, DRhoGEF2, Vav) Family of proteins containing DH domains and with guanine nucleotide exchange factor activity for ***rho*** family of small G-proteins.

DD-PCR See ***differential display PCR***.

deacetylase An enzyme that removes an acetyl group: one of the most active deacetylation reactions is the constant deacetylation (and reacetylation) of lysyl residues in histones (the half life of an acetyl group may be as low as 10min). Acetylation (which removes a positive charge on the lysine ε-amino group) is thought to be increased in active genes, therefore deacetylation would be important in switching off genes.

DEAD-box helicases Family of ATP-dependent DNA or RNA ***helicases*** with a 4 amino acid consensus,-D-E-A-D-, that resembles an ATP binding site. Examples: p68, a human nuclear protein involved in cell growth; vasa, a *Drosophila* protein required for specification of posterior embryonic structures.

DEAD-box proteins Include the ***DEAD-box helicases***; may protect mRNA from degradation by endonucleases.

DEAE- (diethyl-aminoethyl-) Group that is linked to cellulose or Sephadex to give a positive charge and thus to produce an anion exchange matrix for chromatography.

deamination (of nucleic acids) The spontaneous loss of the amino groups of cytosine (yielding uracil), methyl cytosine (yielding thymine) or of adenine (yielding hypoxanthine). It can be argued that the presence of thymine in DNA in place of the uracil of RNA stabilizes genetic information against this lesion, since repair enzymes would restore the GU base pair to GC.

death domain Conserved domain (around 80 amino acids) found in cytoplasmic portion of some receptors (including the TNF receptor), essential for generating signals that often lead to apoptosis.

death receptors Superfamily of ***tumour necrosis factor*** receptors including TNF-R1, CD95, TRAMP, that trigger apoptotic cell death through interaction of various adapter proteins (FADD, TRADD, etc.) with their cytoplasmic ***death domains***. These adaptors then interact with ***caspases*** such as FLICE.

death-effector domain (DED) Domain at the C-terminus of FADD and N-terminus of FLICE. Interaction mediated by these domains leads to the assembly of the death-inducing signalling complex (DISC) which activates other ***caspases***.

debridement A term of French origin for the removal of necrotic, infected or foreign material from a wound.

decapentaplegic (dpp Drosophila gene, product related to ***TGF-α***.

decay-accelerating factor Plasma protein that regulates ***complement*** cascade by blocking the formation of the C3bBb complex (the C3 convertase of the alternate pathway). Widely distributed in tissues but deficient in paroxysmal nocturnal haemoglobinuria.

decidua See ***endometrium***.

deconvolution Process in digital image handling whereby a composite image is formed using information from several separate images taken at different levels (focal planes). The final image can be rotated and viewed from different angles and has usually had noise filtered out so that the image is much clearer and sharper.

decorin A small ***proteoglycan*** (90–140 kD) of the ***extracellular matrix***, so-called because it 'decorates' collagen fibres. The core protein has a mass of approximately 42 kD and is very similar to the core protein of ***biglycan*** and ***fibromodulin***. All three have highly conserved sequences containing 10 internal homologous repeats of approximately 25 amino acids with leucine-rich motifs. Decorin has one ***glycosaminoglycan*** chain, either chondroitin sulphate or dermatan sulphate and N-linked oligosaccharides.

dedifferentiation Loss of differentiated characteristics. In plants, most cells, including the highly differentiated haploid ***microspores*** (immature pollen cells) of angiosperms, can lose their differentiatiated features and give rise to a whole plant; in animals this is less certain, and there is still controversy as to whether the undifferentiated cells of the blastema that forms at the end of an amputated amphibian limb (for example) are derived by dedifferentiation or by proliferation of uncommitted cells. Neither is it clear whether dedifferentiation in animal cells might just be the temporary loss of phenotypic characters, with retention of the ***determination*** to a particular cell type.

deep cells Cells (blastomeres) in the teleost blastula that lie between the outer cell layer and the yolk syncytial layer, and are the cells from which the embryo proper is constructed during gastrulation; much studied in the fish, *Fundulus*.

defective virus A virus genetically deficient in replication, but which may nevertheless be replicated when it co-infects a host cell in the presence of a wild-type 'helper' virus. Most acute transforming ***retroviruses*** are defective, since their acquisition of oncogenes seems to be accompanied by deletion of essential viral genetic information.

defensins Family of small (30–35 residue) cysteine-rich cationic proteins found in vertebrate phagocytes (notably the azurophil granules of neutrophils) and active against bacteria, fungi and enveloped viruses. May constitute up to 5% of the total protein. Insect defensins have some sequence homology with the vertebrate forms.

defined medium Cell culture medium in which all components are known. In practice this means that the serum (that is normally added to culture medium for animal cells) is replaced by insulin, transferrin and possibly specific growth factors such as ***platelet-derived growth factor***.

definitive erythroblast Embryonic ***erythroblast*** found in the liver; smaller than primitive erythroblasts, they lose their nucleus at the end of the maturation cycle and produce ***erythrocytes*** with adult haemoglobin.

degeneracy The coding of a single amino acid by more than one base triplet (***codon***). Of the 64 possible codons, three are used for stop signals, leaving 61 for only 20 amino acids. Since all codons can be assigned to amino acids, it is clear that many amino acids must be coded by several different codons, in some cases as many as six.

degenerate primer A single-stranded synthetic ***oligonucleotide*** designed to hybridize to DNA encoding a particular protein sequence. As the mapping of codons to amino acids is many-to-one, the oligonucleotide must be made as a mixture with several different bases at variable positions. The total number of different oligos in the resulting mixture is known as the degeneracy of the primer. Such primers are widely used in screening a ***genomic library*** or in degenerate ***PCR***, to identify homologues of already known genes.

degenerins Products of *deg*-1, *mec*-4 and *mec*-10 genes in *Caenorhabditis elegans* which turn out to have homology with amiloride-sensitive sodium channels. Mutations cause neuronal degeneration, probably by disrupting ion fluxes. A related protein, the product of *unc*-105, interacts with collagen and may be a stretch-activated channel.

degradosome (RNA degradosome) Multienzyme complex in *E. coli* that contains exoribonuclease, polynucleotide phosphorylase (PNPase), endoribonuclease E (RNAase E), enolase and Rh1B (a member of the DEAD-box family of ATP-dependent RNA helicases).

degranulation Release of secretory granule contents by fusion with the plasma membrane.

dehydration Removal of water as in preparing a specimen for embedding or a histological section for clearing and mounting.

dehydrin Class of plant proteins expressed in response to water shortage, and notable for a run of seven contiguous serines.

dehydrogenase Enzyme that oxidizes a substrate by transferring hydrogen to an acceptor that is either NAD^+/$NADP^+$ or a flavin enzyme.

delayed rectifier channels The potassium-selective ***ion channels*** of ***axons***, so called because they change the potassium conductance with a delay after a voltage step. The name is used to denote any axon-like K channel. They have various roles, eg. regulation of pacemaker potentials, generation of bursts of ***action potentials*** or generation of long plateaux on action potentials.

delayed-type hypersensitivity See ***hypersensitivity***.

deletion mutation A mutation in which one or more (sequential) nucleotides is lost from the genome. If the number lost is not divisible by 3 and is in a coding region, the result is a ***frame-shift mutation***.

Delta (Dl) Neurogenic gene locus in *Drosophila*. Gene product contains nine repeats of the ***EGF-like domain***.

delta cells See ***D-cells***.

delta chains (δ chains) See ***immunoglobulin***. The ***heavy chains*** of mouse and human IgD immunoglobulins.

delta-endotoxin (δ-endotoxin) The toxic glycoprotein produced by sporulating ***Bacillus thuringiensis*** that can kill insects.

delta virus Hepatitis D virus. A defective RNA virus requiring a ***helper virus***, usually hepatitis B virus, for replication. Delta virus infections may exacerbate the clinical effects of Hepatitis B.

dematin Actin microfilament bundling protein (52 kD, but variants of similar molecular weight are reported); contains an ***SH3*** domain and is extensively palmitoylated; associated with membrane of erythrocytes (protein 4.9).

demyelinating diseases Diseases in which the myelin sheath of nerves is destroyed and which often have an autoimmune component. Examples are multiple sclerosis, acute disseminated encephalomyelitis (a complication of acute viral infection), experimental allergic encephalomyelitis and Guillain–Barre syndrome.

denaturation Reversible or irreversible loss of function in proteins and nucleic acids resulting from loss of higher order (secondary, tertiary or quaternary structure) produced by non-physiological conditions of pH, temperature, salt or organic solvents.

dendrite A long, branching outgrowth from a ***neuron***, that carries electrical signals from synapses to the cell body; unlike an axon that carries electrical signals away from the cell body. This classical definition, however, lost some weight with the discovery of axo-axonal and dendro-dendritic synapses.

dendritic cells (1) Follicular dendritic cells, found in germinal centres of spleen and lymph nodes, retain antigen for long periods. (2) Accessory (antigen-presenting) cells, positive for Class II histocompatibility antigens, found in the red and white pulp of the spleen and lymph node cortex and associated with stimulating T-cell proliferation. (3) T-lymphocyte found in epidermis and other epithelial cells involved in antigen recognition expressing predominantly γδ-TCR receptors (dendritic epidermal cells: DECs). (4) ***DOPA***-positive cells derived from neural crest and found in the basal part of epidermis: melanocytes distinct from (3). See also ***Langerhans cells***.

dendritic spines Wine glass- or mushroom-shaped protrusions from dendrites that represent the principal site of termination of excitatory afferent neurons on interneurons, especially in the cortical regions.

dendritic tree Characteristic (tree-like) pattern of outgrowths of neuronal ***dendrites***.

Dendroaspis angusticeps Snake, eastern green mamba. See ***calcicludine***.

Dendroaspis* natriuretic peptide** (DNP) ***Natriuretic peptide (38 residues) found in the venom of the snake *Dendroaspis angusticeps*. Binds to atrial natriuretic peptide receptor A but not to the ANP receptor type B.

Dendroaspis polylepis Snake, Black mamba. See ***calciseptine***.

dendrotoxins Polypeptides (57–60 residues) isolated from *Dendroaspis* (snake) venom that are selective blockers of ***voltage-gated*** K^+ channels in a variety of tissues and cell types. Have sequence similarity with Kunitz-type serine protease inhibitors.

denervation Removal of nerve supply to a tissue, usually by cutting or crushing the ***axons***.

dengue Tropical disease caused by a flavivirus (one of the ***arboviruses***), transmitted by mosquitoes. A more serious complication is dengue shock syndrome, a haemorrhagic fever probably caused by an immune complex hypersensitivity after re-exposure.

dense bodies Areas of electron density associated with the thin filaments in smooth muscle cells. Some are associated with the plasma membrane, others are cytoplasmic.

density-dependent inhibition of growth The phenomenon exhibited by most normal (***anchorage-dependent***) animal cells in culture that stop dividing once a critical cell density is reached. The critical density is considerably higher for most cells than the density at which a monolayer is formed; for this reason, most cell behaviourists prefer the term 'density-dependent inhibition of growth' as this avoids any confusion with contact inhibition of locomotion, a totally different phenomenon that is contact dependent.

density gradient A column of liquid in which the density varies continually with position, usually as a consequence of variation of concentration of a solute. Such gradients may be established by progressive mixing of solutions of different density (as for, eg. sucrose gradients) or by centrifuge-induced redistribution of solute (as for ***caesium chloride*** gradients). Density gradients are widely used for centrifugal and gravity-induced separations of cells, organelles and macromolecules. The separations may exploit density differences between particles, or primarily differences in size, in which latter case the function of the gradient is chiefly to stabilize the liquid column against mixing.

dentate nucleus Nerve cell mass, oval in shape, located in the centre of each of the cerebral hemispheres.

deoxycholate A bile salt. The sodium salt is used as a detergent to make membrane proteins water soluble.

deoxyglucose (2-deoxyglucose) Analogue of glucose in which the hydroxyl on C-2 is replaced by a hydrogen atom. Since it is often taken up by cells but not further metabolized, it can be used to study glucose transport and also to inhibit glucose utilization.

deoxyhaemoglobin Haemoglobin without bound oxygen.

deoxynojirimycin Antibiotic produced by *Bacillus* spp; inhibits α-glucosidases and thus interferes with the ***glycosylation*** of cell surface ***glycoproteins***.

deoxynucleoside of 5-bromo-uracil See ***BUdR***.

deoxyribonuclease (DNAase; DNase) An ***endonuclease*** with preference for DNA.

Pancreatic DNAase I yields di- and oligonucleotide 5′ phosphates, pancreatic DNAase II yields 3′ phosphates. In chromatin, the sensitivity of DNA to digestion by DNAase I depends on its state of organization, transcriptionally active genes being much more sensitive than inactive genes.

deoxyribonucleic acid See ***DNA***.

deoxyribose (2-deoxy-D-ribose) The sugar that when linked by 3′-5′ phosphodiester bonds forms the backbone of DNA.

Dep-1 (CD148; HPTP ε) Density-enhanced phosphatase-1 (220–250 kD). Transmembrane protein with eight extracellular FnIII domains and a single cytoplasmic tyrosine phosphatase domain. Dep-1 is involved in signal transduction in lymphocytes and is also found in smooth muscle cells and tumour cells. When clustered, Dep-1 inhibits Fc γ RII-induced superoxide production. In many tumour cells Dep-1 is associated with a 64 kD serine/threonine kinase that may regulate its activity. Has been speculated to play a role in density-dependent inhibition of growth.

depactin ***Actin*** depolymerizing protein (17.6 kD) originally isolated from echinoderm eggs. Similar to ***actophorin***.

dephosphorylation Removal of a phosphate group.

depolarization A positive shift in a cell's ***resting potential*** (that is normally negative), thus making it numerically smaller and less polarized, eg. –90mV to –50mV. The opposite of ***hyperpolarization***.

depsipeptides Polypeptides that contain ester bonds as well as peptides. Naturally occurring depsipeptides are usually cyclic; they are common metabolic products of microorganisms and often have potent antibiotic activity (eg. ***actinomycin***, enniatins, ***valinomycin***).

depurination (of DNA) The N-glycosidic link between purine bases and deoxyribose in DNA has an appreciable rate of spontaneous cleavage ***in vivo***, a lesion that must be enzymically repaired to ensure stability of the genetic information.

derepression Anything that stops the repression of a gene thereby allowing ***expression*** to occur.

dermal tissue *Bot.* Outer covering of plants, which includes the ***epidermis*** and periderm (non-living bark). Compare ***dermis***.

dermatan sulphate ***Glycosaminoglycan*** (15–40 kD) typical of ***extracellular matrix*** of skin, blood vessels and heart. Repeating units of D-glucuronic acid-N-acetyl-D-galactosamine or L-iduronic acid-N-acetyl-D-galactosamine with 1–2 sulphates per unit. Broken down by L-iduronidase, but accumulates intralysosomally in ***Hurler's disease*** and ***Hunter syndrome***.

dermatitis Inflammation of the ***dermis***, often a result of ***contact sensitivity***.

dermatosparaxis Recessive disorder of cattle in which a procollagen peptidase is absent. In consequence the amino- and carboxy-terminal peptides of procollagen are not removed, the ***collagen*** bundles are disordered, and the dermis is fragile. Similar to Ehlers–Danlos syndrome in humans.

dermis Mesodermally derived ***connective tissue*** underlying the epithelium of the skin.

dermoid cyst Usually benign cyst, the walls of which are of dermal origin. Many ovarian tumours are dermoid cysts.

DeSanctis–Cacchione syndrome A variant of ***xeroderma pigmentosum*** in which a different DNA repair enzyme is involved. Hybrid fibroblasts formed by Sendai virus fusion of the two types show normal repair (complementation).

desensitization In a general sense: see ***adaptation***. Immunologically, the term is used to mean the administration of a graded series of doses of an antigen to which there is an immediate hypersensitivity response. The technique is used in the treatment of allergy and works by inducing the production of blocking antibody (IgG) which inhibits IgE production or blocks IgE binding.

desferrioxamine Iron transporter from *Streptomyces pilosus* that chelates ferric ions. Used clinically to treat acute iron poisoning.

desmid Chlorophyte ***algae*** that are usually freshwater living and unicellular. Their cell wall often has elaborate ornamented shape.

desmin A protein (53 kD) of intermediate filaments, somewhat similar to vimentin, but characteristic of muscle cells. Type III intermediate-filament protein. Co-localizes with ***synemin***, ***paranemin*** and ***plectin*** in the appropriate cell types.

desmocalmin A protein (240 kD) isolated from bovine desmosomes that binds calcium-calmodulin and cytokeratin-type intermediate filaments.

desmocollins ***Glycoproteins*** of 130 and 115 kD (desmocollins I and II) isolated from ***desmosomes***. Antibody fragments directed against desmocollins block desmosome formation, and desmocollins are therefore thought to be involved in the adhesion.

desmoglein Transmembrane ***glycoprotein*** (165 kD) found in desmosomes.

desmoplakins Proteins isolated from ***desmosomes***. Types I (240 kD) and II (210 kD) are long flexible rod-like molecules about 100nm long made of two polypeptide chains in parallel. Desmoplakin III is smaller (81 kD).

desmosine Component of ***elastin***, formed from four side chains of lysine and constituting a crosslinkage.

desmosome (macula adherens junctions; spot desmosomes) Specialized cell junction characteristic of epithelia into which intermediate filaments (tonofilaments of cytokeratin) are inserted. The gap between plasma membranes is of the order of 25–30nm and the intercellular space has a medial band of electron dense material. Desmosomes are particularly conspicuous in tissues such as skin that have to withstand mechanical stress. See also Table D1 (proteins and glycoproteins of desmosomes).

desmotubule Cylindrical membrane-lined channel through a ***plasmodesma***, linking the cisternae of ***endoplasmic reticulum*** in the two cells.

desmoyokin Desmosomal plaque protein (680 kD) from bovine muzzle ***keratinocytes***. Homologous with human AHNAK protein.

desoxy- See ***deoxy-***.

desquamation Shedding of outer layer of skin (squames) or of cells from other epithelia.

Table D1. Proteins and glycoproteins of desmosomes

	Apparent M_r (kD) in gels	Synonyms	Location
Proteins			
dpl	230–250	Band 1, desmoplakin 1	Cytoplasm
dp2	210–220	Band 2, desmoplakin 2	Cytoplasm
dp3	83–90	Band 5, desmoplakin 3, plakoglobin	Cytoplasm and non-desmosomal
Glycoproteins			
dgl	140–160	Band 3, desmoglein 1	Plasma membrane
dg2	110–120	Band 4a, desmoglein 2a, desmocollin 1	Plasma membrane
dg3	97–105	Band 4b, desmoglein 2b, desmocolllin 2	Plasma membrane

(Courtesy of C. Skerrow)

destrin Actin depolymerizing protein (19 kD) from pig, apparently identical to *ADF* and similar to ***cofilin***.

destruxins Cyclic ***depsipeptide*** fungal toxins that suppress the immune response in invertebrates.

desynapsis Separation of the paired homologous chromosomes at the ***diplotene*** stage of meiotic prophase I.

detergents ***Amphipathic***, surface active, molecules with polar (water soluble) and non-polar (hydrophobic) domains. They bind strongly to hydrophobic molecules or molecular domains to confer water-solubility. Examples include sodium dodecyl sulphate, fatty acid salts, the Triton family, octyl glycoside.

determination The commitment of a cell to a particular path of differentiation, even though there may be no morphological features that reveal this determination. Generally irreversible, but in the case of ***imaginal discs*** of *Drosophila* that are maintained by serial passage, ***transdetermination*** may occur.

detoxification reactions Reactions taking place generally in the liver or kidney in order to inactivate toxins, either by degradation or else by conjugation of residues to a hydrophilic moiety to promote excretion.

deuterium oxide See ***heavy water***.

Deuteromycetes Outmoded term for group now reclassified as Deuteromycotina. Includes fungi with no known sexual reproductive stages: the old Fungi Imperfecta.

Deuterostome Embryonic developmental pattern in which the mouth does not form from the blastopore but from a second opening: includes echinoderms and chordates. Contrasts with morphogenesis in protostome phyla which include annelids, molluscs and arthropods. The two groups also differ in many aspects of early development including the pattern of early cleavage and the stage at which blastomeres become committed in differentiation: in deuterostomes the early blastomeres are equipotent whereas in protostomes there is earlier patterning and commitment to form particular cell lineages.

Devoret test Test for potential carcinogens based upon induction of prophage lambda in bacteria (*E. coli* K12 envA uvrB). There is a good correlation between ability of aflatoxins and benzanthracenes to induce lambda and carcinogenicity in rodents.

dexamethasone Steroid analogue (glucocorticoid) used as an anti-inflammatory drug.

dextrans High-molecular weight polysaccharides synthesied by some microorganisms. Consist of D-glucose linked by α-1,6 bonds (and a few α-1,3 and α-1,4 bonds). Dextran 75 (average molecular weight 75 kD) has a colloid osmotic pressure similar to blood plasma, so dextran 75 solutions are used clinically as plasma expanders. They will also cause charge-shielding, and at the right concentrations induce flocculation of red cells, a trick that is used in preparing leucocyte-rich plasma for white cell purification in the laboratory. Crosslinked dextran is the basis for ***Sephadex***. Commercially derived from strains of *Leuconostoc mesenteroides*.

dextrose See ***glucose***.

DH domain (Dbl homology domain) Domain of around 250 amino acids found in ***Dbl***, Vav and a family of other Dbl-family proteins (Lfc, Lsc, Ect2, Dbs, Brx). DH domains are invariably located immediately N-terminal to PH domain and the membrane localization and enzymatic activity of the DH domain requires the PH domain for normal function.

DHEA See ***dihydroepiandosterone***.

DHFR See ***dihydrofolate reductase***.

diabetes insipidus Rare form of ***diabetes*** in which the kidney tubules do not reabsorb enough water. This can be because: the renal tubules have defective receptors for ***antidiuretic hormone***

(ADH; vasopressin); a class of ***aquaporin*** water channel in the collecting duct is defective; or there is inadequate ADH production by the pituitary, leading to the excessive production of dilute urine.

diabetes mellitus Relative or absolute lack of ***insulin*** leading to uncontrolled carbohydrate metabolism. In juvenile-onset diabetes (that may be an autoimmune response to pancreatic B cells) the insulin deficiency tends to be almost total, whereas in adult onset diabetes there seems to be no immunological component but an association with obesity.

diabetes-related peptide See ***islet amyloid peptide***.

diacetoxyscirpenol (DAS) Trichothecene mycotoxin produced by various species of fungi. Cytotoxic for human CFU-GM and BFU-E.

diacylglycerol (DAG) Glycerol substituted on the 1 and 2 hydroxyl groups with long-chain fatty acyl residues. DAG is a normal intermediate in the biosynthesis of phosphatidyl phospholipids and is released from them by phospholipase C activity. DAG from phosphatidyl inositol polyphosphates is important in signal transduction. Elevated levels of DAG in membranes activate protein kinase C by stabilizing its catalytically active complex with membrane-bound phosphatidyl serine and calcium.

diacytosis Discharge of an empty pinocytotic vesicle from a cell. Not commonly used.

diakinesis The final stage of the first ***prophase*** of ***meiosis***. The ***chromosomes*** condense to their greatest extent during this stage and normally the nucleolus disappears and the fragments of the nuclear envelope disperse.

dialysis Separation of molecules on the basis of size through a semi-permeable membrane. Molecules with dimensions greater than the pore diameter are retained inside the dialysis bag or tubing whereas small molecules and ions emerge in the dialysate outside the tubing.

diaminobenzidine (DAB) Peroxidase substrate, but a potent carcinogen.

diaminobenzoic acid (DABA; 3,5-diaminobenzoic acid) Compound used in fluorimetric determination of DNA content: gives fluorescent product when heated in acid solution with aldehydes.

diapedesis Archaic term for the emigration of leucocytes across the endothelium.

diaphorase Any enzyme capable of catalysing oxidation of NAD or NADPH in the presence of an electron acceptor other than oxygen, eg. methylene blue, quinones or cytochromes. Imprecise term.

diatom Algae of the division Bacillariophyta; largely unicellular and characterized by having cell walls of hydrated silica embedded in an organic matrix. The cell walls are formed in two halves that fit together like the lid and base of a pillbox and often have elaborate patterns formed by pores. Diatoms are very abundant in marine and freshwater plankton. Deposits of the cell walls form diatomaceous or siliceous earths.

dibutyryl cyclic AMP An analogue of cyclic AMP that shares some of the pharmacological effects of this nucleotide, but is generally believed to enter cells more readily on account of its greater hydrophobicity.

dicarboxypropyl leucine See ***leucinopine***.

dichlorobenzonitrile (2,6-dichlorobenzonitrile) Inhibitor of cellulose biosynthesis in higher plants.

dichlorophenoxyacetic acid (2,4-dichlorophenoxyacetic acid; 2,4-D) A synthetic ***auxin***, also used as a selective herbicide.

dichroism See ***circular dichroism***.

dickkopf-1 *(dkk-1)* Gene in *Xenopus* that encodes a secreted protein (259 residues, 40 kD) that induces the head region in the developing embryo. Member of a family of genes. Dkk-1 is a potent antagonist of ***Wnt***.

dicotyledonous plants Plants belonging to the large subclass of Angiosperms that have two seed-leaves (cotyledons). Includes the majority of herbaceous flowering plants and most deciduous woody plants of the temperate regions.

dictyosome Organelle found in plant cells and functionally equivalent to the Golgi apparatus of animal cells.

Dictyostelium A genus of the ***Acrasidae***, the cellular slime moulds.

Dictyostelium discoideum The most commonly used member of the ***slime moulds***.

dictyotene Prolonged ***diplotene*** of meiosis: the stage at which oocyte nuclei remain during yolk production.

dideoxy sequencing (Sanger dideoxy sequencing) The most popular method of DNA sequence determination (cf. ***Maxam–Gilbert method***). Starting with single-stranded template DNA, a short complementary primer is annealed, and extended by a DNA polymerase. The reaction is split into four tubes (called 'A, C, G or T') each containing a low concentration of the indicated dideoxynucleotide, in addition to the normal deoxynucleotides. Dideoxynucleotides, once incorporated, block further chain extension, and so each tube accumulates a mixture of chains of lengths determined by the template sequence. The four reactions are denatured and run out on an acrylamide sequencing gel in neighbouring lanes, and the sequence read up the gel according to the order of the bands.

Diels–Alder reaction Reaction used in organic synthesis of six-membered rings.

diencephalon In vertebrate central nervous system, the most rostral part of the ***brainstem***, consisting of the thalamus, hypothalamus, subthalamus and epithalamus. It is a key relay zone for transmitting information about sensation and movement and also contains (in the hypothalamus) important control mechanisms for homeostatic integration.

diethyl-aminoethyl- See ***DEAE-***.

differential adhesion The differential adhesion hypothesis was advanced by Steinberg to explain the mechanism by which heterotypic cells in mixed aggregates sort out into isotypic territories. Quantitative differences in homo- and heterotypic adhesion are supposed to be sufficient to account for the phenomenon without the need to postulate cell type-specific adhesion systems: fairly generally accepted, although some tissue specific ***cell adhesion molecules*** are now known to exist.

differential display PCR Variation of the ***polymerase chain reaction*** used to identify differentially expressed genes. ***mRNA*** from two different tissue samples is reverse transcribed, then amplified using short, intentionally non-specific primers. The array of bands obtained from a series of such amplifications is run on a high-resolution gel, and compared with analogous arrays from different samples. Any bands unique to single samples are considered to be differentially expressed; they can be purified from the gel, and sequenced and used to clone the full-length cDNA. Similar in aim to ***subtractive hybridization***. See also ***differential screening***.

differential interference contrast Method of image formation in the light microscope based on the method proposed by Nomarski (though strictly speaking all forms of optical microscopy rely to a greater or lesser extent on differential interference). The light beam is split by a ***Wollaston prism*** in the condenser, to form slightly divergent beams polarized at right angles. One passes through the specimen (and is retarded if the refractive index is greater), and one through the background nearby: the two are recombined in a second Wollaston prism in the objective and interfere to form an image. The image is spuriously 'three-dimensional': the nucleus, for example, appears to stand out above the cell (or be hollowed out) because it has a higher refractive index than the cytoplasm. The Nomarski system has the advantage that there is no phase-halo, but the contrast is low and image formation with crowded cells is poor because the background does not differ from the specimen.

differential scanning calorimetry (DSC) Form of ***thermal analysis*** in which heat flows to a sample and a standard at the same temperature are compared, as the temperature is changed.

differential screening General term for techniques used to identify genes that are expressed differentially in two different conditions. These are usually based on identifying those mRNAs that are more abundant in one sample of cDNA library than in another. See also ***subtractive hybridization, differential display PCR***.

differentiation Process in development of a multicellular organism by which cells become specialized for particular functions. Requires that there is selective expression of portions of the genome; the fully differentiated state may be preceded by a stage in which the cell is already programmed for differentiation but is not yet expressing the characteristic phenotype ***determination***.

differentiation antigen Any large structural macromolecule that can be detected by immune reagents and that also is associated with the differentiation of a particular cell type or types. Many cells can be identified by their possession of a unique set of differentiation antigens. There should be no implication that the antigens cause differentiation.

diffraction When a wave-train passes an obstacle, secondary waves are set up that interfere with the primary wave and give rise to bands of constructive and destructive interference. Around a point source of light, in consequence, is a series of concentric light and dark bands (coloured bands with white light), a diffraction pattern.

diffusion coefficient (diffusion constant) For the translational diffusion of solutes, diffusion is described by Fick's First Law, which states that the amount of a substance crossing a given area is proportional to the spatial gradient of concentration and the diffusion constant (D), that is related to molecular size and shape. A useful derived relationship is that the mean square distance moved by molecules in time *t* is 6Dt.

diffusion constant See ***diffusion coefficient***.

diffusion limitation The boundary layer hypothesis: that the proliferation of cells in culture is limited by the rate at which some essential component (almost certainly a growth factor) diffuses from the bulk medium into the layer immediately adjacent to the plasma membrane. By spreading out, a cell obtains a suprathreshold level of the factor and can divide; if unable to spread (because of crowding or poor adhesion) then the cell will remain in the G0 stage of the ***cell cycle***.

diffusion potential Potential arising from different rates of diffusion of ions at the interface of two dissimilar fluids; a junction potential.

DiGeorge syndrome Congenital absence of the thymus and parathyroid as a result of which the T-lymphocyte system is absent.

digestive vacuole Intracellular vacuole into which lysosomal enzymes are discharged and digestion of the contents occurs. More commonly referred to as a ***secondary lysosome***.

digitalis General term for pharmacologically active compounds from the foxglove (*Digitalis*). The active substances are the cardiac glycosides, digoxin, digitoxin, strophanthin and ***ouabain***. Causes increased force of contraction of the heart, disturbance of rhythm and reduced beat frequency. Also causes arteriolar constriction, venous dilation, nausea and visual disturbances.

digitonin See ***saponin***.

digoxygenin Small molecule derived from foxgloves that is used for labelling DNA or RNA probes, and subsequent detection by enzymes linked to anti-digoxygenin antibodies. Proprietary to Boehringer-Mannheim.

dihydroepiandrosterone (DHEA) Predominant androgen secreted from the adrenal cortex, an intermediate in androgen and oestrogen biosynthesis. Can be converted to sulphate (DHEA-S) the predominant plasma form. Can be converted to potent androgens and oestrogens. Considered to play an important immunomodulatory role and the decline in DHEA levels with age correlates with reduced immune competence. Administration especially to postmenopausal women is claimed to bring benefit, especially in bone mineralization. Has been shown to have tumour-suppressive and antiproliferative effects in rodents.

dihydrofolate reductase (DHFR) An enzyme (EC 1.5.1.3) involved in the biosynthesis of ***folic acid*** coenzymes that transfers hydrogen from NADP to ***dihydrofolate***, yielding tetrahydrofolic acid, an essential vitamin cofactor in purine, thymidine and methionine synthesis. Inhibitors (eg. aminopterin and amethopterin, components of ***HAT medium***) can be used as anti-microbial and anti-cancer drugs.

dihydropyridines Specific blockers of some types of ***calcium channel***, eg. nifedipine and nitrenidine.

DiI Name used for fluorescent derivatives of indocarbocyanine iodide that have two long alkyl chains and are membrane soluble. Used as general stains for membranes and also as specific probes for membrane fluidity measurements.

dilution cloning Cloning by diluting the cell suspension to the point at which the probability of there being more than one cell in the inoculum volume is small. Inevitably on quite a few occasions there will not be any cells.

dinitrophenol (2,4-dinitrophenol) A small molecule used as an uncoupler of oxidative phosphorylation. Also used after reaction with various proteins to provide a strong and specific identified ***haptenic*** group.

dinoflagellates Photosynthetic organisms of the order Dinoflagellida (for botanists Dinophyceae). They are aquatic and abundant in marine plankton; two ***flagella*** lie in grooves in an often elaborately sculptured shell or ***pellicle*** that is formed from plates of cellulose deposited in membrane vesicles. The pellicle gives some dinoflagellates very bizarre shapes. Their chromosomes lack centromeres and may have little or no protein and may perhaps be intermediate between pro- and eukaryote types; hence the group has been termed ***mesokaryotic***. The nuclear membrane persists during mitosis. *Gymnodinium* and *Gonyaulax*, that causes 'red tide', produce toxins that if accumulated by filter-feeding ***molluscs*** can be fatal. Another common genus is *Peridinium*.

dioecious Flowering plants in which the sexes are separate; each plant is either male or female and flowers have either stamens or pistils but not both.

diphtheria toxin An AB exotoxin (62 kD) coded by β corynephage of virulent *Corynebacterium diphtheriae* strains (that can produce a repressor of toxin production). The B subunit binds to receptors on the surface of the target cell and facilitates the entry of the enzymically active A subunit (21 kD) that ADP-ribosylates ***elongation factor*** 2, thereby halting mRNA translation.

diplococcus Bacterial strain in which two spherical cells (cocci) are joined to form a pair like a dumbbell or figure-of-eight.

diploid A diploid cell has its ***chromosomes*** in homologous pairs, and thus has two copies of each autosomal genetic ***locus***. The diploid number (2n) equals twice the ***haploid*** number and is the characteristic number for most cells other than gametes.

diplornavirus Proposed family of all double-stranded RNA viruses: considered taxonomically unsound by many virologists.

diplotene The final stage of the first prophase of meiosis. All four ***chromatids*** of a ***tetrad*** are fully visible and homologous chromosomes start to move away from one another except at ***chiasmata***.

Diptera Order of insects with one pair of wings, the second pair being modified into balancing organs, the halteres; the mouthparts are modified for sucking or piercing. The insects show complete metamorphosis in that they have larval, pupal and imaginal (imago, adult) stages.

The order includes the flies and mosquitoes; best known genera are ***Anopheles*** and ***Drosophila***.

diptericins Inducible glycine-rich antibacterial peptides (about 8 kD) from Dipteran haemolymph.

direct B-cells Lymphocytes responding to a small range of antigens by antibody production without any requirement for T-cells. The antigens include ***flagellin*** and pokeweed mitogen.

disaccharide Sugar formed from two monosaccharide units linked by a glycosidic bond. The trehalose type are formed from two non-reducing sugars, the maltose type from two reducing sugars.

DISC (death-inducing signalling complex) See ***death-effector domain***.

disc gel Confusingly, nothing to do with shape; gels in which there is a discontinuity in pH, or gel concentration, or buffer composition.

discodermolide Anti-tumour drug (a polyhydroxylated alkatetraene lactone) that, like ***taxol***, promotes formation of stable bundles of microtubules and competes with taxol for binding to polymerized tubulin. Isolated from the marine sponge, *Discodermia dissoluta*.

discoidin A lectin, isolated from the cellular slime mould *Dictyostelium discoideum* (see ***Acrasidae***), that has a binding site for carbohydrate residues related to galactose. The lectin, that consists of two distinct species (discoidins I and II), is synthesized as the cells differentiate from vegetative to aggregation phase, and was originally thought to be involved in intercellular adhesion, but discoidin I is now thought to be involved in adhesion to the substratum by a mechanism resembling that of fibronectin in animals.

disintegrins Peptides found in the venoms of various snakes of the viper family, which inhibit the function of some ***integrins*** of the β1 and β3 classes. They were first identified as inhibitors of ***platelet*** aggregation and were subsequently shown to bind with high affinity to integrins and to block the interaction of integrins with ***RGD***-containing proteins, eg. they block the binding of the platelet integrin αIIb β3 to ***fibrinogen***. Disintegrins are effective inhibitors at molar concentrations 500–2000 times lower than short RGDX peptides. They are cysteine-rich peptides ranging from 45 to 84 amino acids in length and almost all of them have a conserved-RGD-sequence on a β-turn, presumed to be the site that binds to integrins. The assumption is that their biological role in the venom is to inhibit blood clotting. Found in many snake species, where they are called variously albolabrin, applagin, batroxostatin, bitistatin, echistatin, elegantin, flavoridin, halysin, kistrin, triflavin and trigramin.

disjunction mutant *Drosophila* mutant in which chromosomes are partitioned unequally between daughter cells at ***meiosis***, as a result of non-disjunction.

dispase Trade name for a crude protease preparation used for disaggregating tissue in setting up primary cell cultures. Dispase gives less complete disaggregation than trypsin but survival of cells may be better.

dispersion forces Forces of attraction between atoms or non-polar molecules that result from the formation of induced dipoles. Sometimes referred to as London dispersion forces. Important in the ***DLVO*** theory of colloid flocculation and thus in theories of cell adhesion.

disseminated intravascular coagulation Complication of septic shock in which endotoxin (from ***Gram-negative bacteria***) induces systemic clotting of the blood, probably indirectly through the effect of endotoxin on neutrophils. It may also develop in other situations where neutrophils become systemically hyperactivated.

dissociation Any process by which a tissue is separated into single cells. Enzymic dissociation with trypsin or other proteases is often used.

dissociation constant See ***equilibrium constant***.

distemper virus Paramyxovirus of the genus Morbillivirus. Commonest is the canine distemper virus that causes fever, vomiting and diarrhoea; variant that infects seals (Phocavirus) has caused significant mortality in recent years.

disulphide bond The -S-S- linkage. A linkage formed between the SH groups of two ***cysteine*** moieties either within or between peptide chains. Each cysteine then becomes a half-cystine residue. -S-S- linkages stabilize, but do not determine, secondary structure in proteins. They are easily disrupted by -SH groups in an exchange reaction and are not present in cytosolic proteins (cytosol has a high concentration of ***glutathione*** which has a free-SH residue).

dithioerythritol (DTE) Like ***dithiothreitol*** is also referred to as Cleland's reagent and has the same properties.

dithiothreitol (Cleland's reagent) Used to protect sulphydryl groups from oxidation during protein purification procedures or to reduce disulphides to sulphydryl groups.

diurnal Occurring during the day or repeating on a daily basis. Use of ***circadian rhythm*** for the latter avoids ambiguity.

diversity gene segment See ***D-gene segment***.

division septum The cell wall that forms between daughter cells at the end of mitosis in plant cells or just before separation in bacteria.

dizygotic Twins arising as a result of the fertilization of two ova by two spermatozoa and thus genetically non-identical, in contrast to ***monozygotic*** twins.

Dl See ***Delta***.

dl See ***dorsal***.

DLVO theory Theory of colloid flocculation advanced independently by Derjaguin & Landau and by Vervey & Overbeek and subsequently applied to cell adhesion. There exist distances (primary and secondary minima) at which the forces of attraction exceed those of electrostatic repulsion; an adhesion will thus be formed. For cells there is quite good correlation between the calculated separations of primary and secondary minima and the cell separations in tight junctions (1–2nm) and more general cell–cell appositions (12–20nm), respectively, although it is clear that other factors (particularly ***cell adhesion molecules***) also play an important part.

DMARD (disease-modifying anti-rheumatic drug) Drug used for treating rheumatoid arthritis that does more than relieve symptoms. Examples include gold, penicillamine, sulphasalazine and chloroquine, though none are as effective as would be desirable.

DMEM (Dulbecco modified eagle's medium) Very commonly used tissue culture medium for mammalian cells.

DMSO (dimethyl sulphoxide) Much used as a solvent for substances that do not dissolve easily in water and that are to be applied to cells (eg. cytochalasin B, formyl peptides), also as a cryoprotectant when freezing cells for storage. It is used clinically for the treatment of arthritis, although its efficacy is disputed.

DNA (deoxyribonucleic acid) The genetic material of all cells and many viruses. A polymer of ***nucleotides***. The monomer consists of phosphorylated 2-deoxyribose N-glycosidically linked to one of four bases ***adenine***, ***cytosine***, ***guanine*** or ***thymine***. These are linked together by 3′, 5′-phosphodiester bridges. In the Watson-Crick double-helix model, two complementary strands are wound in a right-handed helix and held together by hydrogen bonds between ***complementary base pairs***. The sequence of bases encodes genetic information. Three major conformations exist ***A-DNA***, ***B-DNA*** (which corresponds to the original Watson–Crick model) and ***Z-DNA***.

DNA annealing The reformation of double-stranded DNA from thermally denatured DNA. The rate of reassociation depends upon the degree of repetition, and is slowest for unique sequences (this is the basis of the Cot value; see ***Cot curve***).

DNA-binding proteins Proteins that interact with DNA, typically to pack or modify the DNA, eg. histones, or to regulate gene expression, transcription factors. Among those proteins that recognize specific DNA sequences, there are a number of characteristic conserved 'motifs' believed to be essential for specificity.

DNA fingerprinting See ***restriction fragment length polymorphism***.

DNA footprinting Technique for identifying the recognition site of DNA-binding proteins: see ***footprinting***.

DNA glycosidase See ***DNA glycosylase***.

DNA glycosylase (DNA glycosidase) Class of enzymes involved in ***DNA repair***. They recognize altered bases in DNA and catalyse their removal by cleaving the glycosidic bond between the base and the deoxyribose sugar. At least 20 such enzymes occur in cells.

DNA gyrase A type II ***topoisomerase*** of *Escherichia coli*, that is essential for DNA replication. This enzyme can induce or relax ***supercoiling***.

DNA helicase (unwindase) An enzyme that uses the hydrolysis of ATP to unwind the DNA helix at the ***replication fork***, to allow the resulting single strands to be copied. Two molecules of ATP are required for each nucleotide pair of the duplex. Found in both prokaryotes and eukaryotes.

DNA hybridization See ***hybridization***.

DNA iteron Repeated DNA sequence found near the ***origin of replication*** of some plasmids.

DNA library See ***genomic library***.

DNA ligase Enzyme involved in DNA replication. The DNA ligase of *E. coli* seals nicks in one strand of double-stranded DNA, a reaction required for linking precursor fragments during discontinuous synthesis on the lagging strand. Nicks are breaks in the phosphodiester linkage that leave a free 3′-OH and 5′-phosphate. The ligase from phage T4 has the additional property of joining two DNA molecules having completely base-paired ends. DNA ligases are crucial in joining DNA molecules and preparing radioactive probes (by nick translation) in recombinant DNA technology.

DNA methylation Process by which methyl groups are added to certain nucleotides in genomic DNA. This affects gene expression, as methylated DNA is not easily transcribed. The degree of methylation is passed on to daughter strands at mitosis by maintenance DNA methylases. Accordingly, DNA methylation is thought to play an important developmental role in sequentially restricting the transcribable genes available to distinct cell lineages. In bacteria, methylation plays an important role in the restriction systems, as restriction enzymes cannot cut sequences with certain specific methylations.

DNA polymerases Enzymes (EC 2.7.7.7) involved in template-directed synthesis of DNA from deoxyribonucleotide triphosphates. I, II and III are known in *E. coli*; III appears to be most important in genome replication and I is important for its ability to edit out unpaired bases at the end of growing strands. Animal cells have α, β and γ polymerases, with α apparently responsible for replication of nuclear DNA, and γ for replication of mitochondrial. All these function with a DNA strand as template. Retroviruses possess a unique DNA polymerase (***reverse transcriptase***) that uses an RNA template.

DNA rearrangement Wholesale movement of sequences from one position to another in DNA, such as occur somatically, eg. in the generation of antibody diversity.

DNA renaturation See ***DNA annealing***.

DNA repair Enzymic correction of errors in DNA structure and sequence that protects genetic information against environmental damage and replication errors.

DNA replication The process whereby a copy of a DNA molecule is made, and thus the genetic information it contains is duplicated. The parental double stranded DNA molecule is replicated semi-conservatively, ie. each copy contains one of the original strands paired with a newly synthesized strand that is complementary in terms of AT and GC base pairing. Though in this sense conceptually simple, mechanistically a complex process involving a number of enzymes.

DNA sequence analysis Determination of the nucleotide sequence of a length of DNA. Typically, this is performed by cloning the DNA of interest, so that enough can be prepared to allow the sequence to be determined, usually by the Sanger ***dideoxy sequencing*** method or the ***Maxam–Gilbert method***. The resulting reactions are then run on a large sequencing gel, capable of resolving single nucleotide differences in chain length. Recently, ***PCR***-based methods have obviated the need to clone the DNA under some conditions, and automated DNA sequencing has become widely available.

DNA synthesis The linking together of nucleotides (as deoxyribonucleotide triphosphates) to form DNA. ***In vivo***, most synthesis is ***DNA replication***, but incorporation of precursors also occurs in repair. In the special case of retroviruses, DNA synthesis is directed by an RNA template (see ***reverse transcriptase***).

DNA topoisomerase An enzyme capable of altering the degree of supercoiling of double-stranded DNA molecules. Various topoisomerases can increase or relax supercoiling, convert single-stranded rings to intertwined double-stranded rings, tie and untie knots in single stranded and duplex rings, catenate and decatenate duplex rings. Topoisomerase II of *E. coli* = gyrase.

DNA transfection A technique originally developed to allow viral infection of animal cells by uptake of purified viral DNA rather than by intact virus particles. The term, a hybrid between transformation and infection, is now generally used to describe applications of same methodology to introduction of other kinds of genes or gene fragments into cells as DNA, such as activated oncogenes from tumours into tissue culture cells.

DNA tumour virus Virus with DNA genome that can cause tumours in animals. Examples are ***Papovaviridae***, ***Adenoviridae*** and ***Epstein–Barr virus***.

DNA virus A virus in which the nucleic acid is double- or single-stranded DNA (rather than RNA). Major groups of double-stranded DNA viruses are papovaviruses, adenoviruses, herpesviruses, large bacteriophages, and poxviruses: of single-stranded, parvoviruses and coliphages φ X174 and M13.

***dnaA*, etc.** Genes in *E. coli* that are involved in coding for replication machinery. *dnaA* and *P* produce proteins involved in replication at the chromosome origin; *dnaB*, *C* and *D* are involved in primosome formation; *dnaE* codes for subunits of polymerase II; *dnaF* for ribonucleotide reductase; *dnaG* codes for primase; *dnaH*, *Q*, *X* and *Z* for components of polymerase III; *dnaI* for protein involved at the replication fork; *dnaJ* and *K* are necessary to ensure survival at high temperature, also considered essential for phage lambda replication; *dnaL* and *M* are uncharacterized; dnaT protein interacts with *dnaC* product, *dnaW* codes adenylate kinase.

DNAase See ***deoxyribonuclease***.

DNase See ***deoxyribonuclease***.

docking protein See ***signal recognition particle receptor***.

dolichol Terpenoids with 13–24 isoprene units and a terminal phosphorylated hydroxyl group. Function as transmembrane carriers for glycosyl units in the biosynthesis of glycoproteins and glycolipids.

domain Used to describe a part of a molecule or structure that shares common physicochemical features, eg. hydrophobic, polar, globular, α-helical domains, or properties, eg. DNA-binding domain, ATP-binding domain.

dominant negative A mutation which is capable of exerting an effect even when only one copy is present, as in a ***heterozygote***. Usually explained as a mutation that disrupts one subunit of a multimeric protein thus making the whole complex dysfunctional. Alternatively the mutated protein may compete with the normal protein produced by the other allele so that the overall activity is reduced below a critical threshold level and function is abnormal. The latter explanation only holds if the normal product of two genes is necessary for normal function (haplo-insufficiency) although experimental overexpression of an inactive form may reduce function by more than 50%.

domoic acid An ***excitatory amino acid*** transmitter.

Donnan equilibrium An equilibrium established between a charged, immobile colloid (such as clay, ion exchange resin or cytoplasm) and a solution of electrolyte. Characteristics: ions of like charge to the colloid tend to be excluded; ions of opposite charge tend to be attracted; the colloid compartment is electrically polarized relative to the solution in the same direction as the colloid charges (a 'Donnan potential'); and the osmotic pressure is higher in the colloid compartment.

donor splice junction The junction between an ***exon*** and an ***intron*** at the 5′ end of the intron. When the intron is removed during ***processing*** of ***hnRNA*** the donor junction is spliced to the acceptor junction at the 3′ end of the intron.

DOPA (L-DOPA; levodopa; 3-hydroxytyrosine) Precursor of the neurotransmitter dopamine, made from L-tyrosine by tyrosine 3-monooxygenase and used as a treatment for ***Parkinsonism***.

dopamine A ***catecholamine neurotransmitter*** and ***hormone*** (153 D), formed by decarboxylation of dehydroxyphenylalanine (DOPA). A precursor of ***adrenaline*** and ***noradrenaline***.

dorsal *(dl) Drosophila* polarity gene; homologue of the ***rel proto-oncogene***. See ***tube, pelle*** and ***toll***.

dorsalin-1 Protein that stimulates ***neural crest*** differentiation, ***neural crest*** growth, bone growth and wound healing.

dosage compensation Genetic mechanisms that allow genes to be expressed at a similar level irrespective of the number of copies at which they are present. Usually invoked for genes that lie on sex chromosomes, which are thus present in different copy numbers in males and females.

dot blot Method for detecting a specific protein or message. A spot of solution is dotted onto nitrocellulose paper, a specific antibody or probe is allowed to bind and the presence of bound antibody/probe then shown by using a peroxidase-coupled second antibody, as in ***Western blot*** or by other visualization methods. See also ***slot blot***.

double helix Conformation of a DNA molecule, like a ladder twisted into a helix.

double layer The zone adjacent to a charged particle in which the potential falls effectively to zero. An excess or deficiency of electrons on the surface (charge; not to be confused with the transmembrane potential) leads to an equivalent excess of ions of the opposite charge in the surrounding fluid. For most cells, that have negative charges, there will be an excess of cations immediately adjacent to the plasma membrane, and at physiological ionic strength the double layer is likely to be around 2–3nm thick.

doublet microtubules Microtubules of the axoneme. The outer nine sets are often referred to as doublet microtubules, although only one (the A tubule) is complete and has 13 protofilaments. The B tubule has only 10 or 11 protofilaments, and shares the remainder with the A tubule. A and B tubules differ in their stability and in the other proteins attached periodically to them; it is the ***dynein*** affixed to the A tubule attaching and detaching from the B tubule of the adjacent doublet that generates sliding movement in the ***axoneme***.

doubling time The time taken for a cell to complete the cell cycle.

Down's syndrome Also known as mongolism, most frequently a consequence of trisomy of chromosome 21. Common (1 in 700 live births); incidence increases with maternal age. The cause is usually non-disjunction at meiosis but occasionally a translocation of fused chromosomes 21 and 14.

downregulation Reduction in the responsiveness of a cell to a stimulus following first exposure, often by a reduction in the number of receptors expressed on the surface (as a consequence of reduced recycling). Tends to be used imprecisely.

downstream (1) Portions of DNA or RNA that are more remote from the initiation sites and that will therefore be translated or transcribed later. (2) Shorthand term for things that happen at a late stage in a sequence of reactions, eg. in a signalling cascade.

doxorubicin Cytotoxic antibiotic from *Streptomyces peucetius*. Blocks topoisomerase and reverse transcriptase by intercalating into the DNA. Has been used in clinical oncology.

Dpp Protein product of the *Drosophila* gene *decapentaplegic*, related to TGF.

dpp *See* ***decapentaplegic***.

drebin Drebin is a developmentally regulated actin-binding brain protein which in the chicken has characteristic changes in expression related to developmental stage. Contains a single ***ADF-H domain***. Binds F-actin. Same class as yeast ABP-1.

Drickamer motif Either of the two highly conserved patterns of invariant amino acids found in the carbohdrate recognition domain of C-type and S-type lectins.

Drosophila A genus of small American flies, ***Diptera***. The best-known species is *D. melanogaster*, often called the fruit fly, but more correctly termed the vinegar fly. First investigated by T. H. Morgan and his group, it has been extensively used in genetic studies. More recently it has been used for studies of embryonic development.

drosulphakinins *Drosophila* homologues of the ***gastrin*** family of peptide hormones.

drug resistance factor See ***R plasmid***.

DSE See ***serum response element***.

dual recognition hypothesis An outmoded hypothesis that is known to be incorrect now that the structure of the T-cell receptor is known. The proposal was that viral (and some chemical) antigens were recognized in association with ***histocompatibility antigens*** by separate receptors on the T-cell. The generation of cytotoxic T-cells was by association with Class I MHC antigens, of T-helper cells by association with Class II MHC antigens. See ***altered self-hypothesis***.

Duchenne muscular dystrophy A sex-linked hereditary disease confined to young males and to females with ***Turner's syndrome***. It is characterized by degeneration and ***necrosis*** of skeletal muscle fibres, which are replaced by fat and fibrous tissue. The incidence of this disorder is about 1 in 4000 male births and of these one-third are estimated to be new mutational events. See ***dystrophin***.

ductin Name for the 16 kD transmembrane subunit of the ***V-type ATPase***, reflecting a (controversial) view that it may be a multifunctional transmembrane pore protein, also involved (for example) in gap junction formation. See also ***connexin***.

Duffy ***Blood group system***. Single gene locus.

dunce *(dnc) Drosophila* mutant that is deficient in short-term memory. Gene codes for cAMP-phosphodiesterase and mutation leads to elevated cAMP levels that in turn particularly affect the delayed rectifier potassium currents in neurons of brain centres associated with acquisition and retention. The effect of the mutation does also alter nerve terminal growth and synaptic plasticity. Comparable behavioural defects are associated with ***rutabaga*** (*rut*).

Dupuytren's contracture Fibroma-like lesion of the palm of the hand that causes flexion contracture. Heritable and commoner in men. Never metastasizes.

dura mater Outermost of the three meningeal membranes covering the brain and spinal cord.

dyad symmetry element See ***serum response element***.

dye coupling Measure of intercellular communication, usually through ***gap junctions***. If a fluorescent dye (eg. ***lucifer yellow***) injected into one cell is seen to pass into a neighbouring cell, the presence of junctions at least able to pass solutes of that size can be inferred between the two. See also ***electrical coupling***.

dynactin Dynein activator complex that stimulates vesicle transport. Includes dynactin (160 kD) and polypeptides of 62, 50, 45, 37, and 32, the 45 kD (possibly ***actin-RPV***) being the most abundant. All the subunits co-sediment with antibody to dynactin 160 and the complex behaves as a stable 20S multiprotein assembly. See ***centractin***.

dynamin A protein isolated from microtubule preparations and shown to cause ATP-mediated microtubule sliding toward the plus ends. A GTP-binding protein with classical G-protein motifs and with very high homology to the Mx protein involved in interferon-induced virus resistance. There are tissue-specific and developmentally regulated forms of dynamin in *Drosophila*. Associated with endocytic sorting of proteins.

dynein Large multimeric protein (600–800 kD) with ATPase activity; constitutes the side arms of the outer microtubule doublets in the ciliary axoneme and is responsible for the sliding. Probably (together with ***kinesin***) involved in microtubule-associated movement elsewhere. Cytoplasmic dynein is MAP-Ic.

dynorphin Opiate peptide derived from the hypothalamic precursor prodynorphin (that also contains the neoendorphin sequences). Contains the pentapeptide leu-enkephalin sequence. Its binding affinity is greater for the κ-type than for the μ-type opioid receptor.

dyrk Dual-specificity protein kinase regulated by tyrosine phosphorylation that may regulate cell cycle. Also has nuclear targeting signal, putative leucine zipper and a very conserved 13-histidine repeat sequence. Rat gene *dyrk* is a homologue of *Drosophila minibrain* (*mnb*), a gene involved in postembryonic neurogenesis. Human homologue maps to chromosome 21 and may be involved in pathogenesis of certain phenotypes of ***Down's syndrome***.

dyscrasia Illness as a result of abnormal material in the blood.

dysgenic System of breeding or selection that is genetically deleterious or disadvantageous.

dyskinetoplasty Absence of an organized kinetoplast (and of kinetoplast DNA) from a flagellate protozoan cell.

dysplasia Literally, 'wrong growth'. Usually used to denote early stage of carcinogenesis, marked by abnormal epithelial morphology.

dystroglycan (156DAG) Complex composed of two proteins, α- and β-dystroglycans (formerly known as 156DAG and 43DAG/A3a, respectively) derived from a single precursor by proteolytic cleavage. β-dystroglycan is a transmembrane protein that associates with ***dystrophin*** in the cytoplasm and α-dystroglycan, an extracellular glycoprotein, on the exterior face. α-dystroglycan binds to ***dystrophin***, thus linking actin through dystrophin and β-dystroglycan to the extracellular matrix. Also associates with dystrophin. Dystrophin deficiency leads to a deficiency in the appearance of these proteins on the sarcolemma, even though they are not themselves defective.

dystrophin Protein (400 kD) from skeletal muscle that is missing in ***Duchenne muscular dystrophy***. Its exact role is not yet clear, though it seems to be associated with the cytoplasmic face of the sarcolemma and T-tubules and may form part of the membrane cytoskeleton. There are sequence homologies with non-muscle α-actinin and with spectrin.

E

E See ***glutamic acid***.

E1A ***Oncogene*** from an ***Adenovirus***. Interacts with the *Rb* ***tumour suppressor*** gene product, in a manner similar to ***SV40*** large T-antigen.

E1B ***Oncogene*** from an ***adenovirus***. Interacts with the ***p53*** tumour suppressor gene product.

E2F Family of transcription factors originally identified through their role in transcriptional activation of the adenovirus E2 promoter, subsequently found to bind to promoters for various genes involved in the G1 and S phases of the cell cycle. E2F forms heterodimers with DP-1 to produce an active transcriptional complex. E2F family members are regulated by interaction with ***retinoblastoma*** (Rb) proteins.

E5 ***Oncogene*** from a ***papillomavirus***. Encodes a small protein that binds and blocks the 16 kD ***proteolipid*** of the ***V-type ATPase***, producing abnormal intravesicular processing of growth factor receptors.

E6 ***Oncogene*** from a ***papillomavirus***. Encodes a 16 kD protein of unknown function.

E7 ***Oncogene*** from a ***papillomavirus***. Interacts with the ***Rb tumour suppressor*** gene product, in a manner similar to ***SV40*** large T-antigen.

E_1E_2-ATPase A class of plasma membrane-localized ion-motive pumps that includes ***sodium–potassium ATPase***. The phosphoenzyme has two conformational states, E1 and E2, and ion exchange is inhibited by oligomycin, orthovanadate and ***ouabain***.

E classification Classification of enzymes based on the recommendations of the Committee on Enzyme Nomenclature of the International Union of Biochemistry. The first number indicates the broad type of enzyme (1 = oxidoreductase; 2 = transferase; 3 = hydrolase; 4 = lyase; 5 = isomerase; 6 = ligase (synthetase)). The second and third numbers indicate subsidiary groupings, and the last number, which is unique, is assigned arbitrarily in numerical order by the Committee.

E. coli See ***Escherichia coli***.

E-face In ***freeze fracture*** the plasma membrane cleaves between the acyl tails of membrane phospholipids, leaving a monolayer on each half of the specimen. The E-face is the inner face of the outer lipid monolayer. From within the cell this is the view that you would have of the outer half of the plasma membrane if the inner layer could be removed. The complementary surface is the P-face (the inner surface of the inner leaflet of the bilayer). E stands for ectoplasmic, P for protoplasmic: not terms that are in common usage!

E-rosettes The clustering of sheep erythrocytes (=E) around a leucocyte or other cell. E-rosette formation is used as a marker for T-lymphocytes of humans and most mammals; in this case E are untreated, compared with other rosette tests such as EA where E have antibody bound to their surface.

E-selectin See ***selectin***.

EA-rosettes See ***E-rosettes***. A test for the presence of Fc receptors.

EAA See ***excitatory amino acid***.

EaA cells Insect cell line derived from haemocytes of the salt marsh caterpillar *Estigmene acrea*. An alternative line for baculovirus expression. See ***Sf9 cells***.

EAC-rosettes Rosettes (see ***E-rosettes***) formed from erythrocytes (E) coated with antibody (A) and complement (C). A test for C3b or C3bi receptors (***CR1*** or ***CR3***).

The rosettes form more easily then E or EA rosettes.

Eadie–Hofstee plot Linear transformation of enzyme kinetic data in which the velocity of reaction (v) is plotted on the ordinate, v/S on the abscissa, S being the initial substrate concentration. The intercept on the ordinate is Vmax, the slope is -Km. Preferable to the ***Lineweaver–Burke plot***.

EAE See ***experimental allergic encephalomyelitis***.

early antigens Virus-coded cell surface antigens that appear soon after the infection of a cell by virus, but before virus replication has begun. See ***early gene***.

early gene Genes that are expressed soon after viral infection of a host cell.

early region Part of a viral genome in which ***early genes*** that are transcribed and expressed early during infection of a cell are clustered.

Ebola virus ***Filovirus*** that causes severe fever and bleeding. Outbreaks are usually in Africa.

EC number See ***E classification*** for enzymes

EC_{50} Effective concentration; concentration at which the substance concerned produces a specified effect in 50% of the organisms treated.

ecdysone Family of ***steroid hormones*** found in insects, crustaceans and plants. In insects, α-ecdysone stimulates moulting. The steadily maturing character of the moults is affected by steadily decreasing levels of ***juvenile hormone***. β-ecdysone (ecdysterone) has a slightly different structure and is also found widely. Phytoecdysones are synthesized by some plants.

ECF See ***eosinophil chemotactic peptide***.

ECF of anaphylaxis See ***eosinophil chemotactic peptide***.

ECF-C See ***eosinophil chemotactic peptide***.

ECG (electrocardiograph; electrocardiogram) A recording of the electrical activity of the heart.

echinocytes Erythrocytes that have shrunk (in hypertonic medium) so that the surface is spiky.

Echinodermata Phylum of exclusively marine animals. The phylum is divided into five classes: the Asteroidea (starfish), the Echinoidea (sea urchins), the Ophiuroidea (brittle stars and basket stars), the Holothuroidea (the sea cucumbers) and the Crinoidea (sea lilies and feather stars).

Echinosphaerium Previously *Actinosphaerium*. A ***Heliozoan*** protozoan. The organisms are multinucleate and have a starburst of radiating axopodia, the microtubules of which have been much studied.

echistatin ***Disintegrin*** found in the venom of the saw-scaled viper, *Echis carinatus*.

Echoviruses A group of human ***Picornaviruses***. 'Echo' is derived from enteric cytopathic human orphan, where orphan implies that they are not associated with any disease, though some are now known to cause aseptic meningitis or other disorders.

ECL (1) Electrochemiluminescence. Production of light during an electrochemical reaction, now being applied to various bioassay systems. (2) Enhanced chemiluminescence. Method for enhancing detection of proteins on blots. Involves the use of luminol that is oxidized by peroxidase-coupled antibody used to detect the protein of interest, and the light produced is then detected on film.

eclosion Emergence of an insect from its old cuticle at a moult, particularly from pupa to adult.

ECM See ***extracellular matrix***.

Eco RI Probably the most commonly used type II ***restriction endonuclease*** isolated from ***Escherichia coli***. It cuts the sequence GAATTC between G and A thus generating 5′ ***sticky ends***.

Eco RII Type II ***restriction endonuclease*** isolated from *Escherichia coli*. It cuts the sequence CC(T/A)GG in front of the first C giving 5′ ***sticky ends***.

ecotropic virus *Retrovirus* which can only replicate in its original host species, cf. amphotropic.

Ecstasy (MDMA; 3,4-methylenedioxymethamphetamine) An amphetamine-like drug of abuse.

ectoderm The outer of the three germ layers of the embryo (the other two being mesoderm and endoderm). Ectoderm gives rise to epidermis and neural tissue.

ectoenzyme Enzyme that is secreted from a cell or located on the outer surface of the plasma membrane and therefore able to act on extracellular substrates.

ectopic Misplaced, not in the normal location.

ectoplasm Granule-free cytoplasm of amoeba lying immediately below the plasma membrane.

ectoplasmic tube contraction Model for amoeboid movement in which it is proposed that protrusion of a pseudopod is brought about by contraction of the subplasmalemmal region everywhere else in the cell thus squeezing the central cytoplasm forwards. See ***frontal zone contraction theory***.

ectromelia Congenital absence or gross shortening of long bones of limb or limbs.

ED_{50} Median effective dose, that dose that produces a response in 50% of individuals or 50% of the maximal response.

edema (USA) See *oedema*.

editosome Multiprotein complex (27S) involved in RNA processing.

Edman reagent Phenyl isothiocyanate. The classic method for sequence determination of peptides using sequential cleavage of the N-terminal residue after reaction with Edman reagent. The N-terminal amino acid is removed as a phenylthiohydantoin derivative.

EDRF See ***endothelium-derived relaxation factor***.

EDTA (ethylenediaminetetraacetic acid) Often used as the disodium salt. Chelator of divalent cations; $\log_{10} K_{app}$ for calcium at pH7 is 7.27 (5.37 for magnesium) See ***EGTA***.

EDTA-light chain Myosin light chains (18 kD) from scallop muscle (2 per pair of heavy chains), easily extracted by calcium chelation. Although the EDTA-light chains do not bind calcium they confer calcium sensitivity on the myosin heavy chains.

Edward's syndrome Complex of abnormalities caused by trisomy 18.

EEG (electroencephalograph; electroencephalogram.) Record of electrical activity of the brain obtained using external electrodes.

EF See ***elongation factor***.

EF-G See ***translocase***.

EF-hand A very common calcium-binding motif. A 12 amino acid loop with a 12 amino acid α-helix at either end, providing octahedral coordination for the calcium ion. Members of the family include: ***aequorin, α-actinin, calbindin, calcineurin, calcyphosin, calmodulin, calpain, calcyclin, diacylglycerol*** kinase, ***fimbrin, myosin*** regulatory light chains, ***oncomodulin, osteonectin, spectrin, troponin*** C.

EGF See ***epidermal growth factor, EGF-like domain***.

EGF receptor (HER-1) Receptor tyrosine kinase encoded by c-*erb* B1. Member of the type I family of growth factor receptors that also includes TGF-α receptor, heregulin receptor.

EGF-like domain Region of 30–40 amino acids containing six cysteines found originally in EGF, and also in a range of proteins involved in cell signalling.

Examples: ***TGF-α***, ***amphiregulin***, ***urokinase***, ***tissue plasminogen activator***, ***complement*** C6–C9, ***fibronectin***, ***laminin*** (each subunit at least 13 times), ***nidogen***, ***selectins***. It is also found in the *Drosophila* gene products: Notch (36 times) Delta, Slit, Crumbs, Serrate.

egg-polarity gene A gene whose product distribution in the egg determines the anterior–posterior axis of subsequent development. Best characterized in *Drosophila*: see ***bicoid***, ***maternal-effect gene***.

eglin C A proteinase inhibitor (70 amino acids) from the leech.

EGTA (ethyleneglycol-bis(2-aminoethyl) N, N, N', N',-tetraacetic acid)) Like ***EDTA*** a chelator of divalent cations but with a higher affinity for calcium (log K_{app} 6.68 at pH 7) than magnesium (log K_{app} 1.61 at pH 7). Will also bind other divalent cations. Note: the 'apparent association constant', K_{app}, is used because protons compete for binding and the association constant varies according to pH. Thus, EGTA has $\log_{10} K_{app}$ for calcium of 2.7 at pH 5, 10.23 at pH 9.

Ehlers–Danlos syndrome See ***dermatosparaxis***.

Ehringhaus compensator Device used in ***interference*** or ***polarization microscopy*** to reduce the brightness of the object to zero in order to measure the phase retardation (optical path difference). The compensator consists of a birefringent crystal plate that can be tilted. An alternative to ***Senarmont compensation*** and has the advantage that it can be applied to retardations of more than one wavelength.

EHS cells (Englebreth–Holm–Swarm sarcoma cells) A line of mouse cells that produce large amounts of basement membrane-type ***extracellular matrix*** (ecm), rich in ***laminin***, collagen type IV, ***nidogen*** and heparan sulphate. Often used as a source of these ecm molecules.

eicosanoid Useful generic term for compounds derived from ***arachidonic acid***. Includes ***leukotrienes***, ***prostacyclin***, ***prostaglandins*** and ***thromboxanes***.

Eimeria Coccidian protozoan. All coccidians are intracellular parasites of various vertebrates and invertebrates. *Eimeria tenella* infects chick intestinal epithelial cells and is of veterinary importance. The trophozoites invade host cells, proliferate as merozoites by schizogony which can then infect adjacent cells if released. Merozoites differentiate to male or female gamonts that fuse to form a zygote that undergoes division to form eight zoites that are retained within a zygocyst. If the zygocyst is ingested by a new host the zoites emerge, and reinfect the host as trophozoites.

Eisenberg An algorithm for calculating a ***hydropathy plot***.

ektacytometry Method in which cells (usually erythrocytes) are exposed to increasing shear stress and the laser diffraction pattern through the suspension is recorded; it goes from circular to elliptical as shear increases. From these measurements a deformability index for the cells can be derived.

elaioplast Unpigmented type of ***plastid*** modified as an oil-storage organelle.

ELAM-1 (CD62E; E-selectin) One of the selectin family; upregulated on endothelial cells at sites of inflammation and partly responsible for trapping of neutrophils. The C-type lectin domain binds sialylated Lewis X and a particular glycoform of ESL-1 that is present on myeloid cells.

elastase Serine protease that will digest ***elastin*** and ***collagen*** type IV; inhibited by α-1-protease inhibitor of plasma.

elasticoviscous Alternate form of the commoner term viscoelastic.

elastin Glycoprotein (70 kD) randomly coiled and crosslinked to form elastic fibres that are found in connective tissue. Like collagen, the amino acid composition is unusual with 30% of residues being glycine and with a high proline content. Crosslinking depends upon formation of ***desmosine*** from four lysine side groups. The mechanical properties of elastin are poorer in old animals.

elastonectin Elastin-binding protein (120 kD) found in extracellular matrix, produced by skin fibroblasts.

ELAV proteins (embryonic lethal abnormal visual proteins) RNA-binding proteins that regulate mRNA stability. The ELAV family of RNA-binding proteins is highly conserved in vertebrates, and in humans there are four members: HuR is expressed in all proliferating cells, whereas Hel-N1, HuC and HuD are expressed in terminally differentiated neurons. See ***AREs***.

electrical coupling Of two physically touching cells, denoting the presence of a ***junction*** that allows the passage of electrical current. Usually tested by impaling both cells with ***microelectrodes***, injecting a current into one, and looking for a change in potential in the other. Usually taken as an indication of coupling by ***gap junctions*** or ***electrical synapses***: see also ***dye coupling***. Electrical coupling is not confined to excitable cells: many embryonic and adult ***epithelia*** are coupled, possibly to allow metabolic cooperation.

electrical synapse A connection between two electrically excitable cells, such as neurons or muscle cells, via arrays of ***gap junctions***. This allows ***electrical coupling*** of the cells, and so an action potential in one cell moves directly into the other, without the 1ms delay inherent in ***chemical synapses***. Electrical synapses do not allow modulation of their connection, and so only occur in neuronal circuits where speed of conduction is paramount (eg. the crayfish escape reflex). A few electrical synapses are rectifying, implying a more specialized property than a simple gap junction.

electrochemical potential Defined as the work done in bringing 1 mole of an ion from a standard state (infinitely separated) to a specified concentration and electrical potential. Measured in joules/mole. More commonly used to measure the electrochemical potential difference between two points (eg. either side of a cell membrane), thus sidestepping the rather abstract concept of a standard state. If the molecule is uncharged or the electrical potential difference between two points is zero, the electrochemical potential reduces to the ***chemical potential*** difference of the species. At equilibrium, the electrochemical potential difference (by definition) is zero; the situation can then be described by the ***Nernst equation***.

electrodynamic forces London–Van der Waals forces: see ***DLVO theory***.

electrofocusing Any technique whereby chemical species are concentrated using an applied electric field. See ***isoelectric focusing***.

electrogenic pump Ion pump that generates net charge flow as a result of its activity. The sodium–potassium exchange pump transports two potassium ions inward across the cell membrane for every three sodium ions transported outward. This produces a net outward current that contributes to the internal negativity of the cell.

electron microprobe A technique of elemental analysis in the electron microscope based on spectral analysis of the scattered X-ray emission from the specimen induced by the electron beam. Using this technique it is possible to obtain quantitative data on, for example, the calcium concentration in different parts of a cell, but it is necessary to use ultra-thin frozen sections.

electron microscopy Any form of microscopy in which the interactions of electrons with the specimens are used to provide information about the fine structure of that specimen. In transmission electron microscopy (TEM) the diffraction and adsorption of electrons as the electron beam passes normally through the specimen is imaged to provide information on the specimen. In scanning electron microscopy (SEM) an electron beam falls at a non-normal angle on the specimen and the image is derived from the scattered and reflected electrons. Secondary X-rays generated by the interaction of electrons with various elements in the specimen may be used for ***electron microprobe*** analysis.

electron paramagnetic resonance (EPR; electron spin resonance; ESR) Form of spectroscopy in which the absorption of microwave energy by a specimen in a strong magnetic field is used to study atoms or molecules with unpaired electrons.

electron transport chain A series of compounds that transfer electrons to an eventual donor with concomitant energy conversion. One of the best studied is in the mitochondrial inner membrane, which takes NADH (from the ***tricarboxylic acid cycle***) or FADH and transfers electrons via ***ubiquinone***, cytochromes and various other compounds, to oxygen. Other electron transport chains are involved in ***photosynthesis***.

electrophoresis Separation of molecules based on their mobility in an electric field. High-resolution techniques normally use a gel support for the fluid phase. Examples of gels used are starch, acrylamide, agarose or mixtures of acrylamide and agarose. Frictional resistance produced by the support causes size, rather than charge alone, to become the major determinant of separation. The electrolyte may be continuous (a single buffer), or discontinuous, where a sample is stacked by means of a buffer discontinuity, before it enters the running gel/running buffer. The gel may be a single concentration or gradient in which pore size decreases with migration distance. In ***SDS*** gel electrophoresis of proteins or electrophoresis of polynucleotides, mobility depends primarily on size and is used to determined molecular weight. In pulse-field electrophoresis, two fields are applied alternately at right-angles to each other to minimize diffusion-mediated spread of large linear polymers. See also ***electrofocusing, pulse-field electrophoresis***.

electrophoretogram Result of a zone electrophoresis separation or the analytical record of such a separation.

electroplax A stack of specialized muscle fibres found in electric eels, arranged in series. The fibres have lost the ability to contract; instead they generate extremely high voltages (ca 500V) in response to nervous stimulation. They contain asymmetrically distributed ***sodium–potassium ATPases, acetylcholine receptors*** and ***sodium gates*** at extraordinarily high concentrations.

electroporation Method for temporarily permeabilising cell membranes so as to facilitate the entry of large or hydrophilic molecules (as in ***transfection***). A brief (ca 1ms) electric pulse is given with potential gradients of about 700V/cm.

electroretinogram Record of electrical activity in the retina made with external electrodes.

electrospray mass spectroscopy Method of mass spectroscopy in which the sample is introduced as a fine spray from a highly charged needle so that each droplet has a strong charge. Solvent rapidly evaporates from the droplets leaving the free macromolecule. Beginning to be widely used because of its capacity to identify a wide range of compounds.

electrostatic forces Like charges in close proximity produce forces of repulsion between them. Consequently if two surfaces bear appreciable and approximately equal densities of charged groups on their surfaces appreciable forces of repulsion may occur between them. The range of these forces is determined in the main by the ionic strength of the intervening medium, forces being of minimal range at high ionic strength. The forces are effective over approximately twice the ***double layer*** thickness. See ***DLVO theory***.

elegantin See ***disintegrin***.

eleutherobin Tricyclic compound that, like ***taxol***, will stabilize microtubule bundles. Isolated from a marine soft coral, *Eleutherobia aurea*.

eleuthosides Class of compounds with anti-inflammatory activity.

elicitor Substance that induces the formation of ***phytoalexins*** in higher plants. May be exogenous (often produced by potentially pathogenic microorganisms), or endogenous (possibly cell wall degradation products).

ELISA (enzyme-linked immunosorbent assay) A very sensitive technique for the detection of small amounts of protein or other antigenic substance. The basis of the method is the binding of the antigen by an antibody that is linked to the surface of a plate. Formation of an immune complex is detected by use of peroxidase coupled to antibody, the peroxidase being used to generate an amplifying colour reaction. Various ways of carrying out the assay are possible: if the aim is to detect antibody production from a myeloma clone, for example, then the antigen may be bound to the plate, and the formation of the antibody–antigen complex may be detected using peroxidase coupled to an anti-Ig antibody.

Elixophyllin Proprietary name for elixir (syrup) containing ***theophylline*** as its active ingredient.

Elk proteins (Eph-like kinases) Family of cell surface receptor tyrosine kinases restricted to brain and testis. Not to be confused with ***Elk-1***.

Elk-1 Gene-regulating protein found in lung and testis. Binds to DNA at purine-rich sites. Substrate for ***MAP kinases***; once phosphorylated forms complex with other transcription factors, binds to the serum response element (***SRE***) and induces transcription of ***fos***.

Elk-L Membrane-anchored ligand (38 kD) for ***EPH*** class receptor tyrosine kinases. Becomes tyrosine phosphorylated once bound to receptor (Nuk).

ellipsosome Membrane-bounded compartment containing cytochrome-like pigment and found in the retinal cones of some fish.

elongation factor (EF) Peptidyl transferase components of ribosomes that catalyse formation of the acyl bond between the incoming amino acid residue and the peptide chain. There are three classes of elongation factor: EF-1α (EF-Tu in prokaryotes) binds GTP and aminoacyl-tRNA, delivering it to the A site of ribosomes; EF-1 β (EF-Ts) helps in regeneration of GTP-EF-1 α; EF-2 (EF-G) binds GTP and peptidyl-tRNA and translocates it from the A site to the P site. Diphtheria toxin inhibits protein synthesis in eukaryotes by adding an ADP-ribosyl group to a modified histidine residue (diphthamide) in elongation factor II.

elongation factor G See ***translocase***.

elutriation Separation of particles on the basis of their differential sedimentation rates.

EMA (1) Epithelial membrane antigen; see ***episialin***. (2) E2F-binding site modulating activity: transcriptional repressor (272 residues, 34 kD) that has some similarity with E2F but lacks the activation domain at the carboxy-terminus.

Embden–Meyerhof pathway (glycolysis; Embden–Meyerhof–Parnas pathway) The main pathway for anerobic degradation of carbohydrate. Starch or glycogen is hydrolyzed to glucose-1-phosphate and then through a series of intermediates, yielding two ATP molecules per glucose, and producing either pyruvate (which feeds into the ***tricarboxylic acid cycle***) or lactate.

Embden–Meyerhof–Parnas pathway See ***Embden–Meyerhof pathway***.

embedding Tissue is embedded in wax or plastic in order to prepare sections for microscopical examination. The embedding medium provides mechanical support.

embolus A clot formed by platelets or leucocytes that blocks a blood vessel.

embryo The developmental stages of an animal or, in some cases a plant, during which the developing tissue is effectively isolated from the environment by, eg. egg membranes, foetal membranes and various structures in plants.

embryogenesis The processes leading to the development of an embryo from egg to completion of the embryonic stage.

embryonic induction The induction of differentiation in one tissue as a result of proximity to another tissue arising, for example, during gastrulation. One of the best-known examples is the induction of the neural tube in the ectoderm by the underlying chordo-mesoderm. Although the information to form the tube is present in the competent determined ectoderm, it must be elicited by the inducing tissue. In some cases it is known that cell–cell contact between epithelium and mesenchyme is necessary.

embryonic stem cell (ES cell) Totipotent cell cultured from early embryo. Have the advantage that following modification *in vitro* they can be used to produce chimeric embryos and thus transgenic animals.

emphysema Pulmonary emphysema is associated with chronic bronchitis and may be caused by excessive leucocyte elastase activity in the alveolar walls (possibly as a result of the inactivation of α-1-antiprotease by active oxygen species released by leucocytes in inflammation).

en See ***engrailed***.

enantiomer Either of a pair of stereoisomers of a compound that has chirality.

encephalopathy Any structural degeneration of the brain. See ***spongiform encephalopathy***.

endplate potential Depolarization of the sarcolemma as a result of acetylcholine release from the motoneuron causing an influx of sodium ions. The endplate potential is the sum of quantal ***miniature endplate potentials***. Development of the endplate potential is blocked by curare.

endarteritis Chronic inflammation of the arterial ***intima***, often a late result of syphilis

endergonic An endergonic reaction requires the input of energy.

endocarditis Inflammation of the membrane lining the heart, that over the valves being particularly susceptible. May be caused by viral or bacterial infection, or indirectly as a response to rheumatic fever, scarlet fever or tonsilitis.

endocrine gland Gland that secretes directly into blood and not through a duct. Examples are pituitary, thyroid, parathyroid, adrenal glands, ovary and testis, placenta and B-cells of pancreas.

endocytosis Uptake of material into a cell by the formation of a membrane-bound vesicle.

endocytotic vesicle See ***endocytosis***.

endoderm A germ layer lying remote from the surface of the embryo that gives rise to internal tissues such as gut. Contrast ***mesoderm*** and ***ectoderm***.

endodermis Single layer of cells surrounding the central stele (vascular tissue) in roots. The radial and transverse walls contain the hydrophobic ***Casparian band***, which prevents water flow in or out of the stele through the ***apoplast***. Also present in some stems.

endogenous Product or activity arising in the body or cell, as opposed to agents coming from outside.

endogenous pyrogen Fever-producing substance released by leucocytes (and Kuppfer cells in particular) that acts on the hypothalamic thermoregulatory centre. Now known to be ***interleukin 1***.

endoglin (CD105) Homodimeric glycoprotein (180 kD) with TGF-binding activity, expressed on endothelial cells and pre-B cells.

endoglycosidase Enzyme of the subclass EC 3.2 that has the ability to hydrolyze non-terminal glycosidic bonds in oligosaccharides or polysaccharides. Endoglycosidases F and H are often used as tools to determine the role of carbohydrate moiety on glycoproteins. Endo-F, the product of *Flavobacterium meningosepticum*, cleaves glycans of high mannose and complex type at the link to asparagine in the protein; Endo H is from *Streptomyces* spp. and is an endo-β-N-acetyl-glucosaminidase.

endolyn-78 Glycoprotein (78 kD) present in membranes of endosomes and lysosomes but relatively scarce in other membranes.

endometrium Mucous membrane that lines the uterus and that thickens during the menstrual cycle ready for implantation of the embryo. If implantation does not occur the endometrium returns to its previous state and the excess tissue is shed at menstruation. If implantation does occur the endometrium becomes the decidua and is not shed until after parturition.

endomitosis Chromosome replication without mitosis, leading to polyploidy. Many rounds of endomitosis give rise to the giant polytene chromosomes of Dipteran salivary glands, though in this case the daughter chromosomes remain synapsed.

endomorphins Endogenous peptides (endomorphin-1, YPWF-NH2; endomorphin-2, YPFF-NH2) with high selective affinity for μ-opiate receptor.

endomysium Connective tissue sheath surrounding individual muscle fibres.

endoneurium Connective tissue sheath surrounding individual nerve fibres in a nerve bundle.

endonexin Calcium-dependent membrane-binding protein located on the ***endoplasmic reticulum*** of fibroblasts. Isolated protein will bind to ***liposomes*** if 1–10μM calcium is present but not if the liposomes contain ***sphingomyelin*** or ***cholesterol***. An analogous calcium-dependent membrane-binding protein, ***synexin***, codistributes with endonexin and binds particularly to phosphatidyl serine. Another of the same class is p36, a component of brush-border membrane, a target for the *src* tyrosine kinase, and which binds phosphatidyl serine or phosphatidyl inositol.

endonuclease One of a large group of enzymes that cleave nucleic acids at positions within the chain. Some act on both RNA and DNA (eg. S1 nuclease, EC. 3.1.30.1, that is specific for single-stranded molecules). ***Ribonucleases*** such as pancreatic, T1, etc. are specific for RNA, ***deoxyribonucleases*** for DNA. Bacterial ***restriction endonucleases*** are crucial in recombinant DNA technology for their ability to cleave double-stranded DNA at highly specific sites.

endopeptidase An enzyme that cleaves protein at positions within the chain. Formally, the enzymes are peptidyl-peptide hydrolases, more usually known as ***proteinases*** or ***proteolytic enzymes***.

endoplasm Inner, granule-rich cytoplasm of amoeba.

endoplasmic reticulum (ER) Membrane system that ramifies through the cytoplasm. The membranes of the ER are separated by 50–200nm and the cisternal space thus enclosed constitutes a separate compartment. The Golgi region is composed of flattened sacs of membrane that together with ER and lysosomes constitute the GERL system. See also ***smooth endoplasmic reticulum***, ***rough endoplasmic reticulum***, ***cisterna***.

endoplasmin Most abundant protein in microsomal preparations from mammalian cells (100-fold more concentrated in ER than elsewhere). A glycoprotein (100 kD) with calcium-binding properties. Same as GRP (***glucose-related protein***). A member of the ***hsp90*** family of ***heat-shock proteins***.

endorphins A family of peptide hormones that bind to receptors that mediate the actions of opiates. Released in response to neurotransmitters and rapidly inactived by peptidases. Physiological responses to endorphins include analgesia and sedation.

endosmosis Movement of water into a cell as a result of greater internal osmotic pressure. The ***water potential*** within the vascular sap of a plant cell must be lower than that in the bathing medium or sap of a neighbouring cell.

endosome (1) Endocytotic vesicle derived from the plasma membrane. More specifically, an acidic non-lysosomal compartment in which receptor–ligand complexes dissociate. (2) A chromatinic body near the centre of a vesicular nucleus in some Protozoa.

endosperm Tissue present in the seeds of angiosperms, external to and surrounding the ***embryo***, that it provides with nourishment in the form of ***starch*** or other food reserves. Formed by the division of the ***endosperm mother cell*** after fertilization; may be absorbed by the embryo prior to seed maturation, or may persist in the mature seed.

endosperm mother cell Cell of the higher plant embryo sac. Contains two 'polar nuclei', and fuses with the sperm cell from the pollen grain. Gives rise to the *endosperm*.

endospore (1) An asexual spore formed within a cell. (2) Inner part of the wall of a fungal spore.

endosymbiont hypothesis The hypothesis that semi-autonomous organelles such as mitochondria and chloroplasts were originally endosymbiotic bacteria or cyanobacteria. The arguments are convincing and although the hypothesis cannot be proven it is widely accepted.

endosymbiotic bacteria Bacteria that establish a symbiotic relationship within a eukaryotic cell, eg. the nitrogen-fixing bacteria of legume root nodules. See also ***endosymbiont hypothesis***.

endothelin-converting enzyme (ECE) ECE-1 is an integral membrane protein belonging to the family of metalloproteinases that also includes ECE-2, neprilysin (***endopeptidase*** 24.11) and Kell blood group protein. The catalytic site is in the large extracellular domain and contains a conserved zinc-binding motif.

endothelin receptor There are thought to be two G-protein coupled receptors for endothelin, ET(A) (427 residues) and ET(B) (427 residues in human), present on vascular smooth muscle cells mediating vasoconstriction, and on endothelium mediating ***nitric oxide*** release. ET(A) binds ET-1 preferentially whereas ET(B) binds ET-1, ET-2 and ET-3 with equal affinity.

endothelins (ET-1, ET-2, ET-3) Group of peptide hormones (all 21 residues) released by endothelial cells. All have two disulphide bridges that hold them in a conical spiral shape. They are the most potent vasoconstrictor hormones known. Structurally related to the snake venom ***sarafotoxins***. Pre-pro-endothelin-1 (203 residues) is cleaved to the biologically inactive big endothelin-1 (92 residues) by ***endothelin-converting enzyme*** which will further cleave big endothelin to form active endothelin-1. ET-1, the predominant form, is produced by endothelial cells; ET-2 and ET-3 by various tissues. In addition to their vasoconstrictive properties, endothelins have ***inotropic*** and ***mitogenic*** properties, influence salt and water balance, alter central and peripheral sympathetic activity and stimulate the ***renin–angiotensin***–aldosterone system. Though ET-1 acting through ***endothelin receptor***(A) is vasoconstrictive, it acts through ET(B) to induce the release of ***nitric oxide*** which is a vasodilator.

endothelium Simple ***squamous*** epithelium lining blood vessels, lymphatics and other fluid-filled cavities (such as the anterior chamber of the eye). Mesodermally derived, unlike most epithelia.

endothelium-derived relaxation factor (EDRF) See ***nitric oxide***.

endothermic Process or reaction that absorbs heat and thus requires a source of external energy in order to proceed. The opposite of ***exothermic***.

endotoxin Heat-stable polysaccharide-like toxin bound to a bacterial cell. The term is used more specifically to refer to lipopolysaccharide (LPS) of the outer membrane of ***Gram-negative bacteria***. There are three parts to the molecule, the ***Lipid A*** (six fatty acid chains linked to two glucosamine residues), the core oligosaccharide (branched chain of 10 sugars), and a variable length polysaccharide side chain (up to 40 sugar units in smooth forms) that can be removed without affecting the toxicity (rough LPS). Some endotoxin is probably released into the medium, and endotoxin is responsible for many of the virulent effects of Gram-negative bacteria.

endplate The area of sarcolemma immediately below the synaptic region of the motor neuron in a neuromuscular junction.

Englebreth–Holm–Swarm sarcoma cells See ***EHS cells***.

engrailed *(en) Drosophila* gene that controls segmental polarity. It is the archetype for one of three subfamilies of ***homeobox***-containing genes.

enhancement effect Property of higher plant photosynthesis, discovered by Robert Emerson. The ***quantum yield*** of red light (less than 680nm) and far-red light (700nm), when shone simultaneously on a plant, is greater than the sum of the yields of the light of the two wavelengths separately. This effect provides evidence for the cooperative interaction of two ***photosystems*** in photosynthesis.

enhancer A DNA ***control element*** frequently found 5′ to the start site of a gene, which when bound by a specific transcription factor, enhances the levels of expression of the gene, but is not sufficient alone to cause expression. Distinguished from a ***promoter***, that is alone sufficient to cause expression of the gene when bound; in practice, the two terms merge.

enhancer trap Technique for mapping gene expression patterns, classically in Drosophila. A ***transposon*** element carrying a ***reporter gene*** (usually ***β-galactosidase***), linked to a very weak ***promoter***, is induced to 'jump' within the genome. If the P-element reinserts within the sphere of influence of promoters and ***enhancers*** of some (random) gene, then the reporter gene is also expressed in a similar tissue-specific manner. Usually, many lines of flies carrying such random insertions are studied; if a line shows 'interesting' patterns of expression, it can be possible to clone the gene of interest.

enkephalins Natural ***opiate*** pentapeptides isolated originally from pig brain. Leuenkephalin (YGGFL) and Met-enkephalin (YGGFM) bind particularly strongly to δ-type opiate receptors.

entactin (nidogen) A dumbbell-shaped 150 kD sulphated ***glycoprotein*** that is found in all basement membranes. It binds to ***laminin***, forming a very stable 1:1 complex (K_D = 10nM) and almost all laminin preparations contain entactin. The N-terminal globular domain can self-aggregate, while the C-terminal globular domain binds to the short arm of laminin and to ***collagen*** IV. The connecting rod has 5-6 ***EGF-like domains*** of cysteine-rich repeats, one of which has an ***RGD*** sequence for cellular interaction.

enteric Relating to the intestine.

Enterobacter Genus of enteropathic bacilli of the *Klebsiella* group. Not to be confused with the family ***Enterobacteria*** of which they are members.

Enterobacteriaceae A large family of ***Gram-negative*** bacilli that inhabit the large intestine of mammals. Commonest is *Escherichia coli*; most are harmless commensals but others can cause intestinal disease (*Salmonella, Shigella*).

enterobactin Alternative name for ***enterochelin***.

enterochelin Iron-binding compound (***siderophore***) of *E. coli* and *Salmonella* spp. A cyclic trimer of 2,3-dihydroxybenzoylserine.

enterocytes Cells of the intestinal epithelium.

enterotoxins Group of bacterial ***exotoxins*** produced by enterobacteria and that act on the intestinal mucosa. By perturbing ion and water transport systems they induce diarrhoea. ***Cholera toxin*** is the best-known example.

enterovirus A genus of ***Picornaviridae*** that preferentially replicate in the mammalian intestinal tract. It includes the ***polioviruses*** and ***Coxsackie viruses***.

Entner–Doudoroff pathway Metabolic pathway for degradation of glucose in a wide variety of bacteria. Differs from the Embden–Meyerhoff pathway although the end result is similar.

env Retroviral gene encoding viral envelope glycoproteins.

envelope (1) Lipoprotein outer layer of

some viruses, derived from plasma membrane of the host cell. (2) In bacteriology, the plasma membrane and cell wall complex of a bacterium.

envoplakin Component protein (210 kD) of transglutaminase crosslinked protein layer (cornified envelope) deposited under the plasma membrane of keratinocytes in outer layer of skin. Has sequence homology with ***desmoplakin***, bullous pemphigoid antigen 1, and ***plectin***.

enzyme induction An increase in enzyme secretion in response to an environmental signal. The classic example is the induction of β-galactosidase in *E. coli*.

enzyme-linked immunosorbent assay See ***ELISA***.

eosin Tetrabromofluorescein, a red dye used extensively in histology, for example in the standard H & E (***haematoxylin***–eosin) stain used in routine pathology.

eosinophil Polymorphonuclear leucocyte (granulocyte) of the myeloid series, of which the granules stain red with eosin. Phagocytic, particularly associated with helminth infections and with hypersensitivity.

eosinophil cationic protein Arginine-rich protein (21 kD) in granules of eosinophils, that damages schistosomula ***in vitro***. Not the same as the MBP (major basic protein) of the granules.

eosinophil chemotactic peptide (ECF; ECF of anaphylaxis; ECF-C) Tetrapeptides (of which two are identified: VGSE and AGSE) released by mast cells and that are said to both attract and activate eosinophils.

eosinophilia Condition in which there are unusually large numbers of ***eosinophils*** in the circulation, usually a consequence of helminth parasites or allergy.

eosinophilic (1) Having affinity for the red dye ***eosin***. (2) Inflammatory lesion characterized by large numbers of ***eosinophils***.

eosinophilopoietin Small (1500 D) peptide, possibly released by T-lymphocytes, that regulates ***eosinophil*** development in the bone marrow. Probably interleukin-5.

eotaxin Chemokine with specificity for eosinophils. See Table C4a.

epalons Class of neuroactive steroids. Name derived from epiallopregnanolone, an endogenous metabolite of progesterone that has activity on the GABA-A receptor complex. Have anxiolytic, anticonvulsant and sedative-hypnotic properties.

ependymal cells Cells that line cavities in the central nervous system; considered to be a type of glial cell.

EPH Class of receptor tyrosine kinases (Nuk) implicated in control of axon guidance, regulation of cell migration and defining compartments in embryogenesis. Ligands are generally cell-bound and include Elk-L and Htk-L. Interaction of receptor and ligand lead to tyrosine phosphorylation of both, suggesting that the signalling is bidirectional.

eph kinases Family of protein tyrosine kinases. Named from erythropoietin-producing hepatocellular carcinoma cells in which gene was first identified. Related kinases are elk (EPH-like kinase), eek (eph and elk-like kinase), eck (epithelial cell kinase) and erk (elk-related kinase). If overexpressed *Eph* has tumorigenic effects.

ephedrine ((1R, 2S)-1-phenyl-1-hydroxy-2-methylaminopropane) Alkaloid from plants of genus *Ephedra*. Structural analogue of epinephrine (***adrenaline***) the effects of which it mimics.

Ephestia kuhniella Mediterranean flour moth. Easily maintained in the laboratory.

epi- Prefix indicating something on, above or near. Epi-illumination is from above, epithelia cover (are on top of) other tissues.

epiboly The process in early embryonic development in which a monolayer of dividing cells (blastoderm) spreads over the surface of a large yolk-filled egg (eg. those of teleosts, reptiles and birds).

epicotyl The first shoot of a plant embryo or seedling, above the point of insertion of the cotyledon(s). Can be relatively long in some seedlings showing ***etiolation***.

epidermal cell (1) Cell of epidermis in animals. (2) Plant cell on the surface of a leaf or other young plant tissue, where bark is absent. The exposed surface is covered with a layer of ***cutin***.

epidermal growth factor (EGF) A mitogenic polypeptide (6 kD) initially isolated from male mouse submaxillary gland. The name refers to the early bioassay, but EGF is active on a variety of cell types, especially but not exclusively epithelial. A family of similar growth factors is now recognized. Human equivalent originally named ***urogastrone*** owing to its hormone activity.

epidermis Outer epithelial layer of a plant or animal. May be a single layer that produces an extracellular material (as for example the cuticle of arthropods), or a complex stratified squamous epithelium, as in the case of many vertebrate species.

epididymis Convoluted tubule connecting the vas efferens, that comes from the seminiferous tubules of the mammalian testis, to the vas deferens. Maturation and storage of sperm occur in the epididymis.

epifluorescence Method of fluorescence microscopy in which the excitatory light is transmitted through the objective onto the specimen rather than through the specimen; only reflected excitatory light needs to be filtered out rather than transmitted light which would be of much higher intensity.

epigenesis The theory that development is a process of gradual increase in complexity as opposed to the preformationist view that supposed that mere increase in size was sufficient to produce adult from embryo.

epigenetics The study of mechanisms involved in the production of phenotypic complexity in morphogenesis. According to the epigenetic view of differentiation, the cell makes a series of choices (some of which may have no obvious phenotypic expression, and are spoken of as ***determination*** events) that lead to the eventual differentiated state. Thus, selective gene repression or derepression at an early stage in differentiation will have a wide-ranging consequence in restricting the possible fate of the cell.

epiglycanin Very extensively glycosylated transmembrane glycoprotein found in TA3 Ha mouse mammary carcinoma cells and which may mask histocompatibility antigens. Functionally analogous to ***episialin*** but there is no sequence homology in the protein.

epiligrin Major glycoprotein of epidermal basement membrane, consisting of three disulphide-bonded subunits of 170, 145 and 135 kD. Epiligrin is the major ligand for α_3/β_1 integrin, is particularly prominent in the lamina lucida of the skin and is absent in patients with lethal junctional epidermolysis bullosa.

epimer Diastereomeric monosaccharides that have opposite configurations of a hydroxyl group at only one position, eg. D-glucose and D-mannose.

epimorphosis Pattern of regeneration in which proliferation precedes the development of a new part. Opposite of ***morphallaxis***.

epinasty Asymmetrical growth of a leaf or stem that causes curvature of the structure.

epinemin Intermediate filament-associated protein (44.5 kD monomer) associated with ***vimentin*** in non-neural cells.

epinephrine See ***adrenaline***.

episialin (polymorphic epithelial mucin, PEM; epithelial membrane antigen, EMA) Heavily glycosylated membrane glycoprotein. Encoded by the *MUC*-1 gene; has a molecular weight of around 300 kD, more than half of which is *O*-linked glycan. There is a 69 residue cytoplasmic domain and the extracellular domain may extend hundreds of nanometres beyond the plasma membrane; the increased expression in carcinoma cells may reduce the adhesion and mask antigenic properties of the cells. Similar functions are ascribed to ***ASGP***, ***epiglycanin*** and ***leukosialin***.

episome Piece of hereditary material that can exist as free, autonomously replicating DNA or be attached to and integrated into the chromosome of the cell, in which case it replicates along with the chromosome. Examples of episomes are many ***bacteriophages*** such as lambda and the male sex factor of ***Escherichia coli***.

epistasis Non-reciprocal interaction of non-allelic genes, eg. when the expression of one gene masks the expression of another. Thus a gene that blocks development of an organ will mask the effects of genes that would modify the form of that organ had it been developed.

epitectin Mucin-like glycoprotein found on surface of human tumour cells (also known as CA antigen) but not non-tumorigenic cell lines. It is present on the surface of some specialized cells (sweat glands, type II pneumocytes from lung, bladder epithelium) and may therefore be a normal ***differentiation antigen***. Also present in normal urine.

epithelial membrane antigen (EMA) See ***episialin***.

epithelioid cells In a general sense, a cell that has an appearance that is similar to that of epithelial cells: used specifically of the very flattened macrophages found in granulomas (eg. in tubercular lesions).

epithelium One of the simplest types of tissues. A sheet of cells, one or several layers thick, organized above a basal lamina (see ***basement membrane***), and often specialized for mechanical protection or ***active transport***. Examples include skin, and the lining of lungs, gut and blood vessels.

epitope See ***antigenic determinant***.

epitope library Large collection (hundreds of millions) of peptides each encoded by a randomly mutated piece of DNA in a phage genome and expressed on the surface of that bacteriophage, sometimes as an N-terminal extension of a coat protein. Particular phages can be selected by a binding assay and since the peptide has its encoding DNA associated with it sequencing is straightforward.

epitope tag Short peptide sequence that constitutes an ***epitope*** for an existing antibody. Widely used in molecular biology to 'tag' transgenic proteins (as a translational fusion product) to follow their expression and fate by immunocytochemistry or Western blotting, but without having to raise antibodies against the specific protein. Example: ***myc tag***. See also ***flag tagging***.

epizootic Veterinary equivalent of an epidemic.

EPO See ***erythropoietin***.

epothilones Compounds (epothilone A and B) isolated from myxobacterium *Sorangium cellulosum* Str 90. Cytotoxic to tumour cells as a result of inducing microtubule assembly and stabilization.

Epstein–Barr virus (EBV) Species of Herpetoviridae, that binds ***CR2*** and which causes infective mononucleosis and, in the presence of other factors, tumours such as Burkitt's lymphoma and nasopharyngeal carcinoma.

equatorial plate Region of the mitotic spindle where chromosomes are aligned at metaphase: as its name suggests, it lies midway between the poles of the spindle.

equilibrium constant (equilibrium dissociation constant; dissociation constant) The ratio of the reverse and forward rate constants for a reaction of the type A + B = AB. At equilibrium the equilibrium constant (K) equals the product of the concentrations of reactants divided by the concentration of product, and has dimensions of concentration. K = (concentration A. concentration B)/ (concentration AB). The affinity (association) constant is the reciprocal of the equilibrium constant.

equilibrium dialysis Technique used to measure the binding of a small molecule ligand to a larger binding partner. The macromolecule is contained within a dialysis chamber and the diffusible ligand added to the exterior: once equilibrium is reached an excess of ligand inside the dialysis chamber is evidence of binding and it is possible to calculate the binding affinity from a measurement of the concentrations of ligand and that of the binding macromolecule.

equilibrium dissociation constant See ***equilibrium constant***.

equinatoxins Small peptide toxins (19 kD) from *Actinia equina*. Form cation-selective pores and are cytolytic.

equivalence The situation where two interacting molecular species are present in concentrations just sufficient to produce occupation of all binding sites. Only used to describe high-avidity interactions, especially the antibody–antigen interaction.

ER See ***endoplasmic reticulum***.

ERAB (endoplasmic reticulum-associated binding protein) Protein (262 residues, 27 kD) in the ER that binds amyloid β (A β). Found ubiquitously but more extensively in liver, heart and brain. Overexpressed in brain of patients with ***Alzheimer's disease*** and it may be the complex between A β and ERAB that is cytotoxic. Sequence has similarities with short-chain alcohol dehydrogenases, hydroxysteroid dehydrogenases and acetoacyl CoA reductases.

erabutotoxins Curaremimetic polypeptide toxins (62 residues) from venom of *Laticauda semifasciata*. Bind to nicotinic acetylcholine receptors.

erb Two oncogenes, *erb* A and *erb* B, associated with erythroblastosis virus (an acute transforming retrovirus). The cellular homologue of *erb* B is the structural gene for the cell-surface receptor for epidermal growth factor, and of *erb A* a steroid hormone receptor.

ergastic substances Metabolically inert products of photosynthesis, such as starch grains and fat globules.

ergodic System or process in which the final state is independent of the initial state.

ERKs See ***MAP kinases***.

Erlich ascites cell Commonly used carcinoma ***cell line*** derived from ***ascites*** fluid.

error-prone repair See ***SOS system***.

erucic acid ((Z)-docos-13-enoic acid) Trivial name for 22: 1 fatty acid. Found in rapeseed (canola) oil.

Erwinia chrysanthemi Phytopathogenic bacterium that causes soft-rot. Virulence factors include pectinases coded by *pelB, pelC, pelD, pelE, ogl, kduI* and *kdgT* that degrade the cell walls of the plant being attacked.

Eryf1 See ***erythroid transcription factor***.

erysipelas A spreading infection of the dermis possibly associated with an allergic reaction to products of the causative organism, *Streptococcus pyogenes*.

erythema nodosum Eruption of pink or red nodules, usually on the lower limbs as a result of infection with any of a range of bacteria, viruses or fungi. Often associated with ***IBD*** and in some parts of the world is commonly associated with lepromatous leprosy.

erythroblast General name for a nucleated cell of the bone marrow that gives rise to erythrocytes. See also ***normoblasts, BFU-E, CFU-E, primitive erythroblasts*** and ***definitive erythroblasts***.

erythroblastosis fetalis Severe haemolytic disease of the neonate as a result of transplacental passage of maternal antibodies mainly directed against rhesus blood group antigens.

erythrocyte A red blood cell.

erythrocyte ghost The membrane and cytoskeletal elements of the erythrocyte devoid of cytoplasmic contents, but preserving the original morphology.

erythrogenic toxin Toxin produced by strains of *Streptococcus pyogenes* responsible for

Table E1. Erythrocyte membrane proteins

The mammalian erythrocyte ghost consists of a lipid bilayer linked to a cytoskeletal network. The proteins of the ghost vary across species, but there are some common patterns. Components are identified as far as possible by comparison with the proteins of the human erythrocyte ghost, after electrophoretic separation on SDS polyacrylamide gel, and numbered according to the Steck classification (*Journal of Cell Biology* 1974, 62: 1–29).

Band number after Steck.[a]	MW (kD)	Other name or function
1	240	Spectrin α
2	220	Spectrin β
2.1	200	Ankyrin. Links band 3 to spectrin
3	93	Anion transporter
4.1	82	Links spectrin to glycophorin
4.2	76	
4.5	46	Glucose transporter
4.9	48	
5	43	Actin; forms short oligomers. involved in gelation of spectrin and band 4.1
6	35	Glyceraldehyde 3-phosphate dehydrogenase
7	28	

[a]These bands are visible when the gel is stained with a typical 'protein' dye, eg. Coomassie brilliant blue. Other bands are only detected when stained for carbohydrate with the periodic acid–Schiff reagent (PAS). Four bands are characterized: PAS1, PAS2, PAS3 and PAS4. Of these PAS1 and PAS2 are the glycoprotein glycophorin (55kD) in different oligomeric states. PAS3 and PAS4 are minor components.

scarlet fever. Three antigenic variants of the toxin are known. It is a small protein that is complexed with hyaluronic acid and can intensify the effects of other toxins such as ***endotoxin*** and ***streptolysin O***.

erythroid cell Cell that will give rise to erythrocytes.

erythroid Kruppel-like factor (EKLF) Red cell-specific transcriptional activator essential for establishing high levels of adult β-globin expression.

erythroid transcription factor (Eryf1; GF-1; NF-E1) ***Transcription factor*** that binds to regulatory regions of genes expressed in erythroid cells.

erythroleukaemic cell Abnormal precursor (virally transformed) of mouse erythrocytes that can be grown in culture and induced to differentiate by treatment with, eg. DMSO. See ***Friend murine erythroleukaemia cells***.

erythromycin General name for a variety of wide-spectrum macrolide antibiotics isolated from *Streptomyces erythreus*. Inhibit protein synthesis by binding to the prokaryotic 50S ribosomal subunit and preventing translocation. A variety of proteins will confer resistance to erythromycin, either by degrading the antibiotic, by enhancing its export from the cell, or by causing modification to the RNA so that its affinity for erythromycin is reduced.

erythrophores ***Chromatophores*** that have red pigment.

erythropoiesis Process of production of erythrocytes in the marrow in adult mammals. A pluripotent stem cell (CFU) produces, by a series of divisions, committed stem cells (***BFU-Es***) that give rise to ***CFU-Es***, cells that will divide only a few more times to produce mature erythrocytes. Each stem cell product can give rise to 2^{11} mature red cells.

erythropoietin Glycoprotein (46 kD) produced in the kidney and that regulates the production of red blood cells in the marrow. Higher concentrations are required to stimulate ***BFU-Es*** than ***CFU-Es*** to produce erythrocytes. Recombinant EPO is now being used therapeutically in patients.

ES cells See *embryonic stem cells*.

Escherichia coli The archetypal bacterium for biochemists, used very extensively in experimental work. A rod-shaped ***Gram-negative*** bacillus (0.5 × 3–5μm) abundant in the large intestine (colon) of mammals. Normally non-pathogenic, but the *E. coli* O157 strain, common in the intestines of cattle, has recently caused a number of deaths.

Escherichia coli* hemolysin** (α-hemolysin; HlyA) Exotoxin of the ***RTX family of bacterial cytolysins. Synthesized as an inactive 110 kD precursor that is activated by fatty acid acylation by accessory protein HlyC. Product of many *E. coli* strains responsible for non-intestinal infections.

essential amino acids Those amino acids that cannot be synthesized by an organism and must therefore be present in the diet. The term is often applied anthropocentrically to those amino acids required by humans (Ileu, Leu, Lys, Met, Phe, Thr, Try and Val), though rats need two more (Arg and His).

essential fatty acids The three fatty acids required for growth in mammals, arachidonic, linolenic and linoleic acids. Only linoleic acid needs to be supplied in the diet; the other two can be made from it.

established cell line See *cell line*.

esterase An enzyme that catalyses the hydrolysis of organic esters to release an alcohol or thiol and acid. The term could be applied to enzymes that hydrolyze carboxylate, phosphate and sulphate esters, but is more often restricted to the first class of substrate.

estradiol See *oestradiol*.

ET See *endothelins*.

ET-1, ET-2, ET-3 See *endothelins*.

ethidium bromide A dye that intercalates into DNA and to some extent RNA. Intercalation into linear DNA is easier than into circular DNA and the addition of ethidium bromide to DNA prior to ultracentrifugation on a caesium chloride gradient is much used to separate nuclear and mitochondrial or plasmid DNA for analytical purposes. Because less intercalates into the circular DNA, the density remains higher.

ethylene Plant growth substance (phytohormone, plant hormone), involved in promoting growth, ***epinasty***, fruit ripening, senescence and breaking of dormancy. Its action is closely linked with that of ***auxin***.

ethylenediaminetetraacetic acid See *EDTA*.

ethyleneglycol-bis(2-aminoethyl) N, N, N′, N′,-tetraacetic acid) See *EGTA*.

etiolation Growth habit adopted by germinating seedlings in the dark. Involves rapid extension of shoot and/or hypocotyl and suppression of chlorophyll formation and leaf growth.

etioplast Form of ***plastid*** present in plants grown in the dark. Lacks chlorophyll, but contains chlorophyll precursors and can develop into a functional chloroplast in the light.

etoposide Lignan derivative synthesized from ***podophyllotoxin***. Used as an antitumour drug and probably works by inhibiting topoisomerase II.

ets An ***oncogene*** found in E26 transforming ***retrovirus*** of chickens. Encodes a nuclear protein that regulates the initiation of transcription from a range of cellular and viral promoter and enhancer elements. There is some interaction between ets protein and ***AP-1***. See Table O1.

ETS domain DNA binding domain, formed of 3 ***alpha-helices***. Named after the DNA-binding domain of the human ETS-1 ***transcription factor***.

Eubacteria A major subdivision of the prokaryotes (includes all except ***Archaebacteria***). Most ***Gram-positive bacteria***, cyanobacteria, mycoplasmas, enterobacteria, pseudomonads and chloroplasts are Eubacteria. The cytoplasmic membrane contains ester-linked lipids, there is ***peptidoglycan*** in the cell wall (if present) and no ***introns*** have been discovered.

Eucaryote See ***Eukaryote***.

euchromatin The chromosomal regions that are diffuse during interphase and condensed at the time of nuclear division. They show what is considered to be the normal pattern of staining (eu = true) as opposed to ***heterochromatin***.

Eudorina Simple multicellular alga of the order Volvocida, often quoted as illustrating the path to multicellularity. Small spherical or ovoid colonies of between 4 and 64 flagellated cells coexist within a gelatinous envelope. *Pandorina* and *Volvox* are similar though more complex.

Euglena *Euglena gracilis* and *E. viridis* are phytoflagellate Protozoa of the algal order Euglenophyta (zoological order Euglenida). An elongate cell with two ***flagella***, one emerging from a pocket at the anterior end, the organism exhibits positive ***phototaxis***, determined by a photoreceptive spot on the basal part of the flagellum shaft being shielded by a carotenoid-containing stigma ('eyespot') in the wall of the pocket.

Eukaryote Organism whose cells have (1) chromosomes with nucleosomal structure and separated from the cytoplasm by a two-membrane nuclear envelope. (2) Compartmentalization of a function in distinct cytoplasmic organelles. Contrast ***Prokaryotes*** (bacteria and cyanobacteria).

eumelanin Form of ***melanin*** found in animals, particularly in skin and hair and in pigmented retinal epithelium of the eye.

Eumycetozoida Order of Protozoa, includes true slime moulds (not the cellular slime moulds).

Eumycota Division of fungi having defined cell walls and forming hyphae. The other main group are the Myxomycota.

euploidy Polyploidy in which the chromosome number is an integer multiple of the starting number.

Euplotes Genus of free-living hypotrich Protozoa. Do not have cilia but may have undulating membranes for propulsion.

eve See ***even-skipped***.

even-skipped *(eve)* A ***pair-rule gene*** of *Drosophila*.

evi-1 Putative ***oncogene*** from mouse myeloid leukaemias.

Ewing's sarcoma Sarcoma that develops in bone marrow.

excision repair Mechanism for the repair of environmental damage to one strand of ***DNA*** (loss of ***purines*** due to thermal fluctuations, formation of pyrimidine dimers by UV irradiation). The site of damage is recognized, excised by an ***endonuclease***, the correct sequence is copied from the complementary strand by a ***polymerase*** and the ends of this correct sequence are joined to the rest of the strand by a ***ligase***. The term is sometimes restricted to bacterial systems where the polymerase also acts as endonuclease.

excitable cell A cell in which the membrane response to ***depolarizations*** is non-linear, causing amplification and propagation of the depolarisation (an ***'action potential'***). Apart from neurons and muscle cells, electrical excitability can be observed in fertilized eggs, some plants and glandular tissue. Excitable cells contain ***voltage-gated ion channels***.

excitation–contraction coupling Name given to the chain of processes coupling excitation of a muscle by the arrival of a nervous impulse at the ***motor endplate*** to the contraction of the filaments of the ***sarcomere***. The crucial link is the release of calcium from the sarcoplasmic reticulum, and the analogy is often drawn between this and stimulus–secretion coupling, which also involves calcium release into the cytoplasm.

excitatory amino acid (EAA) The naturally occurring amino acids L-glutamate and L-aspartate and their synthetic analogues, notably ***kainate***, ***quisqualate***, and ***NMDA***. They have the properties of excitatory neurotransmitters in the CNS, may be involved in long-term potentiation, and can act as ***excitotoxins***. At least three classes of EAA receptor have been identified: the agonists of the N-type receptor are L-aspartate, NMDA, and ibotenate; the agonists of the Q-type receptor are L-glutamate and quisqualate; agonists of the K-type are L-glutamate and kainate. All three receptor types are found widely in the CNS, and particularly the telencephalon; N- and Q-type receptors tend to occur together, and may interact; their distribution is complementary to the K-type receptors. The ion fluxes through the Q and K receptors are relatively brief, whereas the flux through the N-type is longer, and carries a significant amount of calcium. Additionally the N-type receptor is blockaded by magnesium near the resting potential, and thus shows ***voltage-gated ion channel*** properties, leading to a regenerative response; this is why N-type receptors have been linked to long-term potentiation. Invertebrate glutamate receptors may not have the same properties as those described above.

excitatory synapse A synapse (either ***chemical*** or ***electrical***) in which an action potential in the presynaptic cell increases the probability of an action potential occurring in the postsynaptic cell. See ***inhibitory synapse***.

excitotoxin Class of substances that damage neurons through paroxysmal overactivity. The best-known excitotoxins are the ***excitatory amino acids***, which can produce lesions in the CNS similar to those of ***Huntington's chorea*** or ***Alzheimer's disease***. Excitotoxicity is thought to contribute to neuronal cell death associated with stroke.

exendin Group of peptide hormones, related to the ***glucagon*** family, found in the Gila monster. Helospectin is exendin-1; helodermin is exendin-2.

exfoliatin Epidermolytic toxin produced by some strains of *Staphylococcus aureus*; causes detachment of outer layer of skin by disrupting desmosomes of the stratum granulosum.

exine External part of pollen wall that is often elaborately sculptured in a fashion characteristic of the plant species. Contains ***sporopollenin***. The term is also used for the outer part of a spore wall.

exocrine Exocrine glands release their secreted products into ducts that open onto epithelial surfaces. See ***endocrine***.

exocytosis Release of material from the cell by fusion of a membrane-bounded vesicle with the plasma membrane.

exocytotic vesicle Vesicle, for example a secretory vesicle or ***zymogen granule***, that can fuse with the plasma membrane to release its contents.

exoenzyme (1) An enzyme attached to the outer surface of a cell (an ectoenzyme) or released from the cell into the extracellular space. (2) An enzyme that only cleaves the terminal residue from a polymer (in contrast to an endoenzyme).

exon The sequences of the RNA ***primary transcript*** (or the DNA that encodes them) that exit the nucleus as part of a ***messenger RNA*** molecule. In the primary transcript neighbouring exons are separated by introns.

exon shuffling Process by which the evolution of proteins with multifunctional domains could be accelerated. If exons each encoded individual functional domains, then ***introns*** would allow their recombination to form new functional proteins with minimal risk of damage to the sequences encoding the functional parts.

exon trapping Technique for identifying regions of a genomic DNA fragment that are part of an expressed gene. The genomic sequence is cloned into an intron, flanked by two exons, in a specialized exon trapping vector, and the construct expressed through a strong promoter. If the genomic fragment contains an exon, it will be spliced into the resulting mRNA, changing its size and allowing its detection.

exonuclease Enzyme that digests the ends of a piece of DNA (cf. ***endonuclease***). The nature of the digestion is usually specified (eg. 5′ or 3′ exonuclease).

exonuclease III (Exo III) Enzyme that degrades DNA from one end. Used to prepare deletions in cloned DNA, or for ***DNA footprinting***.

exopeptidase Peptide hydrolases of the class EC3.4 that cleave the N- or C-terminal amino acid from a peptide.

exothermic Process or reaction in which heat is produced. The opposite of ***endothermic***.

exotoxins Toxins released from ***Gram-positive*** and ***Gram-negative bacteria*** – as opposed to ***endotoxins*** that form part of the cell wall. Examples are ***cholera, pertussis*** and ***diphtheria toxins***. Usually specific and highly toxic. See Table E2.

experimental allergic encephalomyelitis (EAE) An autoimmune disease that can be induced in various experimental animals by the injection of homogenized brain or spinal cord in ***Freund's adjuvant***. The antigen appears to be a basic protein present in myelin, and the response is characterized by focal areas of lymphocyte and macrophage infiltration into the brain, associated with demyelination and destruction of the ***blood–brain barrier***. Sometimes used as a model for demyelinating diseases, although whether this is entirely justifiable is not clear.

expressed sequence tag (EST) DNA sequence derived by sequencing an end of a random cDNA clone from a library of interest. Usually, tens of thousands of such ESTs are generated as part of ***genome projects***. These ESTs provide a rapid way of identifying cDNAs of interest, based on their sequence 'tag'; they can then be purchased cheaply, obviating the need to screen a library.

Table E2. Exotoxins

Name	Source	Target/mode of action
α toxin	*Clostridium perfringens*	Phospholipase C
Anthrax toxin	*Bacillus anthracis*	three components, one a soluble adenyl cyclase
Botulinum toxins	*Clostridium botulinum*	Inhibits acetylcholine release
Cholera toxin	*Vibrio cholerae*	ADP-ribosylation of G_s
Diphtheria toxin	*Corynebacterium diphtheriae*	ADP-ribosylation of EF-2
δ-toxin	*Clostridium perfringens*	Binds to cholesterol
Enterotoxins	*Staphyloccus aureus*	Neurotoxic
	Pseudomonas aeruginosa	Causes diarrhoea
Erythrogenic toxin	*Streptococcus pyogenes*	Skin hypersensitivity
Exfoliatin	*Staphyloccus aureus*	Disrupts desmosomes
Haemolysin	*Staphyloccus aureus*	
α-haemolysin		Unknown mechanism
β-haemolysin		Acts as sphingomyelinase C
γ-haemolysin		Haemolytic
δ-haemolysin		Surfactant, haemolytic
Haemolysin	*Pseudomonas aeruginosa*	Toxic for macrophages
Heat-labile toxin	*Bordetella pertussis*	Dermonecrotic
Heat-labile toxin	*Escherichia coli*	Similar to cholera toxin
Heat-stable enterotoxin	*Escherichia coli*	Analogue of guanylin
Kanagawa haemolysin	*Vibrio haemolytica*	Haemolytic, cardiotoxic
Leucocidin	*Staphyloccus aureus* and *Pseudomonas aeruginosa*	Lyses neutrophils and macrophages
Pertussis toxin	*Bordetella pertussis*	ADP-ribosylates G_i
Pneumolysin	*Streptococcus pneumoniae*	Binds to cholesterol
Stable toxin	*Escherichia coli*	Activates guanylate cyclase
Streptolysin D	*Streptococcus pyogenes*	Binds cholesterol
Streptolysin S	*Streptococcus pyogenes*	Membranolytic
Subtilysin	*Bacillus subtilis*	Haemolytic surfactant
Tetanolysin	*Clostridium tetani*	Binds cholesterol
Tetanus toxin	*Clostridium tetani*	Inhibits glycine release at synapse
Toxin A	*Pseudomonas aeruginosa*	ADP-ribosylates EF-2

G_s, G_i: see GTP-binding proteins.
EF-2: elongation factor 2.

expression cloning Method of ***gene cloning*** based on ***transfection*** of a large number of cells with ***cDNAs*** in an ***expression vector*** (eg. a cDNA library), then screening for a functional property (eg. binding of a radiolabelled hormone to identify receptors, or induction of transforming activity for putative oncogenes).

expression vector A ***vector*** that results in the ***expression*** of inserted DNA sequences when propagated in a suitable host cell, ie. the protein coded for by the DNA is synthesized by the host's system.

extensin Glycoprotein of the plant cell wall, characterized by its high hydroxyproline content. Carbohydrate side chains are composed of simple galactose residues and oligosaccharides containing 1-4 arabinose residues. Part of a larger class of ***hydroxyproline-rich glycoproteins***. Function uncertain.

extinction coefficient Outmoded term for absorption coefficient.

extracellular matrix (ecm; ECM) Any material produced by cells and secreted into the surrounding medium, but usually applied to the non-cellular portion of animal tissues. The ecm of connective tissue is particularly extensive and the properties of the ecm determine the properties of the tissue. In broad terms there are three major components: fibrous elements (particularly ***collagen***, ***elastin***, or ***reticulin***), link proteins (eg. ***fibronectin***, ***laminin***), and space-filling molecules (usually ***glycosaminoglycans***). The matrix may be mineralized to resist compression (as in bone) or dominated by tension-resisting fibres (as in tendon). The basal lamina of epithelial cells is another commonly encountered ecm. Although ecm is produced by cells, it has recently become clear that the ecm can influence the behaviour of cells quite markedly, an important factor to consider when growing cells ***in vitro***: removing cells from their normal environment can have far-reaching effects.

extrachromosomal element Any heritable element not associated with the chromosome(s). It is usually a ***plasmid*** or the DNA of organelles such as mitochondria and chloroplasts.

extremophile An organism that requires an extreme environment in which to flourish, eg. ***thermophiles*** and ***halophiles***.

extrinsic pathway Initiation of blood clotting as a result of factors released from damaged tissue, as opposed to contact with a foreign surface (the intrinsic pathway). Tissue thromboplastin (factor III) in conjunction with factor VII (proconvertin) will activate factor X which, once activated, converts prothrombin to thrombin.

exudate cells Leucocytes that enter tissues (exude from the blood vessels) during an inflammatory response. See also ***peritoneal exudate***.

ezrin Microfilament bundling protein (80 kD) from the core of microvilli. Phosphorylated following stimulation of cells.

F

F See ***phenylalanine***.

f X-174 See ***phi X-174***.

F-actin Filamentous ***actin***.

F-box Motif found in a number of eukaryotic regulatory proteins. Responsible in some cases for ubiquitin-mediated proteolysis.

F-box proteins Adapter proteins that are involved in associating proteins with the ubiquitin-driven proteolytic system. The F-box is a motif originally identified within *Neurospora crassa* negative regulator sulphur controller-2 but subsequently found in a wide variety of proteins including many cell cycle regulatory proteins, though various F-box proteins probably also play a part in regulation of transcription, signal transduction and development. See ***SCF complexes***.

F-factor ***Plasmid*** that confers the ability to conjugate (ie. fertility) on bacterial cells, and carries the *tra* genes; first described in *E. coli*.

f-met-leu-phe (formyl-methionyl-leucyl-phenylalanine; fMLP) See ***formyl peptides***.

F-spondin The F-spondin genes are a family of extracellular matrix molecules united by two conserved domains, FS1 and FS2, at the amino terminus plus a variable number of thrombospondin repeats at the carboxy-terminus. Currently, characterized members include a single gene in *Drosophila* and multiple genes in vertebrates. The vertebrate genes are expressed in the midline of the developing embryo, primarily in the floor plate of the neural tube.

F-type ATPase (F_1F_o ATPase; ATP synthase; F-ATPase) Multi-subunit proton-transporting ATPase, related to the ***V-type ATPase***. Found in the inner membrane of ***mitochondria*** and ***chloroplasts***, and in bacterial ***plasma membranes***. Normally driven in reverse by ***chemiosmosis*** to make ***ATP***, and so also known as ***ATP synthase***.

F1 See ***neuromodulin***.

F1 hybrid First filial generation – product of crossing two dissimilar parents. If the parents are sufficiently dissimilar the hybrid may be sterile (eg. in the crossing of horse and donkey to produce a mule) and the term F1 hybrid generally refers to such sterile hybrids which may, however, show desirable hybrid vigour.

F11 Neural cell recognition molecule with ***immunoglobulin*** type C domains and ***fibronectin*** type III repeats. Its cDNA sequence is almost identical to ***contactin*** except that while F11 is probably attached to the membrane via phosphatidyl inositol and lacks a cytoplasmic domain, contactin is attached to the cytoskeleton. Like contactin, ***neurofascin***, ***TAG-1*** and ***fasciclin*** II, F11 is thought to be associated with the process of ***fasciculation***.

Fab Fragment of immunoglobulin prepared by papain treatment. Fab fragments (45 kD) consist of one light chain linked through a disulphide bond to a portion of the heavy chain, and contain one antigen binding site. They can be considered as univalent antibodies.

Fab(2) The fragment (90 kD) of an immunoglobulin produced by pepsin treatment. These fragments have two antigen combining sites and contain two light chains and two variable region heavy chains plus one constant region domain in each heavy chain. The fragment is divalent but lacks the complement-fixing (Fc) domain.

FABP See ***fatty acid-binding proteins***.

Fabry disease (angiokeratoma) ***Storage disease*** due to deficiency of ceramide trihexosidase.

facilitated diffusion (passive transport) A process by which substances are conveyed across cell membranes faster than would be possible by diffusion alone. This is generally achieved by proteins that provide a hydrophilic environment for polar molecules throughout their passage through the ***plasma membrane***, acting as either shuttles or pores. See ***symport, antiport, uniport***.

facilitation Greater effectiveness of synaptic transmission by successive presynaptic impulses, usually due to increased transmitter release.

facilitator neuron A neuron whose firing enhances the effect of a second neuron on a third. This allows the effects of neuronal activity to be modulated.

FACS See ***fluorescence-activated cell sorter***.

Factor P See ***properdin***.

factor VIII Blot-clotting factor defective or missing in most cases of ***haemophilia***.

Factors I–XII Blood clotting factors, especially from humans. These factors form a cascade in which the activation of the first factor leads to enzymic attack on the next factor and so on, finally resulting in blood clotting. See Table F1.

facultative heterochromatin That ***heterochromatin*** which is condensed in some cells and not in others, presumably representing stable differences in the activity of genes in different cells. The best-known example results from the random inactivation of one of the pair of X chromosomes in the cells of female mammals (***Lyonization***).

FAD (flavin adenine dinucleotide) A prosthetic group of many flavin enzymes. See ***flavin nucleotides***.

FADD Adaptor protein that links ***death receptors*** to ***caspases*** in the signalling pathway that leads to apoptotic cell death.

FAK See ***focal adhesion kinase***.

Falconization Trade name for the treatment of polystyrene to make it appropriate for use in cell culture. The main commercial process is probably corona discharge in air or other gas mixtures at low pressure. Treatment of polystyrene with sulphuric acid will produce the same effect.

familial hypercholesterolemia Excess of ***cholesterol*** in plasma as a result of defects in the recycling process that leads to reduced uptake of LDL (***low density lipoprotein***) into ***coated vesicles***.

famotidine Drug that blocks histamine H2 receptors and is used for treatment of gastric ulcers. A Yamanouchi drug.

Fanconi syndrome Transport disease (recessive defect) in which the renal reabsorption of several substances (phosphate, glucose, amino acids) is impaired.

Table F1. Blood clotting factors

Factor	Name	M_r (kD)	Function
	Fibrinogen	340	Cleaved to form fibrin
II	Prothrombin	70	Converted to thrombin by Factor X
III	Thromboplastin	—	Lipoprotein which acts with VII to activate X
IV	Calcium ions	—	Needed at various stages
V	Proaccelerin	—	Product accelerin, promotes thrombin production
VII	Proconvertin	—	Activated by trauma to tissue
VIII	Antihaemophilic factor	$>10^3$	Acts with IXa to activate X
IX	Christmas factor	55	See VIII
X	Stuart factor	55	When activated converts II to thrombin
XI	Thromboplastin antecedent	124	Converts IX to active form
XII	Hagemann factor	76	Activated by surface contact
XIII	Fibrin-stabilizing factor	350	Transglutaminase which crosslinks fibrin

Fanconi's anaemia Defect in thymine-dimer excision from DNA predisposing to development of leukaemia.

far Western blot Form of ***Western blot*** in which protein–protein interactions are studied. Proteins are run on a gel and transferred to a membrane as in a normal Western blot. The proteins are then allowed to renature, incubated with a candidate protein, and the blot washed. Areas of the blot where the protein has adhered are then detected with an antibody.

FAR1 Yeast gene, induced by α factor, that causes cells to arrest in G1 phase, by interacting with the G1 cyclin, CLN2.

Farber's disease Lipogranulomatosis caused by deficiency of ***ceramide*** degrading enzymes – a storage disease.

Farmer's lung Type III ***hypersensitivity*** response to *Micropolyspora faeni*, a thermophilic bacterium found in mouldy hay. (Conveniently afflicts Joe Grundy in BBC Radio 4's *The Archers*.)

farnesyl transferase Enzyme (EC 2.5.1.21) that adds a farnesyl group to certain intracellular proteins. See ***farnesylation***.

farnesylation The farnesyl group is the linear grouping of three isoprene units. It is specifically attached to proteins that contain the C-terminal motif CAAX by cleavage and addition to the SH group of C; the free carboxylate group is also methylated. Believed to act as a membrane attachment device. See also ***polyisoprenylation***.

Farr-type assay Method of radioimmunoassay in which free antigen remains soluble and antibody–antigen complexes are precipitated.

Fas antigen Cell surface transmembrane protein (35 kD) that mediates ***apoptosis***. Has structural homology with ***TNF***-receptor and ***NGF***-receptor. May play a part in negative selection of autoreactive T-cells in the thymus.

Fas ligand Ligand for the ***Fas antigen*** which is actually a receptor of the TNF receptor family.

fascicle Literally, a bundle. In particular, this is used to describe the tendency of ***neurites*** to grow together (fasciculate).

fasciclins ***Cell adhesion*** molecules of the ***immunoglobulin superfamily*** found in the central nervous system of insects. Involved with ***fasciculation*** of axons and probably in pathfinding during morphogenesis of the nervous system. The sequence of fasciclin II shows that it shares structural motifs with a variety of vertebrate ***CAMs***, See ***contactin, F11, neurofascin, TAG-1,*** *Drosophila* ***neuroglian***.

fascicular cambium Form of ***cambium*** present in the vascular bundles of higher plants.

fasciculation Tendency of developing ***neurites*** to grow along exisiting neurites and hence form bundles or ***fascicles***. Selective fasciculation in developing vertebrate nervous systems is thought to involve the ***axon***-associated ***cell adhesion molecules L1, F11, contactin, neurofascin*** and transient axonal glycoprotein ***TAG-1***, and the ***fasciclins*** in insect nervous systems.

fascin Actin filament-bundling protein (58 kD) from sea urchin eggs.

FAST (forkhead activin signal transducer) DNA-binding component (60 kD) of the ***ARF*** complex, binds to activin response element in *mix* gene promoter. A ***winged helix transcription factor*** of the ***forkhead*** family.

Fast Ponceau 2B See ***Ponceau red***.

fat cell See ***adipocyte***.

fat droplets Micro-aggregates of (mainly) triglycerides visible within cells.

fate map Diagram of an early embryo (usually a ***blastula***) showing which tissues the cells in each region will give rise to (ie. their developmental fate). Fate maps are normally constructed by labelling small groups of cells in the blastula with ***vital dyes*** and seeing which tissues are stained when the embryo develops.

fats A term largely applied to storage lipids in animal tissues. The primary components are triglyceride esters of long-chain fatty acids.

fatty acid-binding proteins (FABP) Group of small cytosolic proteins that bind fatty acids or other organic solutes.

fatty acids Chemically R-COOH where R is an aliphatic moiety. The common fatty acids of biological origin are linear chains with an even number of carbon atoms. Free fatty acids are present in living tissues at low concentrations. The esterified forms are important both as energy storage molecules and structural molecules. See ***triglycerides***, ***phospholipids***.

fatty streak Superficial fatty patch in the artery wall caused by the accumulation of cholesterol and cholesterol oleate in distended ***foam cells***.

Favism Haemolytic anaemia induced in individuals who are glucose 6-phosphate dehydrogenase deficient by eating fava beans (from *Vicia fava*).

Fc That portion of an immunoglobulin molecule (fragment crystallizable) that binds to a cell when the antigen-binding sites (***Fab***) of the antibody are occupied or the antibody is aggregated; the Fc portion is also important in complement activation. The Fc fragment can be separated from the Fab portions by pepsin. Fc moieties from different antibody classes and subclasses have different properties.

Fc receptors Receptors for the Fc portion of immunoglobulins.

FCS See ***foetal calf serum***.

FDA (Food and Drug Authority) American drug regulatory authority responsible for assuring the safety of prescription drugs. FDA approval is essential for a new drug to be launched on the market.

feedback regulation Control mechanism that uses the consequences of a process to regulate the rate at which the process occurs: if, for example, the products of a reaction inhibit the reaction from proceeding (or slow down the rate of the reaction), then there is negative feedback, something that is very common in metabolic pathways. Positive feedback is liable to lead to exponential increase and may be explosively dangerous in some cases. Other examples are the action of voltage-gated ***sodium channels*** in generating action potentials and the activation of blood clotting ***factor V*** and ***factor VIII*** by ***thrombin***. Without damping, feedback can lead to resonance (hunting) and oscillation in the system.

feeder layer In order to culture some cells, particularly at low or clonal density, it is necessary to use a layer of less fastidious cells to condition the medium. Often the cells of the feeder layer are irradiated or otherwise treated so that they will not proliferate. In some cases the feeder layer may be producing growth factors or cytokines.

feline immunodeficiency virus (FIV) Widespread lentivirus (retrovirus) that causes an immunodeficiency in domestic cats. The immunodeficiency may be due to failure to generate an IL-12-dependent type I response. CXC-R4 seems to be the surface receptor for viral binding – CD4 is not required in contrast with HIV infection.

Felis catus Cat that is said to be domesticated.

fermentation Breakdown of organic substances, especially by microorganisms such as bacteria and yeasts, yielding incompletely oxidized products. Some forms can take place in the absence of oxygen, in which case ***ATP*** is generated in reaction pathways in which organic compounds act as both donors and acceptors of electrons. Historically, the production of ethyl alcohol or acetic acid from glucose. Also applied to anaerobic ***glycolysis*** as in ***lactate*** formation in muscle.

ferrichromes (siderochromes) Ligands for iron binding secreted by microorganisms to sequester and transport iron.

ferridoxins Low molecular weight iron–sulphur proteins that transfer electrons from one enzyme system to another without themselves having enzyme activity.

ferritin An iron storage protein of mammals, found in liver, spleen, and bone marrow. Morphologically a shell of apoferritin (protein) with a core of ferrous hydroxide/phosphate. It is much used as an electron-dense label in electron microscopy.

fertilization The essential process in sexual reproduction, involving the union of two specialized ***haploid*** cells, the gametes, to give a diploid cell, the zygote, which then develops to form a new organism.

ferulic acid Phenolic compound present in the plant cell wall that may be involved in crosslinking polysaccharide.

fes (*fps*) An oncogene, identified in avian and feline ***sarcomas***, encoding a tyrosine ***protein kinase***. See also Table O1.

fetuin (α 2-HS-glycoprotein) An α-globulin constituting up to 45% of the total protein in ***foetal calf serum***. Very carbohydrate-rich and a growth factor for many cells. Protein portion has ***cystatin*** features.

Feulgen reaction Specific staining procedure for DNA: mild acid hydrolysis makes the aldehyde group of deoxyribose available to react with Schiff's reagent to give a purple colour.

FGF See ***fibroblast growth factor***.

fgf-5 ***Oncogene*** encoding a member of the ***FGF*** family. See also Table O1.

fgr ***Oncogene*** identified in a feline ***sarcoma***, encoding a tyrosine ***protein kinase***. See also Table O1.

fibre cell Greatly elongated type of plant cell with very thick lignified wall. Usually dead at maturity, this cell type is specialized for the provision of mechanical strength. Fibre cells and ***sclereids*** together make up the tissue known as ***sclerenchyma***.

fibrillar centres Location of the nucleolar ribosomal chromatin at telophase: as the nucleolus becomes active the ribosomal chromatin and associated ribonucleoprotein transcripts compose the more peripherally located dense fibrillar component.

fibrillarin Highly conserved nucleolar protein (34–36 kD) that associates with U3-snoRNP and is found in the ***coiled body*** of the nucleolus. The N-terminus contains a glycine and arginine-rich domain (GAR domain). Yeast homologue is NOP1. Expression of fibrillarin (and ***nucleolin***) is greater in rapidly proliferating cells and in the early stages of lymphocyte activation. Autoantibodies to fibrillarin are found in some patients with scleroderma, systematic sclerosis, ***CREST*** syndrome and other connective tissue diseases.

fibrillin Widely distributed connective tissue protein (350 kD) associated with microfibrils (10nm diameter).

fibrin Monomeric fibrin (323 kD) is produced from ***fibrinogen*** by proteolytic removal of the highly charged (aspartate- and glutamate-rich) ***fibrinopeptides*** by thrombin, in the presence of calcium ions. The monomer readily polymerizes to form long insoluble fibres (23nm periodicity; half-staggered) that are stabilized by covalent crosslinking (by factor XIII, plasma transglutaminase). The fibrin gel acts as a haemostatic plug.

fibrinogen Soluble plasma protein (340 kD; 46nm long), composed of six peptide chains (two each of Aα, Bβ, and γ) and present at about 2–3mg/ml.

fibrinolysis Solubilization of fibrin in blood clots, chiefly by the proteolytic action of ***plasmin***.

fibrinopeptides Very negatively charged peptide fragments cleaved from fibrinogen by thrombin. Two peptides (A and B) are produced from each fibrinogen molecule.

fibroblast Resident cell of connective tissue, mesodermally derived, that secretes fibrillar procollagen, fibronectin and collagenase.

fibroblast growth factor (FGF; α-FGF; β-FGF; HBGF) Also known as heparin-binding growth factor (HBGF). Acidic

FGF (α-FGF, HBGF 1) and basic FGF (β-FGF, HBGF 2) are the two founder members of a family of structurally related *growth factors* for mesodermal and neuroectodermal cells. Both α-FGF and β-FGF lack a signal sequence and the pathway of release is unclear. In addition to their growth promoting activity FGFs play an important part in developmental signalling.

fibroblast growth factor receptor Family of *receptor tyrosine kinases* for *fibroblast growth factor*.

fibroblastic Many types of cultured cell become fibroblastic in appearance – this does not mean that they *are fibroblasts*.

fibroin Structural protein of silk, one of the first to be studied with X-ray diffraction. It has a repeat sequence GSGAGA and is unusual in that it consists almost entirely of stacked antiparallel *beta-pleated sheets*.

fibromodulin A small proteoglycan, around 60 kD, of the extracellular matrix. The core protein has a mass of around 42 kD and is very similar to the core protein of *biglycan* and *decorin*. All three have highly conserved sequences containing 10 internal homologous repeats of around 25 amino acids with leucine-rich motifs. Fibromodulin has four keratan sulphate chains attached to N-linked oligosaccharides.

fibronectin *Glycoprotein* of high molecular weight (two chains each of 250 kD linked by disulphide bonds) that occurs in insoluble fibrillar form in extracellular matrix of animal tissues, and soluble in plasma, the latter previously known as cold-insoluble globulin. The various slightly different forms of fibronectin appear to be generated by tissue-specific differential splicing of fibronectin mRNA, transcribed from a single gene. Fibronectins have multiple domains that confer the ability to interact with many extracellular substances such as collagen, fibrin and heparin, and also with specific membrane receptors on responsive cells. Notable is the *RGD* domain recognized by *integrins*, and two repeats of the *EGF-like domain*. The Fibronectin type III domain (FnIII), about 90 amino acids long, of which there are 15–17 per molecule, is a common motif in many cell surface proteins. Interaction of a cell's fibronectin receptors (members of the *integrin* family) with fibronectin adsorbed to a surface results in adhesion and spreading of the cell.

fibrosarcoma Malignant tumour derived from connective tissue fibroblast.

fibrosis Deposition of avascular collagen-rich matrix (*fibrous tissue*) in a wound, usually as a consequence of slow fibrinolysis or extensive tissue damage as in sites of chronic inflammation.

fibrous lamina Alternative name for the *nuclear lamina*, the region lying just inside the inner nuclear membrane.

fibrous plaque Thickened area of arterial *intima* with accumulation of smooth muscle cells and fibrous tissue (collagen, etc.) produced by the fat-laden smooth muscle cells. Below the thickening may be free extracellular lipid and debris that, if much necrosis is also present, is referred to as an *atheroma*.

fibrous tissue Although most connective tissue has fibrillar elements, the term usually refers to tissue laid down at a wound site – well-vascularized at first (*granulation tissue*) but later avascular and dominated by collagen-rich extracellular matrix, forming a scar. Excessive contraction and hyperplasia leads to formation of a *keloid*.

fibulin Calcium-binding, cysteine-rich glycoprotein found in the extracellular matrix and in plasma. Alternative splicing generates three forms of fibulin with 566, 601 and 683 amino acids respectively. All three forms have three repeated motifs near the N-terminus, with the bulk of the remaining chain formed of nine EGF-like repeats. Fibulin was originally described as a cytoplasmic protein, but this identification was based on fortuitous binding to the integrin β_1 subunit.

Fick's law Equation that describes the process of diffusion. The flux is proportional to the concentration gradient, times the diffusion constant for the molecule in that particular medium. J = –D. dC/dx

Ficoll Synthetic branched copolymer of sucrose and epichlorhydrin. Ficoll solutions have high viscosity and low osmotic pressures. Often used for preparing density gradients for cell separations (sometimes in conjunction with Hypaque for leucocyte separation).

Ficoll-Paque Proprietary name for premixed Ficoll and diatrizoate (Hypaque) with a density of 1.077g/cm^3 used as a cushion for separating lymphocytes (which do not pass through the Ficoll-Paque layer) from other blood cells in a one-step centrifugation method.

field ion microscope Type of microscopy in which the specimen is 'illuminated' with ions, often gallium ions, that are focused electrostatically. The ions remove components of the specimen, lower atomic masses first. These are imaged and provide information on elemental distribution with a resolution of perhaps 30nm.

filaggrins Basic protein components of ***keratohyalin granules*** of the suprabasal cells of the skin. Family of intermediate filament-associated cationic proteins found in mammalian epidermis. Bundle cytokeratin filaments. Various sizes in different species (16 kD bovine, 26 kD mouse, 35 kD human, 45 kD rat).

filamentous phage (Inovirus) Single-stranded DNA ***bacteriophage***s of the genus Inoviridae. Examples that infect *E. coli*: M13, f1.

filaments See ***thick filaments***, ***thin filaments***, ***intermediate filaments***, and ***microfilaments***.

filamin A protein that binds to ***F-actin***, crosslinking it to form an isotropic network; the binding does not require Ca^{2+}. It was originally isolated from smooth muscle and is a homodimer 2 × 250 kD. Similar to actin-binding protein (ABP) from leucocytes.

Filaria Genus of nematode worms causing elephantiasis and filariasis. Transmitted by insects.

filensin Protein (100 kD) of the intermediate filament family found in lens fibre cells. Binds to ***vimentin*** and co-assembles with ***phakinin*** to form the lens-specific intermediate filament system referred to as beaded-chain filaments. Filensin differs from other intermediate filament proteins in having a rather short central rod domain and will not, on its own, assemble to form intermediate filaments.

filiform papillae Curved tapering cone-shaped body on the tongue of rodents, of which the epithelial cell columns have been investigated in detail.

filipin Polyene antibiotic from *Streptomyces filipinensis*. Polymers of filipin associated with cholesterol in the cell membrane form pores which lead to cytolysis (as does ***amphotericin B***)

filopodium (plural, filopodia) A thin protrusion from a cell, usually supported by microfilaments; may be functionally the linear equivalent of the leading lamella.

Filoviridae Family of single-stranded RNA viruses, similar in some respect to rhabdoviruses. Marburg and Ebola viruses are the only two members of the family at present. Filovirus infections seem to cause intrinsic activation of the clotting cascade leading to haemorrhagic complications and high mortality. Morphologically, virions are very long filaments (up to 14mm, 70nm thick), sometimes branched. The RNA is contained within a nucleocapsid that is surrounded by a cell-derived envelope.

filovirus Virus of family Filoviridae. Includes ***Marburg*** and ***Ebola*** viruses. Both cause severe haemorrhagic fevers in humans.

fimbria (singular) See ***fimbriae***.

fimbriae (singular, fimbria) See ***pili***.

fimbrillin Major subunit protein of bacterial ***pili*** (fimbriae). Binds to fibronectin and ***statherin***. Coded by *Fim* genes. In *Porphyromonas* (*Bacteroides*) *gingivalis* fimbrillins are around 43 kD; in *E. coli* around 17 kD. Important as virulence factors.

fimbrin (plastin) Actin-binding protein (68 kD) from the core of epithelial brush-border ***microvilli***. Contains the ***EF-hand*** motif.

finasteride Drug that inhibits 5-α-reductase, the enzyme that converts testosterone to dihydrotestosterone in the prostate. Used to treat benign prostatic hyperplasia and prostatic carcinoma.

fingerprinting The basic principle of the technique is to digest a large molecule with a sequence-specific hydrolase to produce moderate-size fragments that can then be run on an electrophoresis gel. Provided the hydrolase only cleaves at specific sites (eg. between particular amino acids or bases) then the fragments should be characteristic of that molecule. The technique can be used to distinguish strains of virus or to differentiate between similar but non-identical proteins (peptide mapping). Not to be confused with ***footprinting***.

FISH analysis See ***fluorescence in situ hybridization***.

fission yeast See ***Schizosaccharomyces pombe***.

FITC (fluorescein isothiocyanate) FITC is used as a reagent to conjugate ***fluorescein*** to protein. FITC-labelled antibodies are extensively used for fluorescence microscopy: the fluorophore illuminated with UV emits a yellow-green light.

FIV Feline immunodeficiency virus. A lentivirus that, like HIV and SIV, uses ***chemokine receptors*** (CXCR4) on cells as a coreceptors for infection. FIV can use human chemokine receptors and will infect human cells that express CXCR4. SIV-infected cells fuse to form syncytia.

fixation Any chemical or physical treatment of cellular material that tends to result in its insolubilization, thus making it suitable for various types of processing for microscopy, such as ***embedding*** or staining. Typically, fixation involves protein denaturation.

FK 506 Immunosuppressive drug (tacrolimus) that acts in a very similar way to ***cyclosporin***, binding to an immunophilin and affecting calcineurin-mediated activation of the transcription factor NF-AT in T-cells.

FKBP (FK506-binding protein) A family of small intracellular proteins (around 11 kD) that bind the immunosuppressive drug FK506 (tacrolimus), thus are ***immunophilins***. Like cyclophilin have peptidyl-prolyl isomerase activity but are not structurally similar.

flag tagging Molecular biology technique, in which the gene encoding a protein of interest is mutagenized to include an ***epitope*** for which there is a good antibody. The fate of the protein in a transfected cell or transgenic organism can then be followed easily. Popular tags include the ***myc***, ***green fluorescent protein*** or haemagglutinin epitopes. See also ***epitope tag***.

flagella (plural) See ***flagellum***.

flagellin Subunit protein (40 kD) of the bacterial flagellum.

flagellum (plural, flagella) Long thin projection from a cell used in movement. In eukaryotes flagella (like ***cilia***) have a characteristic axial '9 + 2' microtubular array (***axoneme***) and bends are generated along the length of the flagellum by restricted sliding of the nine outer doublets. In prokaryotes the flagellum is made of polymerized ***flagellin*** and is rotated by the basal motor.

flame cells Specialized excretory cells found in Platyhelminthes (flatworms). The basal nucleated cell body has a distal cylindrical extension that surrounds an extracellular cavity lined by cilia. Mode of action unclear.

flanking sequence Short DNA sequences bordering a ***transcription unit***. Often these do not code for proteins.

FLAP (5-lipoxygenase activating protein) Activator of the enzyme responsible for the production of 5-HPETE from arachidonic acid, the first step in leukotriene synthesis.

flare streaming Phenomenon described in isolated cytoplasm of giant amoeba when the medium contains Ca^{2+} and ATP. A loop of cytoplasm flows outward and then returns to the main mass: the appearance is reminiscent of flares around the eclipsed sun.

flat revertant Variant of a malignant-transformed animal tissue cell in which the characteristic high ***saturation density*** and piled-up morphology have reverted to the flatter morphology associated with non-transformed cells.

flavan (2,3-dihydro-2-phenylbenzopyran) Parent ring compound on which flavanols, flavanones, flavones, flavonols and flavonoids are based. Should be distinguished from flavin which shares the yellow colour but not structure.

flavin Group of variously substituted derivatives of 7,8-dimethylisoalloxazine. Yellow coloured. The flavin group is found in FAD, FADH and flavoproteins. Not to be confused with flavan and flavones.

flavin adenine dinucleotide See ***FAD***.

flavin nucleotides General term for flavin adenine dinucleotide (FAD) or flavin mononucleotide (FMN). Act as prosthetic groups (covalently linked cofactors) for flavin enzymes.

Flaviviridae Family of enveloped RNA viruses with spherical particles 40–50nm in diameter. Only genus is Flavivirus. Cause dengue haemorrhagic fever, Japanese encephalitis, tick-borne encephalitis and yellow fever (the latter being the source of the name).

flavone (2-phenylchromen-4-one; 2-phenyl-4H-1-benzopyran-4-one) Specifically the compound and more generally a group of hydroxylated derivatives. Flavone glycosides occur widely as yellow pigments in angiosperms.

flavoproteins Enzymes or proteins that have a ***flavin nucleotide*** as a coenzyme or prosthetic group. Oxidoreductases or electron carriers in the terminal portion of the electron transport chain.

flavoridin See ***disintegrin***.

Flemming-without-acetic An excellent cytoplasmic fixative that contains chromic acid and osmium tetroxide.

FLICE ***Caspase*** 8; see Table C1.

FLIP (FLICE-inhibitory protein) Family of proteins that inhibit the ***caspase***, FLICE, and thus protect from apoptotic death. Viral FLIPs (v-FLIPs) contain two ***death-effector domains*** that interact with ***FADD*** and have been shown to be produced by various herpes viruses and molluscipox virus.

flip-flop A term used to describe the coordinated transfer of two phospholipid molecules from opposite sides of a lipid bilayer membrane. Now used to mean the passage of a phospholipid species from one lamella of a lipid bilayer membrane to the other.

flippase See ***flp-frp recombinase***.

FLIPR (fluorescence imaging plate reader) Machine for fluorescence imaging using a laser that is capable of illuminating a 96-well plate and a means of simultaneously reading each well thus enabling rapid measurements on a large number of samples. Used in high throughput screening.

flk-1 (VEGFR-2; KDR) One of the receptors for ***VEGF***, binds VEGF-121 and VEGF-C. See ***flt*** and ***cloche***.

florigen Hypothetical plant growth substance (hormone) postulated to induce flowering. Existence not proven: recently suggested that it might be an ***oligosaccharin***.

flow cytometry Slightly imprecise but common term for the use of the fluorescence activated cell sorter (FACS). Cells are labelled with fluorescent dye and then passed, in suspending medium, through a narrow dropping nozzle so that each cell is in a small droplet. A laser-based detector system is used to excite fluorescence, and droplets with positively fluorescent cells are given an electric charge. Charged and uncharged droplets are separated as they fall between charged plates, and so collect in different tubes. The machine can be used either as an analytical tool, counting the number of labelled cells in a population, or to separate the cells for subsequent growth of the selected population. Further sophistication can be built into the system by using a second laser system at right angles to the first to look at a second fluorescent label, or to gauge cell size on the basis of light-scatter. The great strength of the system is that it looks at large numbers of individual cells and makes possible the separation of populations with, for example, particular surface properties.

flp-frp recombinase (Pronounced: 'flip-furp') Yeast system for DNA rearrangement. In the presence of 'flippase', a stretch of DNA flanked by matching frp sites is excised, and the ends rejoined. An example of a ***cassette mechanism***.

FLRF-amide Phe-Leu-Arg-Phe-NH2, a tetrapeptide ***neurotransmitter*** found in invertebrates that is a member of a diverse family of RF-amide peptides; all members of which share the same C-terminal RF-amide sequence. See also ***FMRF-amide***.

flt Receptors for VEGF isoforms. Flt-1 is VEGF receptor-1; flt-3 (Flk2; STK-1; CD135) has a ligand of 24 kD similar to c-kit ligand and M-CSF; flt-4 (VEGFR-3) binds VEGF-C and is mainly restricted to lymphatic endothelium during development. All are receptor tyrosine kinases and are involved in regulation of endothelial or haematopoietic cell development.

fluctuation analysis Method used to determine (for example) how many ion channels contribute to the transmembrane current. On the assumption that each channel is either open or shut, the noise in the recorded current can be considered to arise from the statistical fluctuation in the number of channels open, and the magnitude of the fluctuation gives an estimate of the conductance of a single channel.

fluctuation test Test devised by Luria & Delbruck to determine whether genetic variation in a bacterial population arises spontaneously or adaptively. In the original version the statistical variance in the number of bacteriophage-resistant cells in separate cultures of bacteriophage-sensitive cells was compared with variance in replicate samples from bulk culture. The greater variance in the isolated populations indicates that mutation occurs spontaneously before challenge with phage. (The proportion of resistant cells depends upon when after isolation the mutation arises; which will be very different in separate populations.)

fluid bilayer model Generally accepted model for membranes in cells. In its original form, the model held that proteins floated in a sea of phospholipids arranged as a bilayer with a central hydrophobic domain. Although it is now recognized that some proteins are restrained by interactions with cytoskeletal elements, and that the phospholipid annulus around a protein may contain only specific types of lipid, the model is still considered broadly correct.

fluorescein Fluorophore commonly used in microscopy. Fluorescein diacetate can be used as a vital stain, or can be conjugated to proteins (particularly antibodies) using isothiocyanate. Excitation is at 365nm, and the emitted light is green-yellow (450–490nm). The emission spectrum is pH-sensitive and fluorescein can therefore be used to measure pH in intracellular compartments.

fluorescence The emission of one or more photons by a molecule or atom activated by the absorption of a quantum of electromagnetic radiation. Typically the emission, that is of longer wavelength than the excitatory radiation, occurs within 10^{-8} seconds: phosphorescence is a phenomenon with a longer or much longer delay in re-radiation. Note that gamma-rays, X-rays, UV, visible light and IR radiations may all stimulate fluorescence.

fluorescence-activated cell sorter (FACS) See ***flow cytometry***.

fluorescence energy transfer (fluorescence resonant energy transfer) Transfer of energy from one fluorochrome to another. The emission wavelength of the fluorochrome excited by the incident light must approximately match the excitation wavelength of the second fluorochrome. If light at the second emission wavelength is detected, it implies that the two fluorochromes were physically within a few nanometres. Used as a technique to probe protein or cell interactions.

fluorescence *in situ* hybridization (FISH; chromosome painting) Technique of directly mapping the position of a gene or DNA clone within a genome by ***in situ* hybridization** to ***metaphase*** spreads, in which condensed chromosomes are distinguishable by light microscopy. The DNA probe is labelled with a fluorophore, and the hybridization sites visualized as spots of light by ***epifluorescence***. Frequently, several probes can be used at one time, to mark specific chromosomes with different coloured fluorophores ('chromosome painting').

fluorescence microscopy Any type of microscopy in which intrinsic or applied reagents are visualized. Intrinsic fluorescence is often referred to as autofluorescence. The applied reagents typically include fluorescently labelled proteins that are reactive with sites in the specimen. In particular, fluorescently labelled antibodies are widely used to detect particular antigens in biological specimens.

fluorescence recovery after photobleaching (FRAP) Many ***fluorochromes*** are bleached by exposure to exciting light. If, for example, the cell surface is labelled with a fluorescent probe and an area bleached by laser illumination, then the bleached patch that starts off as a dark area will gradually recover fluorescence. The recovery is due to the repopulation of the area by unbleached molecules and diffusion of bleached molecules to other areas. The rate and extent of recovery are a measure of the fluidity of the membrane and the proportion of labelled molecules that are free to exchange with adjacent areas. The technique is usually applied to cell surface fluidity or viscosity measurements, but is also applicable to other structures.

fluorescence resonant energy transfer See ***fluorescence energy transfer***.

fluoride The fluoride ion F^-. Low levels of fluoride in drinking water markedly decrease the incidence of dental caries, probably because bacterial metabolism is much more sensitive to low fluoride levels.

fluorite objective Microscope objective corrected for ***spherical*** and ***chromatic aberration*** at two wavelengths. Better than an ordinary objective corrected at one wavelength but inferior to (and much cheaper than) a planapochromatic objective.

fluorochromes Those molecules that are fluorescent when appropriately excited; fluorochromes such as fluorescein or tetramethyl rhodamine are usually used in their isothiocyanate forms (FITC, TRITC).

fMLP See ***formyl peptides***.

FMN (flavin adenine nucleotide) See ***flavin nucleotides***.

FMRF-amide Phe-Met-Arg-Phe-NH_2, a tetrapeptide ***neurotransmitter***, a member of the same family of RF-amide peptides as ***FLRF-amide***, sharing the same C-terminal RF-amide sequence.

fms An ***oncogene***, identified in a feline ***sarcoma***, encoding a tyrosine ***protein kinase***, as part of a mutant receptor for ***macrophage colony-stimulating factor***. See also Table O1.

Fnr Fnr (fumarate nitrate reductase) protein activates a number of operons in *E. coli* during anaerobic growth and is a transcriptional regulator.

foam cells Lipid-laden macrophages and, to a lesser extent smooth muscle cells, found in ***fatty streaks*** on the arterial wall.

focal adhesion kinase (FAK) Protein kinase which is found at ***focal adhesions*** and is thought to mediate the adhesion or spreading processes.

focal adhesions Areas of close apposition, and thus presumably anchorage points, of the plasma membrane of a fibroblast (for example) to the substratum over which it is moving. Usually 1 × 0.2μm with the long axis parallel to the direction of movement; always associated with a cytoplasmic microfilament bundle that is attached via several proteins to the plasma membrane at an area of high protein concentration (this is noticeably electron-dense in electron micrographs). Focal adhesions tend to be characteristic of slow-moving cells.

focus Group of (frequently ***neoplastic***) cells, identifiable by distinctive morphology or histology.

fodrin Tetrameric protein (α 240 kD, β 235 kD) found in brain: an isoform of ***spectrin***.

foetal calf serum (FCS) Expensive component of standard culture media for many types of animal tissue cells.

folate (tetrahydrofolate) Molecule that acts as a carrier of one-carbon units in intermediary metabolism. It contains residues of *p*-aminobenzoate, ***glutamate***, and a substituted ***pteridine***. The latter cannot be synthesized by mammals, which must obtain tetrahydrofolate as a vitamin or from intestinal microorganisms. One-carbon units are carried at three different levels of oxidation, as methyl-, methylene- or formimino-groups. Important biosyntheses dependent on tetrahydrofolate include those of ***methionine***, ***thymine*** and ***purines***. Analogues of dihydrofolate, such as ***aminopterin*** and ***methotrexate*** block the action of tetrahydrofolate by inhibiting its regeneration from dihydrofolate.

folic acid Pteridine derivative that is abundant in liver and green plants, and is a growth factor for some bacteria. The biochemically active form is tetrahydrofolate (see ***folate***).

follicle Generally a small sac or vesicle. *Bot.* A kind of fruit formed from a single carpel, that splits to release its seeds. *Zool.* Its use includes: hair follicle, an invagination of the epidermis into the dermis surrounding the hair root; ovarian follicle, an oocyte surrounded by one or more layers of ***granulosa*** cells. As the ovarian follicle develops a cavity forms and it is then termed a Graafian follicle.

follicle-stimulating hormone (FSH; follitropin) Pituitary hormone that is an acidic glycoprotein. It induces development of ovarian follicles and stimulates the release of oestrogens.

follistatin Originally identified as an activin-binding protein, follistatin inhibits ***BMP*** activity in early *Xenopus* development.

follitropin See ***follicle-stimulating hormone***.

footprinting A technique used to identify the binding site of, for example, a protein on a nucleic acid sequence. The basic principle is to carry out a very limited hydrolysis of the DNA with or without the protein complexed and then to compare the digestion products. If a cleavage site is masked by the bound protein then the pattern of fragments when protein is present will be different and it is possible to work out, by a series of such procedures, exactly where the protein binds.

Foraminifera Group of Rhizopod Protozoa that secrete a test ('shell') and have slender pseudopods that extend beyond the test and unite to form networks. *Allogromia* is a genus within this group. Extensive remains of Foraminiferan tests are found in sedimentary rocks from the Ordovician to the present.

foreign body giant cell Syncytium formed by the fusion of macrophages in response to an indigestible particle too large to be phagocytosed (eg. talc, silica or asbestos fibres). There may be as many as 100 nuclei randomly distributed: similar cells but with the nuclei more peripherally located (***Langhans' multinucleate giant cells***) are found at the centre of tuberculous lesions.

forkhead *Drosophila* homeotic gene. The *forkhead* gene family of transcription factors belong to the winged helix class of DNA-binding proteins. More than 40 members of the family have been identified and are involved in embryonic development, tumorigenesis and tissue-specific gene expression. The DNA-binding domain is of 100 amino acids and is referred to as the forkhead domain.

formaldehyde Commonly used fixative and antibacterial agent. As a fixative it is cheap and tends to cause less denaturation of proteins than does glutaraldehyde, particularly if used in a well-buffered solution (buffered formalin, formal saline). Old formaldehyde solutions usually contain crosslinking contaminants, and it is therefore often preferable to used a formaldehyde-generating agent such as paraformaldehyde. Formalin fumes, particularly in conjunction with HCl vapour, are potently carcinogenic.

formerly mucopolysaccharides See ***glycosaminoglycans***.

formins A set of protein isoforms encoded by alternatively spliced *ld* locus of the mouse. Mutations in *ld* lead to disruption in pattern formation, small size, fusion of distal bones and digits of limbs and renal aplasia.

formyl peptides Informal term for small peptides with a formylated N-terminal methionine and usually a hydrophobic amino acid at the carboxy-terminal end (fMetLeuPhe is the most commonly used). These peptides stimulate the motor and secretory activities of leucocytes, particularly neutrophils and monocytes, that have a specific receptor (about 60 kD) of high affinity (Kd approximately 10^{-8} M). Leucocytes show chemotaxis towards formyl peptides but the term chemotactic peptides understates the range of activities the molecules will trigger. Thought to be synthetic analogues of bacterial signal sequences, though this is unproven. The leucocytes of many animals (eg. pig, cow, chicken) do not respond.

forskolin (colforsin) Diterpene from the roots of *Coleus forskohlii* that stimulates adenylate cyclase and is often used in conjunction with inhibitors of phosphodiesterase to artificially increase intracellular levels of cAMP.

Forssman antigen A glycolipid heterophil antigen present on tissue cells of many species. It was first described for sheep red cells, and is not present on human, rabbit, rat, porcine or bovine cells.

fos An ***oncogene***, identified in a mouse ***osteosarcoma***, encoding a transcription factor. Fos and ***Jun*** proteins dimerise via a ***leucine zipper*** to form the AP-1 ***transcription factor***. See also Table O1.

Fos-related antigen-1 See ***fra-1***.

founder cell Cell that gives rise to tissue by clonal expansion. For most mammalian tissues there are considerably more than two founder cells, as can be determined by forming chimeras from genetically distinguishable embryos, but single founder cells have been found for the intestine and germ line in *Caenorhabditis elegans*.

four-helix bundle Common protein ***motif*** in which four ***alpha-helices*** bundle closely together to form a hydrophobic core.

Fourier analysis Loosely, the use of Fourier transformations to convert a time-based signal to a frequency spectrum and back, allowing any periodic property of the signal to be identified.

fovea Small pit or depression on the surface of a structure or organ; the fovea centralis is the most cone-rich region of the retina with maximum acuity and colour sensitivity.

FPLC Fast protein liquid chromatography. Chromatographic method for protein purification that is much less commonly used now that recombinant proteins can be purified by affinity methods.

fps See ***fes***.

fra-1 (Fos-related antigen-1) Related to ***fos***.

fractalkine Membrane-bound ***chemokine*** with CX_3C motif. Chemokine domain (76 amino acids) is bound to membrane through mucin-like stalk (241 amino acids) or can be released as a 95 kD glycoprotein. Highly expressed on activated endothelial cells and is both an adhesion molecule and an attractant for T-cells and monocytes.

fraction I protein See ribulose bisphosphate carboxylase/oxidase (***RUBISCO***).

fractionation A term used to describe any method for separating and purifying biological molecules. See also ***cell fractionation***.

fragile X syndrome Most frequent cause of mental retardation. There is an expanded ***trinucleotide repeat***, CGG, in the *fra*(X) gene.

fragmentin-2 See ***granzyme*** B.

fragmin An actin-binding protein (42 kD) from the slime mould *Physarum polycephalum*, that has calcium-sensitive severing and capping properties.

frame-shift mutation Insertion or deletion of a number of bases not divisible by 3 in an open reading frame in a DNA sequence. Such mutations usually result in the generation, downstream, of nonsense, chain-termination codons.

Frankia Genus of ***Actinomycetales*** capable of nitrogen fixation, both independently and in symbiotic association with roots of certain non-leguminous plants, notably alder.

FRAP See ***fluorescence recovery after photobleaching***.

frataxin Product of the X25 gene: deficiency leads to ***Friedreich's ataxia***.

fraxin (7-hydroxy-6-methoxycoumarin 8-glucoside) Coumarinic glucoside from *Fraxinus excelsior* that has anti-inflammatory and anti-metastatic properties, the former probably because of its inhibitory effect on 5-HETE production.

free energy (Gibbs free energy, G) A thermodynamic term used to describe the energy that may be extracted from a system at constant temperature and pressure. In biological systems the most important relationship is: $\Delta G = -RT\ln(K_{eq})$, where K_{eq} is an equilibrium constant.

free radical Highly reactive and usually short-lived molecular fragment with one or more unpaired electrons.

freeze cleavage See ***freeze fracture***.

freeze drying Method commonly adopted to produce a dry and stable form of biological material that has not been seriously denatured. By freezing the specimen, often with liquid nitrogen, and then subliming water from the specimen under vacuum, proteins are left in reasonably native form, and can usually be rehydrated to an active state. Since the freeze-dried material will store without refrigeration for long periods, it is a convenient method for holding back-up or reference material, or for the distribution of antibiotics, vaccines, etc.

freeze etching If a ***freeze-fractured*** specimen is left for any length of time before shadowing, then water will sublime off from the specimen etching (lowering) those surfaces that are not protected by a lipid bilayer. Some etching will take place following any freeze cleavage process; in deep etching the ice surface is substantially lowered to reveal considerable detail of, for example, cytoplasmic filament systems.

freeze fracture Method of specimen preparation for the electron microscope in which rapidly frozen tissue is cracked so as to produce a fracture plane through the specimen. The surface of the fracture plane is then shadowed by heavy metal vapour, strengthened by a carbon film, and the underlying specimen is digested away, leaving a replica that can be picked up on a grid and examined in the transmission electron microscope. The great advantage of the method is that the fracture plane tends to pass along the centre of lipid bilayers, and it is therefore possible to get *en face* views of membranes that reveal the pattern of integral membrane proteins. The ***E-face*** is the outer lamella of the plasma membrane viewed as if from within the cell, the P-face the inner lamella viewed from outside the cell. Fracture planes also often pass along lines of weakness such as the interface between cytoplasm and membrane, so that outer and inner membrane surfaces can be viewed. Further information about the structure can be revealed by ***freeze etching***. Extremely rapid freezing followed by deep etching has allowed the structure of the cytoplasm to be studied without the artefacts that might be introduced by fixation.

French flag problem The French flag (tricolor) is used to illustrate a problem in the determination of pattern in a tissue, that of specifying three sharp bands of cells with discrete properties that do not have blurred edges using, for example, a gradient of a diffusible morphogen.

frequenin Synaptic calcium-binding protein originally found in *Drosophila*. Homologous to ***recoverin*** and ***visinin***.

Freund's adjuvant A water-in-oil emulsion used experimentally for stimulating a vigorous immune response to an antigen (that is in the aqueous phase). Complete Freund's adjuvant contains heat-killed tubercle bacilli; these are omitted from Freund's incomplete adjuvant. Unsuitable for use in humans because it elicits a severe granulomatous reaction.

Friedreich's ataxia Autosomal recessive disorder caused by trinucleotide (GAA) repeats that, unlike those in ***Huntington's chorea*** and ***fragile X syndrome***, are within an intron of the gene 25 that codes for ***frataxin***, the protein deficient in the disease. The intra-intronic repeat may interfere with hnRNA processing and thus lead to a deficiency in frataxin production.

Friend helper virus Mouse (lymphoid) leukaemia virus present in stocks of Friend virus, which was believed at one time to assist its replication. Molecular cloning of Friend virus has since shown that it is non-defective.

Friend murine erythroleukaemia cells Lines of mouse erythroblasts transformed by the Friend virus, that can be induced to differentiate terminally, producing haemoglobin, by various agents such as dimethyl sulphoxide.

Friend murine leukaemia virus Murine leukaemia virus isolated by Charlotte Friend in 1956 while attempting to transmit the Erlich ascites tumour by cell-free extracts. Causes an unusual erythroblastosis-like leukaemia, in which anaemia is accompanied by large numbers of nucleated red cells in blood. Does not carry a host-derived oncogene, but seems to induce tumours by proviral insertion into specific regions of host genome.

Friend spleen focus-forming virus Defective virus found in certain strains of ***Friend helper virus***, detected by its ability to form foci in spleens of mice, and believed to be responsible in those strains for the production of a leukaemia associated with polycythemia rather than anaemia.

fringe (Fng) Protein that regulates the location-specific expression of the ***Notch*** ligands ***serrate protein*** and ***Delta*** protein in the developing *Drosophila* wing.

frizzled *(fz) Drosophila* tissue-polarity gene encoding a serpentine receptor that responds to a polarity signal. Downstream signalling seems to involve JNK/SAPK-like kinases, ***Rho factor*** A and the product of the gene *dishevelled* (*dsh*)

frontal zone contraction theory Model proposed to account for the movement of giant amoebae in which cytoplasmic contraction at the front of the leading pseudopod (fountain zone) pulls viscoelastic cytoplasm forward in the centre of the cell and forms a tube of more rigid cytoplasm immediately below the plasma membrane behind the active region. The peripheral contracted cytoplasm relaxes into a weaker gel at the rear and is pulled forward in its turn. Contrasts with the ***ectoplasmic tube contraction*** model.

frozen stock Because cell lines tend to change their properties with continuous rounds of subculturing, it is common practice to keep stocks of cells frozen (either in liquid nitrogen or at –70° C) and to keep returning to this stock so that experiments are all carried out on cells of comparable passage number. The method also allows strains to be stored for long periods. Cells are usually frozen down in the presence of a cryoprotectant such as DMSO or glycerol. The method is also extensively used for storing semen for artificial insemination.

fructose A 6-carbon sugar (hexose) abundant in plants. Fructose has its reducing group (carbonyl) at C2, and thus is a ketose, in contrast to glucose that has its carbonyl at C1 and thus is an aldose. Sucrose, common table sugar, is the non-reducing disaccharide formed by an α-linkage from C1 of glucose to C2 of fructose (latter in furanose form). Fructose is a component of polysaccharides such as inulin, levan.

frusemide See ***furosemide***.

FSH See ***follicle-stimulating hormone***.

ftsZ Filamentous temperature sensitive gene from *E. coli*, the product of which is a novel GTP-binding protein (43 kD) that may be involved in signalling. The protein has GTPase activity and is widely distributed in bacterial species.

ftz See ***fushi tarazu***.

fuchsin Synthetic rosaniline dye. Used as a red dye (in Schiff's reagent) and as an antifungal agent.

fucose L-fucose (6-deoxy-L-galactose) is found as a constituent of N-glycan chains of glycoproteins; it is the only common L-form of sugar involved. D-fucose is usually encountered as a synthetic galactose analogue.

fucosyl transferase An enzyme catalysing the transfer of fucosyl residues from the nucleotide sugar GDP-fucose.

fucoxanthin ***Carotenoid*** pigment of certain brown algae (***Phaeophyta***) and bacteria; absorbs at 500–580nm.

Fucus Genus of brown algae common on shoreline of Northern seas.

Fugu rubripes Japanese puffer fish. Notorious for the poison (***tetrodotoxin***) found in lethal amounts in the poison gland (which must be removed before the fish can be eaten safely) and at low levels elsewhere. Also of interest and utility because of the very low levels of repetitive DNA found in the genome.

fumagillin Naturally secreted antibiotic from *Aspergillus fumigatus* that inhibits endothelial cell proliferation and is therefore potentially anti-angiogenic.

fumarate A dicarboxylic acid intermediate in the Krebs cycle (***tricarboxylic acid cycle***). Can be derived from aspartate, phenylalanine and tyrosine for input to the Krebs cycle.

functional cloning (expression cloning) Strategy for cloning a desired gene that is based on some property (antigenicity, ligand binding, etc.) of the expressed gene. For example, a cDNA library could be produced in a eukaryotic ***expression vector***, and transfected into a large number of cells. To identify a particular transport protein, a radiolabelled substrate could be added, and cells containing the protein of interest identified by radiography. The plasmid could then be recovered and the gene's sequence determined.

Fundulus heteroclitus The killifish. A teleost much used for the study of early embryonic development because the egg and embryo are transparent.

fura-2 A fluorescent dye, used in measurement of intracellular free calcium levels.

furin ***Subtilisin***-like eukaryotic endopeptidase with substrate specificity for consensus sequence Arg-X-Lys/Arg-Arg at the cleavage site. Furin is known to activate the haemagglutinin of fowl plague virus and will cleave the HIV envelope glycoprotein (gp160) into two portions, gp120 and gp41, a necessary step in making the virus fusion-competent.

furosemide (frusemide) Potent diuretic that increases the excretion of sodium, potassium and chloride ions, and inhibits their resorption in the proximal and distal renal tubules.

furunculosis Disease of fish caused by *Aeromonas salmonicida*. Major problem in fish farms.

Fusarium* mycotoxins** Important fungal mycotoxin contaminants of various food products. Include zearalenone, ***diacetoxyscirpenol, T-2 toxin, neosolaniol monoacetate, deoxynivalenol, nivalenol, fumonisin B1, fumonisin B2, moniliformin, fusarenon-X, HT-2 tioxin and β-zearalenol.

fushi tarazu *(ftz)* (Japanese for 'too few segments'); a ***pair-rule gene*** of *Drosophila*.

fusiform Tapered at both ends, like a spindle, though the current rarity of spindles makes this a somewhat unhelpful description.

fusin Lymphocyte surface protein originally described as being an essential cofactor for HIV bound to CD4 to fuse with and enter the cell, later shown to be a chemokine receptor (CXCR4 in the case of lymphotropic virus strains, CCR5 for myelotropic strains). Since ***FIV*** will infect CD4 cells it is possible that the chemokine receptor is the original binding site and CD4 the coreceptor, rather than the converse.

fusion protein Protein formed by expression of a hybrid gene made by combining two gene sequences. Typically this is accomplished by cloning a cDNA into an ***expression vector*** in-frame with an existing gene, perhaps encoding, eg. β-galactosidase. See ***GST fusion protein***.

futile cycles Any sequence of enzyme-catalysed reactions in which the forward and reverse processes (catalysed by different enzymes) are constitutively active. Frequently used to describe the cycle of phosphorylation and dephosphorylation of phosphatidyl inositol derivatives in cell membranes.

Fx Very small protein (5 kD) from platelets that binds to G-actin rendering it assembly-incompetent.

fyn A non-receptor ***tyrosine kinase***, related to ***src***.

G

G See ***glycine***.

G banding (Giemsa banding) Spreads of metaphase chromosomes, treated briefly with protease then stained with ***Giemsa***, produce characteristic ***banding patterns*** that allow identification of the separate chromosomes. The deeply staining G bands do not coincide with the pattern of quinacrine bands (Q bands).

G-actin Globular ***actin***.

γ-carboxyglutamate See ***carboxyglutamate***.

G-protein (1) See ***GTP-binding proteins***. (2) The spike glycoprotein of vesicular stomatitis virus. This has been an important protein for investigation of membrane transport in eukaryotic cells.

G-protein coupled receptor (GPCR; serpentine receptor; seven-spanners; seven transmembrane receptors) Cell surface receptors that are coupled to heterotrimeric ***G-proteins*** (GTP-binding proteins). All G-protein coupled receptors seem to have seven membrane-spanning domains (are ***serpentine receptors***), and have been divided into two subclasses: those in which the binding site is in the extracellular domain, eg. receptors for glycoprotein hormones, such as ***thyroid-stimulating hormone*** (TSH) and ***follicle-stimulating hormone*** (FSH); and those in which the ligand-binding site is likely to be in the plane of the seven transmembrane domains, eg. ***rhodopsin*** and receptors for small ***neurotransmitters*** and ***hormones***, eg. ***muscarinic acetylcholine receptor***.

G_o See ***GTP-binding protein***.

G1 Phase of the eukaryotic ***cell cycle*** between the end of cell division and the start of DNA synthesis, ***S phase***. G stands for gap.

G2 Phase of the eukaryotic ***cell cycle*** between the end of DNA synthesis and the start of cell division.

gab-1 (Grb2-associated binder-1) Protein (77 kD, 694 residues) that binds to ***Grb-2*** and has homology with IRS-1 (insulin receptor substrate-1), particularly in the ***PH domain*** in the N-terminal region. Gab-1 is a substrate for the EGF-receptor and may integrate signals from different receptors into the control of cellular responses: overexpression of Gab-1 makes cells more responsive to limiting amounts of growth factors. Gab-1 is found in most human tissues.

GABA (gamma-aminobutyric acid) Fast inhibitory ***neurotransmitter*** in the mammalian ***central nervous system***; prevalent in higher regions of the ***neuraxis***. Also mediates peripheral inhibition in crustaceans and in the leech *Hirudo medicinalis*.

GABA receptor Member of a family of receptors for neurotransmitters that includes those for glycine and the nicotinic acetylcholine receptor. Opened by γ-amino butyric acid (***GABA***). There are two main classes; GABA A and GABA C receptors are ***ionotropic***, GABA B receptor is ***metabotropic***. 1. GABA A receptor (ca 250 kD) is a ligand-gated chloride channel specifically blocked by ***bicuculline*** and picrotoxin, a hetero-oligomer with (probably) five subunits, generally two pairs of α and β subunits and a γ subunit, but there are multiple isoform and splicing variants of these and some additional tissue-specific isoforms. The α chains (53 kD) are needed for binding of ***benzodiazepine***, though the site is probably shared with the γ subunit, and the β chains (58 kD) bind GABA. The subunits are thought to form a tight group with the chloride channel in the centre. There is considerable protein sequence similarity between GABA A receptor and the nicotinic acetylcholine receptor. Properties of the receptor can be modified by phosphorylation. Insect GABA receptor resemble vertebrate GABA A but do not bind bicuculline and have significantly different pharmacological profiles. 2. The GABA B receptor (80 kD) is a G-protein-coupled receptor found in the brain and differs from the GABA A receptor both in agonist specificity (baclofen is a specific agonist) and its effects on cells. It is negatively coupled to adenylate cyclase through a G_o-protein and thus acts indirectly on N-type calcium channels. Inhibitory effects mediated through this receptor are due to a reduction in catecholamine release. Has sequence similarity with metabotropic glutamate receptors. 3. The GABA C receptor resembles GABA A but is restricted to the retina.

gadd45 Nuclear protein induced by growth arrest and DNA damage. Level is highest in G1 phase of cell cycle and gadd associates with cyclin dependent kinase inhibitor $p21^{Cip1}$ and with ***PCNA***.

GAG See ***glycosaminoglycan***.

gag-protein (group specific antigen) The protein of the nucleocapsid shell around the RNA of a retrovirus.

gain of function mutation Gene mutation that results in higher than normal levels of activity of the gene product, for example by deletion of a regulatory phosphorylation site on the protein. Examples are ***oncogenic*** mutations in genes involved in growth control.

GAL promoter Inducible promoter region of the yeast operon that encodes, among other things, the enzyme ***beta-galactosidase***. Extensively used, as is the prokaryotic analogue, because the blue colour generated by the action of the enzyme on ***Xgal*** is a convenient marker of colonies containing the vector. The transcription factor that binds to the promoter (*GAL4*) is much used in ***yeast two-hybrid screening***.

GAL4 Yeast ***transcription factor*** that binds the UASG promoter domain. Often used in reporter gene constructs and in ***yeast two-hybrid screening***.

GAL4 enhancer trap Form of ***enhancer trap*** (classically in ***Drosophila***) in which the ***reporter gene*** is the yeast ***transcription factor*** GAL4. The advantage of this system is that such enhancer traps can drive expression of any ***transgene*** under control of the UAS ***promoter*** recognized by GAL4. This thus provides a technique for cell-specific expression of transgenes in an intact organism.

galactan Polymer of galactose.

galactocerebroside (GalC) Surface antigen characteristic of newly differentiated ***oligodendrocytes***. GalC antibody is used to identify this glial cell type in cultures of rat optic nerve and brain.

galactosaemia Inborn disorder in which the enzyme galactose-1-phosphate uridyl transferase, which converts galactose-1-phosphate into glucose-1-phosphate, is absent. Excess galactose-1-phosphate accumulates in the blood and a variety of problems result.

galactose Hexose identical to glucose except that orientation of –H and –OH on carbon 4 are exchanged. A component of ***cerebrosides*** and ***gangliosides***, and glycoproteins. ***Lactose***, the disaccharide of milk, consists of galactose joined to glucose by a β-glycosidic link.

galactose-binding protein A bacterial periplasmic protein, most studied in *E. coli*, that acts both as a sensory element in the detection of galactose as a chemotactic signal, and in the uptake of the sugar.

galactosyl transferase Enzyme catalysing the transfer of galactose units from the sugar-nucleotide, uridine diphosphogalactose (UDP-galactose) to an acceptor, commonly N-acetyl-glucosamine in a glycan chain, forming a glycosidic bond involving C1 of galactose.

galanin Neuropeptide (29 amino acids) isolated from the upper small intestine of pig but subsequently found throughout the central and peripheral nervous system. Regulates gut motility and the activity of endocrine pancreas.

galaptins Soluble ***lectins*** of around 130–140 residues secreted by vertebrates. Developmentally regulated: seem to be important in differentiation of tissues. Larger, related lectin is known as MAC-2 antigen, CBP-35, or IgE-binding protein.

GalC See ***galactocerebroside***.

galectin-1 One of a family of β-galactoside binding proteins that has growth-regulatory and immunomodulatory properties. Galectin-1 mediates cell–cell and cell– substratum adhesion. Recombinant galectin-1 will induce apoptosis in T-cells. Occurs as a homodimer which is cell surface-associated.

galectin-3 (IgE binding protein; Mac-2; ε–BP) One of a family of β-galactoside binding proteins (30 kD) that has growth-regulatory and immunomodulatory properties. Galectin-1 mediates cell–cell and cell–substratum adhesion. Recombinant galectin-1 will induce apoptosis in T-cells. Occurs as a homodimer which is cell-surface associated.

gallic acid (3,4,5-trihydroxybenzoic acid) Phenolic acid, commonly found in flowering plants, usually esterified with tannins.

Galliformes Order of birds that includes chickens, peacocks, grouse, pheasants and turkeys all of which have gizzards.

GALT See ***gut-associated lymphoid tissue***.

galvanotaxis The directed movement of cells induced by an applied voltage. This movement is almost always directed toward the cathode, occurs at fields around 1mV/mm, and is argued to be involved in cell guidance during morphogenesis, and in the repair of wounds. The term 'galvanotropism' is used for neurons, since the cell body remains stationary and the neurites grow toward the cathode. Note that these processes involve cell locomotion, and are distinct from ***cell electrophoresis***.

galvanotropism See ***galvanotaxis***.

gamete Specialized haploid cell produced by meiosis and involved in sexual reproduction. Male gametes are usually small and motile (spermatozoa), whereas female gametes (oocytes) are larger and non-motile.

gametogenesis Process leading to the production of ***gametes***.

gametophyte Haploid stage of life cycle of plants; the major vegetative stage for simple plants like liverworts.

gamma-aminobutyric acid See ***GABA***.

gamma-delta cells (γδ-T-cells) Lineage of T-cells possessing the γδ form of the T-cell receptor. Appear early in development and constitute around 5% of mature T-cells in peripheral lymphoid organs. May be the predominant form at epithelial surfaces. Most have neither CD4 nor CD8.

gamma-toxin Complex toxin (33.4 kD) produced by *Staphylococcus aureus*. Rabbit erythrocytes are particularly sensitive to lysis by the toxin, but the mechanism is unknown.

ganciclovir Antiviral nucleoside analogue: 9-((1,3-dihydroxy-2-propoxy) methyl)-guanine. Used in treatment of cytomegalovirus.

ganglion A physical cluster of ***neurons***. In vertebrates, the ganglia are appendages to the central nervous system; in invertebrates, the majority of neurons are organized as separate ganglia.

ganglion cell A type of ***interneuron*** that conveys information from the retinal ***bipolar***, horizontal and ***amacrine cells*** to the brain.

ganglioside A ***glycosphingolipid*** that contains one or more residues of N-acetyl or other neuraminic acid derivatives. Gangliosides are found in highest concentration in cells of the nervous system, where they can constitute as much as 5% of the lipid.

gangliosidoses Diseases, such as ***Tay–Sachs***, caused by inherited deficiency in enzymes necessary for the breakdown of gangliosides. Cause gross pathological changes in the nervous system, with devastating neurological symptoms.

GAP See ***GTPase-activating protein***.

gap gene ***Segmentation genes*** involved in specifying relatively coarse subdivisions of the embryo. They are expressed sequentially in development between ***egg-polarity genes*** and ***pair-rule genes***. In *Drosophila*, there are at least three such genes, eg. *Kruppel, knirps*.

gap junction A junction between two cells consisting of many pores that allow the passage of molecules up to about 900 D. Each pore is formed by an hexagonal array (connexon) of six transmembrane proteins (connexins) in each plasma membrane: when mated together the pores open, allowing communication and the interchange of metabolites between cells. ***Electrical synapses*** are gap junctions, and metabolic cooperation depends upon the formation of gap junctions.

GAP-43 See ***neuromodulin***.

GAPs See ***growth-associated proteins***.

gar2 Nucleolar protein containing a glycine- and arginine-rich (GAR) domain from *Schizosaccharomyces pombe*. Required for 18S rRNA and 40S ribosomal subunit accumulation and assembly of the preribosomal particles. Reminiscent of ***nucleolin***.

gargoyle cells Fibroblasts with large deposits of mucopolysaccharide, commonly found in storage diseases such as ***Hurler's disease***.

gas motif DNA motif in the Fc ε RI promoter.

gas vacuole A prokaryotic cellular organelle consisting of cylindrical vesicles around 75 × 300nm, often in clusters. The wall of the gas vacuole, which is permeable to gases but not to water, is formed from a monolayer of a single protein. Gas vacuoles are found mainly in planktonic cyanobacteria and their prime function is to make the bacterium buoyant.

gas-1 (growth arrest-specific gene 1) Gene, the product of which has been shown to cause cellular quiescence.

gastric inhibitory polypeptide Peptide ***hormone*** (43 amino acids) that stimulates insulin release and inhibits the release of gastric acid and pepsin. See ***GIP***.

gastrin A group of peptide ***hormones*** secreted by the mucosal gut lining of some mammals in response to mechanical stress or high pH. They stimulate secretion of protons and pancreatic enzymes. Several different gastrins have been identified; human gastrin I has 16 amino acids (2116 D). Gastrin is competitively inhibited by ***cholecystokinin***.

gastrin-releasing peptide (GRP) A regulatory peptide (27 amino acids) thought to be the mammalian equivalent of ***bombesin***. It elicits gastrin release, causes bronchoconstriction and vasodilation in the respiratory tract, and stimulates the growth and mitogenesis of cells in culture.

Gastropoda Class of the phylum Mollusca; snails, slugs, limpets and conches.

gastrula Embryonic stage of an animal when ***gastrulation*** occurs; follows ***blastula*** stage.

gastrulation During embryonic development of most animals a complex and coordinated series of cellular movements occurs at the end of ***cleavage***. The details of these movements, gastrulation, vary from species to species, but usually result in the formation of the three primary germ layers, ***ectoderm***, ***mesoderm*** and ***endoderm***.

gated ion channel Transmembrane proteins of excitable cells, that allow a flux of ions to pass only under defined circumstances. Channels may be either ***voltage-gated***, such as the ***sodium channel*** of neurons, or ***ligand-gated*** such as the ***acetylcholine receptor*** of cholinergic synapses. Channels tend to be relatively ion-specific and allow fluxes of typically 1000 ions to pass in around 1ms; they are thus much faster at moving ions across a membrane than transport ***ATPases***.

gating currents Small currents in the membrane just prior to the increase in ionic permeability, due to the movement of charged particles within the membrane. So called because they open the 'gates' for current flow through ion channels.

Gaucher's disease Familial autosomal recessive defect of glucocerebrosidase (β-glucosidase), most common in Ashkenazi Jews. Associated with hepatosplenomegaly (enlargement of liver and spleen) and, in severe early-onset forms of the disease, with neurological dysfunction.

gax One of the family of growth arrest genes (*gas* and *gadd* genes), a homeobox gene restricted to cardiovascular tissue and downregulated by mitogens.

GC box DNA binding motif (GGGCG) recognized by the mammalian ***transcription factor*** Sp1.

gCAP39 (MCP) Protein (40 kD) that binds to the barbed ends of microfilaments. Has considerable sequence homology with ***gelsolin***, and like gelsolin responds to calcium and to phosphoinositides. Widely distributed in mammalian cells and also secreted into plasma, though does not have a signal sequence. Not the same as ***capZ***, though function similar.

GDGF (glioma-derived growth factor) Growth factor derived from glioma cells. GDGF-I has subsequently shown to be a homodimer of polypeptides immunologically similar to the A-chain of ***PDGF*** and GDGF-II is predominantly a heterodimer containing one peptide similar to A-chain and one similar to B-chain of PDGF with some homodimers of B-chain-like peptides. GDGF-II is more potently mitogenic.

GDP Guanosine diphosphate. Phosphorylation gives ***GTP***.

Gea-1 ***GEF*** (guanine nucleotide exchange factor) for ***ARF*** from *Saccharomyces cerevisiae*. Contains ***Sec7***-like domain. Similar to ***ARNO*** and ***cytohesin***.

Gea-2 Yeast guanine nucleotide exchange factor for ***ARF***. Contains ***Sec7***-like domain.

GEA1, GEA6 Proteins from *Arabidopsis thaliana* that are homologous to the 'early methionine labelled' (Em) proteins of wheat.

GEFs (guanine nucleotide exchange factors.) Family of proteins that facilitate the exchange of bound GDP for GTP on small G-proteins such as ras and rho and thus activate them. Act in the opposite way to GAPs (which promote the hydrolysis of bound GTP, thereby switching the G protein to the inactive form). Family includes ***cytohesin***, ***ARNO***, Gea-1 and 2, kalirin and yeast Sec7.

gel Jelly-like material formed by the coagulation of a colloidal liquid. Many gels have a fibrous matrix and fluid-filled interstices: gels are viscoelastic rather than simply viscous and can resist some mechanical stress without deformation. Examples are the gels formed by large molecules such as collagen (and gelatin), agarose, acrylamide and starch.

gel electrophoresis ***Electrophoresis*** using a gel supporting-phase. Usually applied to systems where the gel is based on polyacrylamide. See ***electrophoresis***.

gel filtration An important method for separating molecules according to molecular size by percolating the solution through beads of solvent-permeated polymer that has pores of similar size to the solvent molecules. Unlike a continous filter that retards flow according to molecular size, separation is achieved because molecules that can enter the beads take a longer path (ie. are retarded) than those that cannot. Typical gels for protein separation are made from polyacrylamide, or from flexible (Sephadex) or rigid (agarose, Sepharose) sugar polymers. The size separation range is determined by the degree of cross-linking of the gel.

gel mobility shift (gel retardation assay) Technique for studying DNA-protein interactions. For example, to study the levels of a particular transcription factor, nuclear protein extracts are incubated with a radiolabelled DNA fragment containing the transcription factor's binding site. When the DNA gel is run, the amount of radiolabel that runs more slowly than the free DNA is directly proportional to the amount of transcription factor present.

gel retardation assay (mobility shift assay) Test for interaction between molecules by looking for a change in gel electrophoretic mobility. For example, to assay for levels of a ***transcription factor***, cell extracts are incubated with a radiolabelled oligonucleotide corresponding to the recognition sequence of the transcription factor, and run on an agarose gel. Most of the radiolabel will run quickly through the gel, but any radioactivity that is retarded is presumably caused by DNA–transcription factor interaction.

gel shift assay See ***bandshift assay***.

gelatin Heat-denatured collagen.

gelatinous lesion A small area of oedema in the arterial intima, possibly a precursor of a ***fibrous plaque***.

gelsolin Actin-binding protein (90 kD) that nucleates actin polymerization, but at high calcium ion concentrations (10^{-6} M) causes severing of filaments.

gemfibrozil Drug that lowers plasma lipoprotein levels. Activates ***PPAR*** α.

gene Originally defined as the physical unit of heredity but the meaning has changed with increasing knowledge. It is probably best defined as the unit of inheritance that occupies a specific locus on a chromosome, the existence of which can be confirmed by the occurrence of different allelic forms. Given the occurrence of ***split genes***, it might be redefined as the set of DNA sequences (***exons***) that are required to produce a single polypeptide.

gene amplification Selective replication of DNA sequence within a cell, producing multiple extra copies of that sequence. The best-known example occurs during the maturation of the oocyte of *Xenopus*, where the set (normally 500 copies) of ribosomal RNA genes is replicated some 4000 times to give about 2 million copies.

gene chip An array of oligonucleotides immobilized on a surface that can be used to screen an RNA sample (after reverse transcription) and thus a method for rapidly determining which genes are being expressed in the cell or tissue from which the RNA came. There are two alternatives for the immobilized oligonucleotides: either a random set of defined sequences or known probes for genes (probes for cytokines, adhesion molecules, etc.). In both cases the position on the chip defines the sequence that hybridizes.

gene cloning The insertion of a DNA sequence into a ***vector*** that can then be propagated in a host organism, generating a large number of copies of the sequence.

gene conversion A phenomenon in which alleles are segregated in a 3:1 not 2:2 ratio in meiosis. May be a result of ***DNA polymerase*** switching templates and copying from the other homologous sequence, or a result of mismatch repair (nucleotides being removed from one strand and replaced by repair synthesis using the other strand as template).

gene dosage Number of copies of a particular ***gene*** locus in the genome; in most cases either one or two

gene duplication A class of DNA rearrangement that generates a supernumerary copy of a gene in the genome. This would allow each gene to evolve independently to produce distinct functions. Such a set of evolutionarily related genes can be called a gene family.

gene expression The full use of the information in a gene via ***transcription*** and ***translation*** leading to production of a protein and hence the appearance of the ***phenotype*** determined by that gene. Gene expression is assumed to be controlled at various points in the sequence leading to protein synthesis and this control is thought to be the major determinant of cellular ***differentiation*** in eukaryotes.

gene family (multigene family) A set of genes coding for diverse proteins which, by virtue of their high degree of sequence similarity, are believed to have evolved from a single ancestral gene. An example is the immunoglobulin family where the characteristic features of the constant-domains are found in various cell surface receptors.

gene knockout See ***knockout***.

gene regulatory protein Any protein that interacts with DNA sequences of a gene and controls its transcription.

gene therapy Treatment of a disease caused by malfunction of a gene, by stable ***transfection*** of the cells of the organism with the normal gene.

gene transfer General term for the insertion of foreign genes into a cell or organism. Synonymous with ***transfection***.

generation time Time taken for a cell population to double in numbers, and thus equivalent to the average length of the cell cycle.

genetic burden See ***genetic load***.

genetic code Relationship between the sequence of bases in ***nucleic acid*** and the order of amino acids in the polypeptide synthesized from it. A sequence of three nucleic acid ***bases*** (a triplet) acts as a 'codeword' (***codon***) for one amino acid. See Table C5.

genetic drift Random change in allele frequency within a population. If the population is isolated and the process continues for long enough it may lead to speciation.

genetic engineering General term covering the use of various experimental techniques to produce molecules of DNA containing new genes or novel combinations of genes, usually for insertion into a host cell for cloning.

genetic linkage The term refers to the fact that certain genes tend to be inherited together, because they are on the same chromosome. Thus parental combinations of characters are found more frequently in offspring than non-parental. Linkage is measured by the percentage recombination between loci, unlinked genes showing 50% recombination. See ***linkage equilibrium, linkage disequilibrium***.

genetic load (genetic burden) In general terms the decrease in fitness of a population (as a result of selection acting on phenotypes) due to deleterious ***mutations*** in the population gene pool. More specifically, the average number of lethal ***recessive mutations***, in the ***heterozygous*** state, estimated to be present in the genome of an individual in a population.

genetic locus The position of a gene in a linkage map or on a chromosome.

genetic recombination Formation of new combinations of alleles in offspring (viruses, cells or organisms) as a result of exchange of DNA sequences between molecules. It occurs naturally, as in ***crossing over*** between homologous chromosomes in meiosis or experimentally, as a result of ***genetic engineering*** techniques.

genetic transformation Genetic change brought about by the introduction of exogenous DNA into a cell. See ***transformation, germ-line transformation, transfection***.

geneticin (Antibiotic G418) Used as a selection agent in transfection. Toxic to bacteria, yeast and mammalian cells. Vector has geneticin-resistance gene from bacteria so that positive transfectants can be selected.

genistein (4,5,7-trihydroxyisoflavone) Inhibitor of protein tyrosine kinases. Competes at ATP-binding site and will inhibit other kinases to some extent.

genome The total set of genes carried by an individual or cell.

genome project Coordinated programme to completely sequence the genomic DNA of an organism. Usually, genomic sequencing is combined with several associated ventures; the ***physical mapping*** of the genome (to allow the genome to be sequenced); the sequencing of ***expressed sequence tags*** (to aid in the identification of transcribed sequences in the genomic DNA sequence); for non-human organisms, a programme of systematic mutagenesis, and genetic mapping of the mutants (to infer function of novel genes by reverse genetics); and an overarching computer database resource, to manage and give access to the data.

genomic imprinting Parent-specific expression or repression of genes or chromosomes in offspring. There are an increasing number of recognized chromosomal imprinting events in pathological conditions, eg. preferential transmission of paternal or maternal predisposition to diabetes or atopy, preferential retention of paternal alleles in ***rhabdomyosarcoma***, ***osteosarcoma***, ***retinoblastoma*** and ***Wilms' tumour***, preferential translocation to the paternal chromosome 9 of a portion of maternal chromosome 22 to form the ***Philadelphia chromosome*** of chronic myeloid leukaemia.

genomic library Type of DNA library in which the cloned DNA is from an organism's genomic DNA. As genome sizes are relatively large compared to individual ***cDNAs***, a different set of ***vectors*** is usually employed in addition to ***plasmid*** and ***phage***; see ***bacterial*** and ***yeast artificial chromosomes***, ***cosmid***.

genotype The genetic constitution of an organism or cell, as distinct from its expressed features or ***phenotype***.

gentamicin A group of aminoglycoside antibiotics produced by *Micromonospora* spp. Members include the closely related gentamycins C1, C2 and C1a, together with gentamycin A. They inhibit protein synthesis on 70S ribosomes by binding to the 23S core protein of the small subunit, which is responsible for binding mRNA. Mode of action similar to that of kanamycin, neomycin, paromomycin, spectinomycin and streptomycin. Active against strains of the bacterium *Pseudomonas aeruginosa*.

geotaxis See ***gravitaxis***. The prefix gravi- is preferable since the gravitational fields used as cues need not necessarily be the Earth's.

geotropism See ***gravitropism***.

gephyrin Peripheral membrane protein of the cytoplasmic face of the glycinergic synapses in the spinal cord. Appears at developing postsynaptic sites before the glycine receptor and may therefore be important in clustering. Thought to interact with the glycine receptors and also with microtubules.

geranyl Prenyl group ((2E)-3,7-dimethyl-2,6-octadien-1-yl). Intermediate in cholesterol synthesis and in production of ***geranyl-geranyl*** group. Can be post-translationally added to proteins, but geranyl-geranyl prenylation is more common.

geranyl transferase (farnesyl pyrophosphate synthetase) Enzyme (EC 2.5.1.10) responsible for the post-translational transfer of geranyl-geranyl residue to protein.

geranyl-geranyl ((2E, 6E, 10E)-3,7,11,15-tetramethyl-2,6,10,14-hexadecatetraen-1-yl group) Prenyl group post-translationally added by geranyl transferase to some cytoplasmic proteins generally at CAAX motif (where X is usually leucine), serves to associate them with membranes. See ***farnesylation***.

geranylation The geranoyl group is a linear sequence of two isoprenyl residues. The term geranylgeranyl is used for the common unit of four residues. See also ***polyisoprenylation***.

GERL The Golgi-endoplasmic reticulum-lysosome system. See individual entries for each of these membranous compartments of the trans-Golgi network.

germ cell Cell specialized to produce ***haploid*** gametes. The germ cell line is often formed very early in embryonic development.

germ layers The main divisions of tissue types in multicellular organisms. Diploblastic organisms (eg. coelenterates) have two layers, ectoderm and endoderm; triploblastic organisms (all higher animal groups) have mesoderm between these two layers. Germ layers become distinguishable during late blastula/ early gastrula stages of embryogenesis, and each gives rise to a characteristic set of tissues, the ectoderm to external epithelia and to the nervous system for example, although some tissues contain elements derived from two layers.

germ-line transformation Micro-injection of foreign DNA into an early embryo, so that it becomes incorporated into the ***germ-line*** of the individual, and thus stably inherited in subsequent generations of ***transgenic*** organisms. Typically, the DNA would be a ***reporter gene*** or ***cDNA*** in a ***vector*** such as a ***transposon***, that might also carry a visible ***marker gene*** (such as eye or coat colour), so that successful transformation could readily be detected.

Gerstmann-Straussler-Scheinker syndrome A familial ***spongiform encephalopathy***. Transgenic mice with a mutant form of the ***PrP*** gene from patients with this syndrome develop degenerative brain disease that is similar, but not identical, to that caused by ***scrapie***.

GF-1 See ***erythroid transcription factor***.

GFAP See ***glial fibrillary acidic protein***.

ghosts See ***erythrocyte ghosts***.

GI Common abbreviation for gastro-intestinal.

G_i See ***GTP-binding protein***.

giant axons Extraordinarily large unmyelinated axons found in invertebrates. Some, like the squid giant axon, can approach 1mm diameter. Large axons have high conduction speeds; the giant axons are invariably involved in panic or escape responses, and may (eg. crayfish) have ***electrical synapses*** to further increase speed. Vertebrate axons with high conduction velocites are much narrower: they have a ***myelin sheath***, allowing ***saltatory conduction***.

Giardia Genus of flagellate protozoans, found as intestinal parasites of vertebrates. The human intestinal parasite is *Giardia lamblia*. The cells have a large disc or 'sucker' on their anterior ventral surfaces, by which they attach to the intestinal mucosa. The attachment of the disc is very strong and can prevent peristaltic clearing. This can result in acute or chronic diarrhoea especially in children. The disease is termed Giardiasis or Lambliasis.

giardin Group of proteins, of 29–38 kD, found in the ventral discs of *Giardia lamblia*.

gibberellic acids Diterpenoid compounds with ***gibberellin*** activity in plants. At least 70 related gibberellic acids have been described and designated as a series GA1, GA2, etc.

gibberellin Plant growth substance (phytohormone) involved in promotion of stem elongation, mobilization of food reserves in seeds, and other processes. Its absence results in the dwarfism of some plant varieties. Chemically all known gibberellins are ***gibberellic acids***.

Gibbs free energy, G See ***free energy***.

Giemsa A Romanovsky-type stain that is often used to stain blood films that are suspected to contain protozoan parasites. Contains both basic and acidic dyes and will therefore differentiate acid and basic granules in granulocytes.

Giemsa banding See ***G banding***.

Ginkgo biloba Ornamental tree originally native to China. Sole surviving member of the family Ginkgoales. Source of various bioactive compounds.

GIP Acronym with too many meanings. (1) ***Gastric inhibitory polypeptide***. (2) Glucose-dependent insulinotropic polypeptide; 42 residue peptide that stimulates insulin release from pancreatic β-cells. (3) Gip or Gip light-regulated inhibitory G-protein from washed microvilli of the photoreceptors of *Octopus*. (4) General insertion protein; Protein involved in the insertion of cytochrome c1 into the mitochondrial matrix. (5) GTPase inhibitory protein. Protein that interferes with the binding of ***GAP*** to *ras*, or enhances nucleotide exchange (a ***GEF***).

gip2 The *gip2* oncogene encodes GTPase-deficient α-subunits of G_s or $G_{i\text{-}2}$ proteins and has been identified in tumours of the ovary and adrenal cortex. It will induce neoplastic transformation of Rat-1 cells but not NIH 3T3 cells and appears to be a tissue-selective oncogene.

GIRKs (G-protein gated inward rectifying potassium channels) A newly identified gene family. The gene products are thought to form functional channels through the assembly of heteromeric subunits. A point mutation in the GIRK2 gene is the cause of the neurological and reproductive defects observed in the ***weaver*** (wv) mutant mouse.

gland Organ specialized for secretion by the infolding of an epithelial sheet. The secretory epithelial cells may either be arranged as an acinus with a duct or as a tubule. Glands from which release occurs to a free epithelial surface are exocrine; those that release product to the circulatory system are endocrine glands.

glandular fever Self-limiting disorder of lymphoid tissue caused by infection with ***Epstein–Barr virus*** (infectious mononucleosis). Characterized by the appearance of many large lymphoblasts in the circulation.

Glanzmann's thrombasthenia Platelet dysfunction in which aggregation is deficient. A specific glycoprotein complex (IIb/IIIa) is absent from the plasma membrane: this seems to be the fibronectin/fibrinogen receptor and is a β_3-***integrin***. See Table I1.

gliadin Group of proline-rich proteins found in cereal seeds constituting the major storage protein. Associate with glutenin to form gluten.

glial cells Specialized cells that surround ***neurons***, providing mechanical and physical support, and electrical insulation between neurons.

glial fibrillary acidic protein (GFAP) Member of the family of ***intermediate filament*** proteins, characteristic of glial cells.

glial filaments ***Intermediate filaments*** of glial cells, made of ***glial fibrillary acidic protein***.

glibenclamide (glyburide) Sulphonyl urea that acts via sulphonyl urea receptor (SUR) to regulate inwardly rectifying K^+-ATP channels (Kir6.1) of pancreatic islet cells thereby increasing insulin release.

glicentin Peptide fragment cleaved from ***glucagon*** by prohormone convertase.

gliding motility Mode of cell motility exhibited by, eg. gregarines. There are no obvious motile appendages, little actin is detectable, and the motor mechanism is poorly understood.

glioblastoma Highly malignant brain tumour derived from glial cells.

gliomas Neuroectodermal tumours of neuroglial origin: include astrocytomas, oligodendroglioma, and ependymoma derived from astrocytes, oligodendrocytes and ependymal cells respectively. All infiltrate the adjacent brain tissue, but they do not metastasize.

gliostatin Cytokine (dimeric, subunits 50 kD) very similar to PD-ECGF. Neurotrophic for cortical neurons and inhibits proliferation of astrocytes (stimulates differentiation). May play a part in aberrant neovascularization of rheumatoid synovium.

Gln See ***glutamine***.

globin The polypeptide moiety of haemoglobin. In the adult human the haemoglobin molecule has two α (141 residues) and two β (146 residues) globin chains.

globoside (cytolipin K) Major neutral glycosphingolipid found in kidney and erythrocytes.

globular protein Any protein that adopts a compact morphology is termed globular. Generally applied to proteins in free solution, but may also be used for compact folded proteins within membranes.

glomerulonephritis Inflammatory response in the kidney glomerulus that often arises because immune complexes cannot pass through the basement membrane of the fenestrated epithelium where plasma filtration occurs. Circulating neutrophils are trapped on the accumulated adhesive immune complexes (which also activate complement). Immune complex tends to be irregularly distributed in contrast to the picture in ***Goodpasture's syndrome***.

Glossina morsitans Tsetse fly (name is onomatopoeic), vector of African trypanosomiasis.

Glu See ***glutamic acid***.

glucagon A polypeptide hormone (3485 D) secreted by the A-***cells*** of the islets of Langerhans in the pancreas in response to a fall in blood sugar levels. Induces ***hyperglycaemia***. A family of structurally related peptides includes glucagon-like peptides 1 and 2 (encoded by the same gene), ***gastric inhibitory polypeptide, secretin, vasoactive intestinal peptide, growth hormone-releasing factor***; ***PACAP***; ***exendins***. See also Table H2.

glucans Glucose-containing polysaccharides, including ***cellulose, callose, laminarin, starch,*** and ***glycogen***.

glucocorticoids Steroid hormones (both natural and synthetic) that promote gluconeogenesis and the formation of glycogen at the expense of lipid and protein synthesis. They also have important anti-inflammatory activity. Type compound is hydrocortisone (cortisol); other common examples are cortisone, prednisone, prednisolone, dexamethasone, betamethasone. See also Table H3.

glucomannan *Hemicellulose*-like plant cell-wall polysaccharide containing glucose and mannose linked by β (1-4)-glycosidic bonds. May contain some side chains of galactose, in which case it may be termed galactoglucomannan. A major polysaccharide of gymnosperm wood (softwood).

gluconeogenesis Synthesis of glucose from non-carbohydrate precursors, such as pyruvate, amino acids and glycerol. Takes place largely in liver, and serves to maintain blood glucose under conditions of starvation or intense exercise.

glucosamine Amino-sugar (2-amino-2-deoxyglucose); component of ***chitin***, heparan sulphate, chondroitin sulphate and many complex polysaccharides. Usually found as β-D-N-acetylglucosamine.

glucosaminoglycan See ***glycosaminoglycan***.

glucose (dextrose) 6-carbon sugar (aldohexose) widely distributed in plants and animals. Breakdown of glucose (***glycolysis***) is a major energy source for metabolic processes. In green plants, glucose is a major product of photosynthesis, and is stored as the polymer ***starch***. In animals it is obtained chiefly from dietary di- and polysaccharides, but also by ***gluconeogenesis***, and is stored as ***glycogen***. Storage polymer in microorganisms is ***dextran***.

glucose transporter (GLUT-1, etc.) Generic name for any protein that transports glucose. In bacteria these may be ***ABC proteins***, in mammals they belong to a family of 12-transmembrane integral transporters. GLUT-4 is found in muscle and is insulin-responsive. Defects in GLUT-4 may be responsible for some forms of ***diabetes***.

glucose-1-phosphate Product of glycogen breakdown by phosphorylase. Converted to glucose-6-phosphate by phosphoglucomutase.

glucose-6-phosphate Phosphomonoester of glucose that is formed by transfer of phosphate from ATP, catalysed by the enzyme ***hexokinase***. It is an intermediate both of the glycolytic pathway (next converted to fructose-6-phosphate), and of the NADPH-generating ***pentose phosphate pathway***.

glucose-related protein (GRP) One of the stress-related proteins: identical to ***endoplasmin***.

glucosylation Transfer of glucose residues, usually from the nucleotide-sugar derivative UDPG. Enzymic glucosylation to generate the glucosyl-galactosyl disaccharide on the hydroxylysine of collagen is a normal process. A recent theory suggests that glucosylation of certain long-lived proteins by a non-enzymic reaction with free glucose may contribute to ageing.

glucuronic acid (GA; GlcA) Uronic acid formed by oxidation of OH group of glucose in position 6. D-glucuronic acid is widely distributed in plants and animals as a subunit of various oligosaccharides.

glucuronoxylan ***Hemicellulose***-like plant cell-wall polysaccharide containing glucuronic acid and xylose as its main constituents. Has a β (1-4)-xylan backbone, with 4-*O*-methylglucuronic acid side chains. Arabinose and acetyl side chains may also be present. Major polysaccharide of angiosperm wood (hardwood).

GLUT See ***glucose transporter***.

glutamate Major fast excitatory ***neurotransmitter*** in the mammalian ***central nervous system***. See ***glutamate receptor***. Also the excitatory neuromuscular transmitter in arthropod skeletal muscles.

glutamate receptor See ***amino acid receptor*** superfamily. Glutamate receptors are implicated in many important brain functions including ***long-term potentiation*** (LTP). At least four major glutamate-gated ***ion channel*** subtypes are at present distinguished on pharmacological grounds, named after their most selective agonists: N-methyl-D-aspartate (NMDA), implicated in memory and learning, neuronal cell death, ischaemia and epilepsy; kainic acid (KA); quisqualate/AMPA; and L-2-amino-4-phosphobutyrate (APB). A fifth subtype (APCD), trans-1-amino-cyclopentane 1,3 dicarboxylate, is a G-protein coupled receptor.

glutamic acid (Glu; E) One of the 20 α-amino acids (147D) commonly found in proteins. Plays a central role in amino acid metabolism, acting as precursor of ***glutamine, proline*** and ***arginine***. Also acts as amino group donor in synthesis by transamination of alanine from pyruvate, and aspartic acid from oxaloacetate. Glutamate is also a neurotransmitter; the product of its decarboxylation is the inhibitory neurotransmitter ***GABA***.

glutamine (Gln; Q) One of the 20 amino acids (146 D) commonly found (and directly coded for) in proteins. It is the amide at the γ-carboxyl of the amino acid ***glutamate***. Glutamine can participate in covalent crosslinking reactions between proteins, by forming peptide-like bonds by a transamidation reaction with lysine residues. This reaction, catalysed by clotting factor XIII stabilizes the aggregates of fibrin formed during blood clotting. Media for culture of animal cells contain some 10 times more glutamine than other amino acids, the excess presumably acting as a carbon source.

glutaraldehyde A dialdehyde used as a fixative, especially for electron microscopy. By its interaction with amino groups (and others) it forms crosslinks between proteins.

glutathione The tripeptide γ-glutamylcysteinylglycine. It contains an unusual peptide linkage between the γ-carboxyl group of the glutamate side chain and the amine group of cysteine. The concentration of glutathione in animal cells is around 5mM and its sulphydryl group is kept largely in the reduced state. This allows it to act as a sulphydryl buffer, reducing any disulphide bonds formed within cytoplasmic proteins to cysteines. Hence, few, if any, cytoplasmic proteins contain disulphide bonds. Glutathione is also important as a cofactor for the enzyme ***glutathione peroxidase***, in the uptake of amino acids and participates in ***leukotriene*** synthesis.

glutathione peroxidase A detoxifying enzyme that eliminates hydrogen peroxide and organic peroxides. Glutathione is an essential cofactor for the enzyme and its reaction involves the oxidation of glutathione (GSH) to glutathione disulphide (GSSG). The GSSG is then reduced to GSH by glutathione reductase. Glutathione peroxidase (GPX) has a selenocysteine residue in its active site. Three forms of the enzyme exist: cytoplasmic GPX; plasma GPX; and phospholipid hydroperoxide GPX.

glutathione reductase An FAD-containing enzyme, a dimer of 50 kD subunits. It catalyses the NADP-dependent reduction of glutathione disulphide (GSSG) to glutathione (GSH). This is an essential reaction that maintains a GSH:GSSG ratio in the cytoplasm of around 500:1.

glutathione S transferase (GST) Enzyme that will couple ***glutathione*** to a xenobiotic as the first step in removal. Now very commonly used as a fusion with a gene of interest that is being expressed in a bacterial system. The fusion construct can be purified easily from lysate by passage down a glutathione affinity column, the purified construct then being eluted with glutathione. The GST can then be cleaved proteolytically from the protein of interest though often the complete fusion protein can be used.

glutelin Group of proteins found in seeds of cereals such as wheat.

gluten Protein-rich fraction from cereal grains, especially wheat. When hydrated forms a sticky mass responsible for the mechanical properties of bread dough. ***Glutelins*** and ***gliadin*** form a substantial component.

glutenin A ***glutelin*** found in endosperm of wheatgrain. Component of gluten and through its tendency to polymerize contributes to properties of dough.

Gly See ***glycine***.

glyceraldehyde-3-phosphate Three-carbon intermediate of the glycolytic pathway formed by the cleavage of fructose 1,6-bisphosphate, catalysed by the enzyme ***aldolase***. Also involved in reversible interchange between ***glycolysis*** and the ***pentose phosphate pathway***.

glyceraldehyde-3-phosphate dehydrogenase (GAPD; GAPDH; G3PD) Glycolytic enzyme (EC 1.2.1.12) that catalyses the reversible oxidative phosphorylation of glyceraldehyde-3-phosphate. Has been shown to interact with various elements of the ***cytoskeleton***, and with the trinucleotide repeat in the ***huntingtin*** gene.

glycerination Permeabilization of the plasma membrane of cells by incubating in aqueous glycerol at low temperature. The technique was first applied to muscle which, once glycerinated, can be made to contract by adding exogenous ATP and calcium.

glycerol A metabolic intermediate, but primarily of interest as the central structural component of the major classes of biological lipids, triglycerides and phosphatidyl phospholipids. Also used as a ***cryoprotectant***.

glycine (Gly; G) The simplest amino acid (75 D). It is a common residue in proteins, especially collagen and elastin, and is not optically active. It is also a major inhibitory ***neurotransmitter*** in spinal cord and brainstem of vertebrate ***central nervous system***.

glycine receptor Chloride channel-forming receptor. One of a family of ***neurotransmitter*** receptors with fast intrinsic ***ion channels***. See ***amino acid receptors***.

glycipan-3 Membrane glycoprotein thought to bind IGF-2. Defective in ***Simpson–Golabi–Behmel syndrome***.

glycocalyx The region, seen by electron microscopy, external to the outer dense line of the ***plasma membrane*** that appears to be rich in glycosidic compounds such as proteoglycans and glycoproteins. Since these molecules are often integral membrane proteins and may be denatured by the processes of fixation for electron microscopy, it might be better to avoid the term or to refer to membrane glycoproteins or to proteoglycans associated with the cell surface.

glycocholate (N-cholyl-glycine) Anion of the bile acid, glycholic acid. Usually found in bile as the sodium salt. Has powerful detergent properties.

glycoconjugate Any biological macromolecule containing a carbohydrate moiety – thus a generic term to cover glycolipids, glycoproteins and proteoglycans.

glycogen Branched polymer of D-glucose (mostly α(1-4)-linked, but some α(1-6) at branch points). Size range very variable, up to 10^5 glucose units. Major short-term storage polymer of animal cells, and is particularly abundant in the liver and to a lesser extent in muscle. In the electron microscope glycogen has a characteristic 'asterisk/star' appearance.

glycolic acid Hydroxyacetic acid; found in young plants and green fruits. Glycolate is formed from ribulose-1,5-bisphosphate in a seemingly wasteful side reaction of photosynthesis, known as photorespiration.

glycolipid Oligosaccharides covalently attached to lipid as in the glycosphingolipids (GSL) found in plasma membranes of all animal and some plant cells. The lipid part of GSLs is sphingosine in which the amino group is acylated by a fatty chain, forming a ***ceramide***. Most of the oligosaccharide chains belong to one of four series, the ganglio-, globo-, lacto-type 1 and lacto-type 2 series. Blood group antigens are GSLs.

glycolysis See ***Embden–Meyerhof pathway***.

glycophorins A class of abundant transmembrane glycoproteins of the human erythrocyte. The major component is a 131-residue peptide chain that is highly O-glycosylated and is rich in terminal sialic acid. The peptide chain carries the ***MN blood group antigens*** at its N-terminus.

glycoprotein Proteins with covalently attached sugar units, either bonded via the OH group of serine or threonine (O-glycosylated) or through the amide NH2 of asparagine (N-glycosylated). Includes most secreted proteins (serum albumin is the major exception) and proteins exposed at the outer surface of the plasma membrane. Sugar residues found include mannose, N-acetyl glucosamine, N-acetyl galactosamine, galactose, fucose and sialic acid.

glycosaminoglycan attachment site In ***proteoglycans*** a number of ***glycosaminoglycan*** chains are attached to a core protein through a xyloside residue which is linked to a serine residue at the consensus pattern S-G-X-G.

glycosaminoglycans (formerly mucopolysaccharides) Polysaccharide side chains of ***proteoglycans*** made up of repeating disaccharide units (more than 100) of amino sugars, at least one of which has a negatively charged side group (carboxylate or sulphate). Commonest are hyaluronate (D-glucuronic acid-N-acetyl-D-glucosamine: MW up to 10 million), chondroitin sulphate (D-glucuronic acid-N-acetyl-D-galactosamine-4 or -6-sulphate), dermatan sulphate (D-glucuronic acid-or L-iduronic acid-N-acetyl-D-galactosamine), keratan sulphate (D-galactose-N-acetyl-D-glucosamine-sulphate) and heparan sulphate (D-glucuronic acid-or L-iduronic acid-N-acetyl-D-glucosamine). Glycosaminoglycan side chains (with the exception of hyaluronate) are covalently attached to a core protein at about every 12 amino acid residues to produce a proteoglycan; these proteoglycans are then non-covalently attached by link proteins to hyaluronate, forming an enormous hydrated space-filling polymer found in ***extracellular matrix***. The extent of sulphation is variable and the structure allows tremendous diversity.

glycosidase (glycosylase) General and imprecise term for an enzyme that degrades linkage between sugar subunits of a polysaccharide. Any of the EC 3.2 class of hydrolases that cleave glycosidic bonds. They may distinguish between α and β links, for example, but are not very substrate specific. See ***endoglycosidase***.

glycosidic bond Bond between anomeric carbon of a sugar and the group to which it is attached (which may be in another sugar or in protein or lipid).

glycosome Microbody containing glycolytic enzymes, found in protozoa of the Kinetoplastida (eg. trypanosomes).

glycosphingolipids ***Ceramide*** derivatives containing more than one sugar residue. If sialic acid is present these are called ***gangliosides***.

glycosyl phosphatidyl inositol (GPI) See ***GPI-anchor***.

glycosyl transferase Enzyme that catalyses the transfer of a sugar (monosaccharide) unit from a sugar nucleotide derivative to a sugar or amino acid acceptor.

glycosylase See ***glycosidase***.

glycosylation The process of adding sugar units such as in the addition of glycan chains to proteins.

glyoxisome Organelle found in plant cells, containing the enzymes of the ***glyoxylate cycle***. Also contains catalase and enzymes for ***beta-oxidation*** of fatty acids. Together with the ***peroxisome*** makes up the class of organelles known as ***microbodies***.

glyoxylate cycle Metabolic pathway present in bacteria and in the ***glyoxisome*** of plants, in which two acetyl-CoA molecules are converted to a four-carbon dicarboxylic acid, initially succinate. Includes two enzymes not found elsewhere: isocitrate lyase and malate synthase. Permits net synthesis of carbohydrates from lipid, and hence is prominent in those seeds in which lipid is the principal food reserve.

glypiation See ***GPI-anchor***.

GM-CSF (granulocyte–macrophage colony stimulating factor) A cytokine that stimulates the formation of granulocyte or macrophage colonies from myeloid stem cells isolated from bone marrow.

Gm-types Genetically determined allotypic antigens found on IgG of some individuals.

gnotobiotic Organism or environment completely or almost completely depleted of all organisms or all other organisms. Animals that are SPF (specific pathogen-free) are gnotobiotic.

GnRH See ***gonadotrophin-releasing hormones***.

goblet cell (1) Cell of the epithelial lining of small intestine that secretes mucus and has a very well-developed Golgi apparatus. (2) Cell type characteristic of larval lepidopteran midgut, containing a potent H^+-***ATPase***, and thought to be involved in maintenance of ion and pH gradients.

Golber–Hogness box See ***TATA box***.

Goldmann equation (Goldmann constant field equation) Equation that describes the electrical potential across a membrane in terms of the distributions and relative permeabilities of the main permeant ions (typically sodium, potassium and chloride). Assumes that the electrical field across the membrane is constant, and that there are no ***active transport*** processes; but none the less gives a reasonable approximation to 'real' membranes.

Golgi apparatus Also known as the Golgi body, Golgi vesicles; in plants, the ***dictyosome***; in flagellate protozoa, the parabasal body. Intracellular stack of membrane-bounded vesicles in which glycosylation and packaging of secreted proteins takes place; part of the ***GERL*** complex. Vesicles from endoplasmic reticulum fuse with the *cis*-Golgi region (the inner concave face) and progress through the vesicular stack to the *trans*-Golgi, whence they move towards the plasma membrane or lysosomes.

Gomori procedure Cytochemical staining procedure used to localize acid phosphatases. Depends upon the production of phosphate ions from organic phosphoesters such as β-glycerophosphate. The phosphate in the presence of lead ions causes the formation of a precipitate of lead salt that is converted to the brown sulphide of lead by the action of yellow ammonium sulphide.

gonadotrophin-releasing hormones (GnRH) Peptide hormones that act on the pituitary to stimulate production of ***luteinizing hormone*** and ***follicle-stimulating hormone***. Decapeptides: consensus QHXSXXXXPG (amidated).

gonadotrophins (gonadotropins) Group of glycoprotein hormones from the anterior lobe of the pituitary gland. They stimulate growth of the gonads and the secretion of sex hormones. Examples: ***follicle-stimulating hormone, luteinizing hormone, chorionic gonadotrophin***.

gonadotropins See ***gonadotrophins***.

Gonyaulax Genus of ***dinoflagellates***. Responsible for 'red tides' and associated shellfish poisoning due to ***saxitoxin***. Some species are bioluminescent.

Goodpasture's syndrome Disease in which there is accumulation of a very uniform layer of autoantibodies to components of basement membrane on the kidney glomerular basement membrane.

gooseberry *(gsb)* A ***segment-polarity gene*** of *Drosophila*. Contains the ***paired box domain***.

GOS19-1 Human ***macrophage inflammatory protein 1*** α.

Gossypium Genus of plants that includes cotton.

gout Recurrent acute arthritis of peripheral joints caused by the accumulation of monosodium urate crytals. Usually due to overproduction of uric acid but may be a result of underexcretion. The problems partly arise because neutrophils release lysosomal enzymes as a result of damage to the phagosome membrane by ingested crystals: colchicine acts to reduce the attack by inhibiting lysosome–phagosome fusion.

Gp-Ib (Glycoprotein Ib) Integral protein of platelets that binds to von Willebrand factor and is involved in thrombus formation. Disulphide-linked heterodimer (α 68 kD; β 22 kD) deficient in ***Bernard–Soulier syndrome***.

GPCR See ***G-protein-coupled receptor***.

GPI-anchor (glypiation) Common modification of the C-terminus of membrane-attached proteins in which a phosphatidyl inositol moiety is linked through glucosamine and mannose to a phosphoryl ethanolamine residue that is linked to the C-terminal amino acid of the protein by its amino group. Glypiation is the sole means of attachment of such proteins to the membrane. The name comes from the addition of glycosyl phosphatidyl inositol (PI).

GPS (1) Gps1 and gps2 suppress *ras* and MAPK-mediated signalling and interfere with JNK activity suggesting that they are signal repressors. (2) gps1-4 are glycoprotein synthesis genes from *Schizosaccharomyces* pombe. (3) GPS1 is GST-P Silencer 1.

Graafian follicle Final stage in the differentiation of follicles in the mammalian ovary. Consists of a spherical fluid-filled blister on the surface of the ovary that bursts at ovulation to release the oocyte.

gradient perception Problem faced by a cell that is to respond directionally to a gradient of, eg. a diffusible attractant chemical. In a spatial mechanism the cell would compare receptor occupancy at different sites on the cell surface; a temporal mechanism would involve comparison of concentrations at different times, the cell moving randomly between readings. In pseudospatial sensing, the cell would detect the gradient as a consequence of positive feedback to protrusive activity if receptor occupancy increased with time as the protrusion moved upgradient. Few cell types have been unambiguously shown to detect gradients.

graft-versus-host response (GVH) When a graft of lymphocytes or a graft containing lymphocytes is made into an animal these may, if appropriately mismatched at MHC Class I to their host, produce lymphocyte clones that will react by a variety of processes against the host and cause damage.

Gram-negative bacteria Bacteria with thin ***peptidoglycan*** walls bounded by an outer membrane containing ***endotoxin*** (lipopolysaccharide). See ***Gram stain***.

Gram-positive bacteria Bacteria with thick cell walls containing ***teichoic*** and ***lipoteichoic acid*** complexed to the ***peptidoglycan***. See ***Gram stain***.

Gram stain A heat-fixed bacterial smear is stained with crystal violet (methyl violet), treated with 3% iodine/potassium iodide solution, washed with alcohol and counterstained. The method differentiates bacteria into two main classes, ***Gram-positive*** and ***Gram-negative***. Certain bacteria, notably mycobacteria, that have walls with high lipid content show acid-fast staining; the stain resists decoloration in strong acid.

Gramicidin A A linear peptide of alternate D- and L-amino acids that acts as a cation ionophore in lipid bilayer membranes. It is proposed that two molecules form a membrane-spanning helix containing a pore lined with polar residues.

Grammostola spatulata Chilean pink tarantula. See ***grammotoxin***.

grammotoxin (ω-grammotoxin SIA) Toxin (peptide, 36 residues) from spider, *Grammostola spatulata*, that inhibits non-L-type voltage-regulated calcium channels, thus resembling ω-conotoxins and ω-***agatoxins***. See ***VSCC***.

grana (plural) See ***granum***.

granins (chromogranins; secretogranins) Family of related acidic proteins (400–600 residues) found in many endocrine cell secretory vesicles. Secretogranin 1 = chromogranin B; secretogranin 2 = chromogranin C.

granular component of nucleolus Area of nucleolus that appears granular in the electron microscope and contains 15nm diameter particles that are maturing ribosomes. In contrast to the pale-staining and ***fibrillar*** areas.

granulation tissue Highly vascularized tissue that replaces the initial fibrin clot in a wound. Vascularization is by ingrowth of capillary endothelium from the surrounding vasculature. The tissue is also rich in fibroblasts (which will eventually produce the ***fibrous tissue***) and leucocytes.

granule cell Type of neuron found in the cerebellum.

granulocyte Leucocyte with conspicuous cytoplasmic granules. In humans the granulocytes are also classified as polymorphonuclear leucocytes and are subdivided according to the staining properties of the granules into ***eosinophils***, ***basophils*** and ***neutrophils*** (using a Romanovsky-type stain); some invertebrate blood cells are also referred to, not very helpfully, as granulocytes.

granulocyte–macrophage colony stimulating factor See ***GM-CSF***.

granulocytopenia Low granulocyte number in circulating blood.

granuloma Chronic inflammatory lesion characterized by large numbers of cells of various types (macrophages, lymphocytes, fibroblasts, giant cells), some degrading and some repairing the tissues.

granulopoiesis The production of granulocytes in the bone marrow.

granum (plural, grana) Stack of ***thylakoids*** in the chloroplast, containing the ***light-harvesting system*** and the enzymes responsible for the ***light-dependent reactions*** of photosynthesis.

granzyme Family of serine proteases found in cytotoxic T-cells and NK cells and involved in ***perforin***-dependent cell killing.

Grave's disease Autoimmune disease characterized by goitre, exophthalmia and thyrotoxicosis. Autoantigen is probably TSH-receptor or closely associated protein in thyroid. In Caucasians is associated with HLA-B8 and DR3.

Graves's disease See ***hyperthyroidism***.

gravitaxis Directed locomotory response to gravity.

gravitropism Directional growth of a plant organ in response to a gravitational field: roots grow downwards, shoots grow upwards. Achieved by differential growth on the sides of the root or shoot. A gravitation field is thought to be sensed by sedimentation of statoliths (starch grains) in root caps.

GRB-2 (growth factor receptor-bound protein 2) Protein that links the cytoplasmic domain of growth factor receptor to ***sos*** and to ***shc*** through its SH2 and SH3 domains, and thus is important in the assembly of the signalling complex. Homologous to yeast sem-5 and *Drosophila* drk.

Greek key Protein ***motif*** in which four ***beta strands*** (in one or two ***beta sheets***) form a twisted arrangement similar to a pattern often seen on Greek vases.

Green algae See ***Chlorophyta***.

green fluorescent protein (GFP) Protein from luminous jellyfish. Excited by blue light (as produced from ***aequorin*** luminescence), it emits green light. The gene has been cloned and mutagenized to give brighter fluorescence and different colour variants. These have become valuable transgenic tools in ***flag tagging*** and as ***reporter gene***s and ***cell lineage*** markers.

gregarine movement Peculiar gliding movement shown by gregarines (Protozoa), the mechanism of which is poorly understood.

grex The multicellular aggregate formed by cellular slime moulds (Acrasidae): the slug-like grex migrates, showing positive phototaxis and negative gravitaxis, until culmination (the formation of a fruiting body) takes place. Coordination of the activities of the hundreds of thousands of individual amoebae that compose the grex may involve pulses of cyclic AMP in *Dictyostelium discoideum*, a species in which cAMP is the chemotactic factor for aggregation.

grey crescent A region near the equator of the surface in the fertilized egg of various amphibia, often of greyish colour, that appears to contain special morphogenetic properties.

GRF See ***growth hormone-releasing factor***.

Grim Regulator of apoptosis in *Drosophila*. Like ***reaper*** and ***hid*** effect is blocked by caspase inhibitors and *Drosophila* homologues of mammalian ***IAPs***.

GRIP (glutamate receptor interacting protein.) Protein (120 kD, 1112 residues) found in postsynaptic terminal, contains seven ***PDZ*** domains and is involved in the clustering of ***AMPA*** receptors with which it interacts through PDZ domains 4 and 5.

griseofulvin Polyketide antibiotic from *Penicillium griseofulvum*. Used therapeutically as an antifungal. Blocks microtubule assembly and thus mitosis.

gRNA (Guide RNA) Small RNA molecules (60–80 nucleotides) that are found in the ***editosome***. Guide RNAs are complementary to edited portions of the mature mRNA and contain poly-U tails that donate the Us added during editing.

gro See ***melanoma growth-stimulatory activity***.

groEL See ***chaperonins***.

groucho *(Gro)* An adaptor molecule that acts as a co-repressor for some negative regulators in *Drosophila* development but not others (does not act on repressor regions of *even-skipped, kruppel*, or *knirps* transcription factors).

ground tissue Plant tissues other than those of the vascular system and the ***dermal tissues***. Composed of relatively undifferentiated cells.

group specific antigen See ***gag-protein***.

growth arrest specific gene 1 See ***gas-1***.

growth cone A specialized region at the tip of a growing ***neurite*** that is responsible for sensing the local environment and moving toward the neuron's target cell. Growth cones are hand-shaped, with several long ***filopodia*** that differentially adhere to surfaces in the embryo. Growth cones can be sensitive to several guidance cues, eg. surface adhesiveness, growth factors, neurotransmitters and electric fields (***galvanotropism***).

growth cone collapse Loss of motile activity and cessation of advance by ***growth cones***. There are now thought to be specific molecules that inhibit the motility of particular growth cones, and which are important in establishing correct pathways in developing nervous systems.

growth control When applied to cells usually means control of growth of the population, ie. of the rate of division rather than of the size of an individual cell.

growth factor receptor-bound protein 2 See ***GRB-2***.

growth factors Polypeptide hormones that regulate the division of cells, eg. EGF (epidermal growth factor), PDGF (platelet-derived GF), FGF (fibroblast GF). Insulin and somatomedin are also growth factors; the status of NGF (***nerve growth factor***) is more uncertain. Perturbation of growth factor production or of the response to growth factor is important in neoplastic transformation.

growth hormone (somatotropin) Polypeptide (191 amino acids) produced by anterior pituitary that stimulates liver to produce ***somatomedins*** 1 and 2.

growth hormone-releasing factor (GRF) Peptide hormone related to the ***glucagon*** family, released from the pituitary, acts on the ***adenohypophysis*** to release ***growth hormone***.

growth substances See ***plant growth substances***.

growth-associated proteins (GAPs) Group of developmentally regulated polypeptides thought to be critical for the formation of neural circuitry. The acidic membrane phosphoprotein GAP-43 is synthesized and transported down regenerating and developing axons; pp46 localized in growth cone membranes during embryogenesis; B-50 in mature presynaptic membranes in the regulation of phosphotidylinositol turnover and F1 in the ***hippocampus*** during ***long-term potentiation***, are now all known to be the same protein.

growth hormone-regulating hormone See Table H2. Hypothalamic hormones that induce (somatoliberin) or inhibit (somatostatin) the release of growth hormone (somatotropin).

growth/differentiation factors Members of the TGF-β family of growth factors. See also ***midkine***.

GRP See ***glucose-related protein*** or ***gastrin-releasing peptide***.

GRP-1 Guanine nucleotide exchange factor with ***PH domain*** that binds PtdIns(3,4,5) P3 and also has domain with homology to yeast Sec7. See ***cytohesin***, ***ARNO***, Gea-1 and Gea-2, other members of the family.

G_s See ***GTP-binding protein***.

gsb See ***gooseberry***.

GSK (glycogen synthase kinase) GSK3 is a serine-threonine protein kinase that plays an important part in various intracellular signalling pathways including the control of glycogen metabolism and protein synthesis by insulin, the modulation of ***AP1*** and ***CREB***, the specification of cell fate in *Drosophila* and the regulation of dorsoventral patterning in *Xenopus*. GSK3 itself is regulated by ***PKB***. GSK3 is identical to tau protein kinase I and may therefore be important in the phosphorylation of ***tau*** that is known to occur in the neurofibrillary tangles of Alzheimer's disease.

GST See ***glutathione S transferase***.

GST fusion protein One way of purifying proteins expressed by a cloned gene is to insert the gene of interest into a vector inframe with a gene encoding glutathione S-transferase; this is then expressed in a cell line to produce a (chimeric) ***fusion protein***. This can be purified on a glutathione affinity column by virtue of the affinity of GST for glutathione, eluted with free glutathione and the GST can subsequently be proteolytically cleaved off and removed by an iteration of the affinity purification, though this is not always necessary.

GST-P Glutathione S-transferase P gene that is strongly and specifically expressed during chemical hepatocarcinogenesis. The promoter region has a silencer (GPS1) to which various transactivators bind.

GTBP (G/T-binding protein) Protein (160 kD) important in mismatch recognition in human cells. The heterodimer of GTBP with hMSH2 (one of the ***MSH*** family) binds mismatches (G paired with T) as a first step in excision repair. Absence of either protein predisposes to tumours. GTBP gene is located near that for hMSH2 and GTBP can be considered one of the MSH family.

GTP (guanosine 5′-triphosphate) Like ***ATP***, a source of phosphorylating potential, but is synthesized separately and takes part in a limited, distinct set of energy-requiring processes. Synthesis is by a substrate-linked phosphorylation involving succinyl coenzyme A, part of the ***tricarboxylic acid cycle***. GTP is required in protein synthesis, the assembly of ***microtubules***, and for the activation of regulatory G-proteins ***GTP-binding proteins***.

GTP-binding proteins (G-proteins) There are two main classes of G-proteins: the heterotrimeric G-proteins that associate with receptors of the seven transmembrane domain superfamily and are involved in signal transduction; and the small cytoplasmic G-proteins. The Gα subunit (39-52 kD) of the heterotrimeric G-proteins dissociates from the βγ subunits (β, 35-36 kD: γ, 6-10 kD) when GTP is bound, and in this state will interact with various second messenger systems, either inhibiting (G_i) or stimulating (G_s). The Gα subunit has slow GTPase activity and once the GTP is hydrolyzed it reassociates with the βγ subunits. There is less diversity among the βγ subunits, but they may have direct activating effects in their own right. Most βγ subunits are post-translationally modified by myristoylation or isoprenylation which may alter their association with membranes. Stimulatory G-proteins are permanently activated by ***cholera toxin***, inhibitory ones by ***pertussis toxin***. ***Transducin*** was one of the first of the heterotrimeric G-proteins to be identified. The small G-proteins are a diverse group of monomeric GTPases that include ***ras***, ***rab***, ***rac*** and ***rho*** and which play an important part in regulating many intracellular processes including cytoskeletal organization and secretion. Their GTPase activity is regulated by activators (GAPs) and inhibitors (GIPs) which determine the duration of the active state. See also ***GEFs***, ***ras-like GTPases***.

GTPase-activating protein (GAP) Originally purified as a 125 kD protein from bovine brain (1044 amino acids); stimulates the GTPase activity of ras-p21 and thereby switches it to the inactive state. GAP may itself be regulated by phospholipids and by phosphorylation on a tyrosine residue by growth factor receptors (PDGF-R, EGF-R). The neurofibromatosis type 1 gene (NF1) codes for a protein homologous to GAP. GAP has both ***SH2*** and ***SH3*** domains. Another example is sar-1 (from yeast).

guanidinium chloride (guanidine hydrochloride) Chloride salt of guanidinium $(C(NH_2)_3)^+$, a powerful chaotropic agent that is used to denature proteins.

guanine (2-amino 6-hydroxy purine) One of the constituent bases of nucleic acids, nucleosides and nucleotides.

guanosine (9-β-D-ribofuranosyl guanine) The nucleoside formed by linking ribose to guanine.

guanosine 5′-triphosphate See ***GTP***.

guanylate cyclase Enzyme catalysing the synthesis of guanosine 3′,5′-cyclic monophosphate from guanosine 5′-triphosphate (cyclic GMP (cGMP) is used as a ***second messenger*** in heart muscle and ***photoreceptor*** cells). The plasma membrane form of guanylate cyclase is an integral membrane protein with an extracellular receptor for peptide hormones, a transmembrane domain, a ***protein kinase***-like domain, and a guanylate cyclase domain. Examples: sea urchin receptors for ***speract*** and ***resact***; and ***atrial natriuretic peptide*** receptors. Two soluble forms of guanylate cyclase are also known, heterodimers of highly related subunits, 70–80 kD.

guanylin Peptide occuring in vertebrate gut that elevates the ***second messenger*** cyclic GMP in a variety of tissues (including gut) via a membrane ***guanylate cyclase***. The receptor is also the target for the *E. coli* heat-stable enterotoxin.

guard cell Plant cells occurring in pairs in the ***epidermis***, flanking each ***stoma***. Changes in ***turgor*** in the guard cells cause the stoma to open and close.

Guarnieri body Acidophilic inclusion body found in cells infected with vaccinia virus; composed of viral particles and proteins, it is the location of virus replication and assembly.

guidance See ***contact guidance***.

Table G1. The properties of heterotrimeric G protein subunits

α subunits

α	MW (kD)	Signal detector	Effector	Toxin	Comments
α_s	44–45	β-adrenergic, glucagon and many other receptors	Activates adenylate cyclase	CT	Stimulatory. At least four splice variants known, and relative concentration varies from one cell type to another
α_{i1}	40.4	α-adrenergic, muscarinic cholinergic, opiate and many other receptors	Inhibits adenylate cyclase and PLC (phosphoinositide hydrolysis)	PT	Inhibitory. The sequence for $G_{i\alpha1}$, and $G_{i\alpha2}$ were derived from two different cDNA clones. The relative importance of the two forms in the cell is not yet clear
α_{i2}	40.5		Inhibits adenylate cyclase, PLC, involved in regulating K^+ -channels, Ca^{2+} channels	PT	
α_{i3}	40.5		Probably same as $\alpha_{i1,2}$	PT	
α_{olf}	44	Seven-membrane spanning	Activates adenylate cyclase	CT	In olfactory neurons
α_{gust}	40.3	Taste sensors	Unclear	Both	Component of Gustducin, restricted to tongue
α_z	40.9	unknown	Inhibits adenylate cyclase	None	In brain and neuronal cells
α_q	42.2		Activates PLCβ (PI hydrolysis)	None	Widely expressed except in T-cells
α_{11-14}	42–45		Activation PLCβ	None	Widely distributed though with variations between tissues
α_o	39	Unknown	Inhibits neuronal calcium channels	PT	'o' stands for other G-protein. It was first detected in large amounts in brain. Splice variants known
α_{t1}	40	Rhodopsin	cGMP-PDE	Both	Subunit of Transducin, found in the rod cells of the retina
α_{t2}	40.4	—	cGMP-PDE	Both	Found in the cone cells of the retina: probably the α subunit of the cone's analogue of Transducin

β and γ subunits

β	MW (kD)	Tissue distribution
β1	37	β subunit in transducin
β2	38	Widely distributed
β3	37	Wide, particularly cone cells of retina
β4	37	Wide, especially brain, eye, lung, heart and testis
γ1	8.4	Rod cells of retina
γ2	7.8	Brain
γ3	8.3	Brain and retina
γ4		Unknown
γ5	7.3	Wide
γ7	7.5	Brain

Based partly upon S. Watson and S Arkinstall (1994) *G-protein Linked rReceptor Facts Book.* Academic Press.

guide RNA Small RNA molecules that hybridize to specific mRNAs and direct their ***RNA editing***.

Guillain–Barre syndrome (Landry–G–B syndrome) Acute infective polyneuritis, associated with cytomegalovirus infection, in which there is cell-mediated immunity to a component of myelin; the disease may be autoimmune in origin. Causes a temporary paralysis, particularly of the extremities.

GUS Glucuronidase; widely used as a ***reporter gene***.

gustducin Taste cell-specific ***GTP-binding protein***. Novel Gα subunit; resembles ***transducin*** more than any other Gα.

gut-associated lymphoid tissue (GALT) Peripheral lymphoid organ consisting of lymphoid tissue associated with the gut (Peyer's patches, tonsils, mesenteric lymph nodes and the appendix).

GVH See ***graft-versus-host response***.

Gymnospermae One of the two major division of seed-bearing vascular plants of which the most common members are conifers. See ***Angiospermae***.

gyrus Any of the ridge-like folds of the cerebral cortex.

H

H400 See ***macrophage inflammatory protein 1*** β.

h See ***hairy***.

H See ***histidine***.

H^+-ATPase See ***proton ATPase***.

H chain Heavy chain of immunoglobulin; see ***IgG***, ***IgM***, etc.

H zone Central portion of the A band of the ***sarcomere***, the region that is not penetrated by thin (actin) filaments when the muscle is only partially contracted. The ***M-line*** is in the centre of the H zone.

H2 antigen An antigen of the H2 region of the major histocompatibility complex of mice. Divided into Class I and Class II antigens.

H2 blocker Antagonist of the histamine type 2 (H2) receptor. Drugs of this type block gastric acid secretion and are therefore clinically useful in treating duodenal ulcers.

H2 complex Mouse equivalent of the human MHC (major histocompatability complex) system, a set of genetic loci coding for Class I and Class II MHC antigens and for complement components. See ***histocompatibility antigen***.

HA tag (haemagglutinin tag; USA, hemagglutinin tag) ***Epitope tag*** (YPYDVPDYA) derived from the haemagglutinin molecule.

ha-ras Harvey-*ras*.

HA1 and HA2 adaptors See ***clathrin adaptor proteins***.

HACBP See ***calreticulin***.

haem (USA, heme) Compounds of iron complexed in a porphyrin (tetrapyrrole) ring that differ in side chain composition. Haems are the prosthetic groups of ***cytochromes*** and are found in most oxygen carrier proteins.

haemagglutination Agglutination of red blood cells, often used to test for the presence of antibodies directed against red cell surface antigens or carbohydrate-binding proteins or viruses in a solution. Requires that the agglutinin has at least two binding sites.

haemagglutinin Substance that will bring about the ***agglutination*** of erythrocytes.

haemangioblast Earliest mesodermal precursor of both blood and vascular endothelial cells. Described in embryonic yolk-sac blood-islands of birds.

Haemanthus katherinae The African blood lily, chiefly known because of classic time-lapse studies done on mitosis in endosperm cells.

haematocrit Relative volume of blood occupied by erythrocytes. An average figure for humans is 45ml per cent, ie. a packed red cell volume of 45ml in 100ml of blood.

haematoma A lump of congealed blood, usually caused by a small haemorrhage, usually under the skin.

haematopoiesis Production of blood cells involving both proliferation and differentiation from stem cells. In adult mammals usually occurs in bone marrow.

haematopoietic stem cell Cell that gives rise to distinct daughter cells, one a replica of the stem cell, one a cell that will further proliferate and differentiate into a mature blood cell. Pluripotent stem cells can give rise to all lineages, committed stem cells (derived from the pluripotent stem cell) only to some.

haematoxylin Basophilic stain that gives a blue colour (to the nucleus of a cell for example), commonly used in conjunction with ***eosin***, which stains the cytoplasm pink/red. Various modifications of haematoxylin have been developed. The histopathologist's 'H & E' is haematoxylin and eosin.

haemocyanin Blue, oxygen-transporting, copper-containing protein found in the blood of molluscs and crustacea. A very large protein with 20–40 subunits and molecular weight of 2–8 million, and having a characteristic cuboidal appearance under the electron microscope. Prior to the introduction of immunogold techniques, it was used for electron-microscopic localization by coupling to antibody. Keyhole limpet haemocyanin (KLH) is widely used as a carrier in the production of antibodies.

haemocytes Blood cells, associated with a haemocoel, particularly those of insects and crustacea. Despite the name they are more leucocyte-like, being phagocytic and involved in defence and clotting of ***haemolymph***, and not involved in transport of oxygen.

haemoglobin four-subunit globular oxygen-carrying protein of vertebrates and some invertebrates. There are two α and two β chains (very similar to myoglobin) in adult humans; the haem moiety (an iron-containing substituted porphyrin) is firmly held in a non-polar crevice in each peptide chain.

haemoglobinopathies Disorders due to abnormalities in the haemoglobin molecule, the best known being sickle-cell anaemia in which there is a single amino acid substitution (valine for glutamate) in position 6 of the β chain. In other cases one of the globin chains is synthesized at a slower rate, despite being normal in structure. See also ***thalassaemia***.

haemolymph Circulating body fluid of invertebrates such as insects that have a haemocoel – sinuses and spaces between organs – rather than a closed circulatory system. Cells in the haemolymph are usually referred to as ***haemocytes***. Unlike vertebrate blood cells, haemocytes do not have an oxygen-carrying function, and subclasses are phagocytic with an immune function, thus resembling the granulocytes of vertebrates.

haemolysins Bacterial ***exotoxins*** that lyse erythrocytes.

haemolysis Leakage of ***haemoglobin*** from erythrocytes due to membrane damage.

haemolytic anaemia *Anaemia* resulting from reduced red cell survival time, either because of an intrinsic defect in the erythrocyte (hereditary spherocytosis or ellipsocytosis, enzyme defects, haemoglobinopathy), or an extrinsic damaging agent, for example autoantibody (autoimmune haemolytic anaemia), isoantibody, parasitic invasion of the cells (***malaria***), bacterial or chemical haemolysins, mechanical damage to erythrocytes.

haemonectin A 60 kD protein found in the bone marrow matrix of mice specifically aiding adhesion of granulocyte-lineage cells.

haemopexin Single-chain haem-binding plasma β_1-glycoprotein (57 kD); unlike ***haptoglobin*** does not bind haemoglobin. Present at around 1mg/ml in plasma. Responsible for transporting haem groups to the liver for breakdown. Structurally related to ***vitronectin*** and some ***collagenases***.

haemophilia Sex-linked congenital deficiency of blood-clotting system, usually of factor VIII.

Haemophilus influenzae Bacterium sometimes associated with influenza virus infections, causes pneumonia and meningitis.

haemorrhagic Related to or causing haemorrhage (bleeding).

haemosiderin A mammalian iron-storage protein related to ferritin but less abundant.

haemostasis Arrest of bleeding through blood clotting and contraction of the blood vessels.

Hagemann factor Plasma β-globulin (110 kD), blood-clotting factor XII, which is activated by contact with surfaces to form factor XIIa, which in turn activates factor XI. Factor XIIa also generates ***plasmin*** from plasminogen and ***kallikrein*** from prekallikrein. Both plasmin and kallikrein activate the complement cascade. Hagemann factor is important both in clotting and activation of the inflammatory process.

Hailey–Hailey disease (familial chronic benign pemphigus) A blistering disease of the skin apparently due to a defect in epidemal cell junctions, even though apparently normal desmosomes and adherens junctions can be assembled. Transmitted as an autosomal dominant.

hair cells (1) Cells found in the epithelial lining of the labyrinth of the inner ear. The hairs are ***stereovilli*** up to 25μm long that restrict the plane in which deformation of the apical membrane of the cell can be brought about by movement of fluid or by sound. Movement of the single stereocilium transduces mechanical movements into electrical receptor potentials. (2) *Bot.* Many plant surfaces are covered with fine hairs (*Tradescantia* stamens are a common source); the hairs are made up of thin-walled cells that are convenient for studying cytoplasmic streaming and for observing mitosis.

hairpin (alpha hairpin; α hairpin; beta hairpin; β hairpin) Protein ***motif*** formed by two adjacent regions of a polypeptide chain that lie antiparallel and alongside each other. Depending on whether the polypeptide is in ***alpha-helix*** or ***beta strand*** configuration, can be described as α hairpin or β hairpin, respectively.

hairy (h) A ***pair-rule gene*** of *Drosophila*.

hairy cell leukaemia (HCL) Clinically associated with severe T-cell dysfunction possibly as a result of defects in the responsiveness to activation although there is also a very restricted repertoire of the T-cell receptor-β family.

half-life ($t_{1/2}$) The period over which the activity or concentration of a specified chemical or element falls to half its original activity or concentration. Typically applied to the half-life of radioactive atoms but also applicable to any other situation where the population is of molecules of diminishing concentration or activity.

halobacteria Bacteria that live in conditions of high salinity.

Halobacterium halobium Photosynthetic (halophilic) bacterium that has patches of purple membrane containing the pigment ***bacteriorhodopsin***.

halophile Literally, salt-loving: organism that tolerates saline conditions, in extreme cases in concentrations considerably in excess of those found in normal sea water such as salt lakes. Some Archaebacteria (eg. ***Halobacterium halobium***) are notable for their ability to survive extremes of salinity.

halophyte Plant that grows in or tolerates salt-rich environments.

halorhodopsin Light-driven chloride ion pump of halobacteria, a retinylidene protein very similar to ***bacteriorhodopsin***.

halothane (2-bromo-2-chloro-1,1,1-trifluorethane) Widely used volatile anaesthetic given by inhalation.

halysin See ***disintegrin***.

hamartin Protein encoded by tumour suppressor gene TSC1. See ***tuberous sclerosis***.

hamartoma Tumour-like but non-neoplastic overgrowth of tissue that is disordered in structure. Examples are haemangiomas (that include the vascular naevus or birthmark) and the pigmented naevus (mole).

Hanks' BSS (HBSS) Balanced salt solution made up according to the recipe given originally by Hanks. Phosphate-buffered to pH 7.0–7.2. Suitable for mammalian and avian cells in temporary culture though is not a growth medium and will not sustain prolonged survival. Usually contains phenol red as an indicator.

Hansen's disease Leprosy.

Hansenula wingei Yeast, used for studies on mating type.

Hanta virus Hantaviruses are responsible for haemorrhagic fevers and exist in various serotypes with different pathogenicity for human beings, varying from asymptomatic infection to highly fatal disease. Human infections arise from inhalation of aerosolized excreta of persistently infected rodents.

haploid Describes a nucleus, cell or organism possessing a single set of unpaired ***chromosomes***. ***Gametes*** are haploid.

haplotype The set, made up of one ***allele*** of each gene, comprising the ***genotype***. Also used to refer to the set of alleles on one ***chromosome*** or a part of a chromosome, ie. one set of alleles of linked genes. Its main current usage is in connection with the linked genes of the ***major histocompatibility complex***.

hapten Could be considered an isolated ***epitope***: although a hapten (by definition) has an antibody directed against it, the hapten alone will not induce an immune response if injected into an animal, it must be conjugated to a carrier (usually a protein). The hapten constitutes a single antigenic determinant; perhaps the best-known example is dinitrophenol (DNP) which can be conjugated to BSA and against which anti-DNP antibodies are produced (antibodies to the BSA can be adsorbed out). Because the hapten is monovalent, immune complex formation will be blocked if the soluble hapten is present as well as the hapten-carrier conjugate (assuming there is more than one hapten per carrier then an immune precipitate can be formed). Competitive inhibition by the soluble small molecule is sometimes referred to as haptenic inhibition, and this term has carried over into lectin-mediated haemagglutination where monosaccharides are added to try to block haemagglutination: the blocking sugar defines the specificity of the lectin.

haptenic inhibition See ***hapten***.

haptoglobin Acid α_2-plasma glycoprotein that binds to oxyhaemoglobin that is free in the plasma, and the complex is then removed in the liver. Tetrameric (two α, two β subunits): the existence of two different α chains in humans means that haptoglobins can exist in three variants in heterozygotes.

haptonema Filament extending between the paired flagella of certain unicellular algae (haptophytes). Supported by six or seven microtubules (not in an axoneme-like array) and apparently used for capturing prey, in a manner analogous to the axopodia of ***heliozoa***.

haptotaxis Strictly speaking, a directed response of cells in a gradient of adhesion, but often loosely applied to situations where an adhesion gradient is thought to exist and local trapping of cells seems to occur.

Hardy–Weinberg law Mathematical formula that gives the relationship between gene frequencies and genotype frequencies in a population. If genotypes are AA, Aa, aa and the frequency of alleles A, a are respectively p, q, then $p + q = 1$ and AA:Aa:aa is $p^2{:}2pq{:}q^2$. Deviation from this equilibrium distribution suggests adverse survival characteristics of organisms with one of the alleles.

Harris–Benedict equation Equation for calculation of basal metabolic rate (BMR).

Hartnup disease Amino acid transport defect that leads to excessive loss of monoamino monocarboxylic acids (cystine, lysine, ornithine, arginine) in the urine, and poor absorption in the gut. See ***iminoglycinuria***.

Harvey sarcoma virus See ***ras***.

Hashimoto's thyroiditis Autoimmune disease in which there is destruction of the thyroid by autoantibodies usually directed against thyroglobulin and a lipoprotein of thyroid cell endosomes.

Hassell's corpuscle (thymic corpuscle) Spherical or ovoid bodies 20–50μm in diameter in the medulla of the thymus, composed of flattened concentrically arranged whorls of keratinized or hyaline cells surrounding dead cells in the core.

HAT medium A selective growth medium for animal tissue cells that contains hypoxanthine, the folate antagonist aminopterin (or amethopterin) and thymine. Used for selection of hybrid somatic cell lines, as in the production of monoclonal antibodies. In HAT medium, cells are forced to use these exogenous bases, via the salvage pathways, as their sole source of purines and pyrimidines. Parental cells lacking enzymes such as HGPRT or TK can be eliminated while hybrids grow.

Hatch–Slack pathway See ***Hatch–Slack–Kortshak pathway***.

Hatch–Slack–Kortshak pathway (Hatch–Slack pathway) Metabolic pathway responsible for primary CO_2 fixation in ***C4 plant*** photosynthesis. The enzymes that are found in ***mesophyll*** chloroplasts include ***PEP carboxylase***, which adds CO_2 to phosphoenolpyruvate to give the four-carbon compound, oxaloacetate. four-carbon compounds are transferred to bundle-sheath chloroplasts, where the CO_2 is liberated and refixed by the ***Calvin–Benson cycle***. The HSK pathway permits efficient photosynthesis under conditions of high light intensity and low CO_2 concentration, avoiding the non-productive effects of photorespiration.

haustorium A projection from a cell or tissue of a fungus or higher plant that penetrates another plant and absorbs nutrients from it. In fungi it is a projection of a ***hypha*** that penetrates into the cytoplasm of a host plant cell; in parasitic angiosperms, it is a modified root.

Hayflick limit See ***cell death, cell line***.

HB-EGF (heparin-binding epidermal growth factor) HB-EGF, like amphiregulin, has a long N-terminal extension that seems to confer the ability to bind to heparin and also to other connective tissue macromolecules (glycosaminoglycans) and cell surface molecules such as CD44. Because it is immobilized the effective local concentration may be much higher and the effects may differ from those of soluble growth factors.

HBGF See ***fibroblast growth factor***.

HCG (hCG; human chorionic gonadotrophin) See ***chorionic gonadotrophin***.

Hck Haematopoietic cell kinase. A protein tyrosine kinase of the ***src family*** found in lymphoid and myeloid cells and is bound to B-cell receptors in unstimulated B-cells. Deletion of hck and src or hck and fgr leads to severe developmental anomalies and impaired immunity in mice.

HCP Histone-rich calcium-binding protein. Major protein of ***sarcoplasmic reticulum***, where it may play a role in sequestering calcium ions. Highly acidic, with multiple repeats of highly conserved domains, believed to be responsible for calcium binding.

HCV ***Hepatitis C*** virus.

HDL See ***high density lipoproteins***.

Heaf test A commonly used tuberculin test in which tuberculin is injected intradermally with a multiple-puncture apparatus. A positive reaction indicates the presence of T-cell reactivity to mycobacterial products.

heart muscle See ***muscle***.

heat-shock factor See ***heat-shock proteins***.

heat-shock proteins (hsp) Families of proteins conserved through pro- and eukaryotic cells, induced in cells as a result of a variety of environmental stresses, though some are constitutively expressed. Some serve to stabilize proteins in abnormal configurations, play a role in folding and unfolding of proteins and the assembly of oligomeric complexes, and may act as ***chaperonins***. Hsp90 complexes with inactive steroid hormone receptor and is displaced upon ligand binding. Four major subclasses are recognized: hsp90, hsp70, hsp60 and small hsps. Hsps have been suggested to act as major immunogens in many infections.

heavy chain In general, the larger polypeptide in a multimeric protein. Thus the immunoglobulin heavy chain is of 50 kD, the light chain of 22 kD, whereas in myosin the heavy chain is very much larger (220 kD) than the light chains (~ 20 kD).

heavy meromyosin See *HMM*.

heavy water (deuterium oxide; D_2O) Most commonly used by cell biologists to stabilize ***microtubules***.

Heidenhain's Azan A particularly beautiful but extremely time-consuming trichrome staining method that, properly carried out, results in chromatin, erythrocytes and neuroglia being stained red, mucus blue, collagen sharp blue and cytoplasmic granules red, yellow or blue. Now very rarely used.

Heidenhain's iron haematoxylin One of many haematoxylin-based staining solutions and one that is particularly good for photography or automatic image processing because of the intensity of black staining that can be achieved. Requires differentiation in iron alum and thus the intensity of staining can be adjusted according to the specimen. Sections stained with Heidenhain are usually counterstained with, eg. eosin or orange G.

Heidenhain's Susa Good general-purpose histological fixative but has the disadvantage of containing mercuric chloride.

HeLa cells An established line of human epithelial cells derived from a cervical carcinoma (said to be from Henrietta Lacks).

Helianthus annuus The sunflower.

helicase See ***DNA helicase***.

Helicobacter pylori S-shaped or curved ***Gram-negative bacteria*** (0.5–0.9 × 3.0μm), non-spore forming, can be flagellate; found in human stomach. Was originally named *Campylobacter pyloridis*. Infection with *H. pylori* is now considered to be a major predisposing cause of gastric ulcers and antibiotic therapy is increasingly used.

helicoidal cell wall Type of plant cell wall in which each wall layer contains parallel ***microfibrils***, but in which the orientation of the microfibrils changes by a fixed angle from one layer to the next. Gives a characteristic 'herringbone' pattern in transmission electron microscopy. A similar architecture of fibrillar material is seen in some insect exoskeletons.

Heliozoa Amoeboid ***Protozoa***, order Heliozoida. They are generally free-floating spherical cells with many straight, slender microtubule-supported ***pseudopods*** radiating from the cell body like a sunburst. These modified pseudopods are termed ***axopodia***. Genera include *Actinophrys* and *Echinosphaerium*.

Helisoma trivolvis Pulmonate mollusc whose relatively simple nervous system contains large identifiable cells and is consequently, like *Hirudo* and *Aplysia*, a favourite preparation for studying neural mechanisms at the cellular level; and in particular for studying isolated neurons in culture.

helix-coil transition See ***random coil***.

helix-destabilizing proteins (single-stranded binding proteins) Proteins involved in DNA replication. They bind cooperatively to single-stranded DNA, preventing the reformation of the duplex and extending the DNA backbone, thus making the exposed bases more accessible for base pairing.

helix-loop-helix (hlh) A motif associated with ***transcription factors***, allowing them to recognize and bind to specific DNA sequences. Two ***alpha-helices*** are separated by a loop. Examples: myoblast *MyoD* 1, c-*myc*, *Drosophila* genes *daughterless, hairy, twist, scute, achaete, asense*. Not the same as ***helix-turn-helix***.

helix-turn-helix A motif associated with ***transcription factors***, allowing them to bind to and recognize specific DNA sequences. Two ***amphipathic*** alpha-helices are separated by a short sequence with a ***beta-pleated sheet***. One helix lies across the major groove of the DNA, while the recognition helix enters the major groove and interacts with specific bases. An example in ***Drosophila*** is the ***homeotic*** gene ***fushi tarazu***, that binds to the sequence TCAATTAAATGA. Not the same as ***helix-loop-helix***.

Heloderma horridum horridum Mexican beaded lizard. See ***helothermine***.

helodermin See ***exendin***.

helospectin See ***exendin***.

helothermine Protein toxin (25 kD) from venom of *Heloderma horridum horridum*. Probably acts by inhibiting the ryanodine-sensitive calcium release channel of sarcoplasmic reticulum.

helper factor A group of factors apparently produced by helper T-lymphocytes that act specifically or non-specifically to transfer T-cell help to other classes of lymphocytes. The existence of specific T-cell helper factor is uncertain.

helper T-cell See ***T-helper cells***.

helper virus A virus that will allow the replication of a coinfecting defective virus by producing the necessary protein.

hema-, hemo- (USA) See ***haema-, haemo-***.

heme (USA) See ***haem***.

hemicellulose Class of plant cell wall polysaccharide that cannot be extracted from the wall by hot water or chelating agents, but can be extracted by aqueous alkali. Includes ***xylan, glucuronoxylan, arabinoxylan, arabinogalactan*** II, ***glucomannan, xyloglucan,*** and galactomannan. Part of the cell wall matrix.

hemidesmosome Specialized junction between an epithelial cell and its basal lamina. Although morphologically similar to half a desmosome (into which intermediate cytokeratin filaments are also inserted) different proteins are involved.

hemimetabolous Of an insect, a species without any marked change in body plan from larval to adult, apart from the development of wings. Examples: grasshoppers and crickets. (cf. ***holometabolous***).

hemizygote Nucleus, cell or organism that has only one of a normally ***diploid*** set of genes. In mammals the male is hemizygous for the ***X chromosome***.

hemocyanin (USA) See ***haemocyanin***.

hemoglobin (USA) See ***haemoglobin***.

hemolytic anemia (USA) See ***haemolytic anaemia***.

Hensen's node (primitive knot) Thickening of the avian blastoderm at the cephalic end of the primitive streak. Presumptive notochord cells become concentrated in this region. May well be a source of retinoic acid that is acting as a morphogen in the developing embryo.

Hep2 cells Line established from human laryngeal carcinoma in 56-year-old Caucasian male. Extensively used in viral studies.

heparan sulphate (glycosaminoglycan) Constituent of membrane-associated ***proteoglycans***. The heparan sulphate-binding domain of ***NCAM*** is proposed to augment NCAM–NCAM interactions, suggesting that cell–cell bonds mediated by NCAM may involve interactions between multiple ligands. The putative heparin-binding site on NCAM is a 28 amino acid peptide shown to bind both heparin and retinal cells, as well as to inhibit retinal cell adhesion to NCAM. This strengthens the argument that this site contributes directly to NCAM-mediated cell–cell adhesion.

heparin Sulphated mucopolysaccharide, found in granules of mast cells, that inhibits the action of thrombin on fibrinogen by potentiating antithrombins, thereby interfering with the blood-clotting cascade. Platelet factor IV will neutralize heparin.

heparin-binding growth factor See ***HBGF***.

hepatitis A Small (27nm diameter) single-stranded RNA virus with some resemblance to ***enteroviruses*** such as polio. Causes 'infectious hepatitis'.

hepatitis B Virion (***Dane particle***), 42nm diameter, with an outer sheath enclosing inner 27nm core particle containing the circular viral DNA. Aggregates of the envelope proteins are found in plasma and are referred to as hepatitis B surface antigen (HBsAg; previously called Australia antigen). Causes 'serum hepatitis'; virus can persist for long periods (and in asymptomatic carriers); association of integrated virus with hepatocellular carcinoma is now well established.

hepatitis C A fairly high proportion of cases of hepatitis of the non-A, non-B, type, but not all, are now recognized to be caused by hepatitis C virus.

hepatitis non-A, non-B A virus somewhat similar in size to hepatitis A but has no antigenic cross-reaction with either A or B.

hepatocarcinoma (hepatocellular carcinoma) Malignant tumour derived from hepatocytes. Associated with ***hepatitis B*** in 80–90% of cases.

hepatocellular carcinoma See ***hepatocarcinoma***.

hepatocyte Epithelial cell of liver. Often considered the paradigm for an unspecialised animal cell. Blood is directly exposed to hepatocytes through fenestrated endothelium, and hepatocytes have receptors for subterminal N-acetylgalactosamine residues on asialo-glycoproteins of plasma.

hepatocyte growth factor (HGF) Polypeptide mitogen originally shown to cause cell division in hepatocytes. In the liver, the main sources of HGF are non-parenchymal cells. It is now clear that HGF is a mitogen for a number of cell types and it is found in many cells outside the liver, including platelets. HGF is synthesized as a single-chain precursor that is proteolytically cleaved to give a heavy chain (70 kD) and a light chain (30 kD) linked by a single disulphide bond. It contains multiple copies of the ***kringle*** domain. However, both the single-chain precursor and the two-chain forms of HGF are biologically active and HGF is generally isolated as a mixture of the two forms. HGF also alters cell motility and is now known to be identical to ***scatter factor***.

hepatoma Carcinoma derived from liver cells: better term is hepatocarcinoma.

hepatopancreas Digestive gland of crustaceans with functions approximately analogous to liver and pancreas of vertebrates: enzyme secretion, food absorption and storage.

HEPES (4-(2-hydroxyethyl)-1-piperazine-ethane-sulphonic acid) Very commonly used buffer for tissue culture medium. Its pKa of 7.5 makes it ideal for most cell culture work. Since related compounds are molluscicides it may be unsuitable for some invertebrate cultures. One of the series of zwitterionic buffers described by Good.

HepG2 cells Cell line derived from hepatic carcinoma. Epithelial in morphology; produce a variety of proteins such as prothrombin, alpha-fetoprotein, C3 activator and fibrinogen.

HER Family of receptors (HER2, erbB2; HER3, erbB3; HER4, erbB4) of the EGF receptor family of receptor ***tyrosine kinases***. Ligands are ***neuregulins***. Overexpression of HER-2, human homologue of erbB2, correlates with poor prognosis in breast carcinoma.

herbimycin A Tyrosine kinase inhibitor from *Streptomyces hygroscopicus*.

herculin Product of the muscle regulatory gene *Myf*-6. Also known as MRF4 (muscle regulatory factor 4).

hereditary angio-oedema Condition in which there seems to be uncontrolled production of ***C2-kinin*** because of a deficiency in C1-inhibitor levels.

Herpesviridae A group of large DNA viruses: *Herpes simplex* causes cold-sores and genital herpes; *Varicella zoster* causes chickenpox and shingles; cytomegalovirus causes congenital abnormalities and is an opportunistic pathogen; ***Epstein–Barr virus*** (EBV) causes glandular fever. *Herpes simplex* type 2 and EBV are associated with human tumours (cervical carcinoma for the former and ***Burkitt's lymphoma*** and nasopharyngeal carcinoma in the case of EBV). *Herpes simplex* establishes a lifelong latent infection of sensory neurons in human dorsal root ganglia and has a tendency to resurgence if the immune system is suppressed (causing shingles).

Herring bodies Granules within axons in the posterior lobe of the pituitary gland. Contain neurosecretory hormones.

Hers disease Glycogen-storage disease in which there is a deficiency of liver phosphorylase.

HETE (hydroxytetraeicosenoic acid) A family of hydroxyeicosenoic acid (C20) derivatives of arachidonic acid produced by the action of lipoxygenase. Potent pharmacological agents with diverse actions. See also ***HPETE***.

hetero- Prefix denoting different or varied.

heterochromatin The chromosomal regions that are condensed during interphase and at the time of nuclear division. They show what is considered an abnormal pattern of staining as opposed to ***euchromatin***. Can be subdivided into constitutive regions (present in all cells) and facultative heterochromatin (present in some cells only). The inactive X chromosome of female mammals is an example of facultative heterochromatin.

heterochrony Lack of synchronization.

heterocyst Specialized cell type found at regular intervals along the filaments of certain ***Cyanobacteria***; site of nitrogen fixation.

heterodimer A dimer in which the two subunits are different. One of the best-known examples is tubulin that is found as an α-tubulin/β-tubulin dimer. Heterodimers are relatively common, and it may be that the arrangement has the advantage that, for example, several different binding subunits may interact with a conserved signalling subunit.

heteroduplex DNA in which the two strands are different, either of different heritable origin, formed ***in vitro*** by annealing similar strands with some complementary sequences, or formed of mRNA and the corresponding DNA strand.

heterogenous nuclear RNA (hnRNA) Originally identified as a class of RNA, found in the nucleus but not the nucleolus, which is rapidly labelled and with a very wide range of sizes, 2–40 kilobases. It represents the primary transcripts of ***RNA polymerase*** II and includes precursors of all ***messenger RNAs*** from which introns are removed by splicing.

heterokaryon Cell that contains two or more genetically different nuclei. Found naturally in many fungi and produced experimentally by cell fusion techniques, eg. ***hybridoma***.

heterologous Derived from the tissues or DNA of a different species. cf. ***autologous, homologous***.

heterolysosome Secondary lysosome formed by fusion of a lysosome with another intracellular vesicle.

heterophile antibody An antibody raised against an antigen from one species that also reacts against antigens from other species. Also used of systems such as the ***Forssman antigen*** where antibody against antigens from a variety of species is present without immunization.

heteroplasia See ***heteroplasmy***.

heteroplasmy (heteroplasia) The occurrence of a tissue in the wrong place in an organism, as a result of inappropriate cellular differentiation.

heterosis Hybrid vigour, the superiority of a heterozygotic organism over the homozygote.

heterospecific Ab Artificially produced antibody in which the two antigen binding sites are for different antigens.

heterospory Condition in vascular plants where the spores are of two different sizes, the smaller producing male prothalli, the larger female prothalli.

heterothallic Situation in some fungi and algae in which there are two mating types and the individual thallus is self-sterile even if hermaphrodite.

heterotroph An organism that requires carbon compounds from other plant or them itself – not an ***autotroph***.

heterotypic Of different types. Thus heterotypic adhesion would be between dissimilar cells, in contrast to homotypic adhesion between cells of the same type.

heterozygosity index Measure of the number of gene loci for which an individual is heterozygous.

heterozygote Nucleus, cell or organism with different ***alleles*** of one or more specific genes. A heterozygous organism will produce unlike gametes and thus will not breed true.

heuristic Proceeding by trial and error rather than according to a planned route.

HEV See ***high endothelial venule***.

hexitol Sugar alcohol with 6 carbon atoms. Natural examples are sorbitol, mannitol.

hexokinase Enzyme that catalyses the transfer of phosphate from ATP to glucose to form glucose-6-phosphate, the first reaction in the metabolism of glucose via the glycolytic pathway.

hexon Subunit of a hexameric structure or with hexameric symmetry, in particular the arrangement of most of the capsomers of ***Adenoviridae*** – one capsomer surrounded by six others to form the hexon.

hexosaminidase Enzyme involved in the metabolism of ***gangliosides***. Deficient in ***Tay–Sachs disease***.

hexose Monosaccharide containing 6 carbon atoms, eg. ***glucose, galactose, mannose***.

Heymann nephritis Rat model of human membranous nephropathy; an autoimmune disease in which the antigen is ***megalin***.

Hfr High frequency, in the sense of bacterial ***conjugation***.

HGF See ***hepatocyte growth factor***.

HGPRT (hypoxanthine-guanine phosphoribosyl transferase) Enzyme that catalyses the first step in the pathway for salvage of the purines hypoxanthine and guanine. The phosphoribosyl moiety is transferred from an activated precursor, 5-phosphoribosyl 1-pyrophosphate. Since animal cells can synthesize purines *de novo*, HGPRT mutants can be selected by their resistance to toxic purine analogues. A genetic lesion in HGPRT in humans underlies the ***Lesch–Nyhan syndrome***. See ***HAT medium***.

hid (head involution defective) *Drosophila* gene that is involved in positive regulation of apoptosis (like ***reaper*** and ***grim***).

high density lipoproteins (HDL) Involved in cholesterol transport in serum. See ***lipoproteins***.

high endothelial venule (HEV) Venules in which the endothelial cells are cuboidal rather than squamous. Found particularly in lymph nodes where there is considerable extravasation of lymphocytes as part of normal traffic. Morphologically similar endothelium is found associated with some chronic inflammatory lesions. Express particular adhesion molecules accounting for preferential lymphocyte adhesion.

high mannose oligosaccharide A subset of the N-glycan chains that are added post-translationally to certain asparagine residues of secreted or membrane proteins in eukaryotic cells; contain 5–9 mannose residues, but lack the sialic acid-terminated antennae of the so-called complex type.

high pressure liquid chromatography See ***HPLC***.

high-energy bond Chemical bonds that release more than 25kJ/mol on hydrolysis: their importance is that the energy can be used to transfer the hydrolyzed residue to another compound. The risk in using the term is that students may think the bond itself is different in some way, whereas it is the compound that matters. Hydrolysis of creatine phosphate yields 42.7kJ/mol; of phosphoenolpyruvate, 53.2; ATP to ADP, 30.5: the latter is important because it shows that energetically the hydrolysis of creatine phosphate will suffice to reconstitute ATP, hence the use of creatine phosphate in muscle.

high-mobility group proteins Family of small, non-***histone***, nuclear proteins. Some appear to be involved in controlling transcription.

high-voltage electron microscopy (HVEM) HVEM has two advantages, the increased voltage shortens the wavelength of the electrons (and therefore increases resolving power) but, more importantly for the biologist, the penetrating power of the beam is increased, and it becomes possible to look at thicker specimens. Thus it is possible, by using stereoscopic views (obtained with a tilting stage) to get a three-dimensional picture of the interior of a cell.

Hill coefficient A measure of cooperativity in a binding process. A Hill coefficient of 1 indicates independent binding, a value of greater than 1 shows positive cooperativity – binding of one ligand facilitates binding of subsequent ligands at other sites on the multimeric receptor complex. Worked out originally for the binding of oxygen to haemoglobin (Hill coefficient of 2.8).

Hill reaction Reaction, first demonstrated by Robert Hill in 1939, in which illuminated chloroplasts evolve oxygen when incubated in the presence of an artificial electron acceptor (eg. ferricyanide). The reaction is a property of ***photosystem II***.

Hind II First type II ***restriction endonuclease*** identified, by Hamilton Smith in 1970. Isolated from *Haemophilus* influenzae Rd, it cleaves the sequence GTPyPuAc between the unspecified pyrimidine and purine, generating 'blunt ends'.

Hind III Commonly used type II ***restriction endonuclease*** isolated from *Haemophilus influenzae* Rd, it cleaves the sequence AAGCTT between the 2 As, generating 'sticky ends'.

hinge region Flexible region of a polypeptide chain, eg. as in immunoglobulins between Fab and Fc regions, and in myosin the S2 portion of heavy meromyosin.

hippocalcin Calcium-binding protein related to recoverin. Found exclusively in pyramidal cells of the hippocampus.

hippocampus Area of mammalian brain, and an important preparation for the study of ***neuronal plasticity***. The hippocampus has been known since the 1950s to be important for long-term memory storage in humans and other mammals; it is essential for initial storing of long-term memory for a period of days to weeks before the memory trace is consolidated elsewhere. Also the site of long-term synaptic plasticity (see ***long-term potentiation***) which is exhibited by defined synaptic pathways in the hippocampus.

Hirano bodies Paracrystalline inclusions found in the brain of patients with neurodegenerative disorders.

Hirschsprung's disease (HSCR) See ***Waardenburg's syndrome***. Aganglionic megacolon: a congenital malformation caused by absence of ganglion cells in myenteric and submucosal neural plexuses of gut. In some cases defect is due to mutation in *RET* receptor tyrosine kinase, in others to mutations in ***endothelin***-3, endothelin-β receptor, ***GDGF***. When associated with Waardenburg's syndrome, the defect is due to mutation in *Sox10*.

Hirudo medicinalis The medicinal leech. The ***central nervous system*** of this annelid contains a relatively small number of large, identifiable cells. This has made the leech, like the molluscs *Aplysia* and *Helisoma*, a chosen preparation for studying nervous system mechanisms at the cellular level. Related species of leeches are the organisms of choice for cellular and molecular genetic studies of early development, since the early embryos also contain identifiable cells.

His See ***histidine***.

his tag (histidine tag) ***Epitope tag*** based on a short stretch (ca 6) of histidine residues.

In addition to detection with anti-polyhistidine antibodies, such tags permit easy protein purification on nickel-based affinity columns.

hisactophilin Actin-binding protein (13.5 kD) from *Dictyostelium discoideum*. Promotes F-actin polymerization and binds to microfilament bundles but is very pH-sensitive as a result of having 31 histidine residues in a total of 118. Structure (though not sequence) very similar to FGF and IL1.

hispid flagella Eukaryotic flagella with two rows of stiff protrusions (mastigonemes) at right-angles to the long axis of the shaft. In hispid flagella, the normal relationship between the direction of flagellar wave propagation and the direction of movement is reversed; a proximal to distal wave pulls the organism forward.

histamine Formed by decarboxylation of histidine. Potent pharmacological agent acting through receptors in smooth muscle and in secretory systems. Stored in mast cells and released by antigen. (See ***hypersensitivity***). Responsible for the early symptoms of anaphylaxis. Also present in some venoms.

histatins Family of small histidine-rich cationic proteins (ranging from 7–8 amino acids in length) secreted into saliva by parotid and submandibular glands. Have potent antifungal activity, particularly against *Candida albicans*, though the mechanism of their fungicidal activity is unclear.

histidine (His; H) An amino acid (155 D) with an imidazole side chain with a pKa of 6–7. Acts as a proton donor or acceptor and has high potential reactivity and diversity of chemical function. Forms part of the catalytic site of many enzymes. See Table A2.

histiocytes Long-lived resident ***macrophages*** found within tissues.

histoblasts Population of small diploid epithelial cells in Dipteran larvae that do not form typical ***imaginal discs***, yet resemble them in some ways.

histochemistry Study of the chemical composition of tissues by means of specific staining reactions.

histocompatibility If tissues of two organisms are histocompatible, then grafts between the organisms will not be rejected. If, however, major histocompatibility antigens are different then an immune response will be mounted against the foreign tissue.

histocompatibility antigen A set of plasmalemmal glycoproteins that are crucial for T-cell recognition of antigens. Particularly the HLA system in humans and the H2 system in mice. There are two classes of histocompatibility antigens: (1) Class I; histocompatibility antigens composed of two glycosylated subunits, a heavy chain of 44 kD and β 2-microglobulin (12 kD). The heavy chain may be coded by K, D or L genes of mouse H2 and A, B or C genes of human HLA complex. Class I antigens are important in T-cell killing and are recognized in conjunction with the foreign cell surface antigens (MHC restriction). (2) Class II antigens; heterodimeric histocompatibility antigens composed of α (32 kD) and β (28 kD) chains. Found mostly on B-lymphocytes, macrophages and accessory cells. The response of T-helper cells requires that the foreign antigen is presented in conjunction with the appropriate Class II antigens. (Murine H2 Ia antigens and human HLA-DR antigens are Class II).

histogenesis The process of formation of a tissue, involving differentiation, morphogenesis, and other processes such as angiogenesis, growth control, cellular infiltration, etc.

histones Proteins found in the nuclei of all eukaryotic cells where they are complexed to DNA in ***chromatin*** and ***chromosomes***. They are of relatively low molecular weight and are basic, having a very high arginine/lysine content. They are highly conserved and can be grouped into five major classes. Two copies of H2A, H2B, H3 and H4 bind to about 200 base pairs of DNA to form the repeating structure of chromatin, the ***nucleosome***, with H1 binding to the linker sequence.

Table H1. Classes of histones

Class	Average M_r (kD)	% arginine	% lysine
H1	23	1.5	29
H2A	14	8	11
H2B	14	5	16
H3	15	13.5	10
H4	11	14	11

They may act as non-specific repressors of gene transcription. See Table H1.

histoplasmin Filtrate of a mycelial culture of *Histoplasma capsulatum*, the causative agent of histoplasmosis. Histoplasmin is intradermally injected in a skin test for the disease.

histotope A site on an MHC Class I or Class II antigen (see ***histocompatibility antigen***) recognized by a T-cell.

HIV (human immunodeficiency virus) Previously known as HTLV-III (human lymphotrophic virus type III) and also referred to as LAV (lymphadenopathy-associated virus), the retrovirus that causes acquired immune deficiency syndrome (***AIDS***) in humans, by killing CD4+-lymphocytes (T-helper cells). There are multiple forms of the virus.

HL60 cells Human promyelocytic cell line that can be induced to differentiate into neutrophil or eosinophil-like cells by various treatments. Have some, but not all, of the features of normal blood cells.

HLA (human leucocyte antigen) Refers to the ***histocompatibility antigens*** found in humans.

hlh See ***helix-loop-helix***.

HMG See ***high-mobility group proteins***.

HMG-CoA reductase (EC 1.1.1.34) Integral membrane protein (97 kD) of ER and peroxisomes that catalyses reaction between hydroxy-methyl-glutaryl-CoA and 2 molecules of NADPH to produce mevalonate, a key starting material for the synthesis of cholesterol and other sterols. Because the enzyme is rate-limiting in synthesis it has made a good target for cholesterol-lowering drugs.

HMM (heavy meromyosin) Soluble tryptic fragment of myosin that retains the ATPase activity and which will bind to F-actin to produce a characteristic ***arrowhead*** pattern (unless ATP is present, in which case it detaches). Papain cleavage of HMM yields S1 and S2 subfragments, the former having the ATPase activity.

hnRNA See ***heterogenous nuclear RNA***.

Hodgkin's disease A human lymphoma that appears to originate in a particular lymph node and later spreads to the spleen, liver and bone marrow. Giant cells, the Sternberg–Reed cells, with mirror-image nuclei are diagnostic. Immunological depletion, caused perhaps by the excessive growth of neoplastic histiocytes, occurs. Four types of the disease are recognized depending on the relative predominance of various neoplastic derivatives of the lymphoid series. Pyrexia is often a feature of the disease. Death often results from generalized immunological inability to respond to infections.

Hoechst 33258 dye A fluorescent dye that is a specific stain for DNA, and can therefore be used to visualize chromosomes, and to monitor animal cell cultures for contamination by microorganisms such as mycoplasma.

Hogness box See ***TATA box***.

holandry Inheritance of characters borne on the male chromosome and therefore only expressed in the male.

holocentric Description of a chromosome in which the centromere is diffuse rather than discrete.

holocrine Form of secretion in which the whole cell is shed from the gland, usually after becoming packed with the main secretory substance. In mammals, sebaceous glands are one of the few examples.

holoenzyme The complete enzyme complex composed of the protein portion (***apoenzyme***) and cofactor or coenzyme.

Hololena curta Funnel web spider. See ***hololena toxin***.

hololena toxin Toxin from the spider *Hololena curta* that irreversibly blocks insect presynaptic calcium channels. A heterodimer with subunits of 7 and 9 kD.

holometabolous Of an insect, a species with a marked change in body plan from larval to pupa to adult. Examples: flies and wasps (cf. ***hemimetabolous***).

Holothuria Class of the phylum Echinodermata. Sea cucumbers are holothurians.

homeobox (homoeobox) Conserved DNA sequence originally detected by DNA–DNA ***hybridization*** in many of the genes that give rise to ***homeotic mutants*** and segmentation mutants in *Drosophila*. The homeobox consists of about 180 nucleotides coding for a sequence of 60 amino acids in a protein, sometimes termed the homeodomain, of which about 80–90% are identical in the various homeodomains identified from *Drosophila*. Homeoboxes have also been detected in the genomes of vertebrates, with about 75% amino acid homology and a similar sequence has been found in the MAT gene of yeast. The homeobox codes for a protein domain that is involved in binding to DNA. Three subfamilies of homeobox-containing proteins can be identified, based on the archetypal *Drosophila* genes *engrailed*, *antennapedia* and *paired*. Interestingly, linear order within genome maps to order of expression in embryo. This may be required for the ***transcriptional silencing*** of certain homeotic genes (see ***Polycomb***).

homeodomain See ***homeobox***.

homeostasis (homoeostasis) The tendency towards a relatively constant state. A variety of homoeostatic mechanisms operate to keep the properties of the internal environment of organisms within fairly well-defined limits.

homeotic gene Gene, containing ***homeobox***, the level of expression of which is set during embryogenesis in response to positional cues, and which then directs the later formation of tissues and appendages appropriate to that part of the organism. Mutation of these genes leads to inappropriate expression of characteristics normally associated with another part of the organism (***homeotic mutants***).

homeotic mutant (homoeotic mutant) A ***mutant*** in which one body part, organ or tissue, is transformed into another part normally associated with another ***segment***. Examples are the ***antennapedia*** and ***bithorax complex*** mutants of *Drosophila*.

Homer Protein (28 kD) found in postsynaptic terminal, contains a single unusual ***PDZ*** domain and is involved in the clustering of a subset of metabotropic glutamate receptors, mGluR1α and mGluR5. Expression level elevated by synaptic activity. See also ***GRIP*** and ***PSD-proteins***.

homocystinuria Recessive condition in which the enzyme (cystathione synthetase) that converts homocysteine and serine into cystathione, a precursor of cysteine, is missing. Deficiency of this enzyme has widespread consequences in connective tissue, circulation and nervous system.

homoeobox See ***homeobox***.

homoeostasis See ***homeostasis***.

homoeotic mutant See ***homeotic mutant***.

homograft Outmoded term for a graft from one of an individual species to another. Includes ***allogeneic*** grafts (allografts) between genetically dissimilar individuals, and syngeneic grafts between identical individuals (eg. twins).

homologous (1) Derived from the tissues or DNA of a member of the same species, cf. ***heterologous, autologous, homologous***

recombination. (2) Of genes, similar in sequence, cf. ***analogous***.

homologous chromosomes Chromosomes that are identical with respect to genetic loci, and which tend to pair or ***synapse*** during mitosis. The two chromosomes (maternal and paternal) of each of the pairs occurring in the nuclei of diploid cells are homologous.

homologous recombination ***Genetic recombination*** involving exchange of homologous loci. Important technique in the generation of null alleles (***knockouts***) in ***transgenic*** mice.

homozygote Nucleus, cell or organism with identical ***alleles*** of one or more specific genes.

hook (1) Basal portion of bacterial flagellum, to which is distally attached the ***flagellin*** filament. Proximally the hook is attached to the rotating spindle of the motor. In some bacteria (Myxobacteria) the rotation of the hook itself (without an attached flagellum) may directly cause forward gliding movement. (2) *Drosophila* gene encoding a large homodimeric protein involved in endocytosis of the ***bride of sevenless/sevenless*** receptor–ligand complexes from the R7 photoreceptor cell.

Hopp Wood An algorithm for calculating a ***hydropathy plot***.

Hordeum vulgare Cultivated barley, very important for the brewing industry.

hormone A substance secreted by specialized cells that affects the metabolism or behaviour of other cells possessing functional receptors for the hormone. Hormones may be hydrophilic, like ***insulin***, in which case the receptors are on the cell surface, or lipophilic, like the ***steroids***, where the receptor can be intracellular. See Tables H2 and H3.

horseradish peroxidase A large enzyme, frequently used in conjunction with diaminobenzidine as an intracellular marker to identify cells both at light- and electron-microscopic levels.

host-range mutant A mutant of phage or animal virus that grows normally in one of its host cells, but has lost the ability to grow in cells of a second host type.

host-versus-graft reaction The normal lymphocyte-mediated reactions of a host against allogeneic or xenogeneic cells acquired as a graft or otherwise, which lead to damage or/and destruction of the grafted cells. The opposite of ***graft-versus-host response***. The common basis of graft rejection.

hotspot A region of DNA that is particularly prone to mutation or transposition.

housekeeping genes Genes that code for proteins or RNAs that are important for all cells and are thus constitutively active. Term used by contrast with '***luxury***' proteins, those which are only produced by differentiated cells.

housekeeping proteins Those sets of proteins involved in the basic functioning of a cell or the set of cells in an organism, eg. enzymes involved in synthesis and processing of DNA, RNA, proteins or the major metabolic pathways. As opposed to ***luxury proteins***.

hox genes ***Homeobox***-containing genes of vertebrates.

HPETE (5-HPETE; 5-hydroperoxy-eicosatetraenoic acid) Intermediate in leukotriene synthesis. Generated from arachidonic acid by 5-lipoxygenase; the starting point for the formation of 5-HETE or of leukotriene A_4. Other related enzymes add the hydroperoxy group in different positions.

HPLC (high pressure liquid chromatography) Chromatographic method in which the sample is forced at high pressure through a tightly packed column of finely divided particles that present a very large surface area. Because HPLC gives gives good separation very rapidly but is expensive, manufacturers tend to speak of 'high performance liquid chromatography' as an encouragement to purchasers.

HPV Human ***Papillomavirus***; causes warts.

Table H2. Polypeptide hormones and growth factors

Name	M_r (Da)	Residues	Source	Actions
Parathyroid hormone (PTH)	9.5k	84	Parathyroid glands	Raises kidney cAMP and blood Ca by stimulating bone release
Calcitonin[a]	4.5k	32	Thyroid, parathyroid	Opposes parathyroid hormone, hypocalcaemic, hypophosphataemic
Growth factors				
Insulin[a]	6k	21+30	β-cells of pancreas	Hypoglycaemic, growth factor
Insulin-like growth factor I (IGF-I, somatomedin C)		70	Liver, kidney	Growth factor, released into plasma
Insulin-like growth factor II (IGF II, multiplication stimulating activity, MSA)		67	Cultured hepatocytes	Mitogen for several cell types
Somatomedin A		50–80	Liver, kidney	Stimulates growth of peripheral nervous system
Epidermal growth factor (urogastrone, EGF)		53	Mouse submaxillary gland, urine	Stimulates epidermal growth, formation of GI tract
Platelet-derived growth factor (PDGF)	3k		Platelets	Stimulate fibroblast proliferation, wound healing
Fibroblast growth factor (FGF)	130k		Brain, pituitary	Stimulates proliferation of fibroblasts, adrenal cells, chondrocytes, endothelia
β-nerve growth factor (NGF)	130k	118	Salivary gland, snake venom	Tropic and trophic effects, mainly on sensory and sympathetic neurons. Peripheral nerve targets
Transforming growth factor (TGF)α	5–20k		Carcinomata	Acts via EGF receptor
TGFβ	2 × 25k	112	Tumour cells	Multifunctional role in tissue damage. Promotes or inhibits proliferation depending on cell type
GM-CSF (MG1, CSFα)	23k	–	Many tissues	Promotes granulocyte or macrophage colony formation
G-CSF (murine; CSFβ in human)	25/30 k	–	Many tissues	Promotes differentiation in myeloleukaemic cells
M-CSF (CSF-1)	40–70 k	–	Mouse tissues	Promotes macrophage differentiation
MultiCSF (IL3)	23–30k	–	Mouse primed T-cells	Proliferation of various haematopoeitic stem cells
Hypothalamic releasing hormones				
Luteinizing hormone-releasing factor[a] (luliberin, LHRH)	10		Hypothalamus	Stimulates release of LH, FSH
Hypothalamic thyrotropic hormone[a]-releasing factor (TRF, TRH)	362	3	Hypothalamus	Stimulates release of thyrotropin from anterior pituitary
Corticotrophin-releasing factor (CRH)[a]		41	Hypothalamus	Stimulates release of corticotrophin
Growth hormone-releasing factor (somatoliberin)	44		Hypothalamus	Stimulates release of growth hormone
Somatostatin (SRIF)[a]		14	Hypothalamus, D-cells of pancreas	Inhibits release of several hormones, including growth hormone

Table H2. (Continued)

Name	M_r (Da)	Residues	Source	Actions
Pituitary gland				
Oxytocin[a]	1007	8	Posterior pituitary	Uterine contraction, lactation
Vasopressin (antidiuretic hormone, ADH)[a]		8	Posterior pituitary	Antidiuretic vasopressor
ACTH-MSH family				
Adrenocorticotrophin (ACTH)[a]	4.5k	39	Anterior pituitary	Stimulates glucocortoid production from adrenals
Melanotropins (melanocyte stimulating hormone, MSH)[a]		11–22	Anterior pituitary	Stimulates darkening of skin
Lipotropins[a]		50–90	Anterior pituitary	Stimulate lipid breakdown
Endorphins[a]		15–31	Pituitary, brain	Opiates
Enkephalins[a]		5	Adrenal medulla, brain, gut	Opiates
Growth hormone (somatotropin, GH)[a]	21.5k	191	Pituitary	Regulates organismal growth
Prolactin	23k	199	Pituitary	Promote mammary growth, lactation
Gonadotropin/thyrotrophin family (glycoproteins of the anterior pituitary with α + β subunits)				
Luteinizing hormone (LH)	30k		Anterior pituitary	Acts on gonads
Follicle stimulating hormone (FSH)	30k		Anterior pituitary	Acts on gonads
Thyroid stimulating hormone (TSH)	30k		Anterior pituitary	Stimulates release of thyroid hormones
Human chorionic gonadotrophin (hCG)	30k		Anterior pituitary, placenta	
Gastrointestinal tract				
Vasoactive intestinal peptide (VIP)[a]		28	Lung, gut H-cells, nervous tissue	Vasodilation, bronchodilation, stimulates insulin, glucagon, prolactin secretion
Gastrin[a]		17 or 34	Gut G-cells	Stimulates acid secretion, muscle contraction
Secretin[a]		27	Gut S-cells	Stimulates alkali secretion from pancreas; decreases gastric acid secretion
Bombesin[a]		14	Skin, gut P-cells, nerves	Acts on CNS, gut smooth muscle, pancreas, pituitary kidney, heart; mitogenic *in vitro*
Motilin[a]		22	Gut enterochromaffin-2	Increases contractile response of stomach muscle cells
Pancreatic polypeptide[a]		36	PP-cells of pancreas;	Released on feeding; alters gut muscle tone, gut mucosa
Pancreozymin-cholecystokinin		8, 33 or 39	Gall bladder, brain, gut I-cells	Secretion of enzymes and electrolytes, secretion of insulin, (PZ-CCK)[a]glucagon, pancreatic polypeptide, contraction of gall bladder

Table H2. (Continued)

Name	M_r (Da)	Residues	Source	Actions
Gastric inhibitory peptide (GIP)		43	Gut	Inhibits gastric acid secretion, stimulates intestinal secretion, stimulates insulin and glucagon secretion
Substance P[a]		11	Gut enterochromaffin-1 cells, nerves	Contracts gut musculature, decreases blood pressure
Neurotensin[a]		13	Gut, hypothalamus	Effects on smooth muscle tone; may also be a neurotransmitter
Plasma kinins				
Bradykinin[a]		9	Formed in plasma	Dilates blood vessels, increases capillary permeability
Kallidin		10	Formed in plasma	Dilates blood vessels, increases capillary permeability
Other				
Renin	40k	gp	Kidney	Cleaves angiotensinogen to angiotensin I in plasma
Angiotensin II[a]		8	Formed from angiotensinogen, *via* angiotensin I	Acts on adrenal gland to stimulate aldosterone release, elevates blood pressure, mitogen for vascular smooth muscle
Atrial natriuretic peptide (ANP, ANF)		21–28	Atria of heart, brain, adrenal glands	Acts on kidney to produce natriuresis and diuresis, relaxes vascular smooth muscle, inhibits catecholamine release from adrenal medulla
Glucagon (HGF)[a]	3550	29	A cells of pancreas	Hyperglycaemic
Placental lactogen		191	Placenta	Promotes lactation
Caerulein	1352	10	Amphibian skin	Hypotensive, stimulates gastric secretion

[a]Neuropeptide. Other neuropeptides include calcitonin gene related peptide, neurokinin, neuromedin, neuropeptide Y, proctolin and carnosine.

Table H3. Steroid hormones

Name	M_r (Da)	Distribution	Actions
Female sex hormones			
β-oestradiol	272	Ovary, placenta, testis	Oestrogen
Progesterone	314	Corpus luteum, adrenal, testis, placenta	Regulates pregnancy, menstrual cycle
Male sex hormones			
Testosterone, dihydrotestosterone	288	Testes	Male secondary sexual characteristics, anabolic effects
Glucocorticoids			
Cortisol	362	Adrenal cortex	Gluconeogenesis, anti-inflammatory
Cortisone	360	Adrenal cortex	Glucocorticoid, anti-inflammatory
Corticosterone	346	Adrenal cortex	
Mineralocorticoids			
Aldosterone	360	Adrenal cortex	Stimulates resorption of Na from kidneys
Thyroid hormones			
Thyroxine (T4)	777	Thyroid	Control basal metabolic rate
Triiodothyronine (T3)	651	Thyroid	
Other			
Vitamin D (calciferol)	384	skin (+ sunlight)	Calcium and phosphate metabolism
Ecdysone	481	—	Moulting hormone of insects and nematodes

HRGP (hydroxyproline-rich glycoprotein) Class of plant glycoproteins and proteoglycans rich in hydroxyproline, that includes ***AGP***, ***extensin*** and certain ***lectins***. Found in the cell wall and are produced in response to injury.

HSK pathway See ***Hatch–Slack–Kortshak pathway***.

hsp See ***heat-shock proteins***.

hsp60 See ***heat-shock proteins***.

hsp70 Widely distributed group of conserved ***heat-shock proteins*** of average weight 70 kD. Possess ATP-binding domains, and may be involved in protein folding or export.

hsp90 Widely distributed group of conserved ***heat-shock proteins*** of average weight 90 kD. Exact function unknown, but are found associated with ***steroid receptors*** and ***tyrosine kinase*** oncogene products. May also bind actin and tubulin.

hst Human ***oncogene*** that encodes a member of the ***FGF*** family.

Htk-L Membrane-anchored ligand (37 kD) for ***EPH*** class receptor tyrosine kinases. Becomes tyrosine phosphorylated once bound to receptor (Nuk).

HTLV-I (human T-cell leukaemia/lymphoma virus type I) A retrovirus causing leukaemia and sometimes a mild immunodeficiency. In addition to *gag*, *pol* and *env*, the virus carries a coding sequence *pX* that does not seem to have a normal genomic homologue and is not a conventional oncogene. The protein product of the *pX* region is a short-lived nuclear protein (around 40 kD). T-cells transformed with HTLV-1 continue to proliferate and are independent of ***interleukin-2***.

HTLV-II (human T-cell leukaemia/lymphoma virus type II) Originally isolated from a T-cell line from a patient with hairy cell leukaemia. It has only partial homology with ***HTLV-I***, and its pathological potential is uncertain.

HTLV-III See *HIV*.

HTRF (homogeneous time-resolved fluorescence) Assay methodology increasingly being used in high throughput screening. An absorbing fluorochrome is coupled to one component, the emitting fluorochrome with slow-release characteristics coupled to the other and if the two components are in proximity (because they bind) then fluorescence resonance energy transfer between absorbing and emitting fluorochromes gives a signal which is analysed in a ***time-resolved fluorescence*** system. No separation step is required – all the reagents are mixed together and inhibitors of binding will reduce the output signal.

human immunodeficiency virus See *HIV*.

human leucocyte antigen See *HLA*.

human T-cell leukaemia/lymphoma virus type I See *HTLV-I*.

human T-cell leukaemia/lymphoma virus type II See *HTLV-II*.

humanized antibody Usually a mouse monoclonal antibody directed against a target of particular therapeutic value that has been modified to have all regions except the antigen-binding portion substituted by human immunoglobulin domains. The procedure should make the antibody minimally antigenic when administered to patients though some anti-idiotype antibodies may still be made to the antigen-binding site.

humoral immune responses Immune responses mediated by antibody.

hunchback (hb) Key regulatory gene in the early segmentation gene hierarchy of *Drosophila*. Codes for a transcription factor of the Cys2-His2 ***zinc finger*** type.

Hunter syndrome Recessive ***mucopolysaccharidosis***, X-linked, in which dermatan and heparan sulphates are not degraded. Because two lysosomal enzymes (heparan sulphate sulphatase and α-iduronidase) are involved in the breakdown of these glycosaminoglycans, fibroblasts from Hunter's syndrome will complement the fibroblasts from Hurler's patients in culture; by recapture of lysosomal enzymes from the medium, both types of cells in mixed culture become competent to digest glycosaminoglycans.

huntingtin (huntingtin-associated protein 1) Protein product of the IT15 gene that has variable numbers of polyglutamine repeats in ***Huntington's chorea***. The IT15 gene is widely expressed and required for nomal development. The polyglutamine repeats (44 in the commonest form of the disease) increase the interaction of huntingtin with huntingtin-associated protein 1 (HAP-1) which is enriched in the brain and may be associated with pathology.

Huntington's chorea (Huntington's disease) Mature onset disease characterized by progressive loss of neuronal functioning. Caused by unstable amplification of a trinucleotide $(CAG)_n$ repeat within the coding region of a gene encoding a 348 kD, widely expressed product.

Huntington's disease See ***Huntington's chorea***.

HuP genes Human equivalents of the *Pax* genes.

Hurler's disease Autosomal mucopolysaccharidosis recessive storage disease in which α-iduronidase is absent, leading to accumulation of heparan and dermatan sulphates. Extensive deposits of mucopolysaccharide are found in ***gargoyle cells***, and in neurons. See ***Hunter syndrome***.

Hurler–Scheie syndrome Although clinically distinct diseases, fibroblasts from patients with Hurler's and with Scheie syndrome do not cross-complement in culture, suggesting that the enzyme defect is the same.

Hurthle cells See *Askenazy cells*.

HUT-78 cells Human T-cell lymphoma from patient with Sezary syndrome. Have features of mature T-cells of helper/inducer phenotype and release IL-2.

HUVEC (human umbilical vein endothelial cells) A convenient source of human endothelial cells are those that line the large vein in the umbilical cord which is usually discarded together with the placenta after childbirth. The cells can be removed as a fairly pure suspension by mild enzymatic treatment of the vein followed by some mechanical distraction and will grow relatively easily in culture, retaining their differentiated characteristics for several passages.

HVEM See ***high-voltage electron microscopy***.

hyaline Clear, transparent, granule-free; as, for example, hyaline cartilage and the hyaline zone at the front of a moving amoeba.

hyaluronic acid Polymer composed of repeating dimeric units of glucuronic acid and N-acetyl-glucosamine. May be of extremely high molecular weight (up to several million daltons) and forms the core of complex proteoglycan aggregates found in extracellular matrix.

hyaluronidase Enzyme that degrades ***hyaluronic acid***; found in lysosomes.

hybrid Ab Artificially produced antibody made by fusing ***hybridomas*** producing two different antibodies; the hybrid cells produce three different antibodies, only one of which is a ***heterophile antibody***. Can also be prepared chemically from two antibodies.

hybrid cells Any cell type containing components from one or more genomes, other than zygotes and their derivatives. Hybrid cells may be formed by ***cell fusion*** or by ***transfection***. See ***heterokaryon***.

hybrid dysgenesis Genetic phenomenon, in which two strains of organism produce offspring at anomalously low rates. Example: the P-M system of *Drosophila*, in which P-strain males (containing multiple ***P elements***) mated to M-strain females produce sterile hybrids.

hybridization (of nucleic acids) Technique in which single-stranded ***nucleic acids*** are allowed to interact so that complexes, or hybrids, are formed by molecules with sufficiently similar, complementary sequences. By this means the degree of sequence identity can be assessed and specific sequences detected. The hybridization can be carried out in solution or with one component immobilized on a gel or, most commonly, nitrocellulose paper. Hybrids are detected by various means: visualization in the electron microscope; by radioactively labelling one component and removing non-complexed DNA; or by washing or digestion with an enzyme that attacks single-stranded nucleic acids and finally estimating the radioactivity bound. Hybridizations are done in all combinations: DNA-DNA (DNA can be rendered single-stranded by heat denaturation), DNA-RNA or RNA-RNA. *In situ* hybridizations involve hybridizing a labelled nucleic acid (often labelled with a fluorescent dye) to suitably prepared cells or histological sections. This is used particularly to look for specific ***transcription*** or localization of genes to specific chromosomes (FISH analysis).

hybridoma A cell hybrid in which a tumour cell forms one of the original source cells. In practice, confined to hybrids between T- or B-lymphocytes and appropriate ***myeloma*** cell lines.

hydatid cyst Large cyst in the viscera of sheep cattle or man following ingestion of eggs of the tapeworm *Echinococcus granulosus*, a cestode. In the normal host the eggs develop into intestinal worms but in abnormal hosts the larvae penetrate through the wall of the intestine and migrate to liver and other organs. The cyst contains many scolices that can form further cysts if liberated: it is thus a metastasizing parasite.

hydatiform mole Abnormal conceptus in which an embryo is absent and there is excessive proliferation of placental villi. In most cases the tissue is diploid XX with both X chromosomes being of paternal origin and thus it is thought that it may arise by fertilization of a dead ovum. May occasionally become invasive though the metastases regress following removal of the mole. Not to be confused with ***hydatid cyst***.

Hydra Genus of freshwater coelenterates (Cnidaria). They are small, solitary and only exist in the polyp form, which is a radially symmetrical cylinder that is attached to the substratum at one end and has a mouth surrounded by tentacles at the other. They have considerable powers of regeneration and have been used in studies on positional information in morphogenesis.

hydraulic motor By altering the internal osmotic pressure within a cell, water will enter and a considerable expansion of the compartment will occur. This has been used as a motor device in plants (turgor pressure), in eversion of ***nematocysts***, and possibly in the production of other cellular protrusions.

hydrocortisone (cortisol; 17-hydroxy-corticosterone) Powerful corticosteroid produced by cells of the zona reticularis in the adrenal gland. Has potent anti-inflammatory effects.

hydrogen peroxide (H_2O_2) Hydrogen peroxide is produced by vertebrate phagocytes and is used in bacterial killing (the ***myeloperoxidase***-halide system).

hydrogenosome Organelle found in certain anaerobic trichomonad and some ciliate protozoa: contains hydrogenase and produces hydrogen from glycolysis.

hydrolase One of a class of enzymes (EC Class 3) catalysing hydrolysis of a variety of bonds, such as esters, glycosides, peptides.

hydrolytic enzymes See ***hydrolase***.

hydropathy plot Approximate way of deducing the higher order structure of a protein, based on the principle that 20–30 consistently hydrophobic residues are necessary to make a membrane-spanning α-helix. For each amino acid residue, a weighted average of the hydrophobicity of the residue and its immediate neighbours are calculated and graphically displayed as a hydropathy plot, with hydrophobic domains plotted as positive numbers. There are several formulae for the calculation, that differ in the calculation of the moving average, the window size, and the scoring system for hydrophobicity of individual residues (eg. Kyte-Doolittle, Hopp Wood, Eisenberg).

hydrophilic group A polar group or one that can take part in hydrogen bond formation, eg. OH, COOH, NH_2. Confers water-solubility, or in lipids and macromolecules causes part of the structure to make close contact with the aqueous phase.

hydrophobic bonding Interaction driven by the exclusion of non-polar residues from water. It is an important determinant of protein conformation and of lipid structures, and is considered to be a consequence of maximizing polar interactions rather than a positive interaction between apolar residues.

hydroxyapatite The calcium phosphate mineral, $Ca_{10}(PO_4)_6(OH)_2$, found both in rocks of non-organic origin and as a component of bone and dentine. Used as column packing for chromatography, particularly for separating double-stranded DNA from mixtures containing single-stranded DNA.

hydroxylysine Post-translationally hydroxylated lysine is found in ***collagen*** and commonly has galactose and then glucose added sequentially by glycosyl transferases. The extent of glycosylation varies with the collagen type.

hydroxyproline Specific proline residues on the amino side of a glycine residue in collagen become hydroxylated at C4, before the polypeptides become helical, by the activity of prolyl hydroxylase. This enzyme has a ferrous ion at the active site, and a reducing agent such as ascorbate is necessary to maintain the iron in the ferrous state. The presence of hydroxyproline is essential to produce stable triple-helical tropocollagen, hence the problems caused by ascorbate deficiency in scurvy. This unusual amino acid is also present in considerable amounts in the major glycoprotein of primary plant cell walls (see ***HRGP***).

hydroxyproline-rich glycoprotein See ***HRGP***.

hydroxytetraecosanoic acid See ***HETE***.

hydroxytetraeicosenoic acid See ***HETE***.

hydroxytryptamine (5-hydroxytryptamine, 5-HT) See ***serotonin***.

hydroxyurea Inhibitor of DNA synthesis (but not repair).

hygromycin B Aminoglycoside antibiotic from *Streptomyces hygroscopicus* that is toxic for both pro- and eukaryotic cells. Used as a selection agent in transfection: the hygromycin-B-phosphotransferase gene, *hph,* in the vector confers resistance.

Hymenolepis Genus of cestode parasites of the gut of mammals. The immunological response to *H. diminuta* infection has been extensively studied.

hyoscyamine Anticholinergic from *Atropa belladona;* when racemized following isolation is known as ***atropine***.

hyper- Prefix denoting more than, bigger than.

hyperaemia An excess of blood in a region of the body.

hypercholesterolaemia (USA, hypercholesterolemia) High serum levels of cholesterol. Can in some cases be caused by a defect in lipoprotein metabolism or, for example, defects in the ***low density lipoprotein*** receptor (familial hypercholesterolaemia).

hyperglycaemia (USA, hyperglycemia) An excess of plasma glucose that can arise through a deficiency in insulin production. See ***diabetes mellitus***, ***diabetes insipidus***.

hyperimmune serum Serum prepared from animals that have recently received repeated injections or applications of a chosen antigen; thus the serum should contain a very high concentration of polyclonal antibodies against that antigen.

hyperlipidaemia (USA, hyperlipidemia) Elevated levels of serum low density lipoprotein, correlated with increased risk of cardiovascular disease. See ***hypercholesterolaemia***.

hyperlipoproteinaemia The same as ***hyperlipidaemia***.

hypermastigote Large multiflagellate symbiotic protozoa found in the gut of termites and wood-eating cockroaches. Most bizarre example of the group is *Mixotricha paradoxica* which actually has few flagella and is propelled by spirochaetes (bacteria) that are attached to special bracket-like regions of the cell wall.

hyperosmotic Of a liquid, having a higher osmotic pressure (usually than the physiological level).

hyperplasia Increase in the size of a tissue as a result of enhanced cell division. Once the stimulus (wound healing, mechanical stress, hormonal overproduction) is removed, the division rate returns to normal (whereas in neoplasia proliferation continues in the absence of a stimulus).

hyperpolarization A negative shift in a cell's ***resting potential*** (which is normally negative), thus making it numerically larger, ie. more polarized. The opposite of ***depolarization***.

hypersensitive response *Bot.* An active response of plant cells to pathogenic attack in which the cell undergoes rapid necrosis and dies. Associated with the production of ***phytoalexins***, ***lignin***, and sometimes ***callose***. The response is thought to prevent a potential pathogen from spreading through the tissues.

hypersensitivity In immunology, a state of excessive and potentially damaging immune responsiveness as a result of previous exposure to antigen. If the hypersensitivity is of the immediate type (antibody mediated), then the response occurs in minutes; in delayed hypersensitivity the response takes much longer (about 24 hours) and is mediated by primed T-cells. Hypersensitivity responses are not simply divisible into the two types, and it is now more common to subdivide immediate responses into types I, II, and III, the delayed response being of type IV. Type I responses involve antigen reacting with IgE fixed to cells (usually mast cells) and are characterized by histamine release; anaphylactic responses and urticaria are of this type. In type II responses circulating antibody reacts with cell surface or cell-bound antigen, and if complement fixation occurs, cytolysis may follow. In type III reactions immune complexes are formed in solution and lead to damage (serum sickness, glomerulonephritis, ***Arthus reaction***). Delayed-type responses of type IV involve primed lymphocytes reacting with antigen and lead to formation of a lymphocyte-macrophage granuloma without involvement of circulating antibody.

hyperthermophile Members of the Archaea that live and thrive in temperatures above 60°C, sometimes above 100°C (cf. thermophiles, which have a tolerance ceiling of about 60°C).

hyperthyroidism (Graves' disease) Excess production of thyroid hormone caused by autoantibodies that bind to the ***thyroid-stimulating hormone*** receptor and induce secretion of ***thyroxine*** by raising cAMP levels in the thyroid cells.

hypertonic Of a fluid, sufficiently concentrated to cause osmotic shrinkage of cells immersed in it. Note that a mildly ***hyperosmotic*** solution is not necessarily hypertonic for viable cells, which are capable of regulating their volumes by ***active transport***. See ***hypotonic, isotonic***.

hypertrophy Increase in size of a tissue or organ as a result of cell growth, rather than an increase of cell number (***hyperplasia***), though often both processes occur.

hypervariable region Those regions of the heavy or light chains of immunoglobulins in which there is considerable sequence diversity within that set of immunoglobulins in a single individual. These regions specify the antigen affinity of each antibody.

hypha (plural, hyphae) Filament of fungal tissue that may or may not be separated into a file of cells by cross-walls (septa). It is the main growth form of filamentous fungi, and is characterized by growth at the tip followed by lateral branching.

hypocotyl Part of the axis of a plant embryo or seedling between the point of insertion of the ***cotyledon(s)*** and the top of the radicle (root). In some ***etiolated*** seedlings, the hypocotyl is greatly extended.

hypogammaglobulinaemia (USA, hypogammaglobulinemia) Syndromes in humans and other vertebrates in which the immunoglobulin level is depressed below the normal range. Congenital, chronic and transient types are known.

hypotonic Of a fluid, having a concentration that will cause osmotic shrinkage of cells immersed in it. Not necessarily hypo-osmotic.

hypoxanthine Purine base present in inosine monophosphate (IMP) from which adenosine monophosphate (AMP) and guanosine monophosphate (GMP) are made. The product of deamination of adenine, 6-hydroxy-purine.

hypoxanthine guanine phosphoribosyl transferase See *HGPRT*.

hypoxia Condition in which there is an abnormally low level of oxygen in the blood and tissues.

I

I See ***isoleucine***.

I-309 (TCA3) Small ***cytokine*** secreted by activated T-lymphocytes that is chemotactic for monocytes.

I band The isotropic band of the sarcomere of striated muscle, where only thin filaments are found. Unlike the ***A band***, the I band can vary in width depending upon the state of contraction of the muscle when fixed.

I-cell disease A human disease in which the lysosomes lack hydrolases but high concentrations of these enzymes are found in the extracellular fluids. The gene defect responsible probably prevents the addition of the lysosome recognition marker (mannose-6-phosphate) to these enzymes so that they are not directed into the lysosomes but are released.

I-domain Binding domain of around 200 amino acids in the N-terminal part of α-subunits of ***integrins***. The I-domain has intrinsic ligand-binding activity that is divalent cation-dependent.

I region (1) The inducible gene region of the genome of *E. coli* involved in the lactose operon. (2) Region of the murine genome coding for products involved in many aspects of the immune response (from Immune Region). Most of the products are Class II histocompatibility antigens.

I region-associated antigens Class II major histocompatibility (MHC) antigens.

Ia antigen Antigens coded for by the I-region of the MHC complex. The majority of, if not all, such antigens are Class II molecules composed of α and β polypeptide chains.

IAA See ***indole acetic acid***.

IAP (inhibitor of apoptosis) One of a family of evolutionarily conserved genes that inhibit apoptosis. Human X-linked IAP directly inhibits two ***caspases***, caspase 3 and caspase 7. Produced by various viruses to allow survival of host cell while virus replicates.

IAPP See ***islet amyloid peptide***.

iatrogenic Iatrogenic disease. Disease caused by attempts at therapy.

IBD (inflammatory bowel disease; identity by descent) (1) General term that covers two different inflammatory disorders of the bowel, ***Crohn's disease*** and ***ulcerative colitis***. (2) Two genes at a locus have identity by descent (IBD) if they were both inherited from a common ancestor. Important in linkage studies for disease susceptibility.

iberiotoxin Peptide toxin (37 residues) from the scorpion *Leiurus quinquestriatus* var *hebraeus* that is selective for the high-conductance calcium-activated K^+ channel. Similar (and highly homologous in sequence) to ***charybdotoxin*** but more selective.

IBMX (isobutylmethylxanthine) Often used as an inhibitor of cAMP phosphodiesterase as a way of raising intracellular cAMP levels for experimental purposes.

ibotenate (ibotenic acid) An ***excitotoxin*** from *Amanita* sp. that acts on the ***NMDA receptor***.

ibotenic acid See ***ibotenate***.

ibuprofen (2-(4-isobutylphenyl) proprionic acid) Non-steroidal anti-inflammatory drug that inhibits cycloxygenases I and II. Related NSAIDs are ketoprofen, flurbiprofen and naproxen.

iC3b Inactivated C3b (C3bi). See ***complement***.

IC50 Concentration of an inhibitor at which 50% inhibition of the response is seen; should only be used of ***in vitro*** test systems. Needs to be used with caution because ED50 (for example) is the concentration that causes an effect in 50% of tests – not necessarily the same thing.

ICaBP See ***intestinal calcium-binding protein***.

ICAM (intercellular adhesion molecule) ICAM-1 is the glycoprotein ligand (80–155 kD) for LFA-1 (CD11a/CD18: β_2-integrin) and to a lesser extent Mac-1 (CD11b/CD18). ICAM-1 is expressed on the luminal surface of endothelial cells and is upregulated in response to IL-1 or TNF treatment. ICAM-1 is also expressed on various haematopoietic cells and is upregulated on activated T- and B-cells. It is a member of the immunoglobulin superfamily and has five C2 domains. Not only is it an important ligand for leucocyte adhesion, but it is also the site to which rhinovirus binds and to which *Plasmodium falciparum*-infected erythrocytes adhere. ICAM-2 (55–68 kD) is constitutively expressed on endothelium and on resting lymphocytes and monocytes and is not upregulated by inflammatory cytokines. It has only two Ig superfamily domains. CD50 has now been identified as ICAM-3 (120 kD) and plays a role in the early stages of the immune response; it is also a ligand for LFA-1. Crosslinking of ICAM-3 seems to induce an increase in intracellular calcium and triggers activation of tyrosine kinases.

iccosomes Immune complex-coated bodies formed when antigen is injected into an immune animal and found in follicular ***dendritic cells***. May serve as a reservoir of antigen to maintain B-cell memory.

ICE See ***interleukin-1 converting enzyme***.

ice nucleation proteins Proteins produced by some ***Gram-negative bacteria*** that promote the nucleation of ice, apparently by aligning water molecules along repeated domains of 48 amino acids, that consist of 16-residue repeats containing the conserved octamer AGYGSTxT. Now finding commercial use in snow-making at ski resorts.

ICK1 Inhibitor of cyclin-dependent kinase identified in *Arabidopsis thaliana*. Has some limited similarity with mammalian $p27^{Kip1}$ kinase inhibitor. Inhibits plant cdc2 kinase but not human $p34^{cdc2}$.

IcsA Actin nucleating protein in the outer membrane of virulent strains of *Shigella flexneri*. Like ***ActA***, IcsA is responsible for unipolar assembly of an F-actin bundle that pushes the bacterium through the cytoplasm.

idazoxan Imidazoline α-2-adrenoreceptor antagonist, also binds to ***imidazoline receptors***.

idiopathic Applied to disease of unknown origin or peculiar to the individual.

idiotope An antigenic determinant (***epitope***) unique to a single clone of cells and located in the variable region of the immunoglobulin product of that clone or to the T-cell receptor. The idiotope forms part of the antigen-binding site. Any single immunoglobulin may have more than one idiotope. Idiotopes are also associated with the antigen-binding sites of T-cell receptors and are the epitopes to which an anti-***idiotype*** antibody or T-cell binds.

idiotype The antigenic specificities defined by the unique sequences (***idiotopes***) of the antigen combining site. Thus anti-idiotype antibodies combine with those specific sequences, may block immunological reactions, and may resemble the epitope to which the first antibody reacts.

idoxuridine Analogue of thymidine that inhibits the replication of DNA viruses. Used in the treatment of *Herpes simplex* and varicella zoster.

iduronic acid (α-L-iduronic acid) A uronic acid, derived from the sugar idose, and bearing one terminal carboxyl group. With N-acetyl-galactosamine-4-sulphate, a component of dermatan sulphate. See ***iduronidase***.

iduronidase (α-1-iduronidase) EC 3.2.1.76. An enzyme (653 amino acids in human) that hydrolyses the bonds between iduronic acid and N-acetylgalactosamine-4-sulphate; a lysosomal enzyme absent in ***Hurler's disease***.

IEG See *immediate early gene*.

IgA Major class of immunoglobulin of external secretions in mammals, also found in serum. In secretions, found as a dimer (400 kD) joined by a short J chain and linked to a secretory piece or transport piece. In serum found as a monomer (170 kD). IgAs are the main means of providing local immunity against infections in the gut or respiratory tract and may act by reducing the binding between an IgA-coated microorganism and a host epithelial cell. Present in human colostrum but not transferred across the placenta. Have α heavy chains.

IgD This immunoglobulin (184 kD) is present at a low level (3–400μg/ml) but is a major immunoglobulin on the surface of B-lymphocytes where it may play a role in antigen recognition. Its structure resembles that of IgG but the heavy chains are of the δ type.

IgE Class of immunoglobulin (188 kD) associated with immediate-type ***hypersensitivity*** reactions and helminth infections. Present in very low amounts in serum and mostly bound to mast cells and basophils that have an IgE-specific Fc receptor (Fc ε R). IgE has a high carbohydrate content and is also present in external secretions. Heavy chain of ε type.

IGF See *insulin-like growth factor*.

IgG The classical immunoglobulin class also known as 7S IgG (150 kD). Composed of two identical light and two identical heavy chains, the constant region sequence of the heavy chains being of the γ type. The molecule can be described in another way as being composed of two ***Fab*** and an ***Fc*** fragment. The Fabs include the antigen combining sites; the Fc region consists of the remaining constant sequence domains of the heavy chains and contains cell-binding and complement-binding sites. IgGs act on pathogens by agglutinating them, by opsonizing them, by activating complement-mediated reactions against cellular pathogens and by neutralizing toxins. They can pass across the placenta to the foetus as maternal antibodies, unlike other Ig classes. In humans four main subclasses are known, IgG2 differs from the rest in not being transferred across the placenta and IgG4 does not fix complement. IgG is present at 8–16mg/ml in serum.

IgM (macroglobulin) An IgM molecule (970 kD) is built up from five IgG type monomers joined together, with the assistance of J chains, to form a cyclic pentamer. IgM binds complement and a single IgM molecule bound to a cell surface can lyse that cell. IgM is usually produced first in an immune response before IgG. The human red cell isoantibodies are IgM antibodies. Heavy chain (μ chain) is rather larger than the heavy chains of other immunoglobulins.

IgX An immunoglobulin class found in Amphibia.

IGF-I See *insulin-like growth factor*-I.

IGIF (interferon-inducing factor) Protein that augments NK cell activity in spleen cells. Precursor has 192 amino acid residues, mature protein 157 amino acids. Has no obvious similarities to other peptides in database. May be involved in development of Th1 cells.

IκB (IkappaB) Protein that inhibits ***NF κ B*** by binding to the p65 subunit. It is thought to prevent NF κ B from entering the nucleus. Two forms have been identified IκBα (37 kD) and IκBβ (43 kD).

IkB See *IκB*.

IL-1, IL-2, etc. See ***interleukins*** and individual interleukin-*n* entries.

IL-6 cytokine family Family of cytokines that all act through receptors sharing a common gp130 subunit. Includes ***interleukin-6, interleukin-11, ciliary neurotrophic factor***, LIF, ***oncostatin M*** and ***cardiotrophin-1***.

ile See *isoleucine*.

imaginal disc Epithelial infoldings in the larvae of holometabolous insects (eg. Lepidoptera, Diptera) that rapidly develop into adult appendages (legs, antennae, wings, etc.) during metamorphosis from larval to adult form. By implanting discs into the haemocoele of an adult insect their differentiation can be blocked, though their ***determination*** remains unchanged except occasionally ***transdetermination*** occurs. The hierarchy of transdetermination has been studied in great detail in *Drosophila*.

imidazoline receptors Receptors for imidazolines such as ***agmatine*** and ***idazoxan*** that modulate sympathetic outflow in the brainstem. I1 is elevated on platelets and in brains of patients with depression. Also expressed on vascular smooth muscle where they may mediate inhibition of proliferation. Candidate protein for I1-receptor is 85 kD and upregulated by idazoxan and agmatine. I2 receptors have been proposed to have a role in modulating pain during acute inflammation.

iminoglycinuria A defect in amino acid transport leading to abnormal excretion of glycine, proline and hydroxyproline in the urine: more seriously, absorption in the intestine may be inadequate. See ***Hartnup disease***.

immaturin Soluble protein produced by *Paramecium caudatum* that represses its mating activity.

immediate early gene (IEG) Class of genes whose expression is low or undetectable in quiescent cells, but whose transcription is activated within minutes after extracellular stimulation such as addition of a ***growth factor***. c-*fos* and c-*myc* proto-oncogenes were among the first IEGs to be identified. Many IEGs encode ***transcription factors*** and therefore have a regulatory function.

immortalization Escape from the normal limitation on growth of a finite number of division cycles (the Hayflick limit), by variants in animal cell cultures, and cells in some tumours. Immortalization in culture may be spontaneous, as happens particularly readily in mouse cells, or induced by mutagens or by ***transfection*** of certain oncogenes.

immotile cilia syndrome Congenital defect in ***dynein*** (either absent or inactive) that leads to male sterility and poor bronchial function. Interestingly, non-ciliated cells show altered locomotion and 50% of patients have ***Kartagener's syndrome***.

immune complex (antigen–antibody complex) Multimolecular antibody–antigen complexes that may be soluble or insoluble depending upon their size and whether or not complement is present. Immune complexes can be filtered from plasma in the kidney, and the deposition of the complexes gives rise to ***glomerulonephritis*** probably because of the trapping of neutrophils via their Fc receptors.

immune complex diseases Diseases characterized by the presence of immune complexes in body fluids. Hypersensitivity of the ***Arthus*** type and serum sickness are examples.

immune deficiency diseases Those diseases in which immune reactions are suppressed or reduced. Reasons may include congenital absence of B- and/or T-lymphocytes, or viral killing of helper lymphocytes (see ***HIV***).

immune response Alteration in the reactivity of an organism's immune system in response to an antigen; in vertebrates, this may involve antibody production, induction of cell-mediated immunity, complement activation or development of immunological tolerance,

immune response gene See ***I-region***.

immune surveillance See ***immunological surveillance***.

immunity A state in which the body responds specifically to antigen and/or in which a protective response is mounted against a pathogenic agent. May be innate or may be induced by infection or vaccination, or by the passive transfer of antibodies of immunocompetent cells.

immunoadsorbent Any insoluble material, eg. cellulose, with either an antigen or an antibody bound to it and that will bind its corresponding antibody or antigen thus removing it from a solution.

immunoblotting Techniques, such as the Western blot, in which very small amounts of protein are transferred from gels to nitrocellulose sheets by electrophoresis and then detected by their antibody binding, usually in combination with peroxidase- or radioactively-labelled IgG. An accurate technique for the specific recognition of very small amounts of protein. See also ***blotting***.

immunoconglutinin Antibodies that react with complement components or their breakdown products. Usually directed against C3b or C4, and found in high titre in patients with rheumatoid arthritis.

immunocytochemistry Techniques for staining cells or tissues using antibodies against the appropriate antigen. Although in principle the first antibody could be labelled, it is more common (and improves the visualization) to use a second antibody directed against the first (an anti-IgG). This second antibody is conjugated either with fluorochromes, or appropriate enzymes for colorimetric reactions, or gold beads (for electron microscopy), or with the biotin–avidin system, so that the location of the primary antibody, and thus the antigen, can be recognized.

immunodeficient See ***immune deficiency diseases***.

immunoelectrophoresis Any form of ***electrophoresis*** in which the molecules separated by electrophoresis are recognized by precipitation with an antibody.

immunofluorescence A test or technique in which one or other component of an immunological reaction is made fluorescent by coupling with a ***fluorochrome*** such as fluorescein, phycoerythrin or rhodamine so that the occurrence of the reaction can be detected as a fluorescing antigen–antibody complex. Used in microscopy to localize small amounts of antigen or specific antibody.

immunogenicity The property of being able to evoke an immune response within an organism. Immunogenicity depends partly upon the size of the substance in question, and partly upon how unlike host molecules it is. Highly conserved proteins tend to have rather low immunogenicity.

immunoglobulin superfamily A large group of proteins with immunoglobulin-like domains. Most are involved with cell surface recognition events. Sequence homology suggests that Igs, MHC molecules, some ***cell adhesion*** molecules and cytokines receptors share close homology, and thus belong to a ***multigene family***.

immunoglobulins See ***IgA***, ***IgD***, ***IgE***, ***IgG***, and ***IgM***.

immunological memory The systems responsible for the situation where reactions to a second or subsequent exposure to an antigen are more extensive than those seen on first exposure (but see also ***immunological tolerance***). The memory is best explained by clonal expansion and persistence of such clones following the first exposure to antigen.

immunological network The concept due to Jerne that the entire specific immune system within an animal is made up of a series of interacting molecules and cell surface receptors, based on the idea that every antibody-combining site carries its own marker antigens or ***idiotypes*** and that these in turn may be recognized by another set of antibody-combining sites, and so on.

immunological surveillance The hypothesis that lymphocyte traffic ensures that all or nearly all parts of the vertebrate body are surveyed by visiting lymphocytes in order to detect any altered-self material, eg. mutant cells.

immunological tolerance Specific unresponsiveness to antigen. Self-tolerance is a process occurring normally early in life due to suppression of self-reactive lymphocyte clones. Tolerance to foreign antigens can be induced in adult life by exposure to antigens under conditions in which specific clones are suppressed. Note that tolerance is not the same as immunological unresponsiveness, since the latter may be very non-specific as in immunodeficiency states.

immunomodulation Modification of the immune response: either activating or, more commonly, suppressing.

immunophilin Generic term for intracellular protein that binds immunosuppressive drugs such as ***cyclosporin***, ***FK 506*** and ***rapamycin***. Both ***cyclophilin*** and the receptor for FK506 are peptidyl prolyl *cis-trans* isomerases (rotamases). Immunophilins are thought to interact with ***calcineurin***.

immunoprecipitation The precipitation of a multivalent antigen by a bivalent antibody, resulting in the formation of a large complex. The antibody and antigen must be soluble. Precipitation usually occurs when there is near equivalence between antibody and antigen concentrations.

immunoregulation The various processes by which antibodies may regulate immune responses. At a simple level, secreted antibody neutralizes the antigen with which it reacts thus preventing further antigenic stimulation of the antibody-producing clone. At a more complex level, anti-idiotype antibodies can be shown to develop against the first antibodies in some cases, and perhaps further anti-idiotype antibodies against them. This is the main concept of the ***immunological network*** theory.

immunostimulatory complexes See ***ISCOMS***.

immunosuppression This occurs when T- and/or B-clones of lymphocytes are depleted in size or suppressed in their reactivity, expansion or differentiation. It may arise from activation of specific or non-specific T-suppressor lymphocytes of either T- or B-clones, or by drugs that have generalized effects on most or all T- or B-lymphocytes. ***Cyclosporin*** A and FK506 act on T-cells, as does antilymphocyte serum; alkylating agents such as cyclophosphamide are less specific in their action and damage DNA replication, while base analogues interfering with guanine metabolism act in a similar way. See ***immunophilin***.

immunotoxins Any toxin that is conjugated to either an immunoglobulin or Fab fragment directed against a specified antigen. Thus if the antigen is borne by a particular type of cell, such as a tumour cell, the toxin may be targeted at the specified cell by the immunological reaction.

IMP See ***intramembranous particles***.

impermeable cell junctions See ***zonula occludens***.

impetigo Contagious skin disease caused by staphylococci or streptococci.

importins Proteins that bind nuclear localization signals on proteins destined for the nucleus and that, in conjunction with ***ran*** and pp15 are involved in transport. Importins α and β form a heterodimer in the cytoplasm that corresponds to the 'NLS-receptor': both have multiple repeated arm-domains. Importin α binds importin β through a NLS-like region on importin β; importin β binds ran-GTP and also several nucleoporins. Since importins α and β recycle from the nucleus to the cytoplasm at different rates it seems that there is a dissociation step in the transport cycle.

imprinting See ***genomic imprinting***.

in silico Term used (rather jokingly) for experiments done using a computer database (ie. on a silicon chip). It is now possible to match a small sequence of nucleotides with a full-length gene by running a search on a database; often it is then possible to order the appropriate cDNA in a vector ready for use. Derived by analogy with ***in vitro*** and ***in vivo***.

in situ Literally, in place. Used particularly in the context of ***in situ hybridization***.

***in situ* hybridization** (1) Technique for revealing patterns of gene expression in a tissue. The tissue is fixed and prepared (usually by sectioning) on a slide, and a labelled DNA or RNA probe is hybridized to the sample. It binds only to complementary mRNA sequences, and so only stains cells that are expressing the gene in question. (2) Use of the technique for identifying the position of a gene on a chromosome by hybridizing a probe to spread chromosomes. Widely used in

Drosophila salivary gland polytene chromosome 'squashes', but now also possible with a variety of eukaryotic nuclei. 'FISH' analysis is the use of ***fluorescence in situ hybridization***.

in vitro Literally, in glass; general term for cells in culture as opposed to in a multicellular organism (***in vivo***).

in vivo Literally, in life; used of cells in their natural multicellular environment or of experiments done on intact organism rather than on isolated cells in culture (***in vitro***)

inactivation For example, of ***voltage-gated*** sodium channels: process by which sodium channels that have been activated or opened by ***depolarization*** subsequently close during the depolarization. Distinguished from activation by its slower kinetics.

inbred strain Any strain of animal or plant obtained by a breeding strategy that tends to lead to homozygosity. Such breeding strategies include brother–sister mating and back-crossing of offspring with parents. See also ***congenic***.

inclusion bodies Nuclear or cytoplasmic structures with characteristic staining properties, usually found at the site of virus multiplication. Semi-crystalline arrays of ***virions***, ***capsids*** or other viral components.

indirect immunofluorescence A method of ***immunofluorescence*** staining in which the first antibody, which is directed against the antigen to be localized, is used unlabelled, and the location of the first antibody is then detected by use of a fluorescently labelled anti-IgG (against IgGs of the species in which the first antibody was raised). The advantage is that there is some amplification, and a well-characterized goat anti-rabbit IgG antibody can, for example, be used against a scarce specific antibody raised in rabbits. The same technique can be used for ultrastructural localization of the first antibody by substituting peroxidase or gold-labelled second antibody.

indole acetic acid (IAA) The most common naturally occurring ***auxin***. Promotes growth in excised plant organs, induces adventitious roots, inhibits axillary bud growth, regulates gravitropism.

indomethacin Non-steroidal anti-inflammatory drug that blocks the production of ***arachidonic acid*** metabolites by inhibiting ***cyclooxygenase***.

inducer cells Cells that induce other nearby cells to differentiate in specified pathways. Perhaps the distinction should be made, as of old, between those cells that evoke a predetermined pathway of differentiation in the target cells and those cells that can actually induce new and unexpected differentiations.

induction See ***embryonic induction*** or ***enzyme induction***.

infarction Death of tissue as a result of loss of blood supply, often as a result of thrombotic occlusion of vessels.

infectious hepatitis See ***hepatitis A***.

infectious mononucleosis See ***glandular fever***.

inflammation Response to injury. ***Acute inflammation*** is dominated by vascular changes and by neutrophil leucocytes in the early stages, mononuclear phagocytes later on. Leucocytes adhere locally and emigrate into the tissue between the endothelial cells lining of the postcapillary venules. Plasma exudation from vessels may lead to tissue swelling, but the early vascular changes are independent of and not essential for the later cellular response. In chronic inflammation, where the stimulus is persistent, the characteristic cells are ***macrophages*** and ***lymphocytes***.

influenza virus Member of the ***Orthomyxoviridae*** that causes influenza in humans. There are three types of influenza virus: type A causes the worldwide epidemics (pandemics) of influenza and can infect other mammals and birds; type B only affects humans; type C causes only a mild infection. Types A and B virus evolve continuously, resulting in changes in the antigenicity of their spike proteins, preventing the development of prolonged immunity to infection. The spike proteins, external haemagglutinin (HA) and ***neuraminidase*** have been studied as models of membrane glycoproteins.

informosome Cytoplasmic complex of mRNA and non-ribosomal protein. May protect the message from degradation during passage from nucleus to cytoplasm.

inhibin Polypeptide hormone secreted by the hypophysis that selectively suppresses the secretion of pituitary FSH (***follicle-stimulating hormone***). The molecule has two subunits (14 and 18 kD) and is a product of the gene family that includes ***TGF*-β**. There are two forms, $\alpha\beta_A$ and $\alpha\beta_B$, the β subunit being shared with ***activin***. Inhibin is now, on the basis of gene knockout experiments, considered to be a tumour-suppressor, the key gene being that for inhibin-α.

inhibitory synapse A synapse in which an action potential in the presynaptic cell reduces the probability of an action potential occurring in the postsynaptic cell. The most common inhibitory neurotransmitter is ***GABA***; this opens channels in the postsynaptic cell which tend to stabilize its ***resting potential***, thus rendering it less likely to fire. See ***excitatory synapse***.

initial cell Actively dividing plant cell in a ***meristem***. At each division one daughter cell remains in the meristem as a new initial cell, and the other is added to the growing plant body. Animal equivalent is a stem cell (potentially confusing terminology in plants).

initiation codon (start codon) The ***codon*** 5′-AUG in mRNA, at which polypeptide synthesis is started. It is recognized by formylmethionyl-tRNA in bacteria and by methionyl-tRNA in eukaryotes.

initiation complex Complex between mRNA, 30S ribosomal subunit and formyl-methionyl-tRNA that requires GTP and ***initiation factors*** to function.

initiation factors (IFs) The set of catalytic proteins required, in addition to mRNA and ribosomes, for protein synthesis to begin. In bacteria three distinct proteins have been identified: IF-1 (8 kD), IF-2 (75 kD) and IF-3 (30 kD). At least 6–8 proteins have been identified in eukaryotes. IFs 1 and 2 enhance the binding of initiator tRNA to the ***initiation complex***.

inner cell mass A group of cells found in the mammalian blastocyst that give rise to the embryo and are potentially capable of forming all tissues, embryonic and extra-embryonic, except the trophoblast.

inner sheath The material that encases the two central microtubules of the ciliary axoneme.

inoculum Cells added to start a culture or, in the case of viruses, viruses added to infect a culture of cells. Also for biological material injected into a human to induce immunity (a vaccine).

inosine The 'fifth base' of nucleic acids. Important because it fails to form specific pair bonds with the other bases. In ***transfer RNAs***, this property is used in the ***anticodon*** to allow matching of a single tRNA to several codons. ***PCR*** performed with primers containing inosine tolerates a limited degree of mismatch between primer and template, useful when trying to clone homologous protein by using degenerate primers.

inositol A cyclic hexahydric alcohol with six possible isomers. The biologically active form is myoinositol.

inositol 3-kinase See ***PI3-kinase***.

inositol phosphates Virtually all of the possible phosphorylated states of inositol have been reported to occur in living tissues. The hexaphosphate, phytic acid, is abundant in many plant tissues and is a powerful calcium chelator. See ***PI3-kinase***.

inositol trisphosphate (IP_3, $InsP_3$) Inositol 1,4,5 trisphosphate is important as a second messenger. It is released from phosphatidyl inositol bisphosphate by the action of a specific phospholipase C enzyme (PLC_γ) and binds to and activates a calcium channel in the ***endoplasmic reticulum***.

inotropic Altering rate of heartbeat, eg. adrenaline has a positive inotropic effect and increases rate of beating.

insect defensins See ***defensins***.

insertin Protein (30 kD) from chicken gizzard smooth muscle. Binds to the barbed ends of actin filaments and apparently allows insertion of further monomers.

insertion sequence (IS elements) Mobile nucleotide sequences that occur naturally in the genomes of bacterial populations. When inserted into bacterial DNA they inactivate the gene concerned; when they are removed the gene regains its activity. Closely related to transposons and range in size from a few hundred to a few thousand bases, but are usually less than 1500 bases.

insertional mutagenesis Generally, ***mutagenesis*** of DNA by the insertion of one or more bases. Specific examples: (1) oncogenesis by insertion of a retrovirus adjacent to a cellular protooncogene; (2) a strategy of mutagenesis with ***transposons***. After a round of transposition, progeny are screened by ***PCR***, with transposon- and gene-specific primers, for the proximity of the transposon sequence to the gene of interest. As PCR can only produce products up to 1–2 kb, a large fraction of progeny identified as positive by PCR will have a transposon close enough to the gene to inactivate or otherwise alter its pattern of expression.

insertosome See ***insertion sequence***.

inside-out patch A variant of the ***patch clamp*** technique, in which a disc of plasma membrane covers the tip of the electrode, with the inner face of the plasma membrane facing outward, to the bath.

inside-out vesicle (IOV) Mechanical disruption of cell membranes gives rise to small closed vesicles surrounded by a bilayer membrane. These may be right-side out (ROV), or IOV if the topography is inverted.

instructive theory Theory of antibody production, now considered untenable, in which antigen acted as template for the production of specific antibody as opposed to the ***clonal selection*** theory in which pre-existing variation occurs and appropriate clones are selectively expanded.

insulin A polypeptide hormone (bovine insulin; 5780 D) found in both vertebrates and invertebrates. Secreted by the B-***cells*** of the pancreas in response to high blood sugar levels, it induces hypoglycaemia. Defective secretion of insulin is the cause of ***diabetes mellitus***. Insulin is also a mitogen, has sequence homologies with other ***growth factors***, and is a frequent addition to cell culture media for demanding cell types.

insulin-like growth factor (IGF) IGFs I and II are polypeptides with considerable sequence similarity to insulin. They are capable of eliciting the same biological responses, including mitogenesis in cell culture. On the cell surface, there are two types of IGF receptor, one of which closely resembles the insulin receptor (which is also present). IGF-I = somatomedin A = somatomedin C; IGF-II = MSA (multiplication-stimulating activity). IGF-I is released from the liver in response to growth hormone.

int-1 ***Oncogene*** from mouse mammary ***carcinoma*** that encodes a secreted protein. Related to *Drosophila* gene *wingless*. See also Table O1.

int-2 ***Oncogene*** from mouse mammary ***carcinoma*** that encodes a member of ***fibroblast growth factor receptor*** family. See also Table O1.

integral membrane protein A protein that is firmly anchored in a membrane (unlike a peripheral membrane protein). Most is known about the integral proteins of the plasma membrane, where important examples include hormone receptors, ion channels, and transport proteins. An integral protein need not cross the entire membrane; those that do are referred to as transmembrane proteins.

integrase protein Enzyme of the bacteriophage lambda (λ) that catalyses the integration of phage DNA into the host DNA.

integration Incorporation of the genetic material of a virus into the host genome.

integrator gene In the Britten and Davidson model for the coordinate expression of unlinked genes in eukaryotes, sensor elements respond to changing conditions by switching on appropriate integrator genes, which then produce transcription factors that activate appropriate subsets of structural genes.

integrins Superfamily of cell surface proteins that are involved in binding to extracellular matrix components in some cases. Most are heterodimeric with a β subunit of 95 kD that is conserved through the superfamily, and a more variable α subunit of 150–170 kD. The first examples described were ***fibronectin*** and ***vitronectin*** receptors of fibroblasts, which bind to an RGD (Arg-Gly-Asp) sequence in the ligand protein, though the 'context' of the RGD seems important and there is also a divalent cation dependence. Subsequently the platelet

Table I1. Vertebrate integrins

Subunits		Ligand	Binding Site	Synonyms[a]
β_1	α_1	Collagens, laminin		VLA-1
	α_2	Collagens, laminin	-DGEA-	VLA-2, Platelet glycoprotein Ia/IIa
	α_3	Collagens, fibronectin, laminin	-RGD-?	VLA-, chick integrin
	α_4	Fibronectin (alternatively spliced domain), VCAM-1	-EILDV-	VLA-4
	α_5	Fibronectin (RGD)	-RGD-	VLA-5 'Fibronectin receptor', Platelet glycoprotein Ic/IIa
	α_6	Laminin		VLA-6, Platelet glycoprotein Ic′/IIa
	α_7	Laminin		
	α_8	?		
	α_V	Vitronectin, fibronectin	-RGD-	
β_2	α_L	ICAM-1, ICAM-2		LFA-1, CD11a
	α_M	C3bi, fibrinogen, factor X, ICAM-1		Mac-1, Mo-1, CR3, CD11b
	a_X	Fibrinogen, C3bi ?	-GPRP-	gp150,95, CR4, CD11c
β_3	α_{IIb}	Fibronectin, fibrinogen, vitronectin, thrombospondin	-RGD-	Platelet glycoprotein IIb/IIIa
		von Willebrand factor	-KQAGDV-	
	α_V	Collagen, fibronectin, fibrinogen, osteopontin, thrombospondin, vitronectin, von Willebrand factor	-RGD-	'Vitronectin receptor'
β_4	α_6	(Laminin ?)		
β_5	α_V	Vitronectin	-RGD-	
β_6	α_V	Fibronectin	-RGD-	
β_7 (? = b_p)	α_4	Fibronectin (alternatively spliced domain), VCAM-1	-EILDV- (in fibronectin only)	LPAM-2
	α_{IEL}	? (expressed on intraepithelial lymphocytes)		
β_8	α_V	?		

[a]CD numbers are only given for the β_2 family of integrins.

IIb/IIIa surface glycoprotein (fibronectin and fibrinogen receptor) and the LFA-1 class of leucocyte surface protein were recognized as integrins, together with the VLA surface protein. The requirement for the RGD sequence in the ligand does not seem to be invariable. See Table I1.

integron Class of DNA element composed of a DNA integrase gene adjacent to a recombination site, at which one or more genes can be found inserted. Frequently, antibiotic resistance genes are found inserted in integron sites in samples of resistant ***Gram-positive bacteria***.

intercalated disc An electron-dense junctional complex, at the end-to-end contacts of cardiac muscle cells, that contains ***gap junctions*** and ***desmosomes***. Most of the disc is formed of a convoluted type of ***adherens junction*** into which the actin filaments of the terminal sarcomeres insert (they are therefore equivalent to half Z bands); desmosomes are also present. The lateral portion of the stepped disc contains gap junctions that couple the cells electrically and thus coordinate the contraction.

intercalation Insertion into a pre-existing structure, eg. nucleotide sequences into DNA (or RNA); molecules into structures such as membranes.

intercellular Between cells: can be used either in the sense of connections between cells (as in intercellular junctions), or as an antonym for intracellular.

intercellular adhesion molecule See ***ICAM***.

interdigitating cells Cells found particularly in thymus-dependent regions of lymph nodes; they have dendritic morphology and ***accessory cell*** function.

interference diffraction patterns The patterns arising from the recombination of beams of light or other waves after they have been split and one set of rays have undergone a phase retardation relative to the other. Such patterns formed by simple objects give information on the correctness of the focus and the presence or absence of optical defects.

interference microscopy Although all image formation depends on interference, the term is generally restricted to systems in which contrast comes from the recombination of a reference beam with light that has been retarded by passing through the object. Because the phase retardation is a consequence of the difference in refractive index between specimen and medium, and because the the refractive increment is almost the same for all biological molecules, it is possible to measure the amount of dry mass per unit area of the specimen by measuring the phase retardation. Quantification of the phase retardation is usually done by using a compensator to reduce the bright object to darkness (see ***Senarmont*** and ***Ehringhaus compensators***). Two major optical systems have been used: the ***Jamin–Lebedeff*** system and the ***Mach-Zehnder*** system. These instruments are often referred to as interferometers, since they are designed for measuring phase retardation. Although their use has passed out of fashion, it may be that they will be employed more frequently in future in conjunction with image analysing systems.

interference reflection microscopy An optical technique for detecting the topography of the side of a cell in contact with a planar substrate, and for providing information on the separation of the plasmalemma from the substrate. Interference between the reflections from the substrate–medium interface and the reflections from the plasmalemma–medium interface generate the image.

interferon-inducing factor See ***IGIF***.

interferon-regulatory factor-1 (IRF-1) Transcription factor. Deletion of the IRF-1 gene in mice leads to severe deficiency in NK cell function. Regulates gene for IL-15.

interferons (IFN) A family of glycoproteins produced in mammals that prevent virus multiplication in cells. IFN-α is made by leucocytes and IFN-β by fibroblasts after viral infection. IFN-γ is produced by immune cells after antigen stimulation. IFN-α and -β are also known as type I interferons. IFN-γ, a type II interferon, is more usually classed as a cytokine.

intergenic suppression Compare ***intragenic suppression***. The situation where a primary gene and the gene that suppresses it do not lie in the same chromosomal locus.

interleukin A variety of substances produced by leucocytes (not necessarily exclusively) and that function during inflammatory responses. (This is the definition recommended by the IUIS-WHO Nomenclature Committee). Interleukins are of the larger class of T-cell products, ***lymphokines***. Now more frequently considered as ***cytokines***.

interleukin-1 (IL-1) Protein (17 kD: 152 amino acids) secreted by ***macrophages*** or ***accessory cells*** involved in the activation of both T- and B-lymphocytes in response to antigens or mitogens, as well as affecting a wide range of other cell types. At least two IL-1 genes are active; α and β forms of IL-1 are recognized. There is an endogenous antagonist, IL-1ra that binds to the receptor but does not elicit effects. IL-1α, IL-1β and IL-1ra are remarkably different is sequence though similar in binding properties. See also ***catabolin***, ***endogenous pyrogen***.

interleukin-1 converting enzyme (ICE) Cytoplasmic cysteine protease that is uniquely responsible for cleaving proIL-1 β (31 or 33 kD) into mature IL-1 β (17.5 kD); the active cytokine is then released by a non-standard mechanism (there is no signal sequence and it does not pass through the Golgi). The enzyme seems to be composed of two non-identical subunits derived from a single proenzyme. The ICE gene has some homology with the *ced-9* gene of *Caenorhabditis elegans*, the product of which is a ***caspase*** involved in mediating cell death by apoptosis.

interleukin-2 (IL-2; T-cell growth factor; thymocyte stimulating factor.) Cytokine (17 kD) released by activated T-cells that causes activation, stimulates and sustains growth of other T-cells independently of the antigen. Blocking production or release of IL-2 would block the production of an immune response

interleukin-3 (IL-3) Product of mitogen-activated T-cells: ***colony-stimulating factor*** for bone marrow stem cells and ***mast cells***.

interleukin-4 (IL-4, B-cell stimulating factor, BSF-1) A ***cytokine***.

interleukin-5 (IL-5) A B-cell growth and differentiation factor; also stimulates eosinophil precursor proliferation and differentiation.

interleukin-6 (IL-6) Cytokine that is coinduced with interferon from fibroblasts, a B-cell differentiation factor, a hybridoma growth factor, an inducer of acute phase proteins, and a colony-stimulating factor acting on mouse bone marrow.

interleukin-7 (IL-7; lymphopoietin 1) Single-chain 25 kD cytokine (20 kD) originally described as a pre-B-cell growth factor but now known to have effects on a range of other cells, including T-cells. Produced by monocytes, T-cells and NK cells.

interleukin-8 (IL-8; neutrophil activating protein, NAP-1) One of the first ***chemokine***s to be isolated; one of the C-X-C family (8 kD). Secreted by a variety of cells and potently chemokinetic and chemotactic for neutrophils and basophils but not monocytes. Receptor is G-protein coupled.

interleukin-9 (IL-9) ***Cytokine*** produced by T-cells, particularly when mitogen stimulated, that stimulates the proliferation of erythroid precursor cells (***BFU-E***). May act synergistically with ***erythropoietin***. Receptor belongs to haematopoietic receptor superfamily.

interleukin-10 (IL-10) ***Cytokine*** produced by Th2 helper T-cells, some B-cells and LPS-activated monocytes. Regulates cytokine production by a range of other cells.

interleukin-11 (IL-11) ***Pleiotropic*** cytokine originally isolated from primate bone marrow ***stromal cell*** line. Stimulates T-cell dependent B-cell maturation, megakaryopoiesis, various stages of myeloid differentiation. Receptor shares gp130 subunit with other members of ***IL-6 cytokine family***.

interleukin-12 (IL-12; NK stimulatory factor; cytotoxic lymphocyte maturation factor) Heterodimeric cytokine (35 kD and 40 kD) that enhances the lytic activity of ***NK*** cells, induces ***interferon***-γ production, stimulates the proliferation of activated T-cells and NK cells. Is secreted by human B-lymphoblastoid cells (NC-37). May play a role in controlling immunoglobulin isotype selection and is known to inhibit IgE production.

interleukin-13 (IL-13) Cytokine (12.4 kD) with anti-inflammatory activity. Produced by activated T-cells; inhibits IL-6 production by monocytes and also the production of other pro-inflammatory cytokines such as TNF-α, IL-1, IL-8. Stimulates B-cells. Gene is located in cluster of genes on human chromosome 5q that also has IL-4 gene.

interleukin-14 (IL-14; high molecular weight B-cell growth factor; HMW-BCGF) Cytokine (53 kD) produced by T-cells that enhances proliferation of activated B-cells and inhibits immunoglobulin synthesis. Unrelated to other cytokines but has homology with complement factor Bb.

interleukin-15 (IL-15) Cytokine that has effects very similar to IL-2 but in addition potently chemotactic for lymphocytes. Levels are elevated in the rheumatoid joint. Receptor shares β and γ subunits with IL-2 receptor but has unique α subunit.

interleukin-16 (IL-16, lymphocyte chemoattractant factor) Secreted from $CD8^+$ cells and will induce migratory responses in $CD4^+$ cells (lymphocytes, monocytes and eosinophils). May bind to CC-CKR-5 and contribute to the blocking of ***HIV*** internalization.

interleukin-17 (IL-17) Pro-inflammatory T-cell product (17 kD) that acts on receptors on a range of cells to activate NFκB. Induces expression of IL-6, IL-8 and ICAM-1 in fibroblasts and enhances T-cell proliferation stimulated by suboptimal levels of ***PHA***. Receptor is a type I transmembrane protein, though a soluble form is also found, and has no homology with other known sequences.

interleukin-18 (interferon-γ-inducing factor; IGIF) First isolated from liver of mice during toxic shock; has sequence homology with IL-1β and IL-1ra and has also been designated IL-1γ.

intermediate filaments A class of cytoplasmic filaments of animal cells so named originally because their diameter (nominally 10nm) in muscle cells was intermediate between thick and thin filaments. Unlike microfilaments and microtubules, the protein subunits of intermediate filaments show considerable diversity and tissue specificity. See ***cytokeratins***, ***desmin***, ***glial fibrillary acidic protein***, ***neurofilament*** proteins, ***nestin*** and ***vimentin***; see also Table I2.

intermembrane space Region between the two membranes of mitochondria and chloroplasts. On the ***endosymbiont hypothesis***, this space would represent the original phagosome.

internal bias Applied to the motile

Table I2. Intermediate filaments and sequence-related proteins

Name of protein	Sequence homology group	M_r (kD)	Cell type
Cytokeratins (19 in humans)	I (acidic)	40–60	Epithelia
	II (neutral-basic)	50–70	
Vimentin	III	53	Many, especially mesenchymal
Desmin	III	52	Muscle
Glial fibrillary acidic protein	III	51	Glial cells. Astrocytes
Neurofilament polypeptides	IV	57–150	Neurons (vertebrates)
		60–120	Neurons (invertebrates)
Nuclear lamins	V	60–70	All eukaryotic cells

behaviour of crawling cells that, in the short term, show ***persistence*** and do not behave as true random walkers. Any intrinsic regulation of the random motile behaviour of the cell could be considered as internal bias.

internal membranes General term for intracellular membrane systems such as endoplasmic reticulum. Not particularly helpful, but has the advantage of being non-committal.

internalin Surface proteins (InlA, InlB) that mediate entry of ***Listeria monocytogenes*** into epithelial cells that express E-cadherin or L-CAM. There appears to be an internalin multigene family (Inl C, D, E, F) although not all the products are involved in bacterial entry into cells.

interneurons Neurons that connect only with other neurons, and not with either sensory cells or muscles. They are thus involved in the intermediate processing of signals.

internexin (α-internexin) Neuronal intermediate filament protein (68 kD). Subunit of type IV filaments found in neurons of CNS.

interphase The stage of the cell or nucleus when it is not in mitosis, hence comprising most of the cell cycle.

interstitial cells (1) Cells lying between but distinct from other cells in a tissue, a good example being the interstitial cells in *Hydra* that serve as stem cells. (2) Cells lying between the testis tubules of vertebrates and that are responsible for the secretion of testosterone.

intervening sequence See ***intron***.

intestinal calcium-binding protein (ICaBP) Calcium-binding proteins containing the ***EF-hand*** motif, induced by vitamin D3.

intestinal epithelium The endodermally derived epithelium of the intestine varies considerably, but the absorptive epithelium of small intestine is usually implied. The apical surfaces of these cells have ***microvilli*** (which increases the absorptive surface, but probably also provides a larger surface area for enzyme activity). The lateral subapical regions have well-developed intercellular ***junctions***.

intima Inner layer of blood vessel, comprising an endothelial monolayer on the luminal face with a subcellular elastic extracellular matrix containing a few smooth muscle cells. Below the intima is the ***media***, then the ***adventitia***. The term may be applied to other organs.

intine Inner layer of the wall of a pollen grain, resembling a ***primary cell wall*** in structure and composition. Also used for the inner wall layer of a ***spore***.

intragenic suppression The situation where a primary gene and the mutated gene that suppresses it lie within the same locus.

intramembranous particles (IMP) Particles (or complementary pits) seen in ***freeze fractured*** membranes. The cleavage plane is through the centre of the bilayer, and the particles are usually assumed to represent ***integral membrane proteins*** (or polymers of such proteins).

intrinsic factor A mucoprotein normally secreted by the epithelium of the stomach and that binds vitamin B12; the intrinsic factor–B12 complex is selectively absorbed by the distal ileum, though only the vitamin is taken into the cell.

intrinsic pathway See ***extrinsic pathway***.

intron (intervening sequence) A non-coding sequence of DNA within a gene (cf. ***exon***), that is transcribed into ***hnRNA*** but is then removed by RNA splicing in the nucleus, leaving a mature mRNA that is then translated in the cytoplasm. Introns are poorly conserved and of variable length, but the regions at the ends are self-complementary, allowing a hairpin structure to form naturally in the hnRNA; this is the cue for removal by RNA splicing. Introns are thought to play an important role in allowing rapid evolution of proteins by ***exon shuffling***. Genes may contain as many as 80 introns.

inulin A polysaccharide of variable molecular weight (around 5 kD), that is a polymer of fructofuranose. Widely used as a marker of extracellular space, an indicator of blood volume in insects (by measuring the dilution of the radiolabel), and in food for diabetics.

invasins Proteins produced by bacterial cells that promote bacterial penetration into mammalian cells. The invasin produced by *Yersinia pseudotuberculosis* seems to bind to the fibronectin receptor (α5-β1 integrin) at a site close to the fibronectin binding site, though the invasin does not have an RGD sequence.

invasion A term that should be used with caution; although most cell biologists would follow Abercrombie in meaning the movement of one cell type into a territory normally occupied by a different cell type, some pathologists might not agree.

invasion index An index devised by Abercrombie and Heaysman as a means to estimate the invasiveness of cells ***in vitro***. The index is derived from measurements on confronted explants of the cells and embryonic chick heart fibroblasts growing in tissue culture: it is the ratio of the estimated movement, had the cells not been hindered, and the actual movement in the zone in which collision occurs.

inverse agonist (reverse antagonist) Any ligand that binds to receptors and reduces the proportion in the active form. Has the opposite effects to an agonist and may actually reduce the background level of activity. Not the same as a partial agonist.

inversion heterozygote Individual in which one chromosome contains an inversion whereas the homologous chromosome does not.

invertase (1) ('sucrase') Enzyme catalysing the hydrolysis of sucrose to glucose and fructose, so called because the sugar solution changes from dextro-rotatory to laevo-rotatory during the course of the reaction. (2) Generally a name for an enzyme that catalyses certain molecular rearrangements. DNA invertases are a class of ***resolvase***.

involucrin Marker protein for ***keratinocyte*** differentiation first appearing in the upper spinous layer of the ***epidermis***. Together with ***trichohyalin***, forms the scaffold for the cell envelope.

involution (1) Restoration of the normal size of an organ. (2) Infolding of the edges of a sheet of cells, as in some developmental processes, notably gastrulation.

ion channel A transmembrane pore that presents a hydrophilic channel for ions to cross a lipid bilayer down their electrochemical gradients. Some degree of ion specificity is usually observed, and typically a million ions per second may flow. Channels may be permanently open, like the potassium leak channel; or they may be ***voltage-gated***, like the ***sodium channel***; or ***ligand-gated*** like the acetylcholine receptor.

ion-exchange chromatography Separation of molecules by absorption and desorption from charged polymers. An important technique for protein purification. For small molecules the support is usually polystyrene, but for macromolecules, cellulose, acrylamide or agarose supports give less non-specific absorption and denaturation. Typical charged residues are CM (carboxymethyl) or DEAE (diethylaminoethyl).

ion-selective electrode An electrode half-cell, with a semi-permeable membrane that is permeable only to a single ion. The electrical potential measured between this and a reference half-cell (eg. a calomel electrode) is thus the Nernst potential for the ion. Given that the solution filling the ion-selective electrode is known, the activity (rather than concentration) of the ion in the unknown solution can be measured. Commercial ion-selective electrodes frequently use a hydrophobic membrane containing an ***ionophore***, such as ***valinomycin*** (for potassium) or ***monensin*** (for sodium). A pH electrode is made with a thin membrane of pH-sensitive (ie. proton-permeable) glass.

ionic coupling The same as ***electrical coupling***.

ionizing radiation ***Radiation*** capable of ionizing, either directly or indirectly, the substances it passes through. α and β radiation are far more effective at producing ionization (and therefore are more likely to cause tissue or cell damage) than γ radiation or neutrons.

ionophore A molecule that allows ions to cross lipid bilayers. There are two classes: carriers and channels. Carriers, like ***valinomycin***, form cage-like structures around specific ions, diffusing freely through the hydrophobic regions of the bilayer. Channels, like ***gramicidin***, form continuous aqueous pores through the bilayer, allowing ions to diffuse through. See ***ion channels*** and Table I3.

ionotropic Allowing the passage of ions (not to be confused with ***inotropic***). Usually used of the class of ***GABA receptors*** that form integral ***ion channels***, cf. ***metabotropic***.

iontophoresis Movement of ions as a result of an applied electric field, eg. the delivery of a charged molecule from the end of a micropipette without hydraulic flow.

IOV See ***inside-out vesicle***.

IP3 See ***inositol trisphosphate***.

IP3, InsP3 See ***inositol trisphosphate***.

IPNS (isopenicillin N synthase) Non-haem iron-dependent oxidase (38 kD) that catalyses the formation of isopenicillin N from α-aminoadipylcysteinyl-valine (ACV).

IPTG (isopropyl β-D-thiogalactoside) Used to trigger gene expression that is under the control of ***gal promoter***, particularly used in expression systems for producing protein.

Ir genes Immune response genes, located within the MHC of vertebrates. Originally recognized as controlling the level of immune response to various synthetic polypeptides, they are now also recognized as mapping within the regions controlling T-cell help and suppression (***I-region***).

Ir-associated antigens Antigens coded for by Ir (immune response) genes or antigens coded for by the genome close to the ***Ir genes***. See also ***Ia antigens***.

Table I3. Ionophores

	M_r (Da)	Ion selectivity	Comments
Neutral			
Valinomycin	1110	Rb>K>Cs>Ag>>NH_4>Na>Li	Depsipeptide, uncoupler
Enniatin A	681	K>Rb≈Na>Cs>>Li	Cyclic hexadepsipeptide
Enniatin B	639	Rb>K>Cs>Na>>Li; Ca>Ba>Sr>Mg	Cyclic hexadepsipeptide
Nonactin	736	NH_4>K≈Rb>Cs>Na	Macrotetralide, product of Actinomyces strains
Monactin	750	NH_4>K>Rb>Cs>Na>Ba	Macrotetralide
Cryptate 211	288	Li>Na>K≈Rb≈Cs; Ca>Sr≈Ba	Amino ether; one of a substantial family
Carboxylic			
Monensin	670	Na>>K>Rb>Li>Cs	Blocks transport through Golgi
Nigericin	724	K>Rb>Na>Cs>>Li	
X-537A (lasalocid)	590	Cs>Rb≈K>Na>Li; Ba>Sr>Ca>Mg	Macrotetralide
A23187	523	Li>Na>K; Mn>Ca>Mg>Sr>Ba	Predominantly selective for divalent cations
Channel-forming			
Gramicidin A	≈1700	H>Cs≈Rb>NH_4>K>Na>Li	Peptide; works as a dimer
Alamethicin	—	K>Rb>Cs>Na	Peptide; voltage-dependent
Monazomycin	≈1422	Cs>Rb>K>Na>Li	Polyene-like; voltage-sensitive

Many uncouplers such as FCCP (carbonyl cyanide-trifluoro-methoxyphenylhydrazone) also act as ionophores. Amphotericin, filipin and nystatin may be anion-specific ionophores.

IRAK (IL-1 receptor associated kinase) Associates with the IL-1R once IL-1 binds. Part of the kinase cascade that eventually leads to NFκB translocation to the nucleus and altered gene expression. Has homology with ***pelle***.

iridoviruses A group of non-occluded viruses of insects; the crystalline array of virus particles in the cytoplasm of epidermal cells gives infected insects an irridescent appearance.

irs-1 (insulin receptor substrate-1) Multisite docking protein (180 kD) that is phosphorylated by insulin and IGF-I receptors following ligand binding. The tyrosine-phosphorylated form of irs-1 will interact with the SH2 domains of p85 of PI-3-kinase, Grb2 and PTP-2.

Is element See ***insertion sequence***.

IS elements See ***insertion sequence***.

ischaemia (USA, ischemia) Inadequate blood flow leading to hypoxia in the tissue.

ISCOMS (immunostimulatory complexes) Small cage-like structures that make it possible to present viral proteins to the immune system in an array, much as they would appear on the virus. Produced by mixing the viral protein with Quill A, a substance isolated from the Amazonian oak, in the presence of detergent. ISCOMS are being used successfully in vaccines.

islet amyloid peptide (IAPP) Peptide of 37 amino acids that selectively inhibits insulin-stimulated glucose uptake in muscle. Structurally related to ***calcitonin gene-related peptide***.

islet cells Cells of the ***islets of Langerhans*** within the pancreas. See ***A-cells, B-cells, D-cells***.

islets of Langerhans Groups of cells found within the pancreas: A-cells and B-cells secrete ***insulin*** and ***glucagon***. See also ***D-cells***.

isoantibody Antibody made in response to antigen from another individual of the same species.

isochores Long stretches of GC- or AT-rich sequences of DNA associated with R and G chromosome bands respectively.

isocitrate An intermediate in the ***tricarboxylic acid cycle*** (citric acid cycle).

isoelectric focusing ***Electrophoresis*** in a stabilized pH gradient. High-resolution method for separating molecules, especially proteins, that carry both positive and negative charges. Molecules migrate to the pH corresponding to their ***isoelectric point***. The gradient is produced by electrophoresis of amphiphiles, heterogeneous molecules giving a continuum of isoelectric points. Resolution is determined by the number of amphiphile species and the evenness of distribution of their isoelectric points.

isoelectric point The pH at which a protein carries no net charge. Below the isoelectric point proteins carry a net positive charge; above it a net negative charge. Due to a preponderance of weakly acid residues in almost all proteins, they are nearly all negatively charged at neutral pH. The isoelectric point is of significance in protein purification because it is the pH at which solubility is often minimal, and at which mobility in an electrofocusing system is zero (and therefore the point at which the protein will accumulate).

isoenzymes Variants of enzymes that catalyse the same reaction, but owing to differences in amino acid sequence can be distinguished by techniques such as electrophoresis or isoelectric focusing. Different tissues often have different isoenzymes. The sequence differences generally confer different enzyme-kinetic parameters that can sometimes be interpreted as fine-tuning to the specific requirements of the cell types in which a particular isoenzyme is found.

isoform A protein having the same function and similar (or identical sequence), but the product of a different gene and (usually) tissue-specific. Rather stronger in implication than ***homologous***.

isohaemagglutinins Natural antibodies that react against normal antigens of other members of the same species.

isoleucine (Ile; I) Hydrophobic amino acid (131 D). See Table A2.

isomers Alternative stereochemical forms of molecules containing the same atoms.

isometric tension Tension generated in a muscle without contraction occurring: cross-bridges are being reformed with the same site on the thin filament, and the tension (in striated muscle) is proportional to the overlap between thick and thin filaments.

isopenicillin N synthase See ***IPNS***.

isoprenaline See ***isoproterenol***.

isoprenaline: isopropyl-noradrenaline See ***isoproterenol***.

isoprenoid Large family of molecules that include carotenoids, phytoids, prenols, steroids, terpenoids and tocopherols. May form only a portion of a molecule being attached to a non-isoprenoid portion. Isoprenoids are synthesized from diphosphate of isopentyl alcohol (isopentenyl diphosphate)

isoprenylation See ***prenylation***.

isoproterenol (isoprenaline: isopropyl-noradrenaline) Synthetic β-adrenergic agonist; causes peripheral vasodilation, bronchodilation and increased cardiac output.

isopycnic Having equal density: thus in equilibrium density gradient centrifugation a particle (molecule) will cease to move when it reaches a level at which it is isopycnic with the medium.

isosbestic Wavelength at which the absorption coefficients of equimolar solutions of two different substances are identical.

isotonic Of a fluid, having a concentration that will not cause osmotic volume changes of cells immersed in it. Note that an isotonic solution is not necessarily isosmotic. See ***hypotonic***, ***hypertonic***.

isotonic contraction Contraction of a muscle, the tension remaining constant. Since the contractile force is proportional to the overlap of the filaments, and the overlap is varying, the numbers of active cross-bridges must be changing.

isotropic environments Environments in which the properties are the same at all points, and there are no vectorial or axial cues.

isotype (1) Applied to a set of macromolecules sharing some features in common. In immunology isotype describes the class, subclass, light chain type and subtype of an immunoglobulin. (2) Antigenic determinant that is uniquely present in individuals of a single species. (3) A conventionalized method for the graphical display of statistical data.

isotype switching The switch of immunoglobulin isotype that occurs, for example, when the immune response progresses (IgM to IgG). The switch from IgM to IgG involves only the constant region of the heavy chains (from μ to γ), the light chain and variable regions of the heavy chain remaining the same, and involves the switch regions, upstream (on the 5′ side) of the constant region genes, at which recombination occurs. Similarly, IgM and IgD with the same variable region of the heavy chain, but with different heavy chain constant regions (μ and δ), seem to coexist on the surface of some lymphocytes.

isotypic variation Variability of antigens common to all members of a species, for example the five classes of immunoglobulins found in humans. See ***idiotype*** and ***allotype***.

isozyme See ***isoenzyme***.

ITAM (immunoreceptor tyrosine-based activation motif) When phosphorylated binds ***zap-70*** and ***Syk*** and initiates T-cell activation. Contrast with ***ITIM***s.

ITIM (immunoreceptor tyrosine-based inhibitory motif) Phosphorylation of the ITIM motif, found in the cytoplasmic tail of some inhibitory receptors (***KIR***s) that bind MHC Class I, leads to the recruitment and activation of a protein tyrosine phosphatase.

Ito cells Hepatic stellate cells that become activated in liver fibrosis due to intoxication or hepatotoxic compounds such as carbon tetrachloride. Activation is associated with expression of a sodium/calcium exchanger.

ivermectin Broad-spectrum anthelminthic drug used to treat eg. parasitic nematode worms such as *Strongylus* and ***Onchocerca***.

J

J774.2 cells Mouse (Balb/c) monocyte/macrophage cells with surface receptors for IgG and complement.

J (1) The joule, SI unit of energy. (2) Used in the single letter code for amino acids to represent trimethyl lysine, eg. in calmodulin.

J chain (J-piece) Polypeptide chain (15 kD), found in IgA and in IgM joining heavy chains (H chains) to each other to form dimers of IgA and pentamers of IgM. Disulphide bonds are formed between the J chain and H chains near the Fc ends of the heavy chains. Despite the similar name, it is not identical with the ***J region*** or coded for by the ***J gene***.

J gene Gene(s) coding for the Joining segment of polypeptide chain which links the V (variable regions) to the C (constant) regions of both light and heavy chains of immunoglobulins. During lymphoid development the DNA is rearranged so that the V genes are linked to the J region sequences.

J-piece See ***J chain***.

J region The polypeptide chains coded for by ***J genes***.

JAB (JAK-binding protein) Cytokine-inducible inhibitor of JAKs, probably negative regulator of cytokine signalling. See ***SOCS***.

JAK (Janus kinase) Family of intracellular tyrosine kinases (120–140kD) that associate with cytokine receptors (particularly but not exclusively interferon receptors) and are involved in the signalling cascade. JAK is so called either from Janus kinase (Janus was the gatekeeper of heaven) or 'just another kinase'. JAK has neither SH2 nor SH3 domains.

Jamin–Lebedeff system ***Interference microscopy*** in which object and reference beams are split and later recombined by birefringent calcite plates, but pass through the same optical components (in contrast to the ***Mach–Zehnder system***).

Janus kinase See ***JAK***.

jaundice Yellowing of the skin (and whites of eyes) by bilirubin, a bile pigment. Frequently because of a liver problem.

JC virus A human ***retrovirus*** similar to ***polyoma*** virus, but which has not been found associated with any human cancer. Member of the ***Papovaviridae***.

JE See ***monocyte chemotactic and activating factor***.

jelly roll Complex protein topology in which four ***Greek key*** motifs form an eight-stranded ***beta sandwich***. So called because the overall structure resembles a swiss (or jelly, USA) roll.

JH See ***juvenile hormone***.

JH1 domain Domain in ***JAKs*** that is probably the binding site for the ***SH2*** domain of ***SOCS***.

Jijoye cells Human lymphoblastic cell line, CD23 positive. Model for B-lymphoctes. Derived from Burkitt's lymphoma.

Jimpy Mouse mutant with reduced lifespan due to a recessive sex-linked defect in PLP gene. Has a severe CNS ***myelin*** deficiency associated with complex abnormalities affecting all glial populations. *Jimpy-J4*, the most severe of the jimpy mutants has virtually no PLP protein. See ***Pelizaeus–Merzbacher disease***.

JNK (Jun kinase; c-jun N-terminal kinase; stress-activated protein kinase; SAPK) Family of kinases involved in intracellular signalling cascades. JNKs are distantly related to ***ERKs*** and are activated by dual phosphorylation on tyrosine and threonine residues. In addition to ***c-jun*** will also phosphorylate p53. JNK1, 46 kD;

JNK2 55 kD. JNK3 (464 residues) is mainly found in neurons and may play a part in regulation of apoptosis.

Job's syndrome Thought to be due to a defect in neutrophil chemotaxis which predisposes to infection by staphylococci, often without the normal signs of inflammation ('cold abscesses'). At one time all patients described were female with red hair and elevated plasma IgE levels, but this is no longer the case.

jumping gene Populist term for ***transposon***.

jun **Oncogene** from an avian ***sarcoma*** virus. Protein product, jun, dimerises with ***fos*** via a ***zipper*** motif to form the ***transcription factor*** AP1. See Table O1.

junction potential Potential difference at the boundary between dissimilar solutions; arises from differences in diffusion constants between ions.

junctional basal lamina Specialized region of the ***extracellular matrix*** surrounding a muscle cell, at the ***neuromuscular junction***. May be responsible for localization of ***acetylcholine receptors*** in the synaptic region, and also binds acetylcholinesterase to this region

junctions See ***adherens junction, desmosome, gap junction, zonula occludens***.

junk DNA Genomic DNA that serves, as yet, no known function.

Jurkat cells Human T-lymphocyte line much used for studies of IL-2 production ***in vitro***. Although a convenient model system they are not identical to real T-cells, particularly in their activation behaviour.

juvenile hormone (JH) A hormone found in insects which affects the balance between mature and juvenile attributes of certain tissues at each moult. In particular, the ***imaginal discs*** of many larval insects only develop into adult wings, sexual organs or limbs when blood juvenile hormone levels fall below a threshold level. There is a complex interaction between juvenile hormone and ***ecdysone***. Synthetic analogues of JH include farnesol and methoprene, which have been tested for insecticide potential (known, with diflubenzuron, as insect growth regulators, IGRs; see also ***chitin***).

juxtacrine activation Activation of target cells by membrane-anchored growth factors; also used for activation of leucocytes by ***PAF*** bound to endothelial cell surface.

K

K See ***lysine***.

K antigen Capsular antigens of bacteria – usually polysaccharide.

K cells See ***killer cells***.

k-ras Kirsten-*ras*; see also ***Kirsten sarcoma virus***.

K_a (1) Acid ***dissociation constant***. Often encountered as pKa (ie. $-\log_{10}$Ka). (2) Association constant (Kass). The equilibrium constant for association, the reciprocal of Kd, with dimensions of litres/mole. Better to use Kd, thereby removing any ambiguity.

kainate An agonist for the K-type ***excitatory amino acid*** receptor. It can act as an ***excitotoxin*** producing symptoms similar to those of ***Huntington's chorea***, and is also used as an anthelminthic drug. Originally isolated from the alga *Digenea simplex*. The receptor is an amino acid-gated ***ion channel***, one of several types gated by the transmitter.

kainic acid See ***kainate***.

kairomone A subclass of ***pheromone***, defined as an interspecific secretion which benefits the receiver. See ***allomone***.

kalanin Protein that provides adhesion between epidemal keratinocytes and dermis. Localizes to anchoring filaments of basement membrane; 400–440 kD with fragments of 165, 155, 130 and 105 when disulphide bonds are reduced. Forms an asymmetric 170nm-long rod with two globules at one end, one at the other. May be the same as ***epiligrin***.

kaliotoxin (KTX) Toxin from the scorpion, *Androctonus mauretanicus m.* (peptide, 4 kD) that blocks some potassium channels. Closely related to ***charybdotoxin*** and ***agitoxins***.

kalirin (P-CIP10) Cytoplasmic protein with spectrin-like and guanine exchange factor (***GEF***) domains that interacts with peptidylglycine α-amidating monooxygenase (peptide processing enzyme). One of the Dbl family. May link secretory pathway to cytosolic regulatory pathways.

kallidin Decapeptide (lysyl-bradykinin; amino acid sequence KRPPGFSPFR) produced in kidney. Like ***bradykinin***, an inflammatory mediator (a ***kinin***); causes dilation of renal blood vessels and increased water excretion.

kallikrein Plasma serine proteases normally present as inactive prekallikreins which are activated by ***Hagemann factor***. Act on ***kininogens*** to produce ***kinins***. Contain an ***apple domain***.

Kanagawa haemolysin See Table E2.

kanamycin Aminoglycoside antibiotic. See Table A4.

Kaposi's sarcoma A sarcoma of spindle cells mixed with angiomatous tissue. Usually classed as an angioblastic tumour. A fairly frequent concomitant to ***HIV*** infection or long-term immunosuppression.

kappa chain (κ-light chains) See ***L chain***.

kappa particle ***Gram-negative*** bacterial endosymbiont of *Paramecium* spp. (*Caedobacter taeniospiralis*) that confers the 'killer' trait; infected *Paramecium* are resistant to the toxin liberated by infected forms. Killing activity is associated with the ***induction*** of defective phage in the endosymbiont, leading to the release of R bodies, coded for by the phage genome and apparently of misassembled phage-coat protein.

kappa toxin (κ-toxin) Exotoxin produced by ***Clostridium***; a collagenase that presumably aids tissue infiltration.

Kar3 Protein of the ***kinesin*** family that, like *NCD*, differs from kinesin in that it moves towards the minus end of the microtubule (like cytoplasmic dynein). Has been implicated in mitotic movement.

Kartagener's syndrome (situs inversus) Condition in which the normal left/right asymmetry of the viscera is reversed. Associated with a ***dynein*** defect (dynein is absent or dysfunctional in some cases) and with ***immotile cilia syndrome***.

karyokinesis (mitosis) Division of the nucleus, whereas cytokinesis is the division of the whole cell.

karyopherins Components of the nuclear pore responsible for regulating, in a ranGTP-dependent manner, transport of proteins with nuclear localization signals.

karyoplast A nucleus isolated from a eukaryotic cell surrounded by a very thin layer of cytoplasm and a plasma membrane. The remainder of the cell is a cytoplast.

karyorrhexis Degeneration of the nucleus of a cell. There is contraction of the chromatin into small pieces, with obliteration of the nuclear boundary.

karyotype The complete set of chromosomes of a cell or organism. Used especially for the display prepared from photographs of mitotic chromosomes arranged in homologous pairs.

karyotyping The production and analysis of a ***karyotype***, usually of a human foetus using a sample obtained from ***amniocentesis***, in antenatal genetic testing.

Kawasaki disease Acute inflammatory disease with systemic angiitis, most commonly occurring in infants and young children. Cause uncertain.

kazal proteins Family of serine ***protease inhibitors***. Includes seminal acrosin inhibitors, pancreatic secretory trypsin inhibitor (PSTI), Bdellin B-3 from leech.

KC See ***melanoma growth-stimulatory activity***.

K_{cat} (catalytic constant) Catalytic constant of an enzyme, also referred to as the turnover number. Represents the number of reactions catalysed per unit time by each active site.

K_d An equilibrium constant for dissociation. Thus, for the reaction: $A + B = C$, at equilibrium $K_d = [A][B]/[C]$. Dimension: moles per litre in this case. K_d is the reciprocal of K_a. In general the concept of K_d is more readily understood than that of K_a; for example, in considering the conversion of A to C by the binding of ligand B, the $K_d = [B]$ when $[A] = [C]$. Thus K_d is equal to the ligand concentration which produces half-maximal conversion (response).

KDEL Single letter code for the C-terminal amino acid consensus, in animals and many plants, for proteins targeted to the ***endoplasmic reticulum***. Other variants in some plants and other phyla include HDEL, DDEL, ADEL and SDEL.

KDR See ***VEGF***.

Kell Blood group system. The K antigen is relatively uncommon (9%) but after the ***rhesus*** antigens is the next most likely cause of haemolytic disease of the newborn.

keloid A bulging scar, the result of excess collagen production. Tendency to produce keloids seems to be heritable (particularly in Negroes) and is associated in some cases with low plasma fibronectin levels.

K_{eq} The equilibrium constant for a reversible reaction. $K_{eq} = [AB]/[A][B]$.

keratan sulphate See ***glycosaminoglycans***.

keratinizing epithelium An epithelium such as vertebrate epidermis in which a keratin-rich layer is formed from intracellular ***cytokeratins*** as the outermost cells die.

keratinocyte Skin cell, of the keratinized layer of epidermis: its characteristic ***intermediate filament*** protein is ***cytokeratin***.

keratinocyte growth factor (KGF) A growth factor structurally related to ***fibroblast growth factor***.

keratins Group of highly insoluble fibrous proteins (of high α-helical content) which are found as constituents of the outer layer of vertebrate skin and of skin-related structures such as hair, wool, hoof and horn, claws, beaks and feathers. Extracellular keratins are derived from cytokeratins, a large and diverse group of intermediate filament proteins.

keratitis Inflammation of the cornea, associated with herpes virus I infection and with congenital syphilis.

keratohyalin granules Granules found in living cells of keratinizing epithelia and which contribute to the keratin content of the dead cornified cells. Some, but not all, contain sulphur-rich keratin.

keratoses (actinic keratoses) Benign but precancerous lesions of skin associated with ultraviolet irradiation.

ketoacidosis Form of ***acidosis*** in which there is excess production of ketone bodies by the ketogenic pathway. Frequently a complication of diabetes, hence the sweet smell said to be characteristic of the breath of diabetics.

ketogenesis Production of ***ketone bodies***. Occurs in mitochondria, mostly in liver.

ketoglutarate (α-ketoglutarate) Intermediate of the ***tricarboxylic acid cycle***, also formed by deamination of ***glutamate***.

ketone body Acetoacetate, β-hydroxybutyrate or acetone. None of these are bodies as defined by morphologists! Ketone bodies accumulate in the body following starvation, in diabetes mellitus, and in some disorders of carbohydrate metabolism.

ketosis Metabolic production of abnormal amounts of ketones. A consequence of diabetes mellitus.

keyhole limpet haemocyanin (KLH) A ***haemocyanin*** from the keyhole limpet. Widely used as a carrier in the production of antibodies; it is chemically coupled to the immunogenic peptide or protein before injection.

KGF See ***keratinocyte growth factor***.

killer cells (K cells) (1) Mammalian cells which can lyse antibody-coated target cells. They have a receptor for the Fc portion of IgG, and are probably of the mononuclear phagocyte lineage, though some may be lymphocytes. Not to be confused with ***cytotoxic T-cells*** (CTL) which recognize targets by other means and are clearly a subset of T-lymphocytes: this confusion exists in the early literature. (2) (NK cells) Natural killer cells are CD3-negative large granular lymphocytes, mediating cytolytic reactions that do not require expression of Class I or II major ***histocompatibility antigens*** on the target cell. (3) (LAK cells) Lymphokine-activated killer cells are NK cells activated by ***interleukin-2***.

killer plasmid These plasmids are found in some strains of *Kluyveromyces marxianus* where the cells contain multiple cytoplasmic copies of dsDNA plasmids. Such cells secrete a glycoprotein toxin. The plasmids and the killer function can be transferred to yeast.

kilobase (kb) One thousand base pairs of DNA. Strictly should probably be kbp (kilobase pairs) but usually truncated.

kinase Widely used abbreviation for ***phosphokinase***, an enzyme catalysing transfer of phosphate from ATP to a second substrate usually specified in less abbreviated name, eg. creatine phosphokinase (***creatine kinase***), ***protein kinase***. Serine/threonine kinases phosphorylate on serine or threonine residues, tyrosine kinases on tyrosines.

kinectin Integral membrane protein (160 kD) of the endoplasmic reticulum and probably other membrane compartments; binds to ***kinesin*** and is the membrane anchor for kinesin-driven vesicle movement. Kinectin has extensive α-helical coiled-coil regions and, like the myosin tail with which it has sequence and structural similarities, may form a very long molecule, possibly 100nm in length when fully extended.

kinesin Cytoplasmic protein (110 kD) that is responsible for moving vesicles and particles towards the distal (plus) end of microtubules. Differs from cytoplasmic ***dynein*** (MAP1C) in the direction in which it moves and its relative insensitivity to vanadate. It has two heavy chains and two light chains. A large number of related gene products are believed to be motor proteins active in mitosis.

kinesis Alteration in the movement of a cell, without any directional bias. Thus speed may increase or decrease (***orthokinesis***) or there may be an alteration in turning behaviour (***klinokinesis***). See ***chemokinesis***.

kinetin (6-furfurylaminopurine) A ***cytokinin*** used as a component of plant tissue culture media. Obtained by heat treatment of DNA, and does not occur naturally in plants.

kinetochore Multilayered structure, a pair of which develop on the mitotic chromosome, adjacent to the ***centromere***, and to which spindle microtubules attach – but not at the end normally associated with a ***microtubule organizing centre***.

kinetodesma (plural, kinetodesmata) Longitudinally oriented cytoplasmic fibrils associated with, and always on the right of, the ***kinetosomes*** of ciliates.

kinetoplast Mass of mitochondrial DNA, usually adjacent to the flagellar ***basal body***, in flagellate protozoa.

kinetosome ***Basal body*** of cilium: used mostly of ciliates.

kinety A row of ***kinetosomes*** and associated ***kinetodesmata*** in a ciliate protozoan.

kininogen Inactive precursor in plasma from which ***kinin*** is produced by proteolytic cleavage.

kinins Inflammatory mediators that cause dilation of blood vessels and altered vascular permeability. Kinins are small peptides produced from ***kininogen*** by ***kallikrein***, and are broken down by kininases. Act on phospholipase and increase arachidonic acid release and thus prostaglandin (PGE2) production. See ***bradykinin, kallidin, C2-kinin***.

KIR (killer cell inhibitory receptor; killer cell immunoglobulin-like receptor) Set of receptors on killer cells. Killer cell immunoglobulin-like receptors are distinct from the CD94/NKG2 KIR (CD94 and NKG2 are C-type lectins) that is HLA-E specific. KIR3D binds to HLA-B, KIR2D to HLA-C. Together these receptors confer tolerance to self by an active signalling in the killer cell.

Kirby–Bauer test (disc diffusion test) Agar disc-diffusion test used to test antibiotics. Bacteria under test are distributed throughout the agar and discs of paper impregnated with different antibiotics are placed on the surface: the extent of the area of growth inhibition around each test disc is a measure of the susceptibility of the organism to the antibiotic.

Kirsten sarcoma virus A murine sarcoma-inducing retrovirus, generated by passaging a murine erythoblastosis virus in newborn rats. Source of the Ki-*ras* oncogene.

kistrin Naturally occurring inhibitor (68 residue peptide) of platelet aggregation found in the venom of Malayan pit viper *Agkistrodon rhodostoma*. Kistrin has an RGD site that competes for the platelet IIb/IIIa ***integrin*** and is therefore one of the ***disintegrins***.

kit (mast cell growth factor) An ***oncogene***, identified in feline ***sarcoma***, encoding a tyrosine ***protein kinase*** that acts on ***stem cell factor***. See Table O1.

Klebsiella Genus of ***Gram-negative bacteria***, non-motile and rod-like, associated with respiratory, intestinal and urinogenital tracts of mammals. *K. pneumoniae* is associated with pneumonia in humans.

Klein–Waardenburg syndrome See ***Waardenburg's syndrome***.

Klenow fragment Larger part of the bacterial DNA polymerase I (76 kD) that remains after treatment with ***subtilisin***; retains some but not all exonuclease and polymerase activity.

Klinefelter's syndrome Human genetic abnormality in which the individual, phenotypically apparently male, has three sex chromosomes (XXY).

klinokinesis ***Kinesis*** in which the frequency or magnitude of turning behaviour is altered. Bacterial chemotaxis can be considered as an adaptive klinokinesis; the probability of turning is a function of the change in concentration of the substance eliciting the response.

Km (Michaelis constant) A kinetic parameter used to characterize an enzyme; defined as the concentration of substrate that permits half-maximal rate of reaction. An analogous constant ***Ka*** is used to describe binding reactions, in which case it is the concentration at which half the receptors are occupied.

km-fibres Bundles of microtubules running longitudinally below and to one side of the bases of cilia in a ***kinety***.

knirps (kni) A *Drosophila* gap gene, asymmetric distribution of which is essential for normal expression of striped patterns of pair-rule genes and thus abdominal segmentation. Encodes a steroid/thyroid orphan receptor-type ***transcription factor***.

knockout (gene knockout) Informal term for the generation of a mutant organism in which the function of a particular gene has been completely eliminated (a 'null allele'). See also ***homologous recombination, transposon***.

knockout mouse Transgenic mouse in which a particular gene has been deleted. Often shows disappointingly little phenotypic change, usually because there are alternative mechanisms or because the right challenge is not being made (some genes are probably unnecessary for the survival of a well-fed laboratory mouse in very well-regulated surroundings).

Koch's postulates The criteria, first advanced by Robert Koch, by which the causative agent of a disease can be unambiguously identified. For an organism to be accepted as the causative agent it must be: (a) present in all cases; (b) isolatable in pure culture; (c) inoculation with the pure isolated organism should cause the disease and; (d) the organism should be observable in the experimentally infected host.

Kohler illumination The recommended type of optical microscope illumination in which the image of the lamp filament is focused in the lower focal plane of the substage condenser. As opposed to collimated illumination in which the light-emitting surface is imaged in the object. Collimated illumination requires even intensity across the light-emitting surface but is preferable for certain types of microscopy. Kohler illumination gives even illumination on the object even if there are irregularities in the brightness of the light-emitting surface.

koilocytes Large cells with cleared cytoplasm and pyknotic nuclei with inconspicuous nucleoli. Koilocytosis is induced by human papilloma virus infection of the superficial epithelial cells of the uterine cervix.

Kostmann's syndrome Autosomal recessive disease characterized by profound ***neutropenia***. It appears that bone marrow precursor cells fail to respond to the endogenous (normal) levels of functional G-CSF though they will respond to pharmacologic doses of G-CSF and the G-CSF receptors seem normal. Defect may be in intracellular signalling downstream of the receptor.

Kozak consensus Consenus for ***translational*** start site of an ***mRNA***. Although the trinucleotide ATG (coding for methionine) is generally considered as the start site, statistical analysis of a large number of mRNAs revealed several conserved residues around this sequence. In eukaryotes, RNNMTGG; in prokaryotes, MAYCATG.

KRAB (Krüppel associated box) Subset of ***zinc finger***-type ***transcription factors***.

Krebs cycle ***Tricarboxylic acid cycle*** or citric acid cycle.

kringle Triple-looped, disulphide-linked protein domains, found in some serine proteases and other plasma proteins, including plasminogen (fives copies), tissue plasminogen activator (two copies), thrombin (two copies), hepatocyte growth factor (four copies) and apolipoprotein A (38 copies).

Krüppel *(Kr)* ***Gap gene*** of *Drosophila*, encoding a ***zinc-finger transcription factor***.

KTX See ***kaliotoxin***.

Kuppfer cell Specialized macrophage of the liver sinusoids; responsible for the removal of particulate matter from the circulating blood (particularly old erythrocytes).

Kurloff cells Cells found in the blood and organs of guinea pigs that contain large secretory granules but are of unknown function.

kuru Degenerative disease of the central nervous system found in members of the Fore tribe of New Guinea: a ***spongiform encephalopathy***.

kwashiokor Form of severe malnutrition of children in the tropics. Generally considered to be due to protein deficiency though it could be due to deficiency in a single essential amino acid. Contrast with ***marasmus***.

Kyte-Doolitle An algorithm for calculating a ***hydropathy plot***.

L

L1 (NgCAM) Neural adhesion molecule with six immunoglobulin-type C2 domains and fibronectin type III repeats, making it another member of the ***immunoglobulin superfamily*** with binding domains similar to ***fibronectin***. The purified molecule, immobilizd on a culture dish, is a potent substrate for neurite outgrowth. See also ***neuroglian*** and ***NCAM***.

L See ***leucine***.

L cells Cell line established by Earle in 1940 from mouse connective tissue. L929 cells are a subclone of this original line.

L chain (light chain) Although ***light chains*** are found in many multimeric proteins, L chain usually refers to the light chains of immunoglobulins. These are of 22 kD and of one of two types: κ or λ. A single immunoglobulin has identical light chains (2 κ or 2 λ). Light chains have one variable and one constant region. There are ***isotype*** variants of both κ and λ.

L-forms Bacteria lacking cell walls, a phenomenon usually induced by inhibition of cell-wall synthesis, sometimes by mutation.

L-myc Relative of the ***myc protooncogene*** overexpressed in lung ***carcinoma***.

L-ring Outermost ring of the basal part of the bacterial flagellum in ***Gram-negative bacteria***. It may serve as a bush to anchor the flagellum relative to the lipopolysaccharide layer.

L-type channels A class of ***voltage-sensitive calcium channels***. L-type channels are found in neurons, neuroendocrine cells, smooth, cardiac and striated muscle, are involved in neurotransmitter release at some synapses and inactivate relatively slowly. They are activated at membrane potentials more positive than –30mV. The long-lasting properties and possible role in ***long-term potentiation*** were the reason for them being designated L-type. They are insensitive to ω-***conotoxin*** but inhibited by dihydropyridines, benzodiazepines and phenylalkylamines.

La protein Protein (45 kD) transiently bound to unprocessed cellular precursor RNAs that have been produced by polymerase III. Mainly located in the nucleus.

LA-PF4 See ***connective tissue-activating peptide III***.

lac operon See ***lactose operon***.

lactacystin Specific inhibitor of ***proteasome***.

lactadherin Mucin-associated glycoprotein (46kD) found in human milk. Binds to rotavirus and inhibits replication.

lactalbumin Milk protein fraction containing β-lactoglobulin and α-lactalbumin. α-lactalbumin is the regulatory subunit of lactose synthetase, thought to be related to lysozyme C.

lactate (2-hydroxypropionic acid) Important as the terminal product of anaerobic glycolysis. Accumulation of lactate in tissues is responsible for the so-called oxygen debt.

lactate dehydrogenase (LDH) The enzyme that catalyses the formation and removal of lactate according to the equation: pyruvate + NADH = lactate + NAD. The appearance of LDH in the medium is often used as an indication of cell death and the release of cytoplasmic constituents.

Lactobacillus Genus of ***Gram-positive*** anaerobic or facultatively aerobic bacilli, product of glucose fermentation is lactate. Important in production of cheese, yoghurt, sauerkraut and silage.

lactoferrin Iron-binding protein of very high affinity (Kd 10^{-19} at pH 6.4, 26-fold greater than that of ***transferrin***) found in milk and in the specific granules of neutrophil leucocytes.

lactoperoxidase Peroxidase enzyme from milk that finds an important use in generating active iodine as a non-permeant radiolabel for membrane proteins.

lactose (4-O-β-D galactopyranosyl-β-D glucose) The major sugar in human and bovine milk. Conversion of lactose to lactic acid by *Lactobacilli*, etc. is important in the production of yoghurt and cheese.

lactose carrier protein The best-known example is the product of the *lacY* gene, coded for in the ***lactose operon*** and responsible for the uptake of lactose by ***E. coli***.

lactose operon Group of adjacent and coordinately controlled genes concerned with the metabolism of lactose in *E. coli*. The *lac* operon was the first example of a group of genes under the control of an ***operator*** region to which a ***lactose repressor*** binds. When the bacteria are transferred to lactose-containing medium, allolactose (which forms by transglycosylation when lactose is present in the cell) binds to the repressor, inhibits the binding of the repressor to the operator, and allows transcription of mRNA for enzymes involved in galactose metabolism and transport across the membrane (β-galactosidase, galactoside permease, and thiogalactoside transacetylase). ***LacZ*** codes for β-galactosidase, *lacY* for the permease, *lacA* for the transacetylase.

lactose repressor Protein (tetramer of 37 kD subunits) that normally binds with very high affinity to the ***operator*** region of the ***lactose operon*** and inhibits transcription of the downstream genes by blocking access of the polymerase to the promoter region. When the lactose repressor binds allolactose, its binding to the operator is reduced and the gene set is derepressed.

lacuna Small cavity or depression, eg. the space in bone where an ***osteoblast*** is found.

LacZ E. coli gene encoding ***beta-galactosidase***. Part of the ***lac operon***.

laddering Apoptotic cells show a regular pattern of oligonucleotide sizes on electrophoretic gels; the ladder-like arrangement is a consequence of the cleavage of the DNA strand between ***nucleosome*** beads by ***endonucleases*** as part of the process by which cell death occurs. See ***apoptosis***.

LAL test See ***Limulus polyphemus***.

lambda bacteriophage (λ phage) Bacterial DNA ***virus***, first isolated from *E. coli*. Its structure is similar to that of the ***T even phages***. It shows a ***lytic*** cycle and a ***lysogenic*** cycle, and studies on the control of these alternative cycles have been very important for our understanding of the regulation of gene ***transcription***. It is used as a cloning vector, accommodating fragments of DNA up to 15 kilobase pairs long. For larger pieces, the ***cosmid*** vector was constructed from its ends.

lambda chain See ***L chain***.

lamellar phase See ***phospholipid bilayer***.

lamellipodium Flattened projection from the surface of a cell, often associated with locomotion of fibroblasts.

lamina Flat sheet; as in ***basal lamina***.

lamina propria Fibrous layer of connective tissue underlying the basal lamina of an epithelium. May contain smooth muscle cells and lymphoid tissue in addition to fibroblasts and extracellular matrix.

laminarin Storage polysaccharide of *Laminaria* and other brown algae; made up of β (1-3)-glucan with some β (1-6) linkages.

laminin Link proteins of basal lamina; consist of an A chain (400 kD) and two B chains (200 kD). Each subunit contains at least 12 repeats of the ***EGF-like domain***. The first laminin studied was from mouse ***EHS cells***, but it is now becoming clear that different forms of laminin occur. In laminin from placenta the A chain is replaced with merosin, in laminin found near the neuromuscular junction the B1 chain is replaced by s-laminin (synapse laminin). Laminin induces adhesion and spreading of many cell types and promotes the outgrowth of neurites in culture.

lamins Proteins that form the nuclear lamina, a polymeric structure intercalated between chromatin and the inner nuclear envelope. Lamins A and C (70 and 60 kD respectively) have C-terminal sequences homologous to the head and tail domains of ***keratins***; their peptide maps are similar, and significantly different from that of lamin B (67 kD), although there are some common epitopes.

LAMP-1 (lysosomal-associated membrane protein 1) Heavily glycosylated protein of lysosomal and plasma membrane. Depending on extent of glycosylation may have molecular weight between 90 and 140 kD; function may be to protect membrane from attack by lysosomal enzymes. Also known as LEP100 (lysosomes, endosomes and plasma membrane, 100 kD) and LGP120.

lampbrush chromosomes Large chromosomes (as long as 1mm), actually meiotic ***bivalents***, seen during prophase of the extended meiosis in the oocytes of some amphibia. Segments of DNA form loops in pairs along the sides of the sister chromosomes, giving them a brush-like appearance. These loops are not permanent structures but are formed by the unwinding of ***chromomeres*** and represent sites of very active RNA synthesis.

lamprey See ***Petromyzon***.

Landry–G–B syndrome See ***Guillain–Barré syndrome***.

Landry–Guillain–Barré syndrome See ***Guillain–Barré syndrome***.

Langendorff perfused heart Classic pharmacological organ preparation in which a rodent heart is maintained ***in vitro*** by perfusion of the aorta with oxygenated fluid so that the fluid passes into the coronary arteries. Can be used to study metabolism of cardiac muscle.

Langerhans See ***islets of Langerhans*** and ***Langerhans cells***.

Langerhans cells Cells of dendritic appearance, strongly MHC Class II positive and weakly phagocytic, found in the basal layers of the epidermis where they serve as ***accessory cells***, responsible for ***antigen processing*** and ***antigen presentation***. Having been exposed to antigen they migrate to the lymph nodes. Are derived from bone marrow and are immature ***dendritic cells***. Part of the immune surveillance system, their location means that they are readily exposed to antigens that penetrate the dermal barrier.

Langhans' multinucleate giant cells Multinucleate cells formed by fusion of epithelioid macrophages and associated with the central part of early tubercular lesions. Similar to ***foreign body giant cells***, but with the nuclei peripherally located.

Langmuir trough A device for studying the properties of lipid monolayers at an air–water interface. A moveable barrier connected to a balance allows measurement of surface pressure.

Langmuir–Blodgett film In biophysics, an ordered monolayer of molecules produced on the surface of water. An ***amphipathic*** molecule is floated at low concentration on the surface of the water and steadily compressed into an ordered surface by moving a barrier across the surface.

lanthanum Lanthanum salts are used as a negative stain in electron microscopy, and as calcium-channel blockers.

LAR (leucocyte antigen-related protein) LAR is the prototype for a family of transmembrane protein tyrosine phosphatases with extracellular domains composed of Ig and fibronectin type III (FnIII) domains and two cytoplasmic catalytic domains, one active, one inactive. LAR-family phosphatases (LAR, PTPδ, PTPσ) play a role in axon guidance, mammary gland development, regulation of insulin action and glucose homeostasis. See ***liprins***.

large extracellular transformation/trypsin sensitive protein See ***LETS***.

large T-antigen See ***T-antigen***.

large-cell lymphoma Highly malignant group of tumours arising from transformed ***lymphocytes*** or ***myeloid*** precursors. Cell of origin often obscure.

Laron dwarfism Human growth defect in which cells do not respond to growth hormone: the defect is due to mutation in the growth hormone receptor gene.

Lassa fever See ***Lassa virus***.

Lassa virus Virulent and highly transmissible member of the ***Arenaviridae*** whose normal host is a rodent (*Mastomys* natalensis); first recorded from Nigeria.

late gene Gene expressed relatively late after infection of a host cell by a virus, usually structural proteins for the viral coat.

latency (1) In electrophysiology, the time between onset of a stimulus and peak of the ensuing ***action potential***. (2) Of an infection, a period in which the infection is present in the host without producing overt symptoms.

latent virus Virus integrated within host genome but inactive: may be reactivated by stress such as ultraviolet irradiation.

lateral diffusion Diffusion in two dimensions, usually referring to movement in the plane of the membrane, such as the motion of fluorescently labelled lipids or proteins measured by the technique of ***fluorescence recovery after photobleaching*** (FRAP).

lateral inhibition A simple form of information processing. The classic example is found in the eye, whereby ganglion cells are stimulated if photoreceptors in a well-defined field are illuminated, but their response is inhibited if neighbouring photoreceptors are excited (an 'on field/off surround' cell) or *vice versa* (an 'off field/on surround' cell). The effect of lateral inhibition is to produce edge- or boundary-sensitive cells, and to reduce the amount of information that is sent to higher centres; a form of peripheral processing.

lathyrism Disorder of collagen crosslinking as a result of copper sequestration by nitriles. (Lysyl oxidase is a copper-containing metalloenzyme). In animals, caused by eating toxic plants of genus *Lathyris*.

Laticauda semifasciata Sea snake. See ***erabutotoxins***.

Latrodectus Genus of spiders, black widows. See ***α-latrotoxin***.

latroinsectotoxin (α-latroinsectotoxin) See ***latrotoxin***.

latrotoxin (α-latrotoxin; α-LTx) Major toxin from *Latrodectus* spp. (1401 residues). Causes release of neurotransmitters from all synapses. An insect-specific toxin, α-latroinsectotoxin, is also present in the venom and has substantial homology and similar mode of action to α-latrotoxin.

lavage Washing out of a cavity (eg. peritoneal cavity) in order to remove loosely adherent cells.

lazy leucocyte syndrome A rare human complaint in which neutrophils display poor locomotion towards sites of infection.

Lbc Oncoprotein of Dbl-like family. See ***Lfc***.

LD50 That dose of a compound which causes death in 50% of the organisms to which it has been administered. Routine use of LD50 tests is now being replaced by more sensitive (and less wasteful) methods.

LDH See ***lactate dehydrogenase***.

LDL See ***low density lipoprotein***.

LE body A globular mass of nuclear material that stains with haematoxylin; associated with lesions of ***systemic lupus erythematosus***.

LE cell Phagocyte that has ingested nuclear material of another cell: characteristic of ***systemic lupus erythematosus***.

leader peptide See ***leader sequence***.

leader sequence In the regulation of gene expression for enzymes concerned with amino acid synthesis in prokaryotes, the leader sequence codes for the leader peptide that contains several residues of the amino acid being regulated. Transcription is closely linked to translation, and if translation is retarded by limited supply of aminoacyl tRNA for the specific amino acid, the mode of transcription of the leader sequence permits full transcription of the operon genes; otherwise complete transcription of the leader sequence prematurely terminates transcription of the ***operon***.

leading lamella Anterior region of a crawling cell, such as a fibroblast, from which most cytoplasmic granules are excluded.

leaky mutation Mutation in which subnormal function exists, eg. if a mutation leads to instability in a protein rather than its complete absence or there is reduced expression of a gene.

LECAM (CD62L; LAM-1; L-selectin; MEL-14 antigen; leu-8) Leucocyte-endothelial cell adhesion molecule (37 kD polypeptide), a ***selectin***, expressed on most haematopoietic cells and on mature monocytes, eosinophils and neutrophils. Important for initial adhesion step (margination) and is then rapidly lost from cell surface through activity of a membrane associated sheddase, a metalloproteinase. Important function is in lymphocyte homing to ***high endothelial venule*** in peripheral lymph nodes, Peyer's patches and areas of inflammation. Has N-terminal ***C-type lectin*** domain and binds particularly to carbohydrates on CD34, CD162, GlyCam and MAdCAM.

lecithin Phospholipids of egg yolk (usually hen's eggs). A mixture of phosphatidyl choline and phosphatidyl ethanolamine, but usually refers to phosphatidyl choline.

lecithinase See ***phospholipases***.

lectin Proteins obtained particularly from the seeds of leguminous plants, but also from many other plant and animal sources, that have binding sites for specific mono- or oligosaccharides. Named originally for the ability of some to selectively agglutinate human red blood cells of particular blood groups. Lectins such as ***concanavalin A*** and ***wheatgerm agglutinin*** are widely used as analytical and preparative agents in the study of glycoproteins. See Table L1.

leghaemoglobin Form of haemoglobin found in the nitrogen-fixing root nodules of legumes. Binds oxygen, and thus protects the nitrogen-fixing enzyme, nitrogenase, which is oxygen sensitive.

Legionella Genus of ***Gram-negative*** asporogenous bacteria. Most species are pathogenic in humans, causing pneumonia-like disease, eg. Legionnaire's disease, named after an outbreak in Philadelphia amongst members of an American Legion reunion.

Table L1. Lectins

Source	Abbreviation	Sugar specificity
Bandieraea simplicifolia	BSL1	α–D-gal > α–D-GalNAc
Concanavalla ensiformis (Jack bean)	ConA	α–D-Man > α–D-Glc > α–D-GlcNAc
Dolichos biflorus	DBA	α–D-GalNAc
Lens culinaris (lentil)	LCA	α–D-Man > α–D-Glc > α–D-GlcNAc
Phaseolus vulgaris (red kidney bean)	PHA	β–D-Gal (1-4)-D-GlcNAc
Arachis hypogaea (peanut)	PNA	β–D-Gal (1-3)-D-GalNAc
Pisum sativum (garden pea)	PSA	α–D-Man > α–D-Glc
Ricinus communis (castor bean)	RCA1	β–D-Gal > α–D-Gal
Sophora japonica	SJA	β–D-GalNAc > β-D-Gal > α–D-Gal
Glycine max (soybean)	SBA	α–D-GalNAc > β–D-GalNAc
Ulex europaeus (common gorse)	UEA1	α–L-fucosyl
Triticum vulgaris (wheatgerm)	WGA	β–D-GlcNAc(1-4) GlcNAc > β–D-GlcNAc(1-4)-β–D-GlcNAc

Gal = galactose; GalNAc = galactosamine; Glc = glucose; GlcNAc = glucosamine; Man = mannose.

Legionnaire's disease See ***Legionella***.

legumin Major storage protein of the seeds of peas and other legumes.

Leidig cells See ***Leydig cells***.

leiomyoma Benign tumour of smooth muscle in which parallel arrays of smooth muscle cells form bundles which are arranged in a whorled pattern. The amount of fibrous connective tissue is very variable. Leiomyoma of the uterus (fibroid) is the commonest form.

leiotonin Smooth-muscle analogue (homologue?) of ***troponin***. Two subunits, leiotonins A and C, the latter similar in size, and homologous, to ***calmodulin*** and ***troponin*** C.

Leishman stain Romanovsky-type stain; a mixture of basic and acid dyes used to stain blood smears and that differentially stains various classes of leucocytes.

leishmaniasis Disease caused by protozoan parasites of the genus *Leishmania*. The parasite lives intracellularly in macrophages. Various forms of the disease are known, depending upon the species of parasite: in particular visceral leishmaniasis (kala-azar), and mucocutaneous leishmaniasis.

Leiurus quinquestriatus hebraeus Scorpion. See ***scyllatoxin*** and ***charybdotoxin***.

lentigo Relatively common pigmented lesion of the skin in which melanocytes replace the basal layer of the epidermis.

Lentivirinae Subfamily of non-oncogenic retroviruses that cause 'slow diseases' that are characterized by horizontal transmission, long incubation periods and chronic progressive phases. Visna virus is in this group, and there are similarities between visna, equine infectious anaemia virus and ***HIV***.

lentoid Spherical cluster of retinal cells, formed by aggregation ***in vitro***, that has a core of lens-like cells inside which accumulate proteins characteristic of normal lens. The cells concerned derive from retinal glial cells.

Lepidoptera Order of insects that comprises butterflies and moths

Lepore haemoglobin Variant haemoglobin in a rare form of ***thalassaemia***: there is a composite δβ chain as a result of an unequal ***crossing-over*** event. The composite chain is functional but synthesized at reduced rate.

leprosy Disease caused by *Mycobacterium leprae*, an obligate intracellular parasite that survives lysosomal enzyme attack by possessing a waxy coat. Leprosy is a chronic disease associated with depressed cellular (but not humoral) immunity; the bacterium requires a lower temperature than 37°C, and thrives particularly in peripheral Schwann cells and macrophages. Only humans and the nine-banded armadillo are susceptible.

leptin Product (16 kD) of the *ob* (obesity) locus. Found in plasma of mouse and man: reduces food uptake and increases energy expenditure.

leptin receptor Receptor for ***leptin***. G-protein coupled, highly expressed in the hypothalamus, the site of appetite regulation in the brain; downstream regulates a K^{+}_{ATP} channel similar to that involved in regulation of insulin release (see ***SUR***).

leptinotarsins (β-leptinotarsins; leptinotoxin) Toxic proteins (45–47 kD) present in the haemolymph of potato beetles (*Leptinotarsa* spp.) Causes release of neurotransmitters from synapses of insect and vertebrates by inducing calcium entry, possibly by acting as a calcium-channel agonist.

leptinotoxin See ***beta-leptinotarsin***.

leptonema See ***leptotene***.

Leptospira Genus of ***spirochaete*** bacteria that cause a mild chronic infection in rats and many domestic animals. The bacteria are excreted continuously in the urine and contact with infected urine or water can result in infection of humans via cuts or breaks in the skin. Infection causes leptospirosis or Weil's disease, a type of jaundice, that is an occupational hazard for sewerage and farm workers.

leptospirosis Weil's disease, caused by infection with ***Leptospira***.

leptotene Classical term for the first stage of ***prophase*** I of ***meiosis***, during which the chromosomes condense and become visible.

Lepus The common European hare.

Lesch–Nyhan syndrome A sex-linked recessive inherited disease in humans that results from mutation in the gene for the purine salvage enzyme HGPRT, located on the X chromosome. Results in severe mental retardation and distressing behavioural abnormalities, such as compulsive self-mutilation.

lethal mutation ***Mutation*** that eventually results in the death of an organism carrying the mutation.

LETS (large extracellular transformation/trypsin-sensitive protein) Originally described as a cell-surface protein that was altered on transformation ***in vitro***: now known to be ***fibronectin***.

leu See ***leucine***.

Leu enkephalin A natural peptide neurotransmitter; see ***enkephalins***.

leucine (leu; L; 2-amino-4-methylpentanoic acid) The most abundant amino acid (131 D) found in proteins. Confers hydrophobicity and has a structural rather than a chemical role. See Table A2.

leucine aminopeptidase An ***exopeptidase*** that removes neutral amino acid residues from the N-terminus of proteins.

leucine zipper Motif found in certain ***DNA-binding proteins***. In a region of around 35 amino acids, every seventh is a leucine. This facilitates dimerization of two such proteins to form a functional ***transcription factor***. Examples of proteins containing leucine zippers are products of the ***protooncogenes myc***, ***fos*** and ***jun***. See also ***AP-1***.

leucine-rich repeat (LRR) Short motif (around 24 residues) with 5–7 leucines generally at positions 2, 5, 7, 12, 21, 24. Forms an amphipathic region and is probably involved in protein–protein interactions.

leucinopine (dicarboxypropyl leucine) An analogue of ***nopaline*** found in crown gall tumours (induced by ***Agrobacterium tumefaciens***) that do not synthesize octopine or nopaline.

leuco- (USA, leuko) Prefix, denoting white or colourless.

leucocidin Exotoxins from staphylococcal and streptococcal species of bacteria that cause leucocyte killing or lysis. There are two subunits, S and F, each inactive alone but synergistically form a pore. Myeloid but not lymphoid cells are affected.

leucocyte (USA, leukocyte) Generic term for a white blood cell. See ***basophil***, ***eosinophil***, ***lymphocyte***, ***monocyte***, ***neutrophil***.

leucocytosis An excess of ***leucocytes*** in the circulation.

leucopenia An abnormally low count of circulating ***leucocytes***.

leucoplast Colourless ***plastid*** that may be an ***etioplast*** or a storage plastid (***amyloplast***, ***elaioplast*** or ***proteinoplast***).

leukaemia (USA, leukemia) Malignant neoplasia of ***leucocytes***. Several different types are recognized according to the stem cell that has been affected, and several virus-induced leukaemias are known (eg. that caused by feline leukaemia virus). Both acute and chronic forms occur: (1) acute lymphoblastic leukaemia (ALL) involves neoplastic proliferation of white cell precursors in which the blood has large numbers of primitive lymphocytes (high nuclear:cytoplasmic ratio characteristic of dividing cells and few specific surface antigens expressed), and tends to be common in the young; (2) acute myeloblastic leukaemia (AML) is more common in adults; the proliferating cells are of the ***myeloid*** haematopoietic series and the cells appearing in the blood are primitive ***granulocytes*** or ***monocytes***; (3) chronic lymphocytic leukaemia (CLL) is a neoplastic disease of middle or old age, characterized by excessive numbers of circulating lymphocytes of normal, mature appearance, usually B-lymphocytes; presumably a neoplastic transformation of lymphoid stem cells; (4) chronic myelogenous leukaemia (CML) involves neoplasia of myeloid stem cells, commonest in middle-aged or elderly people, characterized by excessive numbers of circulating leucocytes, most commonly neutrophils (or precursors), but occasionally eosinophils or basophils.

leukaemia inhibitory factor (LIF) Polypeptide ***growth factor*** or ***cytokine*** with wide range of activities. Regulates growth and differentiation of primordial germ cells and embryonic stem cells but has effects on peripheral neurons, osteoblasts, adipocytes and various cells of the myeloid lineage. Given to adult animals induces weight loss, behavioural disorders and bone abnormalities. Many of the effects of LIF ***in vitro*** can be mimicked by ***interleukin-6, oncostatin M*** and ***ciliary neurotrophic factor***, all of which interact indirectly with gp130, a shared tranducer subunit.

leukosialin (CD43; sialophorin) Widely distributed membrane-associated mucin, the major sialoglycoprotein of thymocytes and mature T-cells. Transmembrane protein with extensive O-linked glycosylation (75–85 oligosaccharides on the 239 residue extracellular domain). Extends at least 45nm beyond plasma membrane. Similar but not homologous to ***episialin***.

leukosis The correct term for an excess of leucocytes in the circulation and other parts of the body, preferable in place of the term leucocytosis.

leukosulfakinin (LSK) Cockroach peptide ***hormones*** that affect gut motility. Related to ***gastrin***.

leukotrienes (LTA_4, LTB_4, LTC_4, LTD_4, LTE_4) A family of hydroxyeicosatetraenoic (HETE) acid derivatives. LTA_4 and LTB_4 are modified lipids; leukotrienes C, D and E have the lipid conjugated to glutathione (LTC_4) or cysteine (LTD_4, LTE_4) to form the peptidyl leukotrienes. A mixture of the latter (LTC_4, LTD_4, LTE_4) constitute SRS-A, the slow reacting substance of anaphylaxis, that has potent bronchoconstrictive effects. LTB_4 is a potent neutrophil chemotactic factor.

leupeptin Family of modified-tripeptide protease inhibitors. Commonest is N-acetyl-Leu-Leu-argininal.

Lewis blood group A pair of blood group activities associated with the A, B, H substances. Lewis Le^a is a separate gene, whereas Le^b arises from the combined activity of the enzymes specified by *Le^a* and *H* genes.

Lewy body Hyaline eosinophilic concentrically laminated inclusions found in the ***substantia nigra*** and locus ceruleus of patients with ***Parkinsonism*** and Lewy body dementia.

LexA *E. coli* ***repressor*** of the ***SOS*** system for response to DNA damage.

Leydig cell Interstitial cells of the mammalian testis, involved in synthesis of testosterone.

LFA-1 (CD11a/CD18; lymphocyte function-related antigen-1) Heterodimeric lymphocyte plasma-membrane protein (α_L 180 kD, β 95 kD) that binds ICAM-1, particularly involved in cytotoxic T-cell killing. One of the ***integrin*** superfamily of adhesion molecules. Deficiency of LFA-1 in leucocyte adhesion deficiency (LAD) syndrome leads to severe impairment of normal defences and poor survival prospects. The related surface adhesion molecules (sometimes referred to as the LFA-1 class of adhesion molecules) are Mac-1 (α_M 170 kD, β 95 kD; CD11b/CD18) and p150,95 (α_X 150 kD, β 95 kD; CD11c/CD18); they are also defective in severe forms of LAD because the β subunit, which is apparently common to all three, is missing. Mac-1 (also known as Mo-1 in earlier literature) is the complement C3bi receptor (CR3) and is present on mononuclear phagocytes and on neutrophils; p150,95 is less well characterized, but is particularly abundant on macrophages.

LFA-3 (lymphocyte function-related antigen-3) Ligand for the CD2 adhesion receptor that is expressed on cytolytic T-cells. LFA-3 is expressed on endothelial cells at low levels. The CD2–LFA-3 complex is an adhesion mechanism distinct from the LFA-1/ICAM-1 system, and binding of erythrocyte LFA-3 to T-lymphocyte CD2 is the basis of formation of ***E-rosettes***.

Lfc Oncoprotein of the Dbl-related family. Contains a Dbl-homology domain in tandem with a ***PH domain*** and is similar to Lsc, Lbc, Tiam-1 and Dbl. Has rho-***GEF*** activity.

LH See ***luteinizing hormone***.

LHRF See ***luteinizing hormone-releasing factor***.

library See ***genomic library***.

lichen A large group of symbiotic associations between fungi and green and occasionally blue-green algae. Several genera of algae and of fungi are involved and the associations are so stable and of such varied but distinct types that the lichens have been classified into genera and species. A variety of incompatibility phenomena are often manifest between individual lichens. Confined to terrestrial habitats and often used as indicators of pollution status of the environment.

lichen planus Rare skin disorder in which there is marked hyperkeratosis and extensive infiltration of lymphocytes into the lower epidermis.

Liddle's disease Hereditable (autosomal dominant) form of salt-sensitive human hypertension caused by mutation in the β or γ subunit of the multisubunit epithelial ***sodium channel*** (ENaC).

lidocaine (lignocaine) Commonly used local anaesthetic.

LIF See ***leukaemia inhibitory factor***.

ligand Any molecule that binds to another; in normal usage a soluble molecule, such as a hormone or neurotransmitter, that binds to a receptor. The decision as to which is the ligand and which the receptor is often a little arbitrary when the broader sense of receptor is used (where there is no implication of transduction of signal). In these cases it is probably a good rule to consider the ligand to be the smaller of the two: thus in a lectin–sugar interaction, the sugar would be the ligand (even though it is attached to a much larger molecule, recognition is of the saccharide).

ligand-gated ion channel A transmembrane ***ion channel*** whose permeability is increased by the binding of a specific ***ligand***, typically a neurotransmitter at a ***chemical synapse***. The permeability change is often drastic; such channels let through effectively no ions when shut, but allow passage at up to 10^7ions/s^{-1} when a ligand is bound. Recently, the receptors for both ***acetylcholine*** and ***GABA*** have been found to share considerable sequence homology, implying that there may be a family of structurally related ligand-gated ion channels.

ligand-induced endocytosis The formation of coated pits and then ***coated vesicles*** as a consequence of the interaction of ligand with receptors, which then interact with ***clathrin*** and associated proteins (coatomers) on the cytoplasmic face of the plasma membrane and come together to form a pit. Not all coated vesicle uptake of receptors requires receptor occupancy.

ligase amplification reaction (LAR) Method for detecting small quantities of a target DNA, with utility similar to ***PCR***. It relies on DNA ligase to join adjacent synthetic oligonucleotides after they have bound the target DNA. Their small size means that they are destabilised by single base mismatches, and so form a sensitive test for the presence of mutations in the target sequence.

ligases (synthetases) Major class of ***enzymes*** that catalyse the linking together of two molecules (category 6 in the ***E classification***), eg. DNA ligases that link two fragments of DNA by forming a ***phosphodiester*** bond.

ligatin Polypeptide (10 kD monomer) that forms 3–4.5nm polymeric fibrils on the outside of chick neural retina cells.

light chain See ***L-chain***.

light-dependent reaction The reaction taking place in the chloroplast in which the absorption of a photon leads to the formation of ATP and NADPH.

Light Green A stain often used for counter-staining cytoplasm following iron haematoxylin; a component of Masson's trichrome.

light-harvesting system Set of photosynthetic pigment molecules that absorb light and channel the energy to the photosynthetic ***reaction centre***, where the light reactions of ***photosynthesis*** occur. In higher plants, contains ***chlorophyll*** and ***carotenoids***, and is present in two slightly different forms in ***photosystem I*** and ***photosystem II***.

light microscopy In contrast to electron microscopy. See ***bright field, phase contrast, interference, interference contrast, interference reflection, dark field, confocal*** and ***fluorescence microscopy***. See also Table L2.

light scattering Particles suspended in a solution will cause scattering of light, and the extent of the scattering is related to the size and shape of the particles (in a somewhat complex relationship).

lignin Complex polymer of phenylpropanoid subunits, laid down in the walls of plant cells such as ***xylem*** vessels and ***sclerenchyma***. Imparts considerable strength to the wall, and also protects it against degradation by microorganisms. It is also laid down as a defence reaction against pathogenic attack, as part of the ***hypersensitive response*** of plants.

LIM domain Domain found in proteins required for developmental decisions. Contain 60-residue conserved, cysteine-rich, repeats. Named after first three genes in group: Lin-11 (*Caenorhabditis elegans*, required for asymmetric division of blast cells), IsI-1 (mammalian insulin gene-binding enhancer protein), mec-3 (*C. elegans*-required for differentiation of a set of sensory neurons).

limb bud The limbs of vertebrates start as outpushings of mesenchyme surrounded by a simple epithelium. The distal region is referred to as the progress zone. There has been extensive study of positional information within the limb-bud that determines, eg. the proximal–distal pattern of bone development and the anterior–posterior specification of digits.

limbic system Those regions of the central nervous system responsible for autonomic functions and emotions. Includes hippocampus, amygdaloid nucleus and portions of the midbrain.

limit of resolution See ***resolving power***.

Limulus polyphemus Now renamed *Xiphosura*, though *Limulus* is still in common usage as a name. The king crab or horseshoe crab, found on the Atlantic coast of North America. It is more closely related to the arachnids than the crustacea, and horseshoe crabs are the only surviving representatives of the subclass Xiphosura. Its compound eyes have been widely used in studies on visual systems, but it is probably better known from the '*Limulus*-amoebocyte lysate' (LAL) test; LAL is very sensitive to small amounts of ***endotoxin***, clotting rapidly to form a gel, and the test is used clinically to test for septicaemia.

lincomycin Antibiotic active against ***Gram-positive*** bacteria. Acts by blocking protein synthesis by binding to the 50S subunit of the ribosome and blocking peptidyl transferase reaction. Clindamycin, a derivative of lincomycin, is used as an antimalarial drug.

linear dichroism See ***circular dichroism***.

Lineweaver–Burke plot A plot of $1/v$ against $1/S$ for an enzyme-catalysed reaction, where v is the initial rate and S the substrate concentration. From the equation: $1/v = 1/V_{max}(1 + K_m/S)$ the parameters V_{max} and K_m can be determined. The equation overweights the contribution of the least accurate points and other methods of analysis are preferred; see ***Eadie–Hofstee plot***.

Table L2. Types of light microscopy

Method		Physical parameter detected
With axial illumination: *without spatial filtration*[a]		
I.	Bright field	Absorption by specimen (May be operated in Visible, UV or IR. regions of the spectrum and in quantitative microspectrophotometric modes)
II.	Interference:	
	Transmitted	Path difference arising in specimen, qualitative or quantitative
	Reflected (interference reflection = IRM)	Path difference in films 1–10 wavelengths thick next to substrate. For cell contacts
III.	Fluorescence	Natural fluorescence or that of probes applied to system
IV.	Dark-field	Refractive index discontinuities revealed by scattered light
V.	Polarisation	Birefringent and/or dichroic properties
With axial illumination: *with spatial filtration*		
VI.	Confocal scanning microscopy	Contrast and resolution enhanced by selection of light paths modified by the object at the back focal plane of the objective. Bright-field or fluorescence modes. Usually combined with video processing of the image
VII.	Phase contrast	Path differences revealed as contrast differences non-quantitatively and non-regularly, using phase plate at back focal plane
VIII.	Differential interference contrast (DIC) = Nomarski	Path difference gradients revealed as contrast or colour differences
IX.	Out-of-focus phase contrast	Path differences revealed as diffraction patterns
B. With anaxial illumination: *with spatial filtration*		
X.	Hoffman modulation contrast	Path differences
XI	Single side band edge enhancement (SEE microscopy)	Path differences from first order diffractions

Nearly all systems can be run in the epi (incident) illumination mode. Video (image) processing can enhance contrast and resolution in images by the application of simple algorithms to expand the grey scale, reduce noise and subtract background. More complex processing is possible, including the extraction of further information by Fourier transforms.

[a]*Spatial filtration.* This is the application of methods to remove those ray paths which have not interacted with the object. This is done at the back focal plane of the objective. It can also be applied to select or remove ray paths that have interacted in some specified way with the object.

lining epithelium An epithelium lining a duct, cavity or vessel, that is not particularly specialized for secretion or as a mechanical barrier. Not a precise classification.

linkage Tendency for certain genes tend to be inherited together, because they are on the same chromosome. Thus parental combinations of characters are found more frequently in offspring than non-parental. Linkage is measured by the percentage recombination between loci.

linkage disequilibrium The occurrence of some genes together, more often than would be expected. Thus, in the HLA system of ***histocompatibility antigens***, HLA-A1 is commonly associated with B8 and DR3, and A2 with B7 and DR2, presumably because the combination confers some selective advantage.

linkage equilibrium Situation that should exist in a population undisturbed by selection, migration, etc., in which all possible combinations of linked genes should be present at equal frequency. The situation is no more common than are such undisturbed populations.

linoleic acid An essential ***fatty acid*** (9, 12, octadecadienoic acid); occurs as a glyceride component in many fats and oils.

linolenic acid An 18-carbon fatty acid with three double bonds (9, 12, 15, octadecatrienoic acid) and α- and γ-isomers. Essential dietary component for mammals. See ***fatty acids***.

lipaemia Presence in the blood of an abnormally large amount of lipid.

lipases Enzymes that break down mono-, di- or triglycerides to release fatty acids and glycerol. Calcium ions are usually required.

lipid A The lipid associated with polysaccharide in the ***lipopolysaccharide*** of ***Gram-negative*** bacterial cell walls.

lipid bilayer See ***phospholipid bilayer***.

lipidosis (plural, lipidoses) ***Storage disease*** in which the missing enzyme is one that degrades sphingolipids (sphingomyelin, ceramides, gangliosides). In ***Tay–Sachs disease*** the lesion is in hexosiminidase A, an enzyme that degrades ganglioside G_{m2}; in Gaucher's disease, glucocerebrosidase; in Niemann–Pick disease, sphingomyelinase.

lipids Biological molecules soluble in apolar solvents, but only very slightly soluble in water. They are a heterogeneous group (being defined only on the basis of solubility) and include fats, waxes and terpenes. See Table L3.

lipoamide The functional form of lipoic acid in which the carboxyl group is attached to protein by an amide linkage to a lysine amino group.

lipoamide dehydrogenase An enzyme that regenerates lipoamide from the reduced form dihydrolipoamide.

lipocalin Family of proteins that transport small, hydrophobic molecules, such as retinol, porphyrins, odorants. Characterized by two orthogonally stranded ***beta sheets***. Examples: α-1-microglobulin, ***purpurin, orosomucoid***.

lipocortin The name given to calcium-binding protein believed to be secreted by macrophages that acts as an inhibitor of phospholipase A2 enzymes and has a possible role in mediating the anti-inflammatory effects of steroids. Lipocortins are identified as proteins of the ***annexin*** class and their extracellular role is in some doubt.

lipofectamine (lipofect amine) Proprietary formulation for lipid-mediated tranfection of cultured cells.

lipofuscin Brown pigment characteristic of ageing. Found in lysosomes and is the product of peroxidation of unsaturated fatty acids and symptomatic, perhaps, of membrane damage rather than being deleterious in its own right.

lipoic acid (thioctic acid; 1,2-dithiolane-3-valeric acid) Regarded as a coenzyme in the oxoglutarate dehydrogenase complex of the ***tricarboxylic acid cycle***. Involved generally in oxidative decarboxylations

Table L3. Lipids

(i) FATTY ACIDS

These are the most important feature of the majority of biological lipids. They occur free in trace quantities and are important metabolic intermediates. They are esterified in the majority of biological lipids. Compounds are included either because they are common components of biological lipids or are used in synthetic 'model' analogues of these lipids. General formula R-COOH. Branched chain compounds are widespread, but are not found in mammalian lipids. All the examples given are straight-chain compounds.

Saturated fatty acids

Number of carbon atoms	Name	M_r (D)
2	Acetic	60
3	Propionic	74.1
4	Butyric	88.1
5	Valeric	102.1
6	Hexanoic (caproic)	116.2
7	Heptanoic	130.2
8	Octanoic (caprylic)	144.2
9	Nonanoic (pelargonic)	158.2
10	Decanoic (capric)	172.2
11	Undecanoic	186.3
12	Lauric	200.3
13	Tridecanoic	214.4
14	Myristic	228.4
15	Pentadecanoic	242.4
16	Palmitic	256.4
17	Margaric	270.7
18	Stearic	284.5
20	Eicosanoic (Arachidic)	312.5
22	Docosanoic (Behenic)	340.6

Unsaturated fatty acids

Designation[a]	Name	Mr (D)
Mono-unsaturated acids		
16: 1 (*cis* 9)	Palmitoleic	254.2
18: 1 (*cis* 9)	Oleic	282.5
18: 1 (*trans* 9)	Elaidic	282.5
18: 1 (*cis* 11)	*cis*-vaccenic	282.5
18: 1 (*trans* 11)	*trans*-vaccenic	282.5
Polyunsaturated acids (all *cis* double bonds)		
18: 2 (9, 12)	Linoleic	280.4
18: 3 (9, 12, 15)	α-linolenic	278.4
18: 3 (6, 9, 12)	γ-linolenic	278.4
20: 4 (5, 8, 11, 14)	Arachidonic (eicosenoic)	304.5
22: 6 (4, 7, 10, 13, 16, 19)	Dodecosahexaenoic acid	328.6

[a]Number of carbon atoms: number of double bonds (configuration and position of bonds).

(ii) ACYL GLYCEROLS

Glycerol esters of fatty acids. Acyl glycerols are the parent compounds of many structural and storage lipids. Diglycerides (DG) may be considered as the parent compounds of the major family of phosphatidyl phospholipids. Triglycerides (TG) are important storage lipids.

Table L3. Lipids (Continued)

Diglycerides
Present as trace components of membranes. They are important metabolites and second messengers in signal–response coupling.

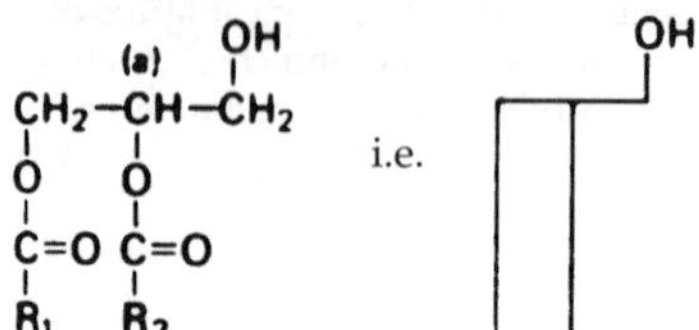

(a)This carbon is asymmetric. See below under phosphatidic acid.

(iii) SPHINGOLIPIDS
Important and widespread classes of phospholipids and glycolipids.
The parent alcohol is SPHINGOSINE:

$$CH_3(CH_2)_{12}CH{=}CH{-}\underset{}{\overset{OH}{\overset{|}{C}H}}{-}\underset{\underset{Y}{|}}{\underset{NH}{\underset{|}{C}H}}{-}CH_2{-}X$$

where X = OH and the primary amino group is free.

SPHINGOSINE is normally substituted at X and Y. When Y is a long-chain unsaturated fatty acyl group the derivative is a CERAMIDE. When the CERAMIDE carries uncharged sugars as the X substituent this is a CEREBROSIDE and where the sugars include sialic acid it is a GANGLIOSIDE.

(iv) PHOSPHOLIPIDS
In animal cell membranes the major class of phospholipids are the phosphatidyl phospholipids for which phosphatidic acid can be considered as the simplest example. These are diacylglycerol (DG) derivatives and in most cases DG is the immediate metabolic precursor.

Outline structure (see diglyceride):

O
|
O=P–O$^-$ (H) pK_a ~6.5
|
O
(b) |
CH_2–CH–CH_2
| |
O O
| |
C=O C=O
| |
R_1 R_2

i.e.

O$^-$
|
$^-$O–P=O
|
O

(b)As in diglycerides, this carbon atom is asymmetric. The biologically important configuration is *syn*. R_1 is usually saturated and R_2 is unsaturated in animal cell membranes.

Phosphatidyl phospholipids
Derived from phosphatidic acid by esterification of the phosphate group.

Base (substituent)	Phospholipid class	Abbreviation	Ionic status
None	Phosphatidic acid	PA	Anionic
Choline	Phosphatidyl choline	PC	Neutral
Ethanolamine	Phosphatidyl ethanolamine	PE	Neutral
Glycerol	Phosphatidyl glycerol	PG	Anionic
Inositol	Phosphatidyl inositol (Ptdyl. Ins.)	PI	Anionic
Inositol 4-monophosphate	Phosphatidyl inositol 4-phosphate (Ptdyl. Ins. 4-phosphate)	PIP	Anionic
Inositol 4,5 diphosphate	Phosphatidyl inositol 4,5-diphosphate (Ptdyl. Ins. 4,5 bisphosphate)	PIP2	Anionic
Phosphatidyl glycerol	Diphosphatidyl glycerol (Cardiolipin)		Anionic
Serine	Phosphatidyl serine	PS	Anionic

Table L3. Lipids (Continued)

Sphingomyelin (SM) is an analogue of phosphatidyl choline in which the diacylglycerol component is replaced by a CERAMIDE. Common variants of these structures are:

Ether phospholipids in which the diacylglycerol structure is modified so that one or both acyl groups are replaced by ether groups.

```
CH-                    CH-
 |                      |
 O      becomes         O
 |                      |
C=O     (acyl)          R        (alkyl)
 |
 R
```

Plasmalogens in which the 1-acyl group is replaced by a 1-alkenyl group. Plasmalogens are abundant lipid components of many membranes.

```
CH-                    CH-
 |                      |
 O                      O
 |      becomes         |
C=O                    CH
 |                      ||
 R                     CH
                        |
```

Phosphonolipids in which the ester linkage between the base (choline or ethanolamine) is replaced by a P-C (phosphono) linkage.

```
     O                              O
     ||                             ||
-O-P-O-CH-     becomes         -O-P-CH-
     |                              |
     O-                             O-
```

Lysophospholipids Derivatives of phosphatidyl phospholipids in which one of the acyl groups has been removed (enzymically). Lysophosphatidyl choline (lysolecithin) is a common, but trace component of membranes.

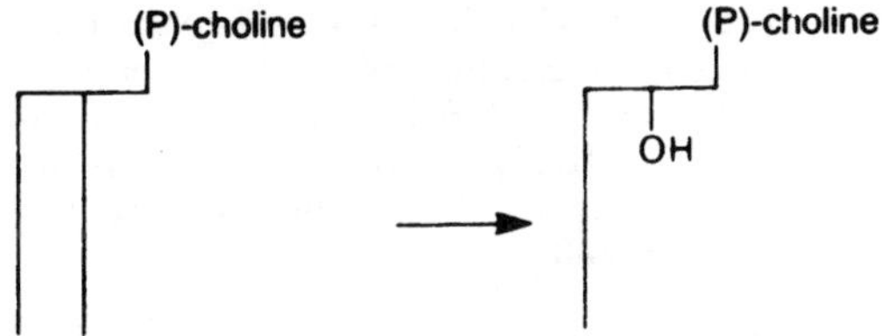

(v) STEROLS. Of this large class of compounds only one member, CHOLESTEROL, is an important structural lipid. It is the single most abundant lipid in the plasma membrane of many animal cell types.

---

of α-keto acids. A growth factor for some organisms.

lipomodulin The name originally given to ***lipocortin*** from neutrophils.

lipophorin A family of high density lipoproteins (6–700 kD) from insect ***haemolymph***, that transport diacyl glycerols. The molecule comprises heavy (250 kD) and light (85 kD) subunits, the remainder of the molecular weight being accounted for by the high lipid content (40–50%, depending on insect species). Lipophorin forms large aggregates during the haemolymph clotting process.

lipopolysaccharide The major constituents of the cell walls of ***Gram-negative bacteria***. Highly immunogenic and stimulates the production of endogenous pyrogen ***interleukin-1*** and ***tumour necrosis factor*** (TNF).

lipoproteins An important class of serum proteins in which a lipid core with a surface coat of phospholipid monolayer is packaged with specific proteins (apolipoproteins). Classified according to density: chylomicrons, large low density particles; very low density (VLDL); low density (LDL) and high density (HDL) species. Important in lipid transport, especially cholesterol transport.

liposome Artificially formed single or multi-layer spherical lipid bilayer structures. Made from solutions of lipids, etc. in organic solvents dispersed in aqueous media. Under appropriate conditions liposomes form spontaneously. Often used as models of the plasma membrane. May also be used experimentally and therapeutically for delivering drugs, etc. to cells since liposomes can fuse with a plasma membrane and deliver their contents to the interior of the cell. Vary in size from submicron diameters to (in a few record-breaking cases) centimetres.

lipoteichoic acid Compounds formed from ***teichoic acid*** linked to glycolipid and found in the walls of most ***Gram-positive bacteria***. The lipoteichoic acid of streptococci may function as an ***adhesin***.

lipotropin (LPH; lipotropic hormone; adipokinetic hormone) Polypeptide hormone (β form: 9894D, 91 residues; γ form has only residues 1–58 of β) from the pituitary hypophysis, which is of particular interest because it is the precursor of ***endorphins***, which are released by proteolysis. Promotes lipolysis and acts through the adenylyl cyclase system. Part of the ACTH group of hormones.

lipoxygenase (5-lipoxygenase; 5-LO) Enzyme that catalyses the addition of a hydroperoxy group to the 5 position of arachidonic acid, the first step in ***leukotriene*** synthesis.

liprins Family of proteins that interact with ***LAR*** family phosphatases. C-terminal portion of liprins binds to membrane-distal phosphatase domain of LAR and N-terminal region may be involved in dimerisation. Some liprins are widely distributed, others are more tissue specific. May affect LAR distribution in the cell, in particular bringing LAR to regions of contact between cell and extracellular matrix.

Listeria monocytogenes Rod-shaped ***Gram-positive*** bacterium. Widespread and can grow over an unusually wide range of temperatures (0–45°C). Normally a ***saprophyte*** but an opportunistic parasite, in that it can survive within cells (particularly leucocytes) and can be transmitted transplacentally. It has caused a number of serious outbreaks of food poisoning with a high mortality rate in recent years.

listeriolysin O ***Cholesterol-binding toxin*** from *Listeria monocytogenes*.

lithium The lightest alkali metal, although it has the largest hydrated cation. Important as an antidepressant and is thought to act by inhibiting the regeneration of ***inositol*** from ***IP3*** and thus reducing the efficiency of the ***phosphatidyl inositol*** signalling pathways.

lithotroph See ***autotroph***.

litorin A peptide that mimics ***bombesin*** in its mitogenic effects, and has a carboxy-terminal octapeptide in common with bombesin.

liver cells Usually implies ***hepatocytes***, even though other cell types are found in the liver (eg. ***Kupffer cells***). Hepatocytes are relatively unspecialised epithelial cells and are the biochemist's 'typical animal cell'.

LMM (light meromyosin) The rod-like portion of the myosin heavy chain (predominantly α-helical) that is involved in lateral interactions with other LMM to form the thick filament of striated muscle, and which is separated from heavy meromyosin (HMM) by cleavage with trypsin.

lobopodia Hemispherical protrusions from the front of a moving tissue cell.

local circuit theory A generally accepted model for neuronal conduction, by which depolarization of a small region of a neuronal plasma membrane produces transmembrane currents in the neighbouring regions, tending to depolarize them. As the ***sodium channels*** are ***voltage-gated***, the depolarization causes further channels to open, thus propagating the action potential.

lock and key models Specific recognition in biological systems might be mediated through interactions that depend upon very precise steric matching between receptor and ligand, or between enzyme and substrate. The commonly used analogy is between lock and key, and implies a precise sterically determined interaction.

locomotion Term used by some authors to distinguish movement of cells from place to place from movements such as flattening, shape-change, ***cytokinesis***, etc.

locus (plural, loci) The site in a linkage map or on a chromosome where the gene for a particular trait is located. Any one of the alleles of a gene may be present at this site.

locus control region (control region; LCR) Region of DNA which contains the ***promoters*** and ***enhancers*** that regulate the expression of a particular gene. Often taken to be a single region 0–2 kb upstream of the transcriptional ***start site***, although there are probably few genes where things are that simple.

Lod score 'Logarithm of the odds' score: a statistical test for the probability that there is linkage. For non-X linked genetic disorders a Lod score of +3 (1000:1) is usually taken to indicate linkage.

Loeffler's medium Suboptimal coagulated-serum medium used to culture ***Corynebacterium*** *diphtheriae* in diagnostic bacteriology.

Loeffler's syndrome Acute but mild and self-limiting eosinophilic pneumonia.

Loligo Squid. Source of giant axons for electrophysiologists.

lomasome Membranous structure, often containing internal membranes, located between the plasma membrane and cell wall of plant cells. Included in the more general term, ***paramural body***.

long-term potentiation (LTP) Increase in the strength of transmission at a ***synapse*** with repetitive use that lasts for more than a few minutes. As a form of long-term ***synaptic plasticity*** it is important as a possible cellular basis of learning and memory storage. It has been studied most extensively at excitatory synapses onto principal neurons of the ***hippocampus*** where it was first demonstrated. Selective inhibition of ***NMDA receptor*** channels has been shown to block LTP, and to block spatial learning.

long-terminal repeat (LTR) Identical DNA sequences, several hundred nucleotides long, found at either end of ***transposons*** and the proviral DNA, formed by reverse transcription of ***retroviral*** RNA. They are thought to have an essential role in integrating the transposon or provirus into the host DNA. LTRs have inverted repeats, that is, sequences close to either end are identical when read in opposite directions. In proviruses the upstream LTR acts as a promoter and enhancer and the downstream LTR as a polyadenylation site.

lophocytes Cells found beneath the dermal membrane of a few species of sponges. Have been postulated to constitute a primitive nervous system though this is uncertain.

Lophophorates Group of minor protostome coelomate phyla. Includes Bryozoa, Phoronida and Brachiopoda.

lophotrichous Cell with flagella arranged as a tuft at one end.

lorica Shell or test secreted by a protozoan; often vase-shaped.

Lou Gehrig's disease See ***amyotrophic lateral sclerosis***.

lovastatin (mevinolin; 6 α-methylcompactin) Fungal metabolite that inhibits hydroxymethylglutaryl-CoA reductase (HMGCoA reductase) and is used as an anti-hypercholesterolemic drug.

low density lipoprotein (LDL) See ***lipoprotein***.

low density lipoprotein receptor (LDL receptor) A cell-surface protein that mediates the endocytosis of LDL by cells. Genetic defects in LDL receptors lead to abnormal serum levels of LDL and hypercholesterolaemia.

low-affinity platelet factor IV See ***connective tissue-activating peptide III***.

Lowry assay One of the most commonly used assays for protein content – the paper describing it is said to be the most frequently cited in the biological literature. Depends upon the interaction of Folin–Ciocalteau reagent with tyrosine or phenylalanine. Proteins that are deficient in these amino acids (eg. collagen) will be underestimated and for this reason the Bradford assay is often preferred.

lox Site in ***bacteriophage*** P1 DNA that is recognized by the ***cre*** recombinase. Now used in vertebrate transgenics: see ***lox-Cre system***.

lox-*Cre* system ***Site-specific recombination*** system from *E. coli* ***bacteriophage*** P1. Now used in transgenic animals to produce conditional mutants. If two ***lox*** sites are introduced into a transgene, the intervening DNA is spliced out if active ***cre*** recombinase is expressed.

loxP Target sequence recognized by the bacterial ***cre*** recombinase.

LPS See ***lipopolysaccharide***.

LRR See ***leucine-rich repeat***.

Lsc One of the Dbl-like oncoproteins. See ***Lfc***.

LTA_4, LTB_4, LTC_4, LTD_4, LTE_4 See ***leukotrienes***.

LTP See ***long-term potentiation***.

Lubrol A non-ionic detergent.

lucifer yellow Bright yellow fluorescent molecule (similar to fluorescein), widely used by ***microinjection*** in developmental biology and neuroscience to study the outline of cells, in ***cell lineage*** studies, or as an indicator of ***dye coupling*** between cells.

luciferase An ***enzyme*** from firefly tails that catalyses the production of light in the reaction between luciferin and ATP. Used by the male firefly for producing light to attract females, and used in the laboratory in a ***chemiluminescence*** bioassay for ATP.

luciferase reporter system ***Reporter genes*** that are based on firefly ***luciferase*** gene offer luminescent detection of reporter activity. As few biological processes emit light, this assay has a very low background.

luciferins Substrates for the enzyme ***luciferase*** that catalyses an oxidative reaction leading to photon emission (bioluminescence).

lucigenin Compound used as a bystander substrate in assaying the ***metabolic burst*** of leucocytes by ***chemiluminescence***. When oxidized by ***superoxide*** it emits light.

Lucké carcinoma A renal carcinoma, caused by a herpesvirus, in frogs; it aroused interest because its abnormal growth appears to be dependent on a restricted temperature range. Nuclei from these cells give rise to normal frogs if transplanted into enucleated eggs, giving support to the ***epigenesis*** theories of neoplasia.

lumen A cavity or space within a tube or sac.

lumican Isoform (37 kD) of corneal keratan sulphate ***proteoglycan*** also found in arterial wall and many other tissues.

lumicolchicine A derivative of ***colchicine*** produced by exposure to ultraviolet light and that does not inhibit tubulin polymerization, although it has many of the non-specific effects of colchicine.

luminol Compound used as a bystander substrate in assaying the ***metabolic burst*** of leucocytes by ***chemiluminescence***. When oxidized by the myeloperoxidase/hydrogen peroxide system, it emits light.

lumirhodopsin Altered form of ***rhodopsin*** produced as a result of illumination.

lumisome Subcellular membrane-enclosed vesicle that is the site of bioluminescence in some marine ***coelenterates***.

lupus erythematosus Skin disease in which there are red scaly patches, especially over the nose and cheeks. May be a symptom of ***systemic lupus erythematosus***.

luteinizing hormone A glycoprotein hormone (26 kD) and ***gonadotropin***. Made up of an α chain (96 amino acids) identical to other gonadotropins, and a hormone-specific β chain. Acts with ***follicle-stimulating hormone*** to stimulate sex hormone release.

luteinizing hormone-releasing factor (LHRF) A decapeptide releasing hormone (1182D) that stimulates release of ***luteinizing hormone***.

lutropin Synonym for ***luteinizing hormone***.

luxury protein A term sometimes used to describe those proteins that are produced specifically for the function of differentiated cells and are not required for general cell maintenance (the so-called 'housekeeping' proteins).

LXR Orphan ***nuclear receptor*** expressed in liver, intestine and adrenal gland, that can complex with retinoid-X receptor (RXR). Ligand may be oxysterol metabolites of cholesterol (particularly 22(R)-hydroxycholesterol) and thus LXR may regulate cholesterol biosynthesis.

lyases Enzymes of the EC Class 4 (see ***E classification***) that catalyse the non-hydrolytic removal of a group from a substrate with the resulting formation of a double bond; or the reverse reaction, in which case the enzyme is acting as a synthetase. Include decarboxylases, aldolases and dehydratases.

Lyb antigen Surface antigens of mouse B-lymphocytes.

lycopene A linear, unsaturated hydrocarbon ***carotenoid*** (536 D); the major red pigment in some fruit.

Lyme disease Disease caused by ***Borrelia burgdorferi***, a tick-borne spirochaete.

lymph Fluid found in the lymphatic vessels that drain tissues of the fluid that filters across the blood vessel walls from blood. Lymph carries lymphocytes that have entered the lymph nodes from the blood.

lymph node (lymph gland) Small organ made up of a loose meshwork of reticular tissue in which are enmeshed large numbers of lymphocytes, macrophages, and accessory cells. Recirculating lymphocytes leave the blood through the specialized high endothelial venules of the lymph node and pass through the node before being returned to the blood through the lymphatic system. Because the lymph nodes act as drainage points for tissue fluids, they are also regions in which foreign antigens present in the tissue fluid are most likely to begin to elicit an immune response.

lymphadenitis Inflammation of ***lymph nodes***.

lymphadenopathy Pathological disorder of ***lymph nodes***. Lymphadenopathy-associated virus (LAV) was the name given to ***HIV*** by the Pasteur Institute group.

lymphoblast Often referred to as a blast cell. Unlike other usages of the suffix blast, a lymphoblast is a further differentiation of a lymphocyte, T- or B-, occasioned by an antigenic stimulus. The lymphoblast usually develops by enlargement of a lymphocyte, active re-entry to the S phase of the cell cycle, mitogenesis and production of much mRNA and ribosomes.

lymphocyte White cells of the blood that are derived from stem cells of the lymphoid series. Two main classes, T- and B-lymphocytes, are recognized, the latter responsible (when activated) for production of antibody, the former subdivided into subsets (helper, suppressor, cytotoxic T-cells), and responsible both for cell-mediated immunity and for stimulating B-lymphocytes.

lymphocyte activation (lymphocyte transformation) The change in morphology and behaviour of lymphocytes exposed to a mitogen or to an antigen to which they have been primed. The result is the production of ***lymphoblasts***, cells that are actively engaged in protein synthesis and that divide to form effector populations. Should not be confused with transformation of the type associated with oncogenic viruses, and 'activation' is therefore perhaps a better term.

lymphocyte transformation See ***lymphocyte activation***.

lymphocytic leukaemia See ***leukaemia***.

lymphoid cell Cells derived from stem cells of the lymphoid lineage: large and small lymphocytes, plasma cells.

lymphoid tissue Tissue that is particularly rich in lymphocytes (and accessory cells such as macrophages and reticular cells), particularly the ***lymph nodes***, spleen, ***thymus***, ***Peyer's patches***, pharyngeal tonsils, adenoids, and (in birds) the ***bursa of Fabricius***.

lymphokine Substance produced by a leucocyte that acts upon another cell. Examples are ***interleukins***, ***interferon*** γ, lymphotoxin (tumour necrosis factor β), granulocyte-monocyte colony-stimulating factor (***GM-CSF***). The term is becoming less common and ***cytokine***, a more general term, is taking over. Cytokines include lymphokines.

lymphoma Malignant neoplastic disorder of lymphoreticular tissue that produces a distinct tumour mass, not a leukaemia (in which the cells are circulating). Includes tumours derived both from the lymphoid lineage and from mononuclear phagocytes; lymphomas arise commonly (but not invariably) in lymph nodes, spleen, or other areas rich in lymphoid tissue. Lymphomas are subclassified as Hodgkin's disease, and non-Hodgkin's lymphomas (eg. Burkitt's lymphoma, large-cell lymphoma, histiocytic lymphoma).

lymphopoietin 1 See ***interleukin-7***.

lymphotoxin Cytotoxic product of T-cells: the term is usually restricted to tumour necrosis factor β which is also known as lymphotoxin.

lymphotropic Having an affinity for lymphocytes, as for example some forms of HIV.

lyn Non-receptor tyrosine ***kinase***, related to ***src***. Plays a critical role in B-cell development and intracellular signalling. Lyn-deficient mice exhibit splenomegaly, elevated serum IgM, production of autoantibody and, later, glomerulonephritis.

Lyon hypothesis Hypothesis, first advanced by Lyon, concerning the random inactivation of one of the two X chromosomes of the cells of female mammals. In consequence females are chimaeric for the products of the X chromosomes, a situation that has been exploited in female Negroes (who are heterotypic for isozymes of glucose-6-phosphate dehydrogenase) as a means to confirm the monoclonal origin of papillomas and of atherosclerotic plaques.

Lyonization See ***Lyon hypothesis***.

lyophilic Characteristic of a material that readily forms a colloidal suspension. Molecules of the solvent form a shell around the particles; if the solvent is water then 'hydrophilic'.

lyophilization Now generally restricted to mean freeze drying: removal of water by sublimation under vacuum.

lyotropic series A listing of anions and cations in order of their effect on protein solubility (tendency to cause salting out). Essentially a competition between the protein and the ion for water molecules for hydration.

Lys See ***lysine***.

lysine (Lys; K) Amino acid (146 D); the only carrier of a side chain primary amino group in proteins. Has important structural and chemical roles in proteins. See Table A2

lysis Rupture of cell membranes and loss of cytoplasm.

lysogenic conversion See ***lysogeny***.

lysogeny The ability of some phages to

survive in a bacterium as a result of the integration of their DNA into the host ***chromosome***. The integrated DNA is termed a prophage. A regulator gene produces a ***repressor protein*** that suppresses the lytic activity of the phage, but various environmental factors, such as ultraviolet irradiation, may prevent synthesis of the repressor, leading to normal phage development and lysis of the bacterium. The best example of this is the ***lambda bacteriophage***.

lysophosphatides Monoacyl derivatives of diacyl phospholipids that are present in membranes as a result of cyclic deacylation and reacylation of phospholipids. Membranolytic in high concentrations, and fusogenic at concentrations that are just sublytic. May have important modulatory roles.

lysosogenic bacteriophage Bacteriophage that can take part in a lysogenic or lytic cycle in its bacterial host. See ***lysogeny***.

lysosomal diseases Diseases (also known as storage diseases) in which a deficiency of a particular lysosomal enzyme leads to accumulation of the undigested substrate for that enzyme within cells. Not immediately fatal, but within a few years lead to serious neurological and skeletal disorders and eventually to death. See the following diseases or syndromes: ***Hurler***, ***Hunter***, ***Sanfillipo***, ***Gaucher's***, ***Niemann–Pick***, ***Pompe's***, ***Tay–Sachs***.

lysosomal enzymes A range of degradative enzymes, most of which operate best at acid pH. The best known marker enzymes are acid phosphatase and β-glucuronidase, but many others are known.

lysosomal-associated membrane protein 1 See ***LAMP-1***.

lysosome Membrane-bounded cytoplasmic organelle containing a variety of hydrolytic enzymes that can be released into a phagosome or to the exterior. Release of lysosomal enzymes in a dead cell leads to autolysis (and is the reason for hanging game, to 'tenderize' the muscle), but it is misleading to refer to lysosomes as 'suicide bags', since this is certainly not their normal function. Part of the GERL complex or trans-Golgi network. Secondary lysosomes are phagocytic vesicles with which primary lysosomes have fused. They often contain undigested material.

lysosome-associated membrane glycoproteins (LAMP-1, LAMP-2) Group of lysosome-specific integral membrane glycoproteins. Long luminal domain, short transmembrane domain, very short cytoplasmic tail. Function not yet clear.

lysosome–phagosome fusion A process that occurs after the internalisation of a primary phagosome. Fusion of the membranes leads to the release of lysosomal enzymes into the phagosome. Some species of intracellular parasite evade immune responses by interfering with this process.

lysosomotropic Having affinity for and thus accumulating in ***lysosomes***.

lysozyme Glycosidase that hydrolyzes the bond between N-acetyl-muramic acid and N-acetyl-glucosamine, thus cleaving an important polymer of the cell wall of many bacteria. Present in tears, saliva, and in the ***lysosomes*** of phagocytic cells, it is an important antibacterial defence, particularly against ***Gram-positive bacteria***.

lysyl oxidase Extracellular enzyme that deaminates lysine and hydroxylysine residues in collagen or elastin to form aldehydes, that then interact with each other or with other lysyl side chains to form crosslinks.

Lyt antigen A set of plasmalemmal surface glycoproteins on mouse T-lymphocytes. Possession of Lyt 1 partly defines a T-helper cell, and of Lyt 2,3 suppressor and cytotoxic cells. Formerly known as Ly antigens; see also ***Lyb antigen***.

lytic complex The large (2000 kD) cytolytic complex formed from complement C5b6789. See ***complement***.

lytic infection The normal cycle of infection of a cell by a ***virus*** or ***bacteriophage***, in which mature virus or phage particles are produced and the cell is then lysed.

lytic vacuole Vacuole found in plant cells that contains hydrolytic enzymes, analogous to the ***lysosome*** of animal cells but differing in morphology, function, enzyme content and mode of origin.

M

m119 See *Mig*.

M See ***methionine***.

M band Central region of the ***A band*** of the ***sarcomere*** in striated muscle.

M-cells Cells found amongst the other cells of the cuboidal surface epithelium of ***gut-associated lymphoid tissue***; have a complex folded surface.

M-channels Voltage-sensitive ***potassium channels*** inactivated by ***acetylcholine***. ACh acting at ***muscarinic acetylcholine receptors*** produces an internal messenger that turns off this class of K channel. A mechanism for regulating the sensitivity of cells to synaptic input.

m-chloro-carbonylcyanide-phenylhydrazine See *CCCP*.

M-current Flow of potassium ions through M-channels.

M line Central part of the ***A band*** of striated muscle (and of the ***M band***): contains M line protein (***myomesin***, 165 kD), ***creatine kinase*** (40 kD) and glycogen phosphorylase b (90 kD). Involved in controlling the spacing between thick filaments.

M phase Mitotic phase of cell cycle of eukaryotic cells, as distinct from the remainder, which is known as interphase (and that can be further subdivided as G1, S and G2). Beginning of M is signalled by separation of centrioles, where present, and by the condensation of chromatin into chromosomes. M phase ends with the establishment of nuclear membranes around the two daughter nuclei, normally followed immediately by cell division (***cytokinesis***).

M phase-promoting factor (MPF) Protein whose levels rise rapidly just before, and fall away just after, ***mitosis***. Thought to be a trigger for mitosis.

M-protein (1) Galactoside carrier in *E. coli*. (2) Cell surface antigen of *Brucella*. (3) Structural protein in the M-line of striated muscle (***myomesin***). (4) Cell wall protein of streptococci: antibody typing of the M-protein is important in identification of different strains of Group A streptococci (at least 55 serotypes are known). The M-protein confers anti-phagocytic properties on the cell and is present as hair-like ***fimbriae*** on the surface. M-protein is an important virulence factor, and antibodies directed against M-protein are essential for phagocytic killing of the bacteria.

M-ring Innermost (motor) ring of the bacterial flagellar base, located in the outer leaflet of the plasma membrane. It is this ring that is linked to the hook region (and thus to the flagellum itself), and that rotates. Composed of 16 or 17 subunits (one more or less than the ***S-ring***).

Mab See ***monoclonal*** antibody.

MAC (membrane attack complex.) See ***complement*** and ***C9***.

Mac-1 (CR3; CD11b/CD18) α **M** β_2 ***integrin*** of leucocytes.

Macaca mulatta Rhesus monkey.

MacConkey's agar Agar-based medium used for isolation of bacteria from faeces, etc. Contains lactose and neutral red as an indicator so that lactose fermenting bacteria will produce red-pink colonies.

Mach–Zehnder system Interferometric system in which the original light beam is divided by a semi-transparent mirror: object and reference beams pass through separate optical systems and are recombined by a second semi-transparent mirror. Interference fringes are displaced if the optical path difference for the reference beam is greater, and this can be compensated with a wedge-shaped auxiliary object. The position of the wedge allows the phase-retardation of the object to be measured. The Mach–Zehnder system was used in a microscope designed by Leitz.

Machupo virus A member of the ***Arenaviridae*** that may cause a severe haemorrhagic fever in humans. The natural hosts are rodents and transmission from human to human is not common.

macrocytes Abnormally large red blood cells, numerous in pernicious anaemia.

macroglobulin See ***IgM***.

macrolides A group of antibiotics produced by various strains of *Streptomyces* that have a complex macrocyclic structure. They inhibit protein synthesis by blocking the 50S ribosomal subunit. Include erythromycin, carbomycin. Used clinically as broad-spectrum antibiotics, particularly against ***Gram-positive bacteria***.

macromolecule Biological term including proteins, nucleic acids and carbohydrates, but probably not phospholipids.

macronucleus The larger nucleus (or sometimes nuclei) in ciliate protozoans. Derived from the micronucleus by a process of DNA polytenization. The DNA in the macronucleus is actively transcribed. The macronucleus degenerates before conjugation.

macrophage Relatively long-lived phagocytic cell of mammalian tissues, derived from blood ***monocyte***. Macrophages from different sites have distinctly different properties. Main types are peritoneal and alveolar macrophages, tissue macrophages (***histiocytes***), ***Kuppfer cells*** of the liver, and osteoclasts. In response to foreign materials may become stimulated or activated. Macrophages play an important role in killing of some bacteria, protozoa and tumour cells, release substances that stimulate other cells of the immune system, and are involved in ***antigen presentation***. May further differentiate within chronic inflammatory lesions to epithelioid cells or may fuse to form ***foreign body giant cells*** or ***Langhans' multinucleate giant cells***.

macrophage colony-stimulating factor See ***MCSF***.

macrophage inflammatory protein 1 (MIP1) Cytokine now recognized to exist in two forms, MIP1 α and MIP1 β (SIS-α, TY-5, L2G25B, 464.1, GOS 19-1). Small ***cytokine*** (monokine) with inflammatory and chemokinetic properties.

macrophage inflammatory protein 2 (MIP2) ***Cytokine*** that is ***chemotactic*** for ***neutrophils***.

macrophage inhibition factor (MIF) A group of lymphokines (including a 14 kD glycoprotein) produced by activated T-lymphocytes that reduce macrophage mobility and probably increases macrophage–macrophage adhesion.

macrophage-stimulating protein (MSP) Ligand for ***STK***. Stimulates motility and phagocytosis of macrophages.

macrosialin (CD68) Mouse homologue of CD68. Transmembrane ***scavenger receptor*** of the mucin-like class that includes ***LAMP***-1 and -2. Expressed specifically on macrophages and related cells, binds oxidized ***LDL***. Predominantly in lysosomal and endosomal membranes and to a lesser extent on cell surface.

macula adherens junctions See ***desmosome***.

macule A spot: only commonly met in the construct 'immaculate' meaning unspotted.

MAD Protein involved in ***Dpp*** signalling (Dpp is TGF β-related). MAD (*mothers against dpp*) contains a specific DNA-binding activity that activates an enhancer in a *Drosophila* wing-patterning gene, *vg* (*vestigial*). See also ***SMAD***.

mad Basic ***helix-loop-helix*** leucine zipper transcription factor (28 kD).

Madin–Darby canine kidney (MDCK) Line of canine epithelial cells that grow readily in culture and form confluent monolayers with relatively low transmonolayer permeability (varies between clones). Often used as a general model for epithelial cells.

MADS Superfamily of transcription factors. See ***mad*** as example.

MAF (macrophage-activating factor) A ***lymphokine***.

MAG protein Myelin-associated glycoprotein; one of the ***immunoglobulin superfamily***.

magainins Peptides of about 20 amino acid residues with antimicrobial activity, found in amphibian skin. Probably have membrane insertion and lytic properties. Sequence related to melittin.

magnesium An essential divalent cation. The major role is as the chelated ion in ATP and presumably other triphosphonucleotides. The Mg^{2+}–ATP complex is the sole biologically active form of ATP. The other essential role of Mg^{2+} is as the central ion of chlorophyll. Cellular concentration is less than 5mM, serum concentration approx. 1mM.

magnetic resonance imaging (MRI) Non-invasive imaging technique that uses nuclear magnetic resonance to look at intact tissues in the body. Particularly valuable for studies on brain and soft tissues.

magnetosome Enveloped compartment in magnetotactic bacteria containing magnetite particles. By using this organelle to detect the vertical component of the Earth's magnetic field, the bacteria swim towards the bottom of the sea.

magnetotaxis Tactic response to magnetic field; in magnetotactic bacteria the Earth's magnetic field is used as a guide to 'up' and 'down' in deep sediment.

magnocellular neuron A neuron in the magnocellular region of the brain. Perhaps the first class of neuron from the central nervous system shown to be sensitive to ***nerve growth factor*** (which had previously been thought only to act at the periphery).

maitotoxin Toxin from ***dinoflagellates***. Activates L-type voltage-sensitive calcium channels (***VSCC***) and mobilizes intracellular calcium stores.

major histocompatibility antigen A set of plasmalemmal glycoprotein antigens involved in rapid (eg. 7 days in the mouse) graft rejection and other immune phenomena. The minor histocompatibility antigens are involved in much slower rejection phenomena. The major antigens show remarkable polymorphism, and occur as Class I and Class II types in mammals; birds may have a Class III molecule as well. See ***histocompatibility antigens***, ***MHC restriction***.

major histocompatibility complex (MHC) The set of gene loci specifying major histocompatibility antigens, eg. HLA in humans, H-2 in mice, RLA in rabbits, RT-1 in rats, DLA in dogs, SLA in pigs, etc.

major intrinsic protein (MIP) Family of structurally related proteins with six transmembrane segments, associated with gap junctions or vacuoles. MIP is found in lens fibre gap junctions. Other members: nodulin-26 (soybean), tonoplast intrinsic protein (TIP) found in plant storage vacuoles, *Drosophila* neurogenic protein 'big brain'.

malabsorption syndrome A variety of conditions in which digestion and absorption in the small intestine are impaired. Multiple causes including lymphoma, amyloid and other infiltrations, ***Crohn's disease***, gluten-sensitive enteropathy and the sprue syndrome in which the villi atrophy for unknown reasons.

malaria In humans, the set of diseases caused by infection by protozoans of the genus ***Plasmodium***. *P. vivax* causes the tertian type, *P. malariae* the quartan type and *P. falciparum* the quotidian or irregular type of disease, the names referring to the frequency of fevers. The fevers occur when the merozoites are released from the erythrocytes. The organisms are transmitted by species of the ***Anopheles*** mosquito.

Malassez cells Cells found in the periodontal ligament as 'epithelial rests of

Malassez'. Malassez cells retain the major characteristics of epithelial cells throughout their differentiation from the root sheath epithelium into the rests of Malassez.

malate The ion from malic acid, a component of the citric acid cycle.

maleate The ion from maleic acid, often used in biological buffers.

malignant As applied to tumours means that the primary tumour has the capacity to show ***metastatic spread*** (metastasize). Implies loss of both ***growth control*** and positional control.

Mallory's one-step stain A modified trichrome method that can give good tissue differentiation using a relatively brief procedure.

malonate The ion from malonic acid, HOOC. CH_2COOH. Malonate is a competitive inhibitor for succinate dehydrogenase in the ***tricarboxylic acid cycle***. Malonyl-SCoA is an important precursor for fatty acid synthesis.

Malpighian tubule Blind-ending tubule opening into the lower intestine of insects and responsible for fluid excretion – the insect equivalent of the kidney.

maltase Enzyme that hydrolyzes maltose (and the glucose trimer maltotriose) to glucose, during the enzymic breakdown of starch.

maltose Disaccharide intermediate of the breakdown of starch, glucose-α (1-4)-glucose. Fermentable substrate in brewing.

maltose-binding protein Protein of the bacterial (*E. coli*) surface that links with ***MCP***-II and is involved in the chemotactic response to maltose; probably derived from a similar protein that links with a transmembrane transport system.

mammalian expression vector In molecular biology, a ***vector*** that will produce large amounts of eukaryotic protein (taxonomy notwithstanding, not necessarily a protein from a mammal).

mammamodulin Protein (52-55 kD) expressed by hormone-independent mammary tumour cells. Affects morphology, motility, growth and hormone-receptor expression.

mammary gland Milk-producing gland of female mammals. An adapted sweat gland, it is made up of milk-producing alveolar cells, surrounded by contractile myoepithelial cells, together with considerable numbers of fat cells. Milk production is hormonally controlled.

mammary tumour virus (Bittner agent) ***Retrovirus*** that induces mammary ***carcinoma*** in mice. Isolated from highly inbred strains that had very high incidence of the tumours, after the discovery that the disease was transmitted in milk by nursing mothers. Endogenous provirus is present in germ-line of all inbred mice. Transcription of the provirus is regulated by a viral promoter that increases transcription in response to glucocorticoid hormones. May transform by proviral insertion activating the cellular ***int-1*** oncogene.

mammary-derived growth inhibitor (MDGI) ***Fatty acid-binding protein*** that inhibits proliferation of mammary carcinoma cells.

Manduca sexta A species of Lepidopteran insect, also known as the tobacco hornworm moth. The caterpillars, which are very large, are used in studies of ion transport, moulting, and as a system for transgenic gene expression (see ***baculovirus***).

manganese An essential trace element. Present in cells as concentrations of around 0.01mM. Activates a wide range of enzymes, eg. ***pyruvate carboxylase*** and one family of ***superoxide dismutases***. Resembles ***magnesium*** and may replace it in many enzymes when it can modify substrate specificities. The addition of manganese salts to buffer solutions will often make cells very adhesive.

mannan Mannose-containing polysaccharides found in plants as storage material, in association with cellulose as hemicellulose. In yeasts a wall constituent.

mannan-binding lectin (MBL; formerly mannan-binding protein, MBP) Plasma protein (32 kD) structurally related to ***complement*** C1 that binds specific carbohydrates (***mannans***) on the surface of various microorganisms including bacteria, yeasts, parasitic protozoa and viruses; activates the complement cascade through ***MASP*** and promotes phagocytosis. Deficiency is associated with frequent infections in childhood.

mannitol (D-mannitol) Hexitol related to D-mannose. Found in plants, particularly fungi and seaweeds.

mannose (D-mannose) Hexose identical to D-glucose except that the orientation of the –H and –OH on carbon 2 are interchanged (ie. the 2-epimer of glucose). Found as constituent of polysaccharides and glycoproteins.

mannose-6-phosphate Mannose derivative, formed by phosphorylation in the ***Golgi apparatus***, of certain mannose residues on N-glycan chains of lysosomal enzymes. Believed to function as targetting signal that causes entry of these enzymes to the lysosomes. The receptor (215 kD) is enriched in specialized pre-lysosomes.

mannosidase Enzyme catalysing hydrolysis of the glycosidic bond between mannose residues and a variety of hydroxyl-containing groups. Alpha-mannosidases in ***rough endoplasmic reticulum*** and *cis*-Golgi are responsible for removing four mannose residues during the synthesis of the complex-type N-linked glycan chains of glycoproteins.

Mantoux test Test for tuberculin reactivity in which tuberculin PPD (purified protein derivative) is injected intracutaneously. The injection site is examined after 2–3 days, a positive reaction, indicating current or previous infection with *Mycobacterium tuberculosis* (in an uninoculated individual), is an oedomatous and reddened area caused by T-cell reactivity.

MAP kinase (mitogen-activated protein kinase; externally regulated kinase, ERK) Serine-threonine kinases that are activated when quiescent cells are treated with mitogens, and which therefore potentially transmit the signal for entry into cell cycle. One target is trancription factor p62TCF. MAP kinase itself can be phosphorylated by MAP kinase kinase and this may in turn be controlled by ***raf*-**1. Confusingly, it also phosphorylates ***microtubule-associated proteins***.

MAP1C Microtubule-associated protein (two heavy chains of 410 kD associated with six or seven light chains of about 50–70 kD), now considered to be the two-headed cytoplasmic equivalent of ciliary dynein and to be responsible for retrograde transport (transport towards the centrosome).

MAPKK ***MAP kinase*** kinase.

mappine See ***bufotenine***.

MAPs (microtubule-associated proteins) May form part of the electron-lucent zone around a microtubule. MAP1A and 1B (approximately 350 kD) from brain, form projections from microtubules; MAP2A and 2B (270 kD) are also from brain microtubules and form projections. MAP3 (180 kD) and MAP4 (220–240 kD) have been described as co-purifying with MAPs 1 and 2. Microtubule-associated protein 2 is associated with axonal processes of vertebrate nerve cells and its presence distinguishes them from dendritic processes. ***MAP1C*** is in a separate class, being a motor molecule.

MAPTAM (1,2-bis-(2-amino-5-methylphenoxy) ethane tetraacetic acid tetraacetoxymethyl ester) Compound that readily enters cells where it is converted to 5-methyl ***BAPTA***, an indicator of calcium concentration.

marasmus Wasting of the body due to a deficiency in energy-giving food, as opposed to protein deficiency (kwashiokor).

Marburg virus A ***filovirus*** that causes Marburg disease, a severe haemorrhagic fever developed in many people who work with African green monkeys.

MARCKS (myristoylated alanine-rich protein kinase C substrate) Membrane associated (through the myristoyl residue) calmodulin- and actin-binding protein that cycles between membrane and cytoplasm depending on its phosphorylation state and on the presence of calcium/calmodulin. As a result of phosphorylation by protein kinase C, MARCKS is displaced from the membrane; dephosphorylation leads to reassociation. Implicated in macrophage activation, neurosecretion and growth factor-dependent mitogenesis.

Marek's disease Infectious cancer of the lymphoid system (lymphomatosis) in chickens, caused by a contagious herpes virus. An effective vaccine is now available.

Marfan syndrome Dominant disorder of connective tissue in which limbs are excessively long and loose jointed. Probably a collagen fibril-assembly disorder since it can be mimicked in mice by aminonitriles that interfere with crosslinking.

margaratoxin ***Charybdotoxin***-related peptide toxin (39 residues) from scorpion *Centruroides margaritatus*. Blocks mammalian voltage-gated potassium channels in neural tissues and lymphocytes. Very similar to ***noxiustoxin*** and ***kaliotoxin***.

marginal band A bundle of equatorially located microtubules that stabilize the biconvex shape of platelets and avian erythrocytes. They are unusual in that they do not derive from the centrosomal *MTOC*.

margination Adhesion of leucocytes to the endothelial lining of blood vessels, particularly postcapillary venules; often, but not always, a prelude to leaving the circulation and entering the tissues.

mariner Group of ***transposons*** with broad phylogenetic distribution (arthropods, nematodes, planaria, humans). Mariner elements consist of a transposase gene flanked by short inverted repeats.

marker gene Gene that confers some readily detectable phenotype on cells carrying the gene, either in culture or in transgenic or chimeric organisms. Gene could be an enzymic ***reporter gene***, a selectable marker conferring antibiotic resistance, or a cell membrane protein with a characteristic ***epitope***.

Markov process A ***stochastic*** process in which the probability of an event in the future is not affected by the past history of events.

Maroteaux–Lamy syndrome ***Mucopolysaccharidosis*** type VI; deficiency of the lysosomal enzyme arylsulphatase B; resembles ***Hurler's disease*** in some respects.

MARPS (microtubule repetitive proteins MARP-1 and MARP-2) Heat-stable high molecular weight proteins (ca 320 kD) with many 38 residue repeats. Isolated from membrane skeleton of ***Trypanosoma*** *brucei*; probably stabilize microtubules – each repeat has tubulin-binding activity.

mas **Oncogene** from brain that encodes a receptor coupled to a ***G-protein*** and to PIP2 turnover. Ligand was originally thought to be ***angiotensin*** II, but this is now less certain. See Table O1.

masked messenger RNA Long-lived and stable mRNA found originally in the oocytes of echinoderms and constituting a store of maternal information for protein synthesis that is unmasked (derepressed) during the early stages of morphogenesis. In these early stages the rate of cell division is so rapid that transcription from the embryonic genome cannot occur. Undoubtedly not restricted to oocytes, and the term can be applied to any mRNA which is present in inactive form.

MASP (MBL-associated serine protease) Proteases (ca 76 kD) in the ***complement*** system, activated by the binding of mannan to ***mannan-binding lectin*** (MBL), (formerly mannose-binding protein, MBP). MASP-1 and MASP-2 are similar to the ***C1*** q-associated proteases, C1r and C1s. All four have a C1r/C1s-like domain, an ***EGF-like domain*** and a second C1r/C1s-like domain, two complement control protein (CCP) domains and a serine protease domain. MASP-1 and MASP-2 are probably responsible for the proteolytic cleavage and activation of C2 and C4.

maspin *Serpin* (42 kD) expressed in normal mammary epithelium but reduced in mammary tumours.

Masson's trichrome stain Trichrome stains are used particularly for connective tissue. Masson's trichrome method uses haemalum, acid fuchsin and methyl blue and has the effect of staining nuclei blue-black, cytoplasm red and collagen blue.

mast cell Resident cell of connective tissue that contains many granules rich in ***histamine*** and heparan sulphate. Release of histamine from mast cells is responsible for the immediate reddening of the skin in a weal-and-flare response. Very similar to ***basophils*** and possibly derived from the same stem cells. Two types of mast cells are now recognized, those from connective tissue and a distinct set of mucosal mast cells; the activities of the latter are T-cell dependent.

mast cell growth factor See ***kit***.

mastigonemes Lateral projections from eukaryotic flagella. May be stiff and alter the hydrodynamics of flagellar propulsion, or flexible and alter the effective diameter of the flagellum (flimmer filaments).

mastocytoma Neoplastic ***mast cells***.

mastoparans Basic peptides from wasp venoms. Analogous to ***melittin*** in honeybee venom they can act as ***phospholipase*** A2 activators, but their relevance to the toxic action of the venoms is not known.

maternal antibody Any antibody transferred from a mammalian mother transplacentally into the foetus. See ***IgG***.

maternal inheritance Inheritance through the maternal cell line, eg. through the oocyte and eggs. Mitochondrial genes are maternally inherited and various other non-Mendelian forms of inheritance may also appear as maternal inheritance.

maternal mRNA Messenger RNA found in oocytes and early embryos that is derived from the maternal genome during oogenesis. See ***masked messenger RNA***.

maternal-effect gene Gene, usually required for early embryonic development, whose product is secreted into the egg by the mother. The phenotype is thus determined by the mother's, rather than the egg's, genotype, cf. ***zygotic-effect gene***. See also ***egg-polarity gene***.

mating-type genes Genes that, in *Saccharomyces cerevisiae,* specify into which of the two mating types (a and α) a particular cell falls. Only unlike mating-type haploids will fuse. The interest derives from the way in which mating type is switched: the existing gene is removed and a new gene, derived from a (silent) master copy elsewhere in the genome is spliced in. Later this gene will in its turn be replaced by a new copy of the old gene, also derived from a silent 'master'. The a- and α-genes code for pheromones that affect cells of the opposite mating type. Similar mating-type genes are known from other yeasts, and the switching mechanism (***cassette mechanism***) may be used more generally.

Matrigel Proprietary name for gel-forming matrix material derived from ***EHS cells***. Cells grown on Matrigel often show morphological characteristics distinct from those seen on a solid tissue-culture substratum, and are probably in a more normal environment both chemically and physically.

matrix Ground substance in which things are embedded or that fills a space (as, for example, the space within the mitochondrion). Most common usage is for a loose meshwork within which cells are embedded (eg. extracellular matrix), although it may also be used of filters or absorbent material.

matrix metalloproteinases (MMP) Proteolytic enzymes that degrade extracellular matrix. Include collagenases and elastases. Inhibitors are predicted to have benefits in arthritis and metastasis though this remains to be proven. See Table M1.

Table M1. Matrix metalloproteinases

Enzyme	Designation	Main substrate
Interstitial collagenase	MMP-1	Fibrillar collagens
Gelatinase A	MMP-2	Fibronectin, type IV collagen
Stromelysin-1	MMP-3	Non-fibrillar collagen, Ln, Fn
Matrilysin	MMP-7	Ln, Fn, non-fibrillar collagen
Neutrophil collagenase	MMP-8	Fibrillar collagens I, II, III
Gelatinase B	MMP-9	Collagen IV, V
Stromelysin-2	MMP-10	Ln, Fn, non-fibrillar collagen
Stromelysin-3	MMP-11	Serpin
Metalloelastase	MMP-12	Elastin
Collagenase-3	MMP-13	Fibrillar collagens
MT-MMP	MMP-14	Progelatinase A
MT2-MMP	MMP-15	Unknown
MT3-MMP	MMP-16	Progelatinase A
MMT4-MMP	MMP-17	Unknown
—	MMP-18	Unknown

Fn = fibronectin; Ln = laminin.

matrix proteins Proteins of the outer layer of the cell wall of ***Gram-negative bacteria***.

maturation-promoting factor See ***cyclin***.

Mauthner neuron Large neuron in the ***mesencephalon*** of fishes and amphibians. A rare example of an individually identifiable neuron in a vertebrate nervous system.

Max Transcription factor: forms homodimers which then interact with CACGTG motif of DNA repressively, but will form heterodimers with ***Myc protein*** that bind the same motif with greater affinity and activate the downstream gene.

Maxam–Gilbert method A method of DNA sequencing based on the controlled degradation of a DNA fragment in a set of independent, nucleotide-specific reactions. The resulting fragments have characteristic sizes depending on the sequence of the template, that can be resolved on a sequencing gel. Although no longer the main protocol, Maxam–Gilbert sequencing still has advantages, eg. for oligonucleotides or covalently modified DNA. See also ***dideoxy sequencing***.

maxiprep Slang, denoting a large scale purification of ***plasmid*** from a bacterial culture. Usually used to describe preparations from 100–500ml culture. See also ***miniprep, midiprep, megaprep***.

MBL See ***mannan-binding lectin***.

MBP See ***myelin basic protein***.

MCAF See ***monocyte chemotactic and activating factor***.

McArdle's disease Glycogen storage disease in which the defective enzyme is muscle phosphorylase.

MCP See ***gCAP39***.

MCP-1 See ***monocyte chemoattractant protein 1***.

MCPs (methyl-accepting chemotaxis proteins) Proteins of the inner cytoplasmic face of the bacterial plasma membane with which the receptors of the outer face interact. Four different MCPs are known in *E. coli*, each with a separate set of receptors. Can be methylated at various sites; methylation is part of the ***adaptation*** to the signal. Although important intermediate signal integration sites, they are not directly connected to the motor.

MCS See ***polycloning site***.

MCSF (CSF-1; macrophage colony-stimulating factor) A 40–76 kD glycoprotein that plays an important role in the activation and proliferation of ***microglial cells*** both ***in vitro*** and in injured neural tissue, is important in adipocyte hyperplasia, in osteoclast differentiation and in one of the early events in atherosclerosis, monocyte to macrophage differentiation in the arterial intima. The receptor for MCSF (c-***fms***) is expressed on the pluripotent precursor and mature osteoclasts and macrophages. Mutation in MCSF leads to osteopetrosis because of osteoclast deficiency.

MDGI See ***mammary-derived growth inhibitor***.

Mdm2 Oncoprotein that inhibits ***p53*** by binding to the transcriptional activator domain of p53 preventing it from regulating its target genes. Mdm2 expression is activated by p53 and Mdm2 promotes degradation of p53 by the proteasome – a complex feedback regulation.

Mdr See ***multidrug transporter***.

Mdx mouse Mouse mutant deficient in ***dystrophin*** and thus a good model system for ***Duchenne muscular dystrophy***.

measles virus ***Paramyxovirus*** that causes the childhood disease measles and is responsible for subacute sclerosing panencephalitis.

media Avascular middle layer of the artery wall, composed of alternating layers of elastic fibres and smooth muscle cells.

medium Shorthand for culture medium or growth medium, the nutrient solution in which cells or organs are grown.

medulla oblongata Region of the brain where the spinal cord tapers into the brain stem. Neurons in this region regulate some very basic functions such as respiration.

MEF2 (myocyte-enhancer factor 2) Group of transcription factors of the ***MADS*** superfamily. MEF2C is expressed in skeletal muscle, spleen, brain and various myeloid cells. In monocytic cells MEF2C has its transactivation activity enhanced by ***LPS*** acting through the MAP kinase ***p38***. The result of activation is increased c-*jun* transcription

megakaryocyte Giant polyploid cell of bone marrow that gives rise to 3–4000 platelets.

megalin (gp330) Epithelial endocytic receptor that internalizes various ligands including apolipoproteins E and B. Major antigen against which there is autoimmune response in ***Heymann nephritis***.

megaprep Slang, denoting a medium-scale purification of ***plasmid*** from a bacterial culture. Usually used to describe preparations from over 500ml culture. See also ***miniprep, midiprep, maxiprep***.

megaspore Haploid spore produced by a plant sporophyte, that develops into a female gametophyte.

meiocytes Cell that will undergo meiosis; a little-used term.

meiosis A specialized form of nuclear division in which there are two successive nuclear divisions (meiosis I and II) without any chromosome replication between them. Each division can be divided into four phases similar to those of mitosis (pro-, meta-, ana- and telophase). Meiosis reduces the starting number of 4n chromosomes in the parent cell to n in each of the four daughter cells. Each cell receives only one of each homologous chromosome pair, with the maternal and paternal chromosomes being distributed randomly between the cells. This is vital for the segregation of genes. During the prophase of meiosis I (classically divided into stages: ***leptotene, zygotene, pachytene, diplotene*** and ***diakinesis***), homologous chromosomes pair to form ***bivalents***, thus allowing ***crossing over***, the physical exchange of chromatid segments. This results in the ***recombination*** of genes. Meiosis occurs during the formation of ***gametes*** in animals, which are thus haploid and fertilization gives a diploid egg. In plants meiosis leads to the formation of the spore by the sporophyte generation.

meiospore Haploid spore formed after meiotic division.

meiotic spindle The meiotic equivalent of the ***mitotic spindle***.

mek (MAPK/ERK kinase) Mitogen-activated protein kinase (Mek1, 45 kD; Mek2, 46 kD).

Mel-14 Antibody that reacts with L-selectin (CD62L). Blocks lymphocyte binding to ***HEV*** both ***in vitro*** and ***in vivo***.

melanin Pigments largely of animal origin. High molecular weight polymers of indole quinone. Colours include black/brown, yellow, red and violet. Found in feathers, cuttle ink, human skin, hair and eyes, and in cellular immune responses and wound-healing in arthropods.

melanocyte Cells that synthesize melanin pigments. The pigments are stored in ***melanosomes*** (***chromatophores***) that can be redistributed in the cytoplasm to change pigment patterns in fish and reptiles.

melanocyte-stimulating hormone (MSH) A releasing hormone produced in the mammalian hypophysis and related structures in lower vertebrates. Made up of α-MSH (1665 D), the same as amino acids 1–13 of ***ACTH***, and β-MSH (18 amino acids, 22 in humans). Causes darkening of the skin by expansion of the ***melanophores*** but its role in mammals is unclear.

melanoma Neoplasia derived from ***melanocytes***; benign forms are moles, but often are highly malignant. Generally the cells contain melanin granules and for this reason they have been used in studies on metastasis because the secondary tumours are easily located in lung.

melanoma growth-stimulatory activity (MGSA, neutrophil-activating protein 3, NAP-3, 'gro', KC, N51, CINC) ***Chemokine*** of the C-X-C subfamily. Potent ***mitogen***. Activates, and is chemotactic for, neutrophils.

melanophore Cell type found in skin of lower vertebrates (amphibian skin, fish scales) that contains granules of the black pigment melanin. The granules can be rapidly redeployed between a dispersed state (which darkens the skin) and concentration at the centre (which lightens it). One of a family of pigmented or light-diffracting, coloured cells, known collectively as ***chromatophores***.

melanosome Membrane-bounded organelle found in ***melanocytes***; when ***melanin*** synthesis is active internal structure is characteristic, containing melanofilaments that have a periodicity of around 9nm and are arranged in parallel arrays. Mature melanosomes, in which the filamentous structure is masked by the dense accumulation of melanin, are transferred to keratinocytes in the skin. Also found in ***pigmented retinal epithelium*** and in some cells of the connective tissue.

melatonin (N-acetyl 5-methoxytryptamine) A hormone secreted by the pineal gland. In lower vertebrates causes aggregation of pigment in melanophores, and thus lightens skin. In humans believed to play a role in establishment of circadian rhythms, and purported to help overcome the effects of jetlag.

melioidosis Fatal bacterial disease of rodents in the tropics, caused by *Pseudomonas pseudomallei*. Affects lymph nodes and viscera.

melittin The major component of bee venom, responsible for the pain of the sting. A 26-amino acid peptide, that has a hydrophobic and a positively charged region. Can lyse cell membranes and activate phospholipase A2 enzymes; it has a very high affinity for ***calmodulin*** but the biological relevance of this is unclear.

mellitose See ***raffinose***.

memantine (1-amino-3,5, dimethyl adamantane) Relatively low affinity non-competitive ***NMDA*** antagonist used in the treatment of ***Parkinsonism*** and some other brain disorders.

membrane Generally, a sheet or skin. In cell biology the term is usually taken to mean a modified lipid bilayer with integral and peripheral proteins, as forms the plasma membrane. Because this usage is so general, it is advisable to avoid other uses where possible, particularly in histology or ultrastructure.

membrane attack complex See ***complement***.

membrane capacitance The electrical capacitance of a membrane. Plasma membranes are excellent insulators and dielectrics: capacitance is the measure of the quantity of charge that must be moved across unit area of the membrane to produce unit change in membrane potential, and is measured in farads (F). Most plasma membranes have a capacitance around 1μF cm^{-2}.

membrane depolarization See ***depolarization***.

membrane fluidity Biological membranes are viscous two-dimensional fluids within their physiological temperature range.

membrane fracture See ***freeze fracture***.

membrane potential More correctly, transmembrane potential difference: the electrical potential difference across a plasma membrane. See ***resting potential***, ***action potential***.

membrane protein A protein with regions permanently attached to a membrane (peripheral membrane protein), or inserted into a membrane (integral membrane protein). Insertion into a membrane implies hydrophobic domains in the protein. All ***transport proteins*** are integral membrane proteins.

membrane recycling The process whereby membrane is internalised, fuses with an internal membranous compartment, and is then reincorporated into the plasma membrane. In cells that are actively secreting by an exocrine method (in which secretory granules fuse with the plasma membrane), it is obviously essential to have some way of reducing the area of the plasma membrane. The membrane can then be used to form new secretory vesicles. The converse is true for phagocytic cells.

membrane transport The transfer of a substance from one side of a plasma membrane to the other, in a specific direction, and at a rate faster than diffusion alone. See ***active transport***.

membrane vesicles Closed unilamellar shells formed from membranes either in physiological transport processes or else when membranes are mechanically disrupted. They form spontaneously when membrane is broken because the free ends of a lipid bilayer are highly unstable.

membrane zippering See ***zippering***.

memory cells Cells of the immune system that 'remember' the first encounter with an antigen and facilitate the more rapid secondary response when the antigen is encountered on a subsequent occasion. The long-lasting immune memory is humoral and resides in B-lymphocytes, although it appears that persistence of the antigen may be essential. T-cell memory is shorter.

menadione (vitamin K_3; 2-methyl-1,4-naphthalene dione.) Synthetic naphthoquinone derivative with properties similar to those of vitamin K. See also Table V1.

Mendelian inheritance Inheritance of characters according to the classical laws formulated by Gregor Mendel, which give the classic ratios of segregation in the F2 generation. In sexually reproducing organisms, any process of heredity explicable in terms of chromosomal segregation, independent assortment and homologous exchange.

meninges Three layers of tissue surrounding the brain: See ***dura mater***, ***pia mater***.

meningitis Inflammation of the meninges of the brain and spinal cord. It can be caused by viral infections, by lymphocytic infiltrations and by various bacteria, but the most serious form is due to infection by *Neisseria meningitidis* with rapidly fatal consequences in up to 70% of untreated cases.

meningococcus *Neisseria meningitidis*; ***Gram-negative*** non-motile pyogenic coccus that is responsible for epidemic bacterial meningitis.

Menke's disease Genetically determined human defect in copper metabolism probably because of an inability to absorb copper from the gut.

mepacrine (quinacrine) Prophylactic antimalarial drug.

mercaptoethanol A pungent water-soluble thiol, not of biological origin. Used in biochemistry to cleave disulphide bonds in proteins or to protect sulphydryl groups from oxidation.

meristem Group of actively dividing plant cells, found as apical meristems at the tips of roots and shoots and as lateral meristems in vascular tissue (vascular ***cambium***) and in cork tissue (***phellogen***). Also found in young leaves, and at the bases of internodes in grasses. Consists of small non-photosynthetic cells, with primary walls and relatively little vacuole.

meristematic Of plant tissue: composed of ***meristem***.

Merkel cell carcinoma Rare and highly malignant skin tumour that arises from neuroendocrine cells with features of epithelial differentiation.

MeroCaM Calcium-sensitive fluorophore that can be used to measure calcium levels within live cells.

merocrine Commonest mode of secretion in which a secretory vesicle fuses with the plasma membrane and releases its contents to the exterior.

meromyosin Fragments of myosin formed by trypsin digestion. Heavy meromyosin (HMM) has the hinge region and ATPase activity, light meromyosin (LMM) is mostly α-helical and is the portion normally laterally associated with other LMM to form the thick filament itself.

merosin See ***laminin***.

merotomy Partial cutting: used in reference to experiments in which protozoa are enucleated and the behaviour of the residual cytoplasm is studied.

merozoite Stage in the life cycle of the malaria parasite (***Plasmodium***): formed during the asexual division of the schizont. Merozoites are released and invade other cells.

merozygote A bacterium that is in part haploid and in part diploid because it has acquired exogenous genetic material, eg. during transduction or conjugation.

mesangial cells Cells found within the glomerular lobules of mammalian kidney, where they serve as structural supports, may regulate blood flow, are phagocytic, and may act as ***accessory cells***, presenting antigen in immune responses.

mesencephalon Region of the brain below the thalamus and above the pons developed from the middle of the three cerebral vesicles of the embryonic nervous system. Includes the superior and inferior colliculi and cerebral peduncles.

mesenchyme Embryonic tissue of mesodermal origin.

Mesocricetus auratus Syrian golden hamster. See ***Cricetulus griseus***.

mesoderm Middle of the three ***germ layers***; gives rise to the musculoskeletal, blood vascular and urinogenital systems, to connective tissue (including that of dermis) and contributes to some glands.

mesokaryotic Those organisms with a cellular organization intermediate between pro- and eukaryotes.

mesophase (smectic mesophase) Arrangement of phospholipids in water where the liquid-crystalline phospholipids form multilayered parallel-plate structures, each layer being a bilayer, the layers separated by aqueous medium.

mesophile Organism that thrives at moderate temperatures (say, 20–40°C).

mesophyll Tissue found in the interior of leaves, made up of photosynthetic (parenchyma) cells, also called ***chlorenchyma*** cells. Consists of relatively large, highly vacuolated cells, with many ***chloroplasts***. Includes ***palisade parenchyma*** and spongy mesophyll.

mesosecrin Glycoprotein (46 kD) secreted by mesothelial cells (including endothelium). In culture, forms a fine coating on the substratum.

mesosome Invagination of the plasma membrane in some bacterial cells, sometimes with additional membranous lamellae inside. May have respiratory or photosynthetic functions.

mesothelioma Malignant tumour of the ***mesothelium***, usually of lung; frequently caused by exposure to asbestos fibres, particularly those of crocidolite, the fibres of which are thin and straight and penetrate to the deep layers of the lung. Because of their shape, the fibres puncture the macrophage phagosome and are released, leading to a chronic inflammatory state that is thought to contribute to development of the tumour.

mesothelium Simple squamous epithelium of mesodermal origin. It lines the peritoneal, pericardial and pleural cavities and the synovial space of joints. The cells may be phagocytic.

mesotocin Peptide hormone secreted by posterior lobe of pituitary; structure and function similar to ***oxytocin***.

messenger RNA (mRNA) Single-stranded RNA molecule that specifies the amino acid sequence of one or more polypeptide chains. This information is translated during protein synthesis when ribosomes bind to the mRNA. In prokaryotes, mRNA is the primary transcript from a DNA sequence and protein synthesis starts while the mRNA is still being synthesized. Prokaryote mRNAs are usually very short-lived (average $t_{1/2}$ 5 min). In contrast, in eukaryotes the primary transcripts (***HnRNA***) are synthesized in the nucleus and they are extensively processed to give the mRNA that is exported to the cytoplasm where protein synthesis takes place. This processing includes the addition of a 5′-5′-linked 7-methyl-guanylate 'cap' at the 5′ end and a sequence of adenylate groups at the 3′ end, the ***poly-A tail***, as well as the removal of any ***introns*** and the splicing together of ***exons***; only 10% of HnRNA leaves the nucleus. Eukaryote mRNAs are comparatively long-lived with $t_{1/2}$ ranging from 30 min to 24 h.

met (1) An ***oncogene***, identified in mouse ***osteosarcoma***. See also Table O1. (2) Abbreviation for ***methionine***.

met repressor–operator complex ***Repressor protein***, 104 residues, product of the *metJ* gene, which regulates methionine biosynthesis in *E. coli*. Dimeric molecules bind to adjacent sites eight base pairs apart on the DNA; sequence recognition is by interaction between antiparallel β-strands of protein and the major groove of the B-form DNA duplex.

met-enkephalin (YGGFM) See ***enkephalins***.

metabolic burst (respiratory burst) Response of phagocytes to particles (particularly after ***opsonization***), and to agonists such as ***formyl peptides*** and ***phorbol esters***; an enhanced uptake of oxygen leads to the production, by an NADH-dependent system, of hydrogen peroxide, superoxide anions and hydroxyl radicals, all of which play a part in bactericidal activity. Defects in the metabolic burst, as in ***chronic granulomatous disease***, predispose to infection particularly with ***catalase***-positive bacteria, and are usually fatal in childhood.

metabolic cooperation Transfer between tissue cells in contact of low molecular weight metabolites such as nucleotides and amino acids. Transfer is via channels constituted by the ***connexons*** of gap junctions, and does not involve exchange with the extracellular medium. First observed in cultures of animal cells in which radiolabelled purines were transferred from wild-type cells to mutants unable to utilize exogenous purines.

metabolic coupling The same as ***metabolic cooperation***.

metabolism Sum of the chemical changes that occur in living organisms.

metabotropic Class of receptor that does not form an integral ion channel, but that acts through a second messenger. Usually used of a class of ***GABA receptors***, cf. ***ionotropic***.

metacentric Descriptive of a chromosome that has its centromere (***kinetochore***) at or near the middle of the chromosome, as opposed to ***acrocentric*** with the centromere near one end.

metachromasia (metachromatic staining) The situation where a stain when applied to cells or tissues gives a colour different from that of the stain solution.

metachronal rhythm See ***metachronism***.

metachronism (metachronal rhythm) Type of synchrony found in the beating of cilia. A metachronal process is one that happens at a later time, and the synchronization is such that the active stroke of an adjacent cilium is slightly delayed so as to minimize the hydrodynamic interference; coordination is by visco-mechanical coupling. Different patterns of metachronal synchronization are recognized: in symplectic m. the wave of activity in the field passes in the same direction as the active stroke of the individual cilium; in antiplectic m. the opposite is true. In dexioplectic and laeoplectic m. the wave of activity in the field is normal to the beat axis. Symplectic and antiplectic m. are considered orthoplectic, the other forms as diaplectic.

metafemales Human females in which there are four X chromosomes in addition to 44 autosomes.

metagon RNA particle found in *Paramecium*, where it behaves as mRNA, and that can behave like a virus if ingested by the protozoan *Didinium*.

metalloenzyme An enzyme that contains a bound metal ion as part of its structure. The metal may be required for enzymic activity, either participating directly in catalysis, or stabilizing the active conformation of the protein.

metalloprotein A protein that contains a bound metal ion as part of its structure.

metalloproteinase (metalloendopeptidase, EC 3.4.24; metallocarboxypeptidase, EC3.4.17) Proteolytic enzymes in which a divalent cation is part of the catalytic mechanism. See ***matrix metalloproteinases***.

metallothioneins Small cysteine-rich metal-binding proteins found in the cytoplasm of many eukaryotes. Synthesis can be induced by heavy metals such as zinc, cadmium, copper and mercury, and metallothioneins probably serve a protective function. Metallothionein gene promoters are used in studies of gene expression.

metamere Unit of ***segmentation*** or metamerism.

metameric *Chem:* having two or more constitutional isomers. *Biol:* having a segmented body form (metamerism).

metamorphosis Change of body form, for example in the development of the adult frog from the tadpole or the butterfly from the caterpillar.

metaphase Classically the second phase of ***mitosis*** or one of the divisions of ***meiosis***. In this phase the ***chromosomes*** are well condensed and aligned along the ***metaphase plate***, making it an ideal time to examine the chromosomes in a cytological preparation (metaphase spread) that flattens the nuclei.

metaphase plate The plane of the ***spindle*** approximately equidistant from the two poles along which the chromosomes are lined up during ***mitosis*** or ***meiosis***. Also termed the equator.

metaplasia Change from one differentiated phenotype to another, for example the change of simple or transitional epithelium to a stratified squamous form as a result of chronic damage.

metastasis Development of secondary tumour(s) at a site remote from the primary; a hallmark of malignant cells.

metastatic spread Process of development of secondary tumours. Involves local invasion (in most cases), passive transport, lodgement and proliferation at a remote site.

metavinculin Splice variant of ***vinculin*** found in smooth and cardiac muscle; has an additional exon 19 that encodes 68 amino acids. In cardiac muscle connects microfilaments to the intercalated disc.

methaemoglobin An oxidized form of haemoglobin containing ferric iron that is produced by the action of oxidizing poisons. Non-functional.

Methanobacteria A genus of strictly anaerobic bacteria that reduce CO_2 using molecular hydrogen, H_2, to give methane. They show a number of features that distinguish them from other bacteria, and are now classified as a separate group, the ***Archaebacteria***. Methanobacteria are found in the anaerobic sediment at the bottom of ponds and marshes (hence marsh gas is the common name for methane) and as part of the microflora of the rumen in cattle and other herbivorous mammals.

methanochondrion A structure of involuted plasma membrane found in many methanogenic bacteria and thought to be an organelle of methane formation.

methicillin Penicillinase-resistant penicillin antibiotic.

methionine (met; M) (149 D) Contains the $-SCH_3$ group that can act as a methyl donor (see ***S-adenosyl methionine***). Common in proteins but at low frequency. The met-x linkage is subject to specific cleavage by cyanogen bromide. See also ***formyl peptides***, and Table A2.

methionine puddle A term used to describe a region of a protein surface composed of a cluster of methionine side chains. Proposed as the active hydrophobic site of calmodulin and also of signal recognition particle (SRP). The concept is of a highly fluid hydrophobic patch.

methisazone (N-methylisatin; β-thiosemicarbazone) Drug that specifically blocks the translation of late viral mRNA in poxvirus infection, and was used prophylactically for smallpox.

methotrexate Analogue of dihydrofolate. Inhibits dihydrofolate reductase and kills rapidly growing cells. Therapeutic agent for leukaemias, but has a low therapeutic ratio.

methyl- ($-CH_3$) Specific reference to the methyl group is made when macromolecules are modified after synthesis by enzymic addition of methyl groups. The group is transferred to nucleic acids and proteins. See also ***methyl transferase*** and ***DNA methylation***.

methyl-accepting chemotaxis proteins See ***MCPs***.

methylcholanthrene (3-methylcholanthrene) Carcinogenic polycyclic hydrocarbon. One of many such substances formed during incomplete combustion of organic material.

methyldopa An antihypertensive drug, preferred in pregnant patients.

Methylene Blue (Swiss blue; Basic Blue 9; tetramethythionine chloride) Water-soluble dye that can be reduced to a colourless form and can be oxidized by atmospheric oxygen. Used as a stain in bacteriology and histology.

methylotroph Yeasts, like *Hansenula polymorpha*, that utilize methanol as an energy source.

methyltransferase Enzyme that transfers a methyl group from S-adenosyl methionine to a substrate. Most commonly encountered in bacterial chemotaxis where the methyl-accepting chemotaxis proteins (***MCPs***) become methylated in the course of adaptation.

methylxanthines Naturally occurring purine alkaloids such as theobromine, theophylline and caffeine (trimethyl-xanthine). They inhibit cAMP phosphodiesterase and thus cause an increase in the intracellular cAMP concentration.

metJ Gene for the repressor of methionine biosynthetic operon. Protein product (12 kD) binds to DNA.

metorphamide Amidated opioid octapeptide from bovine brain. Derived by proteolytic cleavage from proenkephalin.

metrizoate (3-acetimido-5-(N-methyl-acetamido)-triiodobenzoate) Sodium salt of metrizoate is used to produce solutions with high densities suitable for cell density gradient separations.

metronidazole Antiprotozoal and antibacterial drug.

mevalonic acid Key intermediate in polyprenyl biosynthesis and thus cholesterol synthesis. Derived from hydroxymethylglutaryl-CoA (HMG-CoA) – a reaction inhibited by ***mevastatin***.

mevastatin Inhibitor of 3-hydroxy-3-methylglutatyl (HMG)-CoA reductase, and thus of ***mevalonic acid*** production. Induces ***apoptosis***.

mevinolin Intermediate in terpene synthesis; an analogue of compactin, a fungal metabolite that is used to lower plasma ***low density lipoprotein*** levels. It acts as an inhibitor of HMG-CoA-reductase, the rate-controlling enzyme in cholesterol biosynthesis.

MGSA See ***melanoma growth-stimulatory activity***.

MHC See ***major histocompatibility complex***.

MHC restriction Restriction on interaction between cells of the immune system because of the requirement to recognize foreign antigen in association with MHC antigens (major histocompatibility antigens). Thus, cytotoxic T-cells will only kill virally infected cells that have the same Class I antigens as themselves, whereas helper T-cells respond to foreign antigen associated with Class II antigens.

micelle One of the possible ways in which amphipathic molecules may be arranged; a spherical structure in which all the hydrophobic portions of the molecules are inwardly directed, leaving the hydrophilic portions in contact with the surrounding aqueous phase. The converse arrangement will be found if the major phase is hydrophobic.

Michaelis constant See ***K_m***.

Michaelis–Menten equation Equation derived from a simple kinetic model of enzyme action that successfully accounts for the hyperbolic (adsorption–isotherm) relationship between substrate concentration S and reaction rate V. $V = V_{max} \times S/(S + K_m)$, where K_m is the ***Michaelis constant*** and V_{max} is maximum rate approached by very high substrate concentrations.

microangiopathic haemolytic anaemia (USA, microangiopathic hemolytic anemia) Consequence of ***disseminated intravascular coagulation*** (DIC): fragments of red blood cells, damaged by being forced through a fibrin meshwork, are found in the circulation.

microbodies See ***peroxisome***.

microcarrier Microcarriers are small solid, or in some cases immiscible liquid, spheres, on which cells may be grown in suspension culture. They provide a means of obtaining large yields of cells in small volumes. The cells must exhibit anchorage dependence of growth and the dimensions of the carrier bead may be important in controlling growth rate. The term is imprecise and has other potential meanings.

microcentrum Obsolete name for the pericentriolar region.

microcinematography The making of films using a microscope and cine camera.

Micrococcus Genus of ***Gram-positive*** aerobic bacteria, cells around 1–2μm in diameter. *M. lysodeikticus* (now *M. luteus*) was commonly used as the source of bacterial cell wall suspension on which lysozyme activity was measured by a decrease in turbidity.

microcolliculi Broad swellings (0.5μm) on the dorsal surface of a moving epidermal cell in culture, that move rearward as the cell moves forward (as do ruffles on fibroblasts).

microcytes Abnormally small red blood cells, found in some types of anaemia.

microdialysis ***Dialysis*** on a small scale, giving microlitre range samples. Used for example in studies of ***in vivo*** release of transmitters in brain tissue.

microelectrode An electrode, with tip dimensions small enough (less than

1μm) to allow non-destructive puncturing of the plasma membrane. This allows the intracellular recording of ***resting*** and ***action potentials***, the measurement of intracellular ion and pH levels (using ***ion-selective electrodes***), or ***microinjection***. Microelectrodes are generally pulled from glass capillaries, and filled with conducting solutions of potassium chloride or potassium acetate to maximize conductivity near the tip. Electrical contact, if required, is usually made with a silver chloride-coated silver wire.

microfibril Basic structural unit of the plant ***cell wall***, made of ***cellulose*** in higher plants and most algae, ***chitin*** in some fungi, and ***mannan*** or ***xylan*** in a few algae. Higher plant microfibrils are about 10nm in diameter, and extremely long in relation to their width. The cellulose molecules are oriented parallel to the long axis of the microfibril in a paracrystalline array, which provides great tensile strength. The microfibrils are held in place by the wall matrix, and their orientation is closely controlled by the protoplast.

microfilament Cytoplasmic filament, 5–7nm thick, of ***F-actin*** that can be decorated with ***HMM***; may be laterally associated with other proteins (tropomyosin, α-actinin) in some cases, and may be anchored to the membrane. Microfilaments are conspicuous in ***adherens junctions***.

microglial cell Small glial cells of mesodermal origin, with scanty cytoplasm and small spiny processes. Distributed throughout grey and white matter. Derive from monocytes and invade neural tissue just before birth; capable of enlarging to become macrophages.

microglobulin Any small globular plasma protein. See ***beta-2-microglobulin***.

microinjection The insertion of a substance into a cell through a ***microelectrode***. Typical applications include the injection of drugs, histochemical markers (such as ***horseradish peroxidase*** or lucifer yellow) and RNA or DNA in molecular biological studies. To extrude the substances through the very fine electrode tips, either hydrostatic pressure (pressure injection) or electric currents (ionophoresis) is employed.

micronucleus The smaller nucleus in ciliate protozoans, fully active in inheritance and passed after meiosis to conjugating pairs. Gives rise to the macronucleus or macronuclei. Genes in the micronucleus are not actively transcribed.

microperoxidase Part of a cytochrome c molecule that retains haem group and has ***peroxidase*** activity.

microperoxisome Small ***peroxisomes*** of 150–250nm diameter found in most cells.

micropinocytosis Pinocytosis of small vesicles (around 100nm in diameter). Not blocked by cytochalasins.

micropore filters Filters made of a meshwork of cellulose acetate or nitrate and with defined pore size. They can be autoclaved, and the smaller pore sizes (0.22μm, 0.45μm) are used for sterilizing heat-labile materials by filtering out microorganisms. Larger pore size filters are used in setting up ***Boyden chambers***. They are about 150μm thick and should be distinguished from ***Nucleopore filters***. Millipore is a trade name for micropore filters.

microprobe See ***electron microprobe***.

micropyle (1) Small hole or aperture in the protective tissue surrounding a plant ovule, through which the pollen tube enters at fertilization. Develops into a small hole in the seed coat through which, in many cases water enters at germination. (2) Perforation in the shell (chorion) of an insect's egg through which the sperm enters at fertilization.

microsatellites Short sequences of di- or trinucleotide repeats of very variable length distributed widely throughout the genome. Using ***PCR*** primers to the unique sequences upstream and downstream of a microsatellite their location and polymorphism can be determined and the technique is extensively used in investigating genetic associations with disease.

microsequencing Sequencing of very small amounts of protein – often a prelude to producing an oligonucleotide probe, screening a cDNA library and cloning.

microsomal fraction See ***microsomes***.

microsomes Heterogeneous set of vesicles 20–200nm in diameter formed from the endoplasmic reticulum when cells are disrupted.

microspikes Projections from the leading edge of some cells, particularly, but not exclusively, nerve ***growth cones***. They are usually about 100nm diameter, 5–10μm long, and are supported by loosely bundled microfilaments. They are referred to by some authors as filopodia. Functionally a sort of linear version of a ruffle on a leading lamella.

microspore A haploid spore produced by a plant sporophyte that develops into a male gametophyte. In seed plants, it corresponds to the developing pollen grain at the uninucleate stage. The smaller of the spores of a heterosporous species.

microtome A device used for cutting sections from an embedded specimen, either for light or electron microscopy.

microtrabecular network Complex network arrangement seen using the high-voltage electron microscope to look at the cytoplasm of cells prepared by very rapid freezing. The suggestion was that most cytoplasmic proteins are in fact loosely associated with one another in this fibrillar network and are separate from the aqueous phase which contains only small molecules in true solution. If it exists, then it must certainly be very labile in cells where there is cytoplasmic flow and rapid organelle movement; now considered artefactual by most microscopists.

microtubule Cytoplasmic tubule, 25nm outside diameter with a 5nm thick wall. Made of tubulin heterodimers packed in a 3-start helix (or of 13 protofilaments looked at another way), and associated with various other proteins (***MAPs***, ***dynein***, ***kinesin***). Microtubules of the ciliary ***axoneme*** are more permanent than cytoplasmic and spindle microtubules.

microtubule organizing centres (MTOC) Rather amorphous region of cytoplasm from which microtubules radiate. The pattern and number of microtubules is determined by the MTOC. The ***pericentriolar region*** is the major MTOC in animal cells; the basal body of a cilium is another example. Activity of MTOCs can be regulated, but the mechanism is unclear.

microtubule-associated proteins See ***MAPs***.

Microtus Genus of voles and meadow mice.

microvillus (plural, microvilli) Projection from the apical surface of an epithelial cell that is supported by a central core of microfilaments associated with bundling proteins such as ***villin*** and ***fimbrin***. In the intestinal ***brush border*** the microvilli presumably increase absorptive surface area, whereas the stereovilli (***stereocilia***) of the cochlea have a distinct mechanical role in sensory transduction.

Microviridae A diverse group of ssDNA bacteriophages, also known as ϕ X phage group or isometric ssDNA phages.

midbody Dense structure formed during ***cytokinesis*** at the cleavage furrow. It consists of remnants of ***spindle fibres*** and other amorphous material and disappears before cell division is completed.

middle lamella First part of the plant ***cell wall*** to be formed, laid down in the ***phragmoplast*** during cell division as the ***cell plate***. Subsequently makes up the central part of the double cell wall that separates two adjacent cells, cementing together the two primary walls. Rich in ***pectin*** and relatively poor in ***cellulose***.

midiprep Slang, denoting a medium-scale purification of ***plasmid*** from a bacterial culture. Usually used to describe preparations from 10 to 100ml culture. See also ***miniprep***, ***maxiprep***, ***megaprep***.

midkine Heparin-binding growth factor (13 kD) of the TGF-β; superfamily; has 50% sequence identity with heparin-binding growth-associated molecule (HB-GAM). Structurally unrelated to fibroblast growth factor (FGF). Midkine was originally described as associated with tooth morphogenesis induced by epithelial–mesenchyme interactions. ***Nucleolin*** binds midkine.

MIF See ***macrophage inhibition factor***, or ***migration inhibitory factor***.

MIF-related protein 8 See ***cystic fibrosis antigen***.

Mig (m119) Mouse protein induced by IFN-γ (interferon γ). See also ***cytokines***.

migration inhibitory factor Factor that inhibits macrophage movement. Originally defined on basis of inhibition of emigration of mononuclear cells from capillary (haematocrit) tubes; more recently a 13 kD protein with migration inhibitory activity has been isolated.

mil *(raf)* An ***oncogene***, identified in bird and mouse ***sarcomas***, encoding a serine/threonine ***protein kinase***. See Table O1.

Millipore filter Trade name for a well-known brand of ***micropore filters***.

Mimosa pudica The 'sensitive plant' whose leaflets fold inwards very rapidly when touched. A more vigorous stimulus causes the whole leaf to droop, and the stimulus can be transmitted to neighbouring leaves.

mimotope Compound that mimics the structure of a conformational ***epitope*** and which will elicit an identical antibody response (whereas a mimetic would not have the same antigenicity). Mostly used of peptides from ***phage display libraries***; potentially useful as vaccines.

mineralocorticoid Natural or synthetic corticosteroid that acts on water and electrolyte balance by promoting retention of sodium ions and excretion of potassium ions in the kidney. Aldosterone is the most potent natural example and is produced in the outer layer of the adrenal cortex.

miniature endplate potential (MEPP) Small fluctuations (typically 0.5 mV) in the resting potential of postsynaptic cells. They are the same shape as, but much smaller than, the [endplate potentials] caused by stimulation of the presynaptic cell. MEPPs are considered as evidence for the quantal release of ***neurotransmitters*** at ***chemical synapses***, a single MEPP resulting from the release of the contents of a single synaptic vesicle.

minicell Spherical fragment of a bacterium produced by abnormal fission and not containing a bacterial chromosome.

minichromosome (1) Certain viruses complex with the histones of the host eukaryote cells they have infected to form a chromatin structure resembling a small chromosome. (2) A plasmid that contains a chromosomal ***origin of replication***.

minimal medium (minimal essential medium) The simplest tissue culture medium that will support the proliferation of normal cells.

minimyosin Form of ***myosin*** isolated from *Acanthamoeba*; only 180 kD, but capable of binding to actin.

miniprep Slang, denoting a small-scale purification of ***plasmid*** from a bacterial culture. Usually used to describe preparations from 1 to 10ml culture. See also ***midiprep***, ***maxiprep***, ***megaprep***.

minisatellite (variable number tandem repeat; VNTR) Class of highly repetitive ***satellite DNA***, comprising variable (typically 10–20) repeats of short (eg. 64 bases) DNA sequences. The high level of polymorphism of such minisatellites make them very useful in ***physical mapping***.

minisegregant cells Human cells with small amounts of DNA and few chromosomes; obtained experimentally by perturbing cell division. Can readily be fused with whole cells.

minK Widely expressed protein (15 kD) that forms ***potassium channels*** by aggregation with other membrane proteins. A variety of channels have been shown to have minK associated with them and minK seems to be important in regulating structure and activity of the channel.

minute **mutant** A class of recessive lethal mutants of ***Drosophila***. The heterozygotes grow more slowly, are smaller and less fertile than the wild type flies. There are about 40 loci that produce *minute* mutants.

MIP See ***major intrinsic protein*** or ***macrophage inflammatory protein***. Macrophage inflammatory protein is known to have various subclasses, MIP-1α, MIP-1β, MIP-2. See Table C4a.

miranda *Drosophila* gene, the product of which (830 residues) colocalizes with ***Prospero*** in mitotic neuroblasts and apparently directs Prospero exclusively to the ganglion mother cell (GMC) during the asymmetric division that gives rise to another neuroblast and the GMC. Anchors Prospero selectively to the basal side of the cell cortex during mitosis and releases it after cytokinesis. Miranda has two ***leucine zipper*** motifs and eight consensus sites for PKC phosphorylation.

mismatch repair A DNA repair system that detects and replaces wrongly paired, mismatched, bases in newly replicated DNA. *E. coli* has a mismatch correction enzyme coded for by three genes *mutH, mutL* and *mutS*, that is directed to the newly synthesized strand and removes a segment of that strand including the incorrect nucleotide. The gap is then filled by DNA polymerase.

missense mutation A mutation that alters a ***codon*** for a particular amino acid to one specifying a different amino acid.

mitochondrial diseases Illnesses, frequently neurological, which can be ascribed to defects in mitochondrial function. If the defect is in the mitochondrial rather than the nuclear genome unusual patterns of inheritance can be observed.

mitochondrion Highly pleiomorphic organelle of eukaryotic cells that varies from short rod-like structures present in high number to long branched structures. Contains DNA and ***mitoribosomes***. Has a double membrane and the inner membrane may contain numerous folds (cristae). The inner fluid phase has most of the enzymes of the ***tricarboxylic acid cycle*** and some of the urea cycle. The inner membrane contains the components of the ***electron transport chain***. Major function is to regenerate ATP by oxidative phosphorylation (see ***chemiosmotic hypothesis***).

mitogenesis The process of stimulating transit through the cell cycle especially as applied to lymphocytes. Concanavalin A is a mitogen for T-lymphocytes; the best mitogen for B-lymphocytes is Cowan strain *Staphylococcus aureus*.

mitogenic Causing re-entry of cells into the cell cycle, not just into mitosis.

mitogillin Toxin (149 residues) produced by *Aspergillus restrictus*. One of the aspergillins; see ***alpha-sarcin***.

mitomycin C Aziridine antibiotic isolated from *Streptomyces caespitosus*. Inhibits DNA synthesis by crosslinking the strands and is used as an anti-neoplastic agent. Most active in late G1 and early S phase. Mitomycin-treated cells are sometimes used as feeder layers.

mitoplasts Isolated mitochondria without their outer membranes. They have finger-like processes, and retain the capacity for oxidative phosphorylation.

mitoribosomes Mitochondrial ribosomes; these more closely resemble prokaryotic ribosomes than cytoplasmic ribosomes of the cells in which they are found, though they are even smaller and have fewer proteins than bacterial ribosomes.

mitosis The usual process of nuclear division in the somatic cells of ***eukaryotes***. Mitosis is classically divided into four stages. The chromosomes are actually replicated prior to mitosis during the ***S phase*** of the ***cell cycle***. During the first stage, prophase, the chromosomes condense and become visible as double strands (each strand being termed a chromatid) and the ***nuclear envelope*** breaks down. At the same time the mitotic ***spindle*** forms by the polymerization of ***microtubules*** and the chromosomes are attached to spindle fibres at their kinetochores. In metaphase the chromosomes align in a central plane perpendicular to the long axis of the spindle. This is termed the metaphase plate. During anaphase the paired chromatids are apparently pulled to opposite poles of the spindle by means of the spindle fibre microtubules attached to the kinetochore, though the actual mechanism for this movement is still controversial. This separation of chromatids is completed during telophase, when they can be regarded as chromosomes proper. The chromosomes now lengthen and become diffuse and new nuclear envelopes form round the two sets of chromosomes. This is usually followed by cell division or cytokinesis in which the cytoplasm is also divided to give two daughter cells. Mitosis ensures that each daughter cell has a ***diploid*** set of chromosomes that is identical to that of the parent cell.

mitotic apparatus See ***spindle***.

mitotic death Cells fatally damaged by ionizing radiation may not die until the next ***mitosis***, at which point the radiation damage to the DNA becomes evident, particularly when there is fragmentation of chromosomes.

mitotic index The fraction of cells in a sample that are in mitosis. It is a measure of the relative length of the mitotic phase of the ***cell cycle***.

mitotic recombination Somatic crossing over. ***Crossing over*** can occur between ***homologous chromosomes*** during mitosis, but is very rare because the chromosomes do not normally pair. When it occurs it can lead to new combinations of previously linked genes. Although infrequent, mitotic recombination has been utilized for genetic analysis in ***Aspergillus*** and in studies on developmental ***compartments*** in *Drosophila* where the frequency of mitotic recombination can be increased by X-irradiation.

mitotic segregation See ***mitotic recombination***.

mitotic shake-off method A method of collecting cells in mitosis, so that the chromosomes can be examined and the karyotype determined. Many cultured cells round up during mitosis and so become less firmly attached to the culture substratum. Cells in mitosis thus can be removed into suspension by gentle shaking of the culture vessel, leaving the non-mitotic cells still attached. The number of cells that are in mitosis is usually increased by using a drug, such as ***colcemid*** which blocks mitosis at ***metaphase***.

mitotic spindle See ***spindle*** and ***mitosis***.

mix Homeobox genes (*mix1, mix2*), expressed in prospective mesoderm and endoderm after mid-blastula stage, that respond to TGF-β (***transforming growth factor***-β) superfamily signals including ***activin***, TGF-β, Vg-1 and BMP-4, but not to non-TGF inducers.

mixed lymphocyte reaction (MLR) Reaction of mitogenesis produced in T-lymphocytes when allogeneic (ie. mixed) lymphocytes are brought together, provided they are mismatched in histocompatibility loci. Once used as a test for possible graft compatibility in human grafting, it is now known that a negative reaction is a poor predictor of graft acceptance.

MLCK See ***myosin light chain kinase***.

MLEE (multi-locus enzyme electrophoresis) Form of two-dimensional ***electrophoresis*** used to distinguish ***polymorphisms*** between strains or populations.

MMTV Mouse ***mammary tumour virus***.

MN blood group antigens A pair of blood group antigens governed by genes that segregate independently of the ABO locus. The alleles are co-dominant and there are three types: MM, NN, and MN. ***Glycophorin*** has M or N activity and this is associated with oligosaccharides attached to the amino-terminal portion of the molecule. M-type glycophorin differs from N-type in amino acid residues 1 and 5, although the antigenic determinants are associated with the carbohydrate side chains.

Mnt Protein that interacts with ***Max*** and functions as a transcriptional repressor. Not a member of the ***Myc*** or ***Mad***

families. Binds to same site on Max as does *Sin3*.

Mo-1 $\alpha_M\beta_2$ ***integrin*** of leucocytes (CD11b/CD18).

mobile genetic elements See ***transposons***.

mobile ion carrier See ***ionophore***.

modification enzyme Enzyme that introduces minor bases into DNA or RNA or that alters bases already incorporated. Serves to alter the sequence so that ***restriction enzymes*** will not damage the strand.

modulation See ***neuromodulation***.

moesin (membrane-organizing extension spike protein) Isolated from placenta, a member of the ***ezrin***, band 4.1, ***talin*** family of cytoskeleton–membrane link proteins.

molecular clock This term has two separate uses. In one sense it means the rate of fixation of mutations in DNA and thus times the rate of genetic diversification. In the second sense it means a biological system capable of maintaining a timing rhythm or pulse. All such clocks are thought to be entrained by a natural oscillator such as the diurnal rhythm.

molluscan catch muscle Muscle responsible for holding closed the two halves of the shell of bivalves. Specialized to maintain tension with low expenditure of ATP. Rich in ***paramyosin***.

molluscum bodies Intracellular inclusions of poxviruses found in cells of human epidermis; harmless, but contagious, skin lesions (molluscum contagiosum).

Moloney murine leukaemia virus Replication-competent retrovirus (Oncovirinae) that causes leukaemia in mice, isolated by Moloney from cell-free extracts made from a transplantable mouse sarcoma.

Moloney murine sarcoma virus Replication-defective retrovirus, source of the ***oncogene*** *v-mos*, responsible for inducing fibrosarcomas ***in vivo***, and transforming cells in culture.

Moloney test Skin test for immunity to diphtheria in which active toxin is injected into one site and toxoid into another. This is to control for pseudopositive reactions to the toxin.

MOLT-4 cells Suspension culture derived from human male with acute lymphoblastic leukaemia. A stable T-cell leukaemia line.

MOM (mitochondrial outer membrane) Used particularly for mitochondrial outer membrane proteins, in conjunction with the molecular mass in kiloDaltons, eg. MOM19 and MOM72, proteins of 19 and 72 kD respectively.

monellin Basic non-glycosylated heterodimeric protein (44 and 50 residue protomers) that has intensely sweet taste.

monensin A sodium ***ionophore*** (671D) from *Streptomyces cinnamonensis*. Has antibiotic properties, and is used as a feed additive in chickens. Also used in ***ion-selective electrodes***.

Monera Kingdom that contains all prokaryotic organisms (bacteria and cyanobacteria) in the Five Kingdom scheme.

mongolism See ***Down's syndrome***.

monoamine neurotransmitters See ***biogenic amines***.

monoamine oxidase (MAO) Enzyme catalysing breakdown of several ***biogenic amines***, such as serotonin, adrenaline, noradrenaline, dopamine.

monocentric chromosome ***Chromosome*** with a single ***centromere***, ie. most chromosomes.

monocistronic RNA A ***messenger RNA*** that gives a single polypeptide chain when ***translated***. All ***eukaryote*** mRNAs are monocistronic, but some bacterial mRNAs are polycistronic especially those transcribed from ***operons***.

monoclonal Used of a cell line whether within the body or in culture to indicate that it has a single clonal origin. Monoclonal antibodies are produced by a single clone of ***hybridoma*** cells, and are therefore a single species of antibody molecule.

monocotyledonous (monocot) Plants in which the developing plant has only one ***cotyledon***. Grasses are perhaps the commonest examples of the Class (which also contains palms, lilies and orchids).

monocyte Mononuclear phagocyte circulating in blood that will later emigrate into tissue and differentiate into a macrophage.

monocyte chemoattractant protein 1 See ***monocyte chemotactic and activating factor***.

monocyte chemotactic and activating factor (MCAF; JE; monocyte chemoattractant protein 1; MCP-1) ***Cytokine*** of the C-C subfamily, coinduced with ***interleukin-8*** on stimulation of endothelial cells, fibroblasts or monocytes, that activates and is ***chemotactic*** for ***monocytes***. A ***chemokine***.

monokines Soluble factors, derived from macrophages, that act on other cells (eg. ***interleukin-1***). Term becoming unusual – all monokines are ***cytokines*** and that term is more commonly used.

monolayer A single layer of any molecule, but most commonly applied to polar lipids. Can be formed at an air–water interface in experimental systems. The term should not be used to describe one layer of a lipid bilayer, for which the term 'leaflet' is generally used. See also ***monolayering of cells***.

monolayering of cells Tendency of animal tissue cells growing on solid surfaces to cover the surface with a complete layer only one cell thick, before growing on top of each other. This non-random distribution is generated by contact inhibition of locomotion, a phenomenon in which colliding cells change direction rather than move over one another. Of the theories why some (but by no means all) types of cells stop growing when a monolayer is formed, present evidence favours limitation by supply of growth factors from the medium, rather than any inhibitory effect of contact on growth.

mononuclear phagocytes Monocytes and their differentiated products, macrophages. 'Mononuclear cells' are leucocytes other than polymorphonuclear cells and include lymphocytes.

monopodial Adjective describing an ***amoeba*** that has only one ***pseudopod*** (as opposed to polypodial forms).

monosaccharide A simple sugar that cannot be hydrolyzed to smaller units. Empirical formula is $(CH_2O)_n$ and range in size from trioses (n = 3) to heptoses (n = 7).

monosome (1) A single ribosome attached to a strand of mRNA. (2) A ribosome that has dissociated from a polysome. (3) Chromosome in an aneuploid set that does not have a homologue.

monosomy Situation in a normally ***diploid*** cell or organism in which one or more of the ***homologous chromosome*** pairs is represented by only one chromosome of the pair. For example, sex determination in grasshoppers depends on the fact that females are XX and males XO, ie. males have only one sex chromosome and are monosomic for the X chromosome.

monozygotic Twins arising as a result of the cleavage and separate development of a single early embryo derived from a single fertilized ovum. Such siblings are thus genetically identical, in contrast to ***dizygotic*** twins.

MOPS (morpholino-propane sulphonic acid) A 'biological' buffer; a synthetic zwitterionic compound with a pKa of 7.2, that is non-toxic and has a low temperature coefficient. Widely used in biochemical studies, largely as a replacement for phosphate buffers.

Morbilli virus Genus of viruses (of the ***Paramyxoviridae***). Type species is ***measles virus***; other species include canine distemper virus and the related seal virus.

morphallaxis Regenerative process in which part of an organism is transformed directly into a new organism without replication at the cut surface.

morphine An opioid alkaloid, isolated from opium, with a complex ring structure. It is a powerful analgesic with important medical uses, but is highly addictive. Functions by occupying the receptor sites for the natural neurotransmitter peptides, endorphins and enkephalins, but is stable to the peptidases that inactivate these compounds.

morphogen Diffusible substance that carries information relating, for example, to position in the embryo, and thus determines the differentiation that cells perceiving this information will undergo.

morphogenesis The process of 'shape formation': the processes that are responsible for producing the complex shapes of adults from the simple ball of cells that derives from division of the fertilized egg.

morphogenetic movements Movements of cells or of groups of cells in the course of development. Thus the invagination of cells in gastrulation is one of the most dramatic of morphogenetic movements; another much-studied example is the migration of neural crest cells.

morphometry Method that involves measurement of shape. A variety of methods exist to enable one to examine, for example, the distribution of objects in a two-dimensional section of a cell and then to use this to predict the shapes and the distribution of these objects in three dimensions.

mortalin (75kD glucose regulated protein; GRP75; MOT-1; MOT-2) Member of the HSP70 family; functions as a mortality marker in fibroblasts. Mortal fibroblasts have MOT-1 uniformly distributed in the cytoplasm, immortalised fibroblasts have MOT-2 in a juxtanuclear concentration.

morula Stage of development in holoblastic embryos. The morula stage is usually likened to a spherical raspberry, a cluster of blastomeres without a cavity.

mos An ***oncogene*** identified in mouse ***sarcoma***, encoding a serine/threonine ***protein kinase***. See Table O1.

mosaic egg At one time a distinction was drawn between those organisms in which the egg seemed to have a firmly committed fate map built in and 'regulating' embryos. In the former, after the first cleavage one blastomere was committed to produce one set of tissues, the other blastomere a different set, and removal of one blastomere led to the production of an incomplete embryo. This was particularly obvious in mollusc development where one blastomere had the polar lobe material. This early differentiation (or ***determination***) of blastomeres for particular fates was in distinction to 'regulating' embryos in which the removal of one blastomere did not matter, the other blastomere(s) compensating and producing a full set of tissues. The distinction is, however, only based upon the timing of differentiative events, and within a few divisions the regulating embryo also becomes a mosaic of determined cells.

motif A small structural element that is recognizable in several proteins, eg. ***alpha-helix***.

motilin Peptide (22 residues) found in duodenum, pituitary and pineal that stimulates intestinal motility.

motogen Term proposed for substances that stimulate cell motility – by analogy with those that stimulate cell division (***mitogens***). ***Scatter factor*** (hepatocyte growth factor) is an example, though it seems likely that factors may be motogens for some cells and mitogens for others and may be motogens, mitogens or both depending upon the local conditions in which the cell is operating.

motoneuron A ***neuron*** that connects functionally to a ***muscle fibre***.

motor endplate Synonym for ***neuromuscular junction***.

motor neuron Synonym for ***motoneuron***.

motor neuron disease (amyotrophic lateral sclerosis) Degenerative disease of unknown cause that affects predominantly motor neurons of spinal cord, cranial nerve nuclei and motor cortex. There is speculation that deficiency in ***ciliary neurotrophic factor*** may be involved.

motor protein Proteins that bind ATP and are able to move on a suitable substrate with concomitant ATP hydrolysis. Most eukaryotic motor proteins move by binding to a specific site on either actin filaments (myosin) or on microtubules (dynein, kinesin). They are normally elongated molecules with two active binding sites although some kinesin analogues have a single site. The distal end of the molecule normally binds adaptor proteins that enable them to make stable interactions with membranous vesicles or with filamentous structures, which then constitute the 'cargo' to be moved along the substrate filament.

Mott cells Plasma cells containing large eosinophilic inclusions; found in the brain in cases of African ***trypanosomiasis***.

Mounier–Kuhn syndrome Tracheobronchomegaly, a rare disorder of the lower respiratory tract. May be a connective tissue disease.

moxonidine See ***clonidine***.

MPF See ***cyclin***.

MRC-5 cells Cell line established from normal human male foetal lung tissue. Will double 50 to 60 times before showing senescence. Often used as 'normal' cells.

MRF-4 (myf-6) Member of the ***MyoD*** family of muscle regulatory proteins.

mRNA See ***messenger RNA***.

MRP See ***calgranulins***.

MRSA (multidrug-resistant *Staphylococcus aureus)* An increasing problem, particularly in hospitals, suggesting that bacteria are beginning to win the 'arms race' against antibiotics.

msh Multigene family coding for proteins involved in mismatch repair. Homologous to *S. cerevisiae MutS*. Included in family are MSH1, MSH2, hMSH2, hMLH1, hPMS1, hPMS2 and probably ***GTBP***.

MSS4 Ubiquitous protein that binds to a subgroup of ***rab*** proteins.

MTOC See ***microtubule organizing centre***.

MUC-1 See ***episialin***.

mucilage Sticky mixture of carbohydrates in plants.

mucocyst Small membrane-bounded vesicular organelle in ***pellicle*** of ciliate protozoans that will discharge a mucus-like secretion.

mucopeptide Synonym for ***peptidoglycan***.

mucopolysaccharide The polysaccharide components of proteoglycans, now more usually known as ***glycosaminoglycans***.

mucopolysaccharidosis (plural, mucopolysaccharidoses) Inherited disease in humans resulting from inability to break down ***glycosaminoglycans***. ***Hunter syndrome*** and ***Hurler's disease***, for example, result from defects in lysosomal enzymes needed to break down sulphated ***mucopolysaccharides***.

mucous gland A type of ***merocrine*** gland that produces a thick (mucopolysaccharide-rich) secretion (as opposed to a ***serous gland***).

mucus Viscous solution secreted by various membranes; rich in glycoprotein.

Muller cell Supporting cell of the neural retina. Cell body and nucleus lie in the middle of the inner nuclear region, their bases form the internal and external limiting membranes.

multi-locus enzyme electrophoresis See ***MLEE***.

multicopy inhibition Inhibition of translation of the transcript of a transposase gene by a multicopy ***plasmid*** with suitable inhibitory gene. The plasmid inhibits transposition events in the host bacterium.

multidrug transporter (Mdr; P-glycoprotein) Closely related family (***ABC proteins***) of integral membrane glycoproteins that export a variety of solutes from the cytoplasm.

multienzyme complex Cluster of distinct enzymes catalysing consecutive reactions of a metabolic pathway, that remain physically associated through purification procedures. Multifunctional enzymes, found in eukaryotes, are a somewhat different phenomenon, since the several enzymic activities are associated with different domains of a single polypeptide.

multigene family See ***gene family***.

multipain ATP-dependent protease (500 kD) isolated from the cytoplasm of skeletal muscle. May form a complex with the 20S ***proteasome*** to form a 26S-like particle.

multiple cloning site See ***polycloning site***.

multiple myeloma See ***myeloma cell***.

multiple sclerosis Neurodegenerative disease characterized by the gradual accumulation of focal plaques of demyelination particularly in the periventricular areas of the brain. Peripheral nerves are not affected. Onset usually in the third or fourth decade with intermittent progression over an extended period. Cause still uncertain.

multipotent cell Progenitor or precursor cell that can give rise to diverse cell types in response to appropriate environmental cues.

multipotential colony-stimulating factor See ***interleukin-3***.

multivesicular body Secondary ***lysosome*** containing many vesicles of around 50nm diameter.

muramic acid (3-O-α-carboxyethyl-D-glucosamine) Subunit of peptidoglycan of bacterial cell walls.

muramidase See ***lysozyme***.

muramyl dipeptide Fragment of ***peptidoglycan*** from cell wall of mycobacteria that is used as an ***adjuvant***.

murein Crosslinked ***peptidoglycan*** complex from the inner cell wall of all ***Eubacteria***. Constitutes 50% of the cell wall in ***Gram-negative*** and 10% in ***Gram-positive*** organisms, and comprises β (1-4)-linked N-acetyl-glucosamine and N-acetyl-muramic acid extensively crosslinked by peptides.

murine Pertaining to mice.

murine leukaemia virus A group of type C ***Retroviridae*** infecting mice and causing in some strains lymphatic ***leukaemia*** after a long latent period. Nearly all are replication competent and *v-onc* negative. See also ***Abelson leukaemia virus***.

Mus musculus House mouse.

Musca domestica House fly.

muscarine Toxin (alkaloid) from the mushroom *Amanita muscaria* (fly agaric) that binds to (muscarinic) acetylcholine receptors.

muscarinic acetylcholine receptor Distinct from the ***nicotinic acetylcholine receptor*** in having no intrinsic ion channel; a seven membrane-spanning ***G-protein coupled receptor***.

muscle Tissue specialized for contraction. See also ***twitch muscle, catch muscle***.

muscle cell Cell of muscle tissue; in striated (skeletal) muscle it comprises a ***syncytium*** formed by the fusion of embryonic ***myoblasts***, in cardiac muscle a cell linked to the others by specialized junctional complexes (***intercalated discs***), in smooth muscle a single cell with large amounts of actin and myosin capable of contracting to a small fraction of its resting length.

muscle fibre Component of a skeletal muscle comprising a single syncytial cell that contains ***myofibrils***.

muscle spindle A specialized muscle fibre

found in tetrapod vertebrates. A bundle of muscle fibres is innervated by sensory neurons. Stretching the muscle causes the neurons to fire; the muscle spindle thus functions as a stretch receptor.

muscular dystrophy (MD) A group of diseases characterized by progressive degeneration and/or loss of muscle fibres without nervous system involvement. All or nearly all of them have a hereditary origin but details of the type of genetic defect and of the prognosis for the disease vary from type to type. Duchenne MD (pseudohypertrophic MD) is the most common form. It is due to a sex-linked recessive allele and this is expressed as an absence of the protein ***dystrophin***; the disease in boys shows extensive but insufficient muscle fibre reformation from ***satellite cells***.

MuSK Tyrosine kinase localized on the postsynaptic surface of the neuromuscular junction. Mice lacking MuSK fail to form neuromuscular junctions. MuSK is probably involved in ***agrin*** signalling, though it does not interact directly with agrin.

Mustela Genus of mustelids, mink, ferret, stoat, etc.

mut Family of genes products of which are involved in ***mismatch repair***. *MutH* codes for a repair endonuclease (25 kD) specific for unmethylated GATC, *mutL* codes for a protein (68 kD) is involved in the repair process, *mutS* (95 kD) may be involved in the recognition step and has some ATPase activity.

mutagenicity tests See ***Ames test***.

mutagens Agents that cause an increase in the rate of mutation; includes X-rays, ultraviolet irradiation (260nm), and various chemicals.

mutarotation Change in optical rotation with time as an optical isomer in solution converts into other optical isomers.

mutation Usage usually restricted to change in the DNA sequence of an organism, which may arise in any of a variety of different ways. See ***frame-shift mutation***, ***nonsense mutation*** and ***missense mutation***.

mutation rate The frequency with which a particular mutation appears in a population or the frequency with which any mutation appears in the whole genome of a population. Normally the context makes the precise use clear. See ***fluctuation analysis***.

mutein Protein with altered amino acid sequence, usually enough to alter properties.

Mx proteins GTPases (70–100 kD) found in interferon-treated cells. Mx1 is found in the nucleus and determines the resistance of mice to influenza A virus by blocking transcription of the viral RNA genome. Other Mx proteins are cytoplasmic and are related to dynamin.

myalgia Muscle pain.

myasthenia gravis The characteristic feature of the disease is easy fatigue of certain voluntary muscle groups on repeated use. Muscles of the face or upper trunk are especially likely to be affected. In most, and perhaps all cases due to the development of autoantibodies against the ***acetylcholine receptor*** in neuromuscular junctions. Immunization of mice or rats with this receptor protein leads to a disease with the features of myasthenia.

myb An ***oncogene***, identified in avian myeloblastosis, encoding a nuclear protein that binds the DNA sequence YAAC(G/T) G. See Table O1.

myc A ***protooncogene***, identified in several avian tumours, encoding a nuclear protein with a ***leucine zipper*** motif. See Table O1.

Myc proteins Family of proteins involved in control of transcription; have a C-terminal basic helix-loop-helix-zipper domain. Myc-***Max*** heterodimers specifically bind the sequence CACGTG with higher affinity than homodimers of either.

myc tag ***Epitope tag*** frequently expressed as a translational fusion with a transgenic protein of interest. As there are good antibodies to the myc epitope, this allows localization of the fusion gene product by ***immunocytochemistry*** or ***Western blot***, or its immunoaffinity purification.

mycelium Mass of ***hyphae*** that constitutes the vegetative part of a fungus (the conspicuous part in most cases is the fruiting body). Similar, though smaller, structures are found with some saprophytic bacteria such as *Nocardia*.

mycobacteria Bacteria with unusual cell walls that are resistant to digestion, being waxy, very hydrophobic, and rich in lipid, especially esterified ***mycolic acids***. Staining properties differ from those of ***Gram-negative*** and ***Gram-positive*** organisms, being acid fast. Many are intracellular parasites, causing serious diseases such as leprosy and tuberculosis. Cell wall has strong immunostimulating (***adjuvant***) properties due to muramyl dipeptide (MDP).

mycolic acids Saturated fatty acids found in the cell walls of ***mycobacteria, Nocardia*** and ***corynebacteria***. Chain lengths can be as high as 80, and the mycolic acids are found in waxes and in glycolipids.

mycophenolic acid Antibacterial and antitumour compound from *Penicillium brevicompactum*. Inhibits *de novo* nucleotide synthesis.

mycoplasma Prokaryotic microorganisms lacking cell walls, and therefore resistant to many antibiotics. Formerly known as pleuro pneumonia-like organisms (PPLO). *Mycoplasma pneumoniae* is a causative agent of pneumonia in humans and some domestic animals. Troublesome contaminants of animal cell cultures, in which they may grow attached or close to cell surfaces, subtly altering properties of the cells, but escaping detection unless specifically monitored. Similar organisms, spiroplasms, cause various diseases in plants.

mycorhiza Fungi associated with roots of higher plants: relationship is mutually beneficial and in some cases essential to survival of the higher plant.

mycosides Complex glycolipids found in cell walls of ***mycobacteria***. Non-toxic, non-immunogenic molecules that influence the form of the colony and the susceptibility of the bacteria to ***bacteriophages***.

mycosis fungoides A human disease in which a frequent secondary feature is fungal infection of lesions in the skin. Recognized as a tumour of T-lymphocytes that accumulate in the dermis and epidermis and cause loss of the epidermis.

myd A gene that is involved in the determination of muscle cells.

myelin The material making up the ***myelin sheath*** of nerve axons.

myelin basic protein Major component of the ***myelin sheath*** in mammalian CNS. Used as an antigen will induce ***experimental allergic encephalomyelitis***, possibly a model for some neurodegenerative disorders.

myelin figures Structures that form spontaneously when bilayer-forming phospholipids (eg. egg lecithin) are added to water. They are reminiscent of the concentric layer structure of myelin.

myelin sheath An insulating layer surrounding vertebrate peripheral ***neurons***, that dramatically increases the speed of conduction. It is formed by specialized ***Schwann cells***, which can wrap around neurons up to 50 times. Exposed areas are called ***nodes of Ranvier***: they contain very high densities of ***sodium channels***, and ***action potentials*** jump from one node to the next without involving the intermediate axon, a process known as saltatory conduction.

myelodysplastic syndrome (MDS) Haematological disorder that occurs mainly in the elderly as an acquired sporadic disease.

myeloid cells One of the two classes of marrow-derived blood cells; includes ***megakaryocytes***, erythrocyte precursors, ***mononuclear phagocytes***, and all the ***polymorphonuclear leucocytes***. That all these are ultimately derived from one stem cell lineage is shown by the occurrence of the ***Philadelphia chromosome*** in these, but not ***lymphoid*** cells. Most authors tend, however, to restrict the term 'myeloid' to mononuclear phagocytes and granulocytes and commonly distinguish a separate erythroid lineage.

myeloma cell Neoplastic ***plasma cell***. The proliferating plasma cells often replace all the others within the marrow, leading to immune deficiency, and frequently there is destruction of the bone cortex. Because they are monoclonal in origin they secrete a monoclonal immunoglobulin. Bence–Jones proteins are monoclonal immunoglobulin light chains overproduced by myeloma cells and excreted in the urine. Myeloma cell lines are used for producing ***hybridomas*** in raising monoclonal antibodies.

myeloma proteins The immunoglobulins and Bence–Jones proteins secreted by ***myeloma cells***.

myeloperoxidase A ***metalloenzyme*** containing iron, found in the lysosomal granules of myeloid cells, particularly macrophages and neutrophils; responsible for generating potent bacteriocidal activity by the hydrolysis of hydrogen peroxide (produced in the ***metabolic burst***) in the presence of halide ions. Deficiency of myeloperoxidase is not fatal, and it is reportedly entirely absent in chickens.

myf*-5** Member of the ***myoD family of muscle regulatory genes/proteins.

myf*-6** See ***MRF-4.

myoinositol (inositol) 'Muscle sugar' – a name that is only really of historical interest.

myoblast Cell that by fusion with other myoblasts gives rise to ***myotubes*** that eventually develop into skeletal muscle fibres. The term is sometimes used for all the cells recognizable as immediate precursors of skeletal muscle fibres. Alternatively, the term is reserved for those postmitotic cells capable of fusion, others being referred to as presumptive myoblasts.

myocarditis Inflammation of heart muscle usually due to bacterial or viral infection.

myocardium Middle and thickest layer of the wall of the heart, composed of cardiac muscle.

myocyte-enhancer factor 2 See ***MEF2***.

myoD (*MyoD* 1) *MyoD* (myogenic determination) was originally described as a master regulatory gene for the determination of muscle cells, a process now thought to involve a family of genes. It is normally only expressed in myoblasts and skeletal muscle cells, but if transfected into cells will convert many differentiated cells into muscle cells. The *myoD* gene codes for the myoD protein which then switches on the transcription of many muscle specific genes. *MyoD* is a member of a family of genes that all code for nuclear proteins which have a basic DNA-binding motif and a ***helix-loop-helix*** dimerization domain (bHLH or mHLH proteins; b = basic, m = myogenic). They bind to a consensus sequence -CANNTG- and can form homodimers or heterodimers with other members of the bHLH superfamily.

MyoD1 See ***myoD***.

myoepithelial cell (basket cell; basal cell) Cell found between epithelium of exocrine glands (eg. salivary, sweat, mammary, mucous) and their basement membranes, which resembles a smooth muscle cell, and is thought to be contractile.

myofibril Long cylindrical organelle of striated muscle, composed of regular arrays of thick and thin filaments, and constituting the contractile apparatus.

myofibroblasts Histological term for fibroblast-like cells that contain substantial arrays of actin microfilaments, myosin and other muscle proteins arranged in such a way as to suggest that they produce contractile forces. Are commonly described as occurring in granulation tissue (formed during wound healing) and in certain forms of arterial thickening where they are found in the intima. Behave in much the same way as smooth muscle cells and have markers characteristic of these cells.

myogenesis The developmental sequence of events leading to the formation of adult muscle that occurs in the animal and in cultured cells. In vertebrate skeletal muscle the main events are: the fusion of ***myoblasts*** to form ***myotubes*** that increase in size by further fusion to them of myoblasts, the formation of ***myofibrils*** within their cytoplasm, and the establishment of functional ***neuromuscular junctions*** with ***motoneurons***. At this stage they can be regarded as mature muscle fibres.

myogenin Member of the ***MyoD*** family of muscle regulatory genes/proteins. Related to the ***myc*** protooncogene family.

myoglobin Protein (17.5 kD) found in red skeletal muscle. The first protein for which the ***tertiary structure*** was determined by ***X-ray diffraction***, by J. C. Kendrew's group working on sperm whale myoglobin. A single polypeptide chain of 153 amino acids, containing a ***haem*** group bonded via its ferric iron to two histidine residues, that binds oxygen non-cooperatively and has a higher affinity for oxygen than ***haemoglobin*** at all partial pressures. In capillaries oxygen is effectively removed from haemoglobin and diffuses into muscle fibres where it binds to myoglobin which acts as an oxygen store.

myomesin Protein (165kD) found in the ***M-line*** of the ***sarcomere***.

myometrium Uterine ***smooth muscle***.

myoneme Contractile organelle of ciliate protozoans; referred to as M bands in *Stentor*, where they are composed of 8–10nm tubular fibrils. The ***spasmoneme*** of peritrich ciliates was originally called a myoneme.

myosin A family of motor ***ATPases*** that interact with ***F-actin*** filaments. An increasing number of different myosins are being described. Classical striated muscle myosin is myosin II. (See also ***myosin light chains***, ***meromyosin***.) Myosin I is a low molecular weight (111–128 kD) form found in protozoa (*Acanthamoeba* and *Dictyostelium*) that does not self-assemble and is found in the cytoplasm as a globular monomeric molecule that can associate with membranes and transport membrane vesicles along microfilaments.

myosin heavy chain See ***myosin***: do not confuse with ***heavy meromyosin*** which is a subfragment of the heavy chain of myosin II.

myosin light chain kinase (MLCK) ***Calmodulin***-regulated kinase of myosin II light chains: molecular weight varies according to source, 130 kD in non-muscle mammalian cells. May regulate activity of myosin in some cells.

myosin light chains Small subunit proteins (17–22 kD) of ***myosin*** II, all with sequence homology to ***calmodulin***, but not all with calcium-binding activity: two pairs of different light chains are found per myosin. Several types are known: regulatory light chains (LC-2, DNTB light chains) probably regulate the ***ATPase*** activity of the heavy chain directly (through the binding of calcium) or indirectly (activating when they themselves are phosphorylated by ***myosin light chain kinase***), and essential light chains (LC-1, LC-3; alkali light chains), which have a more subtle and apparently non-essential role. In molluscan muscle the EDTA light chains (similar to LC-2 from vertebrate muscle) confer calcium sensitivity on the myosin itself.

myositis Inflammation of muscle. Bacterial myositis can be caused by *Clostridium welchii* (gas gangrene). Viral myositis (epidemic myalgia) is usually due to Coxsackie B virus. Parasitic myositis can be a result of infection with the nematode worm *Trichinella*.

myotonic dystrophy An inherited human neuromuscular disease classed as an autosomal dominant disease in which there is progressive muscle weakening and wasting. Caused by an unstable nucleotide repeat (CTG) in the 3′ untranslated region.

myotoxins Small basic proteins (42–45

amino acids) in rattlesnake venom. Induce rapid necrosis of muscle.

myotube Elongated multinucleate cells (three or more nuclei) that contain some peripherally located ***myofibrils***. They are formed ***in vivo*** or ***in vitro*** by the fusion of ***myoblasts*** and eventually develop into mature muscle fibres that have peripherally located nuclei and most of their cytoplasm filled with myofibrils. In fact, there is no very clear distinction between myotubes and muscle fibres proper.

myristic acid The myristoyl group is one of the less common fatty acyl residues of phospholipids in biological membranes (see Table L3) but is found as an N-terminal modification of a large number of membrane-associated proteins and some cytoplasmic proteins. It is a common modification of viral proteins. In all known examples, the myristoyl residue is attached to the amino group of N-terminal glycine. The specificity of the myristoyl transferase enzymes is extremely high with respect to the fatty acyl residue. For many proteins, the addition of the myristoyl group is essential for membrane association. There is some evidence that myristoylated proteins do not interact with free lipid bilayer, but require a specific receptor protein in the target membrane.

myristoylation Many proteins in eukaryotes are covalently attached to ***myristic acid*** in membranes through amide linkages formed by myristoyl CoA: protein N-myristoyl transferase (NMT), at a glycine, with a consensus site: G-(EDRKHPFYW)-x-x-(STAGCN)-P. This allows a range of enzymes to be concentrated in specific domains within the cell.

Mytilus edulis The edible mussel, a marine bivalve mollusc. The ciliated gills are used for filter feeding and these are utilized in studies on the ***cilium*** and on ***metachronism***.

myxamoebae In the Myxomycetes, such as ***Physarum***, each spore on germination produces two amoeboid cells, myxamoebae, which then transform into flagellated cells.

Myxobacteria Group of ***Gram-negative bacteria***, found mainly in soil. They are non-flagellated with flexible cell walls. They show a gliding motility, moving over solid surfaces leaving a layer of slime (myxo = slime). At some stage in their growth the cells of this group swarm together and form fruiting bodies and spores in a fashion similar to the ***slime moulds***.

myxoedema Severe hypothyroidism usually as a result of autoimmunity to ***thyroglobulin***. A variety of severe physiological problems accompany the reduction in thyroid function.

myxoma virus A poxvirus (see ***Poxviridae***) that causes myxomatosis. Originally isolated from a species of wild rabbit, *Sylvilagus*, in Brazil, in which it causes a mild non-fatal disease, it was found to be 99% fatal in the European rabbit *Oryctylagus*. It causes the characteristic, subcutaneous gelatinous swellings, 'myxomata' and usually kills in 2–5 days. It has been used to control rabbit populations in Australia and Britain, but there are signs that they have developed immunity.

Myxoviridae Single-stranded RNA viruses of animals. Orthomyxoviruses include influenza viruses, Paramyxoviruses include mumps virus.

N

N51 See *melanoma growth-stimulatory activity*.

N See *asparagine*.

N-(3-aminopropyl)-1,4-butanediamine) See *spermidine*.

N. A. See *numerical aperture*.

N, N′ bis(3-aminopropyl)-1,4-butanediamine) See *spermine*.

N-acetyl glucosamine (2-acetamido glucose) A sugar unit found in glycoproteins and various polysaccharides such as ***chitin***, bacterial ***peptidoglycan*** and in *hyaluronic acid*.

N-acetyl muramic acid Sugar unit of bacterial peptidoglycan, consisting of N-acetyl glucosamine bearing an ether-linked lactyl residue on carbon 3. Repeating unit of the cell wall polysaccharide is N-acetyl muramic acid linked to N-acetyl glucosamine via a β (1-4)-glycosidic bond, that can be cleaved by ***lysozyme***.

N-acetyl neuraminic acid A 9-carbon sugar, structurally a condensation product of N-acetyl mannosamine and pyruvate. Also known as sialic acid, but more correctly is a member of the family of ***sialic acids***. Found in ***glycolipids***, especially ***gangliosides***, and in ***glycoproteins***, and therefore in the ***plasma membrane*** of animal cells, to the outer surface of which it contributes negative charge by virtue of its carboxylate group.

N-cadherin See *cadherins*.

N-chimaerin A phorbol ester/diacyl glycerol binding protein found in brain. A GTPase-activating protein for *rac*.

N-CoR (nuclear receptor corepressor) Protein involved in transcriptional repression by thyroid hormone and retinoic acid receptors. The corepressor complex may also contain *Sin3* and histone deacetylase.

N-end rule The N-end rule holds that the ***in vivo half-life*** of a protein is determined by the N-terminal residues.

N-glycanase Enzyme that cleaves asparagine-linked oligosaccharides from glycoproteins.

N-glycosylation site Exposed extracellular asparagine residues are often glycosylated (see ***glycosylation***). The consensus site is N-P-(S/T)-P.

N lines Regions in the sarcomere of striated muscle. The N1 line is in the I band near the Z disc, the N2 line is at the end of the A band. The N lines may represent the location of proteins such as ***nebulin*** that contribute to the stability of the sarcomere.

N-methyl-D-aspartate See *NMDA*.

N-myc ***Oncogene*** related to ***myc***, found in neuroblastomas.

N-protein (1) Anti-terminator protein of the ***lambda bacteriophage*** and other phages that plays a key role in the early stages of infection. During the early phase, only two genes, N and *cro*, are transcribed, by transcription of the DNA in opposite directions. N-protein binds to sites on the DNA (*nut* sites for N-utilization), prevents *rho*-dependent termination and allows transcription of the genes. (2) The name was once used for ***GTP-binding proteins*** (G-proteins); now obsolete and should be avoided because of confusion with N-protein of bacteriophages.

N-type channels A class of ***voltage-sensitive calcium channels***. Restricted to neurons and neuroendocrine cells where they are involved in regulation of neurotransmitter or neurohormone release. Require substantial depolarization to become activated and become inactivated in a time-dependent fashion. Potently inhibited by ω-conotoxin.

Na^+-H^+ exchanger regulatory factor See ***NHE-RF***.

NAA See ***naphthalene acetic acid***.

NAD (nicotinamide adenine dinucleotide; NAD^+; formerly DPN) Coenzyme in which the nicotine ring undergoes cyclic reduction to NADH and oxidation to NAD. Acts as a diffusible substrate for dehydrogenases, etc. $NADH^+$ is one source of reducing equivalents for the electron transport chain. NAD is of special interest as the source of ADP-ribose (see ***ADP-ribosylation***).

NADP (formerly TPN) Analogue of ***NAD***, but NADPH is used for reductive biosynthetic processes (eg. pentose phosphate synthesis) rather than ATP generation.

Naegleria gruberi A normally amoeboid protozoan found in the soil. When it is flooded with water or a solution of low ionic strength it transforms into a swimming form with two ***flagella***.

naevus (USA, nevus) Tumour-like but non-neoplastic ***hamartoma*** of skin. A vascular naevus is a localized capillary-rich area of the skin ('strawberry birthmark'; sometimes the much more extensive 'port-wine stain'). A mole (benign melanoma) is a pigmented naevus, a cluster of melanocytes containing melanin.

nagarse (nagarase) Broad-specificity protease from bacteria.

Nagler's reaction Standard method for identifying *Clostridium perfringens*. When the bacterium is grown on agar containing egg yolk, an opalescent halo is formed around colonies that produce α-toxin (lecithinase).

Naja kaouthia Asian cobra (one of the *Elapidae*). See ***α-cobratoxin***.

nalidixic acid Synthetic antibiotic that interferes with ***DNA gyrase*** and inhibits prokaryotic replication. Often used in selective media.

naloxone An ***alkaloid*** antagonist of ***morphine*** and of the opiate peptides.

Namalwa cells Line of human B-lymphocytes grown in suspension and used to produce interferon (stimulated by Sendai virus infection). Derived from patient with Burkitt's lymphoma.

nanobacteria Nanobacteria are the smallest cell-walled bacteria, only recently discovered in human and cow blood and commercial cell culture serum. They can produce apatite in media mimicking tissue fluids and glomerular filtrate and the aggregates produced closely resemble those found in tissue calcification and kidney stones.

nanovid microscopy Technique of bright-field light microscopy using electronic contrast enhancement and maximum numerical aperture.

NAP-1 See ***interleukin-8***.

NAP-3 See ***melanoma growth-stimulatory activity***.

naphthalene acetic acid (NAA) A synthetic auxin, often used in plant physiology and in plant tissue culture media because it is more stable than ***IAA***.

naphthylamine (β-naphthylamine) Potent carcinogen; used in production of aniline dyes, one of the first chemicals to be associated with a tumour (bladder cancer). The compound itself is not directly carcinogenic; a metabolite produced by hydroxylation (1-hydroxy-2-aminonaphthalene) is detoxified in the liver by conjugation with glucuronic acid, but reactivated by a glucuronidase in the bladder.

napin Angiosperm 2S albumin seed-storage protein from *Brassica napus* (oilseed rape). Consists of two polypeptide chains (3.8 and 8.4 kD) linked by two disulphide bridges. Interacts with calmodulin and has antifungal properties.

napthoquinones Plant pigments derived from napthoquinone.

nastic movement Non-directional movement of part of a plant in response to external stimulus. The tips of growing shoots of plants that twine around supports show nastic movement. See ***epinasty***.

natriuretic Of a substance or hormone, causing natriuresis (elimination of extra sodium in the urine). See ***atrial natriuretic peptide***.

natriuretic peptides Family of four peptides all sharing significant sequence homology and all act through guanylyl cyclase. C-type natriuretic peptide (CNP-22) is a vasodilator and plays an important part in regulating blood pressure, renal function and volume homeostasis. It is produced by endothelial and renal cells and is considered an autocrine regulator of endothelium as well as a neuropeptide. Effects may oppose those of ***atrial natriuretic peptide***. Receptor is natriuretic peptide receptor B. See also ***brain natriuretic peptide*** and ***dendroaspis natriuretic peptide***.

natural killer cells (NK cells) See ***killer cells***.

natural selection The hypothesis that genotype–environment interactions occurring at the phenotypic level lead to differential reproductive success of individuals and hence to modification of the gene pool of a population.

NBQX (6-nitro-7-sulphamoylbenzo[f] quinoxaline-2,3-dione) Blocker of ***AMPA receptors***.

NBT (nitroblue tetrazolium) See ***nitroblue tetrazolium reduction***.

NCAM (neural cell adhesion molecule) One of the first of the ***CAM***s to be isolated from chick brain. Part of the ***immunoglobulin superfamily***, as is NgCAM (neural–glial CAM). Initially defined by adhesion-blocking antiserum. Thought to be important in divalent cation-independent (***L1***) intercellular adhesion of neural and some embryonic cells. See also ***neuroglian***.

NCD Protein of the ***kinesin*** family that differs from kinesin in that it moves towards the minus end of the microtubule (like cytoplasmic dynein). Implicated in spindle organization. See also ***Kar3***.

Nck Small adaptor protein with SH2 and SH3 domains. Similar to ***Crk*** and ***Grb2*** but of uncertain function.

Ndk (nucleoside diphosphate kinase) Enzyme that generates nucleoside triphosphates or their deoxy derivatives by terminal phosphotransfer from ATP or GTP.

nebulin Family of large matrix proteins (600–900 kD) found in the ***N line*** of the ***sarcomere*** of striated muscle. Consist of many (more than 200) repeats of conserved actin-binding motifs; bind to ***F-actin*** and may serve as templates for assembly of the sarcomere.

necrosis Death of some or all cells in a tissue as a result of injury, infection or loss of blood supply.

nectin (1) Another name for SAM (substrate adhesion molecule), eg. fibronectin. (2) A protein forming the stalk of mitochondrial ATPase.

nedd2 (caspase-2; Ich-1) See ***caspases***.

nef ***HIV*** protein that is important for pathogenesis, enhances infectivity and regulates the sorting of at least two cellular transmembrane proteins, CD4 and MHC Class I. Has a proline-rich sequence that interacts with ***Hck***-SH3 domain and activates kinase activity of Hck; also interacts with a serine/threonine kinase of 62 kD. Nef itself has a sorting signal (ENTSLL) that functions as an endocytosis marker and it has some amino acid homology with α-scorpion toxins that bind to potassium channels.

negative feedback This occurs where the products of a process can act at an earlier stage in the process to inhibit their own formation. The term was first used widely in conjunction with electrical amplifiers where negative feedback was applied to limit distortion of the signal by the amplification mechanism. Tends to stabilize the process. In contrast to ***positive feedback***.

negative regulation Negative feedback in biological systems mediated by allosteric regulatory enzymes.

negative staining Microscopic technique in which the object stands out against a dark background of stain. For electron microscopy the sample is suspended in a solution of an electron-dense stain such as sodium phosphotungstate and then sprayed onto a support grid. The stain dries as structureless solid and fills all crevices in the sample. When examined in the electron microscope the sample appears as a light object against a dark background. Quite fine structural detail can be observed using negative staining and it has been used extensively to study the structure of viruses and other particulate samples.

negative-stranded RNA virus Class V ***viruses*** that have an RNA genome that is complementary to the mRNA, the positive strand. They also carry the virus-specific ***RNA polymerase*** necessary for the synthesis of the mRNA. Includes ***Rhabdoviridae, Paramyxoviridae*** and Myoviridae (eg. the T-even phages).

Negri body Acidophilic cytoplasmic inclusion (mass of ***nucleocapsids***) characteristic of rabies virus infection.

Neisseria ***Gram-negative*** non-motile pyogenic cocci. Two species are serious pathogens, *N. meningitidis* (see ***meningitis***) and *N. gonorrhoeae*. The latter associates specifically with urinogenital epithelium through surface ***pili***. Both species seem to evade the normal consequences of attack by phagocytes.

nematocyst (cnidocyst) Stinging mechanism used for defence and prey capture by ***Hydra*** and other members of the Cnidaria (Coelenterata). It is located within a specialized cell, the ***nematocyte***, and consists of a capsule containing a coiled tube. When the nematocyte is triggered, the wall of the capsule changes its water permeability and the inrush of water causes the tube to evert explosively ejecting the nematocyst from the cell. The tube is commonly armed with barbs and may also contain toxin.

nematocyte (cnidoblast) Stinging cells found in ***Hydra***, used for capturing prey and for defence. There are four major types, containing different sorts of ***nematocysts***: stenoteles (60%), desmonemes, holotrichous isorhizas and atrichous isorhizas. They differentiate from interstitial cells and are almost all found in the tentacles.

nematode sperm The nematode *Caenorhabditis elegans* has an unusual amoeboid spermatozoon that is actively motile yet appears to lack both actin and tubulin.

nematosome Cytoplasmic inclusion in some neurons.

neoantigen Antigen acquired after a cell has been transformed by an oncogenic virus.

neoendorphin Opioid peptide (***endorphin***) cleaved from pro-dynorphin.

neointima The new intima laid down in a vessel that has been dilated by ***angioplasty***; often hyperplastic and the cause of restenosis.

neomycin Either of two aminoglycosides (B and C) produced by *Streptomyces fradiae* that have generalized antibiotic activity. Neomycin A (Ineamine) contains 2-deoxy-1,3-diamino-inositol combined with the aminoglycoside.

neoplasia Literally, new growth, usually refers to abnormal new growth and thus means the same as ***tumour***, which may be benign or malignant. Unlike ***hyperplasia***, neoplastic proliferation persists even in the absence of the original stimulus.

neoplastic Adjective, describing cells that exhibit ***neoplasia***.

neoteny The persistence in the reproductively mature adult of characters usually associated with the immature organism.

neoxanthin A ***xanthophyll carotenoid*** pigment, found in higher plant chloroplasts as part of the ***light-harvesting system***.

NEP (neutral endopeptidase) Cell surface zinc endopeptidase (EC 3.4.24.11) that hydrolyzes regulatory peptides such as ***ANP***. Spontaneously hypertensive hamsters have elevated levels of NEP in two organs that contribute appreciably to vascular resistance, skeletal muscle and kidney.

nephelometry Any method for estimating the concentration of cells or particles in a suspension by measuring the intensity of scattered light, often at right-angles to the incident beam. Light scattering depends upon number, size and surface characteristics of the particles.

nephron The structural and functional unit of the vertebrate kidney. It is made up of the glomerulus, Bowman's capsule and the convoluted tubule.

neprilysin (EC 3.4.24.11) Neuropeptide-degrading neutral endopeptidase, a zinc metalloproteinase similar to bacterial thermolysin with some homology to endothelin-converting enzyme.

Nernst equation A basic equation of biophysics that describes the relationship between the equilibrium potential difference across a ***semi-permeable membrane***, and the equilibrium distribution of the ionic permeant species. It is described by: $E = (RT/zF). \ln([C1]/[C2])$, where E is the potential on side 2 relative to side 1 (in volts), R is the gas constant (8.314 J K^{-1} mol^{-1}), *T* is the absolute temperature, *z* is the charge on the permeant ion, F is the Faraday constant (96 500°C mol^{-1}) and C1 and C2 are the concentrations (more correctly activities) of the ions on sides 1 and 2 of the membrane. It can be seen that this equation is a solution of the more general equation of ***electrochemical potential***, for the special case of equilibrium. The equation described the voltage generated by ion-selective electrodes, like the laboratory pH electrode; and approximates the behaviour of the resting plasma membrane (see ***resting potential***).

Nernst potential See ***Nernst equation*** and ***ion-selective electrodes***.

nerve cell See ***neuron***.

nerve ending See ***synapse***.

nerve growth cone See ***growth cone***.

nerve growth factor (NGF) A peptide (13.26 kD) of 118 amino acids (usually dimeric) with both chemotropic and chemotrophic properties for ***sympathetic*** and ***sensory neurons***. Found in a variety of peripheral tissues, NGF attracts ***neurites*** to the tissues by chemotropism, where they form synapses. The successful neurons are then 'protected' from neuronal death by continuing supplies of NGF. It is also found at exceptionally high levels in snake venom and male mouse submaxillary salivary glands, from which it is commercially extracted. NGF was the first of a family of nerve tropic factors to be discovered. Amino acids 1–81 show homology with proinsulin. Besides its peripheral actions, NGF selectively enhances the growth of ***cholinergic neurons*** that project to the forebrain and that degenerate in ***Alzheimer's disease***.

nerve impulse An ***action potential***.

nesidioblast Precursor cell of pancreatic ***B-cells***.

nested PCR Variety of ***polymerase chain reaction***, in which specificity is improved by using two sets of primers sequentially. An initial PCR is performed with the 'outer' primer pairs, then a small aliquot is used as a template for a second round of PCR with the 'inner' primer pair.

nested primers Sets of primers for PCR so arranged that the second set to be used lie within the sequence amplified by the first set of primers and so on.

nestin (neural stem cell protein) Large (200 kD) intermediate filament protein found in developing rat brain. Functionally similar to other intermediate filament proteins but the sequence is very different. Forms class VI of the ***intermediate filaments***.

netrin Genes identified in studies of vertebrate neuronal development. Netrins are ***chemotropic*** for embryonic commissural neurons: netrin-1 is secreted by the floorplate, whereas netrin-2 is distributed ventrally except for the floorplate. The netrins are homologous to the product of *unc-6*, a gene identified in studies of neuronal development of the worm ***Caenorhabditis elegans***.

netropsin Basic peptide antibiotic from *Streptomyces*. Binds selectively in minor groove of ***B-DNA*** and will induce A to B transition.

neu (erb-B2) ***Oncogene*** originally identified in a ***neuroblastoma*** that encodes a receptor ***tyrosine kinase*** of the EGF-receptor family. Ligand is ***neuregulin***. See Table O1.

neu differentiation factor See ***neuregulin***.

neural cell adhesion molecule See *NCAM*.

neural crest A group of embryonic cells that separate from the ***neural plate*** during ***neurulation*** and migrate to give several different lineages of adult cells: the spinal and autonomic ganglia, the ***glial cells*** of the peripheral nervous system, and non-neuronal cells, such as ***chromaffin cells***, ***melanocytes*** and some haematopoietic cells.

neural fold A crease that forms in the ***neural plate*** during ***neurulation***.

neural induction In vertebrates the formation of the nervous system from the ***ectoderm*** of the early embryo as a result of a signal from the underlying ***mesoderm*** of the archenteron roof; also known as primary neural induction. The mechanism of neural induction is not yet clear.

neural plate A region of embryonic ectodermal cells, called neuroectoderm, that lie directly above the ***notochord***. During ***neurulation*** they change shape so as to produce an infolding of the neural plate (the neural fold) that then seals to form the neural tube.

neural retina Layer of nerve cells in the retina, embryologically part of the brain. The incoming light passes through nerve fibres and intermediary nerve cells of the neural retina, before encountering the light-sensitive rods and cones at the interface between neural retina and the pigmented retinal epithelium.

neural tube The progenitor of the central nervous system. See ***neural plate***, ***neurulation***.

neuraminic acid See ***N-acetyl neuraminic acid***.

neuraminidase (sialidase) Enzyme catalysing cleavage of ***neuraminic acid*** residues from oligosaccharide chains of glycoproteins and glycolipids. Since these residues are usually terminal, neuraminidases are generally exo-enzymes, although an endoneuraminidase is known. For use as a laboratory reagent, common sources are from bacteria such as *Vibrio* or *Clostridium*. A neuraminidase is one of the transmembrane proteins of the envelope of influenza virus.

neuraxin Protein associated with neuronal ***microtubules***. Structurally related to ***MAP***-1B.

neuraxis The neural axis of the body, the brain and spinal cord.

neuregulin (neu differentiation factor) Growth factor of the ***epidermal growth factor*** family that induces growth and differentiation of epithelial, glial and muscle cells in culture. Receptor is a tyrosine kinase: will bind erbB3 which will heterodimerise with erbB2. Gene disruption is lethal during embryogenesis with heart malformation and defects in Schwann cells and neural ganglia. Neuregulin 1 and neuregulin 2 show distinct expression patterns and mediate distinct biological processes.

neurexin Synaptic cell surface proteins related to ***α-latrotoxin*** receptor, laminin and agrin. At least 180 transcripts from at least two genes. Cell recognition molecules at nerve terminal.

neurite A process growing out of a neuron. As it is hard to distinguish a ***dendrite*** from an ***axon*** in culture, the term neurite is is used for both.

neuroblast Cells arising by division of precursor cells in neural ectoderm (***neurectoderm***) that subsequently differentiate to become neurons.

neuroblastoma Malignant tumour derived from primitive ganglion cells. Mainly a tumour of childhood. Commonest sites are adrenal medulla and retroperitoneal tissue. The cells may partially differentiate into cells having the appearance of immature neurons.

neurocalcin Calcium-binding protein (23 kD) related to ***recoverin***. Abundant in CNS.

neuroectoderm (neurectoderm) ***Ectoderm*** on the dorsal surface of the early vertebrate embryo that gives rise to the cells (neurons and glia) of the nervous system. Also called the ***neural plate***.

neuroendocrine cell See ***neurohormone***.

neuroepithelium See ***neuroectoderm***.

neurofascin Axon-associated adhesion molecule of the vertebrate nervous system. Contains six Ig-like motifs and four fibronectin type II repeats. Related to Ng-CAM, Nr-CAM.

neurofibrillary tangle A characteristic pathological feature of the brain of patients with ***Alzheimer's disease*** is the presence of tangles of coarse ***neurofibrils*** within large neurons of the cerebral cortex. Whether this causes neuronal degeneration or is a secondary consequence remains contentious.

neurofibrils Filaments found in neurons; not necessarily ***neurofilaments*** in all cases, and in the older literature 'fibrils' are composed of both microtubules and neurofilaments. Originally used by light microscopists to describe much larger fibrils seen particularly well with silver-staining methods.

neurofibromatosis Tumours of neuronal sheath; the most common genetic disease. Type 1 neurofibromatosis is associated with the the von Recklinghausen Neurofibromatosis locus that encodes the NF-1 protein, a ***GTPase-activating protein*** which interacts with the ***ras*** proteins.

neurofilament Member of the class of ***intermediate filaments*** found in axons of nerve cells. In vertebrates assembled from three distinct protein subunits (NF-L 68 kD; NF-M 160 kD; NF-H 200 kD) these proteins, if introduced into fibroblasts, will incorporate into the vimentin filament system.

neurogenesis Differentiation of the nervous system from the ***ectoderm*** of the early embryo. There are major differences between neurogenesis in vertebrates and invertebrates.

neurogenic gene Best described in *Drosophila*, genes that are required to determine a neuronal fate. Examples: ***Notch, Delta***.

neuroglia See ***glial cells***.

neuroglian Protein isolated from *Drosophila* nervous system that is a member of the ***immunoglobulin superfamily***. It contains six immunoglobulin-like domains and five fibronectin type III domains and has strong sequence homology to mouse ***NCAM*** and ***L1***. Two different forms of neuroglian arise by differential splicing. These have identical extracellular domains but differ in the size of the cytoplasmic domains: the long form is restricted to neurons in central and peripheral nervous systems of embryos and larvae.

neurohormone A hormone secreted by specialized ***neurons*** (neuroendocrine cells), eg. releasing hormones.

neurokinin See ***tachykinins***.

neuroleptic drugs (antischizophrenic drugs; antipsychotic drugs; tranquillisers) Literally 'nerve-seizing': used of chlorpromazine-like drugs. Antagonise the effects of ***dopamine***.

neuromeres Alternate swellings and constrictions seen along the ***neuraxis*** at early stages of ***neural tube*** development, thought to be evidence of intrinsic segmentation in the central nervous system. Neuromeres or segments in the hindbrain region are called ***rhombomeres*** and have been shown to be lineage-restriction units, each constructing a defined piece of hindbrain.

neuromodulation Alteration in the effectiveness of ***voltage-gated*** or ***ligand-gated ion channels*** by changing the characteristics of current flow through the channels. The mechanism is thought to involve ***second messenger*** systems.

neuromodulin (GAP-43; pp46; B-50; F1; P 57) Protein associated with actively growing axons, especially in the ***growth cone***. Binds ***calmodulin***, is phosphorylated by ***protein kinase C***.

neuromuscular junction A ***chemical synapse*** between a motoneuron and a muscle fibre. Also known as a motor endplate.

neuron (neurone; nerve cell) An ***excitable cell*** specialized for the transmission of electrical signals over long distances. Neurons receive input from sensory cells or other neurons, and send output to muscles or other neurons. Neurons with sensory input are called 'sensory neurons', neurons with muscle outputs are called 'motoneurons'; neurons that connect only with other neurons are called 'interneurons'. Neurons connect with each other via ***synapses***. Neurons can be the longest cells known; a single ***axon*** can be several metres in length. Although signals are usually sent via ***action potentials***, some neurons are ***non-spiking***.

neuronal differentiation Acquisition during development of specific biochemical, physiological and morphological properties by nerve cells.

neuronal guidance See ***axonal guidance***.

neuronal plasticity Ability of nerve cells to change their properties, eg. by sprouting new processes, making new synapses or altering the strength of existing synapses. See ***long-term potentiation*** and ***synaptic plasticity***.

neuronal polarity Distribution of specific functions to discrete cellular domains, eg. axons and dendrites that have different molecular composition, morphology and ultrastructure and perform different functions.

neuropeptide Y (NPY; melanostatin) Peptide neurotransmitter (36 residues) found in adrenals, heart and brain. Potent stimulator of feeding and regulates secretion of gonadotrophin-releasing hormone. ***Leptin*** inhibits NPY gene expression and release. Receptors are G-protein coupled.

neuropeptides Peptides with direct synaptic effects (peptide neurotransmitters) or

Table N1. Neurotransmitters

Transmitter	Peripheral nervous system	Central nervous system
Noradrenaline	Some postganglionic sympathetic neurons	Diverse pathways especially in arousal and blood pressure control
Dopamine	Sympathetic ganglia	Diverse; perturbed in Parkinsonism and schizophrenia
Serotonin	Neurons in myenteric plexus	Distribution very similar to that of noradrenergic neurons Lysergic acid (LSD) may antagonize
Acetylcholine	Neuromuscular junctions (nmj). All postganglionic parasympathetic and most postganglionic sympathetic neurons	Widely distributed, usually excitatory. Possibly antagonizes dopaminergic neurons
Amino acids		
GABA	Inhibitory at nmj of arthropods	Inhibitory in many pathways
Glutamate	Excitatory at nmj of arthropods	Widely distributed; excitatory
Glycine	—	Diverse, particularly in grey matter of spinal cord
Aspartate	Locust nmj	—
Neuropeptides	Diverse actions in both peripheral and central nervous systems; see Table H2	
Histamine	—	Minor role
Purines	Particularly neurons controlling blood vessels	Mostly inhibitory
Octopamine	Invertebrate nmj	—
Substance P	Sensory neurons of vertebrates	Sensory neurons

Other substances known, or proposed to have neurotransmitter function are: adrenaline, β-alanine, taurine, proctolin and cysteine.

indirect modulatory effects on the nervous system (peptide neuromodulators). See Table N1.

neurophysin Carrier protein (10 kD, 90–97 amino acids) that transports neurohypophysial hormones along axons, from the hypothalamus to the posterior lobe of the pituitary. See also ***brain***.

neurosecretory cells Cells that have properties both of electrical activity, carrying impulses, and a secretory function, releasing hormones into the bloodstream. In a sense, they are behaving in the same way as any chemically signalling neuron, except that the target is the blood (and remote tissues), not another nerve or postsynaptic region.

Neurospora An Ascomycete fungus, haploid and grows as a ***mycelium***. There are two mating types, and fusion of nuclei of two opposite types leads to meiosis followed by mitosis. The resulting eight nuclei generate eight ascospores, arranged linearly in an ordered fashion in a pod-like ***ascus*** so that the various products of meiotic division can be identified and isolated. Because of this, *Neurospora crassa* is one of the classic organisms for genetic research; studies on biochemical mutants led Beadle and Tatum to propose the seminal 'one gene-one enzyme' hypothesis.

neurosteroids Steroids synthesized in the brain that have effects on neuronal excitability. The ***epalons*** may regulate type A ***GABA receptors*** by allosteric potentiation.

neurotactin Membrane-anchored chemokine (395 amino acid residues in mouse, 397 in man) with unique CXXXC pattern (unlike α-chemokines, CXC; β-chemokines, CC; and γ-chemokines, C). Neurotactin is predominantly expressed in brain and is upregulated on capillary vessels and microglia in LPS-induced inflammation and ***EAE***. Unlike other chemokines, gene is on Chromosome 8 (mouse) and 16q (man). Proteolytically released soluble active fragments may be generated. Chemotactic for neutrophils.

neurotensin Tridecapeptide hormone (sequence: ELYENKPRRPYIL) of gastrointestinal tract: has general vascular and neuroendocrine actions.

neurotoxin A substance, often exquisitely toxic, that inhibits neuronal function. Neurotoxins act typically against the ***sodium channel*** (eg. ***TTX***) or block or enhance ***synaptic transmission*** (***curare, bungarotoxin***).

neurotransmitter A substance found in ***chemical synapses*** that is released from the presynaptic terminal in response to depolarization by an action potential, diffuses across the synaptic cleft, and binds a ligand-gated ion channel on the postsynaptic cell. This alters the resting potential of the postsynaptic cell, and thus its excitability. Examples: ***acetylcholine, GABA, noradrenaline, serotonin, dopamine***. See Table N1.

neurotrophic Involved in the nutrition (or maintenance) of neural tissue. Classic example is ***nerve growth factor***.

neurotrophin-3 (NT-3; hippocampal-derived neurotrophic factor or NGF-2) Member of the family of neurotrophic factors or ***neurotrophins*** that also includes ***nerve growth factor*** and ***brain-derived neurotrophic factor*** (BDNF) that have about 50% amino acid sequence identity. NT-3 was the first member of the family to have its sequence determined by molecular techniques without the need for prior purification of the protein from natural sources. NT-3 shows strong similarities to NGF and BDNF (including strictly conserved domains that contain six cysteine residues) but has a different pattern of neuronal specificity and regional expression.

neurotrophins Molecules with closely related structures that are known to support the survival of different classes of embryonic neurons. See ***nerve growth factor*** (NGF), ***brain-derived neurotrophic factor*** (BDNF), ***neurotrophin-3*** (NT-3), ***GDGF*** and ***ciliary neurotrophic factor***.

neurotropic Having an affinity for, or growing towards, neural tissue. Rabies virus, which localizes in neurons, is referred to as neurotropic; can also be used to refer to chemicals.

neurotubules A term for ***microtubules*** in a neuron.

neurula The stage in vertebrate embryogenesis during which the neural plate closes to form the central nervous system.

neurulation The embryonic formation of the ***neural tube*** by closure of the ***neural plate,*** directed by the underlying notochord.

neutral mutation A mutation that has no selective advantage or disadvantage. Considerable controversy surrounds the question of whether such mutations can exist.

neutral protease Protease that is optimally active at neutral pH: see ***proteases***.

neutropenia Condition in which the number of ***neutrophils*** circulating in the blood is below normal.

neutrophil (neutrophil granulocyte; polymorphonuclear leucocyte; PMN or PMNL) Commonest (2500–7500/mm^3) blood leucocyte; a short-lived phagocytic cell of the ***myeloid*** series, which is responsible for the primary cellular response to an acute inflammatory episode, and for general tissue homeostasis by removal of damaged material. Adheres to endothelium (***margination***) and then migrates into tissue, possibly responding to chemotactic signals. Contain ***specific*** and ***azurophil granules***.

neutrophil-activating protein See ***interleukin-8***.

neutrophil-activating protein 3 See ***melanoma growth-stimulatory activity***.

neutrophilin Neutrophil-derived platelet activator, probably a serine protease.

nevus See ***naevus***.

Newcastle disease virus A paramyxovirus that causes the fatal disease, fowl-pest, in poultry.

nexin Protein (165 kD) that links the adjacent microtubule doublets of the ***axoneme*** in a ***cilium***.

nexus A connection or link.

Nezelof syndrome Congenital T-cell deficiency associated with thymic hypoplasia.

NF-1 (1) Nuclear factor 1; see ***CTF***. (2) ***Neurofibromatosis***-related protein 1.

NF-E1 See ***erythroid transcription factor***.

NFAT (nuclear factor of activated T-cells) Transcription factor involved in regulation of IL-2 and IL-4 gene transcription (in concert with other transcription factors). NFAT is cytoplasmic until dephosphorylated by ***calcineurin,*** a step that is inhibited by cyclosporin and FK506, then translocates to the nucleus.

NFκB (NFkappaB) A ***transcription factor*** (originally found to switch on transcription of genes for the kappa class of immunoglobulins in B-lymphocytes). It is involved in activating the transcription of more than 20 genes in a variety of cells and tissues. NFκB is found in the cytoplasm in an inactive form, bound to the protein IκB. A variety of stimuli, such as tumour necrosis factor, phorbol esters and bacterial lipopolysaccharide activate it, by releasing it from IκB, allowing it to enter the nucleus and bind to DNA. It has two subunits p50 and p65 that bind DNA as a heterodimer. The dimerization and DNA-binding activity are located in N-terminal regions of 300 amino acid that are similar to regions in the Rel and *dorsal* transcription factors.

NgCAM See ***L1***.

NGF Abbreviation for ***nerve growth factor***.

NHE-RF (Na^+-H^+ exchanger regulatory factor) Cytoplasmic phosphoprotein involved in ***protein kinase A*** (PKA) mediated regulation of ion transport. Contains 2 ***PDZ domains*** and will bind to C-terminus of ***CFTR***.

niacin (nicotinic acid) One of the B vitamins. See Table V1.

niasomes Multilamellate ***liposomes*** made from non-ionic lipids and used for drug delivery.

nibrin Protein thought to be defective in ***Nijmegen breakage syndrome***. Has two modules found in cell cycle checkpoint proteins and a *forkhead*-associated domain.

nicardipine Calcium channel blocker used to treat hypertension.

nick A point in a double-stranded DNA molecule where there is no ***phosphodiester bond*** between adjacent nucleotides of one strand typically through damage or enzyme action.

nick translation A technique used to radioactively label DNA. *E.coli* DNA polymerase I will add a nucleotide, copying the complementary strand, to the free 3′-OH group at a nick, at the same time its exonuclease activity removes the 5′-terminus. The enzyme then adds a nucleotide at the new 3′-OH and removes the new 5′-terminus. In this way one strand of the DNA is replaced starting at a nick, which effectively moves along the strand. Nick translation refers to this translation or movement and not to protein synthesis. In practice, DNA is mixed with trace amounts of ***DNAase*** I to generate nicks, ***DNA polymerase*** I and labelled nucleotides. Because the nicks are generated randomly the DNA preparation can be uniformly labelled and to a high degree of specific activity.

nicotinamide adenine dinucleotide See *NAD*.

nicotinamide adenine dinucleotide phosphate See *NADP*.

nicotine A plant alkaloid from tobacco; blocks transmission at nicotinic synapses.

nicotinic acetylcholine receptor (nAChR) Integral membrane protein of the postsynaptic membrane to which ***acetylcholine*** binds. The receptor contains an integral ***ion channel***; as a result of binding of acetylcholine, ion channels in the subsynaptic membrane are opened. At the ***neuromuscular junction***, the nicotinic acetylcholine receptor initiates muscle contraction. Currently the best characterized ion channel protein: made of a heteropentamer of related subunits, although a homopentamer is functional in insects. Structural studies show that the acetylcholine binding site and the ionic channel are part of the same macromolecular unit. The nAChR mediates rapid transduction events (1ms) whereas receptors activating ***G-protein*** coupled receptors operate on slower time scales (millisecond to second range).

nicotinic acid (pyridine 3-carboxylic acid) A precursor of ***NAD*** that is a product of the oxidation of nicotine.

NIDDM (non-insulin dependent diabetes mellitus.) Type II, maturity (adult) onset diabetes. Can usually be treated by regulating sugar intake.

nidogen See ***entactin***.

Niemann–Pick disease Severe lysosomal storage disease caused by deficiency in sphingomyelinase; excess sphingomyelin is stored in 'foam' cells (macrophages) in spleen, bone marrow and lymphoid tissue. More common in Ashkenazi Jews than in other groups.

nif genes The complex of genes in nitrogen-fixing bacteria that code for the proteins required for ***nitrogen fixation***, particularly the ***nitrogenase***. Present as an operon in ***Klebsiella*** and carried on plasmid in ***Rhizobium***.

nifedipine (BAYa1041; Nifedin; Procardia) A calcium channel blocker (346 D) used experimentally and as a coronary vasodilator.

niflumic acid ((2,(3-(trifluoromethyl)-anilino) nicotinic acid)) A rather non-specific inhibitor of chloride channels.

nigericin An ionophore capable of acting as a carrier for K^+ or Rb^+ or as an exchange carrier for H^+ with K^+. Originally used as an antibiotic. Has been used in investigating chemiosmosis and other transport systems.

NIH 3T3 cells Very widely used mouse fibroblast cell line; 3T3 cells have been derived from different mouse strains and it is therefore important to define the particular cell line. NIH strain were from the National Institute of Health in the USA.

Nijmegen breakage syndrome (NBS) Autosomal recessive chromosomal instability syndrome characterized by microencephaly, growth retardation, immunodeficiency and predisposition to tumours. Cells from patients are hypersensitive to ionizing radiation in the same way as cells from ***ataxia telangiectasia***. A novel protein, ***nibrin*** has been implicated in the syndrome.

ninhydrin (triketohydrindene hydrate) Pale yellow substance used to detect amino acids and proteins (compounds containing free amino or imino groups) with which it forms a deeply coloured purple-blue compound.

Nissl granules Discrete clumps of material seen by phase contrast microscopy in the perikaryon of some neurons, particularly motor neurons. They are basophilic and contain much RNA, and are regions very rich in ***rough endoplasmic reticulum***. Their reaction following damage to neurons is characteristic; they disperse through the cytoplasm giving a general basophilia to the whole cell body.

Nitella Characean alga that has giant, multinucleate internodal cells. These show ***cytoplasmic streaming*** at rates of up to 100μm/s and have been used as models for motile phenomena in cells, and in studies on ionic movement.

nitric oxide (endothelium-derived relaxation factor; NO) Gas produced from L-arginine by the enzyme ***nitric oxide synthase***. Acts as an intracellular and intercellular messenger in a wide range of processes, in the vascular and nervous systems. The intracellular 'receptor' is a soluble (cytoplasmic) ***guanylate cyclase***. In the immune system, large amounts can be generated as a cytotoxic attack mechanism. NO signalling is phylogenetically widespread, suggesting it is an ancient mechanism.

nitric oxide synthase (NO synthase) Enzyme that produces the vasorelaxant ***nitric oxide*** (endothelium-derived relaxation factor) from L-arginine. There are two isoforms, one constitutive and calmodulin dependent, the other inducible (iNOS) and calcium independent.

nitroblue tetrazolium reduction Nitroblue tetrazolium, a yellow dye, is taken up by phagocytosing neutrophils and reduced to insoluble formazan, which is deep blue, if the ***metabolic burst*** is normal. Reduction does not take place in ***chronic granulomatous disease***.

nitrocellulose paper Paper with a high non-specific absorbing power for biological macromolecules. Very important as a receptor in blot-transfer methods. Bands are transferred from a chromatogram or electropherogram either by blotting on nitrocellulose sheets or by electrophoretic transfer. The replica can then be used for sensitive analytical detection methods.

nitrogen fixation The incorporation of atmospheric nitrogen into ammonia by various bacteria, catalysed by ***nitrogenase***. This is an essential stage in the nitrogen cycle and is the ultimate source of all nitrogen in living organisms. In the sea, the main nitrogen fixers are ***Cyanobacteria***. There are several free-living bacteria in soil that fix nitrogen including species of *Azotobacter*, *Clostridium* and *Klebsiella*. *Rhizobium* only fixes nitrogen when in symbiotic association, in root nodules, with leguminous plants. The oxygen-sensitive nitrogenase is protected by plant-produced leghaemoglobin and the plant obtains fixed nitrogen from the bacteria. See ***Frankia***.

nitrogen mustards A series of tertiary amine compounds having vesicant properties similar to those of mustard gas. They have the general formula $RN(CH_2CH_2Cl)_2$. They can alkylate compounds such as DNA and are used as the basis of cytostatic drugs for cancer chemotherapy.

nitrogenase Nitrogenases are enzymes found in nitrogen-fixing bacteria that reduce nitrogen to ammonia (also ethylene to acetylene).

nitrosamines These molecules contain the N–N=O group (N-nitrosamines): many are carcinogens or suspected carcinogens.

NK cell (natural killer cell) See ***null cell***.

NK stimulatory factor See ***interleukin-12***.

NMDA (N-methyl-D-aspartic acid) A powerful agonist for a class of receptor ***NMDA receptor*** found on some vertebrate nerve cells involved in synaptic transmission.

NMDA receptor Glutamate receptor subtype (see ***excitatory amino acids***). NMDA channels seem to be potentiated by intracellular ***arachidonic acid***.

NO See ***nitric oxide***.

NO synthase See ***nitric oxide synthase***.

Nocardia Genus of ***Gram-positive bacteria*** that form a ***mycelium*** that may fragment into rod- or coccoid-shaped cells. They are very common ***saprophytes*** in soil but some are opportunistic pathogens of humans, causing nocardiosis. This is characterized by abscesses, particularly of the jaw, which if untreated may invade the surrounding bone.

nociception Detection of pain. See ***capsaicin***.

nociceptor Pain receptor. Many nociceptors respond to ***capsaicin***.

nocodazole Microtubule blocker that binds to tubulin heterodimer rendering it assembly-incompetent.

Noctiluca A bioluminescent dinoflagellate. Responsible for many instances of marine phosphorescence.

NOD mice (non-obese diabetic mice) Have unique histocompatibility antigens; pancreatic B-cells are destroyed by an autoimmune response.

node A point in a plant stem at which one or more leaves are attached.

nodes of Ranvier Regions of exposed neuronal plasma membrane in a myelinated axon. Nodes contain very high concentrations of ***voltage-gated ion channels***, and are the site of propagation of action potentials by saltatory conduction.

nodulin Plant protein. Soybean nodulin-24 is closely related to ***major intrinsic protein***.

noggin Dorsalizing factor (26 kD) from ***Spemann's organizer*** region of amphibian embryo.

nojirimycin Antibiotic produced by *Streptomyces* strains; inhibits α-glucosidases and prevents normal glycosylation of proteins by interfering with the early pruning down to the core carbohydrate that is normally followed by addition of specific sugar residues.

Nomarski differential interference contrast See ***differential interference contrast***.

non-coding DNA DNA that does not code for part of a polypeptide chain or RNA. This includes ***introns*** and ***pseudogenes***. In eukaryotes the majority of the DNA is non-coding. Non-coding strand refers to the so-called nonsense strand, as opposed to the sense strand which is actually translated into mRNA.

non-competitive inhibitor Reversible inhibition of an enzyme by a compound that binds at a site other than the substrate-binding site.

non-cyclic photophosphorylation Process by which light energy absorbed by ***photosystems*** in chloroplasts is used to generate ATP (and also NADPH). Involves photolysis of water by photosystem II, passage of electrons along the photosynthetic electron transport chain with concomitant phosphorylation of ADP, and reduction of $NADP^+$ using energy derived from photosystem I.

non-disjunction Failure of homologous chromosomes or sister ***chromatids*** to separate at meiosis or mitosis respectively. It results in aneuploid cells. Non-disjunction of the X chromosome in *Drosophila* allowed Bridges to confirm the theory of chromosomal inheritance.

non-equivalence Term used in cell determination for cells that will give rise to the same sorts of differentiated tissues but that have different positional values (eg. cells of forelimb and hindlimb buds).

non-histone chromosomal proteins ***Chromatin*** consists of DNA, ***histones*** and a very heterogeneous group of other proteins that includes DNA polymerases, regulator proteins, etc. They are often generically referred to as non-histone proteins or acidic proteins, to distinguish them from the basic histones.

non-ionic detergent Detergent in which the hydrophilic head-group is uncharged. In practice hydrophilicity is usually conferred by –OH groups. Examples are the polyoxyethylene p-t-octyl phenols known as Tritons, and octyl glucoside. Non-ionic detergents can be used to solubilize intrinsic membrane proteins with less tendency to denature them than charged detergents. They do not usually cause disassembly of structures such as microfilaments and microtubules that depend on protein–protein interactions.

non-Mendelian inheritance In eukaryotes, patterns of gene transmission not explicable in terms of segregation, independent assortment and linkage. May be due to ***cytoplasmic inheritance***, ***gene conversion***, meiotic drive, etc.

non-Newtonian fluid A fluid in which the viscosity varies depending upon the shear stress. The effect can arise because of alignment of non-spherical molecules as flow is established or because of suspended deformable particles as in blood.

non-receptor protein-tyrosine kinase See ***tyrosine kinase***.

non-reciprocal contact inhibition Collision behaviour between different cell types in which one cell shows contact inhibition of locomotion, and the other does not. An example is the interaction between sarcoma cells and fibroblasts (the former not being inhibited).

Nonidet Trade name for non-ionic detergents, usually octyl or nonyl-phenoxy-polyethoxy-ethanols.

non-permissive cell Originally a cell of a tissue type or species that does not permit replication of a particular virus. Early stages of the virus cycle may be possible in such a cell, that in the case of tumour viruses may become transformed. Now used in a more general sense, of agents and treatments other than viruses.

non-polar group (hydrophobic group) Group in which the electronic charge density is essentially uniform, and which cannot therefore interact with other groups by forming hydrogen bonds, or by strong dipole–dipole interactions. In an aqueous environment, nonpolar groups tend to cluster together, providing a major force for the folding of macromolecules and formation of membranes. Clusters are formed chiefly because they cause a smaller increase in water structure (decrease in entropy) than dispersed groups. (Non-polar groups interact with each other only by the relatively weak London-van der Waals forces).

nonsense codon (nonsense triplet) The three ***codons***, UAA (known as ochre), UAG (amber) and UGA (opal), that do not code for an amino acid but act as signals for the termination of protein synthesis. Any mutation that causes a base change which produces a nonsense codon results in premature termination of protein synthesis and probably a non-functional or nonsense protein. See also Table C5.

nonsense mutation Mutation in coding DNA that prevents the protein from being synthesized.

nonsense strand See ***non-coding DNA***.

non-spiking neuron A neuron that can convey information without generating action potentials. As passive electrical potentials are attenuated over distances greater than the space constant for a neuron (typically 1mm), this implies that most non-spiking neurons are involved in signalling over relatively short distances. Typical examples are invertebrate stretch receptors and ***interneurons*** in the central nervous system.

nopaline (N-alpha-(1,3-dicarboxypropyl)-L-arginine) An ***opine***. The gene for nopaline synthase is carried on the T-DNA of the ***Ti plasmid***.

noradrenaline (norepinephrine; arterenol) Catecholamine neurohormone, the neurotransmitter of most of the sympathetic nervous system (of so-called adrenergic neurons): binds more strongly to α-adrenergic receptor than β-adrenergic receptor. Stored and released from ***chromaffin cells*** of the adrenal medulla.

norepinephrine See ***noradrenaline***.

norleucine (Ahx; Nle; 2-aminohexanoic acid) Non-protein amino acid. Formyl-Norleucyl-Leucyl-Phenylalanine has been used as a substitute for fMLP in studies on neutrophil chemotaxis since it is not so susceptible to oxidation.

normoblast Nucleated cell of the ***myeloid cell*** series found in bone marrow that gives rise to red blood cells. See ***erythroblast***.

normocyte Erythrocyte of normal size and shape.

Northern blot An electroblotting method in which RNA is transferred to a filter and detected by hybridization to ^{32}P-labelled RNA or DNA. See ***blots***.

Northwestern blot Technique for identifying protein–RNA interactions in which protein is run on a gel, blotted, and probed with a labelled RNA of interest. Interactions are detected as hot-spots on the filter. So-called because it involves both RNA (***Northern blot***) and protein (***Western blot***).

Norwalk virus Unclassified single-stranded RNA virus causing common acute infectious gastroenteritis.

nosocomial infections Hospital-acquired infections: commonest are due to *Staphylococcus aureus*, *Pseudomonas aeruginosa*, *E. coli*, *Klebsiella pneumoniae*, *Serratia marcescens*, and *Proteus mirabilis*.

Notch Family of large transmembrane receptor proteins (350 kD) that mediate developmental cell-fate decisions; Notch contains 36 repeats of the ***EGF-like domain***. Mammalian *Notch* gene mutations have been associated with leukaemia, breast cancer, stroke and dementia (see ***CADASIL***). In *Drosophila* wing development, Notch receptor is activated at the dorsal/ventral boundary and is important in growth and patterning. Notch binds transmembrane ligands encoded by *Ser* (serrate protein) and *Dl* (Delta protein). *Fringe* (*fng*) is also involved in Notch signalling, encoding a pioneer protein.

notexins Notexins Np and Ns are phospholipase A2 isoforms found in the venom of *Notexis scutatus scutatus*. They block acetylcholine release at neuromuscular junction.

Notexis scutatus scutatus Tiger snake. Venom contains a range of toxins including ***notexins***.

notochord An axial mesodermal tissue found in embryonic stages of all chordates and protochordates, often regressing as maturity is approached. Typically a rod-shaped mass of vacuolated cells immediately below the nerve cord, and may provide mechanical strength to the embryo.

notoplate Region of the ***neural plate*** overlying the ***notochord***.

noxiustoxin ***Charybdotoxin***-related peptide toxin (39 residues) from scorpion *Centruroides noxius*. Blocks mammalian voltage-gated potassium channels and high-conductance calcium-activated potassium channels.

Nramp (natural resistance associated macrophage protein; Nramp1, Nramp2) Nramp1 codes for an integral membrane protein (65 kD) expressed only in macrophages/monocytes and PMNs. Localized to endosomal/lysosomal compartment and rapidly recruited to the phagosome membrane following phagocytosis. Mutations in Nramp1 seem to abrogate the ability of macrophages to kill intracellular parasites such as *Mycobacterium tuberculosis* and to be associated with onset of rheumatoid arthritis. Nramp2 is very similar to Nramp1 but is expressed in more tissues and is known to be an iron transporter. Yeast homologues Smf1 and Smf2 transport divalent cations.

NSAID (non-steroidal anti-inflammatory drug) Range of anti-inflammatory drugs that include aspirin, ibuprofen and a wide range of derivatives. Mostly act on the production of early low molecular weight mediators of the acute inflammatory response. Are particularly good at inhibiting swelling, but often have undesirable side-effects.

NSF (N-ethyl-maleimide sensitive factor) Homotetrameric protein (76 kD) involved, together with three ***SNAPs***, in mediating vesicle traffic between medial and *trans*-Golgi compartments.

NSO cells Murine myeloma cell line (plasmacytoma).

Ntk Protein tyrosine kinase (56 kD) similar to ***Csk*** but found in nervous tissue and T-cells. Has SH2 and SH3 domains but lacks consensus tyrosine phosphorylation and myristoylation sites of src. Ntk levels drop when T-cells are activated

nuclear actin-binding protein (NAB) Nuclear protein, dimer of 34 kD subunits. Binds actin with Kd of around 25μM.

nuclear envelope Membrane system that surrounds the nucleus of eukaryotic cells. Consists of inner and outer membranes separated by perinuclear space and perforated by nuclear pores. The term should be used in preference to the term 'nuclear membrane' which is potentially very confusing.

nuclear factor 1 See ***CTF***.

nuclear lamina A fibrous protein network lining the inner surface of the nuclear envelope. The extent to which this system also provides a scaffold within the nucleus is controversial. Proteins of the lamina are ***lamins*** A, B and C, which have sequence homology to proteins of ***intermediate filaments***.

nuclear localization signal (NLS) In eukaryotes, peptide signal sequence that identifies a protein as being destined for the nucleus (see ***importins***). Frequently the signal sequence is a collection of basic amino acids downstream of a helix-breaking proline, eg. SV40 T (Pro-Lys-Lys-Lys-Arg-Lys-Val).

nuclear magnetic resonance (NMR) Biophysical technique which allows the spectroscopy or imaging of molecules containing at least one paramagnetic atom (eg. ^{13}C, ^{31}P). Although non-invasive, the scale of the equipment needed to generate the radiofrequency electromagnetic and magnetic fields, and the computer power needed to analyse the results, are non-trivial. Widely used as a medical imaging technology.

nuclear matrix Protein latticework filling the nucleus that anchors ***DNA replication*** and ***transcription*** complexes.

nuclear membrane See ***nuclear envelope***.

nuclear pore Openings in the nuclear envelope, diameter about 10nm, through which molecules such as nuclear proteins (synthesized in the cytoplasm) and mRNA must pass. Pores are generated by a large protein assembly.

nuclear receptor Receptor for a diffusible signal molecule that can enter the nucleus, particularly receptors for ***steroid hormones***.

nuclear RNA The nucleus contains RNA that has just been synthesized, but in addition there is some that seems not to be released, or is only released after further processing, the heterogeneous nuclear RNA (***hnRNA***) and small RNA molecules associated with protein to form ***snRNPs*** (small nuclear ribonucleoproteins).

nuclear run-on (nuclear run-off) Strictly different, the two terms tend to be used interchangeably. A nuclear run-on assay is intended to identify the genes that were being transcribed at a particular instant; nuclei are rapidly isolated from cells, and incubated with labelled nucleotides. This gives a population of labelled RNAs that were being transcribed immediately before isolation. These can be studied directly; or (more commonly) used as a probe to identify corresponding cDNAs.

nuclear transplantation Experimental approach in study of nucleo-cytoplasmic interactions, in which a nucleus is transferred from one cell to the cytoplasm (which may be anucleate) of a second.

nuclear transport Passage of molecules in and out of the nucleus, presumably via nuclear pores. Passage of proteins into the nucleus may depend on possession of a nuclear location sequence containing five consecutive positively-charged residues (PKKKRKV).

nuclease An enzyme capable of cleaving the ***phosphodiester bonds*** between nucleotide subunits of ***nucleic acids***.

nucleation A general term used in polymerization or assembly reactions where the first steps are energetically less favoured than the continuation of growth. Polymerization is much faster if a preformed seed is used to nucleate growth (eg. microtubule growth is nucleated from the ***microtubule organizing centre***, although the nature of this nucleation is not known).

nucleic acids Linear polymers of nucleotides, linked by 3′,5′ phosphodiester linkages. In DNA, deoxyribonucleic acid, the sugar group is deoxyribose, and the bases of the nucleotides adenine, guanine, thymine and cytosine. RNA, ribonucleic acid, has ribose as the sugar, and uracil replaces thymine. DNA functions as a stable repository of genetic information in the form of base sequence. RNA has a similar function in some viruses but more usually serves as an informational intermediate (mRNA), a transporter of amino acids (tRNA), in a structural capacity or, in some newly discovered instances, as an enzyme.

nucleocapsid The coat (***capsid***) of a virus plus the enclosed nucleic acid genome.

nucleocytoplasmic transport Transport of molecules from the nucleus to the cytoplasm.

nucleoid Region of cell in a bacterium that contains the DNA.

nucleolar organizer Loop of DNA that has multiple copies of rRNA genes. See ***nucleolus***.

nucleolin A major nucleolar protein (100 kD) that functions as a shuttle protein between nucleus and cytoplasm and is also found on the cell surface. Nucleolin binds ***midkine*** and heparin-binding growth associated molecule (HB-GAM).

nucleolus A small dense body (sub-organelle) within the nucleus of eukaryotic cells, visible by phase contrast and interference microscopy in live cells throughout interphase. Contains RNA and protein, and is the site of synthesis of ribosomal RNA. The nucleolus surrounds a region of one or more chromosomes (the nucleolar organizer) in which are repeated copies of the DNA coding for ribosomal RNA.

nucleoplasm By analogy with cytoplasm, that part of the nuclear contents other than the nucleolus.

nucleoplasmin First protein to be described as a molecular ***chaperone***; its major function seems to be in assembly of ***nucleosomes***.

Nucleopore filter Filter of defined pore size made by etching a polycarbonate filter that has been bombarded by neutrons, the extent of etching determining the pore size. Very thin, with neat circular holes going right through the membrane, not a complex meshwork like micropore filters.

nucleoporins Proteins that make up the nuclear pore complex that regulates the traffic of proteins and nucleic acids into and out of the nucleus. Many contain N-acetyl-glucosamine residues.

nucleoproteins Structures containing both nucleic acid and protein. Examples are chromatin, ribosomes, certain virus particles.

nucleoside Purine or pyrimidine base linked glycosidically to ribose or deoxyribose, but lacking the phosphate residues that would make it a nucleotide. Ribonucleosides are adenosine, guanosine, cytidine and uridine. Deoxyribosides are deoxyadenosine, deoxyguanosine, deoxycytidine and deoxythymidine (the latter is almost universally referred to as thymidine).

nucleoskeletal DNA DNA that is proposed to exist mostly to maintain nuclear volume and not for coding protein.

Table N2. Nucleotides

Phosphate esters of nucleosides, which are themselves conjugates between the biological bases and sugars, either ribose or 2-deoxyribose.
Nucleosides are derived from the bases by the addition of a sugar in the position indicated (H).

Adenine Cytosine Guanine Uracil Thymine

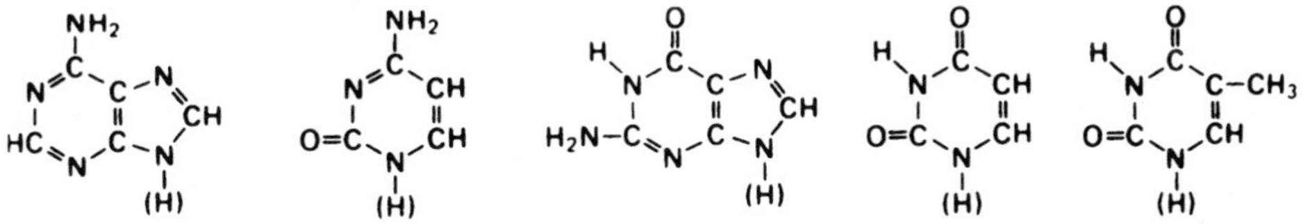

The structure of nucleotides, exemplified by adenine derivatives is:

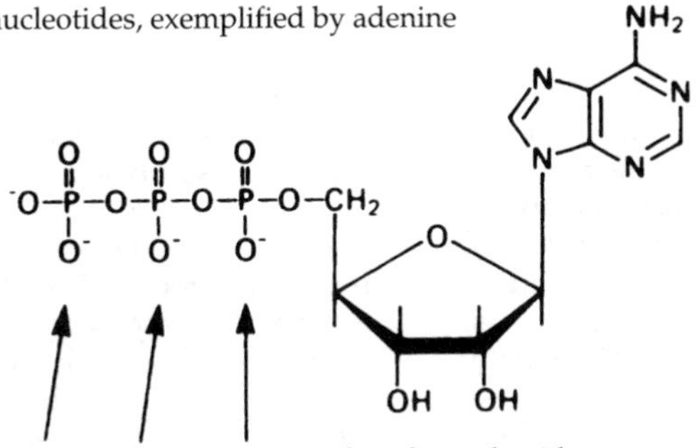

i.e. tri di- mono- phosphonucleotides
e.g. ATP ADP AMP

The phosphate may also be a cyclic diester involving two hydroxyl groups of the sugar. eg. 3′,5′ cyclic AMP

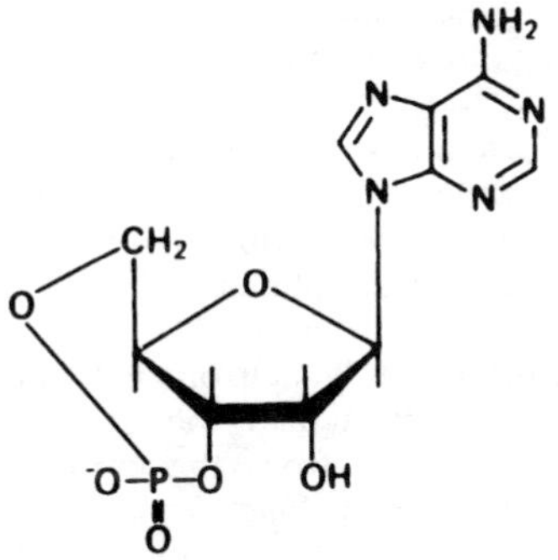

Ribonucleotides are precursors of RNA and also common metabolic intermediates and regulators; examples of the shorthand nomenclature are given.

	Adenine	Cytosine	Guanine	Uracil
Mononucleotide	AMP	CMP	GMP	UMP
Dinucleotide	ADP	CDP	GDP	UDP
Trinucleotide	ATP	CTP	GTP	UTP
Cyclic nucleotide	3′,5′ cyclic AMP		3′,5′ cyclic GMP	

Deoxyribonucleotides, required for the synthesis of DNA, are made by the biological reduction of the corresponding ribose dinucleotides and the deoxyribonucleotides are phosphorylated to give the triphosphonucleotides. dTMP is made by methylation of dUMP, which is then phosphorylated to give dTTP.

	Adenine	Cytosine	Guanine	Thymidine
Dinucleotide	dADP	dCDP	dGDP	
Trinucleotide	dATP	dCTP	dGTP	dTTP

Nucleotides occur as part of other biological molecules, eg. NAD is the ADP-ribose derivative of nicotinamide. Nucleotide adducts are important intermediates in anabolic processes. CDP derivatives occur in the biosynthesis of lipids. UDP and TDP derivatives are important in sugar metabolism.

nucleosome Repeating units of organization of chromatin fibres in chromosomes, consisting of around 200 base pairs, and two molecules each of the ***histones*** H2A, H2B, H3 and H4. Most of the DNA (around 140 base pairs) is believed to be wound around a core formed by the histones, the remainder joins adjacent nucleosomes, thus forming a structure reminiscent of a string of beads.

nucleotidase (5′-nucleotidase) Enzyme that cleaves the 5′ monoester linkage of nucleotides, and thus converts them to the corresponding nucleoside.

nucleotide Phosphate esters of ***nucleosides***. The metabolic precursors of nucleic acids are monoesters with phosphate on carbon 5 of the pentose (known as 5′ to distinguish sugar from base numbering). However many other structures, such as adenosine 3′5′-cyclic monophosphate (cAMP), and molecules with two or three phosphates are also known as nucleotides. See Table N2.

nucleotide-binding fold Protein motif consisting of a fold or pocket with certain conserved residues, required for the binding of nucleotides.

nucleus The major organelle of eukaryotic cells, in which the chromosomes are separated from the cytoplasm by the ***nuclear envelope***.

nude mice Strains of athymic mice bearing the recessive allele *nu/nu* which are largely hairless and lack all or most of the T-cell population. Show no rejection of either allografts or xenografts; *nu/nu* alleles on some backgrounds have near normal numbers of T-cells.

null cell Lymphocytes lacking typical markers of T- or B-cells capable of lysing a variety of tumour or virus-infected cells without obvious antigenic stimulation, also effect ***antibody-dependent cell-mediated cytotoxicity***, and in humans carry CD16 marker.

null mutant Mutation in which there is no gene product.

numerical aperture (N. A.) For a lens the resolving power depends upon the wavelength of light being used and inversely upon the numerical aperture. The N. A. is the product of the refractive index of the medium (1 for air, 1.5 for immersion oil) and the sine of the angle, i, the semi-angle of the cone formed by joining objects to the perimeter of the lens. The larger the value of N. A., the better the resolving power of the lens; most objectives have their N. A. value engraved on the barrel and this should be quoted when describing an optical system.

nurse cells Cells accessory to egg and/or sperm formation in a wide variety of organisms. Usually thought to synthesize special substances and to export these to the developing gamete.

nystatin A polyene antibiotic active against fungi. The name is derived from 'New York State Health Department' where it was discovered as a product of *Streptomyces noursei*.

O-2A progenitor Bipotential progenitor cells in rat optic nerve that give rise initially to oligodendrocytes and then to type-2 astrocytes. Production of type-2 astrocytes from O-2A progenitor cells ***in vitro*** is triggered by ***ciliary neurotrophic factor*** (CNTF).

O-antigens Tetra- and pentasaccharide repeat units of the cell walls of ***Gram-negative bacteria***. They are a component of ***lipopolysaccharide***.

O-hydroxycinnamic acid See ***coumarin***.

oat cell carcinoma Form of carcinoma of the lung in which the cells are small, spindle-shaped and dark-staining. May derive from argyrophilic APUD cells of the mucosa and certainly tends to be associated with endocrine symptoms.

obelin Calcium-activated photoprotein in the photocyte of the colonial hydroid coelenterate, *Obelia geniculata*.

occludens junction Tight junction. See ***zonula occludens***.

ochre codon The ***codon*** UAA, one of the three that causes termination of protein synthesis. The most frequent termination codon in *E. coli* genes.

ochre mutation Mutation that changes any codon to the ***termination*** codon UAA.

ochre suppressor A gene that codes for an altered tRNA so that its ***anticodon*** can recognize the ***ochre codon*** and thus allows the continuation of protein synthesis. A suppressor of an ***ochre mutation*** is a tRNA that is charged with the amino acid corresponding to the original codon or a neutral substitute. Ochre suppressors will also suppress ***amber codons***.

ochronosis Deposition of dark brown pigment in cartilage, joint capsules and other tissues, usually as a result of ***alkaptonuria***.

oct Family of genes for transcription factors that act as RNA Polymerae II promoters. Protein products contain a ***POU domain*** and are ***leucine zipper*** proteins that bind to ***octamer*** sequences.

octamer (1) 8-base sequence motif common in eukaryotic promoters. Consensus is ATTTGCAT; binds various transcription factors. (2) Assembly of 8 histone proteins (two each of H2A, H2B, H3 and H4) that forms core of nucleosome.

octamer-binding protein ***Transcription factor*** that binds to the ***octamer*** motif. Examples: mammalian proteins Oct-1, Oct-2.

octamer motif A DNA motif found in certain promoters that can produce B-cell specific gene expression. Sequence: ATGCAAAT.

octopaline (N-α-(D-1-carboxyethyl)-L-arginine) An ***opine***.

octopamine A ***biogenic amine*** found in both vertebrates and invertebrates (identified first in the salivary gland of *Octopus*). Octopamine can have properties of both a hormone and a neurotransmitter, and acts as an adrenergic agonist.

octyl glucoside A biological detergent characterized by its ease of removal from hydrophobic proteins. Used to solubilize membrane proteins.

odontoblasts Columnar cells derived from the dental papilla after ***ameloblasts*** have differentiated, and that give rise to the dentine matrix that underlies the enamel of a tooth.

odontogenic epithelial cells Epithelial layer that will give rise to teeth.

oedema (USA, edema) Swelling of tissue: can result from increased permeability of vascular endothelium.

oesophageal dysmotility See ***CREST***.

oestradiol (USA, estradiol; follicular hormone) A hormone (272 D) synthesized mainly in the ovary, but also in the placenta, testis, and possibly adrenal cortex. A potent ***oestrogen***.

oestrogen (USA, estrogen) A type of hormone that induces oestrus ('heat') in female animals. It controls changes in the uterus that precede ovulation, and is responsible for development of secondary sexual characteristics in pubescent girls. Some tumours are sensitive to oestrogens. See ***oestradiol***.

OFAGE (orthogonal field alternation gel electrophoresis) Electrophoresis in which macromolecules are electrophoresed in a gel using electric fields applied alternately at right-angles to each other.

okadaic acid Derived from a dinoflagellate toxin. This compound is a powerful inhibitor of serine–threonine-specific protein ***phosphatases*** 1 and 2A. Also can act as a tumour promoter.

Okazaki fragments Short fragments of newly synthesized DNA strands produced during DNA replication. All the known ***DNA polymerases*** can only synthesis DNA in one direction, the 5′ to 3′ direction. However as the strands separate, replication forks will be moving along one parental strand in the 3′ to 5′ direction and 5′ to 3′ on the other parental strand. On the former, the leading strand, DNA can be synthesized continuously in the 5′ to 3′ direction. On the other, the lagging strand, DNA synthesis can only occur when a stretch of single-stranded DNA has been exposed and proceeds in the direction opposite to the movement of the replication fork (still 5′ to 3′). It is thus discontinuous and the series of fragments are then covalently linked by ***ligases*** to give a continuous strand. Such fragments were first observed by Okazaki using pulse-labelling with radioactive thymidine. In eukaryotes, Okazaki fragments are typically a few hundred nucleotides long, whereas in prokaryotes they may contain several thousands of nucleotides.

oleic acid See ***fatty acids*** and Table L3.

oleosome Plant spherosome rich in lipid that serves as a storage granule in seeds and fruits. There are none of the enzymes characteristic of lysosomes.

olfactory epithelium The ***epithelium*** lining the nose. Has the diverse G protein-coupled receptors responsible for the sense of smell.

olfactory neuron ***Sensory neuron*** from the lining of the nose. They are the only neurons that continue to divide and differentiate throughout an organism's life.

oligodendrocyte Neuroglial cell of the central nervous system in vertebrates whose function is to myelinate CNS axons. See ***neuroglia***.

oligomycin A bacterial toxin inhibitor of oxidative phosphorylation that acts on a small subunit of the ***F-type ATPase***.

oligomycin sensitivity conferral protein (OSCP) The δ subunit of the ***ATP synthase***, believed to link the F_1 catalytic segment to the F_0 proton-conduction segment. Binds the toxin ***oligomycin***.

oligonucleotide Linear sequence of up to 20 nucleotides joined by phosphodiester bonds. Above this length the term polynucleotide begins to be used.

oligopeptide A peptide of a small number of component amino acids as opposed to a polypeptide. The exact size range is a matter of opinion but peptides from 3 to about 40 member amino acids might be so described.

oligosaccharide A saccharide of a small number of component sugars, either O- or N-linked to the next sugar. Number of component sugars not rigorously defined.

oligosaccharin An oligosaccharide derived from the plant cell wall that in small quantities induces a physiological response in a nearby cell of the same or a different plant, and thus acts as a molecular signal. Sometimes considered to be a plant hormone or plant growth substance. The best authenticated examples are involved in host–pathogen interactions and in the control of plant cell expansion.

oligotroph Organism that can grow in an environment poor in nutrients.

olomoucine Inhibitor of ***cdk*** 5.

omega-oxidation Minor metabolic pathway for medium chain length fatty acids

omeprazole Non-proprietary name for gastric proton pump (H^+/K transport-ATPase) inhibitor, much prescribed for gastric ulcers since it inhibits acid secretion by parietal cells.

ommatidium (plural, ommatidia) Single facet of an insect compound eye. Composed of a set of photoreceptor cells, overlain by a crystalline lens.

Onchocerca Genus of filarial nematode parasites that cause river blindness.

oncocytes See ***Askenazy cells***.

oncogen Synonym for ***carcinogen***, an agent causing cancer.

oncogene Mutated and/or overexpressed version of a normal gene of animal cells (the ***protooncogene***) that in a dominant fashion can release the cell from normal restraints on growth, and thus alone, or in concert with other changes, convert a cell into a tumour cell. See Table O1.

oncogenic virus A virus capable of causing cancer in animals or in humans. These include DNA viruses, ranging in size from Papova viruses to Herpes viruses, and the RNA-containing retroviruses. See ***Oncovirinae***.

oncomodulin Calcium-binding proteins containing the ***EF-hand*** motif. Found only in tumours, and related to panalbumin.

oncoprotein 18 See ***stathmin***.

oncostatin M Multifunctional cytokine (28 kD) of the ***IL-6 cytokine family***. Produced by activated T-cells; inhibits tumour cell growth and induces IL-6 production by endothelial cells via the tyrosine kinase $p62^{yes}$.

Oncovirinae The family of retroviruses (Retroviridae) that can cause tumours. They are enveloped by membrane derived from the plasma membrane of the host cell, from which they are released by budding without lysing the cell. Within each virion is a pair of single-stranded RNA molecules. Replication involves a DNA intermediate made on an RNA template by the enzyme ***reverse transcriptase***.

ontogeny The total of the stages of an organism's life history.

oocyte The developing female gamete before maturation and release.

oocyte expression Technique whereby the cellular translational machinery of an oocyte (typically *Xenopus*) is utilized to generate functional protein from microinjected mRNA or to produce protein encoded by an introduced expression vector.

oogenesis The process of egg formation.

oogonium Female sexual structure in certain algae and fungi, containing one or more gametes. After fertilization the oogonium contains the oospore.

oomycetes Group of fungi in which the mycelium is non-septate, ie. lacks cross-walls, and the nuclei are diploid. Sexual reproduction is oogamous.

opal codon The codon UGA, one of the three that causes termination of protein synthesis.

opal mutation Mutation that changes any codon to the termination codon UGA.

opal suppressor A gene that codes for an altered tRNA so that its anticodon can recognize the ***opal codon*** and thus allows the continuation of protein synthesis. A suppressor of an ***opal mutation*** is a tRNA that is charged with the amino acid corresponding to the original codon or a

Table O1. Oncogenes and tumour viruses

Acronym	Virus	Species	Tumour origin	Comments
abl	Abelson leukaemia	mouse	Chronic myelogenous leukaemia	TyrPK(src)
erbA	Erythroblastosis	chicken		Homology to human glucocorticoid receptor
erbB	Erythroblastosis	chicken		TyrPK EGF/TGFc receptor
ets	E26 myeloblastosis	chicken		Nuclear
fes(fps)[a]	Snyder–Theilen sarcoma, Gardner–Arnstein sarcoma	cat		TyrPK (src)
fgr	Gardner–Rasheed sarcoma	cat		TyrPK(src)
fms	McDonough sarcoma	cat		TyrPK CSF-1 receptor
fps(fes)[a]	Fujinami sarcoma	chicken		TyrPK(src)
fos	FBJ osteosarcoma	mouse		Nuclear, TR
hst	NVT	human	Stomach tumour	FGF homologue
intl	NVT	mouse	MMTV-induced carcinoma	Nuclear, TR
int2	NVT	mouse	MMTV-induced carcinoma	FGF homologue
jun	ASV17 sarcoma	chicken		Nuclear, TR
hit	Hardy–Zuckerman 4 sarcoma	cat		TyrPK GFR L?
B-1ym	NVT	chicken	Bursal lymphoma	
mas	NVT	human	Epidermoid carcinoma	Potentiates response to angiotensin II?
met	NVT	mouse	Osteosarcoma	TyrPK GFR L?
mil (raf)	Mill Hill 2 acute leukaemia	chicken		Ser/ThrPK
mos	Moloney sarcoma	mouse		Ser/ThrPK
myb	Myeloblastosis	chicken	Leukaemia	Nuclear, TR
myc	MC29 Myelocytomatosis	chicken	Lymphomas	Nuclear TR?
N-myc	NVT	human	Neuroblastomas	Nuclear
neu(ErbB2)	NVT	rat	Neuroblastoma	TyrPK GFR L?
ral(mil)[b]	3611 sarcoma	mouse		Ser/ThrPK
Ha-ras	Harvey murine sarcoma	rat	Bladder, mammary and skin carcinomas	GTP-binding
Ki-ras	Kirsten murine sarcoma	rat	Lung, colon carcinomas	GTP-binding
N-ras	NVT	human	Neuroblastomas, leukaemias	GTP-binding
rel	Reticuloendotheliosis	turkey		
ros	UR2	chicken		TyrPK GFR L?
sis	Simian sarcoma	monkey		One chain of PDGF
src	Rous sarcoma	chicken		TyrPK
ski	SKV770	chicken		Nuclear
trk	NVT	human	Colon carcinoma	TyrPK GFR L?
yes	Y73, Esh sarcoma	chicken		TyrPK(src)

[a]fps/fes are species equivalents.
[b]mil/raf are species equivalents.
GFR L? = from sequence, a growth-factor receptor for unknown ligand.
MMTV = mouse mammary tumour virus.
NVT = isolated from non-retroviral tumour. In most cases detected by transfection of 3T3 cells.
Ser/ThrPK = serine, threonine protein kinase; TyrPK = tyrosine protein kinase.
TR = transcriptional regulator.

neutral substitute. Some eukaryote cells normally synthesize opal suppressor tRNAs. The function of these is not clear and they usually do not prevent normal termination of protein synthesis at an opal codon.

Opalina A genus of parasitic protozoans found in the guts of frogs and toads. They look superficially like ciliates, but are classified in a separate group as they have a number of similar nuclei.

open reading frame (ORF) A possible ***reading frame*** of DNA which is capable of being translated into protein, ie. is not punctuated by stop codons. (This capacity does not indicate *per se* that the ORF is translated.)

operator The site on DNA to which a specific ***repressor protein*** binds and prevents the initiation of transcription at the adjacent ***promoter***.

operon Groups of bacterial genes with a common ***promotor***, that are controlled as a unit and produce mRNA as a single piece, ***polycistronic*** messenger. An operon consists of two or more structural genes, which usually code for proteins with related metabolic functions, and associated control elements that regulate the transcription of the structural genes. The first described example was the ***lac operon***.

opiates Naturally occuring basic (alkaloid) molecules with a complex fused ring structure. Have high pharmacological activity. See ***morphine***.

opines Carbon compounds produced by crown galls and hairy roots induced by ***Agrobacterium tumefaciens*** and *A. rhizogenes* respectively. They are utilized as nutritional sources by *Agrobacterium* strains that induced the growth and are, in some cases, chemoattractants. Chemotactic activity seems to be specific for plasmids carrying the relevant opine synthase gene. Octopine, ***nopaline***, mannopine and agrocinopines A and B are examples.

opioid receptors Membrane proteins, widely distributed in animal cells, but especially in the brain (enkephalin receptors) and gut. The natural ligands are the opiate peptide neurotransmitters, but the name is given because opiates are potent agonists that occupy the receptors and mimic the action of the natural transmitters.

opportunistic pathogen (secondary pathogen) Pathogenic organism that is often normally a commensal, but which gives rise to infection in immunocompromised hosts.

opsin General term for the apoproteins of the ***rhodopsin*** family.

opsonin Substance that binds to the surface of a particle and enhances the uptake of the particle by a phagocyte. Probably the most important in mammals derive from ***complement*** (C3b or C3bi) or immunoglobulins (which are bound through the ***Fc receptor***).

opsonization Process of coating with an ***opsonin***. Often done simply by incubating particles (eg. ***zymosan***) with fresh serum.

optic nerve Projection from the vertebrate retina to the midbrain. Embryologically, a CNS tract rather than a peripheral nerve. Popular experimental preparation for studies of regeneration of ***retino-tectal connections*** in lower vertebrates and also for studies of glial cell lineage in CNS.

optic tectum A region of the midbrain in which input from the optic nerve is processed. Because the retinally derived neurons of the optic nerve 'map' onto the optic tectum in a defined way, the question of how this specificity is determined has been a long-standing problem in cell biology. Although there is some evidence for adhesion gradients and for some adhesion specificity, the problem is unresolved.

optical diffraction A technique used to obtain information about repeating patterns. Diffraction of visible light can be used to calculate spacings in the object.

optical isomers Isomers (stereoisomers) differing only in the spatial arrangement of groups around a central atom. Optical isomers rotate the plane of polarized light in different directions. For most biological molecules in which the possibility of optical isomerism exists, only one of the isomers is functional, although there are some exceptions. For example, both glucose and mannose are used and D-amino acids are found in the cell walls of some bacteria (whereas L-forms are universal in proteins).

optical tweezers (laser tweezers; optical trap) By focusing a beam of light on a particle it is possible to trap the particle as a result of the forces due to radiation pressure (the forces involved are of the order of pico-newtons). Laser beams exert sufficient force for it to be possible to move small organelles around under the microscope or to measure the forces that motor molecules are exerting by measuring the force needed to oppose their activity.

orbivirus Genus of ***Reoviridae*** that infects a wide range of vertebrates and insects.

organ culture Culture ***in vitro*** of pieces of tissue (as opposed to single cells) in such a way as to maintain some normal spatial relationships between cells and some normal function. Contrast with ***tissue culture***.

organelle A structurally discrete component of a cell.

organizing centre See ***microtubule organizing centre***.

organogenesis The process of formation of specific organs in a plant or animal involving morphogenesis and differentiatio.

Oriental sore Skin disease caused by the flagellate protozoan, *Leishmania tropica*.

orientation chamber Chamber designed by Zigmond in which to test the ability of cells (***neutrophils***) to orient in a gradient of chemoattractant. The chamber is similar to a haemocytometer, but with a depth of only about 20μm. The gradient is set up by diffusion from one well to the other, and the orientation of cells towards the well containing chemoattractant is scored on the basis of their morphology or by filming their movement.

origin of replication Regions of DNA that are necessary for its replication to begin, such as pBR322 *ori*, required for plasmid replication.

ornithine decarboxylase The enzyme that converts ornithine to putrescine (dibasic amine) by decarboxylation. Rate limiting in the synthesis of the polyamines spermidine and spermine that regulate DNA synthesis.

ornithine transcarbamylase deficiency X-linked disorder, the most common cause of inherited urea cycle disorders.

orosomucoid (α-1-seromucoid; α-1-acid glycoprotein) Plasma protein of mammals and birds, 38% carbohydrate. In humans a single-chain glycoprotein of 39 kD. Increased levels are associated with inflammation, pregnancy and various diseases.

orotic acid (orotate) Intermediate in the *de novo* synthesis of pyrimidines. Linked glycosidically to ribose 5′-phosphate, orotate forms the pyrimidine nucleotide orotidylate, that on decarboxylation at position 5 of the pyrimidine ring yields the major nucleotide uridylate (uridine 5′-phosphate).

orthogonal arrays Arrays that are at (approximately) right angles to one another. Confluent fibroblasts often become organized into such arrays; other examples are the packing of collagen fibres in the cornea, and cellulose fibrils in the plant cell wall.

orthograde transport Axonal transport from the cell body of the neuron towards the synaptic terminal. Opposite of retrograde transport and probably dependent on a different mechanochemical protein (almost definitely ***kinesin***) interacting with microtubules.

orthokinesis ***Kinesis*** in which the speed or frequency of movement is increased (positive orthokinesis) or decreased.

orthologous genes Genes related by common phylogenetic descent. Contrast with ***paralogous genes***.

Orthomyxoviridae Class V ***viruses***. The genome consists of a single negative strand of RNA that is present as several separate segments each of which acts as a template for a single mRNA. The ***nucleocapsid*** is helical and has a viral-specific RNA polymerase for the synthesis of the mRNAs. They leave cells by budding out of the plasma membrane and are thus enveloped. They usually have two classes of spike protein in the envelope. One has ***haemagglutinin*** activity and the other acts as a ***neuraminidase*** and both are important in the invasion of cells by the virus. The major viruses of this group are the influenza viruses.

Orthopoxviridae Genus of double-stranded DNA viruses (250–390 × 200–260nm) that preferentially infect epithelial cells. Includes variola (smallpox) and vaccinia.

Oryctolagus cuniculus Rabbit.

Oryza sativa Rice.

oscillator Something that changes regularly or cyclically. Examples: oscillator neurons, which generate regular breathing or locomotory rhythms; slime moulds which secrete cyclic AMP in regular pulses.

Oscillatoria princeps Large cyanobacterium that exhibits gliding movements, possibly involving the activity of helically arranged cytoplasmic fibrils of 6–9nm diameter.

oscillin Soluble protein (oligomeric: 33 kD subunits) from mammalian sperm that is involved with the oscillations in calcium concentration that occur in the egg following fertilization. Has sequence similarity with prokaryote hexose phosphate isomerase.

OSCP See ***oligomycin sensitivity conferral protein***.

osk See ***oskar***.

oskar *(osk)* An ***egg-polarity gene*** in ***Drosophila***, concentrated at the posterior pole of the egg, and required for subsequent posterior structures. A ***maternal-effect gene***.

osmiophilic Having an affinity for ***osmium tetroxide***.

osmium tetroxide (OsO_4) Used as a post-fixative/stain in electron microscopy. Membranes in particular are osmiophilic, ie. bind osmium tetroxide.

osmoregulation Processes by which a cell regulates its internal osmotic pressure. These may include water transport, ion accumulation or loss, synthesis of osmotically active substances such as glycerol in the alga *Dunaliella*, activation of membrane ATPases, etc.

osmosis The movement of solvent through a membrane impermeable to solute, in order to balance the chemical potential due to the concentration differences on each side of the membrane. Frequently misused in the popular press.

osmotic pressure See ***osmosis***. The pressure required to prevent osmotic flow across a semi-permeable membrane separating two solutions of different solute concentration. Equal to the pressure that can be set up by osmotic flow in this system.

osmotic shock Passage of solvent into a membrane-bound structure due to osmosis, causing rupture of the membrane. A method of lysing cells or organelles.

osteoarthritis Disease of joints due to mechanical trauma. There is major disturbance in homeostasis of extracellular matrix with cartilage degradation (involving ***matrix metalloproteinases***) and loss of normal joint function. Unlike ***rheumatoid arthritis*** there is not an autoimmune element and, though there is inflammation, it is generally considered to be secondary rather than causative.

osteoblast Mesodermal cell that gives rise to bone.

osteocalcin (bone γ-carboxyglutamic acid protein: BGP) Polypeptide of 50 residues formed from a 76–77 amino acid precursor, and found in the extracellular matrix of bone. Binds hydroxyapatite. Has limited homology of its leader sequence with that of other vitamin K-dependent proteins such as ***prothrombin***, factors IX and X and ***protein C***.

osteoclast Large multinucleate cell formed from differentiated ***macrophage***, responsible for breakdown of bone.

osteocyte ***Osteoblast*** that is embedded in bony tissue and which is relatively inactive.

osteogenesis Production of bone.

osteogenesis imperfecta Heterogeneous group of human genetic disorders that affect connective tissue in bone, cartilage and tendon. Bones are very brittle and fracture-prone.

osteogenin Bone-inducing protein (less than 50 kD) associated with extracellular matrix. Binds heparin. See ***bone morphogenetic protein***.

osteoid Uncalcified bone matrix, the product of osteoblasts. Consists mainly of collagen, but has ***osteonectin*** present.

osteoma Benign tumour of bone.

osteomalacia Softening of bone caused by vitamin D deficiency: adult equivalent of rickets.

osteomyelitis Inflammation of bone marrow caused by infection.

osteonectin (basement membrane protein BM-40; SPARC) Calcium-binding protein of bone, containing the ***EF-hand*** motif. Binds to both collagen and ***hydroxyapatite***.

osteopetrosis The formation of abnormally dense bone, as opposed to osteoporosis.

osteophytes Small dense areas seen in X-ray pictures of bone around margin of joints with osteoarthritis. Sometimes larger, more porotic protrusions are also seen.

osteopontin Bone-specific sialoprotein (57 kD: probably two similar peptides) that links cells and the hydroxyapatite of mineralised matrix; has ***RGD*** sequence. Found only in calcified bone, probably produced by osteoblasts.

osteoporosis Loss of bony tissue; associated with low levels of ***oestrogen*** in older women.

osteosarcoma Malignant tumour of bone (probably neoplasia of ***osteocytes***).

ouabain (Strophanthin G) A plant alkaloid from *Strophantus gratus*, that specifically binds to and inhibits the ***sodium–potassium ATPase***. Related to ***digitalis***.

Ouchterlony assay Immunological test for antigen–antibody reactions in which diffusion of soluble antigen and antibody in a gel leads to precipitation of an antigen–antibody complex, visible usually as a whitish band. The system has the advantage that, because of radial diffusion of the reagents, a very wide range of ratios of antigen to antibody concentration develop; thus it is likely that precipitation will occur somewhere in the gel even when no care is taken with quantitation of the system.

outron Found at the 5′ end of pre-mRNAs, that are to be trans-spliced: contains an intron-like sequence, followed by a splice acceptor.

outside-out patch A variant of ***patch clamp*** technique, in which a disc of plasma membrane covers the tip of the electrode, with the outer face of the plasma membrane facing outward, to the bath.

ovalbumin A major protein constituent of egg-white. A phosphoprotein of 386 amino acids (44 kD) with one N-linked oligosaccharide chain. Synthesis is stimulated by oestrogen. The gene, of which there is only one in the chicken genome, has 8 exons and is of 7.8 kbase; it was one of the first genes to be studied in this sort of detail.

ovalocytosis Hereditary disorder of erythrocytes relatively common in areas where malaria is endemic. Not only are the erythrocytes more rigid, but there is a mutation in ***band III***, the anion transporter.

ovarian follicle In mammals the group of cells around the primary ***oocyte*** proliferate and form a surrounding non-cellular layer. A space opens up in the follicle cells and the whole structure is then the ovarian (Graafian) follicle.

ovarian granulosa cells During oogenesis in mammals the ovarian (Graafian) follicle, in which the developing ovum lies, is lined with follicle cells; the peripheral follicle cells form the stratum granulosum or ovarian granulosa.

overlap index A measure of the extent to which a population of cells in culture forms multilayers. The predicted amount of overlapping is calculated knowing the cell density, the projected area of the nucleus (usually), and assuming a Poisson distribution. The actual overlap is measured on fixed and stained preparations and the ratio of actual:predicted is derived. A value of 1 implies a random distribution with no constraint on overlapping; normal fibroblasts may have values as low as 0.05. Although a useful measure it does not unambiguously indicate the reason for the effect, which may be ***contact inhibition of locomotion*** or differential adhesion of cells between substratum and other cells.

overlapping In cell locomotion, the situation in which the ***leading lamella*** of one cell moves actively over the dorsal surface of another cell; should be distinguished from ***underlapping***.

overlapping genes Different genes whose nucleotide coding sequences overlap to some extent. The common nucleotide sequence is read in two or three different reading frames thus specifying different polypeptides.

oviduct The tubular tract in female animals through which eggs are discharged either to the exterior or, in mammals, to the uterus.

Ovis aries Domestic sheep (eg. 'Dolly').

ovomucoid Egg-white protein produced in tubular gland cells in the epithelium of the chicken oviduct in response to progesterone or oestrogen.

ovotransferrin See ***conalbumin***.

ovum (plural, ova) An egg cell.

owl-eye cells Enlarged cells infected with ***Cytomegalovirus*** that contain large inclusion bodies surrounded by a halo, hence the name.

oxacladiellanes (1,4-oxacladiellanes) Class of compounds that includes ***sarcodictyin A*** and ***eleutherobin*** that stabilize microtubules, and the ***valvidones*** and ***eleuthosides*** that are anti-inflammatory.

oxalic acid Occurs in plants, and is toxic to higher animals by virtue of its calcium-binding properties; it causes the precipitation of calcium oxalate in the kidneys, prevents calcium uptake in the gut and is not metabolized.

oxaloacetate Metabolic intermediate. Couples with acetyl CoA to form citrate, ie. the entry point of the ***tricarboxylic acid cycle***. Formed from aspartic acid by transamination.

oxidation Occurs when a compound donates electrons to an oxidizing agent. Also combination with oxygen or removal of hydrogen in reactions where there is no overt passage of electrons from one species to another.

oxidation-reduction potential See ***redox potential***.

oxidative metabolism Respiration in the biochemical sense.

oxidative phosphorylation The phosphorylation of ATP coupled to the ***respiratory chain***.

oxidoreductase An oxidase that uses molecular oxygen as the electron acceptor.

oxygen electrode A sensitive method to detect oxygen consumption; involves a PTFE (Teflon) membrane.

oxygen radical Any oxygen species that carries an unpaired electron (except free oxygen). Examples are .OH, the hydroxyl radical and O_2^-, the superoxide anion. These radicals are very powerful oxidizing agents and cause structural damage to proteins and nucleic acids. They mediate the damaging effects of ionizing radiation.

oxygen-dependent killing One of the most important bactericidal mechanisms of mammalian phagocytes involves the production of various toxic oxygen species (hydrogen peroxide, superoxide, singlet oxygen, hydroxyl radicals) through the ***metabolic burst***. Although anaerobic killing is possible, the oxygen-dependent mechanism is crucial for normal resistance to infection, and a defect in this system is usually fatal within the first decade of life (***chronic granulomatous disease***). See ***myeloperoxidase, chemiluminescence***.

oxygenase Enzyme catalysing the incorporation of the oxygen of molecular oxygen into organic substrates. Dioxygenases (oxygen transferases) catalyse introduction of both atoms of molecular oxygen; monoxygenases (mixed function oxygenases) introduce one atom, the other becomes reduced to water, so that these enzymes require a second substrate, acting as oxygen donor. Both types are used by bacteria in degradation of aromatic compounds. Dioxygenases all contain iron, eg. tryp-2,3 dioxygenase. Examples of monooxygenases are the enzymes that hydroxylate proline and lysine of collagen, using α-ketoglutarate.

oxyntic cell (parietal cell) Cell of the gastric epithelium that secretes hydrochloric acid.

oxyntomodulin Peptide produced by cleavage of proglucagon by prohormone convertases. Cleavage also produced proglucagon in addition to glucagon itself.

oxyphil cells See ***Askenazy cells***.

oxysome Multimolecular array that acts as a unit in oxidative phosphorylation.

oxytocin A peptide hormone (1007 D) from hypothalamus: transported to the posterior lobe of the pituitary (see ***neurophysin***). Induces smooth muscle contraction in uterus and mammary glands. Related to ***vasopressin***.

Oxyuranus scutelatus scutelatus The Australian taipan snake. See ***taicatoxin***.

P

p21 (Waf1; Cip1) Cyclin dependent kinase inhibitor (21 kD) that inhibits multiple cdks.

p35, P36 See ***annexin***.

p38 Serine/threonine protein kinase activated by MAP kinase kinase (MKK6b) that acts in the signalling cascade downstream of various inflammatory cytokines such as IL-1 and TNFα. Homologous to the yeast HOG protein. See ***MEF2***.

p53 A 393 residue (in humans) phosphoprotein that is a common tumour antigen, expressed in many transformed cells. However, it is believed to be the product of a ***tumour suppressor*** gene, rather than an ***oncogene***, as it is frequently inactivated or mutated in tumours. See Table O1.

P57 See ***neuromodulin***.

p57^{KIP2} ***Cyclin-dependent kinase inhibitor*** of the p21^{CIP1}, p27^{KIP1} family. Mice deficient in this inhibitor have major developmental defects similar to those seen in ***Beckwith–Wiedemann syndrome*** as a result of altered cell proliferation and differentiation. Inhibits G1/S phase ***cdks***. Encoded by a maternally imprinted gene in both human and mouse.

p65 Protein from human T-cells that has sequences in common with ***plastin***. Major target for IL-2 stimulated phosphorylation. On basis of sequence has two calcium binding sites, a calmodulin binding site and two actin binding sites.

p68 See ***annexin, DEAD-box helicases***.

p70 See ***annexin***.

p107 Protein (107 kD) with many similarities to ***retinoblastoma*** gene product. Binds to ***E2F*** and is found in the ***cyclin***–E2F complex together with p33cdk2.

p130^{cas} Substrate for ***src family*** kinases. Once phosphorylated can act as docking protein for proteins with ***SH2*** domains. Related protein is Sin/Efs.

P680 Form of chlorophyll that has its absorption maximum at 680nm. See ***photosystem II***.

P700 Form of chlorophyll that has its absorption maximum at 700nm. See ***photosystem I***.

p21-activated kinases (PAK-1, 2, 3) Serine/threonine protein kinases of the STE20 subfamily. Form an activated complex with GTP-bound ras-like proteins (p21, cdc2 and rac1). Activated, autophosphorylated PAK acts on a variety of targets, shows highly specific binding to the SH3 domains of phospholipase C γ and of adapter protein ***nck***, and regulates various morphological and cytoskeletal changes. ***Caspase***-mediated cleavage of 62 kD PAK2 generates a constitutively active catalytic fragment (34 kD) and induces apoptosis in ***Jurkat*** cells. Its activity is apparently essential for the formation of apoptotic bodies.

p34 kinase See ***cyclin***.

P388 D1 cells Clonal derivative of P388 cells, a mouse macrophage line that produces a large amount of ***interleukin-1***.

p56lck (lck) Lymphoid isoform of src-family tyrosine kinase. Expressed predominantly in thymocytes and peripheral T-cells. Associates with cytoplasmic domains of CD4 and CD8, and with the β chain of the IL-2 receptor.

P400 protein Inositol 1,4,5-triphosphate receptor (InsP3-R). Key protein to understanding mechanisms of InsP3-mediated Ca^{2+} mobilization. Originally found as a cerebellar glycoprotein of 250 kD and subsequently shown to be identical to the InsP3 receptor protein.

P See ***proline***.

P antigen Antigenic determinant on the surface of human red blood cells to which the Donath–Landsteiner antibody reacts. This antibody binds in the cold (a 'cold IgG'), but elutes from red cells at 37°C, is particularly associated with tertiary syphylis, and its binding causes paroxysmal nocturnal haemoglobinuria. (See also ***decay-accelerating factor***).

P domain See ***trefoil motif***.

P element A class of *Drosophila* ***transposon***, widely used as a vector for reporter genes, for efficient germ-line transformation, and for ***enhancer trap*** or ***insertional mutagenesis*** studies.

P-face See ***freeze fracture***.

P(GAL4) Synthetic ***P element*** of ***Drosophila*** *melanogaster*, comprising ***long-terminal repeats*** flanking a mini-*white* gene to mark flies carrying the P element by their red eye colour, ***Bluescript*** to allow plasmid rescue of DNA flanking the genomic insertion site, and a gene encoding the yeast transcription factor ***GAL4***, downstream of a weak (permissive) promoter. Although itself unable to move within the genome, P(GAL4) mobilization can be induced by crossing in a source of transposase. Patterns of expression of neighbouring genes can be detected (see ***enhancer trapping***) by crossing in a reporter gene (eg. ***lacZ***, ***green fluorescent protein*** (GFP)) under the control of the UAS promoter recognized by GAL4.

P glycoprotein See ***multidrug transporter***.

P light chain (DNTB light chain) Myosin light chain that can be phosphorylated by ***myosin light chain kinase***; as a result of phosphorylation, the myosin is activated.

P-loop See ***ATP-binding site***.

P-protein Protein found in large amounts in phloem sieve tubes. Appears as thin strands when seen in the electron microscope.

P-ring One of the bushes at the base of the flagellum of ***Gram-negative bacteria***, anchoring it in the peptidoglycan layer of the cells wall. Lies below the ***L-ring***.

P-selectin See ***selectins***.

P-site The peptidyl-tRNA binding site on the ***ribosome***, the one to which the growing chain is attached; the incoming ***aminoacyl tRNA*** attaches to the A site.

P-type ATPase One of three major classes of ion transport ***ATPases***, characterized by vanadate sensitivity and a phosphorylated intermediate. The archetype is the ***sodium pump***. See ***F-type ATPase***, ***V-type ATPase***.

P-type channels A class of ***voltage-sensitive calcium channels***. Found in various neurons but particularly in Purkinje cells of the cerebellum (hence the name). Involved in induction of long-term depression. Require substantial depolarization to activate and inactivate slowly. Inhibited by ***agatoxin*** and a polyamine FTX, through a G-protein coupled mechanism.

PACAP (pituitary adenylate cyclase-activating peptide) Genes for PACAP and its receptor are widely expressed in the mouse neural tube at day 10 of development and may play a role in patterning and regulated gene expression.

pachynema See ***pachytene***.

pachytene Classical term for the third stage of prophase I of meiosis, during which the homologous chromosomes are closely paired and ***crossing over*** takes place.

packaging Of a virus, the process by which the genetic material is encapsulated by the coat proteins.

paclitaxel Proprietary name for ***taxol***.

pactamycin Antibiotic that inhibits translation by blocking the binding of initiator tRNA to the initiator complex.

PADGEM (CD62P; P-selectin; GMP-140) One of the ***selectin*** family, present on megakaryocytes, activated platelets and activated endothelial cells; rapidly upregulated in platelets and endothelial cells and only transiently expressed. Mediates rolling of neutrophils, platelets and some T-cell subsets along luminal surface of vessels. Although knockout is defective in cellular infiltration into inflammatory sites it is necessary to inhibit both P-selectin and E-selectin to see total blockade.

PAGE See ***polyacrylamide gel electrophoresis***.

Paget's disease Breast carcinoma characterized by large cells with clear cytoplasm.

pagoda cells Ganglion cells, from the central nervous system of a leech, with a spontaneous firing pattern that can look a little like a pagoda on an oscilloscope.

PAI-1 (plasminogen activator inhibitor 1) PAI-1 and PAI-2 are plasma serpins that inhibit ***plasminogen***.

pair-rule gene A ***segmentation gene***, expressed sequentially between ***gap genes*** and ***segment-polarity genes***. In development of *Drosophila*, a set of about eight genes that are expressed only in alternate segments (odd or even) of the developing embryo. Loss-of-function mutants thus lack alternate segments. Examples: *even-skipped* (*eve*), *fushi tarazu* (*ftz*), *hairy*.

paired *(prd)* Developmentally regulated gene in *Drosophila* that contains the ***paired box domain***.

paired box domain Conserved domain of 128 amino acids, found in several developmentally regulated proteins in *Drosophila* (eg. paired, gooseberry, Pox), mouse and human (eg. Pax, HuP1, HuP48).

PAK-1, 2, 3 See ***p21-activated kinases***.

PAL See ***phenylalanine ammonia lyase***.

Palade pathway The routing of protein(s) from the site of their synthesis to the final cellular or secreted position. Several different pathways are known and others suspected. Glycosylation of the proteins may provide specific 'address labels' for the proteins.

palindromic sequence Nucleic acid sequence that is identical to its complementary strand when each is read in the correct direction (eg. TGGCCA). Palindromic sequences are often the recognition sites for ***restriction enzymes***. Degenerate palindromes with internal mismatching can lead to loops or hairpins being formed (as in tRNA).

palisade parenchyma Tissue found in the upper layers of the leaf ***mesophyll***, consisting of regularly shaped, elongated parenchyma cells, orientated perpendicular to the leaf surface, which are active in ***photosynthesis***.

Pallister–Killian syndrome Rare disorder with multiple congenital abnormalities, seizures and mental retardation. Cause is an extra metacentric chromosome 12p only in skin fibroblasts, so that the body is a tissue-specific mosaic.

palmitic acid (n-hexadecanoic acid) One of the most widely distributed of fatty acids. The palmitoyl residue is one of the common acyl residues of membrane phospholipids. It is also found as a thioester attached to cystein residues on some membrane proteins. The proteins so modified are often transmembrane proteins and the modified residue is on the cytoplasmic surface of the membrane. The specificity of the transferase for the acyl residue is not high and both stearoyl and oleoyl residues can replace the palmitoyl residue. (cf. ***myristoylation***).

palmitoylation See ***palmitic acid***.

palytoxin Linear peptide (2670 D) from corals of *Palythoa* spp. that binds to Na^+/K^+ATPase at a site overlapping that of ***ouabain*** and converts it into a channel. Extremely toxic and said to be the most potent animal-derived toxin.

pancreatic acinar cells Cells of the pancreas that secrete digestive enzymes; the archetypal secretory cell upon which much of the early work on the sequence of events in the secretory process was done.

pancytopenia Simultaneous decrease in the numbers of all blood cells: can be caused by aplastic anaemia, hypersplenism, or tumours of the marrow.

Pandorina Colonial phytomonad in which the cells are held together in a gelatinous matrix. More complex than ***Eudorina***, less complex than *Volvox*.

Paneth cells Coarsely granular secretory cells found in the basal regions of crypts in the small intestine.

panmictic A panmictic population is one in which there is random mating.

panning Method in which cells are added to a dish with a particular surface coat or a layer of other cells and the non-adherent cells are then washed off. Those that remain are expressing particular surface adhesive properties and can be cloned, or in the case of an expression library, the identity of the adhesion molecule can be determined.

pannus (1) Vascularized granulation tissue rich in fibroblasts, lymphocytes, and macrophages, derived from synovial tissue; overgrows the bearing surface of the joint in rheumatoid arthritis and is associated with the breakdown of the articular surface. (2) Granulation tissue that invades the cornea from the conjunctiva in response to inflammation.

pantonematic flagella ***Flagella*** without ***mastigonemes***; cf. ***hispid flagella***.

pantophysin Ubiquitously expressed ***synaptophysin*** homologue (29 kD) found in cells of non-neuroendocrine origin. May be a marker for small cytoplasmic transport vesicles.

pantothenic acid Vitamin of the B2 group. See Table V1.

PAP technique (1) Colloquial abbreviation for ***Papanicolaou's stain***; (2) Peroxidase–antiperoxidase method for obtaining an enhanced peroxidase reaction to indicate antibody binding to antigen. In the first stage the material, eg. a section, is reacted with a specific antiserum (say rat) against the antigen. In the next stage a large excess of (say) rabbit anti-rat immunoglobulin is applied so that only one of the binding sites is bound to the first antibody. Then a rat antiperoxidase antiserum is bound to the second antibody's unfilled sites and finally peroxidase is added and binds to the third antiserum before the peroxidase is used to develop a colour reaction.

papain Thiol protease (EC 3.4.22.2) from *Carica papaya* (pawpaw). Thermostable and will act in the presence of denaturing agents. Although it will cleave a variety of peptide bonds there is greatest activity one residue towards the C-terminus from a phenylalanine.

Papanicolaou's stain A complex stain for detecting malignant cells in cervical smears. Contains in separate staining stages (a) haematoxylin; (b) Orange-G phosphotungstic acid; (c) Light green, Bismarck Brown, Eosin and phosphotungstic acid.

papaverine Constituent of opium that acts as a smooth muscle relaxant probably by blocking membrane calcium channels and inhibiting phosphodiesterase.

paper chromatography Separation method in which filter paper is used as the support. Not a very sensitive method, but historically important as one of the first methods available for separating natural compounds.

Papilio glaucus Swallowtail butterfly.

papilla (1) A projection occurring in various animal tissues and organs. (2) A small blunt hair on plants.

papilloma Benign tumour of epithelium. Warts (caused by papilloma virus) are the most familiar example, and each is a clone derived from a single infected cell.

papillomaviruses Genus of ***Papovaviridae***. See ***papilloma, Shope papilloma virus***.

Papovaviridae (papovavirus) Family of oncogenic ***DNA viruses*** including papilloma, polyoma and simian vacuolating virus (SV40). Non-enveloped small viruses that mainly infect mammals.

papule Small raised spot on skin (as in the rash of chickenpox).

PAR See ***protease-activated receptor***.

PAR1, 2, 3 See ***protease-activated receptor***.

parabiosis Surgical linkage of two organisms so that their circulatory systems interconnect.

Paracentrotus lividus Species of sea urchin commonly used in developmental biology.

paracortex Midcortical region of lymph node; area that is particularly depleted of T-lymphocytes in thymectomized animals, and is referred to as the thymus-dependent area.

paracrine Form of signalling in which the target cell is close to the signal-releasing cell. Neurotransmitters and neurohormones are usually considered to fall into this category.

parainfluenza virus Species of the ***Paramyxoviridae***; there are four types: type 1 is also known as Sendai virus or haemagglutinating virus of Japan (HVJ), and the inactivated form is used to bring about cell fusion; types 2–4 cause mild respiratory infections in humans.

paralogous genes Genes that result from duplication of existing genes and then divergence of function. Contrast with ***orthologous genes***.

Paramecium *(Paramoecium)* Genus of ciliate protozoans. The 'slipper animalcule' is cigar-shaped, covered in rows of cilia and about 250μm long. Free-swimming, common in freshwater ponds, feeds on bacteria and other particles. Reproduces asexually by binary fission, and sexually by conjugation involving the exchange of micronuclei. See ***kappa particle***.

Paramoecium See ***Paramecium***.

paramural body Membranous structure located between the plasma membrane and cell wall of plant cells. If it contains internal membranes, it may be called a ***lomasome***; if not, it may be termed a plasmalemmasome.

paramylon Storage polysaccharide of ***Euglena*** and related algae, present as a discrete granule in the cytoplasm and consisting of β(1-3)-glucan.

paramyosin Protein (200–220 kD) that forms a core in the thick filaments of invertebrate muscles. The molecule is rather like the rod part of myosin and has a two-chain coiled-coil α-helical structure, 130 × 2nm. Paramyosin is present in particularly high concentration in the 'catch' muscle of bivalve molluscs, where it forms the almost crystalline core of the thick filaments.

Paramyxoviridae (paramyxovirus) Class V viruses of vertebrates. The genome consists of a single negative strand of RNA as one piece. The helical nucleocapsid has a virus-specific RNA polymerase (transcriptase) associated with it. They are enveloped viruses: main members are ***Newcastle disease virus, measles virus***, and ***parainfluenza virus***.

paranemin Developmentally regulated protein (280 kD) associated with desmin and vimentin filaments.

paraoxonase (PON1) Human serum protein (45 kD glycoprotein) located on high density ***lipoprotein*** (HDL) and that has been implicated in detoxification of organophosphates and possibly preventing the oxidation of LDL. Mice lacking serum paraoxonase are susceptible to toxicity and to atherosclerosis.

parasegment In development of *Drosophila*, the genetic boundaries between developing segments are thought to lie along the middle of each visible segment. To distinguish them from the segments in everyday use, these compartments are called 'parasegments'.

parasitaemia Infection of a host by a parasite or the level of infection by the parasite, depending upon context.

parasympathetic nervous system One of the two divisions of the vertebrate ***autonomic nervous system***. Parasympathetic nerves emerge cranially as preganglionic fibres from oculomotor, facial, glossopharyngeal and vagus, and from the sacral region of the spinal cord. Most neurons are cholinergic and responses are mediated by ***muscarinic acetylcholine receptors***. The parasympathetic system innervates, for example, salivary glands, thoracic and abdominal viscera, bladder and genitalia. cf. ***sympathetic nervous system***.

parathormone (parathyrin; parathyroid hormone) A peptide hormone of 84 amino acids (9402 D). Stimulates osteoclasts to increase blood calcium levels, the opposite effect to ***calcitonin***.

parathyroid hormone See ***parathormone***.

paratope In immune network theory, an ***idiotope***; an antigenic site of an antibody that is responsible for that antibody binding to an antigenic determinant (epitope). Also used of the site on a ligand molecule to which a cell-surface receptor binds.

paraxial Lying along an axis: commonest use is in reference to paraxial mesoderm, the mesoderm that forms somites as opposed to the axial mesoderm that forms notochord.

Pardachirus marmoratus Red sea flatfish. See ***pardaxin***.

pardaxin Polypeptide (33 residues) from toxin gland of *Pardachirus marmoratus* that forms an eight-subunit voltage-dependent pore that will induce neurotransmitter release.

parenchyma Type of unspecialized cell making up the ground tissue of plants. The cells are large and usually highly vacuolated, with thin, unlignified walls. They are often photosynthetic, in which case they may be termed ***chlorenchyma***.

parenteral Administration of a substance to an animal by any route other than the alimentary canal.

parenthesome Structure shaped rather like a parenthesis '(', found on either side of pores in the septum of a basidiomycete fungus. More logically called septal pore caps.

parfocal Microscope objectives that are mounted in such a way that changing objectives does not put the specimen out of focus are parfocal.

parkin Gene found to be mutated in unusual form of ***Parkinsonism*** (autosomal recessive juvenile Parkinsonism). Located on long arm of chromosome 6. Gene is large (500 kb), very active in ***substantia nigra***, and codes for large protein that has one portion resembling ***ubiquitin*** and also has a ***zinc finger***.

Parkinsonism (paralysis agitans) Disease (Parkinson's disease) characterized by tremor and associated with the underproduction of L-DOPA (dihydroxyphenylalanine) by dopaminergic neurons and their death, particularly in the substantia nigra of the brain. Can be treated quite successfully in many cases by administration of L-DOPA.

paroral membrane In ciliates the cilia in the region of the mouth may be fused into a paroral membrane.

paroxysmal nocturnal haemoglobinuria Disease in which there is haemolysis by complement as a result of deficiency in ***decay-accelerating factor***.

PARP See ***poly(ADP-ribose) polymerase***.

parthenocarpy Fruit formation without fertilization. Occurs spontaneously in some plants, eg. banana, and in other plants can be induced by application of auxin. Results in seedless fruits.

parthenogenesis Development of an ovum without fusion of its nucleus with a male pronucleus to form a zygote.

partial agonist Agonist for a receptor population that is unable to produce a maximal response even if all the receptors are occupied.

partition coefficient Equilibrium constant for the partitioning of a molecule between hydrophobic (oil) and hydrophilic (water) phases. A measure of the affinity of the molecule for hydrophobic environments, and thus, for example, a rough guide to the ease with which a molecule will cross the plasma membrane.

parvalbumins Calcium-binding proteins (12 kD), found in teleost and amphibian muscle, with sequence homology to ***calmodulin*** but only two ***EF-hand*** calcium-binding sites.

Parvoviridae Class II viruses. The genome of these simple viruses is single-stranded DNA and they have an icosahedral nucleocapsid. The autonomous parvoviruses have a negative strand DNA and include viruses of vertebrates and arthropods. The defective adeno-associated viruses cannot replicate in the absence of helper adenoviruses and have both positive and negative stranded genomes, but packaged in separate virions.

PAS (1) See ***periodic acid–Schiff reaction***. (2) p-amino-salicylic acid.

PAS genes Genes essential for the biogenesis and proliferation of peroxisomes in yeast (*S. cerevisiae*). PAS1 codes for a rather hydrophilic 117 kD protein with two ATP-binding sites and similarity with some ATPases, PAS2 codes for a 183 residue polypeptide that seems to be a member of the ubiquitin-conjugating protein family, PAS3 codes for a 48 kD integral membrane protein that may be part of the import machinery.

pasin Proteins of unknown function bound, on the cytoplasmic face, to the sodium–potassium ATPase. Two pasins have been identified, pasin 1 (77 kD) and pasin 2 (73 kD). The name is derived from ATPase-associated.

passage Term that derives originally from maintenance of, for example, a parasite by serially infecting host animals, passaging the parasite each time. Subsequently also used to describe the subculture of cells in culture, and therefore not equivalent to cell division number.

passive immunity Immunity acquired by the transfer from another animal of antibody or sensitized lymphocytes. Passive transfer of antibody from mother to offspring is important for immune defence during the perinatal period.

passive transport The movement of a substance, usually across a plasma membrane, by a mechanism that does not require metabolic energy. See ***active transport, transport protein, facilitated diffusion, ion channels***.

Pasteur effect Decrease in the rate of carbohydrate breakdown that occurs in yeast and other cells when switched from anaerobic to aerobic conditions. Results from a relatively slow flux of material through the biochemical pathways of respiration compared with those of fermentation.

Yersinia pestis **(**Pasteurella pestis**)** Causative agent of black death plague.

Patau syndrome (trisomy 13) Set of congenital defects in man caused by presence of an extra chromosome 13.

patch clamp A specialized and powerful variant of ***voltage clamp*** method, in which a patch electrode of relatively large tip diameter (5μm) is pressed tightly against the plasma membrane of a cell, forming an electrically tight, 'gigohm' seal. The current flowing through individual ***ion channels*** can then be measured. Different variants on this technique allow different surfaces of the plasma membrane to be exposed to the bathing medium: the contact just described is a 'cell-attached patch'. If the electrode is pulled away, leaving just a small disc of plasma membrane occluding the tip of the electrode, it is called an 'inside-out patch'. If suction is applied to a cell-attached patch, bursting the plasma membrane under the electrode, a 'whole cell patch' (similar to an intracellular recording) is formed. If the electrode is withdrawn from the whole cell patch, the membrane fragments adhering to the electrode reform a seal across the tip, forming an 'outside-out patch'.

patching Passive process in which integral membrane components become clustered following crosslinking by an external or internal polyvalent ligand. See ***capping***.

pathogenic Capable of causing disease.

pathognomic A sign or symptom that is diagnostic of a disease.

pattern formation One of the classic problems in developmental biology is the way in which complex patterns are formed from an apparently uniform field of cells. Various hypotheses have been put forward, and there is now evidence for the existence of gradients of diffusible substances (morphogens) specifying the differentiative pathway that should be followed according to the concentration of the ***morphogen*** around the cell.

pavementing Term used to describe the ***margination*** of leucocytes on the endothelium near a site of damage.

pawn Mutant of *Paramecium* that, like the chess piece, can only move forward and is unable to reverse to escape noxious stimuli. Defect is apparently in the voltage-sensitive calcium channel of the ciliary membrane.

***Pax* genes** Mouse genes that contain a DNA-binding domain similar to one in the paired genes of *Drosophila*. Eight *Pax* genes have been identified, and most of them are expressed in the nervous system during development. A number of mouse mutations have been found to map to *Pax* genes. For example, *undulated*, which causes distortions of the vertebral column and sternum results from a point mutations of *Pax*-1, and is expressed in the sclerotome.

paxillin Cytoskeletal protein (68 kD) that localizes, like ***talin***, to focal adhesions, to dense plaques in smooth muscle, and to the myotendonous and neuromuscular junctions of skeletal muscle. Binds to ***vinculin***.

PBMC (peripheral blood mononuclear cells) A mixture of monocytes and lymphocytes; blood leucocytes from which granulocytes have been separated and removed.

PBP See ***platelet basic protein***.

Pc See ***Polycomb***.

PC12 A rat ***phaeochromocytoma*** cell line from adrenal medulla. Widely used in the study of stimulus–secretion coupling, and because it differentiates to resemble sympathetic neurons on application of nerve growth factor.

pCEF-4 See ***9E3***.

PCNA (proliferating cell nuclear antigen) Commonly used marker for proliferating cells, a 35 kD protein that associates as a trimer, and as a trimer interacts with DNA polymerases δ and ε; acts as an auxiliary factor for DNA repair and replication. Transcription of PCNA is modulated by p53.

PCP See ***Angel dust***.

PCR See ***polymerase chain reaction***.

PCR *in situ* hybridization New technique for detection of very rare mRNA or viral transcripts in a tissue. Tissue sections are subjected to ***PCR***, usually in a temperature-cycling oven, before detection of the (hugely amplified) transcript.

PD-ECGF (platelet-derived endothelial cell growth factor) Cytokine (471 residues) produced by platelets, fibroblasts and smooth muscle cells. Stimulates endothelial proliferation ***in vitro*** and angiogenesis ***in vivo***. Also promotes survival and differentiation of neurons. ***Gliostatin*** is related.

PDE (phosphodiesterase) Any enzyme (in EC 3.1 class) that catalyses the hydrolysis of one of the two ester linkages in a phosphodiester. PDE-I (EC 3.1.4.1) catalyses removal of 5′-nucleotides from the 3′ end of an oligonucleotide. PDE-II (EC 3.1.16.1) catalyses removal of 3′-nucleotides from the 5′-end of a nucleic acid. Often the name is used loosely when cAMP-phosphodiesterase is meant. See cyclic ***nucleotide phosphodiesterases***.

PDGF See ***platelet-derived growth factor***.

PDZ domain Domains found in various intracellular signalling proteins associated with the plasma membrane; named for the postsynaptic density, Discs-large, ZO-1 proteins in which they were first described. May mediate formation of membrane-bound macromolecular complexes, eg. of receptors and channels, by homotypic interaction, also of cell–cell junctions. Usually bind to short linear C-terminal sequences in the protein with which they interact.

PEA-15 (phosphoprotein enriched in astrocytes, 15 kD) PEA-15 is an acidic serine-phosphorylated protein highly expressed in the CNS, where it can play a protective role against cytokine-induced apoptosis. PEA-15 is phosphorylated in astrocytes by CaMKII (or a related kinase) and by protein kinase C in response to endothelin

peanut agglutinin ***Lectin*** from *Arachis hypogaea* that binds to ***glycoproteins*** containing β-D-gal (1-3) D-galNAc in membranes; used to investigate differential adhesiveness in developing systems.

Pecten (scallop) A bivalve mollusc. The adductor muscle, a ***catch muscle***, has been a favourite with muscle physiologists and biochemists as well as with gourmets.

pectin Class of plant cell wall polysaccharide, soluble in hot aqueous solutions of chelating agents or in hot dilute acid. Includes polysaccharides rich in galacturonic acid, rhamnose, arabinose and galactose, eg. the polygalacturonans, rhamnogalacturonans, and some arabinans, galactans and arabinogalactans. Prominent in the ***middle-lamella*** and ***primary cell wall***.

pedicels See ***podocytes***.

PEG See ***polyethylene glycol***.

pelB The 18-residue N-terminal leader sequence of pelB is commonly used in various vector constructs. The gene from which it comes, *pelB*, codes for pectin lyase B, one of the many virulence factors of ***Erwinia chrysanthemi***, pectinases that degrade cell walls of plants.

Pelizaeus–Merzbacher disease Dysmyelinating disease resulting from defects in PLP (proteolipid protein) gene. Mouse model is ***jimpy***.

pelle *Drosophila* protein kinase that is involved in the activation of ***dorsal*** (NFκB homologue).

pellicle The outer covering of a ***protozoan***: the plasma membrane plus underlying reinforcing structures, for example the membrane-bounded spaces (alveoli) just below the plasma membrane in ciliates.

PEM (polymorphic epithelial mucin) See ***episialin***.

pemphigus A group of dermatological diseases characterized by the production of bullae (blisters).

penetrance The proportion of individuals with a specific genotype who express that character in the phenotype.

penicillamine (dimethyl cysteine) Product of acid hydrolysis of ***penicillin*** that chelates heavy metals (lead, copper, mercury) and assists in their excretion in cases of poisoning. Also used in treatment of rheumatoid arthritis although its mode of action as an anti-rheumatic drug is not clear.

penicillin Probably the best known of the antibiotics, derived from the mould *Penicillium notatum*. It blocks the crosslinking reaction in ***peptidoglycan*** synthesis, and therefore destroys the bacterial cell wall making the bacterium very susceptible to damage.

pentobarbital An anticonvulsant and anaesthetic, usually used as the sodium or calcium salt.

pentosan Glycan that, when hydrolyzed, yields only pentoses.

pentose phosphate pathway (pentose shunt; hexose monophosphate pathway; phosphogluconate oxidative pathway) Alternative metabolic route to ***Embden–Meyerhof pathway*** for breakdown of glucose. Diverges from this when ***glucose-6-phosphate*** is oxidized to ribose 5-phosphate by the enzyme glucose-6-phosphate dehydrogenase. This step reduces ***NADP*** to NADPH, generating a source of reducing power in cells for use in reductive biosyntheses. In plants, part of the pathway functions in the formation of hexoses from CO_2 in photosynthesis. Also important as source of pentoses, eg. for nucleic acid biosynthesis. This pathway is the main metabolic pathway in neutrophil leucocytes; congenital deficiency in the pathway produces sensitivity to infection.

pentoses Sugars (monosaccharide) with 5 carbon atoms. Include ***ribose*** and ***deoxyribose*** of nucleic acids, and many others such as the aldoses ***arabinose*** and ***xylose***, and the ketoses ***ribulose*** and ***xylulose***.

pentraxins Family of proteins that share a discoid arrangement of five non-covalently linked subunits. Includes ***CRP*** and ***serum amyloid P***.

PEP See ***phosphoenolpyruvate***.

PEP carboxylase Enzyme responsible for the primary fixation of CO_2 in ***C4 plants***. Carboxylates PEP (***phosphoenolpyruvate***) to give oxaloacetate. Also important in ***crassulacean acid metabolism***, since it is responsible for CO_2 fixation in the dark.

pepsin Acid protease (EC 3.4.23.1) from stomach of vertebrates. Cleaves preferentially between two hydrophobic amino acids (eg. F-L, F-Y,), and will attack most proteins except protamines, keratin and highly glycosylated proteins. A single-chain phosphoprotein (327 amino acids; 34.5 kD) released from the enzymatically inactive zymogen, pepsinogen, by autocatalysis at acid pH in the presence of HCl. One of the peptides cleaved off in this process is a pepsin inhibitor and has to be further degraded to allow the pepsin to have full activity.

pepsinogen The inactive precursor (42.5 kD) of ***pepsin***.

pepstatin Peptide from *Streptomyces* spp. that inhibits pepsin and other aspartic proteases, eg. cathepsin D and renin.

peptidase Alternative name for a ***protease***.

peptide bond The amide linkage between the carboxyl group of one amino acid and the amino group of another. The linkage does not allow free rotation and can occur in *cis* or *trans* configuration, the latter the most common in natural peptides, except for links to the amino group of proline, which are always *cis*.

peptide map Proteases will produce fragments of a characteristic size from a protein, and this can be used as a test for the identity or otherwise of two similar-sized proteins. It is possible to produce a peptide fragment map from a single gel band.

peptide neurotransmitter Small peptides used as primary or co-transmitters in nerve cells, eg. ***FMRF-amide***, ***FLRF-amide***.

peptide nucleic acid (PNA) Synthetic ***nucleic acid*** mimic, in which the sugar-phosphate backbone is replaced by a peptide-like polyamide. Instead of 5′ and 3′ ends, PNAs have N- and C-termini. Their resistance to both ***nucleases*** and ***proteinases***, and their ability to bind closely to complementary DNA or RNA sequences, have made them promising candidates in ***antisense*** and ***gene theapy*** technologies.

peptide receptor Specific receptor for ***peptide neurotransmitters***.

peptidoglycan (murein) Crosslinked polysaccharide–peptide complex of indefinite size found in the inner cell wall of all bacteria (50% of the wall in ***Gram-negative***, 10% in ***Gram-positive***). Consists of chains of approximately 20 residues of β (1-4)-linked N-acetyl glucosamine and N-acetyl muramic acid crosslinked by small peptides (4–10 residues).

peptidyl transferase (EC 2.3.2.12) Integral enzymic activity of the large subunit of a ribosome, catalysing the formation of a peptide bond between the carboxy-terminus of the nascent chain and the amino group of an arriving tRNA-associated amino acid.

peptidyl-arginine deiminase (PAD) EC 3.5.3.15 Enzyme responsible for formation of protein-bound citrulline, a major amino acid in the inner root sheath and medulla of the hair follicle. Substrate is ***trichohyalin*** and postsynthetic modification of trichohyalin by PAD alters its properties so that it is able to act as a rigid matrix component.

peptidyl-prolyl *cis-trans* isomerase See ***PPIase*** and ***immunophilin***.

per See ***period***.

Percoll Trademark for colloidal silica coated with polyvinylpyrrolidone that is used for density gradients. Inert and will form a good gradient rapidly when centrifuged. Useful for the separation of cells, viruses, and subcellular organelles.

perforins Perforins 1 and 2 form tubular transmembrane complexes (16nm diameter) at the sites of target cell lysis by ***NK cells*** and ***cytotoxic T-cells***.

perfringolysin O (theta toxin; θ-toxin) ***Cholesterol-binding toxin*** from *Clostridium perfringens*. Shares with other ***thiol-activated haemolysins*** a highly conserved sequence (ECTGLAWEWWR) near the C-terminus.

periaxin Protein originally suggested to initiate myelin deposition in peripheral nerves. Two isoforms exist coded by a single gene. L-periaxin (147 kD) is localized to plasma membrane of ***Schwann cells***, S-periaxin (16 kD) is diffusely cytoplasmic. Both possess ***PDZ domains***.

peribacteroid membrane Membrane derived from the plasma membrane of a plant cell and that surrounds the nitrogen-fixing bacteroids in legume root nodules. Has a high lipid content and may regulate the passage of material from the plant cell cytoplasm to the symbiotic bacterial cell. The idea that it restricts ***leghaemoglobin*** to the peribacteroid space seems untenable since leghaemoglobin is found in the cytoplasm of some cells.

pericanicular dense bodies Electrodense membrane-bounded cytoplasmic organelles found near the canaliculi in liver cells: lysosomes.

pericarp That part of a fruit that is produced by thickening of the ovary wall. Composed of three layers, epicarp (skin), mesocarp (often fleshy) and endocarp (membranous or stony in the case of, eg. plum).

pericentric inversion Chromosomal inversion in which the region that is inverted includes the kinetochore.

pericentrin Conserved protein (200–220 kD) of the pericentriolar region involved in organization of microtubules during meiosis and mitosis; concentration highest at metaphase, lowest at telophase.

pericentriolar region Rather amorphous region of electron-dense material surrounding the centriole in animal cells: the major ***microtubule organizing centre*** of the cell.

perichondrial cell Cell of the perichondrium, the fibrous connective tissue surrounding cartilage.

pericyte Cell associated with the walls of small blood vessels: neither a smooth muscle cell, nor an endothelial cell.

periderm The outer cork layer of a plant that replaces the epidermis of primary tissues. Cells have their walls impregnated with ***cutin*** and ***suberin***.

perikaryon Cell body surrounding nucleus of a neuron; does not include axonal and dendritic processes.

perinuclear space Gap 10–40nm wide separating the two membranes of the ***nuclear envelope***.

period *(per) Drosophila* gene regulating circadian rhythm. Expressed in ***CNS***, ***Malpighian tubules***, and a number of other tissues. Per contains a PAS structural domain, a nuclear localization sequence and a cytoplasmic localization domain that restricts it to the cytoplasm in the absence of Tim (product of ***timeless***) with which it forms a heterodimer.

periodic acid–Schiff reaction (PAS) A method for staining carbohydrates: adjacent hydroxyl groups are oxidized to form aldehydes by periodic acid (HIO_4) and these aldehyde groups react with Schiff's reagent (basic fuchsin decolorized by sulphurous acid) to give a purple colour. Used in histochemistry and in staining gels on which glycoproteins have been run.

periodontal Adjective describing the region around the teeth, ie. gums and gingival crevice. Periodontal disease is a common consequence of inadequate phagocyte function.

periosteum Dense fibrous ***connective tissue*** covering all bones. Contains ***osteoblasts***, and has the capacity to generate new bone.

peripheral lymphoid tissue Secondary lymphoid tissue, not necessarily located peripherally. See ***lymphoid tissue***.

peripheral membrane protein Membrane proteins that are bound to the surface of the membrane and not integrated into the hydrophobic region. Usually soluble and were originally thought to bind to integral proteins by ionic and other weak forces (and could therefore be removed by high ionic strength, for example). However, it is now clear that some peripheral membrane proteins are covalently linked to molecules that are part of the membrane bilayer (see ***glypiation***), and that there are others that fit the original definition but are perhaps more appropriately considered proteins of the cytoskeleton (eg. Band 4.1 and ***spectrin***) or extracellular matrix (eg. ***fibronectin***).

peripherin (1) Type III intermediate filament protein (57–58 kD) coexpressed with ***neurofilament*** triplet proteins. (2) Photoreceptor-specific glycoprotein found on the rim region of rod outer segment disk membranes. Thought to be essential for assembly, orientation and physical stability of outer segment disks. Predicted sequence 346 residues, highly conserved between rodents, man and cattle. Mutations in the gene (RDS gene: retinal degeneration, slow) can cause retinitis pigmentosa or macular dystrophy.

Periplaneta Genus of insects that includes *P. americana*, the American cockroach, a favourite experimental animal.

periplasmic binding proteins Transport proteins located within the ***periplasmic space***. Some act as receptors for bacterial chemotaxis, interacting with ***MCPs***. Their mode of action is unclear.

periplasmic space Structureless region between the plasma membrane and the cell wall of ***Gram-negative bacteria***.

periseptal annulus Organelle associated with cell division in Gram negative bacteria. There are two circumferential zones of cell envelope in which membranous elements of the envelope are closely associated with ***murein***. The annuli appear early in division and in the region between them, the periseptal compartment, the division septum is formed.

peritoneal exudate A term most commonly used to describe the fluid drained from the peritoneal cavity some time after the injection of an irritant solution. For example, a standard method for obtaining neutrophil leucocytes is to inject intraperitoneally saline with glycogen (to activate complement) and drain off the leucocyte-rich peritoneal exudate some hours later.

permease General term for a membrane protein that increases the permeability of the plasma membrane to a particular molecule, by a process not requiring metabolic energy. See ***facilitated diffusion***.

permissive cells Cells of a type or species in which a particular virus can complete its replication cycle.

permissive temperature Of a temperature-sensitive mutation, a temperature at which the mutated gene product behaves normally, and so the cell or organism survives as if wild-type (compare with restrictive temperature, at which the gene product takes on a mutant phenotype).

Peromyscus Genus of mice native to Central and North America.

peroxidase A ***haem*** enzyme that catalyses reduction of hydrogen peroxide by a substrate that loses two hydrogen atoms. Within cells, may be localized in peroxisomes. Coloured reaction products allow detection of the enzyme with high sensitivity, so peroxidase-coupled antibodies are widely used in microscopy and ***ELISA***. ***Lactoperoxidase*** is used in the catalytic surface-labelling of cells by radioactive iodine.

peroxisome Organelle containing peroxidase and catalase, sometimes as a large crystal. A site of oxygen utilization, but not of ATP synthesis. In plants, associated with ***chloroplasts*** in ***photorespiration*** and considered to be part of a larger group of organelles, the ***microbodies***.

persistence (1) The tendency of a cell to continue moving in one direction: an internal bias on the random walk behaviour that cells exhibit in isotropic environments. (2) Of viruses that persist in a cell population, animal, plant or population for long periods often in a non-replicating form, by such strategies as integration into host DNA, immunological suppression, or mutation into forms with slow replication.

pertussis toxin Protein complex (ca 117 kD). An ***AB toxin***, the active subunit is a single polypeptide (28 kD), the binding subunit a pentamer (two heterodimers, 23 + 11.7 kD, 11.7 + 22 kD, and a monomer (9.3 kD) that binds the heterodimers). The active subunit ADP-ribosylates the α subunit of the inhibitory ***GTP-binding protein*** (Gi). Crucial to the pathogenicity of *Bordetella pertussis*, the causative agent of whooping cough. See ***ADP-ribosylation***.

PEST sequence (Pro-Glu-Ser-Thr) Amino acid motif that is thought to target cytoplasmic proteins for rapid proteolytic degradation.

petechia (plural, petechiae) Small round red-purple spot, not raised and caused by intradermal haemorrhage.

petite mutants A class of yeast mutants, most studied in *Saccharomyces cerevisiae*. Mutants grow slowly and rely on anaerobic respiration: mitochondria, although present, have reduced cristae and are functionally defective (termed promitochondria). There are three types of petite mutant: (i) segregational mutants that show Mendelian behaviour and result from mutations in mitochondrial genes located in the nucleus; (ii) neutral petites, which are recessive genotypes and result from the complete absence of mitochondrial DNA; and (iii) suppressive petites, in which most of the mitochondrial DNA is lost (60–99%), though what remains is often amplified.

Petromyzon (lamprey) Primitive marine vertebrate (Class Agnatha) with eel-like body and lacking true jaws. Their relatively simple nervous system has been studied in some detail.

Peyer's patches Lymphoid organs located in the submucosal tissue of the mammalian gut containing very high proportions of IgA-secreting precursor cells. The patches have B- and T-dependent regions and germinal centres. A specialized epithelium lies between the patch and the intestine. Involved in gut-associated immunity.

PF4 See ***cytokine***.

Pfr The form of ***phytochrome*** that absorbs light in the far red region, 730nm, and is thus converted to ***Pr***. It slowly and spontaneously converts to Pr in the dark.

PGs (prostaglandins; PGA, PGB, PGD, PGE, PGF, PGG, PGH, PGI) PGA is prostaglandin A, etc. PGI is more commonly known as prostacyclin.

pH ($-\log_{10}$ [H^+]) A logarithmic scale for the measurement of the acidity or alkalinity of an aqueous solution. Neutrality corresponds to pH 7, whereas a 1 molar solution of a strong acid would approach pH 0, and a 1 molar solution of a strong alkali would approach pH 14.

PH-30 Heterodimeric sperm surface transmembrane protein involved in sperm–egg fusion. The α subunit has some similarities to viral fusion proteins, and the β subunit has a domain similar to soluble integrin ligands (***disintegrins***).

PH domain (pleckstrin homology domain) Domain found in various intracellular signalling cascade proteins (eg. ***tec family kinases***). Seem to be involved in interactions with phospholipids, particularly PIP3. At one stage it was suggested that they were involved in the interaction with heterotrimeric G-proteins.

PHA See ***phytohaemagglutinin***.

phaeochromocytoma (USA, pheochromocytoma) A normally benign neoplasia (***neuroblastoma***) of the ***chromaffin tissue*** of the adrenal medulla. In culture, the cells secrete enormous quantities of ***catecholamines***, and can be induced to form neuron-like cells on addition of, eg. cyclic AMP or nerve growth factor. Excessive production of adrenaline and noradrenaline leads to secondary hypertension, sometimes paroxysmal.

Phaeophyta (USA, Pheophyta); brown algae) Division of algae, generally brown in colour, with multicellular, branched thalluses. Includes large seaweeds such as *Laminaria* and *Fucus*. The brown colour is due to the ***xanthophylls***, fucoxanthin and lutein. Many have ***laminarin*** as a food reserve and alginic acid as a wall component.

phage See ***bacteriophage***.

phage display library Phage library in which the insert is expressed as a translational fusion with a phage coat protein. This makes it particularly easy to screen with antibodies. Widely used to identify epitopes recognized by some particular antibody; commercial phage display libraries randomly encoding all possible six or seven residue peptides can be screened with the antibody, and the inserts of bound phage sequenced to build up a picture of the binding profile of the antibody.

phage integrase family See ***recombinases, site-specific recombination***.

phage typing Bacteria may be typed by their susceptibility to a range of bacteriophages though confusion may arise if the bacteria carry plasmids encoding ***restriction endonucleases***.

phagemid ***Bacteriophage*** whose genome contains a ***plasmid*** that can be excised by coinfection of the host with a Helper phage. Useful as ***vectors*** for library production, as the library can be amplified and screened as phage, but the inserts of selected ***plaques*** can readily be prepared as plasmids without subcloning. Example of a commercial phagemid: λ Zap, from which pBluescript can be excised with helper phage.

phages See ***bacteriophages***.

phagocyte A cell that is capable of phagocytosis. The main mammalian phagocytes are ***neutrophils*** and ***macrophages***.

phagocytic vesicle Membrane-bounded vesicle enclosing a particle internalised by a phagocyte. The primary phagocytic vesicle (phagosome) will subsequently fuse with ***lysosomes*** to form a secondary phagosome in which digestion will occur.

phagocytosis Uptake of particulate material by a cell (endocytosis). See ***opsonization, phagocyte***.

phakinin Eye-lens specific protein, 47 kD, that coassembles with ***filensin*** (three phakinin molecules per filensin) to form beaded-chain ***intermediate filaments***. Phakinin has very strong sequence homology with ***cytokeratins*** but lacks the rod domains that are involved in filament formation. It is effectively a tailless intermediate filament protein.

phalangeal cells Cells of the organ of Corti (in the inner ear).

phalloidin Cyclic peptide (789D) from the death cap fungus (*Amanita phalloides*) that binds to and stabilizes ***F-actin***. Fluorescent derivatives are used to stain actin in fixed and permeabilized cells, although there is some uptake by live cells.

pharate Of an insect, having its new cuticle formed beneath its present cuticle, and thus ready for its next moult.

pharmacodynamics The study of how drugs affect the body: contrast with ***pharmacokinetics***.

pharmacokinetics The study of what the body does to drugs, in contrast to ***pharmacodynamics***.

phase contrast microscopy A simple non-quantitative form of ***interference microscopy*** of great utility in visualizing live cells. Small differences in optical path length due to differences in refractive index and thickness of structures are visualized as differences in light intensity.

phase separation The separation of fluid phases that contain different concentrations of common components. Occurs with partially miscible solvents used in many biochemical separation methods. Also temperature-dependent phase separation occurs with some detergent solutions. With reference to membranes means the segregation of lipid components into 'domains' that have different chemical composition.

phase variation Alteration in the expression of surface antigens by bacteria. For example, ***Salmonella*** can express either of two forms of ***flagellin***, H1 and H2, that are coded by different genes. Control of which form is expressed is brought about by inversion of the promoter for the H2 gene, which if functional (non-inverted) is associated with the expression of H2 and the production of a repressor of the H1 gene. Inversion occurs about every 1000 bacterial divisions, and is under the control of another gene, *hin*, which is within the invertable sequence.

phaseollin A ***phytoalexin*** produced by *Phaseolus* (bean) plants in response to pathogenic attack or other stress.

phasic See ***adaptation***.

phasing of nucleosomes A non-random arrangement of ***nucleosomes*** on DNA, in which, at certain segments of the genome, nucleosomes are positioned in the same way relative to the nucleotide sequence in all cells. Most nucleosomes are arranged randomly, but phasing has been detected in some genes.

phasmid Hybrid phage/plasmid formed by integration of plasmid containing the *att* site, and lambda phage, mediated by phage integrase site-specific recombination.

phelloderm Tissue containing parenchyma-like cells, in the bark of tree roots and shoots. Produced by cell division in the ***phellogen***.

phellogen ***Meristematic*** tissue in plants, giving rise to cork (phellem) and ***phelloderm*** cells. Also termed 'cork cambium'.

phencyclidine (1-(1-phenylcyclohexyl) piperidine; angel dust; PCP) Anaesthetic and drug of a kind that can produce marked behavioural effects. Interacts with the ***NMDA receptor***.

phenocopy An environmentally produced phenotype simulating the effect of a particular genotype.

phenol red Dye used as pH indicator: changes from yellow to red in range 6.8–8.4. Very commonly used in tissue culture medium though it can interfere with luminescence assays.

phenolphthalein Indicator dye that is colourless at neutral pH and red-pink in slightly alkaline solutions. Also used as a laxative.

phenome Phenotypic equivalent of the genome: the sum of all phenotypic characters.

phenothiazines A group of antipsychotic drugs, thought to act by blocking dopaminergic transmission in the brain. Examples are ***chlorpromazine*** and ***trifluoperazine***. Trifluoperazine binds to and inhibits ***calmodulin*** and has been used experimentally to block calcium/calmodulin-controlled reactions.

phenotype The characteristics displayed by an organism under a particular set of environmental factors, regardless of the actual genotype of the organism.

phenyalanine (Phe; F) An essential amino acid with an aromatic side chain (165 D). See Table A2.

phenylalanine ammonia lyase (PAL) Enzyme involved in the synthesis of ***lignin*** and other phenolic compounds from phenylalanine. Used as an enzymic marker for lignification and other developmental processes in plant cells.

phenylephrine hydrochloride An α1-adrenergic agonist (204 D).

phenylketonuria (PKU) Congenital absence of phenylalanine hydroxylase (an enzyme that converts phenylalanine into tyrosine). Phenylalanine accumulates in blood and seriously impairs early neuronal development. The defect can be controlled by diet and is not serious if treated in this way. Incidence highest in Caucasians.

pheromone A volatile hormone or behaviour modifying agent. Normally used to describe sex attractants (eg. bombesin for the moth *Bombyx*) but includes volatile aggression stimulating agents (eg. isoamyl acetate in honeybees).

phi X-174 (ϕ X-174) Bacteriophage of *E. coli* with a single-stranded DNA genome and an icosahedral shell. This was the first DNA phage to be fully sequenced: the genome consists of 10 genes, some of which are ***overlapping genes***.

Philadelphia chromosome Characteristic chromosomal abnormality of chronic myelogenous ***leukaemia*** in which a portion of chromosome 22 is translocated to chromosome 9.

phl See ***polehole***.

phlebotomy The cutting of veins, a fancy name for taking blood by venepuncture, usually with a needle, not a knife.

phloem Tissue forming part of the plant vascular system, responsible for the transport of organic materials, especially sucrose, from the leaves to the rest of the plant. Consists of ***sieve tubes***, ***companion cells***, ***fibre cells***, and ***parenchyma***.

phorbol esters Polycyclic compounds isolated from croton oil in which two hydroxyl groups on neighbouring carbon atoms are esterified to fatty acids. The commonest of these derivatives is phorbol myristoyl acetate (PMA). Potent co-carcinogens or tumour promotors, they are diacyl glycerol analogues and activate protein kinase C irreversibly.

phormicin Insect ***defensin*** produced by the blowfly, *Phormia terranovae*.

phosducin Protein (33 kD) that inhibits Gs-GTPase activity (a ***GIP***). Isolated from bovine brain and found in retina, pineal gland and many other tissues. Activity of phosducin is inhibited if phosphorylated by a cAMP-dependent protein kinase.

phosphatases Enzymes that hydrolyze phosphomonoesters. Acid phosphatases are specific for the single-charged phosphate group and alkaline phosphatases for the double-charged group. These specificities do not overlap. The phosphatases comprise a very wide range of enzymes including broad- and narrow-specificity members. Phosphoprotein phosphatases specifically dephosphorylate a particular protein and are essential if phosphorylation is to be used as a reversible control system; they are also specific for phosphoserine/threonine or phosphotyrosine residues within the target protein.

phosphatides The family of phospholipids based on 1, 2 diacyl 3-phosphoglyceric acid. See ***phospholipids***.

phosphatidic acid (PA; diacyl glycerol 3-phosphate) The 'parent' structure for phosphatidyl phospholipids, present in low concentrations in membranes. The acyl groups are derived from long-chain fatty acids. An intermediate in the synthesis of diacyl glycerol, the immediate precursor of most of the phosphatidyl phospholipids (except phosphatidyl inositol) and of triacyl glycerols.

phosphatidyl choline (PC) The major phospholipid of most mammalian cell membranes where the 1-acyl residue is normally saturated and the 2-acyl residue unsaturated. Choline is attached to phosphatidic acid by a phosphodiester linkage. Major synthetic route is from diacyl glycerol and CDP-choline. Forms monolayers at an air–water interface, and forms bilayer structures (liposomes) if dispersed in aqueous medium. A zwitterion over a wide pH range. Readily hydrolyzed in dilute alkali.

phosphatidyl ethanolamine (PE) A major structural phospholipid in mammalian systems. Tends to be more abundant than phosphatidyl choline in the internal membranes of the cell, and is an abudant component of prokaryotic membranes. Ethanolamine is attached to phosphatidic acid by a phosphodiester linkage. Synthesis from diacyl glycerol and CDP-ethanolamine.

phosphatidyl inositol (PI) Very important minor phospholipid in eukaryotes, involved in signal transduction processes. Contains ***myo-inositol*** linked through the 1-hydroxyl group to phosphatidic acid. The 4-phosphate (PIP) and 4,5 bisphosphate derivatives (PIP2) are formed and broken down in membranes by the action of specific kinases and phosphatases (futile cycles). Signal-sensitive phospholipase C enzymes remove the inositol moiety, in particular from 1,4,5 trisphosphate (PIP2) as inositol 1,4,5-triphosphate (InsP3: IP3). Both the diacyl glycerol and inositol phosphate products act as ***second messengers***.

phosphatidyl serine (PS) An important 'minor' species of phospholipid in membranes. Serine is attached to phosphatidic acid by a phosphodiester linkage. Synthesis is from phosphatidyl ethanolamine by exchange of ethanolamine for serine. Distribution is asymmetric, as the molecule is only present on the cytoplasmic side of cellular membranes. It is negatively charged at physiological pH and interacts with divalent cations; involved in calcium-dependent interactions of proteins with membranes (eg. protein kinase C).

phosphocreatine (creatine phosphate) Present in high concentration (about 20mM) in striated muscle, and is synthesized and broken down by creatine phosphokinase to buffer ATP concentration. It acts as an immediate energy reserve for muscle.

phosphodiester bond Not a precise term. Refers to any molecule in which two parts are joined through a phosphate group. Examples are found in RNA, DNA, phospholipids, cyclic nucleotides, nucleotide diphosphates and triphosphates.

phosphodiesterase An enzyme that cleaves phosphodiesters to give a phosphomonoester and a free hydroxyl group. Examples include RNAase, DNAase, phospholipases C and D and the enzymes that convert cyclic nucleotides to the monoester forms. In casual usage the cAMP phosphodiesterase is usually meant.

phosphoenolpyruvate (PEP) An important metabolic intermediate. The enol (less stable) form of pyruvic acid is trapped as its phosphate ester, giving the molecule a high phosphate transfer potential. Formed from 2-phosphoglycerate by the action of enolase.

phosphofructokinase The pacemaker enzyme of glycolysis. Converts fructose 6-phosphate to fructose 1,6-bisphosphate. A tetrameric allosteric enzyme that is sensitive to the ATP: ADP ratio.

phosphoglycerate The molecules 2-phosphoglycerate and 3-phosphoglycerate are intermediates in glycolysis. 3-phosphoglycerate is the precursor for synthesis of phosphatidic acid and diacyl glycerol, hence of phosphatidyl phospholipids.

phospholipases (PLA; PLB; PLC; PLD) Enzymes that hydrolyze ester bonds in phospholipids. They comprise two types: aliphatic esterases (phospholipase A1, A2 and B) that release fatty acids, and phosphodiesterases (types C and D) that release diacyl glycerol or phosphatidic acid respectively. Type A2 is widely distributed in venoms and digestive secretions. Types A1, A2 and C (the latter specific for phosphatidyl inositol) are present in all mammalian tissues. Type C is also found as a highly toxic secretion product of pathogenic bacteria. Type B attacks monoacyl phospholipids and is poorly characterized. Type D is largely of plant origin. PLA2 type II (a secreted enzyme, but not the same as the type I digestive pancreatic enzyme) is probably very important in inflammation because its action can release arachidonic acid, the starting point for ***eicosanoid*** synthesis. Phosphatidylinositol bisphosphate-specific phospholipase C is important in generating ***diacylglycerol*** and ***inositol trisphosphate***, both ***second messengers***.

phospholipid The major structural lipid of most cellular membranes (except the chloroplast, which has galactolipids). Contain phosphate, usually as a diester.

Examples include phosphatidyl phospholipids, plasmalogens and sphingomyelins. See Table L3.

phospholipid bilayer A lamellar organization of phospholipids that are packed as a bilayer with hydrophobic acyl tails inwardly directed and polar head groups on the outside surfaces. It is this bilayer that forms the basis of membranes in cells, though in most cellular membranes a very substantial proportion of the area may be occupied by integral proteins. The triple-layered appearance of membranes seen in electron microscopy is thought to arise because the ***osmium tetroxide*** binds to the polar regions leaving a central, unstained, hydrophobic region.

phospholipid transfer protein Cytoplasmic proteins that bind phospholipids and facilitate their transfer between cellular membranes. May also cause net transfer from the site of synthesis.

phosphomannose See ***mannose-6-phosphate***.

phosphoprotein Proteins that contain phosphate groups esterified to serine, threonine or tyrosine (S, T or Y). The phosphate group usually regulates protein function.

phosphoramidite Nucleotide derivative used in oligonucleotide synthesis.

phosphorescence (1) Emission of light following absorption of radiation. Emitted light is of longer wavelength than the exciting radiation and is a result of decay of electrons from the triplet to the ground state. Lasts longer than fluorescence (electron decay from singlet to ground state) and occurs after a longer delay. (2) Popularly misused as a term for biological luminescence, eg. by fireflies.

phosphorylase (glycogen phosphorylase) Enzyme that catalyses the sequential removal of glycosyl residues from glycogen to yield one glucose-1-phosphate per reaction. Its activity is controlled by phosphorylation (by ***phosphorylase kinase***).

phosphorylase kinase The enzyme that regulates the activity of phosphorylase and glycogen synthetase by addition of phosphate groups. A large and complex enzyme, itself regulated by phosphorylation. Integrates the hormonal and calcium signals in muscle.

phosphorylation of proteins Addition of phosphate groups to hydroxyl groups on proteins (to the side chains of serine, threonine or tyrosine) catalysed by a protein kinase (often specific) with ATP as phosphate donor. Activity of proteins is often regulated by phosphorylation.

phosphotransferase An enzyme that transfers a phosphate group from a donor to an acceptor. Very important in metabolism.

phosphotyrosine Strictly speaking, tyrosine phosphate, but normally refers to the phosphate ester of a protein tyrosine residue. Present in very small amounts in tissues, but believed to be important in systems that regulate growth control, and is therefore of interest in studies of malignancy. The *src* gene product (pp60src) was one of the first kinases shown to phosphorylate at a tyrosine residue.

photoaffinity labelling A technique for covalently attaching a label or marker molecule onto another molecule such as a protein. The label, which is often fluorescent or radioactive, contains a group that becomes chemically reactive when illuminated (usually with ultraviolet light) and will form a covalent linkage with an appropriate group on the molecule to be labelled: proximity is essential. The most important class of photoreactive groups used are the aryl azides, which form short-lived but highly reactive nitrenes when illuminated.

photobleaching Light-induced change in a ***chromophore***, resulting in the loss of its absorption of light of a particular wavelength. A problem in fluorescence microscopy where prolonged illumination leads to progressive fading of the emitted light because less of the exciting wavelength is being absorbed.

photodynamic therapy Therapeutic approach in which a light-sensitive pro-drug is given and then the target area (usually a tumour) is illuminated to generate the active drug in the right place.

photodynesis Initiation of cytoplasmic streaming by light. Uncommon usage.

photolyase (DNA photolyase) Family of ubiqitous enzymes found in bacteria, archaebacteria and eukaryotes that can repair UV-induced DNA damage. The protein (between 454 to 614 residues) is associated with two prosthetic groups, FADH and a light-harvesting cofactor, MTHF (5,10-methenyltetrahydofolyl polyglutamate). Light is needed for the repair step.

photolysis Light-induced cleavage of a chemical bond, as in the process of photosynthesis.

photoperiodism Events triggered by duration of illumination or pattern of light/dark cycles: often the wavelength of the illuminating light is important, as for example in control of circadian rhythm in plants. See ***phytochromes***.

photophosphorylation The synthesis of ATP that takes place during photosynthesis. In non-cyclic photophosphorylation the photolysis of water produces electrons that generate a [proton motive force] which is used to produce ATP, the electrons finally being used to reduce $NADP^+$ to NADPH. When the cellular ratio of reduced to non-reduced NADP is high, ***cyclic photophosphorylation*** occurs and the electrons pass down an electron transport system and generate additional ATP, but no NADPH.

photopigment Pigment involved in ***photosynthesis*** in plants. Includes ***chlorophyll, carotenoids*** and ***phycobilins***.

photoreceptor A specialized cell type in a multicellular organism that is sensitive to light. This definition excludes single-celled organisms, but includes non-eye receptors, such as snake infra-red (heat) detectors or photosensitive pineal gland cells. See ***retinal rods, retinal cones***.

photorespiration Increased respiration that occurs in photosynthetic cells in the light, due to the ability of ***RuDP carboxylase*** to react with oxygen as well as carbon dioxide. Reduces the photosynthetic efficiency of ***C3 plants***.

photosynthesis Process by which green plants, algae, and some bacteria absorb light energy and use it to synthesize organic compounds (initially carbohydrates). In green plants, occurs in ***chloroplasts***, that contain the photosynthetic pigments. Occurs by slightly different processes in ***C3*** and ***C4 plants***. See also ***Z scheme of photosynthesis*** and contrast with ***chemosynthesis***.

photosynthetic bacteria Bacteria that are able to carry out ***photosynthesis***. Light is absorbed by ***bacteriochlorophyll*** and ***carotenoids***. Two principal classes are the green bacteria and the purple bacteria.

photosynthetic unit Group of photosynthetic pigment molecules (***chlorophylls*** and ***carotenoids***) that supply light to one ***reaction centre*** in ***photosystem I*** or ***photosystem II***.

photosystem I Photosynthetic system in ***chloroplasts*** in which light of up to 700nm is absorbed and its energy used to bring about charge separation in the ***thylakoid*** membrane. The electrons are passed to ferredoxin and then used to reduce NADP+ to NADPH (non-cyclic electron flow) or to provide energy for the phosphorylation of ADP to ATP (cyclic photophosphorylation).

photosystem II Photosynthetic system in ***chloroplasts*** in which light of up to 680nm is absorbed and its energy used to split water molecules, giving rise to a high energy reductant, Q–, and oxygen. The reductant is the starting point for an electron transport chain that leads to ***photosystem I*** and which is coupled to the phosphorylation of ADP to ATP.

phototaxis Movement of a cell or organism towards (positive phototaxis) or away from (negative phototaxis) a source of light.

phototransduction The transformation by photoreceptors (eg. ***retinal rods*** and ***cones***) of light energy into an electrical potential change.

phototrophic Any organism that can utilize light as a source of energy.

phototropism Movement or growth of part of an organism (eg. a plant shoot) towards (positive phototropism) a source of light, without overall movement of the whole organism.

phox (phox47; phox67) Components of the NADPH oxidase system in phagocytes, the system responsible for generating an oxidative burst and thus bacterial killing. Phox47 and phox67 are cytoplasmic and only associate with the integral membrane component following activation. Both contain ***SH3*** domains. Deletion or mutation in either leads to a form of chronic granulomatous disease.

phragmoplast Central region of mitotic spindle of a plant cell at telophase, in which vesicles gather and fuse to form the ***cell plate***, apparently guided by spindle microtubules.

phragmosome In plant cells, the region of the cytoplasm in which the nucleus is located during nuclear division. Can also refer to ***microbodies*** associated with the developing ***cell plate*** after nuclear division.

phycobilins Photosynthetic pigments found in certain algae, especially red algae (Rhodophyta) and ***cyanobacteria***.

phycobilisome An accessory light energy harvesting structure in ***cyanobacteria***. They have cores of allophycocyanin with radiating rods composed of discs of ***phycocyanin*** and ***phycoerythrin***. Linker polypeptides attach the core to the ***thylakoid*** membranes. These structures, 20–70nm across, contain the pigments named above that transfer light energy to chlorophyll *a*. The pigments are extracted and used as fluorochromes for labelling various probe reagents.

phycocyanin Blue ***phycobilin*** found in some algae, and especially in ***cyanobacteria***.

phycoerythrin Red ***phycobilins*** found in some algae, especially red algae (***Rhodophyta***).

Phycomycetes A group of fungi possessing hyphae that are usually non-septate (without cross walls).

physaliphorous cells Cells of chordoma (tumour derived from notochordal remnants) that appear vacuolated because they contain large intracytoplasmic droplets of mucoid material.

Physarum A member of the Myxomycetes or acellular slime moulds. Normally exists as a multinucleate ***plasmodium*** that may be many centimetres across, but if starved and stimulated by light will produce spores that later germinate to produce amoeboid cells, myxamoebae, which may transform into flagellated swarm cells. Either of these cell types may fuse to produce a zygote that forms the plasmodium by synchronous nuclear division. Easily grown in the laboratory and much used for studies on cytoplasmic streaming and on the cell cycle (because they show synchronous DNA synthesis and nuclear division).

physical mapping The process of assembling genomic DNA clones that completely cover a genetic locus. In ***genome projects***, this is an essential prerequisite for sequencing; in ***positional cloning***, it assists in designing a strategy to identify the gene of interest. The procedure is to screen candidate clones for a series of characteristic marker sequences, based either on ***satellite DNA*** or on ***PCR***-derived sequence tagged sites. Clones that share particular markers are assumed to overlap in that region, and computer analysis is used to identify the smallest set of clones that completely cover the region.

phytic acid Inositol hexaphosphate, found in plant cells, especially in seeds, where it acts as a storage compound for phosphate groups.

phytoalexins Toxic compounds produced by higher plants in response to attack by pathogens and to other stresses. Sometimes referred to as plant antibiotics, but rather non-specific, having a general fungicidal and bactericidal action. Production is triggered by ***elicitors***. Examples: ***pisatin, phaseollin***.

phytochelatins ***Metallothionein***-type peptides of plants that bind heavy metals such as cadmium, zinc, lead, mercury and copper. General form is (γ-glutamyl-cysteinyl) n-glycine where n is from 2 to 11. Involved in the detoxification of heavy metals and the homeostasis of non-essential metals.

phytochrome Plant pigment protein that absorbs red light and then initiates physiological responses governing light-sensitive processes such as germination, growth and flowering. Exists in two forms, ***Pr*** and ***Pfr***, that are interconverted by light.

phytohaemagglutinin (PHA) Sometimes used as synonym for ***lectins*** in general, but more usually refers to lectin from seeds of the red kidney bean *Phaseolus vulgaris*. Binds to oligosaccharide containing N-acetyl galactosyl residues. Binds to both B- and T-lymphocytes, but acts as a ***mitogen*** only for T-cells.

phytohormones See ***plant growth substances***.

phytol Long-chain fatty alcohol (C20) forming part of ***chlorophyll***, attached to the protoporphyrin ring by an ester linkage.

pi protein (π protein) Polypeptide (35 kD) that is required for the initiation of DNA replication in the R6K antibiotic-resistance plasmid, of which there are 12–18 copy equivalents in the *E. coli* chromosome.

PI3 kinase (phosphatidyl inositol-3-kinase; PI kinase) Lipid kinase that phosphorylates phosphatidylinositol phosphate on the 3 position. Now recognised to be key enzymes acting downstream of many receptors, particularly ***receptor tyrosine kinases*** such as ***PDGF***-receptor (in the case of Class Ia). Classical form has p85 regulatory subunit and p110 enzymatic subunit. The p85 adaptor associates with the cytoplasmic domain of various ***growth factor receptors*** through SH2 domains that bind to phosphotyrosine residues in the ligated (phosphorylated) receptor and with the catalytic subunit. An increasing family is being identified, some of which are regulated by ***G-proteins*** or calcium. Most are inhibited by ***wortmannin***.

pia mater Innermost of the three meningeal membranes that surround the ***brain***, lying between the dura mater and the arachnoid layer. Contains a plexus of small blood vessels.

Pick's disease Rare neurodegenerative disease similar in clinical symptoms to ***Alzheimer's disease***. Affects mostly frontal and temporal lobes.

Picornaviridae (picornavirus) Class IV viruses, with a single positive strand of RNA and an icosahedral capsid. There are two main classes: enteroviruses, which infect the gut and include poliovirus: and rhinoviruses which infect the upper respiratory tract (common cold virus, Coxsackie A and B, foot-and-mouth disease virus and hepatitis A).

pigment cells Cells that contain pigment: see ***melanocytes, chromatophores***.

pigmented retinal epithelium (PRE; retinal pigmented epithelium, RPE) Layer of unusual phagocytic epithelial cells lying below the photoreceptors of the vertebrate eye. The dorsal surface of the PRE cell is closely apposed to the ends of the rods, and as discs are shed from the rod outer segment they are internalized and digested by the PRE. Do not have ***desmosomes*** or ***cytokeratins*** in some species.

pigtail One name for the covalent assembly of sugars linked to phosphatidyl inositol joined to the C-terminal residue of many proteins by a modified ethanolamine residue. Also called a greasy foot. Another term for this modification is glypiation. The function of the pigtail is to act as the sole anchor of the protein to the external surface of the lipid bilayer. The moiety is added to the protein during co-translational insertion into the ER membrane on the luminal side. The addition is synchronized with the removal of a large C-terminal polypeptide sequence that is usually hydrophobic and could itself have formed a membrane anchor. The surface proteins of many unicellular protozoa very commonly have this modification, the best known being the variable surface glycoprotein of ***trypanosomes*** and of ***malaria*** parasites. Examples are probably present in all eukaryotic plasma membranes.

PIIF (proteinase inhibitor-inducing factor) Factor produced by a plant in response to attack by insects. Induces the formation of a substance that inhibits the proteinase that the insect secretes to digest plant tissues. May be mobile within the plant, thus inducing inhibitor formation away from the site of original attack.

pilin (1) General term for the protein subunit of ***pilus***. (2) Protein subunit (7.2 kD) of F-pili, ***sex pili*** coded for by the F-plasmid.

pilocarpine Alkaloid with muscarinic cholinomimetic activity isolated from *Pilocarpus jaborandi*.

pilus (plural, pili); fimbrium; plural, fimbria) Hair-like projection from the surface of some bacteria. Involved in adhesion to surfaces (may be important in virulence), and specialized ***sex pili*** are involved in conjugation with other bacteria. Major constituent is a protein, ***pilin***.

pim-1 ***Oncogene*** from murine T-cell ***lymphomas***, encoding a serine/threonine ***protein kinase***. See also Table O1.

pinacocyte Flattened polygonal cell that lines ostia and form the epidermis of sponges. Capable of synthesizing collagen.

pinocytosis Uptake of fluid-filled vesicles into cells (endocytosis). Macro-pinocytosis and micro-pinocytosis are distinct processes, the latter being energy independent and involving the formation of receptor–ligand clusters on the outside of the plasma membrane, and ***clathrin*** on the cytoplasmic face.

pinocytotic vesicle Fluid-filled endocytotic vesicle, usually less than 150μm diameter. Micropinocytotic vesicles are around 70nm diameter.

pinosome A ***pinocytotic vesicle***.

PIP2 (phosphatidyl inositol 4,5, bisphosphate) Formed by linked 'futile cycles' from ***phosphatidyl inositol*** via phosphatidyl inositol phosphate (PIP). Chiefly important because a ligand-activated PIP2-specific phosphodiesterase (phospholipase Cγ) breaks down PIP2 to form diacyl glycerol, which stimulates protein kinase C, and inositol 1,4,5 trisphosphate (InsP3), which releases calcium from the endoplasmic store.

piroplasm Class of Protista, Phylum Apicomplexa (Sporozoa or Telosporidea), which includes the tick-transmitted parasite, *Babesia*.

Pisaster Echinoderm of the class Asteroidea, a starfish.

pisatin ***Phytoalexin*** produced by peas.

pit Region of the plant cell wall in which the ***secondary wall*** is interrupted, exposing the underlying ***primary cell wall***. One or more ***plasmodesmata*** are usually present in the primary wall, communicating with the other half of a pit pair. May be simple or bordered; in the latter case, the secondary wall overarches the pit field. Do not confuse with ***coated pits***.

pituicytes Dominant intrinsic cells of the neural lobe of the hypophysis. Have long branching processes and resemble ***neuroglia***: secrete ***antidiuretic hormone***.

pituitary (hypophysis) Possibly the most important of the vertebrate endocrine glands. Located below the ***brain*** to which it is attached by a stalk. Has two lobes, the anterior adenohypophysis and posterior neurohypophysis. Secretes a wide range of hormones including ***somatotropin***, ***follicle-stimulating hormone***, ***gonadotrophins***, ***thyroid-stimulating hormone***, ***lipotropin*** and many others. See Table H2.

Pitx2 Bicoid-type homeobox gene that is expressed asymmetrically in the left lateral plate mesoderm and may be involved in determining left–right asymmetry in mouse and chick.

pK_a See ***association constant***.

PLA2 (phospholipase A2) See ***phospholipases***.

placental calcium-binding protein (18a2; nerve growth factor-induced protein 42a; pE2-9 δ; calvasculin p9k) Calcium-binding protein of placenta, uterus and vasculature containing the ***EF-hand*** motif.

placode Area of thickened ectoderm in the embryo from which a nerve ganglion, or a sense organ will develop.

plakalbumin Fragment of ovalbumin produced by ***subtilisin*** cleavage: more soluble than ovalbumin itself.

plakoglobin Polypeptide (83 kD) present at cell–cell but not cell–substratum contacts. Associated with desmosomes and with ***adherens junctions***: soluble 7S form present in cytoplasm.

planapochromat Expensive microscope objective that is corrected for ***spherical aberration*** and ***chromatic aberration*** at three wavelengths.

plant growth substances Substances that, at low concentration, influence plant growth and differentiation. Formerly referred to as plant hormones or phytohormones, these terms are now suspect because some aspects of the 'hormone concept', notably action at a distance from the site of synthesis, do not necessarily apply in plants. Also known as 'plant growth regulators'. The major classes are ***abscisic acid***, ***auxin***, ***cytokinin***, ***ethylene*** and ***gibberellin***; others include steroid and phenol derivatives.

plaque assay (1) Assay for virus in which a dilute solution of the virus is applied to a culture dish containing a layer of the host cells; convective spread is prevented by making the medium very viscous. After incubation the 'plaques', areas in which cells have been killed (or transformed), can be recognized, and the number of infective virus particles in the original suspension estimated. (2) Assay for cells producing antibody against erythrocytes or against antigen that has been bound to the erythrocytes. The cell is surrounded by a clear plaque of haemolysis. Basic principle behind the assay is the same as for the virus plaque assay.

plaque-forming cell Antibody-secreting cell detected in a ***plaque assay***.

plaque-forming unit (pfu) Number of Ig-producing cells or infectious virus particles per unit volume. Of a virus like bacteriophage λ, the number of viable viral particles, established by counting the number of plaques formed by serial dilution of the library. For example, a cDNA library might have a titre of 50 000 pfu/μl of library.

plasma Acellular fluid in which blood cells are suspended. Serum obtained by defibrinating plasma (plasma-derived serum) lacks platelet-released factors and is less suitable to support the growth of cells in culture.

plasma cell A terminally differentiated antibody-forming, and usually antibody-secreting, cell of the B-cell lineage.

plasma kallikrein A plasma serine protease with an ***apple domain***.

plasma membrane The external, limiting lipid bilayer membrane of cells.

plasmacytoma Malignant tumour of ***plasma cells***, very similar to a ***myeloma*** (plasmacytomas usually develop into multiple myeloma). Can easily be induced in rodents by the injection of complete Freund's adjuvant. Plasmacytoma cells are fused with primed lymphocytes in the production of monoclonal antibodies.

plasmal reaction Long-chain aliphatic aldehydes occurring in ***plasmalogens*** react with ***Schiff's reagent*** in the so-called plasmal reaction to form, eg. palmitaldehyde, stearaldehyde.

plasmalemma Archaic name for the plasma membrane of a cell (the term often included the cortical cytoplasmic region). Adjectival derivative (plasmalemmal) still current.

plasmalogens A group of glycerol-based phospholipids in which the aliphatic side chains are not attached by ester linkages. Widespread distribution. Less easily studied than the acyl phospholipids.

plasmid (episome) A small, independently replicating piece of cytoplasmic DNA that can be transferred from one organism to another. Linear or circular DNA molecules found in both pro- and eukaryotes capable of autonomous replication. 'Stringent' plasmids occur at low copy number in cells, 'relaxed' plasmids at high copy number, ca 10–30. Plasmids can become incorporated into the genome of the host, or can remain independent. An example is the F-factor of *E.coli*. May transfer genes, and plasmids carrying antibiotic-resistant genes can spread this trait rapidly through the population. Described largely from bacteria and protozoa. Widely used in genetic engineering as vectors of genes (cloning vectors).

plasmid prep Generic term for the isolation of ***recombinant*** plasmids from liquid bacterial culture, usually by alkaline/detergent lysis, selective precipitation of other components, and affinity purification of plasmid. As this is the most exciting thing most molecular biologists ever do, there is an informal shorthand for the scale of the preparation based on the size of the overnight culture: see ***miniprep***, ***midiprep***, ***maxiprep*** and ***megaprep***.

plasmin (fibrinolysin) Trypsin-like serine protease that is responsible for digesting fibrin in blood clots. Generated from plasminogen by the action of another protease, plasminogen activator. Also acts on activated ***Hagemann factor*** and on complement.

plasminogen Inactive precursor of ***plasmin***; occurs at 200mg/l in blood plasma. Contains multiple copies of the ***kringle*** domain.

plasminogen activator ***Serine protease*** that acts on ***plasminogen*** to generate ***plasmin***. Has also been implicated in invasiveness, and is produced by many normal and invasive cells. The vascular form (tPA; 55 kD) is very similar to tissue plasminogen activator (uPA; 70 kD) and to ***streptokinase*** and ***urokinase***.

plasmodesma (plural, plasmodesmata) Narrow tube of cytoplasm penetrating the plant cell wall, linking the protoplasts of two adjacent cells. A desmotubule runs down the centre of the tube, which is lined by plasma membrane.

plasmodesmata See ***cytoplasmic bridge***.

Plasmodium Genus of parasitic protozoa that cause ***malaria***. The life-cycle is complex, involving several changes in cellular morphology and behaviour. Intermediate host is female mosquito (***Anopheles***) that infects vertebrate host when taking a blood meal. Predominant form of the organism in humans is the intracellular parasite (the merozoite) in the erythrocyte, where it undergoes a form of multiple cell division termed schizogony. As a result the erythrocyte bursts and the progeny infect other erythrocytes. Eventually some cells develop into gametes that, when ingested by a female mosquito, will fuse in her gut to form a zygote (ookinete). Multiple cell division within the resultant oocyte, attached to the gut wall, gives rise to infective sporozoites; these migrate to the salivary glands and are ejected with the saliva the next time the mosquito takes a blood meal.

plasmodium Multinucleate mass of protoplasm bounded only by a plasma membrane; the main vegetative form of acellular slime moulds (eg. ***Physarum***).

plasmolysis Process by which the plant cell protoplast shrinks, so that the plasma membrane becomes partly detached from the wall. Occurs in solutions of high osmotic potential, due to water moving out of the protoplast by osmosis.

plastid Type of plant cell organelle, surrounded by a double membrane and often containing elaborate internal membrane systems. Partially autonomous, containing some DNA, RNA and ribosomes, and reproducing itself by binary fission. Includes ***amyloplasts, chloroplasts, chromoplasts, etioplasts, leucoplasts, proteinoplasts,*** and ***elaioplasts.*** Develop from ***proplastids.***

plastin See ***fimbrin.***

plastocyanin An electron-carrying protein present in chloroplasts, forming part of the electron transport chain. Contains two copper atoms per molecule. Associated with ***photosystem*** I.

plastoglobuli Globules found in plastids, containing principally lipid, including ***plastoquinone.***

plastoquinone A ***quinone*** present in chloroplasts, forming part of the photosynthetic electron transport chain. Closely associated with ***photosystem II.*** May be stored in ***plastoglobuli.***

platelet Anucleate discoid cell (3μm diameter) found in large numbers in blood; important for blood coagulation and for haemostasis. Platelet α-granules contain lysosomal enzymes; dense granules contain ADP (a potent platelet aggregating factor), and ***serotonin*** (a vasoactive amine). They also release ***platelet-derived growth factor*** which presumably contributes to later repair processes by stimulating fibroblast proliferation.

platelet activating factor (PAF; PAFacether; 1-0-hexadecyl-2-acetyl-sn-glycero-3-phosphorylcholine) Potent activator of many leucocyte functions, not just platelet activation.

platelet basic protein (PBP) Protein that is the precursor of ***connective tissue-activating peptide III*** and ***beta- thromboglobulin.***

platelet factor 3 Phospholipid associated with the platelet plasma membrane that contributes to the blood clotting cascade by forming a complex (thromboplastin) with other plasma proteins and activating ***prothrombin.***

platelet factor 4 Platelet-released protein that promotes blood clotting by neutralizing ***heparin.***

platelet-derived growth factor (PDGF) The major mitogen in serum for growth in culture of cells of connective tissue origin. It consists of two different but homologous polypeptides A and B (~ 30 000 D) linked by disulphide bonds. Believed to play a role in wound healing. The B chain is almost identical in sequence to $p28^{sis}$, the transforming protein of simian sarcoma virus, which can transform only those cells that express receptors for PDGF, suggesting that transformation is caused by ***autocrine*** stimulation. Receptor is a ***tyrosine kinase.***

PLCPI (porcine leucocyte cysteine protease inhibitor) ***Stefin*** type protease inhibitor (103 residues, 11 kD) that co-purifies with ***cathelin.*** Inhibits papain and cathepsins L and S by forming a tight complex.

pleckstrin Protein of 47 kD, the major substrate for protein kinase C in platelets. Pleckstrin homology domains (***PH domains***) are being identified in a number of proteins.

plectin Abundant protein of cytomatrix (apparent 300 kD but 466 kD on basis of cDNA sequence). Co-localizes with various intermediate filament proteins and may be involved in their crosslinking or anchoring.

pleiomorphic Having more than one body shape during the life cycle or having the ability to change shape or to adopt a variety of shapes.

pleiotropic Having multiple effects. For example, the cyclic-AMP concentration in a cell will have a variety of effects because the cAMP acts to control a protein kinase that in turn affects a variety of proteins.

Pleurobrachia Small free-swimming marine organism, member of the phylum Ctenophora. Roughly spherical and transparent with most of the body made up from transparent jelly-like material. The animal has two long tentacles for catching prey, and swims by means of eight rows of ***comb plates*** (made of fused cilia) that run along the body.

Pleurodeles Genus of salamanders.

pleuropneumonia-like organism (PPLO) See ***mycoplasmas***.

PLGF (placental growth factor) Growth factor similar in activity to ***VEGF***.

pluripotent stem cell Cells in a stem cell line capable of differentiating into several different final differentiated types, eg. there may be a pluripotent stem cell line for erythrocytes, granulocytes and megakaryocytes.

pluteus Free-swimming ciliated larval stage of some echinoderms.

PMA (TPA; tumour promotor activity) Phorbol myristate acetate, a ***phorbol ester***.

PMF (proton motive force) The proton gradient across a prokaryote membrane that provides the coupling between oxidation and ATP synthesis, and is used to drive the flagellar motor.

PMN (PMNL) Polymorphonuclear leucocyte: could be an ***eosinophil***, ***basophil*** or ***neutrophil granulocyte***, but usually intended to mean the latter (an idle habit).

PMSF (phenylmethylsulphonyl fluoride) Broad spectrum protease inhibitor.

pneumococci ***Gram-positive*** pyogenic organisms (about 1μm diameter), usually encapsulated, closely related to streptococci; associated with diseases of the lung.

Pneumocystis carinii Organism that commonly causes pneumonia in immunocompromised patients (eg. with AIDS). Apparently most closely related to ustomycetous yeasts.

pneumolysin ***Cholesterol binding toxin*** from *Streptococcus pneumoniae*.

PNMT (phenylethanolamine N-methyl transferase) Terminal enzyme in the catcholamine biosynthetic pathway; converts noradrenaline to adrenaline.

podocalyxin Major sialoprotein (140 kD) of renal glomerular epithelial cells (podocytes).

podocytes Cells of the visceral epithelium that closely invest the network of glomerular capillaries in the kidney. Most of the cell body is not in contact with the ***basal lamina***, but is separated from it by trabeculae that branch to give rise to club-shaped protrusions, known as pedicels, interdigitating with similar processes on adjacent cells. The complex interdigitation of these cells produces thin filtration slits that seem to be bridged by a layer of material (of unknown composition), that acts as a filter for large macromolecules.

podophyllotoxin Toxin (414D) that binds to tubulin and prevents microtubule assembly.

podosomes Punctate substratum-adhesion complexes in osteoclasts. Contain ***vinculin***, ***talin***, ***fimbrin*** and ***F-actin***. Podosomes form a broad ring of contacts with the underlying bone and the enclosed area below the cell is then absorbed.

poikilocytosis Irregularity of red cell shape.

poikilotherm Organism whose body temperature varies with environment, opposite of homeotherm. Though poikilothermic animals are often referred to as cold-blooded, this is not necessarily true.

point mutation ***Mutation*** that causes the replacement of a single base pair with another pair.

pokeweed mitogen Any of the ***lectins*** derived from the pokeweed, *Phytolacca americana*, all of which will stimulate T-cells. Binds β-D-acetylglucosamine.

pol* genes** Genes coding for ***DNA polymerases of which there are three in *E. coli*, *polA*, *polB*, and *polC* coding for polymerases I, II, and III respectively. *Pol* genes in ***oncogenic*** retroviruses code for ***reverse transcriptase***.

polar body In animals each meiotic division of the oocyte leads to the formation of one large cell (the egg) and a small polar body as the other cell. Polar body formation is a consequence of the very eccentric position of the nucleus and the spindle.

polar granules Granules containing a basic protein found in insect eggs that induce the formation of and become incorporated into germ cells.

polar group Any chemical grouping in which the distribution of electrons is uneven enabling it to take part in electrostatic interactions.

polar lobe In some molluscs a polar lobe appears as a clear protrusion close to the vegetal pole of the cell prior to the first cleavage, and becomes associated with only one of the daughter cells. Removal of the first polar lobe, or of any polar lobe that forms at a subsequent mitosis, leads to defects in the embryo; it seems that the polar lobe contains special morphogenetic factors.

polar plasm Differentiated cytoplasm associated with the animal or vegetal pole of an oocyte, egg or early embryo.

polarity Literally 'having poles' (like a magnet), but used to describe cells that have one or more axes of symmetry. In epithelial cells, the polarity meant is between apical and basolateral regions; in moving cells, having a distinct front and rear. Some cells seem to show multiple axes of polarity (which will hinder forward movement).

polarization microscopy Any form of microscopy capable of detecting birefringent objects. Usually performed with a polarizing element below the stage to produce plane-polarized light, and an analyser that is set to give total extinction of the background, and thus to detect any birefringence.

pole cell A cell at or near the animal or vegetal pole of an embryo.

pole fibres Microtubules inserted into the pole regions of the mitotic spindle (each pole is the product of the division of the centrioles and constitutes a ***microtubule organizing centre***).

polehole (phl) Drosophila homologue of the [***raf***] ***oncogene***.

poliovirus A member of the enterovirus group of ***Picornaviridae*** that causes poliomyelitis.

pollen mother cell A diploid plant cell that forms four ***microspores*** by meiosis; the microspores give rise to pollen grains in seed plants.

poly(ADP-ribose) polymerase (PARP) (EC 2.4.2.30) An abundant nuclear protein activated by DNA nicks and important in DNA repair. PolyADP ***ribosylation***, brought about by ADP-ribosyl protein ligase, is a post-transcriptional modification of proteins and ***p53*** is one of the proteins that can be modified in this way. PARP knockout mice show defects in fibroblast proliferation and impaired capacity to handle radiation-induced damage. One of the earliest proteins cleaved by ***caspase*** 3 in apoptosis.

poly-A See ***polyadenylic acid***.

poly-A tail Polyadenylic acid sequence of varying length found at the 3′ end of most eukaryotic mRNAs. Histone mRNAs do not have poly-A tail. The poly-A tail is added post-transcriptionally to the primary transcript as part of the nuclear processing of RNA yielding ***hnRNAs*** with 60–200 adenylate residues in the tail. In the cytoplasm the poly-A tail on mRNAs is gradually reduced in length. The function of the poly-A tail is not clear but it is the basis of a useful technique for the isolation of eukaryotic mRNAs. The technique uses an ***affinity chromatography*** column with oligo(U) or oligo(dT) immobilized on a solid support. If cytoplasmic RNA is applied to such a column, poly-A-rich RNA (mRNA) will be retained.

polyacrylamide gel electrophoresis (PAGE) Analytical and separative technique in which molecules, particularly proteins, are separated by their different electrophoretic mobilities in a hydrated gel. The gel suppresses convective mixing of the fluid phase through which the electrophoresis takes place, and contributes molecular sieving. Commonly carried out in the presence of the anionic detergent sodium dodecylsulphate (SDS). SDS denatures proteins so that non-covalently associating subunit polypeptides migrate independently, and by binding to the proteins confers a net negative charge roughly proportional to the chain weight. See also ***SDS-PAGE***.

polyadenylic acid ***Polynucleotide*** chain consisting entirely of residues of adenylic acid (ie. the base sequence is AAAA . . . . AAAA). Polyadenylic chains of various lengths are found at the 3′ end of most eukaryotic mRNAs, the ***poly-A tail***.

polyamine Polycations at physiological pH, polyamines can bind and interact with various other molecules within the cell. In particular interact with DNA but also may modulate ion channels and act as growth factors. ***Spermine*** has four positive charges, ***spermidine*** has three. The precursor of both, ***putrescine***, has two.

polyanion Macromolecule carrying many negative charges. The commonest in cell biological systems is nucleic acid.

polycation Macromolecule with many positively charged groups. At physiological pH the most commonly used in cell biology is poly-L-lysine; this is often used to coat surfaces thereby increasing the adhesion of cells (which have net negative surface charge). See also ***cationized ferritin***.

polycistronic mRNA A single ***mRNA*** molecule that is the product of the ***transcription*** of several tandemly arranged genes; typically the mRNA transcribed from an ***operon***.

polyclonal antibody An antibody produced by several clones of B-lymphocytes as would be the case in a whole animal. refers to antibodies raised in immunized animals, whereas a ***monoclonal*** antibody is the product of a single clone of B-lymphocytes, usually maintained ***in vitro***.

polyclonal compartment When the progeny of several cells occupy an area or volume with a defined boundary, it is referred to as a polyclonal compartment, eg. clones lying close to the midline of the wing of *Drosophila*.

polycloning site (multiple cloning site; MCS) Region of a phage or plasmid vector that has been engineered to contain a series of restriction sites that are usually unique within the entire vector. This makes it particularly easy to insert or excise (subclone) DNA fragments.

Polycomb (*Pc*) *Drosophila* gene, which when mutated leads to extra sex combs on the legs of male flies, suggesting that the posterior legs have become anterior legs. There are at least 10 genes in the *Polycomb* group; they are thought to act by ***transcriptional silencing*** of ***homeotic genes***.

polycythemia Increase in the haemoglobin content of the blood, either because of a reduction in plasma volume or an increase in red cell numbers. The latter may be a result of abnormal proliferation of red cell precursors (polycythemia vera, Vaquez–Osler disease).

polyelectrolyte An ion with multiple charged groups.

polyendocrine syndrome Autoimmune disorder (the antigen to which the response is mounted is in the ***B-cells*** of the pancreas) in which there is involvement of several organ systems.

polyethylene glycol (PEG) A hydrophilic polymer that interacts with cell membranes and promotes fusion of cells to produce viable hybrids. Often used in producing ***hybridomas***.

polygalacturonan Plant cell wall polysaccharide consisting predominantly of galacturonic acid. May also contain some rhamnose, arabinose and galactose. Those with significant amounts of rhamnose are termed ***rhamnogalacturonans***. Found in the ***pectin*** fraction of the wall.

polygalacturonase Enzyme that degrades ***polygalacturonan*** by hydrolysis of the glycosidic bonds that link galacturonic acid residues. Important in fruit ripening and in fungal and bacterial attack on plants.

polygenic Something that is controlled or caused by the action of many genes. Thus many of the major non-infectious diseases (eg. arthritis, cardiovascular disease, asthma, diabetes) are likely to be caused by the interaction of many genes; no single gene mutation is responsible, rather the coincidence of polymorphic variants that together contribute risk factors that predispose an individual to the disease.

polyisoprenylation See ***geranylation***.

polylysine A polymer of ***lysine***, it carries multiple positive charges and is used to mediate adhesion of living cells to synthetic culture substrates, or of fixed cells to glass slides (eg. observation by fluorescence microscopy).

polymer A macromolecule made of repeating (monomer) units or ***protomers***.

polymerase chain reaction (PCR) The first practical system for ***in vitro*** amplification of DNA, and as such one of the most important recent developments in molecular biology. Two synthetic oligonucleotide primers, which are complementary to two regions of the target DNA (one for each strand) to be amplified, are added to the target DNA (that need not be pure), in the presence of excess deoxynucleotides and ***Taq polymerase***, a heat-stable DNA polymerase. In a series (typically 30) of temperature cycles, the target DNA is repeatedly denatured (around 90°C), annealed to the primers (typically at 50–60°C) and a daughter strand extended from the primers (72°C). As the daughter strands themselves act as templates for subsequent cycles, DNA fragments matching both primers are amplified exponentially, rather than linearly. The original DNA need thus be neither pure nor abundant, and the PCR reaction has accordingly become widely used not only in research, but in clinical diagnostics and forensic science.

polymerization The process of polymer formation. In many cases this requires ***nucleation*** and will only occur above a certain critical concentration.

polymorphic epithelial mucin See ***episialin***.

polymorphism (1) The existence, in a population, of two or more alleles of a gene, where the frequency of the rarer alleles is greater than can be explained by recurrent mutation alone (typically greater than 1%). HLA alleles of the ***major histocompatibility complex*** are very polymorphic. (2) The differentiation of various parts of the units of colonial animals into different types of unit specialized for different purposes, eg. as in the colonial hydroid *Obelia*.

polymorphonuclear leucocyte (PMNL; PMN) Mammalian blood leucocyte (granulocyte) of myeloid series in distinction to mononuclear leucocytes: see ***neutrophil, eosinophil, basophil***.

polymyxins Group of peptide antibiotics produced by *Bacillus* spp. Molecular weights are around 1000-1200 D and the molecules are cyclic. Act against many ***Gram-negative bacteria***, working apparently by increasing membrane permeability.

polynucleotide Linear sequences of ***nucleotides***, in which the 5′-linked phosphate on one sugar group is linked to the 3′ position on the adjacent sugars. In the polynucleotide DNA the sugar is ***deoxyribose*** and in RNA, ***ribose***. They may be double-stranded or single-stranded with varying amounts of internal folding.

polyomavirus A DNA tumour virus with very small genome (of the ***Papovaviridae***). Polyoma was isolated from mice, in which it causes no obvious disease, but when injected at high titre into baby rodents, including mice, causes tumours of a wide variety of histological types (hence poly-oma). ***In vitro***, infected mouse cells are permissive for virus replication, and thus are killed, while hamster cells undergo ***abortive infection***, and at a low frequency become transformed.

polyp (1) Growth, usually benign, protruding from a mucous membrane. (2) The sessile stage of the Cnidarian (***coelenterate***) life-cycle; the cylindrical body is attached to the substratum at its lower end, and has a mouth surrounded by tentacles bearing ***nematocysts*** at the upper end; *Hydra* and the feeding polyps of the colonial *Obelia* are examples.

polypeptide Chains of ***α-amino acids*** joined by peptide bonds. Distinction between peptides, oligopeptides and polypeptides is arbitrarily by length; a polypeptide is perhaps more than 10 residues.

polyploid Of a nucleus, cell or organism that has more than two ***haploid*** sets of ***chromosomes***. A cell with three haploid sets (3n) is termed triploid, four sets (4n) tetraploid, and so on.

polypodial Adjective describing an amoeba with several pseudopods.

polyposis coli (adenomatous polyposis coli; Familial adenomatous polyposis; FAP) Hereditary disorder (Mendelian dominant) characterized by the development of hundreds of adenomatous ***polyps*** in the large intestine, which show a tendency to progress to malignancy. The APC gene has also been implicated in a chromosome 5 gastric and pancreatic cancer.

polyprotein Protein that, after synthesis, is cleaved to produce several functionally distinct polypeptides. Some viruses produce such proteins, and some polypeptide hormones seem to be cleaved from a single precursor polyprotein (eg. ***pro-opiomelanocortin***).

polyribosome Functional unit of protein synthesis consisting of several ***ribosomes*** attached along the length of a single molecule of mRNA.

polysaccharide Polymers of (arbitrarily) more than about 10 monosaccharide residues linked glycosidically in branched or unbranched chains.

polysialic acid (PSA) Potential regulator of cell–cell interactions. Polysialic acid chains in ***glycoprotein*** may have negative regulatory effects on cell–cell contact. Thus the low PSA form of NCAM is thought to promote cell-cell contact and enhance ***fasciculation*** whereas NCAM with a high PSA content is thought to prevent close membrane–membrane apposition.

polysome See ***polyribosome***.

polysomy Situation in which all chromosomes are present, and some are present in greater than the diploid number, eg. trisomy 21.

polyspermy Penetration of more than one spermatozoon into an ovum at time of fertilization. Occurs as normal event in very yolky eggs (eg. bird), but then only one sperm fuses with egg nucleus. Many other eggs have mechanisms to block polyspermy.

Polysphondylium A genus of ***Acrasidae***, the cellular slime moulds.

polytene chromosomes Giant ***chromosomes*** produced by the successive replication of homologous pairs of chromosomes, joined together (synapsed) without chromosome separation or nuclear division. They thus consist of many (up to 1000) identical chromosomes (strictly chromatids) running parallel and in strict register. The chromosomes remain visible during interphase and are found in some ciliates, ovule cells in angiosperms, and in larval Dipteran tissue. The best known polytene chromosomes are those of the salivary gland of the larvae of ***Drosophila*** *melanogaster* which appear as a series of dense bands interspersed by light interbands, in a pattern characteristic for each chromosome. The bands, of which there are about 5 000 in *D.melanogaster*, contain most of the DNA (ca 95%) of the chromosomes, and each band roughly represents one gene. The banding pattern of polytene chromosomes provides a visible map to compare with the linkage map determined by genetic studies. Some segments of polytene chromosome show chromosome ***puffs***, areas of high transcription.

polyuridylic acid Homopolymer of uridylic acid. Historically, was used as an artificial mRNA in cell-free ***translation*** systems, where it coded for polyphenylalanine; thus began the deciphering of the genetic code.

polyvinylpyrrolidone Polymer used to bind phenols in plant homogenates, and hence to protect other molecules, especially enzymes, from inactivation by phenols. Also occasionally used to produce viscous media for gradient centrifugation.

Pompe's disease Severe glycogen ***storage disease*** caused by deficiency in α (1-4)-glucosidase, the lysosomal enzyme responsible for glycogen hydrolysis. Even though the non-lysosomal glycogenolytic system is normal, glycogen still accumulates in the lysosomes.

Ponceau red (Ponceau S; Fast Ponceau 2B) Dye used to stain proteins.

Ponceau S See *Ponceau red*.

ponticulin Developmentally regulated 17 kD transmembrane glycoprotein from *Dictyostelium* that regulates actin binding and nucleation. Preferentially located at actin-rich regions such as sites of cell adhesion. Analogue found in human neutrophils.

population diffusion coefficient Coefficient that describes the tendency of a population of motile cells to diffuse through the environment. Its use presupposes that the cells move in a random-walk.

porins Transmembrane matrix proteins (37 kD) found in the outer membranes of ***Gram-positive bacteria***. Associate as trimers to form channels (1nm diameter, ca 10^5 per bacterium) through which hydrophilic molecules of up to 600 D can pass. Similar porins are also found in outer mitochondrial membranes (VDAC, voltage-dependent anion-selective channel).

porphyria Any of a group of disorders in which there is excessive excretion of porphyrins or their precursors.

porphyrins Pigments derived from porphin: all are chelates with metals (Fe, Mg, Co, Zn, Cu, Ni). Constituents of haemoglobin, chlorophyll, cytochromes.

position effect Effect on the expression of a gene depending upon its position relative to other genes on the chromosome. Moving (transposing) a gene from an inactive region to an active region can alter expression markedly, sometimes with unfortunate consequences as with the ***Philadelphia chromosome*** abnormality that leads to CML.

positional cloning Identification of a gene based on its location in the genome. Typically, this will result from ***linkage*** analysis based on a mutation in the target gene, followed by a ***chromosome walk*** from the nearest known sequence.

positional information The instructions that are interpreted by cells to determine their differentiation in respect of their position relative to other parts of the organism, eg. digit formation in the limb buds of vertebrates.

positive control Mechanism for gene regulation that requires that a regulatory protein must interact with some region of the gene before transcription can be activated.

positive feedback See ***feedback***.

positive strand RNA viruses Class IV and VI viruses that have a single-stranded RNA ***genome*** that can act as mRNA (plus strand) and in which the virus RNA is itself infectious. Includes ***Picornaviridae***, ***Togaviridae*** and ***Retroviridae***.

post-translational modification Changes that occur to proteins after peptide bond formation has occurred. Examples include glycosylation, acylation, limited proteolysis, phosphorylation, isoprenylation.

postcapillary venule That portion of the blood circulation immediately downstream of the capillary network; the region having the lowest wall-shear stress, and the most common site of leucocytic margination and endothelial transmigration (diapedesis).

postsynaptic cell In a chemical ***synapse***, the cell that receives a signal (binds neurotransmitter) from the presynaptic cell and responds with depolarization. In an electrical synapse, the postsynaptic cell would just be downstream, but since many electrical synapses are ***rectifying***, one of the two cells involved will always be postsynaptic.

postsynaptic membrane In a ***chemical synapse***, the membrane of the ***postsynaptic cell*** specialised to detect ***neurotransmitter*** release, cf. ***presynaptic membrane***.

postsynaptic potential In a synapse, a change in the ***resting potential*** of a postsynaptic cell following stimulation of the presynaptic cell. For example, in a cholinergic synapse, the release of acetylcholine from the presynaptic cell causes channels to open in the postsynaptic cell. Each channel opening causes a small depolarization, known as a ***miniature endplate potential*** (MEPP); these sum to produce an excitatory postsynaptic potential.

postsynaptic protein (43 kD postsynaptic protein) A peripheral membrane protein thought to help anchor ***acetylcholine receptors*** in the postsynaptic membrane. Highly conserved.

potassium channel Ion channel selective for potassium ions. There are diverse types with different functions, eg. ***delayed rectifier channels, M-channels***, A-channels, inward rectifier channels, Ca-dependent K^+ channels.

potato lectin ***Lectin*** from the potato, *Solanum tuberosum*. Binds to N-acetyl glucosaminyl residues.

potentiation Increase in quantal release at a synapse following repetitive stimulation. Whereas ***facilitation*** at synapses lasts a few hundred milliseconds, potentiation may last minutes to hours.

potocytosis Transport of small molecules across membrane using ***caveolae*** rather than ***coated vesicles***.

POU domain A conserved protein domain of around 150 amino acids, composed of a 20 amino acid ***homeobox*** domain and a larger POU-specific domain, and so is the target of some ***transcription factors***. Named POU (Pit-Oct-Unc) after three such proteins: Pit-1 that regulates expression of certain pituitary genes, Oct-1 and 2, that bind an octamer sequence in the promoters of histone H2A and some immunoglobulin genes, and Unc-86, involved in nematode sensory neuron development.

Poxviridae Class I viruses with double-stranded DNA genome that codes for more than 30 polypeptides. They are the largest viruses and their shell is complex, consisting of many layers, and includes lipids and enzymes, amongst which is a DNA-dependent RNA polymerase. Uniquely among the DNA viruses they multiply in the cytoplasm of the cell, establishing what is virtually a second nucleus. The most important poxviruses are ***vaccinia, variola*** (smallpox), and ***myxoma virus***.

pp46 See ***neuromodulin***.

pp60src The phosphoprotein (60 kD) encoded by the ***src*** oncogene. A tyrosine kinase.

PPAR (peroxisome proliferator-activated receptors) PPARα stimulates β-oxidative degradation of fatty acids, PPARγ promotes lipid storage by regulating adipocyte differentiation. Are implicated in metabolic disorders predisposing to atherosclerosis and inflammation. PPARα-deficient mice show prolonged response to inflammatory stimuli. PPARα is activated by gemfibrozil and other fibrate drugs.

PPD (purified protein derivative) Protein purified from the culture supernatant of tubercle bacteria (*Mycobacterium tuberculosis*) and used as a test antigen in ***Heaf*** and ***Mantoux tests***.

PPIase (peptidyl-prolyl *cis-trans* isomerase) Enzymes that accelerate protein folding by catalysing *cis-trans* isomerizations. ***Immunophilins*** are PPIases though their enzymic activity may not be essential for their immunosuppressive effects.

PPLO Pleuropneumonia-like organisms. See ***mycoplasma***.

Pr The form of ***phytochrome*** that absorbs light in the red region (660nm), and is thus converted to ***Pfr***. In the dark the equilibrium between Pr and Pfr favours Pr, which is therefore more abundant.

Prader–Willi syndrome Syndrome in which there is an absence of paternal chromosome 15q11q13. Short stature, obesity and mild mental retardation are features of the syndrome. Uniparental disomy leads to differences between this and ***Angelman syndrome*** where it is the equivalent maternal region that is deleted. See ***imprinting***.

prazosin Antagonist of α -adrenergic receptors.

prd See ***paired***.

pre-pro-protein A pre-protein is a form that contains a signal sequence that specifies its insertion into or through membranes. A pro-protein is one that is inactive; the full function is only present when an inhibitory sequence has been removed by proteolysis. A pre-pro-protein has both sequences still present. Pre-pro-proteins usually only accumulate as products of ***in vitro*** protein synthesis.

pre-prophase band Band of microtubules 1–3μm wide that appears just below the plasma membrane of a plant cell before the start of mitosis. The position of the pre-prophase band determines the plane of cytokinesis and of the cell plate that will eventually separate the two cells.

precipitin Any antibody that forms a precipitating complex (a precipitin line) with an appropriate multivalent antigen. The term is now outmoded.

prednisolone (1,4-pregnadiene-11 β, 17 α, 21-triol-3,20-dione) Steroid with glucocorticoid action, very similar to prednisone. An effective anti-inflammatory drug but with serious side-effects.

prednisone (1,4-pregnadiene-17 α, 21-diol-3,11,20-trione) Synthetic steroid that acts as a glucocorticoid, with powerful anti-inflammatory and anti-allergic activity.

preleptonema Rarely used term that designates an extra stage in the prophase of meiosis I. Usually lumped in with ***leptotene***.

prenylation Post-translational addition of prenyl groups to a protein. Farnesyl, geranyl or geranyl-geranyl groups may be added. Consequence is usually to promote membrane association.

preprophase Rarely used term to designate an extra stage of mitosis, normally included as part of prophase.

presecretory granules Vesicles near the maturation face of the Golgi. Also known as Golgi condensing vacuoles.

presenilins Multi-pass transmembrane proteins, PS1 and PS2, found in Golgi. Mutations in genes for PS1 are associated with 25% of early-onset ***Alzheimer's disease*** and altered amyloid β protein (***β -amyloid precursor protein***) processing. PS1 is a functional homologue of SEL-12, a protein found in *Caenorhabditis elegans* that facilitates signalling mediated by ***Notch***, and the expression of PS1 seems to be essential for the spatio-temporal expression of Notch1 and Dll1 (δ-like gene 1) during embryogenesis. PS1 and PS2 are also similar to *C. elegans Spe-1*, a gene involved in protein trafficking in the Golgi during spermatogenesis.

prespore cells Cells in the rear portion of the migrating slug (grex) of a cellular slime mould, which will later differentiate into spore cells. Can be recognized as having different proteins by immunocytochemical methods. See also ***Acrasidae***.

prestalk cells Cells at the front of the migrating grex of cellular slime moulds that will form the stalk upon which the ***sorocarp*** containing the spores is borne. See ***prespore cells***.

presynaptic cell In a chemical synapse, the cell that releases neurotransmitter that will stimulate the ***postsynaptic cell*** In an electrically synapsed system, the cell that has the first action potential, but since synapses are rectifying, one of the two cells involved is always presynaptic.

presynaptic membrane In a ***chemical synapse***, the membrane of the ***presynaptic cell*** from which ***neurotransmitter*** is released, cf. ***postsynaptic membrane***.

presynaptic receptor Receptors located on presynaptic terminals at ***synapses***.

prezygonema Rarely used term to designate an extra stage in the prophase of meiosis I. Usually lumped in with ***zygotene***.

Pribnow box See *promoter*.

prickle cell Large flattened polygonal cells of the stratum germinosum of the epidermis (just above the basal stem cells), that appear in the light microscope to have fine spines projecting from their surfaces; these terminate in desmosomes that link the cells together and have many tonofilaments of ***cytokeratin*** within them.

primaquine An 8-aminoquinoline drug used to treat malaria. Affects the mitochondria of the exo-erythrocytic stages (see ***Plasmodium***), but the mechanism is not understood. The most effective drug at preventing spread of all four species of human malaria.

primary cell culture Of animal cells, the cells taken from a tissue source and their progeny grown in culture before subdivision and transfer to a subculture. See also Table C3.

primary cell wall A plant cell wall that is still able to expand, permitting cell growth. Growth is normally prevented when a ***secondary wall*** has formed. Primary cell walls contain more ***pectin*** than secondary walls, and no lignin is present until a secondary wall has formed on top of them.

primary immune response The immune response to the first challenge by a particular antigen. Usually less extensive than the secondary immune response, being slower and shorter-lived with smaller amounts of lower affinity antibody being produced.

primary lymphoid tissue See *lymphoid tissues*.

primary lysosome A lysosome before it has fused with a vesicle or vacuole.

primary meristem See *apical meristem*.

primary oocyte The enlarging ovum before maturity is reached, as opposed to the secondary oocyte or polar body.

primary spermatocyte A stage in the differentiation of the male germ cells. Spermatogonia differentiate into primary spermatocytes, showing a considerable increase in size in doing so; primary spermatocytes divide into secondary spermatocytes.

primary transcript RNA transcript immediately after transcription in the nucleus, before ***RNA splicing*** or polyadenylation to form the mature ***mRNA***.

primary tumour The mass of tumour cells at the original site of the neoplastic event: from the primary tumour metastasis will lead to the establishment of secondary tumours.

primase The enzyme that polymerizes nucleotide triphosphates to form oligoribonucleotides in a 5′ to 3′ direction. The enzyme synthesizes the RNA for RNA-DNA sequences that later become ***Okazaki fragment***s and also RNA primers for some types of phage using an sDNA template.

primer extension Technique for finding the ***transcriptiona***l start site of a gene. ***mRNA***s cannot be relied on to be complete at the 5′ end, so a labelled antisense oligonucleotide ***primer*** is designed to complement the putative mRNA near its 5′ end, and used to prime a ***reverse transcription*** reaction. The products are run on a sequencing gel, and the lengths of products allow the putative start sites to be deduced.

priming Treatment that does not in itself elicit a response from a system but that induces a greater capacity to respond to a second stimulus.

primitive erythroblast Large cell with euchromatic nucleus found in mammalian embryos. In the mouse, the cells are located in the yolk sac and are responsible for early production of erythrocytes with foetal haemoglobin.

primitive knot See ***Hensen's node***.

primitive streak Thickened streak of cells in early mammalian and avian embryos: marks location of embryonic axis. Hensen's node is at the end of the primitive streak until the cellular movements of gastrulation cause it to regress caudally.

primordial germ cells Germ cells at the earliest stage of development. Since germ cells may originate in the embryo at some distance from the gonads they then have to migrate to the gonadal primordia, a process that may involve chemotaxis or, more probably, random movement with trapping.

primosome Complex of proteins involved in the synthesis of the RNA primer sequences used in DNA replication. Main components are primase and DNA helicase that move as a unit with the replication fork.

prions See ***PrP***, ***Gerstmann–Straussler–Scheinker syndrome***. Suggested as the causative agents of several infectious diseases such as scrapie (in sheep), kuru and ***Creutzfeldt–Jakob disease*** in humans. Prions (proteinaceous infective particles) apparently contain no nucleic acid. The 27 kD protein of scrapie is related to a normal cell protein and may possibly cause its overproduction.

pristane (2,6,10,14-tetramethylpentadecane.) Extracted from shark liver. Will induce a lupus-like syndrome in non-autoimmune mice and a form of experimental arthritis.

PRK (PKC related kinase; PKN) Serine/threonine kinases (120 kD). PRK-1 is found in hippocampus and is activated by ***phospholipids*** and ***arachidonic acid***, binds to rho-GTP and possibly regulates cytoskeletal changes. PRK-2 (PKN-2; PAK-2) was isolated from U937 cells and foetal brain and is activated by cardiolipin and acidic phospholipids. Binds ***rac*** and ***rho*** which will activate its kinase activity, and interacts with SH3 domain of ***Nck*** and ***PLC*** γ.

pro See ***proline***.

pro-enzyme Enzyme that does not have full (or any) function until an inhibitory sequence has been removed by limited proteolysis. See also ***zymogen***.

pro-opimelanocortin ***Polyprotein*** produced by the anterior pituitary that is cleaved to yield ***adrenocorticotrophin***, α, β and γ ***melanocyte-stimulating hormones***, lipotropic hormones, β-***endorphin***, and other fragments.

proband First patient to present with a disorder, usually heritable, and from whom the descent can be traced.

probe General term for a piece of DNA or RNA corresponding to a gene or sequence of interest, that has been labelled either radioactively, or with some other detectable molecule, such as ***biotin***, ***digoxygenin*** or ***fluorescein***. As stretches of DNA or RNA with complementary sequences will hybridize, a probe will label viral ***plaques***, bacterial colonies or bands on a gel that contain the gene of interest. See also ***Northern blot***, ***Southern blot***.

procaine Organic base (234 D). Procaine butyrate, borate and hydrochloride are used as local anaesthetics.

procambium Plant ***meristem*** that gives rise to the primary vascular system.

procardia See ***nifedipine***.

procaryote See ***prokaryote***.

procentriole The forming centriole composed of microtubules. Multiple procentrioles are present in some cells as a structure called the blepharoplast.

procollagen Triple-helical trimer of collagen molecules in which the terminal extension peptides are linked by disulphide bridges; the terminal peptides are later removed by specific proteases to produce a ***tropocollagen*** molecule.

procollagen peptidases The proteases that remove the terminal extension peptides of ***procollagen***; deficiency of these enzymes leads to ***dermatosparaxis*** or ***Ehlers–Danlos syndrome***.

prodigiosin See *Serratia marcescens*.

prodromal A prodromal sign is an early indication of a disease, often before classical symptoms appear.

profilin Actin-binding protein (15 kD) that forms a complex with ***G-actin*** rendering it incompetent to nucleate ***F-actin*** formation. The profilin–actin complex seems to interact with inositol phospholipids that may regulate the availability of nucleation-competent G-actin.

progenitor cell In development a 'parent' cell that gives rise to a distinct ***cell lineage*** by a series of cell divisions.

progeria Accelerated ageing syndrome in which most of the characteristic stages of human senescence are compressed into less than a decade. Defect probably in DNA repair.

progesterone (luteohormone) Hormone (314 D) produced in the corpus luteum, as an antagonist of ***oestrogens***. Promotes proliferation of uterine mucosa and the implantation of the blastocyst; prevents further follicular development. See Table H3.

programmed cell death A form of cell death, best documented in development, in which activation of the death mechanism requires protein synthesis. Morphologically, the cell appears to die by ***apoptosis*** though this is not necessarily the case. Presumably requires some form of genetic code that determines that certain cells are to die at specific stages and specific sites during development. Classic example is the death of cells in the spaces between the developing digits of vertebrates, thus dividing them.

progress zone An undifferentiated population of mesenchyme cells beneath the apical ectodermal ridge of the chick limb bud from which the successive parts of the limb are laid down in a proximodistal sequence.

prohormone A protein hormone before processing to remove parts of its sequence and thus make it active.

Prokaryotes Organisms, namely bacteria and cyanobacteria (formerly known as blue-green algae), characterized by the possession of a simple naked DNA chromosome, occasionally two such chromosomes, usually of circular structure, without a nuclear membrane and possessing a very small range of organelles, generally only a plasma membrane and ribosomes.

prolactin Pituitary lactogenic hormone (23 kD), synthesized on ER-bound ribosomes as preprolactin that has an N-terminal signal peptide that is cleaved from the mature form. The conversion of preprolactin to prolactin has been much used as an assay for membrane insertion.

prolamellar body The disorganized membrane aggregations in chloroplasts that have been deprived of light (***etioplasts***).

proliferating cell nuclear antigen (PCNA) Commonly used marker for proliferating cells, a 35 kD protein that associates as a trimer and as a trimer interacts with DNA polymerases δ and ε; acts as an auxiliary factor for DNA repair and replication. Transcription of PCNA is modulated by p53.

proliferative unit (of epidermis) The basal layer of the mammalian epidermis contains cells that undergo repeated divisions. The cells outwards from a particular basal cell are often derived from this cell or a nearby one so that columns of cells exist running outwards from the stem cell in the basal layer from which they were derived. Such columns of cells are referred to as proliferative units.

proliferin A hormone, related to ***prolactin***, associated with the induction of cell division that is triggered by serum.

proline (pro; P) One of the 20 amino acids directly coded for in proteins (115 D). Structure differs from all the others, in that its side chain is bonded to the nitrogen of the α-amino group, as well as the α-carbon. This makes the amino group a secondary amine, and so proline is described as an imino acid. Has strong influence on secondary structure of proteins and is much more abundant in collagens than in other proteins, occurring especially in the sequence glycine-proline ***hydroxyproline***. A proline-rich region seems to characterize the binding site of ***SH3*** domains. See Table A2.

prolyl hydroxylase See ***hydroxyproline***.

prometaphase Rarely used term that designates an extra stage in mitosis, starting with the breakdown of the nuclear envelope. Usually lumped in with ***metaphase***.

promoter A region of DNA to which RNA polymerase binds before initiating the transcription of DNA into RNA. The nucleotide at which transcription starts is designated +1 and nucleotides are numbered from this with negative numbers indicating upstream nucleotides and positive downstream nucleotides. Most bacterial promoters contain two ***consensus sequences*** that seem to be essential for the binding of the polymerase. The first, the Pribnow box, is at about –10 and has the consensus sequence 5′-TATAAT-3′. The second, the –35 sequence, is centred about –35 and has the consensus sequence 5′-TTGACA-3′. Most factors that regulate gene transcription do so by binding at or near the promoter and affecting the initiation of transcription. Much less is known about eukaryote promoters; each of the three RNA polymerases has a different promoter. RNA polymerase I recognizes a single promoter for the precursor of rRNA. RNA polymerase II, that transcribes all genes coding for polypeptides, recognizes many thousands of promoters. Most have the Goldberg–Hogness or ***TATA box*** that is centred around position –25 and has the consensus sequence 5′-TATAAAA-3′. Several promoters have a CAAT box around –90 with the consensus sequence 5′-GGCCAATCT-3′. There is increasing evidence that all promoters for genes for [***'housekeeping' proteins***] contain multiple copies of a GC-rich element that includes the sequence 5′-GGGCGG-3′. Transcription by polymerase II is also affected by more distant elements known as enhancers. RNA polymerase III synthesizes 5s ribosomal RNA, all tRNAs, and a number of small RNAs. The promoter for RNA polymerase III is located within the gene either as a single sequence, as in the 5s RNA gene, or as two blocks, as in all tRNA genes.

promoter insertion Activation of a gene by the nearby ***integration*** of a virus. The ***long-terminal repeat*** acts as a promoter for the host gene. A form of ***insertional mutagenesis***.

promyelocytes Cells of the bone marrow that derive from myeloblasts and will give rise to myelocytes; precursors of ***myeloid cells*** and neutrophil granulocytes.

pronase Mixture of proteolytic enzymes from *Streptomyces griseus*. At least four enzymes are present, including trypsin and chymotrypsin-like proteases.

pronucleus Haploid nucleus resulting from meiosis. In animals the female pronucleus is the nucleus of the ovum before fusion with the male pronucleus. The male pronucleus is the sperm nucleus after it has entered the ovum at fertilization but before fusion with the female pronucleus. In plants the pronuclei are the two male nuclei found in the pollen tube.

properdin (factor P) Component of the alternative pathway for ***complement*** activation: complexes with C3b and stabilizes the alternative pathway C3 convertase (C3bBbP) that cleaves C3.

prophage The genome of a ***lysogenic*** bacteriophage when it is integrated into the chromosome of the host bacterium. The prophage is replicated as part of the host chromosome.

prophase Classical term for the first phase of mitosis or of one of the divisions of meiosis. During this phase the chromosomes condense and become visible.

prophylaxis Preventative action that will, for example, prevent infection; thus vaccination is a prophylactic treatment.

propidium iodide Used as a fluorescent stain for DNA and also for detecting dead cells which are permeable and therefore stain.

Propionibacterium Genus of bacteria that will ferment glucose to propionic acid or acetic acid.

proplastid Small, colourless ***plastid*** precursor, capable of division. It can develop into a chloroplast or other form of plastid, and has little internal structure. Found in cambial and other young cells.

propranolol Potent adrenergic antagonist acting at β1- and β2-adrenergic receptors.

prorenin Inactive precursor of ***renin***.

Prosite Searchable dictionary of conserved protein domains. Useful in inferring likely function of novel proteins.

prosome Raspberry-shaped ribonucleoprotein particle (19S) composed of small cytoplasmic RNA (15%) and heat-shock proteins, thought to be involved in post-transcriptional repression of mRNA translation: found in both nucleus and cytoplasm.

prospero *Drosophila* gene, product of which is asymmetrically distributed in the division of neural stem cells (neuroblasts) and is not present in one daughter (pluripotent neuroblast) but is retained in the ganglion mother cell (which has more restricted developmental potential). Prospero protein, once released from interaction with ***Miranda***, translocates to the nucleus and causes differential gene expression.

prospherosome Proposed stage in the development of ***spherosomes*** in plant cells. There is an accumulation of lipid in the prospherosome that is mobilized at a later stage.

prostacyclin (PGI_2) Unstable ***prostaglandin*** released by mast cells and endothelium, a potent inhibitor of platelet aggregation; also causes vasodilation and increased vascular permeability. Release enhanced by ***bradykinin***.

prostaglandins (PGs) Group of compounds derived from arachidonic acid by the action of ***cyclooxygenase*** that produces cyclic endoperoxides (PGG_2 and PGH_2) which can give rise to ***prostacyclin*** or ***thromboxanes*** as well as prostaglandins. Were originally purified from prostate (hence the name), but are now known to be ubiquitous in tissues. PGs have a variety of important roles in regulating cellular activities, especially in the inflammatory response where they may act as vasodilators in the vascular system, cause vasoconstriction or vasodilation together with bronchodilation in the lung, and act as hyperalgesics. Prostaglandins are rapidly degraded in the lungs, and will not therefore persist in the circulation. Prostaglandin E_2 (PGE_2) acts on ***adenylate cyclase*** to enhance the production of ***cyclic AMP***.

prostanoids Collective term for ***prostaglandins***, ***prostacyclins*** and ***thromboxanes***: slightly narrower than ***eicosanoids***.

prosthetic group A tightly bound non-polypeptide structure required for the activity of an enzyme or other protein, eg. the ***haem*** of ***haemoglobin***.

protamine Highly basic (arginine-rich) protein that replaces ***histone*** in sperm heads, enabling DNA to pack in an extremely compacted form, eg. clupein, iridin (4 kD). See also ***transition proteins***.

protease See also ***peptidase***. The term is normally reserved for endopeptidases that have very broad specificity and will cleave most proteins into small fragments. These are usually the digestive enzymes, eg. trypsin, pepsin, etc., or enzymes of plant origin (eg. ficin, papain) or bacterial origin (eg. pronase, proteinase K). Proteases are widely used for peptide mapping and for structural studies. See Table P1.

protease nexin Specific inhibitor of ***urokinase***, ***thrombin*** and ***plasmin***. Reported to influence neurite outgrowth by regulating the degree of proteolytic activity and thereby preventing excess degradation of substrate macromolecules and promoting neurite adhesion.

protease-activated receptor (PAR1, 2, 3) PAR1 is the human ***thrombin*** receptor, PAR2 is a possible trypsin receptor and PAR3 is similar to PAR1. All are G-protein coupled receptors activated by cleavage of part of the extracellular domain at K38/T39; the cleaved fragment

Table P1. Proteases[a]

Proteolytic enzymes (proteases; proteinases) can be divided into 'mechanistic' sets according to their mode of action. Most inhibitors tend to be specific for one set alone (the important exception being the plasma inhibitor α_2-macroglobulin). Alternatively, proteases can be classified simply according to whether they act on terminal amino acids (exopeptidases; aminopeptidases act at the N-terminal, carboxypeptidases at the C-terminus) or on peptide bonds within the chain (endopeptidases).

Set	Feature	Inhibitors	Examples
Serine	Serine at active site	Organic phosphate esters (DFP, PMSF)	Trypsin, chymotrypsin, thrombin, plasmin, elastase, subtilisin
Metallo-exopeptidase	Metal ion, often zinc	o-phenanthroline, EDTA	Carboxypeptidase
Metallo-endopeptidase	Metal ion, often zinc	o-phenanthroline, EDTA	Collagenase, thermolysin
Sulphydryl (Thiol)	CysSH at active site	Iodoacetate	Papain, cathepsin B, bromelain
Acid	Acid pH optimum	Diazoketones	Pepsin

[a]Based on Walsh (1975) *Cold Spring Harbor Conferences on Cell Proliferation*, Vol. 2.

then acts as a ligand and stimulates phosphoinositide hydrolysis. Found on platelets and megakaryocytes.

protegrins Family of ***cathelin***-associated antimicrobial peptides found in mammalian leucocytes. Will kill a range of bacteria including ***Gram-positive bacteria*** and are active against multidrug resistant strains.

protein A linear polymer of amino acids joined by peptide bonds in a specific sequence.

protein A Protein obtained from ***Staphylococcus aureus*** that binds immunoglobulin molecules without interfering with their binding to antigen. Widely used in purification of immunoglobulins, and in antigen detection, eg. by ***immunoprecipitation***. A very effective B-cell mitogen.

protein B Cell surface protein of group B streptococci that, like ***protein A***, will bind Fc region of immunoglobulin, but preferentially IgA.

protein C Vitamin K-dependent glycoprotein (62 kD) that is the zymogen of a serine endopeptidase (activated protein C; EC 3.4.21.69) found in plasma. Activated protein C in combination with ***protein S*** will hydrolyze blood-clotting factors Va and VIIIa, thereby inhibiting blood coagulation.

protein engineering Normally means the use of recombinant DNA technology to produce proteins with desired modifications in the primary sequence. See ***site-specific mutagenesis***.

protein G Protein from group C streptococci that binds the Fc portion of IgG; less species-specific than ***protein A***.

protein kinase Enzyme catalysing transfer of phosphate from ***ATP*** to hydroxyl side chains on proteins, causing changes in function. Most phosphate on proteins of animal cells is on ***serine*** residues, less on ***threonine***, with a very small amount on ***tyrosine*** residues. Tyrosine kinases phosphorylate proteins on tyrosine; serine/threonine kinases on serine or threonine.

protein kinase A (PKA) Cyclic AMP-dependent protein kinase.

protein kinase B (PKB) See **Alet**.

protein kinase C (PKC) Family of protein serine/threonine kinases activated by phospholipids that play an important part in intracellular signalling. The classical PKCs (α, β 1, β 2, γ) are also calcium dependent and can be activated by diacyl glycerol, one of the products of phospholipase C or, non-physiologically, by phorbol esters. A growing set of non-classical calcium independent isoforms are known. The catalytic domain is highly conserved and specific properties are conferred by a variety of regulatory domains including a pseudo substrate region which is displaced upon activation. The specific physiological substrates for these enzymes are not yet well defined.

protein kinase C phosphorylation site Protein kinase C tends to phosphorylate serine or threonine residues near a C-terminal basic residue, with the consensus pattern: [ST]-x-[RK].

protein kinase IV A calcium/calmodulin-dependent protein kinase (53 kD) found in brain, T-cells and postmeiotic male germ cells. Present in nucleus where it phosphorylates and activates ***CREB*** and CREM-tau. See ***calspermin*** and ***reticalmin***. May be important in preventing apoptosis during T-cell development and during activation of T-cells in response to mitogens.

protein kinase N (PKN; PRK1) Serine/threonine kinase (100 kD) probably regulated by ***rho***-dependent phosphorylation. Kinase activity resides in the C-terminal region (which has high homology with the catalytic domain of PKC). The N terminal region, through which regulation occurs, has a polybasic region and a ***leucine zipper*** domain. PKN has some sequence homology with ***rhophilin***.

protein S Single-chain glycoprotein (69 kD) that promotes binding of ***protein C*** to platelets; a vitamin K-dependent cofactor.

protein tyrosine phosphatase (PTP) A phosphatase that specifically cleaves the phosphate from a tyrosine residue in a protein, thus reversing the action of a ***tyrosine kinase***. Examples include CD45, ***shp, dep, lar***.

protein Z Major protein (43 kD) of barley endosperm, structurally similar to a ***serpin***.

protein zero (P ϕ) The major ***glycoprotein*** of peripheral nerve ***myelin***, an integral transmembrane protein (28 kD), synthesized by ***Schwann cells***.

proteinase-inhibitor inducing factor See ***PIIF***.

proteinoid droplets Membrane-bounded droplets supposed to have been formed in 'primaeval soup' as an early stages in the evolution of cells.

proteinoplast (proteoplast) Form of ***plastid*** adapted as a protein storage organelle; the protein may be crystalline.

proteoglycan A high molecular weight complex of protein and polysaccharide, characteristic of structural tissues of vertebrates, such as bone and ***cartilage***, but also present on cell surfaces. Important in determining viscoelastic properties of joints and other structures subject to mechanical deformation. ***Glycosaminoglycans*** (GAGs), the polysaccharide units in proteoglycans, are polymers of acidic disaccharides containing derivatives of the amino sugars glucosamine or galactosamine.

proteoheparan sulphate A ***proteoglycan*** containing as its ***glycosaminoglycan*** heparan sulphate whose constituent N-acetyl glucosamine is often sulphated. Hence highly negatively charged. ***Syndecan*** is one.

proteolipid Obsolete term for hydrophobic integral membrane proteins.

proteolipid protein (PLP) Highly conserved membrane protein (30 kD) in myelin. Cellular function obscure but mutations lethal, eg. ***jimpy*** mouse and Pelizaeus–Merzbacher disease of humans.

proteolysis Cleavage of proteins by proteases. Limited proteolysis occurs where proteins are functionally modified (activated in the case of zymogens) by highly specific proteases.

proteolytic enzyme See ***protease*** or peptidase.

proteome Protein equivalent of the genome: the next great challenge!

proteosome The 20S proteosome has 28 protein subunits arranged as an (α1–α7, β1–β7)$_2$ complex in four stacked rings. The interior of the complex has the active sites. The β-type subunits are synthesized as pro-proteins and are proteolytically cleaved before assembly. Nomenclature is complex: see Table P2.

Proteus (1) Genus of highly motile ***Gram-negative bacteria***. They are found largely in soil but are also found in the intestine of humans. They are opportunistic pathogens; *P. mirabilis* is a major cause of urnary tract infections. (2) An urodele amphibian. It is a cave dweller and is blind, has external gills and lacks any pigment.

prothallus Independent gametophyte phase of horsetail or fern.

prothrombin Inactive precursor of ***thrombin***, found in blood plasma.

protirelin See ***thyrotropin-releasing hormone***.

Protista The kingdom of eukaryotic unicellular organisms. It includes the ***Protozoa***, unicellular eukaryotic algae and some fungi (myxomycetes, acrasiales and oomycetes).

protochlorophyllide Precursor of chlorophyll, found in ***proplastids*** and ***etioplasts***. Lacks the phytol side chain of chlorophyll.

protofilaments One way of viewing microtubule structure is to consider it to be built of (usually) 13 protofilaments arranged in parallel.

protolignin An immature form of lignin that can be extracted from the plant cell wall with ethanol or dioxane.

protolysosome Primary lysosome that has not been involved in fusion with another vesicle or in digestive activity.

protomers Subunits from which a larger structure is built. Thus the tubulin

Table P2. Proteosome subunits

Systematic name	Traditional name (*S. cerevisiae*)	Human equivalent
α1.sc	C7/PRS2	ι
α2.sc	Y7	C3
α3.sc	Y13	C9
α4.sc	PRE6	C6
α5.sc	PUP2	ζ
α6.sc	PRE5	C2
α7.sc	C1/PRS1	C8
β1.sc	PRE3	Y/δ
β1i.hs	—	LMP2
β2.sc	PUP1	Z
β2i.hs	—	MECL1
β3.sc	PUP3	C10
β4.sc	C11/PRE1	C7
β5.sc	PRE2	X/MB1
β5i.hs	—	LMP7
β6.sc	C5/PRS3	C5
β7.sc	PRE4	N3/β

sc = in *Saccharomyces cerevisiae*; hs = *Homo sapiens*. β1i, β2i and β5i are exchangeable subunits in the immunoproteosome.
Based upon Table 2 in M Groll, L Ditzel, J Löwe, D Stock, M Bochtler, HD Bartunik and R Huber (1997) *Nature* 386: 463.

heterodimer is the protomer for microtubule assembly, G-actin the protomer for F-actin. Because it avoids the difficulty that arises with, for example, dimers that serve as subunits for assembly, it is a useful term that deserves wider currency.

protooncogene The normal, cellular equivalent of an ***oncogene***; thus usually a gene involved in the signalling or regulation of cell growth. In general, cellular protooncogenes are prefixed with a 'c', rather than their abnormal viral counterparts, which are prefixed with a 'v', eg. c-*myc* and v-*myc*.

proton ATPase (H^+ -ATPase) An ion pump that actively transports hydrogen ions across lipid bilayers in exchange for ATP. Major groups are the ***F-type ATPases***, that run in reverse to synthesize ATP in bacterial, mitochondrial and chloroplast membranes ('ATP synthase'); and the ***V-type ATPases*** found in intracellular vesicles with an acidic lumen, and on certain epithelial cells (eg. kidney intercalated cells). Gastric H^+/K^+ATPase is a proton ATPase.

proton motive force See ***PMF***.

protonophore ***Ionophore*** that carries protons. Many ***uncoupling agents*** are protonophores.

protoplast A bacterial cell deprived of its cell wall, eg. by growth in an isotonic medium in the presence of antibiotics that block synthesis of the wall ***peptidoglycan***. Alternatively, a plant cell similarly deprived by enzymic treatment.

protoporphyrin Porphyrin ring structure lacking metal ions. The most abundant is protoporphyrin IX, the immediate precursor of haem.

protostome Invertebrate phylum in which the mouth forms from the embryonic blastopore. Major protostome phyla are Annelida, Mollusca and Arthropoda. See ***deuterostome***.

Protozoa A very diverse group comprising some 50 000 eukaryotic organisms that consist of one cell. Because most of them are motile and heterotrophic, the Protozoa were originally regarded as a phylum of the animal kingdom. However it is now clear that they have only one common characteristic, they are not multicellular, and Protozoa are now usually classed as a subkingdom of the kingdom ***Protista***. On this classification the Protozoa are grouped into several phyla, the main ones being the Sarcomastigophora (flagellates, heliozoans and amoeboid-like protozoa), the Ciliophora (ciliates) and the Apicomplexa (sporozoan parasites such as ***Plasmodium***).

provacuoles In plant cells provacuoles are budded directly from the ***rough endoplasmic reticulum*** and fuse with other provacuoles to form vacuoles. Since vacuoles may contain hydrolytic enzymes, it is therefore possible to consider them as analogues of primary lysosomes in animal cells.

provirus The genome of a virus when it is integrated into the host cell DNA. In the case of the retroviruses, their RNA genome has first to be transcribed to DNA by ***reverse transcriptase***. The genes of the provirus may be transcribed and expressed, or the provirus may be maintained in a latent condition. The integration of the ***oncogenic*** viruses, such as ***Papovaviridae*** and retroviruses can lead to cell transformation.

Prozac (fluoxetine) Proprietary name for a serotonin uptake inhibitor used extensively as an antidepressant.

prozone Prozone phenomena occur in immunological reactions when the concentrations of antibody or other active immune agent are so high that the optimum concentration for maximal reaction with antigen is exceeded. Immunological phenomena in the prozone region may show partial or total inhibition.

prozymogen granule (condensing vacuole) Stage in the development of a mature ***secretory vesicle*** (zymogen granule).

PrP PrPc is a normal protein anchored to the outer surface of neurons and, to a lesser extent, the surfaces of other cells, including lymphocytes. The ***prion***, PrPs, thought to be responsible for ***scrapie*** and other ***spongiform encephalopathies*** is hypothesized to be a modified form of PrPc.

PSA (prostate specific antigen) Antigen in serum that seems to be a relatively reliable marker for prostatic hyperplasia/ carcinoma. Antigen is a serine endopeptidase related to kallikreins.

PSD-proteins Family of postsynaptic density proteins containing ***PDZ*** domains and apparently responsible for the clustering of receptors. The PSD-95 protein is responsible for the clustering of NMDA receptors and K^+ channels; a similar protein is responsible for the formation of synaptic complexes.

pseudogene Non-functional DNA sequences that are very similar to the sequences of known genes. Examples are those found in the β-like globin gene cluster. Some probably result from gene duplications that become non-functional because of the loss of promoters, accumulation of stop codons, mutations that prevent correct processing, etc. Some pseudogenes contain a ***poly-A tail*** suggesting that an mRNA, at some point, was copied into DNA which was then integrated into the genome.

Pseudomonas Genus of ***Gram-negative bacteria***. They are rod-shaped and motile, possessing one or more polar ***flagella***. Several species produce characteristic water-soluble fluorescent pigments. They are found in soil and water. *P. syringae* is a plant pathogen causing leaf spot and wilt. *P. aeruginosa*, normally a soil bacterium, is an opportunistic pathogen of humans who are immunocompromised. It can infect the wounds of victims with severe burns, causing the formation of blue pus.

Pseudonaja textilis textilis Australian brown snake. See ***textilotoxin***.

pseudopod Blunt-ended projection from a cell: usually applied to cells that have an amoeboid pattern of movement.

pseudopterosins Class of natural compounds (diterpene-pentose glycosides) isolated from the soft coral *Pseudopterogorgonia elisabethae*, and that interfere with arachidonic acid metabolism. Have anti-inflammatory and analgesic properties.

pseudospatial gradient sensing Mechanism for sensing a gradient of a diffusible chemical in which the cell sends protrusions out at random; upgradient protrusions are stabilized by positive feedback (because receptor occupancy is rising with time) and others are transitory because of adaptation. Possibly the mechanism by which neutrophils sense chemotactic gradients.

pseudouridine (5-β -D-ribofuranosyluracil) Unusual nucleotide found in some tRNA: glycosidic bond is associated with position 5 of ***uracil***, not position 1.

PSGL-1 (CD162) P-selectin glycoprotein ligand; expressed on neutrophils, monocytes and most lymphocytes. An extended mucin-like transmembrane glycoprotein expressed as a homodimer. See ***PADGEM***.

psoralens Drugs capable of forming photoadducts with nucleic acids if ultraviolet-irradiated.

psoriasis Chronic inflammatory skin disease characterized by epidermal hyperplasia. Lesions may be limited or widespread and in the latter case the disease can be life-threatening. Unlike many chronic inflammatory conditions it seems to be T-cell mediated. There is a fairly strong genetic predisposition.

psychrophile Organism that grows best at low temperatures.

PTEN (phosphatase and tensin homolog deleted on chromosome 10; MMAC1/ PTEN; TEP-1) A protein tyrosine phosphatase with homology to ***tensin***, is a tumour-suppressor gene on chromosome 10q23. Somatic mutations in PTEN occur in multiple tumours, most markedly glioblastomas. Germ-line mutations in PTEN are responsible for Cowden disease (CD), a rare autosomal dominant multiple-hamartoma syndrome. Mutated in MMAC1/ PTEN (multiple advanced cancers 1/ phosphatase and tensin) homologue.

pteridine (pyrazino [2,3-d] pyrimidine) Nitrogen-containing compound composed of two 6-membered rings (pyrazine and pyrimidine rings). Structural component of ***folic acid*** and ***riboflavin*** and parent compound of pterins such as xanthopterin, a yellow pigment found in the wings of some butterflies.

Pteridophyta Division of the plant kingdom that includes ferns, horsetails and clubmosses.

PtK2 cells Cell line from *Potorous tridactylis* (potoroo or kangaroo rat) kidney. Often used in studies on mitosis because there are only a few large chromosomes and the cells remain flattened during mitosis.

ptosis Drooping of the upper eyelid for any one of a number of causes. May be a result of damage to the third cranial nerve, to ***myasthenia gravis***, to Horner's syndrome or simply be an isolated congenital feature.

PTP (protein tyrosine phosphatase) A ***phosphatase*** that reverses the effect of a tyrosine kinase.

ptyalin Common name for α-amylase found in saliva.

PU box Purine rich sequence recognized by the product of the *Sp-1* oncogene.

pUC9 *E. coli* ***phagemid vector***, derived from pUC19 and M13MP9.

PUFA (polyunsaturated fatty acid) Increasing the ratio of unsaturated to saturated fatty acids can alter the behaviour of cells, probably by altering physical characteristics of membranes and thus influencing the behaviour of integral membrane proteins. Whether this is true for whole organisms is a matter for debate.

puffs Expanded areas of a ***polytene chromosome***. At these areas the chromatin becomes less condensed and the fibres unwind, though they remain continuous with the fibres in the chromosome axis. A puff usually involves unwinding at a single band, though they can include many bands as in ***Balbiani rings***. Puffs represent sites of active RNA transcription. The pattern of puffing observed in the larvae of *Drosophila*, in different cells, and at different times in development provides possibly the best evidence that differentiation is controlled at the level of transcription.

pulse-chase An experimental protocol used to determine cellular pathways, such as precursor–product relationships. A sample (organism, cell or cellular organelle), is exposed for a relatively brief time to a radioactively labelled molecule, the pulse. It is then replaced with an excess of the unlabelled molecule, the chase (cold chase). The sample is then examined at various later times to determine the fate of radioactivity incorporated during the pulse.

pulse-field electrophoresis A method used for high-resolution electrophoretic separation of very large (megabase) fragments of DNA. Electric fields 100° apart (the angle may vary) are applied to the separation gel alternately. The continuous change of direction prevents the molecules aligning in the electric field and greatly improves resolution on the axis between the two fields.

Punnett's square The checkerboard (matrix) method used to determine the types of zygotes produced by fusion of gametes from parents of defined genotype.

purine A heterocyclic compound with a fused pyrimidine/imidazole ring. Planar and aromatic in character. The parent compound for the purine bases of nucleic acids.

purinergic receptors Receptors that use purine nucleotides (eg. ATP) as ligands.

Purkinje cell A class of ***neuron*** in the ***cerebellum***; the only neurons that convey signals away from the cerebellum.

Purkinje fibres Specialized cardiac muscle cells that conduct electrical impulses through the heart and are involved in regulating the beat.

puromycin An antibiotic that acts as an ***aminoacyl tRNA*** analogue. Binds to the A-site on the ***ribosome***, forms a peptide linkage with the growing chain and then causes premature termination.

purple membrane Plasma membrane of *Halobacterium* and *Halococcus*, that contains a protein-bound carotenoid pigment that absorbs light and uses the energy to translocate protons from the cytoplasm to the exterior. The proton gradient then provides energy for ATP synthesis. The binding protein is called ***bacteriorhodopsin***, or purple membrane protein.

purpurin Heparin-binding protein (20 kD) released by chick neural retina cells in culture.

putrescine A dibasic amine associated with putrifying tissue. Associates strongly with DNA. Has been suggested as a growth factor for mammalian cells in culture. Metabolic precursor of the polyamines ***spermine*** and ***spermidine***.

PVP (polyvinyl pyrrolidone) Water-soluble white compound that when dissolved makes a very viscous solution.

pyaemia Invasion of bloodstream by pyogenic organisms.

pyknosis Contraction of nuclear contents to a deep-staining irregular mass; sign of cell death.

pyocins ***Bacteriocins*** produced by bacteria of the genus *Pseudomonas*.

pyocyanin Blue-green phenazine pigment produced by *Pseudomonas aeruginosa*; has antibiotic properties.

pyogenic Causing the formation of pus, a thick yellow or greenish liquid formed at a site of infection, and that contains dead leucocytes, bacteria and tissue debris.

pyramidal cells Commonest nerve cells of the cerebral cortex.

pyranose Sugar structure in which the carbonyl carbon is condensed with a hydroxyl group (ie. in a hemi-acetal link), forming a ring of five carbons and one oxygen. Most hexoses exist in this form, although in sucrose, fructose is found with the smaller (four carbon) furanose ring.

pyrenoid Small body found within some chloroplasts that may contain protein. In green algae may be involved in starch synthesis.

pyrethrum (pyrethroid) Toxic hydrocarbon, originally from chrysanthemum flowers, that now forms the basis for a wide range of 'natural' synthetic pyrethroid insecticides.

pyridoxal phosphate The coenzyme derivative of vitamin B6. Forms ***Schiff bases*** of substrate amino acids during catalysis of transamination, decarboxylation and racemisation reactions.

pyrimidine (1,3 diazine) A heterocyclic 6-membered ring, planar and aromatic in character. The parent compound of the pyrimidine bases of nucleic acid.

pyrogen Substance or agent that produces fever. The major endogenous pyrogen in mammals is probably ***interleukin-1***.

pyrogenic Causing fever. See ***pyrogen***.

pyrophosphate Two phosphate groups linked by esterification. Released in many of the synthetic steps involving nucleotide triphosphates (eg. protein and nucleic acid elongation). Rapid cleavage by enzymes that have high substrate affinity ensures that these reactions are essentially irreversible.

pyrrole ring A heterocyclic ring structure, found in many important biological pigments and structures that involve an activated metal ion, eg. chlorophyll, haem.

pyruvate (2-oxopropanoate; $CH_3 COCOO^-$) Important intermediate in many metabolic pathways, particularly of glucose metabolism and the synthesis of many amino acids.

pyruvate carboxylase An enzyme that catalyses the formation of oxaloacetate from pyruvate, CO_2 and ATP in gluconeogenesis.

pyruvate dehydrogenase A complex multienzyme system that catalyses the conversion of (pyruvate + CoA + NAD^+) to (acetyl CoA + CO_2 + NAD).

Q

Q10 Ratio of the velocity of reaction at one temperature and that at a temperature 10°C lower. Usually around two for biological reactions.

Q See ***glutamine***.

Q banding See ***banding patterns*** and ***quinacrine***.

Q beta (Q β) An RNA virus that infects *E. coli*. Genome circular, single stranded and acts both as template for replication of a complementary strand and as messenger RNA.

Q box Glutamine-rich sequence found in some transcription factors.

Q enzyme Converts amylose to amylopectin.

Q fever Typhus-like illness caused by rickettsia, *Coxiella burneti*. Mainly a disease of domestic animals but can be caught by humans.

Q-type channels A class of ***voltage-sensitive calcium channels***. May be identical or very similar to ***P-type channels***. Inhibited by neurotransmitters that act through G-protein coupled receptors, high concentrations of ω-***conotoxin*** and ω-***agatoxin***.

QH2-cytochrome c reductase Membrane-bound complex in the mitochondrial inner membrane, responsible for electron transfer from reduced coenzyme Q to cytochrome c. Contains cytochromes b and c1, and iron–sulphur proteins.

qmf1, qmf2, qmf3 ***Quail*** homologues of MyoD, myogenin and myf-5, respectively.

QSAR (quantitative structure–activity relationship) Relationship between the structure of a compound and its activity in binding or inhibiting something, based upon computed parameters of structure. Computational chemists rely upon complex computational methods to derive appropriate paramters of shape and electronic distribution.

quail Small galliform bird. Quail embryos are often use in developmental studies because quail cells can be distinguished from chicken cells, yet the two are sufficiently closely related that it is possible to graft embryonic tissue from one to the other.

quantal mitosis A controversial concept in cellular differentiation proposed by H. Holtzer and defined by him as a mitosis 'that yields daughter cells with metabolic options very different from those of the mother cell as opposed to proliferative mitoses in which the daughter cells are identical to the mother cell'. Implicit in this is the idea that the changes in cell ***determination*** that occur during development take place at these special quantal mitoses.

quantasome Smallest structural unit of photosynthesis, a particulate component of the ***thylakoid*** membrane containing chlorophyll and cytochromes.

quantum yield (quantum requirement) The number of photons required for the formation of one oxygen molecule in photosynthesis. Varies from 8 to 14 depending on the system used to measure it.

quasi-equivalence Term used to refer to the way in which subunits pack into a quasi-crystalline array as, for example, in viral coat assembly. There is usually some strain in the packing.

quaternary structure Fourth order level of structural organization of proteins. Tertiary structure defines the shape of single protein molecules; quaternary structure the way in which dimers or multimers are arranged.

Quellung reaction Swelling of the capsule surrounding a bacterium as a result of interaction with anticapsular antibody; consequently the capsule becomes more refractile and conspicuous.

quercetin Mutagenic flavonol pigment found in many plants. Inhibits ***F-type ATPases***.

quiescence 'Quietness'. In cells, the state of not dividing; in neurons, the state of not firing.

quiescent stem cell A stem cell that is not at that time undergoing repeated cell cycles but that might be stimulated so to do later. For example, the satellite cells in the skeletal muscles of mammals that are quiescent myoblasts that will proliferate after wounding and give rise to more muscle cells by fusion.

quin2 A fluorescent calcium indicator. Resembles the chelator ***EGTA*** in ability to bind calcium much more tightly than magnesium. Binding of calcium causes large changes in ultraviolet absorption and fluorescence.

quinacrine A fluorescent dye that intercalates into DNA helices. Chromosomes stained with quinacrine show typical banding patterns of fluorescence at specific locations, Q bands, that can be used to recognize chromosomes and their abnormalities.

quinate: NAD oxidoreductase A plant enzyme converting hydroquinic acid (a derivative of the shikimate pathway) to quinic acid. The enzyme is activated by a calcium- and calmodulin-dependent phosphorylation.

quinine An alkaloid isolated from cinchona bark. Used as an antimalarial. It is believed to act by raising the pH of endocytotic vesicles and inhibiting internal membrane fusion processes.

quinone Aromatic dicarbonyl compound derived from a dihydroxy aromatic compound. Ubiquinone (coenzyme Q) is a dimethoxy-dicarbonyl derivative of benzene involved in electron transport. Other quinones may act as tanning agents.

quinone reductase Enzymes that reduce quinones to phenols usually using NADH or NADPH as a source of reductant.

quisqualate An agonist of the Q-type ***excitatory amino acid*** receptor. See ***kainate***.

QX-222 Open-channel blocker at ***nicotinic acetylcholine receptors***; a quaternary ammonium derivative of the local anaesthetic Lidocaine. In mouse muscle AChR it blocks by interacting with adjacent turns of the M2 (membrane spanning region) helix.

R

R17 bacteriophage Bacteriophage with RNA genome that codes for the enzyme ***RNA synthetase*** and for the coat protein, a protein to which the RNA is attached and that is involved in attachment to the bacterium.

R7G See ***G-protein coupled receptors***.

R See ***arginine***.

R body A protein structure, visible by optical microscopy, found in various bacteria, probably related to plasmid presence. Found both in free-living pseudomonads and in various bacteria endosymbiotic in *Paramecium*. Has toxic activity against *Paramecium* and confers killer characteristics on *Paramecium* that ingest bacteria containing the structure.

R factor See ***R plasmid***.

R loop A single-stranded loop section of DNA formed by the association of a section of ssRNA with the other strand of the DNA in this region whereby one DNA strand is displaced as the loop. Mature mRNA can be used to form loops from exons with the intervening double-stranded linear regions being introns.

R plasmid (R factor; drug resistance factor) A plasmid that confers resistance to one or more antibiotics or other poisonous compounds in a bacterium.

R point of cell cycle See ***restriction point***.

R-type channels A class of neuronal ***voltage-sensitive calcium channels*** that are unaffected by dihydropyridines, phenylalkylamines and conotoxins. Thought to carry much of the current, stimulated by glutamate release in response to ischaemia, that induces neuronal death.

rab genes (1) One of the three main groups of ***ras***-like genes specifying small GTP-binding proteins (the others are ***ras*** and ***rho***). Rab proteins are involved in vesicular traffic and seem to control translocation from donor to acceptor membranes. (2) Gene family in plants 'responsive to abscisic acid': encode proteins of 15–17 kD.

rabaptin Rabaptin-5 (117 kD) interacts with rab5-GTP and is essential for rab-mediated endosomal fusion. Both ends of rabaptin have coiled-coil domains characteristic of vesicular transport proteins.

rabies virus Species of the ***Rhabdoviridae*** that causes rabies in humans. The virus infects the cells in the brain, causing a fatal encephalomyelitis. It is found all over the world, but strict quarantine regulations have excluded it from Britain and Australia. The virus infects a number of domestic and wild mammals, whose saliva is infective. Some bats and small mammals can carry the virus without showing any symptoms of disease.

rabin3 Protein (50 kD) that interacts with rab3A and rab3D but not with other small GTPases.

rabphilin Receptor (704 amino acid residues) for the small GTP-binding protein rab3A that is implicated in regulated secretion, particularly of neurotransmitters. The N-terminal region interacts with rab3A, the C terminal domain interacts with calcium and phospholipid, and rabphilin is found in association with sites of exocytosis in neurites. There are indications that rabphilin may inhibit ***GAP***-activated GTPase activity of rab3A.

rac Small GTP-binding protein involved in regulating actin cytoskeleton; the activated form of rac seems to induce membrane ruffling (whereas ***rho*** acts on stress fibres). Rac may be activated by specific ***GAPs*** such as ***bcr*** and ***N-chimaerin***.

RACE See ***rapid amplification of DNA ends***.

racemic mixture (racemate) A mixture containing equimolar amounts of two enantiomers (D and L forms) of a chiral molecule.

rachitic To do with rickets, a disease caused by vitamin D deficiency.

RACKs (receptors for activated C kinase) Proteins, usually anchored to specific areas of the cell, which selectively bind activated protein kinase C and thus control the regions of the cell on which it acts.

rad (1) Abbreviation for radian. (2) Unit of radiation 1rad = 0.01Gy. (3) *rad1* is a *Schizosaccharomyces pombe* checkpoint control gene important in both DNA damage-dependent and replication-dependent cycle control; various *rad* genes are of comparable function in other organisms (*Hrad1* from man, *Mrad1* from mouse, *RAD17* from *Saccharomyces cerevisiae*)

rad23 Conserved protein that is important in nucleotide excision repair. N-terminal domain has similarity with ***ubiquitin*** and links rad23 to the proteosome. Binds to rad14 and TFIIH and forms a stable complex with rad4.

rad proteins (rad 50, 51, 52, 54, 55, 57, 59; MRE11, XRS2) Yeast proteins coded by genes originally identified as being particularly sensitive to X-rays. Required for spontaneous and induced mitotic recombination, meiotic recombination and mating-type switching. Human homologues of many of these proteins have been identified. rad51 (37 kD) is the functional homologue of ***recA*** and promotes ATP-dependent homologous pairing and DNA strand exchange; binds Rad52.

radial glial cell A type of glial cell, organized as parallel fibres joining the inner and outer surfaces of the developing cortex. They are thought to play a role in ***neuronal guidance*** in development. See ***contact guidance***.

radial spoke The structure that links the outer microtubule doublet of the ciliary axoneme with the sheath that surrounds the central pair of microtubules. The spokes are arranged periodically along the axoneme every 29nm, have a stalk about 32nm long and a bulbous region adjacent to the sheath. At least 17 different polypeptides are associated with the spokes. Spokes are thought to restrict the sliding of doublets relative to one another; digestion of the radial spokes will allow sliding apart of the doublets.

radiation inactivation The technique of inactivating proteins in freeze-dried (lyophilized) preparations using high-energy particles (eg. electrons). One high-energy particle can apparently inactivate all of the components of a multisubunit polypeptide; the method is therefore used to determine the molecular weight of functional oligomers.

radioautography See ***autoradiography***.

radioimmunoassay Any system for testing antigen–antibody reactions in which use is made of radioactive labelling of antigen or antibody to detect the extent of the reaction.

radioisotope Form of a chemical element with unstable neutron number, so that it undergoes spontaneous nuclear disintegration. Major use in biology is to trace the fate of atoms or molecules that follow the same metabolic pathway or enzymic fate as the normal stable isotope, but which can be detected with high sensitivity by their emission of radiation. Also used to locate the position of the radioactive metabolite, as in ***autoradiography***, and to measure relative rates of synthesis of compounds from radioactive precursors.

Radiolaria Subclass of the ***Sarcodina***. Marine protozoans with silicaceous exoskeleton and radiating filopodia.

radixin Barbed-end capping actin-modulating protein (82 kD) found in ***adherens junctions*** and in the cleavage furrow of many cells.

raf Serine/threonine protein kinase implicated in signal–reponse transduction pathways involving tyrosine-kinases. Apparently raf1 is downstream of ras1 in the signalling cascade.

raf See *mil*.

raffinose (mellitose) A non-reducing trisaccharide found in sugar beet and many seeds, consisting of the disaccharide ***sucrose*** bearing a D-galactosyl residue linked α (1-6) to its glucose group.

Raji cell binding test A test for the detection of soluble IgG–antigen complexes. Raji cells are a line of EBV-transformed lymphocytes with surface Fc receptors. Complexes are detected by their ability to compete with a radiolabelled aggregated IgG for binding to the cells.

Raji cells Cell line that grows in suspension derived from patient with Burkitt's lymphoma.

rak Tyrosine kinase (54 kD) found in nucleus. Has N-terminal ***SH2*** and ***SH3*** domains and has similarities with src. Binds to ***Rb*** protein and leads to growth suppression.

ral Oncogene related to ***ras***. Protein product is a ras-like GTPase.

Raman spectroscopy Method for measuring the Raman spectrum, the plot of Raman scattering of light that produces weak radiation at frequencies not present in the incident radiation. The spectrum is characteristic of the compound and independent of the wavelength of the incident light.

RAMPs (receptor activity modifying proteins) RAMPs are type I transmembrane proteins that associate with and regulate properties of receptors. If the calcitonin receptor-like receptor is transported to the membrane by RAMP1 it behaves as a ***calcitonin gene-related peptide*** receptor; if associated with RAMP2 its glycosylation pattern is different and it acts as an ***adrenomedullin*** receptor.

ran Small G-protein (GTPase) required, together with ***importins*** α and β and pp15, for protein transport into the nucleus. The only known nucleotide exchange factor for ran is nuclear (RCC1), whereas the only known activating factor is cytoplasmic. This would provide a mechanism for vectorial transport. Ran-GTP binds importin and may cause dissociation of the transport complex.

Rana pipiens Common European frog.

random amplification of polymorphic DNA See ***RAPD***.

random coil A term originally invented by polymer chemists to describe a disordered tangle of a linear polymer chain with curved sections. In DNA parlance the random coil refers to the structure that results from melting or other forms of separation of the double helix, ie. helix–coil transition.

random priming Method of labelling a DNA probe for use in hybridization. Double-stranded DNA is denatured to form a single stranded template. Random oligonucleotide primers (usually hexamers) are allowed to anneal, nucleotides and DNA polymerase added, and new DNA fragments synthesized in the presence of trace amounts of radioactive or non-radioactive label. The result is a population of short, labelled DNA molecules of indeterminate length that represent the whole length of the template DNA.

random walk A description of the path followed by a cell or particle when there is no bias in movement. The direction of movement at any instant is not influenced by the direction of travel in the preceding period. If changes of direction are very frequent, then the displacement will be small, unless the speed is very great, and the object will appear to vibrate on the spot. Although the behaviour of moving cells in a uniform environment can be described as a random walk in the long term, this is not true in the short term because of ***persistence***.

RANTES (regulated upon activation normal T-expressed and secreted) ***Cytokine*** of the C-C subfamily, produced by T-cells, and chemotactic for monocytes, memory T-cells and eosinophils. Uniquely among the ***chemokines***, it is down-regulated when the secreting cells are activated.

rap Oncogene related to ***ras***.

rap1 (1) Small GTP-binding protein which seems to antagonize ***ras*** activity. Has antimitogenic activity, seems to be involved in NGF-induced neuronal differentiation and T-cell activation. Activated by ***CRK*** adaptor proteins and rap-specific ***GEF***s. See ***tuberous sclerosis***. (2) Rhoptry-associated protein 1 of *Plasmodium falciparum*, a potential component of an antimalarial vaccine.

RAP74 Large subunit of the transcription initiation factor (TFIIF).

rapamycin Immunosuppressive macrolide antibiotic with structural similarity to ***FK 506***; inhibits T-and B-cell proliferation but at a much later stage than FK 506, despite binding to the same ***immunophilin***. Inhibits TOR (target of rapamycin) in the Ras/MAP kinase signalling pathway.

RAPD Variation of the ***polymerase chain reaction*** used to identify differentially expressed genes. ***mRNA*** from two different tissue samples is reverse transcribed, then amplified using short, intentionally non-specific primers. The array of bands obtained from a series of such amplifications is run on a high-resolution gel and compared with analogous arrays from different samples. Any bands unique to single samples are considered to be differentially expressed; they can be purified from the gel, and sequenced and used to clone the full-length cDNA. Similar in aim to ***subtractive hybridization***. See also ***differential display PCR***.

raphidosome Rod-shaped particle found in bacterial cell near DNA-rich region.

rapid amplification of DNA ends (RACE; 3′ RACE; 5′ RACE) Techniques, based on the ***polymerase chain reaction***, for amplifying either the 5′ end (5′ RACE) or 3′ end (3′ RACE) of a cDNA molecule, given that some of the sequence in the middle is already known. The two procedures differ slightly; in the more straightforward 3′ RACE, first strand cDNA is prepared by reverse transcription with an oligo dT primer (to match the poly-A tail), from an mRNA population believed to contain the target. PCR then proceeds with a gene-specific, forward-facing primer and an oligo-dT reverse facing primer. 5′ RACE is an example of ***anchored PCR***; the first-strand cDNA population is tailed with a known sequence, either by homopolymer tailing (eg. with dA) or by ligation of a known sequence. PCR then proceeds as before with a primer specific for the gene, and one specific for the added tail.

ras One of a family of ***oncogenes***, first identified as transforming genes of ***Harvey*** and ***Kirsten*** murine ***sarcoma viruses***. ('ras' from rat sarcoma because Harvey virus, though a mouse virus, obtained its transforming gene during passage in a rat). Transforming protein coded is p21ras, a GTP-binding protein with GTPase activity, that resembles regulatory G-proteins.

ras-like GTPases Family of small G-proteins (rac, rab, ran, rad, rho, gem, kir, ric. rin, rit, Ypt). The ***rab*** subfamily is required for membrane traffic in eukaryotic cells, ral has been associated with growth factor-induced DNA synthesis and oncogenic transformation. ***Ran*** is highly conserved and found in the nucleus. ***Rho*** and ***rac*** are involved in cytoskeletal control. Rin, ric and rit lack prenylation sequences and are well conserved between *Drosophila* and humans; rin is confined to neuronal cells. Ypts are the yeast homologues, of which 11 are known. The Rad subfamily – rad (ras associated with diabetes), gem (immediate early gene expressed in mitogen-stimulated T-cells) and kir (tyrosine kinase-inducible ras-like) – bind calmodulin in a calcium-dependent manner via a C-terminal extension, and also have various serine phosphorylation sites so their activity may be regulated by kinases including CaMKII, PKA, PKC and CKII.

ratio-imaging fluorescence microscopy A method of measurement of intracellular pH or intracellular calcium levels, using a fluorescent probe molecule (see ***fura-2***), in which the two different excitation wavelengths are used, and the emitted light levels compared. If emission at one wavelength is sensitive to the intracellular ion level, and emission at the other wavelength is not, then standardization for intracellular probe concentration, efficiency of light collection, inactivation of probe and thickness of cytoplasm can all be performed automatically.

RAW 264 cells Murine monocyte/ macrophage line derived from ascitic tumour induced with Abelson leukaemia virus.

Rb ***Tumour suppressor*** gene encoding a nuclear protein that, if inactivated, enormously raises the chances of development of cancer, classically ***retinoblastoma***, but also other ***sarcomas*** and ***carcinomas***.

rbc Red blood cell or erythrocyte

RBL-1 cells Rat basophilic leukaemia cell line: shows wide variation but can be used as a model for basophils.

RCA See ***Ricinus communis agglutinin***.

rDNA DNA that codes for ribosomal RNA.

reaction centre The site in the chloroplast that receives the energy trapped by chlorophyll and accessory pigments, and initiates the electron transfer process.

reactive oxygen species (ROS) Oxygen-containing radical or reactive ions such as superoxide, singlet oxygen and hydroxyl radicals, the product of the ***respiratory burst*** in phagocytes and responsible for bacterial killing as well as incidental damage to surrounding tissue.

reading frame One of the three possible ways of reading a nucleotide sequence. As the genetic code is read in non-overlapping triplets (***codons***) there are three possible ways of ***translation*** of a sequence of nucleotides into a protein, each with a different starting point. For example, given the nucleotide sequence: AGCAGCAGC, the three reading frames are: AGC AGC AGC, GCA GCA, CAG CAG.

reagin Reaginic antibodies; an outmoded term for ***IgE***.

reannealing Renaturation of a DNA sample that has been dissociated by heating. In reannealing the two strands that recombine to form a double-stranded molecule are from the same source. Differences in the rate of reannealing led to the early recognition of repetitive sequences, which rapidly recombine (have low values on the ***Cot curve***).

reaper *(rpr)* Regulator of apoptosis in *Drosophila*. Has no known homology with vertebrate proteins but reaper-induced apoptosis is blocked by ***caspase*** inhibitors and human ***IAPs***. See also ***grim*** and ***hid***.

recA protein Protein (40 kD) product of the *rec* (recombination) gene, that catalyses the pairing of a single-stranded piece of DNA with its complementary sequence, displacing a loop of single stranded DNA (***D loop***).

recapitulation The outmoded theory that the stages of development (ontogeny) recapitulated the evolutionary stages through which an organism had passed (phylogeny); thus the primitive mammalian embryo was supposed to go through fish-like and amphibian-like stages before gaining mammal-like features.

recB protein Protein (140 kD); one subunit of nuclease that unwinds double-stranded DNA and fragments the strands sequentially; the other subunit is recC (128 kD)

receiver cell Cells in the photosynthetic tissues of plants into which the solutes from xylem are pumped.

receptor In general terms, a membrane-bound or membrane-enclosed molecule that binds to, or responds to something more mobile (the ligand), with high specificity. Examples: ***acetylcholine receptor***, ***photoreceptors***.

receptor downregulation A phenomenon observed in many cells: following stimulation with a ligand the number of receptors for that ligand on the cell surface diminishes because internalization exceeds replenishment. Often used very loosely, thus destroying the utility of the term.

receptor potential The transmembrane potential difference of a sensory cell. Such cells are not generally excitable, but their response to stimulation is a gradual change in their ***resting potential***.

receptor tyrosine kinase Class of membrane receptors that phosphorylate tyrosine residues. Many play significant roles in development or cell division. Examples: insulin receptor family, c-*ros* receptor, *Drosophila sevenless*, trk family.

receptor-mediated endocytosis Endocytosis of molecules by means of a specific receptor protein that normally resides in a ***coated pit***, but may enter this structure after complex formation occurs. The structure then forms a coated vesicle that delivers its contents to the endosome whence it may enter the cytoplasm or the lysosomal compartment. Many bacterial toxins and viruses enter cells by this route.

receptors for activated C kinase See ***RACK***s.

receptosome Synonym for ***endosome***.

recessive An ***allele*** or ***mutation*** that is only expressed phenotypically when it is present in the homozygous form. In the heterozygote it is obscured by dominant alleles.

recombinant DNA Spliced DNA formed from two or more different sources that have been cleaved by ***restriction enzymes*** and joined by ***ligases***.

recombinase Enzymes that mediate ***site-specific recombination*** in prokaryotes. They fall into two families, 'phage integrases' and 'resolvases'.

recombination The creation, by a process of intermolecular exchange, of chromosomes combining genetic information from different sources, typically two genomes of a given species. Site-specific, homologous, transpositional and non-homologous (illegitimate) types of recombination are known. Recombination can be intragenic, between two alleles of a gene (***cistron***), or intergenic where there is information exchange between non-allelic genes.

recombination nodule Protein-containing assemblies of about 90nm diameter placed at intervals in the ***synaptonemal complexes*** that develop between homologous chromosomes at the zygotene stage of meiosis. Some nodules may be associated with the site of ***recombination***.

recon Unit of genetic ***recombination***, the smallest section of a chromosome that is capable of recombination.

recoverin Calcium-binding protein containing three ***EF-hand*** motifs. No longer thought to be activator of photoreceptor guanylate cyclase. Related to visinin, P26, 23 kD protein, S-modulin, also to 21 kD CaBP and neurocalcin.

recruitment zone Region of cytoplasm in the rear third of a moving amoeba where endoplasm is recruited from ectoplasm.

rectifying synapse An ***electrical synapse*** at which current flow can only occur in one direction.

red blood cell (erythrocyte) Cell specialized for oxygen transport, having a high concentration of ***haemoglobin*** in the cytoplasm (and little else). Biconcave, anucleate discs, ca 7μm diameter in mammals; nucleus contracted and chromatin condensed in other vertebrates.

red drop effect Experimental observation that the photosynthetic efficiency of monochromatic light is greatly reduced above 680nm, even though chlorophyll absorbs well up to 700nm. Led to the discovery of the two light reactions of photosynthesis; see ***photosystems I*** and ***II***.

red tide Phenomenon in which the sea appears red as a result of massive increase in the population of a dinoflagellate (*Gymnopodium*); unfortunately the dinoflagellate contains a ***saxitoxin*** that is fatal; shellfish that have filtered out the dinoflagellates in large numbers become poisonous. Other algal blooms do occur, though the effect may only be to asphyxiate fish by clogging gills.

redox potential The reducing/oxidizing power of a system measured by the potential at a hydrogen electrode.

Reed–Sternberg cell Giant histiocytic cells, a common feature of ***Hodgkin's disease***.

reeler Mouse autosomal recessive mutant, deficient in ***reelin*** and with disruption in large areas of the brain. Reelin, a large extracellular protein, is secreted by Cajal–Retzius cells in the forebrain and by granule neurons in the cerebellum.

reelin Protein product (99 kD) of mouse *releer* gene, an extracellular matrix component produced by pioneer neurons and important in cortical neuronal migration.

refractile Adjective usually used in describing granules within cells that scatter (refract) light. Not to be confused with refractory.

refractory period Most commonly used in reference to the interval (typically 1ms) after the passage of an ***action potential*** during which an axon is incapable of responding to another. This is caused by inactivation of the sodium channels after opening. The maximum frequency at which neurons can fire is thus limited to a few hundred Hertz. An analogous refractory period occurs in individuals of *Dictyostelium discoideum*, which are insensitive to extracellular cyclic AMP immediately after a pulse of cAMP has been secreted. The term can be applied to any system where a similar insensitive period follows stimulation.

REG proteins (REGα = PA28α; REGβ; REGγ (Ki antigen)) Components of the 11S ***proteosome*** activator that stimulates peptidase activity and enhance the processing of antigens for presentation. All three proteins share substantial sequence homology. Purified REGα (27.8 kD) forms a heptamer in solution with a 20–30Å cone-shaped pore, but will preferentially form a heteromeric complex with REGβ.

Reg-1 Islet cell mitogen, the product of the pancreatic regulating (*reg1*) gene. Expression of *reg1* inversely correlates with cell differentiation and can be modulated by glucocorticoid receptor but not gastrointestinal hormones. Reg-1 is associated with pancreatic islet regeneration and recombinant rat reg-1 will stimulate pancreatic ***B-cell*** replication ***in vitro*** and *in vivo*. Also expressed in developing and Alzheimer's disease-affected cerebral cortex.

Reg-2 Potent Schwann cell mitogen, a secreted protein of 16 kD. Produced by motor and sensory neurons during development, production possibly regulated by ***LIF/CNTF***.

regeneration Processes of repair or replacement of missing structures.

regulatory sequence (control element) DNA sequence to which regulatory molecules such as ***promotors*** or ***enhancers*** bind, thereby altering the expression of the adjacent gene.

regulatory T-cell Vague term for any class of T-lymphocyte not directly involved in the effector side of immunity, but involved in controlling responses and actions of other cells; especially T-helper and T-suppressor cells.

regulon A situation in which two or more spatially separated genes are regulated in a coordinated fashion by a common regulator molecule.

rejection Usually used of grafts. Any process leading to the destruction or detachment of a graft or other specified structure.

Rel Protein that acts as a transcription factor. It was first identified as the oncogene product of the lethal avian retrovirus Rev-T. It has a N-terminal region of 300 amino acids that is similar to the N-terminal regions of NF κ B subunits.

rel An ***oncogene***, identified in avian endotheliosis, encoding a nuclear gene. See Table O1.

relaxation time Time taken for a system to return to the resting or ground state or a new equilibrium state following perturbation. Often used in context of receptor systems that have a refractory period after responding and then relax to a competent state. Can be used more precisely to mean the time for a system to change from its original equilibrium value to 1/ e of this original value.

relaxin Polypeptide hormone produced by corpus luteum and found in the blood of pregnant animals. Acts, as its name suggests, to cause muscle relaxation during parturition. Human relaxin has an A chain of 24 amino acids and a B chain of 29. Has structural similarity to ***insulin***.

relaxosome Complex multisubunit structure forming at the plasmid origin of replication which nicks supercoiled DNA.

release factor A component of the specialized transport system involved in the transport of cobalamin (vitamin B12) across the wall of the intestine. Dissociates the complex between cobalamin and the extracellular cobalamin-binding glycoprotein known as ***intrinsic factor***.

renal Associated with the kidney.

renaturation The conversion of denatured protein or DNA to its native configuration. This is rare for proteins. However, if DNA is denatured by heating, the two strands separate and if the heat-denatured DNA is then cooled slowly the double stranded helix reforms. This renaturation is also termed reannealing.

renin An acid protease released from the walls of afferent arterioles in the kidney when blood flow is reduced, plasma sodium levels drop, or plasma volume diminishes. Catalyses splitting of ***angiotensin*** I from ***angiotensinogen***, an α_2-globulin of plasma.

Reoviridae Class III viruses, with a segmented double-stranded RNA genome; there are about 8–10 segments each coding for a different polypeptide and only one strand of the RNA (minus strand) acts as template for ***mRNA*** (plus strand). Icosahedral capsid, and the ***virion*** includes all the enzymes needed to synthesize mRNA. The viruses originally included in this group do not seem to cause any disease in humans, though they have been isolated from the respiratory tract and gut of patients with a variety of diseases; the name is derived from 'respiratory, enteric, orphan viruses'. Several pathogenic viruses are now classed as reoviruses including orbivirus, a tick-borne virus that causes Colorado tick fever, and ***rotavirus***.

repair nucleases Class of enzymes involved in DNA repair. It includes ***endonucleases*** that recognize a site of damage or an incorrect base pairing and cut it out, and exonucleases that remove neighbouring nucleotides on one strand. These are then replaced by a ***DNA polymerase***.

repellant guiding molecule Specific molecules that inhibit the activity of growth cones and are thought to be important in establishing axon pathways during nervous system development. See ***growth cone collapse***.

reperfusion injury Damage that occurs to tissue when blood flow is restored after a period of ischaemia. The damage is caused by neutrophils that adhere to the microvasculature and release various inflammatory mediators, hydrolytic enzymes, etc. as a response to damage. Can exacerbate the effects of stroke or cardiac ischaemia and also can follow from extended periods of arterial occlusion during surgical operations.

repetitive DNA Nucleotide sequences in DNA that are present in the genome as numerous copies, originally identified by the value on the ***Cot curve*** derived from kinetic studies of DNA renaturation. These sequences are not thought to code for polypeptides. One class of repetitive DNA, termed highly repetitive DNA, is found as short sequences, 5–100 nucleotides, repeated thousands of times in a single long stretch. It typically comprises 3–10% of the genomic DNA and is predominantly ***satellite DNA***. Another class, which comprises 25–40% of the DNA and is termed moderately repetitive DNA, usually consists of sequences about 150–300 nucleotides in length dispersed evenly throughout the genome, and includes ***Alu*** sequences and ***transposons***.

replica methods Methods in the preparation of specimens for transmission electron microscopy. The specimen (eg. a piece of ***freeze-fractured*** tissue) is shadowed with metal and coated with carbon and then the tissue is digested away. The replica is then picked up on a grid and it is the replica that is examined in the microscope.

replica plating Technique for testing the genetic characteristics of bacterial colonies. A dilute suspension of bacteria is first spread, in a petri dish, on agar containing a medium expected to support the growth of all bacteria, the master plate. Each bacterial cell in the suspension is expected to give rise to a colony. A sterile velvet pad, the same size as the petri dish, is then pressed onto it, picking up a sample of each colony. The bacteria can then be 'stamped' onto new sterile petri dishes, plates, in the identical arrangement. The media in the new plates can be made up to lack specific nutritional requirements or to contain antibiotics. Thus colonies can be identified that cannot grow without specific nutrients or which are antibiotic-resistant, and cells with mutations in particular genes can be isolated.

replicase Generic (and rather unhelpful) term for an enzyme that duplicates a polynucleotide sequence (either RNA or DNA). The term is more usefully restricted to the enzyme involved in the replication of certain viral RNA molecules.

replication Copying, but usually the production of daughter strands of nucleic acid from the parental template.

replication factor A (replication protein A; RPA) Protein that associates, together with the PIK3-kinase-like kinase (***ATM***), at sites where homologous regions of DNA interact during meiotic prophase and at breaks associated with meiotic recombination after ***synapsis***.

replication fork Point at which DNA strands are separated in preparation for ***replication***. Replication forks thus move along the DNA as replication proceeds.

replication protein A See ***replication factor A***.

replicative intermediate Intermediate stage(s) in the replication of a RNA virus; a copy of the original RNA strand, or of a single-strand copy of the first replicative intermediate. Essentially an amplification strategy.

replicons Tandem regions of replication in a chromosome, each about 30μm long, derived from an ***origin of replication***. By definition a replicon must contain an origin of replication.

replisome Complex of proteins involved in the replication (elongation) of DNA that moves along as the new complementary strand is synthesized. On this basis a minimum content would be DNA polymerase III and a ***primosome***. An RNA replisome has been proposed as a putative ancestor of the ribosome.

reporter gene A gene that encodes an easily assayed product (eg. ***CAT***) that is coupled to the upstream sequence of another gene and transfected into cells. The reporter gene can then be used to see which factors activate response elements in the upstream region of the gene of interest.

repressor protein A protein that binds to an ***operator*** of a gene preventing the ***transcription*** of the gene. The binding affinity of repressors for the operator may be affected by other molecules. Inducers bind to repressors and decrease their binding to the operator, while corepressors increase the binding. The paradigm of repressor proteins is the lactose repressor protein that acts on the ***lac operon*** and for which the inducers are β-galactosides such as ***lactose***; it is a polypeptide of 360 amino acids that is active as a tetramer. Other examples are the λ repressor protein of lambda bacteriophage that prevents the transcription of the genes required for the lytic cycle leading to ***lysogeny***, and the *cro* protein, also of lambda, which represses the transcription of the λ repressor protein establishing the lytic cycle. Both of these are active as dimers and have a common structural feature, the ***helix-turn-helix*** motif, which is thought to bind to DNA with the helices fitting into adjacent major grooves.

reproduction Propagation of organisms. The act of producing new organisms. May be asexual or sexual.

resact Sea urchin peptide hormone, affecting motility and metabolism. Receptor is a plasma-membrane ***guanylate cyclase***.

resealed ghosts Membrane shells formed by lysis of erythrocytes resealed by adjusting the cation composition of the medium. Relatively impermeable, although more permeable than the original membrane.

reserpine ***Alkaloid*** derived from *Rauwolfia*; blocks the packaging of noradrenaline in to presynaptic vesicles. Useful experimental tool to determine the involvement of sympathetic innervation.

residual body (1) ***Secondary lysosomes*** containing material that cannot be digested. (2) The surplus cytoplasm shed by spermatids during their differentiation to spermatozoa. Usually the cytoplasm from several spermatids connected by ***cytoplasmic bridges***. (3) Surplus cytoplasm containing pigment and left over after production of merozoites during schizogony of ***malaria*** parasites.

resilin Amorphous rubber-like protein found in insect cuticle: similar to ***elastin***, though there is no fibre formation.

resiniferatoxin Potent analogue of ***capsaicin*** from *Euphorbia resinifera* (flowering cactus). An agonist at ***vanilloid receptor-1***.

resolution Complete return to normal structure and function: used, for example, of an inflammatory lesion, or of a disease. See also ***resolving power***.

resolvase See ***recombinase, site-specific recombination***.

resolving power (1) The resolution of an optical system defines the closest proximity of two objects that can be seen as two distinct regions of the image. This limit depends upon the ***numerical aperture*** (N. A.) of the optical system, the contrast step between objects and background and the shape of the objects. The often quoted Airy limit applies only to self-luminous discs. (2) In genetics, the smallest map distance measurable by an experiment involving a certain number of classified recombinant progency.

resonance energy transfer See ***fluorescence energy transfer***.

resorufin Pink fluorescent dye; a caged form of resorufin (non-fluorescent unless activated – released – by irradiation with UV) coupled to ***G-actin*** and microinjected has been used as a marker for microfilaments in the leading lamella of moving cells.

respiration Term used by physiologists to describe the process of breathing and by biochemists to describe the intracellular oxidation of substrates coupled with production of ATP and oxidized coenzymes (NAD^+ and FAD). This form of respiration may be anaerobic as in glycolysis, or aerobic in the case of oxidations operating via the ***tricarboxylic acid cycle*** and the ***electron transport chain***.

respiratory burst See ***metabolic burst***.

respiratory chain The mitochondrial ***electron transport chain***.

respiratory enzyme complex The enzymes that make up the ***respiratory chain***: NADH-Q reductase, succinate-Q reductase, cytochrome reductase, cytochrome C and cytochrome oxidase.

respiratory quotient Molar ratio of carbon dioxide production to oxygen consumption.

restenosis Reocclusion of coronary arteries after ***angioplasty*** (PTCA) or after replacement with blood vessels from elsewhere. Probably due to excessive proliferation of vascular smooth muscle that inappropriately thickens the intima and narrows the lumen.

resting potential The electrical potential of the inside of a cell, relative to its surroundings. Almost all animal cells are negative inside; resting potentials are in the range –20 to –100mV, –70mV typical.

Resting potentials reflect the action of the ***sodium pump*** only indirectly; they are mainly caused by the subsequent diffusion of potassium out of the cell through potassium leak channels. The resting potential is thus close to the ***Nernst potential*** for potassium. See ***action potential***.

restriction endonuclease (restriction enzyme) Class of bacterial enzymes that cut DNA at specific sites. In bacteria their function is to destroy foreign DNA, such as that of ***bacteriophages*** (host DNA is specifically modified at these sites). Type I restriction endonucleases occur as a complex with the methylase and a polypeptide that binds to the recognition site on DNA. They are often not very specific and cut at a remote site. Type II restriction endonucleases are the classic experimental tools. They have very specific recognition and cutting sites. The recognition sites are short, 4–8 nucleotides, and are usually ***palindromic sequences***. Because both strands have the same sequence running in opposite directions the enzymes make double-stranded breaks, which, if the site of cleavage is off-centre, generates fragments with short single-stranded tails; these can hybridize to the tails of other fragments and are called sticky ends. They are generally named according to the bacterium from which they were isolated (first letter of genus name and the first two letters of the specific name). The bacterial strain is identified next and multiple enzymes are given Roman numerals. For example, the two enzymes isolated from the R strain of *E. coli* are designated *Eco* RI and *Eco* RII. The more commonly used restriction endonucleases are shown in Table R1.

restriction enzyme See ***restriction endonuclease***.

restriction fragment length polymorphism (RFLP) Technique, also known as DNA fingerprinting, that allows familial relationships to be established by comparing the characteristic polymorphic patterns that are obtained when certain regions of genomic DNA are amplified (typically by PCR) and cut with certain restriction enzymes. In principle, an individual can be identified unambiguously by RFLP (hence the use of RFLP in forensic analysis of blood, hair or semen). Similarly, if a polymorphism can be identified close to the locus of a genetic defect, it provides a valuable marker for tracing the inheritance of the defect.

restriction fragments The fragments of DNA generated by digesting DNA with a specific ***restriction endonuclease***. Each of the fragments ends in a site recognized by that specific enzyme.

restriction map Map of DNA showing the position of sites recognized and cut by various ***restriction endonucleases***.

restriction nucleases See ***restriction endonuclease***.

restriction point (of cell cycle) A point, late in ***G1***, after which the cell must (normally) proceed through to division at its standard rate.

restriction site Any site in DNA that can be cut by a ***restriction enzyme***.

restrictive temperature See ***permissive temperature***.

restrictocin Toxin (149 residues) produced by *Aspergillus restrictus*. One of the aspergillins; see ***alpha-sarcin***.

ret A human ***oncogene***, encoding a receptor ***tyrosine kinase***. See Table O1.

retention signal The sequence of amino acids on proteins that indicates that they are to be retained in the secretory processing system, for example the ***Golgi apparatus***, and not passed on and released. Can be applied to any similar situation in which sorting of macromolecules into different compartments occurs.

reticalmin Calcium/calmodulin-dependent protein kinase IV-like protein (35 kD) found in rat retina, mainly in outer segment of photoreceptors and dendrites of inner plexiform layers. See also ***calspermin***.

reticular fibres Fine fibres (of ***reticulin***) found in extracellular matrix, particularly

Table R1. Recognition sequences of various type II restriction endonucleases

<table>
<tr><th>Enzyme</th><th>Bacterium from which enzyme is derived</th><th>Recognition sequence[a]</th></tr>
<tr><td>Aha III</td><td>Aphanothece halophytica</td><td>↓
TTT | AAA</td></tr>
<tr><td>Alu I</td><td>Arthrobacter luteus</td><td>↓
AG | CT</td></tr>
<tr><td>Ava I</td><td>Anabaena variabilis</td><td>↓
C PyC | GPuG</td></tr>
<tr><td>Bam Hl</td><td>Bacillus amyloliquefaciens H</td><td>↓ m
G GA | TCC</td></tr>
<tr><td>Bst Ell</td><td>Bacillus stearothermophilus ET</td><td>↓
G GTNACC</td></tr>
<tr><td>Cla I</td><td>Caryphanon latum</td><td>↓
AT C | GAT</td></tr>
<tr><td>Dde I</td><td>Desulfovibrio desulfuricans</td><td>↓
C TNAG</td></tr>
<tr><td>Eco Rl</td><td>Escherichia coli</td><td>↓ m
G AA | TTC</td></tr>
<tr><td>Eco Rll</td><td>Escherichia coli</td><td>↓ m
CC(AT)TT</td></tr>
<tr><td>Eco RV</td><td>Escherichia coli</td><td>↓
GAT | ATC</td></tr>
<tr><td>Hae ll</td><td>Haemophilus aegyptius</td><td>↓
PuGC | CG Py</td></tr>
<tr><td>Hae 111</td><td>Haemophilus aegyptius</td><td>↓m
GG | CC</td></tr>
<tr><td>Hha I</td><td>Haemophilus haemolyticus</td><td>m ↓
GC | GC</td></tr>
<tr><td>Hin dl ll</td><td>Haemophilus influenzae R_d</td><td>m ↓
AA G | CTT</td></tr>
<tr><td>Hin f l</td><td>Haemophilus influenzae R_f</td><td>↓
G ANTC</td></tr>
<tr><td>Hpa I</td><td>Haemophilus parainfluenzae</td><td>↓
GTT | AAC</td></tr>
<tr><td>Hpa ll</td><td>Haemophilus parainfluenzae</td><td>↓ m
CC | GG</td></tr>
<tr><td>Kpn I</td><td>Klebsiella pneumoniae OK8</td><td>↓
GGT | AC C</td></tr>
<tr><td>Mbo I</td><td>Moraxella bovis</td><td>↓
GA | TC</td></tr>
<tr><td>Msp I</td><td>Moraxella species</td><td>↓
C C | GG</td></tr>
<tr><td>Mst I</td><td>Microcoelus species</td><td>↓
TGC | GCA</td></tr>
<tr><td>Pst I</td><td>Providencia stuartii</td><td>↓
CTG | CA G</td></tr>
<tr><td>Pvu I</td><td>Proteus vulgaris</td><td>↓
CGA | T CG</td></tr>
<tr><td>Sac I</td><td>Streptomyces achromogenes</td><td>↓
GAG | CT C</td></tr>
<tr><td>Sma I</td><td>Serratia marcescens</td><td>↓
CCC | GGG</td></tr>
<tr><td>Xba I</td><td>Xanthomonas badrii</td><td>↓
T CT | AGA</td></tr>
<tr><td>Xho I</td><td>Xanthomonas holcicola</td><td>↓
C TC | GAG</td></tr>
</table>

↓
5′-XXX X | XXXX-3′ (↓: Cleavage site |: Axis of symmetry).
Pu = purine, ie. A or G are recognized; Py = pyrimidine, ie. C or T are recognized; N = any base.
m
X = base methylated by corresponding methylase, where known, to give N6-methyladenosine or 5-methylcytosine.

in lymph nodes, spleen, liver, kidneys and muscles.

reticular lamina The lower region of extracellular matrix underlying an epithelial monolayer, separated from the basal surface of the epithelial cells by the basal lamina. The reticular lamina contains fibrillar elements (collagen, elastin, etc.) and is probably secreted by fibroblasts of the underlying connective tissues. The reticular lamina and the basal lamina together form what older textbooks referred to as the ***basement membrane***.

reticulin Constituent protein of ***reticular fibres***: collagen type III.

reticulocyte Immature red blood cells found in the bone marrow, and in very small numbers in the circulation.

reticulocyte lysate Cell lysate produced from ***reticulocytes***; used as an ***in vitro*** translation sytem.

reticuloendothelial system The phagocytic system of the body, including the fixed macrophages of tissues, liver and spleen. Rather old-fashioned term that is coming back into use; 'mononuclear phagocyte system' is probably better when only phagocytes are meant.

reticulum cells Cells of the reticuloendothelial system, found particularly in lymph nodes, bone marrow, and spleen. In lymph nodes they are stromal cells and probably not reticuloendothelial cells in the current sense of that term.

retina Light-sensitive layer of the eye. In vertebrates, looking from outside, there are four major cell layers: (i) the outer neural retina, which contains neurons (ganglion cells, ***amacrine cells***, ***bipolar cells***) as well as blood vessels; (ii) the photoreceptor layer, a single layer of rods and cones; (iii) the ***pigmented retinal epithelium*** (PRE or RPE); (iv) the choroid, composed of connective tissue, fibroblasts, and including a well-vascularized layer, the chorio capillaris, underlying the basal lamina of the PRE. Behind the choroid is the sclera, a thick organ capsule. See ***retinal rods***, ***retinal cones***, ***rhodopsin***. In molluscs (especially cephalopods such as the squid) the retina has the light-sensitive cells as the outer layer with the neural and supporting tissues below.

retinal Aldehyde of ***retinoic acid*** (vitamin A); complexed with opsin forms ***rhodopsin***. Photosensitive component of all known visual systems. Absorption of light causes retinal to shift from the 11-*cis* form to the all-*trans* configuration, and through a complex cascade of reactions excites activity in the neurons synapsed with the rod cell.

retinal cone The other light-sensitive cell type of the retina, that, unlike ***retinal rods***, is differentially sensitive to particular wavelengths of light, and is important for colour vision. There are three types of cones, each type sensitive to red, green or blue. Present in large numbers in the ***fovea***.

retinal ganglion cell See ***ganglion cell***.

retinal pigmented epithelial cell See [pigmented retinal epithelium] and [retina].

retinal rod Major photoreceptor cell of vertebrate retina (about 125 million in a human eye). Columnar cells (about 40μm long, 1μm diameter) having three distinct regions: a region adjacent to, and synapsed with, the neural layer of the ***retina*** contains the nucleus and other cytoplasmic organelles, below this is the inner segment, rich in mitochondria, that is connected through a thin 'neck' (in which is located a ***ciliary body***) to the outer segment. The outer segment largely consists of a stack of discs (membrane infoldings that are incompletely separated in cones) that are continually replenished near the inner segment, shed from the distal end and phagocytosed by the pigmented epithelium. The membranes of the discs are rich in ***rhodopsin***, the pigment that absorbs light.

retinitis pigmentosa Disease caused by overactivity of the pigmented retinal epithelial cells, leading to damage and occlusion of photoreceptors and blindness.

retino-tectal connection A problem that has exercised developmental biologists is the way in which nerve fibres from the developing retina are 'mapped' on to the tectum of the brain. There seems to be a good positioning system in operation, and a variety of mechanisms probably operate, including control of the fasciculation of fibres in the optic nerve, and some specific recognition of the correct target area by the nerve growth cone.

retinoblastoma Malignant tumour of the retina, usually arising in the inner nuclear layer of the neural retina. Retinoblastoma is unusual in being caused by an autosomal dominant mutation in some cases (about 6%), in which case it may be bilateral. The gene product of the retinoblastoma gene is a tumour suppressor that interacts with transcription factors such as ***E2F*** to block transcription of growth-regulating genes. The Rb gene plays a role in normal development, not just that of the retina.

retinoic acid (vitamin A) The aldehyde (***retinal***) has long been known to be involved in photoreception, but retinoic acid has other roles. There are cytoplasmic retinoic acid binding proteins and retinoic acid response elements that regulate gene transcription. Retinoic acid is thought to be a morphogen in chick limb-bud development and in early development of the chick (see ***Hensen's node***) which probably accounts for its potent teratogenic action. See also Table V1.

retraction fibres Thin projections from crawling cells associated with areas where the cell body is becoming detached from the substratum, but ***focal adhesions*** persist. Usually contain a bundle of microfilaments that are under tension.

retrograde axonal transport The transport of vesicles from the synaptic region of an axon towards the cell body: involves the interaction of ***MAP1C*** with microtubules.

retrotransposon ***Transposable element*** with a transpositional mechanism requiring ***reverse transcriptase*** in a manner reminiscent of ***retroviruses***, to which they may be related.

retroviral vector See ***Retroviridae***. Retroviral vectors are used in the genetic modification of cells as a means of introducing foreign DNA into the genome. For example, RVs encoding histochemical markers (***reporter genes***) are used in the study of neural cell lineage in vertebrates. RVs may contain the bacterial ***lacZ*** gene that encodes for the enzyme β-galactosidase. When the retrovirally infected cells divide, they replicate the foreign DNA. Progeny of infected cells will therefore express the protein and can then be detected histochemically.

Retroviridae (retrovirus) Viruses with a single-stranded RNA genome (class VI). On infecting a cell the virus generates a DNA replica by action of its virally coded ***reverse transcriptase***. ***Oncovirinae*** are one of three subclasses of retroviruses, the others being ***Lentivirinae*** and Spumavirinae. See ***retroviral vector***.

retrovirus See ***Retroviridae***.

reverse genetics The technique of determining a gene's function by first sequencing it, then mutating it, and then trying to identify the nature of the change in the phenotype.

reverse passive haemagglutination If antibodies are bonded to the surface of red blood cells haemagglutination will occur if the appropriate bi- or multivalent antigen is added in soluble or microparticulate form. Used as a test for, eg. ***Hepatitis B*** virus in the serum.

reverse transcriptase RNA-directed DNA polymerase. Enzyme first discovered in retroviruses that can construct double-stranded DNA molecules from the single-stranded RNA templates of their genomes. Reverse transcription now appears also to be involved in movement of certain mobile genetic elements, such as the Ty plasmid in yeast, in the replication of other viruses such as ***Hepatitis B***, and possibly in the generation of mammalian ***pseudogenes***.

reversion Reversion of a mutation occurs when a second mutation restores the function that was lost as a result of the first mutation. The second mutation causes a change in the DNA that either reverses the original alteration or compensates for it.

Reynaud's phenomenon See ***CREST***.

Reynold's number A constant without dimensions that relates the inertial and viscous drag acting to hinder a body moving through fluid medium. For cells the Reynold's number is very small; viscous drag is dominant, and inertial resistance can be neglected.

RF (release factor; RF-1, RF-2, RF-3, eRF-1) Proteins that are involved in the release of the nascent polypeptide from the ***ribosome***. In bacteria RF-1 (40 kD) is specific for UAG/UAA codons and RF-2 (41 kD) specific for UGA/UAA. Act on the ribosomal A site and are assisted by RF-3 which is not codon-specific. Eukaryotic equivalents (eRF-1, etc.) have also been identified.

RFLP See ***restriction fragment length polymorphism***.

RGD A domain found in ***fibronectin*** and related proteins, recognized by ***integrins***. In most cases, the consensus is -R-G-D-S- (arginine-glycine-aspartic acid-serine).

RH factor (Rh factor) Rhesus factor: see ***rhesus blood group***.

Rhabdocoela Order of aquatic turbellaria (flatworms). Superceded as a classification.

rhabdomyosarcoma Malignant tumour (sarcoma) derived from striated muscle.

Rhabdoviridae (rhabdovirus) Class V viruses with a single negative strand RNA genome and an associated virus-specific RNA polymerase. The capsid is bullet shaped and enveloped by a membrane that is formed when the virus buds out of the plasma membrane of infected cells. The budded membrane contains host lipids but only glycoproteins coded for by the virus, of which there are usually 1–3 species. In the electron microscope these appear as regularly arranged spikes about 10nm long and are called spike glycoproteins. This group includes ***rabies virus***, ***vesicular stomatitis virus*** and a number of plant viruses.

rhamnogalacturonan Plant cell wall polysaccharide consisting principally of rhamnose and galacturonic acid. Present as a major part of the pectin of the primary cell wall. Two types known: rhamnogalacturonan I (RG-I), the major component, which contains rhamnose, galacturonic acid, arabinose and galactose, and rhamnogalacturonan II (RG-II), containing at least four different sugars in addition to galacturonic acid and rhamnose.

rhamnose (6-deoxy L-mannose) A sugar found in plant glycosides.

rheotaxis A ***taxis*** in response to the direction of flow of a fluid.

Rhesus blood group Human blood group system with allelic red cell antigens C, D and E. The D antigen is the strongest. Red cells from a Rhesus-positive foetus cross the placenta and can sensitize a Rhesus-negative mother, expecially at parturition. The mother's antibody may then, in a subsequent pregnancy, cause haemolytic disease of the newborn if the foetus is Rhesus-positive. The disease can be prevented by giving anti-D IgG during the first 72 hours after parturition to mop up D^+ red cells in the maternal circulation.

rheumatic fever Disease involving inflammation of joints and damage to heart valves that follows streptococcal infection and is believed to be autoimmune, ie. antibodies to streptococcal components cross-react with host tissue antigens.

rheumatoid arthritis Chronic inflammatory disease in which there is destruction of joints. Considered by some to be an autoimmune disorder in which immune complexes are formed in joints and excite an inflammatory response (complex-mediated ***hypersensitivity***). Cell-mediated (type IV) hypersensitivity also occurs, and macrophages accumulate. This in turn leads to the destruction of the synovial lining (see ***pannus***).

rheumatoid factor Complex of IgG and anti-IgG formed in joints in ***rheumatoid***

arthritis. Serum rheumatoid factors are more usually formed from IgM antibodies directed against IgG.

rhinovirus ***Picornaviridae*** that largely infect the upper respiratory tract. Include the common cold virus and foot-and-mouth disease virus.

Rhizobium ***Gram-negative bacterium*** that fixes nitrogen in association with roots of some higher plants, notably legumes. Forms root nodules, in which it is converted to the nitrogen-fixing ***bacteroid*** form.

rhizoid Portion of a cell or organism that serves as a basal anchor to the substratum.

rhizoplast Striated contractile structure attached to the basal region of the cilium in a variety of ciliates and flagellates. May regulate the flagellar beat pattern, and is sensitive to calcium concentration. Composed of a 20 kD protein rather similar to ***spasmin***.

Rhizopoda Phylum that includes single celled amoebae such as *Amoeba proteus*.

rho factors (ρ factors) Protein factors found in prokaryotes, especially *E. coli*, involved in the termination of transcription. Mutations in *rho* may cause the RNA polymerase to read through from one ***operon*** to the next. Not to be confused with rho (small G-protein) in eukaryotic cells.

rho genes Genes coding for small GTP-binding proteins; implicated in actin organization and the interaction of the cytoskeleton with intracellular membranes. See also ***ras*** and ***rab***.

rhodamines A group of triphenylmethane-derived dyes are referred to as rhodamines, lissamines, etc. Many are fluorescent and are used as fluorochromes in labelling proteins and membrane probes.

Rhodnius prolixus Reduviid blood-sucking bug, vector of *Trypanosoma cruzi* in South America. Much used by insect physiologists because the the transition from one instar to the next is triggered by a blood meal. The blood meal, which may exceed substantially the body weight of the insect, triggers complex changes in the mechanical properties of abdominal cuticle allowing the insect to swell, and activates excretory mechanisms. The haemolymph contains relatively few ***haemocytes***, which made some of the classical parabiosis experiments much easier.

Rhodophyta (red algae) Division of algae, many of which have branching filamentous forms and red coloration. The latter is due to the presence of ***phycoerythrin***. The food reserve is floridean (starch), found outside the plastid. The walls contain sulphated galactans such as ***carrageenan*** and ***agar***.

rhodopsin (visual purple) Light-sensitive pigment formed from ***retinal*** linked through a ***Schiff base*** to opsin: rhodopsin is an integral membrane protein found in the discs of ***retinal rods*** and cones, comprising some 40% of the membrane. Vertebrate opsins are proteins of 38 kD. See also ***bacteriorhodopsin***.

Rhodospirillum rubrum A purple non-sulphur bacterium with a spiral shape; contains the pigment ***bacteriochlorophyll*** and under anaerobic conditions undergoes ***photosynthesis*** using organic compounds as electron donors for the reduction of carbon dioxide. The purple colour results from the presence of ***carotenoids***, though the bacteria are often more red or brown.

rhombomere *Neuromeres* or segments in the hindbrain region that are of developmental significance. Shown to be lineage restriction units in that cells of adjacent rhombomeres do not mix with each other. Regulatory genes have been shown to be expressed in patterns in the developing hindbrain that relate to the neuromeric or ***segmentation*** pattern.

rhophilin Protein (71 kD, 643 residues) that interacts with ***rho***-GTP. Has sequence homology with the N terminal region of ***protein kinase N*** though possesses no catalytic activity.

rhoptry Electron-opaque dense body found in the apical complex of parasitic protozoa of the phylum Apicomplexa.

ribavirin Ribavirin is an orally active guanoside analogue that inhibits the replication of a variety of RNA viruses. It inhibits inosine monophosphate dehydrogenase.

ribbon synapse Ultrastructurally distinct type of synapse found in a variety of sensory receptor cells such as retinal ***photoreceptor*** cells, ***cochlear hair cells*** and vestibular organ receptors, as well as in a non-sensory neuron, the retinal bipolar cell. Unlike most neurons, these cells do not use regenerative action potentials but release transmitter in response to small graded potential changes. Ribbon synapses have different exocytotic machinery from conventional synapses in containing dense bars or ribbons anchored to the presynaptic membrane covered with a layer of synaptic vesices. The ribbons have been proposed to shuttle synaptic vesicles to exocytotic sites.

riboflavin (vitamin B2) Ribose attached to a flavin moiety that becomes part of FAD and FMN. See also Table V1.

ribonuclease (RNAse, RNAase) Widely distributed type of enzyme that cleaves RNA. May act as endonucleases or exonucleases depending upon the type of enzyme. Generally recognize target by tertiary structure rather than sequence. Ribonuclease E is an RNAase involved in the formation of 5S ribosomal RNA from pre-rRNA. F is stimulated by interferons and cleaves viral and host RNAs and thus inhibits protein synthesis. H specifically cleaves an RNA base-paired to a complementary DNA strand. P is an endonuclease that generates t-RNAs from their precursor transcripts. T is an endonuclease that removes the terminal AMP from the 3′ CCA end of a non-aminoacylated tRNA. RNAase T1 cleaves RNA specifically at guanosine residues. RNAase III cleaves double-stranded regions of RNA molecules.

ribonucleic acid See ***RNA***.

ribonucleoprotein (RNP) Complexes of RNA and protein involved in a wide range of cellular processes. Besides ***ribosomes*** (with which RNP was originally almost synonymous), in eukaryotic cells both initial RNA transcripts in the nucleus (***hnRNA***) and cytoplasmic ***mRNAs*** exist as complexes with specific sets of proteins. Processing (splicing) of the former is carried out by small nuclear RNPs (***snRNPs***). Other examples are the ***signal recognition particle*** responsible for targetting proteins to endoplasmic reticulum and a complex involved in termination of transcription.

ribophorin Glycoproteins of the endoplasmic reticulum that interact with ribosomes whilst cotranslational insertion of membrane or secreted proteins is taking place. Ribophorins may form a pore through which the nascent polypetide chain passes.

riboprobe Somewhat casual term for an RNA segment used to probe for a complementary nucleotide sequence, either in the mRNA pool or in the DNA of a cell.

ribose (D-ribose) A monosaccharide pentose of widespread occurrence in biological molecules, eg. RNA.

ribose-binding protein Periplasmic binding proteins of bacteria that interact either with the ribose transport system or with the methyl-accepting chemotaxis protein, ***MCP*** III (*trg*).

ribosomal protein Proteins present within the ribosomal subunits. In prokaryotes there are 31 proteins in the large subunit and 21 in the small subunit. Eukaryotic subunits have 50 (large subunit) and 33 (small subunit) proteins.

ribosomal RNA See ***RNA***.

ribosome A heterodimeric multisubunit enzyme composed of ribonucleoprotein and protein subunits. Interacts with aminoacylated tRNAs, and mRNAs and translates protein coding sequences from messenger RNA. During protein elongation, the nascent protein is held at the P site (peptidyl-tRNA complex), while aminoacyl-tRNAs bearing new amino-acids are bound at the A site. Similar ribosomes are found in all living organisms, all composed of large and small subunits, as well as chloroplasts and mitochondria. Differences are apparent between prokaryotic and eukaryotic ribosomes.

ribotype The RNA complement of a cell – by analogy with phenotype or genotype.

ribozyme RNA with catalytic capacity – an enzyme made of nucleic acid not protein. It is of particular interest because of the implications for self-replicating systems in the earliest stages of the evolution of (terrestrial) life.

ribulose 1,5-bisphosphate An intermediate in the ***Calvin–Benson cycle*** of photosynthesis.

ribulose bisphosphate carboxylase (RUBISCO) Enzyme responsible for CO_2 fixation in photosynthesis. Carbon dioxide is combined with ribulose diphosphate to give two molecules of 3-phosphoglycerate, as part of the ***Calvin–Benson cycle***. It is the sole CO_2-fixing enzyme in C3 plants, and collaborates with ***PEP carboxylase*** in CO_2 fixation in C4 plants. In the presence of oxygen the products of the reaction are 1 molecule of phosphoglyceric acid and 1 molecule of phosphoglycolic acid. The latter is the initial substrate for photorespiration and this oxygenase function occurs in C3 plants where the enzyme is not protected from ambient oxygen; in C4 plants the enzyme acts exclusively as a carboxylase since it is protected from oxygen. Also known as fraction 1 protein, the major protein of leaves.

ribulose diphosphate carboxylase (RuDPC; RuDP carboxylase) See ***ribulose bisphosphate carboxylase***.

ricin Highly toxic ***lectin*** (66 kD) from seeds of the castor bean, *Ricinus communis*. Has toxic A subunit (32 kD), carbohydrate-binding B subunit (34 kD). Toxic subunit inactivates ribosomes, and the binding subunit is specific for β-galactosyl residues.

Ricinus communis agglutinin (RCA) ***Lectin*** (120 kD) from castor bean, with specificity similar to ***ricin***, but much less toxic.

Rickettsia Genus of ***Gram-negative bacteria*** responsible for a number of insect-borne diseases of humans (including scrub typus and ***Rocky Mountain spotted fever***). Obligate intracellular parasites.

rifampicin Semi-synthetic member of the ***rifamycin*** group of antibiotics.

rifamycin Antibiotic produced by *Streptomyces mediterranei* that acts by inhibiting prokaryotic, but not eukaryotic, DNA-dependent RNA synthesis. Blocks initiation, but not elongation of transcripts.

rigor Stiffening of muscle as a result of high calcium levels and ATP depletion, so that actin–myosin links are made, but not broken.

rilmenidine See ***clonidine***.

RING finger motif A ***zinc finger*** motif found in various nuclear proteins and in some receptor-associated proteins. The RING finger, or Cys3HisCys4, family of zinc binding proteins play important roles in differentiation, oncogenesis and signal transduction.

Ringer's solution Isotonic salt solution used for mammalian tissues; original version (for frog tissues) much modified and often used loosely to mean any physiological saline.

ristocetin Mixture of ristocetins A and B: isolated from actinomycete, *Nocardia lurida*. Induces platelet aggregation.

RMP pathway (ribulose monophosphate pathway; allulose phosphate pathway) A metabolic pathway used by methylotropic bacteria for the conversion of formaldehyde to hexose sugars, etc. In the first stage ribulose-5-phosphate is condensed with HCHO.

RNA (ribonucleic acid) This molecular species has an informational role, a structural role and an enzymic role and is thus used in a more versatile way than either DNA or proteins. Considered by many to be the earliest macromolecule of living systems. The structure is of ribose units joined in the 3′ and 5′ positions through a phosphodiester linkage with a purine or pyrimidine base attached to the 1′ position. All RNA species are synthesized by transcription of DNA sequences, but may involve post-transcriptional modification.

RNA (catalytic) Species of RNA that catalyse cleavage or transesterification of the phosphodiester linkage. Operates in the self-splicing of group I and group II introns and in the maturation of various tRNA species.

RNA editing A process responsible for changes in the final sequence of mRNA that are not coded in the DNA template. Excludes mRNA ***splicing*** and modifications to tRNA. Various kinds of editing are known, the commonest being cytidine (C) to uridine (U) substitution, though polyadenylation of mitochondrial mRNA and guanosine (G) insertion in paramyxoviruses are other examples of editing. The process involves guide RNA (gRNA) in some cases but not all. Though the commonest examples are in organelles, the process of editing does also occur in nuclear transcripts.

RNA plasmid dsRNA found in yeasts, also known as killer factors. Their nomenclature is uncertain and some scientists consider them viruses.

RNA polymerases Enzymes that polymerize ribonucleotides in accordance with the information present in DNA. Prokaryotes have a single enzyme for the three RNA types that is subject to stringent regulatory mechanisms. Eukaryotes have type I that synthesizes all ***rRNA*** except the 5S component, type II that synthesizes ***mRNA*** and ***hnRNA*** and type III that synthesizes ***tRNA*** and the 5S component of rRNA.

RNA primase An RNA polymerase that synthesizes a short RNA primer sequence to initiate DNA replication.

RNA primer The primer sequence synthesized by RNA primase.

RNA processing Modifications of primary RNA trancripts including splicing, cleavage, base modification, capping and the addition of ***poly-A tails***. See also ***RNA editing***.

RNA splicing The removal of ***introns*** from primary RNA transcripts. See ***alternative splicing***.

RNA tumour virus See ***Oncovirinae***.

RNAase protection assay Sensitive and quantitative alternative to ***Northern blots*** for the measurement of gene expression levels. Labelled antisense cRNA is transcribed from a DNA clone in an appropriate vector, hybridized with an mRNA sample, and single stranded RNA digested away with RNAase, then run out on a gel. The amount of labelled RNA surviving is directly proportional to the amount of target mRNA present in the sample.

RNases E, F, H, P, T, III See ***ribonucleases***.

RNP See ***ribonucleoprotein***.

Robertsonian translocation A special type of non-reciprocal translocation in chromosomes whereby the long arms of two non-homologous acrocentric chromosomes are attached to a single centromere. The short arms become attached to form a reciprocal structure that, however, often disappears some divisions after its formation.

Rock (Rock-1, ROK-β; Rock-2, ROK-α) Serine threonine kinases, putative targets for ***rho*** activation. Rock-2 may act as an effector for GTP-Rho-A in inducing cytoskeletal rearrangement.

Rocky Mountain spotted fever Acute infectious tick-borne rickettsial disease caused by *Rickettsia rickettsii*.

rod cell See ***retinal rod***.

rod outer segment See ***retinal rod***.

rolipram Inhibitor of cAMP-specific phosphodiesterase IV.

rolling circle mechanism A mechanism of DNA replication in many viral DNAs, in bacterial F factors during mating, and of certain DNAs in gene amplification in eukaryotes. DNA synthesis starts with a cut in the + strand at the replication origin, the 5′ end rolls out and replication starts at the 3′ side of the cut around the intact circular DNA strand. Replication of the 5′ end (tail) takes place by the formation of ***Okazaki fragments***.

Romanovsky-type stain Composite histological stains including ***Methylene Blue***, Azure A or B and ***eosin***, sometimes with other stains. Examples are ***Giemsa***, Wrights, and ***Leishman stain***.

root cap Tissue found at the apex of roots, overlying the root apical meristem and protecting it from friction as the root grows through the soil. Secretes a glycoprotein mucilage as a lubricant.

root hair cell Root epidermal cell, part of which projects from the root surface as a thin tube, thus increasing the root surface area and promoting absorption of water and ions.

root nodule Globular structure formed on the roots of certain plants, notably legumes and alder, by symbiotic association between the plant and a nitrogen-fixing microorganism (*Rhizobium* in the case of legumes and *Frankia* in the case of alder and a variety of other plants).

rootlet system Microtubules associated with the base of the flagellum in ciliates and flagellates. Also associated with this region is the ***rhizoplast***.

ros An ***oncogene***, identified in bird ***sarcoma***, encoding a receptor ***tyrosine kinase***. See Table O1.

rotamase Prokaryotic peptidyl-prolyl *cis-trans* isomerase, homologue of ***immunophilins*** but not inhibited by ***cyclosporin***. Located in the periplasm.

rotamer A rotational isomer, conformationally different by rotation at a single bond.

Rotavirus Genus of the Reoviridae having a double-layered capsid and 11 double-stranded RNA molecules in the genome. They have a wheel-like appearance in the electron microscope, and cause acute diarrhoeal disease in their mammalian and avian hosts.

rotenone An inhibitor of the ***electron transport chain*** that blocks transfer of reducing equivalents from NADH dehydrogenase to coenzyme Q. A very potent poison for fish and for insects.

rough endoplasmic reticulum (rough ER; RER) Membrane organelle of eukaryotes that forms sheets and tubules. Contains the receptor for the signal receptor particle and binds ribosomes engaged in translating mRNA for secreted proteins and the majority of transmembrane proteins. Also a site of membrane lipid synthesis. The membrane is very similar to the nuclear outer membrane. The lumen contains a number of proteins that possess the C-terminal signal ***KDEL***.

rough microsome Small vesicles obtained by sonicating cells and which are derived from the ***rough endoplasmic reticulum***. Have bound ribosomes and can be used to study protein syntheis.

rough strain Bacterial strains that have altered outer cell wall carbohydrate chains causing colonies on agar to change their appearance from smooth to dull. In Streptococci the smooth strains are virulent whereas the rough strains are not. This is partly because the rough strains are much more readily phagocytosed.

rouleaux Cylindrical masses of red blood cells. Horse blood will spontaneously form rouleaux, in other species it can be induced by reducing the repulsion forces between erythrocytes.

Rous sarcoma virus (RSV) The virus responsible for the classic first cell-free transmission of a solid tumour, the chicken ***sarcoma***, first reported by Rous in 1911. An avian ***C-type*** oncorna virus, original source of the *src* oncogene.

royalisin Insect ***defensin*** found in honeybee royal jelly.

RPA See ***replication factor A***.

RPMI Rich medium for mammalian cell culture originally developed at Roswell Park Memorial Institute.

rRNA (ribosomal RNA) Structural RNA components of the ribosome. Prokaryotes have 5S and 23S species in the large subunit and a 16S species in the small subunit. Eukaryotes have a 5S, 5.8S and 28S species in the large subunit and an 18S species in the small subunit.

RS domain Region of arginine/serine repeats found at C-terminus of SR proteins (splicing factors).

RSC complex RSC, an abundant, essential chromatin-remodelling complex, related to the ***SWI/SNF complex,*** binds nucleosomes and naked DNA with comparable affinities. The RSC complex of *Saccharomyces cerevisiae* is closely related to the SWI/SNF complex. Both complexes are involved in remodelling chromatin structure and they share conserved components. The RSC proteins Sth1, Rsc8/Swh3, Sfh1 and Rsc6 are homologues of the SWI/SNF proteins Swi2/Snf2, Swi3, Snf5 and Swp73 respectively.

RSV See ***Rous sarcoma virus***.

RT-PCR (reverse transcriptase polymerase chain reaction; reverse transcription PCR) PCR in which the starting template is RNA, implying the need for an initial reverse transcriptase step to make a DNA template. Some thermostable polymerases have appreciable reverse transcriptase activity; however, it is more common to perform an explicit reverse transcription, inactivate the reverse transcriptase or purify the product, and proceed to a separate conventional PCR.

RTF (resistance transfer factor) The part of a conjugative R plasmid, usually self-repressed, that specifies conjugation.

RTX family of toxins A group of related cytolysins and cytotoxins produced by ***Gram-negative bacteria*** including *E. coli, Proteus vulgaris* (hemolysin), *Pasteurella haemolytica* (leukotoxin) and *Bordetella pertussis* (adenylate cyclase-hemolysin). Characteristically contain a repeat domain (hence the designation, repeats in toxins) with glycine and aspartate-rich motifs repeated within the domain. All are produced in inactive pro-form that must be post-translationally modified to generate an active toxin and are calcium-dependent pore-forming toxins. See ***Escherichia coli hemolysin***.

rubidium One of the alkali earth metals, can be used to substitute for potassium in some ion flux experiments.

RUBISCO See ***ribulose bisphosphate carboxylase***.

RUDP carboxylase See ***ribulose bisphosphate carboxylase***.

Ruffinini's corpuscles Ovoid encapsulated sensory nerve ending in subcutaneous tissue. Probably mechanosensors.

ruffles Projections at the leading edge of a crawling cell. In time-lapse films the active edge appears to 'ruffle'. The protrusions are apparently supported by a microfilament meshwork, and can move centripetally over the dorsal surface of a cell in culture.

rutabaga** (rut) Drosophila* memory mutant; gene codes for calcium/calmodulin-responsive adenylyl cyclase; net result is elevated cAMP levels and comparable behavioural defect to ***dunce.

ruthenium red A stain used in electron microscopy for acid mucopolysaccharides on the outer surfaces of cells.

ryanodine Drug that blocks the release of calcium from the sarcoplasmic reticulum of skeletal muscle. Ryanodine-binding proteins have also been found in the CNS.

S

S1 Soluble fragment (102 kD) of ***heavy meromyosin*** that is produced by papain cleavage: retains the ATPase, actin-binding activity and motor function, and can be used to decorate actin filaments for identification by electron microscopy.

S2 Fibrous fragment of heavy meromyosin (***HMM***). Links the ***S1*** head to the light meromyosin (***LMM***) region that lies in the body of the thick filament and acts as a flexible hinge.

S100 Family of calcium-binding proteins containing an ***EF-hand***. Found on glial cell surfaces. S100β has neurotrophic and mitogenic activity through the ability to cause a rise in intracellular calcium. The family includes calcyclin, MRP8 and MRP14, ***calmodulin***, ***troponin***, and ***calgranulins***.

S180 (sarcoma 180) Highly malignant mouse sarcoma cells, often passaged in ***ascites*** form. Used in some of the classical studies on contact inhibition of locomotion.

S6 kinase A serine/threonine ***kinase***, activated by ***MAP kinase***; phosphorylates ribosomal protein S6 to elevate protein production in cells stimulated by a ***mitogen***.

S See ***serine***.

S-adenosyl methionine (S-(5′-deoxyadenosine-5′)-methionine) An activated derivative of ***methionine***, that functions as a methyl group donor, in (for example) phospholipid methylation and bacterial ***chemotaxis***.

S-antigen (1) Abundant protein of the retina and pineal gland that elicits experimental autoimmune uveitis; now known to be ***arrestin***. (2) Soluble heat-stable antigens (195 kD) on the surface of ***Plasmodium*** *falciparum* that are responsible for antigenic heterogeneity.

S-gene complex Genes coding for molecular components of the pollen–stigma recognition system in the cabbage genus (*Brassica*). The gene products govern the ***self-incompatibility*** response and include a glycoprotein found on the stigma surface and a lectin on the pollen grain surface that binds to the stigma glycoprotein.

S layer A continuous layer of glycoprotein or protein repeating units forming usually the outermost layer of several species of archae- and eubacteria. About 10nm thick.

S phase The phase of the cell cycle during which DNA replication takes place.

S region The non-MHC gene in the midst of the H-2 ***major histocompatibility complex*** of the mouse genome that codes for complement component C4. Sometimes confusingly known as the gene for the type III MHC product in mice.

S-ring The static part of the bacterial motor: a ring of 15 or 17 subunits (one less or one more than the ***M-ring***), anchored to the inner surface of the cell wall.

S-type lectins One of two classes of ***lectin*** produced by animal cells. The classification of animal lectins into two classes, the other being the C-type, was originally proposed by K. Drickamer. The carbohydrate-binding activity of the S-type lectins requires their cysteines to have free thiols and does not need divalent cations (cf. ***C-type lectins***). They mostly have molecular masses in the range 14–16 kD and often form dimers and higher oligomers. The carbohydrate recognition domain contains a number of critically conserved amino acids and largely binds to β-galactosides. S-type lectins certainly occur as cytoplasmic proteins but the existence of extracellular S-type lectins is still a matter of debate.

S value Svedberg unit. See ***sedimentation coefficient***.

Sab Protein (425 residues) that binds selectively to SH3 domain of ***btk***.

Saccharomyces Genus of Ascomycetes; yeasts. Normally haploid unicellular fungi that reproduce asexually by budding. Also have a sexual cycle in which cells of different mating types fuse to form a diploid ***zygote***. Economically important in brewing and baking, and are also suitable eukaryotic cells for the processes of genetic engineering, and for the analysis of, for example, cell division cycle control by selecting for mutants (see ***cdc genes***). *S. cerevisiae* is baker's yeast; *S. carlsbergensis* is now the major brewer's yeast. See also ***Schizosaccharomyces pombe***

Salmonella Genus of Enterobacteriaceae; motile, ***Gram-negative***. Enteric organisms that if invasive cause enteric fevers (eg. typhoid, caused by *S. typhi*), food-poisoning (usually *S. typhimurium* or *S. enteridis*, the latter notorious for contamination of poultry), and occasionally septicaemia in non-intestinal tissues.

saltatory conduction A method of neuronal transmission in vertebrate nerves, where only specialized ***nodes of Ranvier*** participate in excitation. This reduces the capacitance of the neuron, allowing much faster transmission. See ***myelin, Schwann cells***.

saltatory movements Abrupt jumping movements of the sort shown by some intracellular particles. Mechanism unclear.

saltatory replication The sudden amplification of a DNA sequence to generate many copies in a tandem arrangement. Possible mechanism for the origin of ***satellite DNA***.

salvage pathways Metabolic pathways that allow synthesis of important intermediates from materials that would otherwise be waste products. An experimentally important pathway is that from hypoxanthine to nucleotides. See ***HGPRT***.

Sam68 (Src-associated in mitosis, 68 kD) Has KH RNA-binding domain, SH2 and SH3 domains. Interacts with RNA, src-family kinases, grb2 and PLCγ. Important for mitosis, and inhibition of phosphorylation of Sam68 by radicicol will block exit from mitosis.

Sanfillipo syndrome ***Lysosomal disease*** in which either keratan sulphate sulphatase or N-acetyl-α-D-glucosaminidase is defective: cross correction (complementation) of co-cultured fibroblasts from apparently clinically identical patients can therefore occur if a different enzyme is missing in each.

Sanger dideoxy sequencing See ***dideoxy sequencing***.

Sanger–Coulson method See ***dideoxy sequencing***.

SAP (1) See ***serum amyloid P-component***. (2) SLAM-associated protein: a T-cell-specific protein (128 residues) that interacts with the cytoplasmic tail of ***SLAM*** and blocks recruitment of ***Shp-2***, thus blocking activation. SAP contains an SH2 domain. Mutations in SAP have been found in patients with X-linked lymphoproliferative disease (XLP).

sapecin Insect ***defensin*** produced by the flesh fly, *Sarcophaga peregrina*.

saponin Glycosidic surfactants produced by plant cells. Used to solubilize membrane proteins, etc.

saprophyte Organism that feeds on complex organic materials, often the dead and decaying bodies of other organisms. Many fungi are saprophytic

SAR See ***structure–activity analysis***.

sarafotoxin Group of snake cardiotoxic venoms from *Atractaspis engaddensis*. Structurally related to the ***endothelins***.

sarcodictyin A Tricyclic compound that, like ***taxol***, will stabilize microtubule bundles.

Sarcodina Group of aquatic protozoa that includes Amoebae, Foraminifera and Radiolaria.

sarcoglycan (adhalin) Complex of α-, β- and γ-sarcoglycans. α-sarcoglycan (50DAG, A2, adhalin), β-sarcoglycan (43DAG; A3b) and γ-sarcoglycan (35DAG; A4) are all transmembrane glycoproteins that associate with dystroglycan in the ***sarcolemma*** (approximate molecular weights are indicated by the old names). Defects in sarcoglycans have been shown to be associated with autosomally inherited ***muscular dystrophy***.

sarcoidosis Disease of unknown aetiology in which there are chronic inflammatory granulomatous lesions in lymph nodes and other organs.

sarcolemma Plasma membrane of a striated muscle fibre.

sarcoma cells Cells of a malignant tumour derived from connective tissue. Often given a prefix denoting tissue of origin, eg. osteosarcoma (from bone).

sarcoma growth factor Polypeptide released by ***sarcoma*** cells that promotes the growth of cells by binding to a cell surface receptor; the sarcoma cell is therefore self-sufficient and independent of normal growth control. See ***growth factors***. The name is no longer commonly used.

sarcoma virus Virus that causes tumours originating from cells of connective tissue such as ***fibroblasts***. See ***Rous sarcoma virus***, and ***src gene***.

Sarcomastigophora Phylum of unicellular protozoa with pseudopodia or flagella or both.

sarcomere Repeating subunit from which the ***myofibrils*** of striated muscle are built. Has ***A*** and ***I bands***, the I band being subdivided by the ***Z disc***, and the A band being split by the ***M line*** and the H zone.

sarcoplasm Cytoplasm of striated muscle fibre.

sarcoplasmic reticulum Endoplasmic reticulum of striated muscle, specialized for the sequestration of calcium ions that are released upon receipt of a signal relayed by the ***T-tubules*** from the neuromuscular junction.

sarin Nerve gas; inhibitor of acetylcholine esterase.

satellite cell (1) Sparse population of mononucleate cells found in close contact with muscle fibres in vertebrate skeletal muscle. Seem normally to be inactive, but may be important in regeneration after damage. May be considered quiescent stem cell. (2) An alternative name for glial cell.

satellite DNA (minisatellite; microsatellite) DNA, usually containing highly repetitive sequences, that has a base composition (and thus density) sufficiently different from that of normal DNA that it sediments as a distinct band in caesium chloride density gradients. Typically 10% of mammalian, and 50% of insect genomes are composed of satellites. As satellites are dispersed widely in the genome, are easily detectable (with a highly repetitive probe) and are frequently polymorphic in length, they are ideal markers for ***linkage*** studies of disease or inheritance, and for genomics.

satellite virus A term used in plant virology for a virus associated functionally, at least for the purpose of its own replication, with another virus.

saturated fatty acids In eukaryotic membranes refers to stearic, palmitic and myristic acids, that are linear aliphatic chains with no double bonds. Prokaryotes have numerous branched-chain saturated fatty acids.

saturation density The maximal population density achieved by a cell type grown under particular ***in vitro*** culture conditions. Although transformed cells generally grow to a higher saturation density than normal cells this is not necessarily the case. Many factors affect the final density achieved by a cell population; the critical factor may be availability of surface upon which to spread or the serum concentration in the medium. Population densities in culture never approach those found in whole organisms.

saturation of receptors Saturation, the state in which all receptors are effectively occupied all the time, can be said to occur in a simple binding equilibrium when the concentration of ligand is more than five times the Kd value, although strictly this will only be true at infinite ligand concentration.

sauvagine Peptide (40 amino acids) originally isolated from the skin of the frog, *Phyllomedusa sauvagei*, which is closely related to ***corticotrophin releasing factor*** and to urotensin I.

saxitoxin (STX) ***Neurotoxin*** produced by the 'red tide' dinoflagellates, *Gonyaulax catenella* and *G. tamarensis*. It binds to the ***sodium channel***, blocking the passage of action potentials. Its action closely resembles that of ***tetrodotoxin***. The toxin was originally isolated from the clam, *Saxidomus giganteus*.

scaffold proteins Proteins that remain when chromosomes are digested with DNAase. Many antigenic species have been identified.

scallop See ***Pecten***.

SCAMP (secretory carrier membrane protein) Integral membrane proteins of secretory and transport vesicles.

scanning electron microscopy (SEM) Technique of electron microscopy in which the specimen is coated with heavy metal, and then scanned by an electron beam. The image is built up on a monitor screen (in the same way as the raster builds a conventional television image). The resolution is not so great as with transmission electron microscopy, but preparation is easier (often by fixation followed by ***critical point drying***), the depth of focus is relatively enormous, the surface of a specimen can be seen (though not the interior unless the specimen is cracked open), and the image is aesthetically pleasing.

scanning probe microscopy (SPM) Methods for visualizing surfaces at microscopic scale that rely on moving a tiny probe over a surface (usually in an x-y scan), and recording some property of interest (current, force) at each coordinate. These techniques have the ability to resolve detail down to single atoms. See also ***scanning tunnelling microscopy*** and ***atomic force microscopy***.

scanning transmission electron microscopy (STEM) Method of electron microscopy in which image formation depends upon analysis of the pattern of energies of electrons that pass through the specimen. Has comparable resolving power to conventional transmission EM.

scanning tunnelling microscopy (STM) A form of ultra-high resolution microscopy of a surface in which a very small current is passed through a surface and is detected by a microprobe of atomic dimensions at its tip that scans the surface by use of a piezodrive. In the simplest form the current transferred to the probe is recorded as an indication of the contours of molecules on the surface above the local plane. In more complex forms feedback is used to hold the probe at a constant difference and the signal in the feedback loop indicates the contours of the molecule. Capable of resolving single atoms and known to work for non-conducting molecules as well as conducting ones.

Scatchard plot A method for analysing data for freely reversible ligand–receptor binding interactions. The graphical plot is: [bound ligand/free ligand] against [bound ligand]; the slope gives the negative reciprocal of the binding affinity, the intercept on the x-axis the number of receptors (bound/free becomes zero at infinite ligand concentration). The Scatchard plot is preferable to the ***Eadie–Hofstee plot*** for binding data because it is more dependent upon the values at high ligand concentration, which will be the most reliable values. A non-linear Scatchard plot is often taken to indicate heterogeneity of receptors, although this is not the only explanation possible.

scatter factor A motility factor (***motogen***) isolated from conditioned medium in which human fibroblasts have been grown. It causes colonies of epithelial and endothelial cells in culture to separate into single cells that move apart, ie. they scatter. It has been shown to be identical to human ***hepatocyte growth factor***, but it is not mitogenic for all cell types.

scavenger receptor Structurally diverse family of receptors on macrophages that are involved in the uptake of modified ***LDL*** and have been implicated in development of atherosclerotic lesions. Six classes are recognized with different binding preferences. Macrophage scavenger receptors Class A bind a wide range of ligands, including bacteria and it is speculated that scavenger receptors may be important in recognizing apoptotic cells.

SCF See ***stem cell factor***.

SCF complexes Class of E3 ubiquitin protein ligases that are composed of a Skp1p-cdc53p-F-box protein complex and play a role in regulation of cell division. Substrate specificity for targeting for ubiquitinylation is conferred by the F-box component. See ***skp*** and ***cullin***.

SCG10 gene A neural-specific gene that encodes a growth-associated protein expressed early in the development of neuronal derivatives of the neural crest. The 22 kD intracellular protein is associated with the membranous organelles that accumulate in growth cones.

Scheie syndrome Mucopolysaccharidosis (***lysosomal disease***) in which there is a defect in α-L-iduronidase. Fibroblasts from Scheie syndrome patients do not cross-correct fibroblasts from ***Hurler's disease***, although the two conditions are clinically distinct.

Schiff base The reaction of a primary amine with an aldehyde or ketone yields an imine sometimes called a Schiff base. When an arylamine is used the Schiff base may form an intermediate in a staining reaction, eg. for polysaccharides.

Schiff's reagent See ***periodic acid–Schiff reaction***.

schistocytes Fragments of red blood cells found in the circulation.

schistosomiasis Disease (bilharzia) caused by trematode worms (flukes). Three main species, *Schistosoma haematobium*, *S. japonicum* and *S. mansoni*, cause disease in man. Larval forms of the parasite live in freshwater snails; cercariae liberated from the snail burrow into skin, transform to the schistosomulum stage, and migrate to the urinary tract (*S. haematobium*), liver or intestine (*S. japonicum*, *S. mansoni*) where the adult worms develop. Eggs are shed into the urinary tract or the intestine and hatch to form miracidia which then infect snails, completing the life cycle. Adult worms cause substantial damage to tissue and seem to resist immune damage by mechanisms that are not fully understood.

schistosomulum (plural, schistosomula) See ***schistosomiasis***.

schizogony The division of cells, especially of protozoans, in non-sexual stages of the life history of the organism.

Schizosaccharomyces pombe Species of fission yeast commonly used for studies on cell cycle control because there is a distinct G2 phase to the cycle. Only distantly related to the budding yeast ***Saccharomyces*** *cerevisiae*. A further advantage is that some mammalian introns are processed correctly.

Schultz–Charlton test Test for scarlet fever in which antitoxin to ***erythrogenic toxin*** of *Streptococcus pyogenes* is injected subcutaneously.

Schwann cell A specialized glial cell that wraps around vertebrate ***axons*** providing extremely good electrical insulation. Separated by ***nodes of Ranvier*** about once every millimetre, at which the axon surface is exposed to the environment. See ***saltatory conduction***, ***myelin***.

Schwannoma-derived growth factor (SDGF) A ***growth factor*** containing an ***EGF-like domain***, mitogenic for ***astrocytes***, ***Schwann cells*** and ***fibroblasts***.

Schwartzmann reaction Misspelling of ***Shwartzman*** reaction.

SCID (severe combined (or congenital) immunodeficiency disease). A heterogeneous group of inherited disorders characterized by gross functional impairment of the immune system. About half the patients with recessive autosomal form have adenosine deaminase (ADA) deficiency. Another form is caused by lack of a transcription factor required for expression of HLA class II genes.

scinderin Protein (80 kD) of the ***gelsolin*** family isolated from vertebrate neural and secretory tissue. Subcortical scinderin is redistributed into patches following stimulation of ***chromaffin cells*** through nicotinic receptors. Similar to [adseverin].

scintillation counting Technique for measuring quantity of a radioactive isotope present in a sample. In biology, liquid scintillation counting is mainly used for β emitters such as ^{14}C, ^{35}S and ^{32}P and particularly for the low energy β emission of ^{3}H. Gamma emissions are often measured by counting the scintillations that they cause in a crystal. Autoradiographic images can be enhanced by using a screen of scintillant behind the film.

scintillation proximity assay Assay system in which antibody or receptor molecule is bound to a bead that will emit light when β emission from an isotope occurs in close proximity, ie. from a radioactively labelled ligand. Avoids the need for scintillant in order to measure the amount of bound isotope, and thus the amount of antigen or ligand present.

scirrhous carcinoma Carcinoma having a hard structure because of excessive production of dense connective tissue.

sclereid Type of ***sclerenchyma*** cell that differs from the ***fibre cell*** by not being greatly elongated. Often occurs singly (an idioblast) or in small groups, giving rise to a gritty texture in, for instance, the pear fruit, where it is known as a 'stone cell'. May also occur in layers, eg. in hard seed coats.

sclerenchyma Plant cell type with thick lignified walls, normally dead at maturity and specialized for structural strength. Includes ***fibre cells*** which are greatly elongated, and ***sclereids***, that are more isodiametric. Intermediate types exist.

sclerodactyly See ***CREST***.

scleroderma Hardening of skin.

sclerosis Pathological hardening of tissue.

scoliosis Abnormal lateral curvature of the spine.

scopolamine (hyoscine) Alkaloid found in thorn apple (*Datura stramonium*). Related to atropine both in effects and structure and acts as a ***muscarinic acetylcholine receptor*** antagonist.

scorpion toxins Polypeptide toxins (7 kD) with four disulphide bridges. The α toxins are found in venom of Old World scorpions, β toxins in those of the New World. Bind with high affinity to the voltage-sensitive ***sodium channel*** of nerve and muscle (α and β toxins bind to different sites).

scotophobin Peptide (15 residues) isolated from brains of rats trained to avoid the dark that will transfer this aversion to naive animals.

scrapie A chronic neurological disease of sheep and goats, similar to other ***spongiform encephalopathies*** and much used as a model for studying the diseases. Controversy still surrounds the nature of the transmissible agent; slow viruses are proposed by some workers, ***prions*** by an increasing number of others.

scRNP Small cytoplasmic ribonucleoprotein.

scruin Actin-binding protein found associated with the acrosomal process of *Limulus polyphemus* sperm. Scruin holds the microfilaments of the core process in a strained configuration so that the process is coiled. The myosin binding sites on the microfilaments are blocked so HMM decoration is impossible, indicating that there is an unusual packing conformation; when the scruin–actin binding is released the process straightens, the conformation of the actin changes and myosin binding is possible.

scurfy The murine X-linked lymphoproliferative disease scurfy is similar to the ***Wiskott-Aldrich syndrome*** in humans. Disease in scurfy (sf) mice is mediated by CD4 T-cells but there is general overproduction of various cytokines.

scurvy Disease caused by vitamin C deficiency. The effects are due to a failure of the hydroxylation of proline residues in collagen synthesis, and the consequent failure of fibroblasts to produce mature collagen. See ***hydroxyproline***.

scutellum Part of the embryo in seeds of the Poaceae (grasses). Can be considered equivalent to the cotyledon of other monocotyledenous seeds. During germination, absorbs degraded storage material from the endosperm and transfers it to the growing axis.

scyllatoxin Toxin from the scorpion *Leiurus quinquestriatus hebraeus*. Peptide of 31 residues that specifically blocks low conductance calcium-dependent potassium channels that are also a target for ***apamin***.

SDGF See ***Schwannoma-derived growth factor***.

SDS (sodium dodecyl sulphate; sodium lauryl sulphate) Anionic detergent that at millimolar concentrations will bind to and denature proteins, forming an SDS–protein complex. The amount of SDS bound is proportional to the molecular weight of the protein, and each SDS molecule, bound by its hydrophobic domain, contributes one negative charge to the protein thus swamping its intrinsic charge. This property is exploited in the separation of proteins by ***SDS-PAGE***.

SDS-PAGE ***Polyacrylamide gel electrophoresis*** (PAGE) in which the charge on the proteins results from their binding of SDS. Since the charge is proportional to the surface area of the protein, and the resistance to movement proportional to diameter, small proteins migrate further.

sea An ***oncogene***, identified in bird ***sarcoma***, encoding a receptor ***tyrosine kinase***. See Table O1.

sea hare See ***Aplysia***.

SEC (serpin–enzyme complex) Receptor mediates catabolism of α1-antitrypsin–elastase complexes and elevates α1-antitrypsin synthesis. Also implicated in neutrophil chemotaxis and neurotoxicity of amyloid β-peptide.

sec65 Gene of *Saccharomyces cerevisiae* that encodes a protein very similar to the SRP19 subunit of the mammalian ***signal recognition particle***.

sec7 (ySec7p) Protein from *Saccharomyces cerevisiae* that plays an important part in the secretory pathway. Mutations in sec7 lead to accumulation of Golgi cisternae and loss of secretory granules. Sec7 contains a domain of around 200 amino acids that is found in several ***GEF***s for ADP ribosylation factors (***ARF***s).

second messenger In many hormone-sensitive systems the systemic hormone does not enter the target cell but binds to a receptor and indirectly affects the production of another molecule within the cell; this diffuses intracellularly to the target enzymes or intracellular receptor to produce the response. This intracellular mediator is called the second messenger. Examples include cyclic AMP, cyclic GMP, IP3 and diacylglycerol.

secondary immune response The response of the immune system to the second or subsequent occasion on which it encounters a specific antigen.

secondary lymphoid tissue See ***lymphoid tissue***.

secondary lysosome Term used to describe intracellular vacuoles formed by the fusion of lysosomes with organelles (***autosomes***) or with primary phagosomes. ***Residual bodies*** are the remnants of secondary lysosomes containing indigestible material.

secondary product End product of plant cell metabolism, which accumulates in, or is secreted from, the cell. Includes ***anthocyanins***, ***alkaloids***, etc. Some are of major economic importance, eg. as drugs. In contrast to a primary product that is involved in the vital metabolism of the plant.

secondary structure Structures produced in polypeptide chains involving interactions between amino acids within the chain. Especially α-helical and ***beta pleated sheet*** structures. Also applies to the complex folding of nucleic acids as, for example, the clover-leaf structure of tRNA.

secondary wall (of plants) That part of the plant cell wall which is laid down on top of the ***primary cell wall*** after the wall has ceased to increase in surface area. Only occurs in certain cell types, eg. tracheids, vessel elements and sclerenchyma. Differs from the primary wall both in composition and structure, and is often diagnostic for a particular cell type.

secretagogue Substance that induces secretion from cells; originally applied to peptides inducing gastric and pancreatic secretion.

secretin Peptide hormone of gastrointestinal tract (27 residues) found in the mucosal cells of duodenum. Stimulates pancreatic, pepsin and bile secretion, inhibits gastric acid secretion. Considerable homology with ***gastric inhibitory polypeptide***, ***vasoactive intestinal peptide*** and ***glucagon***.

secretion Release of synthesized product from cells. Release may be of membrane-bounded vesicles (merocrine secretion) or of vesicle content following fusion of the vesicle with the plasma membrane (apocrine secretion). In holocrine secretion whole cells are released.

secretogranins See ***granins***.

secretory carrier membrane protein See ***SCAMP***.

secretory cells Cells specialized for secretion, usually epithelial. Those that secrete proteins characteristically have well developed ***rough endoplasmic reticulum***, whereas conspicuous ***smooth endoplasmic reticulum*** is typical of cells that secrete lipid or lipid-derived products (eg. ***steroids***).

secretory component of IgA A polypeptide chain of about 60 kD that aids secretion of the IgA; a portion of the IgA receptor on the plasmalemma of the inner side of the epithelial cells lining the gut, which is proteolysed when the IgA receptor complex has travelled through the cell after receptor-mediated endocytosis at the inner face, to the outer (luminal) face.

secretory proteins In eukaryotes, proteins synthesized on ***rough endoplasmic reticulum*** and destined for export. Nearly all proteins secreted from cells are glycosylated (in the ***Golgi apparatus***, although there are exceptions i.e. ***albumin***). In prokaryotes, secreted proteins may be synthesized on ribosomes associated with the plasma membrane or exported post-translation.

secretory vesicle Membrane-bounded vesicle derived from the ***Golgi apparatus*** and containing material that is to be released from the cell. The contents may be densely packed, often in an inactive precursor form (***zymogen***).

sedimentation Settling of a component of a mixture under the influence of gravity (natural or artificial) so that the mixture separates into two or more phases or zones.

sedimentation coefficient The ratio of the velocity of sedimentation of a molecule to the centrifugal force required to produce this sedimentation. It is a constant for a particular species of molecule, and the value is given in Svedberg units (S) that, it should be noted, are non-additive.

sedoheptulose Seven-carbon sugar, whose phosphate derivatives are involved in the ***pentose phosphate pathway*** and the ***Calvin–Benson cycle***.

segment long-spacing collagen See ***SLS collagen***.

segment-polarity gene A ***segmentation gene***, responsible for specifying anterior– posterior polarity within individual embryonic segments. In ***Drosophila***, there are at least 10 such genes, eg. *gooseberry*.

segmentation Organization of the body into repeating units called segments is a common feature of several phyla, eg. arthropods and annelids, although the segments arise by very different mechanisms. Segmentation also occurs during embryonic development in vertebrates, eg. partition of the mesoderm into ***somites***, and is a feature of early ***CNS*** development. See ***rhombomeres, neuromeres***.

segmentation gene Genes required for the establishment of ***segmentation*** in the embryo. In ***Drosophila*** about 20 such genes are required.

segregation of chromosomes The separation of pairs of ***homologous chromosomes*** that occurs at meiosis so that only one chromosome from each pair is present in any single gamete.

selectin Group of cell adhesion molecules that bid to carbohydrates via a ***lectin***-like domain. The name is derived from select and lectin. They are integral membrane glycoproteins with an N-terminal, C-type lectin domain, followed by an EGF-like domain, a variable number of repeats of the short consensus sequence of complement regulatory proteins and a single transmembrane domain. Three selectins have been identified and are distinguished by capital letters based on the source of the original identification, ie. E-, L- and P-selectin. Examples: ELAM-1, GMP-140, LECAM. See Table S1.

selector genes A group of genes that determines which part of a developmental pattern cells will be allocated within a developmental segment.

selenium Essential trace element that must be provided as a supplement in serum-free culture media for most animal cells.

selenocysteine (sec) An unusual amino acid of proteins, the selenium analogue of ***cysteine***, in which a selenium atom replaces sulphur. Involved in the catalytic mechanism of seleno-enzymes such as formate dehydrogenase of *E. coli*, and mammalian glutathione peroxidase. May be cotranslationally coded by a special ***opal suppressor*** tRNAase that recognizes certain UGA nonsense codons.

self-antigens (autoantigen) The antigens of an individual's body have the potential to be self-antigens for the immune system and unless clones of immune cells reactive with self-antigens are eliminated an autoimmune response can be initiated.

self-assembly The property of forming structures from subunits (protomers) without any external source of information about the structure to be formed such as priming structure or template.

self-cloning Any system in which inappropriate cell types or organisms are eliminated because they possess some character that allows them to die or to remove themselves from the system. Thus a transfected cell with genetic material including a drug resistance marker will be self-cloning in the presence of the drug and non-transfected cells will die.

self-incompatibility Inability of pollen grains to fertilize flowers of the same plant or its close relatives. Acts as a mechanism to ensure outbreeding within some plant species, eg. in the case of the ***S-gene complex*** in Brassicas.

self-replicating Literally, replication of a system by itself without outside intervention. In practice often taken to refer to systems that replicate without the contribution of any information from outside the system.

self-splicing Self-catalysed removal of group 5 **introns**, mediated by six paired conserved regions.

sem-5 Cell-signalling gene of *Caenorhabditis elegans* that encodes a protein (228 residues) with SH2 and SH3 domains and that acts in vulval development and sex myoblast migration.

semaphorins Family of proteins that mediate neuronal guidance by inhibiting ***nerve growth cone*** movement. Both transmembrane and secreted proteins are included and many domains of the proteins are highly conserved between invertebrates and vertebrates. Most are around 750 residues with a conserved 'sema' domain of up to 500 residues extracellularly with a single immunoglobulin C2-type domain C-terminally to this. ***Collapsin***, responsible for the collapse of nerve growth cones of chick sensory

Table S1. Selectins

Selectin	Synonyms	Cellular expression	Domain structure	Role in adhesion
E-selectin	ELAM-1	Activated endothelial cells	1 C-lectin 1 EGF 6 CRP (90 kD)	Binding of leucocytes in the inflammatory response
L-selectin	LECAM-1, LAM-1, gp90[MEL-14], LHR, lymphocyte homing receptor	Lymphocytes, polymorphonuclear neutrophils	1 C-lectin 1 EGF 2 CRP (115 kD)	Homing of lymphocytes to peripheral lymph nodes
P-selectin	CD62, GMP-140, PADGEM	α-granules of platelets and Weibel–Palade granules of endothelial cells. Plasma membranes of both cells after activation	1 C-lectin 1 EGF 9 CRP (140 kD)	Binding of cells to monocytes and leucocytes?

neurites in culture following contact with retinal axons was one of the first semaphorins described.

semiautonomous Of systems or processes that are not wholly independent of other systems or processes.

semiconservative replication The system of replication of DNA found in all cells in which each daughter cell receives one old strand of DNA and one strand newly synthesized at the preceding ***S phase***. The existence of semiconservative replication was demonstrated by the Meselson–Stahl experiment and implies the double or multistrandedness of DNA.

semipermeable membrane A membrane that is selectively permeable to only one (or a few) solutes. The potential developed across a membrane permeable to only one ionic species is given by the ***Nernst equation*** for the species: this is the basis for the operation of ***ion-selective electrodes***.

Semliki forest virus Enveloped virus of the alphavirus group of ***Togaviridae***. First isolated from mosquitoes in the Semliki Forest in Uganda; not known to cause any illness. The synthesis and export of its three spike glycoproteins, via the endoplasmic reticulum and Golgi complex, have been used as a model for the synthesis and export of plasma membrane proteins.

Senarmont compensation In interference microscopy, compensation for the phase difference introduced by the object, measured by introducing a quarter-wavelength plate and rotating the analyser: the angle of rotation is proportional to the optical path difference.

Sendai virus Parainfluenza virus type 1 (Paramyxoviridae). Can cause fatal pneumonia in mice, and may cause respiratory disease in humans. The ability of ultraviolet-inactivated virus to fuse mammalian cells has been extensively used in the study of ***heterokaryons*** and ***hybrid cell*** lines.

senescent cell antigen An antigen (62 kD) that appears on the surface of senescent erythrocytes and is immunologically cross-reactive with isolated ***band III***. Seems to be recognized by an autoantibody, and the immunoglobulin-coated erythrocyte is then removed from circulation by cells such as Kuppfer cells of the liver that have Fc receptors. Intracellular cleavage of intact band III by a calcium-activated protease, ***calpain***, may reveal the antigen *in situ*.

senile plaque Characteristic feature of the brains of ***Alzheimer's disease*** patients and aged monkeys, consisting of a core of amyloid fibrils surrounded by dystrophic neurites. The principal component of amyloid fibrils in senile plaques is B/A4, a peptide of about 4 kD that is derived from the larger ***amyloid precursor protein*** (APP). The B/A4 sequence is located near the C-terminus of APP.

sensitization A state of heightened responsiveness, usually referring to the state of an animal after primary challenge with an antigen. The term is frequently used in the context of ***hypersensitivity***.

sensory neuron (1) A ***neuron*** that receives input from sensory cells. (2) Sensory cells such as cutaneous mechanoreceptors and muscle receptors.

Sephacryl Trade name for a covalently crosslinked allyl dextrose gel formed into beads. Used in ***gel filtration*** columns for separating molecules in the size range 5 kD to 1.5 million D.

Sephadex Trade name for a crosslinked dextran gel in bead form used for ***gel filtration*** columns: by varying the degree of crosslinking the effective fractionation range of the gel can be altered.

Sepharose Trade name for a gel of agarose in bead form from which charged polysaccharides have been removed. Used in ***gel filtration*** columns.

septate junction An intercellular junction found in invertebrate epithelia that is characterized by a ladder-like appearance in electron micrographs. Thought to provide structural strength and to provide a barrier to diffusion of solutes through the intercellular space. Occurs widely in transporting epithelia, and is controversially considered analogous to tight junctions (***zonula occludens***).

septic shock Condition of clinical shock caused by ***endotoxin*** in the blood. A serious complication of severe burns and abdominal wounds, frequently fatal. Part of the problem seems to be due to increased leucocyte adhesiveness, which leads to massive sequestration of neutrophils in the lung, increased vascular permeability, and acute (adult) respiratory distress syndrome.

septicaemia See ***bacteraemia***.

septin Family of homologous proteins (around 40 kD) first identified in *Saccharomyces cerevisiae* where they associated with cytokinesis and septum formation. Encoded by CDC3, CDC10, CDC11 and CDC12 genes in *S. cerevisiae*: seem to form 10nm filaments that form a ring around the plasma membrane in the mother-bud neck. Homologous proteins, associated with cleavage furrows, are reported from *Drosophila*, amphibians and mammals.

septum Literally, a separating wall. Mainly applied to the structure composed of plasmalemmae and cell wall material formed in cell division in prokaryotes and fungi. Also applied to the sealing layers in various packages of sterile fluids, or barriers through which injections, needles, etc. may be passed.

sequence homology Strictly, refers to the situation where nucleic acid or protein sequences are similar because they have a common evolutionary origin. Often used loosely to indicate that sequences are very similar. Sequence similarity is observable; homology is an hypothesis based on observation.

SER See ***smooth endoplasmic reticulum***.

ser See ***serine***.

Ser See ***Serrate***.

serglycin An intracellular ***proteoglycan***, found particularly in the storage granules of connective tissue mast cells. The core protein consists of 153 amino acids with 24 serine-glycine repeats between amino acids 89 and 137, hence the name. The serine-glycine repeats are the linkage sites for around 15 ***glycosaminoglycan***

chains that are either heparin or highly sulphated chondroitin sulphate. These negatively charged chains are thought to concentrate positively charged proteases, histamine and other molecules within the storage granules.

sericin Protein found in silk. Very serine-rich: 30% of the residues are serine.

serine (Ser; S:) (105 D) One of the amino acids found in proteins and that can be phosphorylated. See Table A2.

serine protease One of a group of endoproteases from both animal and bacterial sources that share a common reaction mechanism based on formation of an acyl-enzyme intermediate on a specific active serine residue. Serine proteases are all irreversibly inactivated by a series of organophosphorus esters, such as di-isopropylfluorophosphate (DFP) and by naturally-occurring inhibitors (***serpins***). Examples are ***trypsin***, ***chymotrypsin***, and the bacterial enzyme ***subtilisin***.

serosa A serous epithelium, having ***serous glands*** or cells, as opposed to a mucous membrane.

serotonin (5-hydroxytryptamine; 5-HT) A ***neurotransmitter*** and ***hormone*** (176 D), found in vertebrates, invertebrates and plants.

serotype The genotype of a unicellular organism as defined by antisera against antigenic determinants expressed on the surface.

serous gland An exocrine gland that produces a watery, protein-rich secretion, as opposed to a carbohydrate-rich mucous secretion.

serpins Superfamily of proteins, mostly ***serine protease*** inhibitors, that includes ***ovalbumin***, ***α1-antitrypsin***, ***antithrombin***.

Serrate *(Ser) Drosophila* locus. Gene product contains 14 repeats of the ***EGF-like domain***.

serrate protein Transmembrane ligand for ***Notch***, expressed on dorsal cells of *Drosophila* wing, activates Notch on ventral cells and induces the expression of ***Delta*** protein. Serrate protein expression is reciprocally induced by Delta and modulated by ***fringe***.

Serratia marcescens A ***Gram-negative bacterium*** that is very common in soil and water; most strains produce a characteristic pigment, prodigiosin. Opportunistic human pathogens, infecting mainly hospital patients.

Sertoli cell Tall columnar cells found in the mammalian testis closely associated with developing spermatocytes and spermatids. Probably provide appropriate microenvironment for sperm differentiation and phagocytose degenerate sperm.

serum Fluid that is left when blood clots; the cells are enmeshed in ***fibrin*** and the clot retracts because of the contraction of platelets. It differs from plasma in having lost various proteins involved in clot formation (***fibrinogen***, ***prothrombin***, various blood-clotting factors such as ***Hagemann factor***, ***factor VIII*** etc.) and in containing various platelet-released factors, notably ***platelet-derived growth factor***. For this reason serum is a better supplement for cell culture medium than defibrinated plasma (plasma-derived serum).

serum amyloid (SAA) In secondary amyloidosis the fibrils deposited in tissues are unrelated to immunoglobulin light chains (in contrast to the situation in primary amyloidosis) and are made of amyloid A protein (AA protein). This is derived from serum amyloid A (SAA) that is the apolipoprotein of a high-density lipoprotein and an ***acute-phase protein***. Partial proteolysis converts SAA into the ***beta-pleated sheet*** configuration of the amyloid fibrils. Amyloid P protein is also found as a minor component of the fibrils (in both primary and secondary amyloidosis) and is derived from serum amyloid P that has similarity to ***C-reactive protein***. The physiological role remains obscure.

serum amyloid P component (SAP) Precursor of amyloid component P, found in basement membrane. Member of the ***pentraxin*** family. See ***serum amyloid***.

serum hepatitis See ***hepatitis B***.

serum requirement The amount of serum that must be added to culture medium to permit growth of an animal cell in culture. Transformed cells frequently have less stringent serum requirements than their normal counterparts.

serum response element (SRE) DNA motif found, eg. in the c-fos ***promoter***, which is bound by the ***serum response factor***.

serum response factor (SRF; p67SRF) ***Transcription factor*** that interacts with Elk-1 (p62TCF) to bind the ***serum response element*** promoter found in many growth-related genes.

serum sickness A ***hypersensitivity*** response (type III) to the injection of large amounts of antigen, as might happen when large amounts of antiserum are given in a passive immunization. The effects are caused by the presence of soluble immune complexes in the tissues.

sev See ***Sevenless***.

seven-membrane spanning receptors See ***serpentine receptors***.

Sevenless *(sev) Drosophila* gene that is required for development of the R7 cell in each ***ommatidium*** in the eye. Gene product is a ***receptor tyrosine kinase***, related to the insulin receptor. Ligand is the product of the ***bride of sevenless*** gene. In the downstream signalling cascade ***son of sevenless*** plays an important part.

severe combined immunodeficiency syndrome See ***combined immunodeficiency***.

severin Calcium-dependent F-actin cleaving protein (40 kD) isolated from the slime mould *Dictyostelium discoideum* that binds irreversibly to the barbed ends of the microfilament; apparently not essential for movement.

sex chromatin Condensed chromatin of the inactivated X chromosome in female mammals (***Barr body***).

sex chromosome Chromosome that determines the sex of an animal. In humans, where the two sex chromosomes (X and Y) are dissimilar, the female has two X chromosomes and the male is heterogametic (XY). In birds, the opposite is the case, the male being XX and the female XY; in many organisms, there is only one sex chromosome, and one sex is XX, the other X0. A portion of the X and Y chromosomes is similar and is known as the pseudoautosomal region.

sex hormone Hormone that is secreted by gonads, or that influences gonadal development. Examples are ***oestrogen***, ***testosterone***, ***gonadotrophins***.

sex pili Fine filamentous projections (***pili***) on the surface of a bacterium that are important in conjugation. Often seem to be coded for by plasmids that confer conjugative potential on the host; in the case of the F-plasmid, the F-pili are 8–9nm diameter and several microns long, composed of ***pilin***. Whether the pili merely serve to establish and maintain adhesive contact between the partners in conjugation, or whether DNA is actually transferred through the central core of the pilus is still unresolved, although a simple adhesion role is more generally accepted.

sex-duction The transfer of genes from one bacterium to another by the process of conjugation. May involve one bacterium with an F′-plasmid, in which case the process is called F-duction.

sex-linked disorder A genetic defect, usually due to a gene on the unpaired portion of the X chromosome. Recessive X-linked alleles are fully expressed in the heterogametic sex because they can have only one copy of the gene. Thus X-linked mutant disorders are more common in human males than in females.

sex-related gene on Y See ***sry***.

Sf9 cells Insect cell line derived from *Spodoptera frugiperda* much used for production of recombinant protein. Gene is incorporated into ***baculovirus*** vector which is then used to infect the cells.

SH domains Src homology domains: domains of protein that, from their homology with src are involved in the interaction with phosphorylated tyrosine residues on other proteins (SH2 domains) or with proline-rich sections of other proteins (SH3 domains).

SH2 See ***SH domains***.

SH3 See ***SH domains***.

shadowing Procedure much used in electron microscopy, in which a thin layer of material, usually heavy metal or carbon, is deposited onto a surface from one side, in such a way as to cast 'shadows'. Deposition is usually done by vaporizing the metal on an electrode under vacuum.

shaker *Drosophila* gene encoding a potassium channel. Related genes, *shab*, *shal* and *shaw* are known in flies and humans. The mutation is so-called because the fly's legs shake under ether anaesthesia.

shc Gene family identified by presence of ***SH2*** domains. Shc also has a tyrosine motif that, when phosphorylated, will bind the SH2 of the adaptor protein ***grb2*** and may link receptor tyrosine kinases with the ras signalling pathway. Overexpression of *shc* will transform fibroblasts.

shear stress response element (SSRE) Various cells can be stimulated to divide if subject to fluid shear stress. This is particularly interesting in the case of endothelial and vascular smooth muscle cells and efforts have been made to identify the response element that activates gene expression in response to shear. There are suggestions, however, that the signalling is through the ras pathway.

shibire *Drosophila* gene that encodes ***dynamin***. *Shibire* is temperature-sensitive and in affected flies synaptic vesicles are depleted at high temperatures but are restored in nerve terminals when endocytosis resumes at lower temperatures.

shiga toxin (verotoxin) Bacterial toxin from *Shigella dysenteriae* that blocks eukaryotic protein synthesis. See ***Shiga-like toxins***.

shiga-like toxins (SLT) Group of structurally related toxins that block eukaryotic protein synthesis by cleaving a single residue from the 28S rRNA subunit of ribosomes thus blocking interaction with elongation factors eEF-1 and eEF-2. Examples: ***Shiga toxin***, Shiga-like toxins SLT-1 and SLT-2 of *Escherichia coli*.

Shigella Genus of non-motile ***Gram-negative*** enterobacteria (Escherichiae group): cause dysentery. See ***Shiga toxin***.

shikimic acid pathway Metabolic pathway in plants and microorganisms, by which the aromatic amino acids (phenylalanine, tyrosine and tryptophan) are formed from phosphoenolpyruvate and erythrose-4-phosphate via shikimic acid. The aromatic amino acids in turn serve as precursors for the formation of lignin and other phenolic compounds in plants. Inhibitors of this pathway are used as herbicides.

Shine–Dalgarno region A polypurine sequence found in bacterial mRNA about seven nucleotides in front of the ***initiation codon***, AUG. The complete sequence is 5′-AGGAGG-3′ and almost all messengers contain at least half of this sequence. It is complementary to a highly conserved sequence at the 3′ end of 16s ribosomal RNA, 3′-UCCUCC-5′, and is thought to be involved in the binding of the mRNA to the ribosome.

shingles Disease in adults caused by ***Varicella zoster*** virus (Herpetoviridae), which in children causes chickenpox. Disease arises by reactivation (usually associated with a decline in cell-mediated immunity) of latent virus that persists in spinal or cranial sensory nerve ganglia.

SHIP Lipid phosphatase containing an SH2 domain; dephosphorylates 5′-inositol phosphate. Important in regulation of mast cell degranulation and cytokine signal transduction in lymphoid and myeloid cells generally. SHIP also modulates PI3-kinase signalling downstream of growth factor and insulin receptors. Negative signalling through SHIP appears to inhibit the ras pathway by competition with ***grb2*** and ***shc*** for SH2 domain binding

ShK toxin A potassium channel-blocking polypeptide (35 residues) from the sea anemone *Stichodactyla helianthus*.

shock Condition associated with circulatory collapse: a result of blood loss, bacteraemia, an anaphylactic reaction or emotional stress.

Shope fibroma virus Poxvirus associated with the production of benign skin tumours in rabbits.

Shope papilloma virus ***Papovavirus*** that produces ***papillomas*** (warts) in rabbits.

Shp (Shp-1, Shp-2) Protein tyrosine phosphatases with SH2 domains. Are recruited to ***ITIM*** motif of receptor tyrosine kinases and play an important role in the control of cytokine signalling. Shp-1 is important in regulating antigen responses in T-cells and the mouse mutant (*motheaten*) is immunosuppressed. Shp-2 is more ubiquitously expressed and functions downstream of a variety of growth factor receptors and has a role in cell spreading and migration; the homozygous mouse knockout is embryonic lethal.

shuttle flow See ***cytoplasmic streaming***.

shuttle vector ***Cloning vector*** that replicates in cells of more than one organism, eg. *E. coli* and yeast. This combination allows DNA from yeast to be grown in *E. coli* and tested directly for ***complementation*** in yeast. Shuttle vectors are constructed so that they have the origins of replication of the various hosts.

Shwartzman reaction Reaction that occurs when two injections of ***endotoxin*** are given to the same animal, particularly rabbits, 24 hours apart. In the local Shwartzman reaction the first injection is given intradermally, the second intravenously, and a haemorrhagic reaction develops at the dermal site. If both injections are intravenous the result is a generalized Shwartzman reaction, often accompanied by ***disseminated intravascular coagulation***. The reaction depends upon the response of platelets and neutrophils to endotoxin.

sialic acid See ***neuraminic acid***.

sialidase See ***neuraminidase***.

sialoglycoprotein Glycoprotein of which the N- or O-glycan chains include residues of ***neuraminic acid***.

sialophorin See ***leukosialin***.

sialyl Lewis^x (sLe-x; CD15s) Sialylated form of CD15, the ligand for E-, P- and L-selectins. Expressed on neutrophils, basophils and monocytes and only some lymphocytes. Also present on some ***HEV***. Deficiency in sialyl Lewis^x will cause leucocyte adhesion deficiency type II.

SIC1 S phase ***cyclin-dependent kinase inhibitor*** from *Saccharomyces cerevisiae*. See ***Skp***.

sickle cell anaemia Disease common in races of people from areas in which malaria is endemic. The cause is a point mutation in haemoglobin (valine instead of glutamic acid at position 6), and the altered haemoglobin (HbS) crystallizes readily at low oxygen tension. In consequence, erythrocytes from homozygotes change from the normal discoid shape to a sickled shape when the oxygen tension is low, and these sickled cells become trapped in capillaries or damaged in transit, leading to severe anaemia. In heterozygotes, the disadvantages of the abnormal haemoglobin are apparently outweighed by increased resistance to *Plasmodium falciparum* malaria, probably because parasitized cells tend to sickle and are then removed from circulation.

sideramines Naturally occurring iron-binding compounds, hydroxamic acids.

sideroblasts Red blood cells containing Pappenheimer bodies: small, deeply basophilic granules that contain ferric iron.

siderochromes See ***ferrichromes***.

sideromycins Non-chelating antibiotic analogues produced by some enteric bacteria; interfere with the uptake of sideramine-ferric ion complexes.

siderophilins Family of non-haem iron chelating proteins (about 80 kD) found in vertebrates. Examples are ***lactoferrin*** and ***transferrin***.

siderophores Natural iron-binding compounds that chelate ferric ions (which form insoluble colloidal hydroxides at neutral pH and are then inaccessible) and are then taken up together with the metal ion. See ***sideramines***.

sieve plate Perforated end walls separating the component cells (sieve elements) that make up the phloem ***sieve tubes*** in vascular plants. The perforations permit the flow of water and dissolved organic solutes along the tube, and are lined with ***callose***. The plates are readily blocked by further deposition of callose when the sieve tube is stressed or damaged.

sieve tube The structure within the phloem of higher plants that is responsible for transporting organic material (sucrose, raffinose, amino acids, etc.) from the photosynthetic tissues (eg. leaves) to other parts of the plant. Made up of a column of cells (sieve elements) connected by ***sieve plates***.

sigma factor (σ factor) ***Initiation factor*** (86 kD) that binds to *E. coli* DNA-dependent ***RNA polymerase*** and promotes attachment to specific initiation sites on DNA. Following attachment, the sigma factor is released.

signal peptidase See ***signal sequence***.

signal peptidase complex (SPC) See ***signal sequence***.

signal peptide See ***signal sequence***.

signal recognition particle (SRP) A complex between a 7S RNA and six proteins. SRP binds to the nascent polypeptide chain of eukaryotic proteins with a ***signal sequence*** and halts further translation until the ribosome becomes associated with the ***rough endoplasmic reticulum***. One of the SRP proteins (srp54) binds GTP and in association with 7SRNA and srp19 has GTPase activity.

signal recognition particle receptor Receptor for the ***signal recognition particle*** (SRP) found in the membrane of the ***endoplasmic reticulum*** (ER). Also known as docking protein. Heterodimeric, both protomers having GTP-binding capacity, though dissimilar binding sites. Not until the complex of SRP, ribosome, message, and nascent polypeptide chain binds to the SRP receptor is the block to further chain elongation released – and concurrently the SRP is released, leaving the ribosome attached to the ER membrane. ***Cotranslational transport*** of the polypeptide delivers it into the lumen of the ER.

signal sequence A peptide present on proteins that are destined either to be secreted or to be membrane components. It is usually at the N-terminus and normally absent from the mature protein. Normally refers to the sequence (ca 20 amino acids) that interacts with ***signal recognition particle*** and directs the ribosome to the ***endoplasmic reticulum*** where cotranslational insertion takes place. Could also refer to sequences that direct post-translational uptake by organelles. Signal peptides are highly hydrophobic but with some positively charged residues. The signal sequence is normally removed from the growing peptide chain by signal peptidase, a specific protease located on the cisternal face of the endoplasmic reticulum. See ***signal recognition particle***.

signal transduction The cascade of processes by which an extracellular signal (typically a hormone or neurotransmitter) interacts with a receptor at the cell surface, causing a change in the level of a ***second messenger*** (eg. calcium or cyclic AMP) and ultimately effects a change in the cell's functioning (eg. triggering glucose uptake, or initiating cell division). Can also be applied to sensory signal transduction, eg. of light at photoreceptors.

signal–response coupling See ***signal transduction***.

signet-ring cell Cell (adipocyte) with a large central fat-filled vacuole that pushes the nucleus to one side to give an appearance reminiscent of a signet ring.

silanizing Conversion of active silanol (–SiOH) groups on surface of, eg. glass into less polar silyl ethers (–SiOR), thereby making the surface less adhesive. See ***siliconization***.

silent mutation Mutations that have no effect on ***phenotype*** because they do not affect the activity of the product of the gene, usually because of codon ambiguity.

siliconization Non-covalent coating of surface with a layer of silicone oil making it less adhesive or reactive. See ***silanizing***.

silicosis Inflammation of the lung caused by foreign bodies (inhaled particles of silica): leads to fibrosis but unlike ***asbestosis*** does not predispose to neoplasia.

simian virus 40 See ***SV40***.

simple epithelium An epithelial layer composed of a single layer of cells all of which are in contact with the basal lamina (see ***basement membrane***). May be cuboidal, columnar, squamous or pseudostratified; though the last of these appears superficially to be multilayered, all cells are in contact with the basal lamina.

Simpson–Golabi–Behmel syndrome (SGBS) Human disorder leading to overgrowth. Arises through mutation in ***glycipan-3*** gene on X chromosome, probably reducing the extent to which IGF-2 is bound and unavailable for growth stimulation.

Sin3 Component of the multiprotein complex involved in repression of transcription. Sin3 is a corepressor and forms a complex with Rpd3 histone deacetylase; the complex then interacts with DNA-binding proteins. Yeast SIN3 is involved in the repression of a diverse range of genes. Sin3 does not itself bind to DNA.

Sindbis virus Enveloped virus of the alphavirus group of ***Togaviridae***. It is thought to be an infection of birds spread by fleas, and there is little evidence that it causes any serious infection in humans. The synthesis and export of the spike proteins, via the ***endoplasmic reticulum*** and Golgi complex, have been used as a model for the synthesis and export of plasma membrane proteins.

single cell protein Protein(s) produced by single cells in culture, especially *Candida* species. Of possible commercial importance in providing food sources from biotechnological processes.

single-stranded DNA (ssDNA) DNA that consists of only one chain of nucleotides rather than the two ***base-pairing*** strands found in DNA in the double-helix form. ***Parvoviridae*** have a single-stranded DNA genome. Single-stranded DNA can be produced experimentally by rapidly cooling heat-denatured DNA. Heating causes the strands to separate and rapid cooling prevents ***renaturation***.

single-channel recording Variant of ***patch clamp*** technique.

single-stranded binding proteins See ***helix-destabilizing proteins***.

single-stranded conformational polymorphism (SSCP) Technique for detecting point mutations in genes by amplifying a region of genomic DNA (using asymmetric ***PCR***) and running the resulting product on a high-quality gel. Single base substitutions can alter the secondary structure of the fragment in the gel, producing a visible shift in its mobility.

singlet oxygen $^1O_2$ An energized but uncharged form of oxygen that is produced in the ***metabolic burst*** of leucocytes and which can be toxic to cells.

sis An ***oncogene***, identified in monkey ***sarcoma***, encoding a B-chain of ***PDGF***. See Table O1.

SIS See ***macrophage inflammatory protein 1*** α and β.

sister chromatid One of the two ***chromatids*** making up a ***bivalent***. Both are semi-conservative copies of the original chromatid.

site-directed mutagenesis See ***site-specific mutagenesis***.

site-specific mutagenesis An *in vitro*

technique in which an alteration is made at a specific site in a DNA molecule, which is then reintroduced into a cell. Various techniques are used; for the cell biologist, a very powerful approach to determining which parts of a protein or nucleotide sequence are critical to function.

site-specific recombination A type of ***recombination*** that occurs between two specific short DNA sequences present in the same or in different molecules. An example is the integration and excision of lambda prophage.

situs inversus See ***Kartagener's syndrome***.

SIV (simian immunodeficiency virus) Very similar to ***HIV*** and used extensively as an animal model.

Sjögren's syndrome One of the so-called connective tissue diseases that also include rheumatoid arthritis, systemic lupus erythematosus and rheumatic fever. Characterized by inflammation of conjunctiva and cornea.

skeletal muscle A rather non-specific term usually applied to the striated muscle of vertebrates that is under voluntary control. The muscle fibres are syncytial and contain myofibrils, tandem arrays of ***sarcomeres***.

ski An ***oncogene***, identified in avian ***carcinoma***, encoding a nuclear protein. See Table O1.

Skn-1 Maternally expressed transcription factor that specifies the fate of certain blastomeres during early development of *Caenorhabditis elegans*. Binds DNA with high affinity as a monomer even though it has a basic region similar to basic leucine zipper (bZIP) proteins that only bind as dimers.

skp (S-phase kinase-associated proteins; Skp1, Skp2) Component of the ***SCF complex***. The SCF (Skp1-cullin-F-box complex) ubiquitin protein ligase of *Saccharomyces cerevisiae* triggers DNA replication by catalysing ubiquitinylation of the S phase cyclin-dependent kinase inhibitor, SIC1.

skyllocytosis Phagocytic process in *Allogromia*.

Sl locus (*Steel* locus) Mouse mutant; see ***stem cell factor***.

SL1 Transcription factor composed of four proteins including ***TBP***, required for activity of RNA polymerase I, resembles in some respect bacterial sigma factor.

SLAM (signalling lymphocyte-activation molecule; CDw150) Glycosylated type-I transmembrane protein (70 kD) present on T- and B-cell surfaces that is a high-affinity self-ligand. Triggering of SLAM co-activates T-or B-cell responses.

sleeping sickness See ***Trypanosoma***.

sliding filament model Generally accepted model for the way in which contraction occurs in the sarcomere of striated muscle, by the sliding of the thick filaments relative to the thin filaments.

slime moulds Two distinct groups of fungi, the cellular slime moulds or ***Acrasidae*** that include *Dictyostelium*, and the ***acellular slime moulds*** or Myxomycetes that include *Physarum*.

slot blot A ***dot blot*** in which samples are placed on a membrane through a series of rectangular slots in a template. This is slightly advantageous because hybridization artefacts are usually circular.

slow muscle Striated muscle used for long-term activity (eg. postural support). Depend therefore on oxidative metabolism and have many mitochondria and abundant ***myoglobin***.

slow reacting substance of anaphylaxis (SRS-A) Potent bronchoconstrictor and inflammatory agent released by mast cells; an important mediator of allergic bronchial asthma. A mixture of three ***leukotrienes*** (LTC_4 mainly, LTD_4 and LTE_4)

slow virus (1) Specifically one of the ***Lentivirinae***. (2) Any virus causing a disease that has a very slow onset. Diseases such as subacute ***spongiform encephalopathy***, Aleutian disease of mink,

scrapie, ***kuru***, and ***Creutzfeldt–Jacob disease*** may be caused by slow viruses. See also ***prion***.

SLS collagen (segment long spacing collagen) Abnormal packing pattern of collagen molecules formed if ATP is added to acidic collagen solutions, in which lateral aggregates of molecules are produced. Each aggregate is 300nm long, and the molecules are all in register. If SLS aggregates are overlapped with a quarter-stagger, the 67nm banding pattern of normal fibrils is reconstituted.

Smad proteins Intracellular proteins that mediate signalling from receptors for extracellular TGF-β related factors. Smad2 is essential for embryonic mesoderm formation and establishment of anterior-posterior patterning. Smad4 is important in gastrulation. Smads1 and 5 are activated (serine/threonine phosphorylated) by ***BMP*** receptors, Smad2 and 3 by ***activin*** and TGF-β receptors. Smads activated by occupied receptors then form complexes with Smad4/DPC4 and move into the nucleus where they regulate gene expression. Interact with ***FAST*** 2.

small acid-soluble spore proteins (SASP) DNA-binding proteins in the spores of some bacteria, thought to stabilize the DNA in the ***A-DNA*** configuration, so protecting it from cleavage by enzymes or UV light.

small cell carcinoma Common malignant neoplasm of bronchus. Cells of the tumour have endocrine-like characteristics and may secrete one or more of a wide range of hormones, especially regulatory peptides like ***bombesin***.

small nuclear RNA (snRNA) Abundant class of RNA found in the nucleus of eukaryotes, usually including those RNAs with sedimentation coefficients of 7s or less. They are about 100–300 nucleotides long. Although 5S rRNA and tRNA are of a similar size, they are not normally regarded as snRNAs. Most are found in complexes with proteins (see ***ribonucleoprotein*** particles, SnRNPs) and at least some have a role in processing hnRNA.

SMARC (BRG1-associated factors) The ***SWI/SNF complex***-related, matrix-associated, actin-dependent regulators of chromatin (SMARC), also called BRG1-associated factors, are components of human SWI/SNF-like chromatin-remodelling protein complexes.

Smith–Watermann alignment Algorithm for detecting sequence similarities when searching a genomic database.

smooth endoplasmic reticulum (SER; smooth ER) An internal membrane structure of the eukaryotic cell. Biochemically similar to the ***rough endoplasmic reticulum*** (RER), but lacks the ribosome-binding function. Tends to be tubular rather than sheet-like, may be separate from the RER or may be an extension of it. Abundant in cells concerned with lipid metabolism and proliferates in hepatocytes when animals are challenged with lipophilic drugs.

smooth microsome Fraction produced by ultracentrifugation of a cellular homogenate. It consists of membrane vesicles derived largely from the ***smooth endoplasmic reticulum***.

smooth muscle Muscle tissue in vertebrates made up from long tapering cells that may be anything from 20–500μm long. Smooth muscle is generally involuntary, and differs from striated muscle in the much higher actin:myosin ratio, the absence of conspicuous sarcomeres, and the ability to contract to a much smaller fraction of its resting length. Smooth muscle cells are found particularly in blood vessel walls, surrounding the intestine (particularly the gizzard in birds) and in the uterus. The contractile system and its control resemble those of motile tissue cells (eg. fibroblasts, leucocytes), and antibodies against smooth muscle myosin will cross-react with myosin from tissue cells, whereas antibodies against skeletal muscle myosin will not. See also ***dense bodies***.

smooth strain See ***rough strain***.

SMRT Silencing mediator of retinoic acid and thyroid hormone receptors. Part of a corepressor complex.

SNAP (1) S-nitroso-N-acetyl penicillamine. (2) Soluble ***NSF*** attachment (accessory) protein (25 kD), involved in the control of vesicle transport. α and γ SNAPs are found in a wide range of tissues. βSNAP is a brain-specific isoform of αSNAP. SNAPs bind, together with NSF, to ***SNAREs***.

SNAREs Receptors for ***SNAPs***. The neuronal receptor for vesicle SNAPs, v-SNARE, is ***synaptobrevin***, also known as VAMP-2. The target (t-SNARE) associated with the plasma membrane of the axonal terminal is ***syntaxin***. The SNAP–SNARE complex is apparently responsible for regulating vesicle targeting: neurotoxins such as ***tetanus toxin*** and ***botulinum toxin*** selectively cleave SNAREs or SNAPs. See also ***cellubrevin***.

snoRNA Small nucleolar ribonucleic acid. See ***fibrillarin***.

snRNA See ***small nuclear RNA***.

snRNP Small nuclear ***ribonucleoprotein***.

SOCS (suppressor of cytokine signalling) Family of proteins (SOCS 1-3 and CIS; 211 amino acids) rapidly induced in response to IL-6 and other cytokines, thought to act as negative feedback regulator of ***JAKs*** in the intracellular signalling pathway. All contain a central ***SH2*** domain. Similar, possibly identical, to JAB and SSI-1.

sodium channel (sodium gate) The protein responsible for electrical excitability of neurons. A multi-subunit transmembrane ***ion channel***, containing an aqueous pore around 0.4nm diameter, with a negatively charged region internally (the 'selectivity filter') to block passage of anions. The channel is ***voltage-gated***: it opens in response to a small depolarization of the cell (usually caused by an approaching action potential), by a multistep process. Around 1000 sodium ions pass in the next millisecond, before the channel spontaneously closes (an event with single-step kinetics). The channel is then refractory to further depolarizations until returned to near the ***resting potential***. There are around 100 channels per μm^2 in unmyelinated axons; in myelinated axons, they are concentrated at the ***nodes of Ranvier***. The sodium channel is the target of many of the deadliest ***neurotoxins***.

sodium dodecyl sulphate See ***SDS***.

sodium pump See ***sodium-potassium ATPase***.

sodium-potassium ATPase (Na^+,K^+ATPase, sodium pump) A major transport protein of the plasma membrane that moves three sodium ions out of the cell and two potassium ions in for each ATP hydrolyzed. A member of the ***P-type ATPases***. The sodium gradient established is used for several purposes (see ***facilitated diffusion***, ***action potential***), while the potassium gradient is dissipated through the potassium leak channel. Must not be confused with a ***sodium channel***.

soft agar Semi-solid agar used to gelate medium for culture of animal cells. Placed in such a medium, over a denser agar layer, the cells are denied access to a solid substratum on which to spread, so that only cells that do not show ***anchorage dependence*** (usually transformed cells) are able to grow.

sol-gel transformation Transition between more fluid cytoplasm (***endoplasm***) and stiffer gel-like ***ectoplasm*** proposed as a mechanism for amoeboid locomotion: since the endoplasm cannot really be considered a simple fluid and has viscoelastic properties like a gel, the term is misleading.

Solanum tuberosum The potato.

solenoid See ***beta helix***.

somatic cell Usually any cell of a multicellular organism that will not contribute to the production of gametes, ie. most cells of which an organism is made: not a ***germ cell***. Notice, however, the alternative use in ***somatic mesoderm***.

somatic cell genetics Method for identifying the chromosomal location of a particular gene without sexual crossing. Unstable ***heterokaryons*** are made between the cell of interest and another cell with identifiably different characteristics (or without the gene in question), and a series of clones isolated. By correlating retention of gene expression with the remaining chromosomes, it is possible to deduce which chromosome must carry the gene. Human–mouse heterokaryons have been extensively used in this sort of work.

somatic hybrid ***Heterokaryon*** formed between two somatic cells, usually from different species. See ***somatic cell genetics***.

somatic mesoderm The portion of the embryonic mesoderm that is associated with the body wall and is divided from the splanchnic (visceral) mesoderm by the coelomic cavity.

somatic mutation ***Mutation*** that occurs in the somatic tissues of an organism, and that will not, therefore, be heritable, since it is not present in the ***germ cells***. Some neoplasia is due to somatic mutation; a more conspicuous example is the reversion of some branches of variegated shrubs to the wild-type (completely green) phenotype. Somatic mutation is probably also important in generating diversity in V-gene regions of immunoglobulins.

somatic recombination One of the mechanisms used to generate diversity in antibody production is to rearrange the DNA in B-lymphocytes during their differentiation, a process that involves cutting and splicing the immunoglobulin genes. Somatic recombination via homologous crossing over occurs at a low frequency in *Aspergillus, Drosophila* and *Saccharomyces* and in mammalian cells in culture. It may be detected through the production of homozygous patches or sectors after mitosis of cells heterozygous for suitable marker genes.

somatocrinin Peptide (44 residues) with high growth hormone releasing activity. Can be isolated from rat hypothalamus and some human pancreatic tumours. Acts on adenylate cyclase.

somatomedin Generic term for ***insulin-like growth factors*** (IGFs) produced in the liver and released in response to ***somatotropin***. Somatomedins stimulate the growth of bone and muscle, and also influence calcium, phosphate, carbohydrate and lipid metabolism. Somatomedin A is IGF-II, somatomedin C is IGF-I. Somatomedin B is a serum factor of uncertain function.

somatostatin Gastrointestinal and hypothalamic peptide hormone (two forms: 14 and 28 residues); found in gastric mucosa, pancreatic islets, nerves of the gastrointestinal tract, in posterior pituitary and in the central nervous system. Inhibits gastric secretion and motility: in hypothalamus/pituitary inhibits somatotropin release. See also Table H2.

somatotropin See ***growth hormone***.

somites Segmentally arranged blocks of mesoderm lying on either side of the notochord and neural tube during development of the vertebrate embryo. Somites are formed sequentially, starting at the head. Each somite will give rise to muscle (from the myotome region), spinal column (from the sclerotome) and dermis (from dermatome).

son of sevenless *Drosophila* ras-GRF (GDF releasing factor), mammalian homologues of which (sos1, sos2) play an important part in intracellular signalling. See ***sos***.

sonic hedgehog (Shh) Secreted protein that is involved in organization and patterning of several vertebrate tissues during development. The zone of polarizing activity (***ZPA***) that determines anterior–posterior patterning of the limb expresses sonic hedgehog.

sorbitol (glucitol) The polyol (polyhydric alcohol) corresponding to glucose. Occurs naturally in some plants, is used as a growth substrate in some tests for bacteria, and is sometimes used to maintain the tonicity of low ionic strength media.

sorbose A monosaccharide hexose: L-sorbose is an intermediate in the commercial synthesis of ***ascorbic acid***.

sorocarp Fruiting body formed by some

cellular ***slime moulds***; has both stalk and spore-mass.

sorting out Phenomenon observed to occur when mixed aggregates of dissimilar embryonic cell types are formed ***in vitro***. The original aggregate sorts out so that similar cells come together into homotypic domains, usually with one cell type sorting out to form a central mass that is surrounded by the other cell type. Much controversy has arisen over the years as to the underlying mechanism, whether there is specificity in the adhesive interactions (which would imply tissue-specific receptor–ligand interactions), or whether it is sufficient to suppose that there are quantitative differences in homo-and hetero-typic adhesion (the ***differential adhesion*** hypothesis). With the exception perhaps of the main protagonists, most cell biologists consider that there are probably elements both of tissue specificity (***CAMs***) and of quantitative adhesive differences involved.

sorus A group of ***sporangia*** or spore cases, eg. on the underside of fern leaves.

sos Guanine nucleotide-releasing factor (155kD), the mammalian homologue of ***son of sevenless***. The proline-rich region of sos binds to the ***SH3*** domain of ***GRB2***. Has homology with CDC-25, the yeast GTP-releasing factor for ras. A family of related proteins are now known and include vav, ***C3G***, Ost, NET1, Ect2, RCC1, tiam, RalGDS and ***Dbl***.

SOS system The DNA repair system also known as error-prone repair in which apurinic DNA molecules are repaired by incorporation of a base which may be the wrong base but that permits replication. RecA protein is required for this type of repair. SOS genes function in control of the cell cycle in pro- and eu-karyotes.

Southern blots See ***blotting***. Originally developed by Dr Ed Southern, hence the name.

sox (1) SoxR: Redox sensory protein in *E. coli*. (2) SOX syndrome: Sialadenitis, osteoarthritis and xerostomia syndrome. (3) *Sox* genes: Gene family involved in many developmental processes. *Sox*-2 regulates transcription of FGF-4 gene, *sox*-3 is involved in neural tube closure and lens specification, *sox*-9 is related to ***sry*** and found in mouse testis, *sox*-10 is important in neural crest development.

soybean trypsin inhibitor (STI; SBTI) Single polypeptide (21 kD; 181 amino acids) that forms a stable, stoichiometric, enzymically inactive complex with trypsin.

spacer DNA The DNA sequence between genes. In bacteria, only a few nucleotides long. In eukaryotes, can be extensive and include ***repetitive DNA***, comprising the majority of the DNA of the genome. The term is used particularly for the spacer DNA between the many tandemly repeated copies of the ribosomal RNA genes.

SPARC See ***osteonectin***.

sparsomycin Antibiotic that inhibits peptidyl transferase in both prokaryotes and eukaryotes.

spasmin Protein (20 kD) that forms the ***spasmoneme***. Thought to change its shape when the calcium ion concentration rises, and to revert when the calcium concentration falls: the reversible shape change is used as a motor mechanism. Contraction does not require ATP, relaxation does, probably to pump calcium ions back into the ***smooth endoplasmic reticulum***.

spasmoneme Contractile organelle found in *Vorticella* and related ciliate protozoans. Capable of shortening faster than any actin–myosin system, and of expanding actively. See ***spasmin***.

spatial sensing Mechanism of sensing a gradient in which the signal is compared at different points on the cell surface and cell movement directed accordingly. Translocation of all or part of the cell is not required. See ***temporal*** and ***pseudospatial gradient sensing***.

specific activity The number of activity units (whatever is appropriate) per unit of mass, volume or molarity. Perhaps most often encountered in the context of radiochemicals, the number of microcuries per micromole.

specific granules (secondary granules) One of the two main classes of granules found in neutrophils: contain lactoferrin, lysozyme, vitamin B12-binding protein and elastase. Are released more readily than the ***azurophil*** (primary) granules which have typical lysosmal contents.

spectinomycin Aminocyclitol antibiotic: acts on ribosome, but is bacteriostatic rather than bactericidal.

spectrin Membrane-associated dimeric protein (240 and 220 kD) of erythrocytes. Forms a complex with ***ankyrin, actin,*** and probably other components of the 'membrane cytoskeleton', so that there is a meshwork of proteins underlying the plasma membrane, potentially restricting the lateral mobility of integral proteins. Isoforms have been described from other tissues (***fodrin, TW-240/260 kD protein***), where they are assumed to play a similar role. Contains the ***EF-hand*** motif.

spectrophotometry Quantitative measurements of concentrations of reagents made by measuring the absorption of visible, ultraviolet or infrared light.

Spemann's organizer Signalling region located on the dorsal lip of the blastopore in the early embryo, essential for defining the main body axis.

speract Sea urchin peptide hormone, from the jelly coat of the eggs of the sea-urchins *Strongylocentrotus purpuratus* and *Hemicentrotus pulcherrinus*, affecting motility and metabolism. Receptor is a plasma-membrane ***guanylate cyclase***.

spermatids The haploid products of the second meiotic division in spermatogenesis. Differentiate into mature spermatozoa.

spermatocytes Cells of the male reproductive system that undergo two meiotic divisions to give haploid spermatids.

spermatogenesis The process whereby primordial germ cells form mature spermatozoa.

spermatogonium Plant gonad cell that undergoes repeated mitoses, leading to the production of spermatocytes.

Spermatophyte Division of the plant kingdom, consisting of plants that reproduce by means of seeds.

spermatozoon Mature sperm cell (male gamete).

spermidine (N-(3-aminopropyl)-1,4-butanediamine)) A polybasic amine (polyamine); see ***spermine***.

spermine (N, N' bis(3-aminopropyl)-1, 4-butanediamine)) Polybasic amine (polyamine). Found in human sperm, in ribosomes and in some viruses. Involved in nucleic acid packaging. Synthesis is regulated by ornithine decarboxylase which plays a key role in control of DNA replication.

SPF (specific pathogen free) Animals that have been raised in carefully controlled conditions so that they are not infected with any known pathogens. May require that they are delivered by Caesarean section and raised in strict quarantine.

spherical aberration Deficiency in simple lenses in which the image is sharp in the centre but out of focus at the periphery of the field, more a problem when taking photographs than when observing directly. Lenses compensated for this defect are referred to as plan-lenses (eg. ***planapochromat***).

spherocytosis A condition in which erythrocytes lose their biconcave shape and become spherical. It occurs as cells age, and is also found in individuals with abnormal cytoskeletal proteins (hereditary spherocytosis, a disorder that leads to haemolytic anaemia).

spheroplast Bacterium from which the cell wall has been removed but that has not lysed.

spherosome Lysosome-like compartment in plants that derives from the ***endoplasmic reticulum*** and is a site for lipid storage.

sphingolipid Structural lipid of which the parent structure is sphingosine rather than glycerol. Synthesized in the Golgi complex.

sphingomyelin A sphingolipid in which the head group is phosphoryl choline. A close analogue of phosphatidyl choline. In many cells the concentration of sphingomyelin and phosphatidyl choline in the plasma membrane seems to bear a reciprocal relationship.

sphingosine Long-chain amino alcohol that bears an approximate similarity to glycerol with a hydrophobic chain attached to the 3-carbon. Forms the class of sphingolipids when it carries an acyl group joined by an amide link to the nitrogen. Forms sphingomyelin when phosphoryl choline is attached to the 1-hydroxyl group. Gives rise to the cerebroside and ganglioside classes of glycolipids when oligosaccharides are attached to the 1-hydroxyl group. Not found in the free form.

Spi-1 Protooncogene encoding a transcription factor (PU1) that binds to purine-rich sequences (PU boxes) expressed in haematopoietic cells.

spin labelling The technique of introducing a grouping with an unpaired electron to act as an electron spin resonance (ESR) reporter species. This is almost invariably a nitroxide compound (-N-O) in which the nitrogen forms part of a sterically hindered ring.

spinal cord Elongated, approximately cylindrical part of the central nervous system of vertebrates that lies in the vertebral canal, and from which the spinal nerves emerge.

spinal ganglion (dorsal root ganglion) Enlargement of the dorsal root of the spinal cord containing cell bodies of afferent spinal neurons. Neural outgrowth from dorsal root ganglia has been studied extensively ***in vitro***.

spindle See ***mitosis***.

spindle fibres Microtubules of the spindle that interdigitate at the equatorial plane with microtubules of the opposite polarity derived from the opposite pole ***microtubule organizing centre***. Usually distinguished from kinetochore fibres which are microtubules that link the poles with the kinetochore, although these could be included in a broader use of the term.

spinner culture Method for growing large numbers of cells in suspension by continuously rotating the culture vessel.

spiral cleavage Pattern of early cleavage found in molluscs and annelids (both ***mosaic eggs***). The animal pole blastomeres are rotated with respect to those of the vegetal pole. The handedness of the spiral twist shows ***maternal inheritance***.

spirillum A fairly rigid helically twisted bacterial cell often, but not necessarily, a member of the genus *Spirillum*.

spirochaete (USA, spirochete) An elongated, spiral-shaped bacterium, eg. the organism responsible for syphilis.

Spirogyra Genus of green filamentous algae found in freshwater ponds. Contain helically disposed ribbon-like chloroplasts.

spironolactone ***Aldosterone*** antagonist: diuretic; used to treat low-***renin*** hypertension and Conn's syndrome (in which there is overproduction of aldesterone).

Spirostomum Genus of large free-living ciliate protozoans with an elongated body.

splanchnic Relating to the viscera. See also ***splanchnic mesoderm***.

splanchnic mesoderm That portion of the embryonic ***mesoderm*** that is associated with the inner (endodermal) part of the body, in contrast to ***somatic mesoderm*** which is associated with the body wall. The two mesodermal regions are separated by the ***coelom***.

splenocytes Phagocytic cells (macrophages) of the spleen are usually meant by this vague term.

spliceosome A complex of small nuclear RNA/protein particles (snRNPs- 'snurps') that participate in hnRNA splicing.

splicing The process by which introns are removed from hnRNA to produce mature messenger RNA that contains only exons. Alternative splicing seems to occur in many proteins and by alternative exon usage a set of related proteins can be generated from one gene, often in a tissue or developmental stage-specific manner (***alternative splicing***).

split gene See ***introns***.

split ratio The fraction of the cells in a fully grown culture of animal cells that should be used to start a subsequent culture. Minimum may be dictated by inadequacies of the medium that result in poor growth of some cells at high dilution.

spokein Constituent protein of the ***radial spokes*** of the ciliary ***axoneme***. Since a number of complementary spoke mutants are known to occur in *Chlamydomonas*, and one mutant lacks 17 proteins, it seems likely that spokein is a complex mixture.

spongiform encephalopathy Any disease characterized by long incubation and fatal progressive course with characteristic spongiform degeneration of grey matter of the cortex. The two main human diseases are ***kuru*** and ***Creutzfeldt–Jakob disease***. Diseases such as ***scrapie***, mink encephalopathy and bovine spongiform encephalopathy (BSE) are considered to be similar. Controversy still surround the causative agent; the two main theories being slow viruses or ***prions***. See also ***Gerstmann–Straussler– Scheinker syndrome***.

spongioblast Cell found in developing nervous system: gives rise to ***astrocytes*** and ***oligodendrocytes***.

spongiocytes Lipid droplet-rich cells from the middle region of the cortex of the adrenal gland.

spongy parenchyma Tissue usually found in the lower part of the leaf ***mesophyll***. Consists of irregularly shaped, photosynthetic parenchyma cells, separated by large air spaces.

spontaneous transformation ***Transformation*** of a cultured cell that occurs without the deliberate addition of a transforming agent. Cells from some species, especially rodents, are particularly prone to such spontaneous transformation.

sporadic Of a ***tumour*** or genetic disease, a novel occurrence without any previous family history of the disease (cf. inherited). Examples of diseases with both sporadic and inherited forms: ***retinoblastoma, Wilms' tumour***.

sporangium (plural, sporangia) Spore case within which asexual spores are produced.

spore Highly resistant dehydrated form of reproductive cell produced under conditions of environmental stress. Usually have very resistant cell walls (integument) and low metabolic rate until activated. Bacterial spores may survive quite extraordinary extremes of temperature, dehydration or chemical insult. Gives rise to a new individual without fusion with another cell.

sporocarp Multicellular structure in fungi, lichens, ferns or other plants. Location of spore formation.

sporophyte Spore-producing plant generation. The dominant generation in pteridophytes and higher plants, and alternates with the gametophyte generation.

sporopollenin Polymer of ***carotenoids***, found in the exine of the pollen wall. Extremely resistant to chemical or enzymic degradation.

Sporozoa Class of spore-forming parasitic protozoa without cilia, flagella or pseudopodia.

spot desmosome Macula adherens: see ***desmosome***.

sprouting Production of new processes (outgrowths) by nerve cells, eg. by embryonic neurons undergoing primary differentiation, by adult neurons in response to nervous system damage, or by dissociated neurons redifferentiating in culture.

squames Flat, keratinized, dead cells shed

from the outermost layer of a squamous *stratified epithelium*.

squamous epithelium An epithelium in which the cells are flattened. May be a *simple epithelium* (eg. *endothelium*) or a *stratified epithelium* (eg. *epidermis*).

squamous cell carcinoma Slow-growing carcinoma that develops from the squamous layer of the epithelium.

squid giant axon Large axons, up to 1mm in diameter, that innervate the mantle of the squid. Because of their large size, many of the pioneering investigations of the mechanisms underlying resting and action potentials in excitable cells were done on these fibres.

squidulin Calcium-binding protein from the optic lobe of squid, which contains the *EF-hand* motif.

SR proteins Family of splicing factors, highly conserved in metazoans. Contain one or two RNA-binding domains and region enriched in arginine/serine repeats (RS domain) at C terminus.

SRBC Sheep red blood cells.

src family Family of protein tyrosine kinases of which *src* was the first example. Includes Fyn, Yes, Fgr, Lyn, Hck, Lck, Blk and Yrk. All cells studied so far have one at least of these kinases which act in cellular control. Family members all have characteristic src-homology (SH) domain structure, a kinase domain (SH1), SH2 and SH3 domains, and a domain (SH4) which has myristoylation and membrane-localization sites. Interdomain interactions, themselves regulated by phosphorylation (see *Csk*), regulate the activity of the kinase.

src **gene** The transforming (sarcoma-inducing) gene of Rous sarcoma virus. Protein product is pp60vsrc, a cytoplasmic protein with tyrosine-specific *protein kinase* activity, that associates with the cytoplasmic face of the plasma membrane.

SRE See *serum response element*.

SRF See *serum response factor*.

SRP See *signal recognition particle*.

SRS-A See *slow reacting substance of anaphylaxis*.

SRTX See *sarafotoxin*.

sry (sex-related gene on Y) Family of high-mobility group (*HMG*) proteins that bind to a subset of sequences recognized by C/EBP family of DNA-binding proteins. *Sry* itself is the primary testis-determining gene and located, as its name suggests, on the Y chromosome. *Sox*-10 is related to Sry.

SSCP See *single-stranded conformational polymorphism*.

ssDNA phage Single-stranded DNA phages such as MS2, FX174, as opposed to double-stranded DNA phages or RNA phages.

SSI-1 (STAT-induced STAT inhibitor 1) Protein induced by cytokines (through *STAT* pathway) that binds to and inhibits JAK2 and Tyk2 thereby acting as a negative feedback signal. See *SOCS*.

stable transfection (stable expression) In *transfection* of animal cells, a clone of cells in which the *transgene* has been physically incorporated into the genome. It thus provides stable, long-term expression at the cost of being more difficult to produce.

stachyose Digalactosyl-sucrose, a compound involved in carbohydrate transport in the phloem of many plants, and also in carbohydrate storage in some seeds.

stanniocalcin Glycoprotein hormone secreted by the corpuscle of Stannius, an endocrine gland in teleosts. Prevents hypercalcaemia and inhibits calcium uptake through gills. Mammalian homologues are reported.

Staphylococcal α toxin (leucocidin) Pore-forming exotoxin (33 kD) secreted by *Staphylococcus aureus*. Protein (monomer) has two domains connected by flexible hinge region: oligomerizes in the plasma membrane by lateral mobility to form a hexameric oligomer (220 kD) that has a pore approximately 1nm in diameter with some anion selectivity. Osmotic lysis leads to death of the cell. Has been used to selectively permeabilize cells to small molecules.

Staphylococcal δ toxin Small peptide exotoxin (26 residues) secreted by *Staphylococcus aureus*. Binds to membranes and a range of cellular components. Very amphipathic and surface active. Properties, though not sequence, very similar to ***melittin***.

staphylococcins ***Bacteriocins*** produced by staphylococci.

Staphylococcus Genus of non-motile ***Gram-positive bacteria*** that are found in clusters, and which produce important exotoxins (see Table E2). *Staphylococcus aureus* (*S. pyogenes*) is pyogenic, an opportunistic pathogen, and responsible for a range of infections. It has ***protein A*** on the surface of the cell wall. Coagulase production correlates with virulence: hyaluronidase, lipase and ***staphylokinase*** are released in addition to the toxins.

staphylokinase Enzyme released by ***Staphylococcus*** *aureus* that acts as a ***plasminogen activator***.

starch Storage carbohydrate of plants, consisting of ***amylose*** (a linear α (1–4)-glucan) and ***amylopectin*** (an α (1–4)-glucan with α (1–6)-branch points). Present as starch grains in plastids, especially in ***amyloplasts*** and ***chloroplasts***.

start codon See ***initiation codon***.

start site Vague, as can refer to either a ***transcriptional*** start site or a ***translational*** start site.

statherin A low molecular weight (5380 D, 43 amino acid residues) acidic tyrosine-rich phosphoprotein secreted mainly by salivary glands. Binds ***fimbrillin***.

stathmin (oncoprotein 18; Op18) Ubiquitous coiled-coil cytosolic phosphoprotein (19 kD) that interacts with tubulin heterodimers and increases the rate of rapid ('catastrophic') disassembly of microtubules. Overexpressed in some tumours and probably regulated by phosphorylation.

statocyst An organ for the perception of gravity and thus body orientation, found in many invertebrate animals; a cavity lined with sensory cells and containing a ***statolith***.

statocyte A root-tip cell containing one or more ***statoliths***, involved in the detection of gravity in geotropism.

statolith (1) *Bot.* A type of ***amyloplast*** found in root-tip cells of higher plants. It can sediment within the cell under the influence of gravity, and is thought to be involved in the detection of gravity in geotropism. (2) *Zool.* A sand-grain or a structure of calcium carbonate or other hard secreted substance, found in the cavity of a ***statocyst***. It stimulates sensory cells lining the cavity with which it comes in contact under the influence of gravity.

STATs (signal transducers and activators of transcription) Contain ***SH2*** domains that allow them to interact with phosphotyrosine residues in receptors, particularly cytokine-type receptors; they are then phosphorylated by ***JAKs***, dimerize and translocate to the nucleus where they act as transcription factors. Many STATs are known; some are relatively receptor-specific, others more promiscuous, so that a wide range of responses is possible with some STATs being activated by several different receptors, sometimes acting synergistically with other STATs.

staurosporine Inhibitor of PKC-like protein kinases derived from *Streptomyces* sp. Has a rather broad inhibitory spectrum and cannot be used to ascribe a specific role for protein kinase C in a signalling pathway.

stearic acid (n-octadecanoic acid) See ***fatty acids***.

steatoblasts Cells that give rise to fat cells (adipocytes).

steel factor Murine equivalent of ***stem cell factor***.

Steel locus See ***Sl locus***.

stefin (cystatin) Family of cysteine proteinase inhibitors. Stefin A has 98 residues and inhibits cathepsins D, B, H and L.

stem cell (1) Cell that gives rise to a lineage of cells. (2) More commonly used of a cell that, upon division, produces dissimilar daughters, one replacing the original stem cell, the other differentiating further (eg. stem cells in basal layers of skin, in haematopoietic tissue and in meristems).

stem cell factor (SCF; Steel factor in mice; mast cell growth factor; c-kit ligand) Haematopoietic growth factor, 18.6 kD from sequence; found as dimer (35 kD protein, 53 kD in its glycosylated form).

stem-and-loop structure The structure of tRNAs is so termed because it has four base-paired stems and three loops (not base-paired), one of which contains the anticodon.

stenohaline Descriptive of an organism that is unable to tolerate a range of salinities.

stereocilium (plural, stereocilia) Microfilament bundle-supported projection, several microns long, from the apical surface of sensory epithelial cells (***hair cells***) in inner ear: like a ***microvillus***, but larger. It is stiff and may act as a transducer directly, or merely restrict the movement of the sensory cilium (which does have an axoneme). Also described on cells of pseudo-***stratified epithelium*** of the epididymal duct. Recently, stereocilia have been referred to as stereovilli, a much better and less confusing name.

stereovillus (plural, stereovilli) Better name for ***stereocilium***.

steroid finger motif See ***steroid receptor***.

steroid hormone A group of structurally related hormones, based on the cholesterol molecule. They control sex and growth characteristics, are highly lipophilic, and are unique in that their receptors are in the nucleus, rather than on the plasma membrane. Examples: ***testosterone***, ***oestrogen***.

steroid receptor Family of nuclear ***transcription factors***, most of which are receptors for hormones of the steroid family, for example androgen, oestrogen, glucocorticoid, mineralocorticoid, progesterone, retinoic acid, ecdysone, thyroid hormone; and the *Drosophila* transcription factors knirps, ultraspiracle and seven-up. This family contains a conserved domain (the 'steroid finger' motif) containing 2 C4-type ***zinc fingers***.

steroid response element DNA sequence that is recognized and bound by a ***steroid receptor***.

sterols Molecules that have a 17-carbon steroid structure, but with additional alcohol groups and side chains. Commonest example is ***cholesterol***.

sticky ends The short stretches of single-stranded DNA produced by cutting DNA with ***restriction endonucleases*** whose site of cleavage is not at the axis of symmetry. The cut generates two complementary sequences that will hybridize (stick) to one another or to the sequences on other DNA fragments produced by the same restriction endonuclease.

stimulus–secretion coupling A term used to describe the events that link receipt of a stimulus with the release of materials from membrane-bounded vesicles (the analogy is with excitation–contraction coupling in the control of muscle contraction). A classical example is the link between membrane depolarization at the presynaptic terminal and the release of neurotransmitter into the synaptic cleft.

sting cells *Nematocysts* of coelenterates.

stipe A stalk, especially of fungal fruiting bodies or of large brown algae.

STK (1) Stem cell-derived tyrosine kinase, one of the hepatocyte growth factor receptor family and the murine homologue of the human RON receptor tyrosine kinase. Expressed on macrophages. Ligand is macrophage-stimulating protein (***MSP***), a serum protein activated by the coagulation cascade. (2) Soluble tyrosine kinase: STK-1 is a soluble form of c-*src* underphosphorylated on C-terminal tyrosine residues; STK-2 is a 48 kD protein tyrosine kinase molecularly and functionally related to ***Csk***. (3) ***Streptokinase***.

stochastic Random or probabilistic event.

stoichiometry Ratio of the participating molecules in a reaction; in the case of an enzyme–substrate or a receptor–ligand interaction, it should be a small integer.

Stoke's radius Stoke's law of viscosity defines the frictional coefficient for a particle moving through a fluid, a coefficient that depends upon the viscosity of the fluid and the radius of the particle. The apparent radius of a molecule sedimenting under centrifugal force calculated from this law (the Stoke's radius) is a feature of the tertiary structure and thus informative about the molecule in question.

stoma (plural, stomata) Pore in the epidermis of leaves and some stems, which permits gas exchange through the epidermis. Can be open or closed, depending upon the physiological state of the plant. Flanked by stomatal ***guard cells***.

stone cell See ***sclereid***.

stop codon See ***termination codon***.

STOPS (stable tubulin-only proteins) A family of calcium calmodulin regulated microtubule-stabilizing proteins that confer cold-stability on microtubules. Subunit seems to be 145kD but may bind as multimer.

storage diseases Another name for ***lysosomal diseases***.

storage granules (1) Membrane-bounded vesicles containing condensed secretory materials (often in an inactive, zymogen, form). Otherwise known as zymogen granules or condensing vacuoles. (2) Granules found in plastids, or in cytoplasm; assumed to be 'food reserves', often of glycogen or other carbohydrate polymer.

strain birefringence See ***birefringence***.

stratified epithelium An epithelium composed of multiple layers of cells, only the basal layer being in contact with the basal lamina (see ***basement membrane***). The basal layer is of ***stem cells*** that divide to produce the cells of the upper layers; in skin, these become heavily keratinized before dying and being shed as squames. Stratified epithelia usually have a mechanical/protective role.

stratum corneum Outermost layer of skin, composed of clear, dead, scale-like cells with little remaining except ***keratin***.

streptavidin Analogue of avidin. A protein isolated from *Streptomycetes avidinii* that has a high affinity for biotin. Used to detect biotin markers.

streptococcal toxins Group of haemolytic ***exotoxins*** released by ***Streptococcus*** spp. α-haemolysin: 26–39 Kd (four types), forms ring-like structures in membranes (see ***streptolysin O***). Lipid target unclear. β-haemolysin: a hot–cold ***haemolysin*** with sphingomyelinase C activity. γ-haemolysin: complex of two proteins (29 and 26 kD) that act synergistically; rabbit erythrocytes particularly sensitive. δ-toxin: heat-stable peptide (5 kD) with high proportion of hydrophobic amino acids. Seems to act in a detergent-like manner (cf. ***subtilysin***), but may form hydrophilic transmembrane pores by cooperative interaction with other δ-toxin molecules. Leucocidin (Panton–Valentine leucocidin) has two components: F (fast migration on CM-cellulose column: 32 kD); and S (slow: 38 kD). Mode of action contentious. See also ***Streptococcus, streptolysins O and S, erythrogenic toxin***.

streptococcins ***Bacteriocins*** released by ***streptococci***.

Streptococcus Genus of ***Gram-positive*** cocci that grow in chains. Some species (*Strep. pyogenes* in particular) are responsible for important diseases in humans (pharyngitis, scarlet fever, rheumatic fever): *Strep. pneumoniae* is the main culprit in lobar- and bronchopneumonia. Streptococci have anti-phagocytic components (hyaluronic acid-rich capsule and ***M-protein***), and release various toxins (***streptolysins*** O and S, ***erythrogenic toxin***) and enzymes (***streptokinase, streptodornase***, hyaluronidase and proteinase). α-haemolytic streptococci (viridans streptococci) produce limited haemolysis on blood agar; they include *S. mutans, S. salivarius, S. pneumoniae*. β-haemolytic streptococci, of which *S. pyogenes* is the only species, though there are many serotypes, produce a broad zone of almost complete haemolysis on blood agar as a result of streptolysin O and S release. γ-streptococci are non-haemolytic (eg. *S. faecalis*).

streptodornase Mixture of four DNAases released by streptococci. By digesting DNA released from dead cells the enzyme reduces the viscosity of pus and allows the organism greater motility.

streptokinase ***Plasminogen activator*** released by *Streptococcus pyogenes*. Occurs in two forms, A and B.

streptolydigin Antibiotic that blocks peptide chain elongation by binding to the polymerase.

streptolysin O Oxygen-labile thiol-activated haemolysin (native toxin is 61 kD and is cleaved to form 55 kD fragment that retains activity). Haemolysis is inhibited by cholesterol, and only cells with cholesterol in their membranes are susceptible. Toxin aggregates are linked to cholesterol to form a channel 30nm diameter in the membrane, and non-osmotic lysis follows. Markedly inhibits neutrophil movement and stimulates secretion, but has little effect on monocytes.

streptolysin S Thought to be a peptide of 28 residues: causes zone of β-haemolysis around streptococcal colonies on blood agar. Mechanism of haemolysis unclear. Toxic to leucocytes, platelets and several cell lines.

Streptomyces Genus of ***Gram-positive*** spore-forming bacteria that grow slowly in soil or water as a branching filamentous mycelium similar to that of fungi. Important as the source of many antibiotics, eg. ***streptomycin, tetracycline, chloramphenicol, macrolides***.

streptomycin Commonly used antibiotic in cell culture media: acts only on prokaryotes, and blocks transition from ***initiation complex*** to chain-elongating ribosome. Isolated originally from a soil streptomycete.

streptovaricins Antibiotics that block initiation of transcription in prokaryotes. (cf. ***rifamycins*** and ***rifampicin***).

streptozotocin Methyl nitroso-urea with a two-substituted glucose, used as an antibiotic (effective against growing ***Gram-positive*** and ***Gram-negative*** organisms), and also to induce a form of diabetes in experimental animals (rapidly induces pancreatic ***B-cell*** necrosis if given in high dose). By using multiple low doses in a particular strain of mice, it is possible to produce insulitis followed later by diabetes, a model for juvenile-onset diabetes in humans.

stress-fibres Bundle of microfilaments and other proteins found in fibroblasts, particularly slow-moving fibroblasts cultured on rigid substrata. Shown to be contractile; have a periodicity reminiscent of the ***sarcomere***. Anchored at one end to a ***focal adhesion***, although sometimes seem to stretch between two focal adhesions.

stress-induced proteins Alternative and preferable name for ***heat-shock proteins*** of eukaryotic cells, which emphasizes that the same small group of proteins is stimulated both by heat and various other stresses.

striated border Obsolete term for the apical surface of an epithelium with microvilli.

striated muscle Muscle in which the repeating units (***sarcomeres***) of the contractile ***myofibrils*** are arranged in register throughout the cell, resulting in transverse or oblique striations observable at the level of the light microscope, eg. the voluntary (skeletal) and cardiac muscle of vertebrates.

stringency (low stringency; high stringency; stringency wash) In nucleic acid ***hybridization***, the labelled ***probe*** is used to label matching sequences by base-pairing. Unbound probe is removed though a series of stringency washes. Low stringency washing (low temperature, high ionic strength), allows some mismatching of probe and target, and thus allows the detection of similar sequences at some cost in specificity. By contrast, high stringency conditions allow only closely matching sequence to remain base-paired.

stroma (1) The soluble, aqueous phase within the chloroplast, containing water-soluble enzymes such as those of the ***Calvin–Benson cycle***. The site of the ***dark reaction*** of photosynthesis. (2) Loose connective tissue with few cells.

stromal cell Resident cell of loose fibrous connective tissue. Relatively non-committal term.

stromelysin Metalloproteinase involved in breaking down the extracellular matrix.

strong promoter Promoter which when bound by a ***transcription factor***, strongly activates expression of the associated gene. The term is widely used, but lacks precision.

Strongylocentrotus purpuratus Common sea urchin. Echinoderms are popular tools for developmental biology because the early embryo is transparent and cell movements can easily be observed. Sea urchin eggs are available in large numbers and were a convenient material for early biochemical studies on histones and mRNA. Because they are relatively large they have also been convenient for various electrophysiological studies.

strophanthin Mixture of glycosides from *Strophanthus kombe* that has properties similar to digoxin and ***ouabain***. See ***digitalis***.

Strophanthin G See ***ouabain***.

structural gene A gene that codes for a product (eg. an enzyme, structural protein, tRNA), as opposed to a gene that serves a regulatory role.

structure–activity analysis (structure activity relationship; SAR) Study in which systematic variation in the structure of a compound is correlated with its activity, in an attempt to determine the characteristics of the (receptor) site at which it acts.

strychnine ***Alkaloid*** obtained from the Indian tree *Strychnos nux-vomica*; specific blocking agent for the action of the amino acid transmitter glycine. Convulsive effects of strychnine are probably due to its blockage of inhibitory synapses onto spinal cord motoneurons.

STX See ***saxitoxin***.

Stylonychia mytilus Large ciliate protozoan of the order Hypotrichida, which has compound cilia (cirri) that can be used for walking or swimming.

subacute Description of a disease that progresses more rapidly than a chronic disease and more slowly than an acute one.

subacute sclerosing panencephalitis (SSPE) Chronic progressive illness seen in children a few years after measles infection, and involving demyelination of the cerebral cortex. Virus apparently persists in brain cells: usually considered a ***slow virus*** disease.

suberin Fatty substance, containing long-chain fatty acids and fatty esters, found in the cell walls of cork cells (phellem) in higher plants. Also found in the ***Casparian band***. Renders the cell wall impervious to water.

subfragment 1 of myosin See ***S1***.

submitochondrial particle Formed by sonicating mitochondria. Small vesicles in which the inner mitochondrial membrane is inverted to expose the innermost surface.

substance P A ***vasoactive intestinal peptide*** (1348 D) found in the brain, spinal ganglia and intestine of vertebrates. Induces vasodilation, salivation and increases capillary permeability. Sequence: RPKPQFFGLM.

substantia nigra Area of darkly pigmented dopaminergic neurons in the ventral midbrain thought to control movement and damaged in ***Parkinsonism***.

substrate Substance that is acted upon by an enzyme: one can also speak of a suitable substrate for maintaining a species of bacterium – the compound is one that can support cell growth.

substratum The solid surface over which a cell moves, or upon which a cell grows: should be used in this sense in preference to ***substrate***, to avoid confusion.

subtilisin Extracellular serine protease produced by *Bacillus* spp.

subtilysin Haemolytic surfactant produced by *Bacillus subtilis*; hexapeptide linked to a long-chain fatty acid.

subtraction cloning See ***subtractive hybridization***.

subtractive hybridization (subtraction cloning) Technique used to identify genes expressed differentially between two tissue samples. A large excess of ***mRNA*** from one sample is hybridized to cDNA from the other, and the double stranded hybrids removed by physical means. Remaining cDNAs are those not represented as RNA in the first sample, and thus presumably expressed uniquely in the second. To improve specificity, the process is often repeated several times. See also ***differential screening***.

subunits Components from which a structure is built; thus myosin has six subunits, microtubules are built of tubulin subunits. In some cases it may be more informative to speak of ***protomers***.

succinate (ethane dicarboxylic acid) Intermediate of the ***tricarboxylic acid cycle*** and ***glyoxylate cycle***.

succinyl CoA An intermediate product in the ***tricarboxylic acid cycle***.

succinylcholine Cholinergic antagonist and therefore a skeletal muscle relaxant.

sucrase See ***invertase***.

sucrose (table sugar) Non-reducing disaccharide, α-D-glucopyranosyl-β-D-fructofuranose.

Sudan stains Histochemical stains used for lipids.

sugars See separate entries, and Table S2.

sulfur, sulfo- The British spelling, sulphur, sulpho-, is used throughout.

sulphatase An ***esterase*** in which one of the substituents of the substrate is a sulphate group.

sulphinpyrazone Pyrazole compound related to phenylbutazone, but without anti-inflammatory activity. Has no effect on platelet aggregation ***in vitro***, but inhibits platelet adhesion and release reactions. Inhibits uric acid resorption in the proximal convoluted tubule of the kidney.

sulpholipids Lipids in which the polar head group contains sulphate species. Synthesized in the Golgi complex.

sulphonamides Group of drugs derived from sulphanilamide (a red dye): act by blocking folic acid synthesis from p-aminobenzoic acid (PABA), because they are competitive analogues.

sulphydryl reagent Compounds that bind to SH groups. Include p-chlormercuribenzoate, N-ethyl maleimide, iodoacetamide. Very important in studies of protein structure.

superantigen Antigens, mostly of microbial origin, that activate all T-lymphocytes that have a T-cell receptor with a particular Vβ sequence; as a consequence superantigens activate large numbers of T-cells. Are presented on MHC Class II but are not processed and though they bind with high affinity, not in the groove of the MHC molecule where peptides are normally bound. Presentation is not MHC-restricted. *Staphylococcal* enterotoxins are the best known superantigens and stimulate $CD4^+$ T-cells in humans. The *Mls* gene product in mice can act as a self-superantigen.

supercoiling In circular DNA or closed loops of DNA, twisting of the DNA about

Table S2. Sugars

The list includes only the most common compounds found in metabolic pathways and in structural molecules. The structures are presented as Haworth models and it should be noted the configuration at the carbon which carries the carbonyl oxygen is not determined unless the hydroxyl group takes part in a glycosidic linkage, which it always does in higher oligomers. The convention for depicting glycosidic linkages is:

glycosyl carbon → acceptor hydroxyl.

▨ Configuration not defined in free molecule.

Monosaccharides

Pentoses

L-arabinose

D-ribose

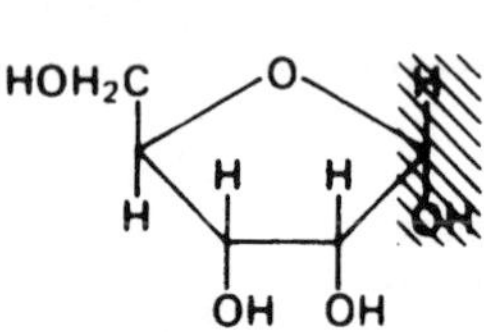

D-xylose

2-deoxy-D-ribose MW 134 1

Hexoses

D-fructose

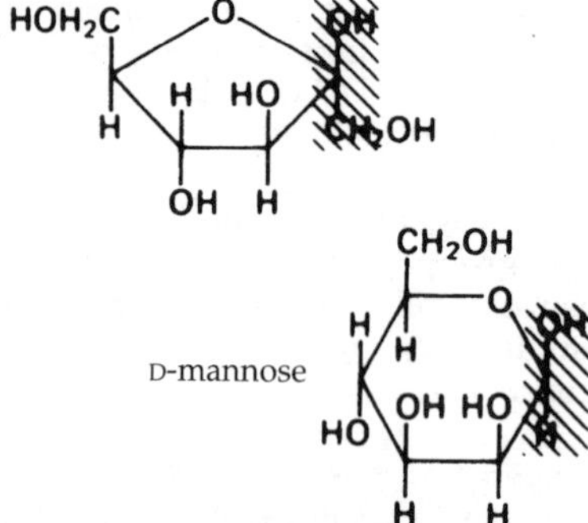

D-galactose

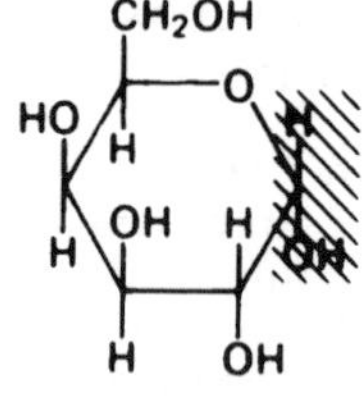

D-glucose

Free amino sugars are not found in structural oligosaccharides but N-acetyl aminohexoses are widely distributed. Most common are:

N-acetylgalactosamine

N-acetylglucosamine

Table S2. (Continued)

Monosaccharides (continued)
Other common components of structural oligosaccharides are:

fucose

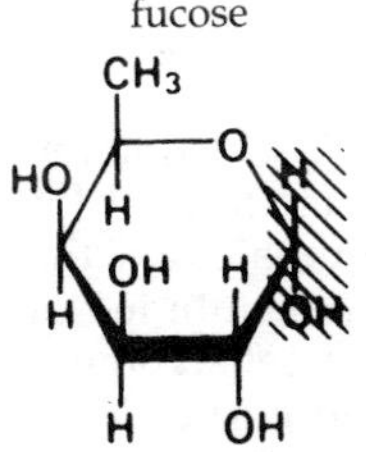

sialic acids (*N*-acetyl-neuraminic acid)

Hexose derivatives found in proteoglycans also include:

D-glucuronic acid

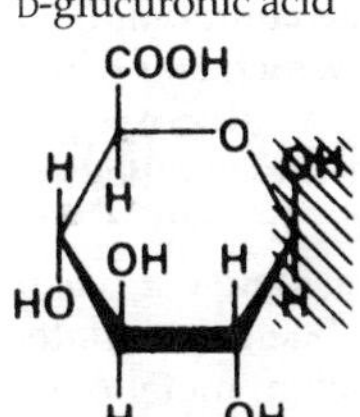

muramic acid

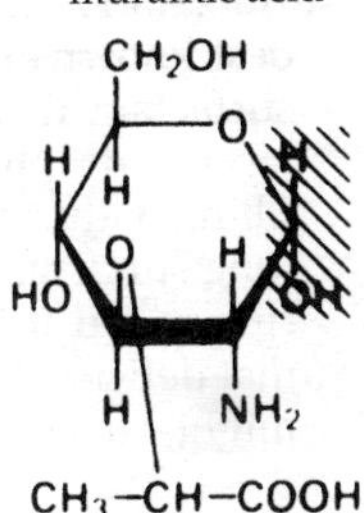

Sulphated derivatives of N-acetyl aminohexoses are also widespread and include the 4- and 6-sulphate esters of N-acetyl glucosamine and N-acetyl galactosamine.

Disaccharides and polysaccharides
These are fully specified by the residue names, sequence, bond direction, and the position numbers of the carbon atoms giving rise to the linkage. The configuration around the glycosidic carbon is also specified as α or β.

its own axis changes the number of turns of the double helix. If twisting is in the opposite direction to the turns of the double helix, ie. anticlockwise, the DNA strands will either have to unwind or the whole structure will twist or supercoil – termed negative supercoiling. If twisting is in the same direction as the helix, clockwise, which winds the DNA up more tightly, positive supercoiling is generated. DNA that shows no supercoiling is said to be relaxed. Supercoiling in circular DNA can be detected by electrophoresis because supercoiled DNA migrates faster than relaxed DNA. Circular DNA is commonly negatively supercoiled and the DNA of eukaryotes largely exists as supercoils associated with protein in the ***nucleosome***. The degree of supercoiling can be altered by ***topoisomerases***.

superoxide (superoxide radical) Term used interchangeably for the superoxide anion $^{\cdot}O_2^{-}$, or the weak acid $HO_2^{\cdot}$. Superoxide is generated both by prokaryotes and eukaryotes, and is an important product of the ***metabolic burst*** of neutrophil leucocytes. A very active oxygen species, it can cause substantial damage, and may be responsible for the inactivation of plasma antiproteases that contributes to the pathogenesis of emphysema.

superoxide dismutase (SOD) Any of a range of metalloenzymes (EC 1.15.1.1) that catalyses the formation of hydrogen peroxide and oxygen from superoxide, and thus protects against superoxide-induced damage. Usually has either iron or manganese as the metal cation in prokaryotes, copper or zinc in eukaryotes.

supershift Phenomenon in ***bandshift assays*** where the reduction in mobility on a gel induced by a binding interaction with a protein is enhanced by the addition of an antibody to the protein (or another interacting protein). Net result is that the mobility of the band of interest is further decreased (shifted).

suppressor factor (1) Factors released by T-suppressor cells. (2) See ***suppressor mutation*** and ***ochre suppressor, opal suppressor***.

suppressor mutation Mutation that alleviates the effect of a primary mutation at a different locus. May be through almost any mechanism that can give a primary mutation, but perhaps the most interesting class is the ***amber*** and ***ochre supressors***, where the anticodon of the tRNA is altered so that it misreads the termination codon and inserts an amino acid, preventing premature termination of the peptide chain.

suppressor T-cells (T-suppressor cell) Ill-defined class of T-cells that suppress T- or B-antigen dependent responses.

SUR (sulphonylurea receptor; SUR1, SUR2) ***ABC protein*** that interacts with K-ATP (Kir6.1 and Kir6.2) channels and regulates the response of the cell to glucose levels by sensing intracellular ATP concentration. The channel is formed by four SUR and four Kir subunits; the presence of both is essential for function.

suramin Compound that uncouples G-proteins from receptors, inhibits phospholipase D and inhibits binding of EGF, PDGF to cell surface receptors.

surface envelope model A way of treating the hydrodynamics of a ciliary field – by considering the whole surface of the ciliate to have an undulating surface. The undulations arise because of ***metachronism***.

surface plasmon resonance Alteration in light reflectance as a result of binding of molecules to a surface from which total internal reflection is occurring. Used in the Biacore (Pharmacia trademark) machine that detects the binding of ligand to surface-immobilized receptor or antibody.

surface potential The electrostatic potential due to surface charged groups and adsorbed ions at a surface. It is usually measured as the zeta potential at the Helmholtz slipping plane outside the surface.

surface-active compound Usually, in biological systems, means a detergent-like molecule that is amphipathic and that will bind to the plasma membrane, or to a surface with which cells come in contact, altering its properties from hydrophobic to hydrophilic, or *vice versa*.

surfactant A ***surface-active compound***; the best-known example of which is the lung surfactant that renders the alveolar surfaces hydrophobic and prevents the lung filling with water by capillary action. The lung surfactant is produced just at parturition, and it has often been speculated that deficiencies in surfactant metabolism might cause cot death.

survivin Protein in tumour cells that blocks apoptosis, possibly by inhibiting ***caspases***. Related to IAPs.

Sus scrofa Domestic pig.

suspensor cell Plant cell linking the growing embryo to the wall of the embryo sac in developing seeds.

suxamethonium A depolarizing neuromuscular blocking agent, which resembles acetylcholine in structure and binds to acetylcholine receptors, acting as an agonist. Unlike acetylcholine it persists for long enough to cause the loss of electrical excitability.

SV3T3 ***Swiss 3T3 cells*** transformed with ***SV40***.

SV40 (simian virus 40) A small DNA ***tumour virus***, a member of the ***Papovaviridae***. Isolated from monkey cells, which were being used for the preparation of ***poliovirus*** vaccine, and originally named 'vacuolating agent' owing to a cytopathic effect observed in infected cells. Found to induce tumours in newborn hamsters. In culture, transforms the cells of many non- and semi-permissive species, including mouse and human. See also ***T-antigen***.

Svedberg unit The unit applied to the sedimentation coefficient of a particle in a high-speed or ultracentrifuge. The unit S is calculated as follows, S = rate of sedimentation × $1/\rho^2 r$, where ρ is the speed of rotation in radians per second and r is the radius to a chosen point in the centrifuge tube. One Svedberg unit is defined as a velocity gradient of 10–13 seconds. Named after a pioneer of the ultracentrifuge. The units are non-additive: a particle formed from two 5S particles will not have a sedimentation coefficient of 10S.

swainsonine Fungal alkaloid that inhibits the mannosidase in the Golgi that is involved in processing the oligosaccharide chains of glycoproteins.

SWI– SNF complex The SWI–SNF complex remodels nucleosome structure in an ATP-dependent manner. In yeast the SWI–SNF chromatin remodelling complex is comprised of 11 tightly associated polypeptides (SWI1, SWI2, SWI3, SNF5, SNF6, SNF11, SWP82, SWP73, SWP59, SWP61 and SWP29). SWP59 and SWP61 are encoded by the ARP9 and ARP7 genes, respectively, which encode members of the actin-related protein (ARP) family. The similarity of ARP7 and ARP9 to the ***heat-shock protein*** and HSC family of ATPases suggests the possibility that chromatin remodelling by SWI–SNF may involve chaperone-like activities.

Swiss 3T3 cells An immortal line of fibroblast-like cells established from whole trypsinized embryos of Swiss mice (not an inbred stock) under conditions that favour establishment of cells with low saturation density in culture.

Swiss blue See ***Methylene Blue***.

switch regions The nucleotide sequences in heavy chain immunoglobulin genes located in the introns at the 5′ end of each CH locus concerned with DNA recombination events that lead to changes in the type of heavy chain produced by a B-cell, eg. IgM to IgG switching. These regions are highly conserved sequences. See ***isotype switching***.

syk Tyrosine kinase (72 kD), an effector of the B-cell receptor signalling pathway. Contains 2 tandem SH2 domains through which it interacts with ***ITAM*** motif. More widely distributed than ***zap70*** and important in signalling in both myeloid and lymphoid cells.

symbiont One of the partners in a ***symbiosis***.

symbiosis Of two organisms of different species, living together for mutual benefit.

symbiotic algae Algae (often *Chlorella* spp.) that live intracellularly in animal cells (eg. endoderm of ***Hydra** viridis*). The relationship is complex, because lysosomes do not fuse with the vacuoles containing the algae, and the growth rates of both cells are regulated to maintain the symbiosis. There is considerable strain specificity. The term is imprecise, since there are many other symbiotic algae (as in lichens) where the relationship is different.

sympathetic nervous system One of the two divisions of the vertebrate ***autonomic nervous system*** (the other being the ***parasympathetic nervous system***). The sympathetic preganglionic neurons have their cell bodies in the thoracic and lumbar regions of the spinal cord, and connect to the paravertebral chain of sympathetic ganglia. Innervate heart and blood vessels, sweat glands, viscera and the adrenal medulla. Most sympathetic neurons, but not all, use noradrenaline as a postganglionic neurotransmitter.

symplast The intracellular compartment of plants, consisting of the cytosol of a large number of cells connected by ***plasmodesmata***.

symplectic metachronism See ***metachronism***.

symport A mechanism of transport across a membrane in which two different molecules move in the same direction. Often, one molecule can move up an electrochemical gradient because the movement of the other molecule is more favourable (see ***facilitated diffusion***). Example:

the sodium/glucose cotransport. See ***antiport, uniport***.

synapse A connection between ***excitable cells***, by which an excitation is conveyed from one to the other. (1) Chemical synapse: one in which an ***action potential*** causes the exocytosis of neurotransmitter from the presynaptic cell, which diffuses across the synaptic cleft and binds to ***ligand-gated ion channels*** on the postsynaptic cell. These ion channels then affect the resting potential of the postsynaptic cell. (2) Electrical synapse: one in which electrical connection is made directly through the cytoplasm, via ***gap junctions***. (3) Rectifying synapse: one in which action potentials can only pass across the synapse in one direction (all chemical and some electrical synapses). (4) Excitatory synapse: one in which the firing of the presynaptic cell increases the probability of firing of the postsynaptic cell. (5) Inhibitory synapse: one in which the firing of the presynaptic cell reduces the probability of firing of the postsynaptic cell.

synapsins Family of phosphoproteins associated with synaptic vesicles and implicated in control of release. Synapsin Ia (84 kD) and Ib (80 kD) are alternatively spliced variants as are synapsins IIa (74 kD) and IIb (55 kD). Can be phosphorylated by several ***protein kinases***. Thought to be involved in regulation of neurotransmitter release at ***synapses***.

synapsis The specific pairing of the chromatids of homologous chromosomes during ***prophase*** I of meiosis. It allows ***crossing-over*** to take place.

synaptic cleft The narrow space between the presynaptic cell and the postsynaptic cell in a chemical ***synapse***, across which the ***neurotransmitter*** diffuses.

synaptic plasticity Change in the properties of a synapse, usually in the context of learning and memory. Very few synapses provide simple 1:1 transfer of ***action potentials***, and very small changes in the efficiency of a synapse (usually mediated by changes in either the pre-or postsynaptic membrane) can have profound influences on the electrical properties of a neuronal circuit. See also ***neuronal plasticity***.

synaptic transmission The process of propagating a signal from one cell to another via a ***synapse***.

synaptic vesicle Intracellular vesicles found in the presynaptic terminals of chemical synapses, which contain ***neurotransmitter***.

synaptobrevin (vSNARE; VAMP-2) Small integral membrane proteins (16.7 kD) of synaptic vesicles. Two isoforms, VAMP-1 and VAMP-2 are known. They bind SNAPs and also interact with target-SNARE (syntaxin). Cleaved by clostridial toxins encoding zinc endopeptidases, such as tetanus toxin and botulinum toxin, blocking synaptic release.

synaptogenesis Formation of a ***synapse***.

synaptogyrin Integral component (29 kD) of synaptic vesicle with some similarity to synaptophysin. Has four transmembrane domains.

synaptojanin Protein of the vertebrate nerve terminal (145 kD) that seems to participate with ***dynamin*** in the process of vesicle recycling. Has phosphatase activity and is a member of the inositol-5-phosphatase family. Amino-terminal region has homology with yeast Sac1 (involved in phospholipid metabolism) and C-terminal region has proline-rich sequences that probably interact with ***SH3*** domains of ***amphiphysin*** and ***GRB 2***.

synaptonemal complex Structure, identified by electron microscopy, lying between chromosomes during ***synapsis***; consists of two lateral plates closely apposed to the chromosomes and connected to a central plate by filaments. It appears to act as a scaffold, and is essential for ***crossing over***.

synaptophysin Abundant glycoprotein component of synaptic vesicle membranes composed of a 38 kD subunit that spans the membrane four times and has both its N- and C-termini located cytoplasmically. Its transmembrane organization and putative quaternary structure resemble the molecular topology of ***gap junction*** proteins, ***connexins***.

synaptoporin Putative channel protein of synaptic vesicles, and a member of the ***synaptophysin/connexin*** superfamily. It has 58% amino acid identity to synaptophysin, with highly conserved transmembrane segments but a divergent cytoplasmic tail.

synaptosome A subcellular fraction prepared from tissues rich in chemical ***synapses***, used in biochemical studies. Consists mainly of vesicles from presynaptic terminals.

synaptotagmin (p65) Calcium-binding synaptic vesicle protein that binds acidic phospholipids and recognizes the cytoplasmic domain of the ***neurexins***. May be involved in vesicle docking.

synchronous cell population A culture of cells that all divide in synchrony. Particularly useful for certain studies of the cell cycle, cells can be made synchronous by depriving them of essential molecules, which are then restored. Synchronization breaks down after a few cycles, however, as individual cells have unique division rates.

syncolin Microtubule-associated protein (280 kD) found in chicken erythrocytes. Has some similarities with MAP-2, but thought to be distinct.

syncytiotrophoblast Syncytial layer that forms the outermost foetal layer in the placenta and is thus the interface with maternal tissue. Has invasive capacity, though in a regulated manner.

syncytium An ***epithelium*** or tissue in which there is cytoplasmic continuity between the constituent cells.

syndecan An integral membrane proteoglycan (250–300 kD) associated largely with epithelial cells. The core protein of 294 amino acids has an extracellular domain of 235 amino acids and a single transmembrane domain of 25 amino acids. The extracellular domain has up to three heparan sulphate and two chondroitin or dermatan sulphate chains plus an N-linked oligosaccharide. The heparan sulphate chains bind to several proteins of the extracellular matrix, including collagens, fibronectin and ***tenascin***. The cytoplasmic domain is thought to interact with actin filaments. Its name is derived from the Greek *syndein*, to bind together. Ligation of N-syndecan (syndecan-3) by heparin-binding growth-associated molecule increases phosphorylation of c-src and ***cortactin***, and N-syndecan may act as a neurite outgrowth receptor.

synemin An intermediate filament-associated protein (230 kD) isolated from avian smooth muscle, but homologue also found in mammalian muscle. Colocalizes with ***desmin*** near myofibrillar ***Z discs***.

synexin Annexin VII. See Table A3.

syngamy Fusion of two haploid gametic nuclei to form the diploid nucleus of the zygote.

syngeneic Organisms that are antigenically identical: monozygotic twins or highly inbred strains of animals. Thus cells injected into a syngeneic host will not be rejected because of a general lack of histocompatibility.

synkaryon A somatic hybrid cell in which chromosomes from two different parental cells are enveloped in a single nucleus.

synomone See ***allomone***.

synovium Connective tissue that forms the bearing surface of the joint and that is eroded in arthritis.

syntaxin (t-SNARE) Integral membrane protein of presynaptic membrane. Has a long cytoplasmic domain involved in the targeting of vesicles.

syntenic Syntenic genes lie on the same chromosome. Some loci are syntenic in both man and mouse, others are not.

synthetase Enzymes of class 6 in the ***E classification***; catalyse synthesis of molecules, their activity being coupled to the breakdown of a nucleotide triphosphate.

Syp (PTP1D; Shp-2) An adaptor molecule mediating ***GRB-2***/ras signalling. See ***Shp***.

syringyl alcohol (sinapyl alcohol) A phenylpropanoid alcohol, one of the three precursors of lignin.

systemic lupus erythematosus (SLE) Disease of humans, probably autoimmune with antinuclear and other antibodies in plasma. Immune complex deposition in the glomerular capillaries is a particular problem.

syzygy In some parasitic protozoa the pairing of gamonts prior to sexual fusion, in gregarines the end-to-end attachment of the sporonts, in some crinoids the fusion of organs or skeletal elements.

T

t$_{1/2}$ See ***half-life***.

T7 (bacteriophage T7) A T-odd phage.

T See ***threonine***.

t-antigen The small ***T-antigen*** of polyoma virus.

T-antigen Proteins coded by viral genes that are expressed early in the replication cycle of papovaviruses such as SV40 and polyoma. Essential for normal viral replication, they are also expressed in non-permissive cells transformed by these viruses. Originally detected as tumour-antigens by immunofluorescence with antisera from tumour-bearing animals. SV40 has two, large T and small t; polyoma has three, large, middle and small. Appear to be collectively responsible for transformation by these viruses.

T-box genes The T-box gene family codes for transcription factors (and putative transcription factors) that share a unique DNA-binding domain, the T-domain. In all metazoans studied from *Caenorhabditis elegans* to man, they are found as a small, highly conserved group of genes; mutations are associated with developmental defects. See ***brachyury***.

T-cell A class of lymphocytes, so called because they are of thymic origin and have been through thymic processing. Involved primarily in cell-mediated immune reactions and in the control of ***B-cell*** development. They bear T-cell antigen receptors (CD3) and lack Fc or C3b receptors. Major T-cell subsets are $CD4^+$ (mainly helper cells) and $CD8^+$ (mostly cytotoxic or suppressor T-cells).

T-cell factor (TCF) Transcription factor, one of the high-mobility group domain proteins, activated by wnt/wingless signalling and repressed by ***CREB-binding protein***. Coactivator is β-catenin. Activation of TCF in colonic epithelium and other cells leads to tumours.

T-cell growth factor See ***interleukin-2***.

T-cell leukaemia/lymphoma viruses See ***HTLV-I, HTLV-II***.

T-cell receptor The antigen-recognizing receptor on the surface of ***T-cells***. Heterodimeric (disulphide linked), one of the immunoglobulin superfamily of proteins; binds antigen in association with the ***major histocompatibility complex*** (MHC), leading to the activation of the cell. There are two subunits (α and β, 42–44 kD in mice, 50–40 kD in humans), each with variable and constant regions, that are associated non-covalently with T3 (20–30 kD). A second heterodimer on $CD3^+$ cells with γ (35 kD in mice, 55 kD in humans) and δ (45 kD in mice, 40 kD in humans) chains is a second T-cell antigen receptor that is not ***MHC-restricted***. The γδ T-cell receptors (TCRs) are formed on very early T-cells in the thymus.

T even phage A group of dsDNA bacteriophages of enterobacteria including T2, T4, T6 as opposed to T odd phage (T1,3,5 and 7)

T-helper cells (T_H cells; Th1; Th2) There are now recognized to be two subclasses of $CD4^+$ T-helper cells, Th1 and Th2. Th1 cells produce IL-2, IFN-γ and TNF-α and do not produce IL-4, IL-5 and IL-10. They are associated with cell-mediated immunity. Selective activation of Th1 cells is promoted by IFN-γ and IL-12 and inhibited by IL-4 and IL-10, the products of Th2 cells. Th2 cells are involved with the humoral immune response, produce IL-4, IL-5 and IL-10 and promote antibody production; IL-4 is essential for growth and differentiation of Th2 cells. There is cross-inhibition between the 2 classes, if one subclass is activated it will inhibit the activity of the other so that the response is polarized.

T loop of RNA (thymine pseudo-uracil loop; T ψ loop) The T loop of tRNA is the region of the molecule that is responsible for ribosome recognition.

T-lymphocyte See ***T-cell.***

T-suppressor cell Set of ***T-cells*** (usually $CD8^+$) specifically involved in suppressing ***B-cell*** differentiation into antibody-secreting cells. There may also be T-suppressors of T-cell functions. Still controversial.

T-tubule (transverse tubule) Invagination of the plasma membrane (sarcolemma) of striated muscle that lies between two tubular portions of the endoplasmic (sarcoplasmic) reticulum to form a triad of membrane profiles adjacent to the A band/I band junction in some cases, in other cases to the ***Z-disc***, of the resting sarcomere. Depolarization of the T tubule membrane triggers the release of calcium from the sarcoplasmic reticulum and eventually muscle contraction.

T-type channels A class of ***voltage-sensitive calcium channels*** that open transiently in response to relatively small depolarizations of the neuronal membrane. May have a role in repetitive firing. No selective inhibitors are known.

TA cloning Cloning strategy for ***PCR*** products that relies on the tendency of ***Taq polymerase*** to add an extra dA at the 3′ end of newly synthesized DNA strands, thus leaving a single base 3′ overhang. Vectors are accordingly prepared with single base dT 3′ overhangs, allowing ligation of ***sticky ends***.

Tacaribe complex Group of 8 ***Arenaviridae*** isolated in South America from bats.

tachykinins A group of neuropeptide hormones including ***substance P***, substance K (neurokinin A) and neurokinin B in mammals, eledoisin from *Octopus* and physalaemin (amphibian). All have 10 or 11 residues with a common -FXGLM-NH2 ending. Elicit a wide range of responses from neurons, smooth muscle, endothelium, exocrine glands and cells of the immune system; effects similar in many ways to ***bradykinin*** and ***serotonin***.

tachyphylaxis A decrease in the response to an agonist following repeated exposure. Can arise through a variety of mechanisms.

TAG-1 (transient axonal glycoprotein) A 135 kD surface glycoprotein that is expressed transiently on commissural and ***motoneurons*** in developing vertebrate nervous system. TAG-1 and ***L1*** have been shown to be on different segments of the same embryonic spinal axons. See ***axonin*** and ***tax-1***.

taicatoxin Complex oligomeric protein toxin from *Oxyuranus scutelatus scutelatus*. Blocks high- but not low-threshold calcium channels of heart muscle. The oligomer contains a neurotoxin-like peptide (8 kD), a phospholipase (16 kD) and a serine-protease inhibitor (7 kD).

taipoxin Heterotrimeric toxin from *Oxyuranus scutelatus scutelatus*. All three subunits (α, β, γ) have homology with pancreatic phospholipase A2. Blocks transmission at the neuromuscular junction.

talin Protein (215 kD) that binds to ***vinculin***, but not to actin, and is associated with the subplasmalemmal cytoskeleton.

Talon resin Proprietary name for immobilized nickel-beads. Used to purify recombinant proteins containing ***his tags***.

Tamiami virus Arenavirus of the ***Tacaribe complex***.

tamoxifen Synthetic anti-oestrogen used in chemotherapy of breast carcinoma. Probably has other effects, including inhibition of chloride channel conductance.

tandem repeats Copies of genes repeated one after another along a chromosome, eg. the 40S-rRNA genes in somatic cells of toads, of which there are about 500 copies.

tannic acid Penta-(*m*-digalloyl)-glucose, or any soluble tannin; used in electron microscopy to enhance the contrast. Addition of tannic acid to fixatives greatly improves, eg. the image obtained of tubulin subunits in the microtubule, or the ***HMM*** decoration of microfilaments.

tannins Complex phenolic compounds found in the vacuoles of certain plant cells, eg. in bark. They are strongly astringent and are used in tanning and dyeing.

tapasin Accessory protein required for the interaction of MHC Class I with ***TAPs*** thus ensuring efficient peptide binding. Tapasin is related to the immunoglobulin superfamily and has an endoplasmic reticulum retention signal.

tapetum (1) Layer of reflective tissue just behind the pigmented retinal epithelium of many vertebrate eyes. May consist of either a layer of guanine crystals, or a layer of connective tissue. In bovine eyes reflects a blue-green iridescent colour. (2) Layer of cells in the sporangium of a vascular plant that nourishes the developing spores.

TAPs Transporters associated with antigen processing: ***ABC proteins*** involved in transporting protein fragments across ER membranes during antigen processing. See ***tapasin***.

Taq polymerase A heat-stable ***DNA polymerase*** that is normally used in the ***polymerase chain reaction***. It was isolated from *Thermus aquaticus*.

TAR RNA (transactivating-response RNA) RNA structure at the extreme 5′ terminus of virion RNA.

target regulation General term for an interaction between neurons and their targets by which target-derived signals influence the differentiation of the innervating neurons.

targeting signal Peptide sequence within a protein that determines where it will be located. Thus there are targeting signals for proteins that accumulate in the nucleus, others for endoplasmic reticulum, lysosomes, etc.

Tat protein Transactivator protein from lentiviruses, notably ***HIV***; sequence-specific RNA-binding protein that recognizes ***TAR RNA***. Will induce endothelial cell migration and invasion ***in vitro*** and rapid angiogenesis ***in vivo***. Peptides from this protein are potent neurotoxins, implying a possible route for HIV-mediated toxicity.

TATA box (Goldber–Hogness box) A consensus sequence found in the promoter region of most genes transcribed by eukaryotic ***RNA polymerase*** II. Found about 25 nucleotides before the site of initiation of transcription and has the consensus sequence: 5′-TATAAAA-3′. This sequence seems to be important in determining accurately the position at which transcription is initiated.

tau protein (*tau* factor) Protein (60–70 kD) that copurifies with ***tubulin*** through cycles of assembly and disassembly, and the first microtubule-associated protein to be characterized. Tau proteins are a family made by alternative splicing of a single gene. It has tandem repeats of a tubulin-binding domain and promotes tubulin assembly. Although tau proteins are found in all cells they are major components of neurons where they are predominantly associated with microtubules of the axon. See ***MAPs***.

taurine (2-aminoethanesulphonic acid) Compound derived from cysteine by oxidation of the sulphydryl group and decarboxylation. Present in the cytoplasm of some cells (particularly neutrophils) at high concentration.

taurocholate Major bile salt (derived from taurocholic acid) with strong detergent activity. Formed by conjugation of taurine with cholate.

tautomerism Form of isomerism in which there are two or more arrangements usually of hydrogens bonded to oxygen. Keto-enol tautomerism is one common example. The balance between two coexisting tautomers may shift with time or as a result of changes in conditions.

tautomycin Antibiotic, inhibitor of type 1 and type 2a protein phosphatases.

tax-1 Axonal surface glycoprotein (135 kD), the human homologue of rat ***TAG-1*** and chicken ***axonin***-1. GPI-linked to neuronal plasma membrane and is involved in adhesion. There are six Ig-like and four Fibronectin III-like domains. Will support neurite outgrowth **in vitro**.

taxis A response in which the direction of movement is affected by an environmental cue. Should be clearly distinguished from a ***kinesis***.

taxol Drug isolated from yew (*Taxus brevifolis*) that stabilizes microtubules: analogous in this respect to ***phalloidin*** which stabilizes microfilaments.

Tay-Sachs disease Lysosomal disease (***lipidosis***) in which hexosaminidase A, an enzyme that degrades ***ganglioside*** GM2, is absent. A lethal autosomal recessive; mostly affects brain, where ganglion cells become swollen and die.

TBP (TATA-binding protein) A 30 kD component of TFIIIB and of ***SL1***, responsible for positioning the polymerase. Also involved in positioning RNA polymerase II in which case it binds directly to the ***TATA box***.

TCA cycle See ***tricarboxylic acid cycle***.

TCA3 See ***I-309***.

TDG (thymine-DNA glycosylase.) Enzyme responsible for repair of G/T mispairings.

Tec family kinase Family of intracellular protein ***tyrosine kinases*** involved in signalling. Includes ***Btk***. Unlike ***src family*** they are not regulated by C-terminal phosphorylation. Have ***PH*** and ***TH domains*** in N-terminal region.

teichoic acid Acidic polymers (glycerol or ribitol linked by phosphodiester bridges) found in cell wall of ***Gram positive bacteria***. May constitute 10–50% of wall dry weight and are crosslinked to peptidoglycan. Related to ***lipoteichoic acid***.

tektins Family of filamentous proteins (A, 55 kD; B, 51 kD; C, 47 kD) associated with some microtubules in ciliary and flagellar axonemes. Have homology with some intermediate filament proteins (keratins and lamins).

telangielactasia See ***CREST***.

teleost melanophores Large stellate cells found in the epidermis of fish. Cytoplasmic pigment granules (containing ***melanin***) can be centrally located, or rapidly dispersed, using a microtubule-associated system. Altering the granule distribution changes the colour of the skin.

telocentric chromosome Chromosome with the centromere located at one end.

telokin Acidic protein (24 kD) found in some muscle tissue, identical to the C-terminal 155 residues of smooth muscle ***myosin light chain kinase*** (MLCK) and independently expressed.

telomerase (telomere terminal transferase) A DNA polymerase with rather unusual properties that will only elongate oligonucleotides from the telomere and not other sequences. The enzyme contains an essential 159 residue RNA sequence that provides a template for the replication of the G-rich telomere sequences (so that the enzyme could in fact be considered a ***reverse transcriptase***).

telomerase repeat binding factor 1 See [TRF1].

telomere The end of a chromosome.

telopeptides Portions of the amino acid sequence of a protein that are removed in maturation of the protein. Best examples are the N- and C-terminal telopeptides of procollagen that are involved in development of the quaternary structure and are then proteolytically removed by ***procollagen peptidases***.

telophase The final stage of mitosis or meiosis, when chromosome separation is completed.

temperate phage A bacteriophage that integrates its DNA into that of the host (***lysogeny***) as opposed to virulent phages that lyse the host.

temperature-sensitive mutation (ts mutation) A type of conditional mutation in organism, somatic cell or virus that makes it possible to study genes whose total inactivation would be lethal. Such ts mutations can also make possible studies of the effect of reversible switching (by temperature changes) in expression of the mutated gene. The usual mechanism of temperature sensitivity is that the mutated gene codes for a protein with a temperature-dependent conformational instability, so that it possesses normal activity at one temperature (the permissive temperature), but is inactive at a second (non-permissive) temperature.

template A structure that in some direct physical process can cause the patterning of a second structure, usually complementary to it in some sense. In current biology it is almost exclusively to refer to a nucleotide sequence that directs the synthesis of a sequence complementary to it by the rules of Watson–Crick base pairing.

temporal sensing Mechanism of gradient sensing in which the value of some environmental property is compared with the value at some previous time, the cell having moved position between the two samplings. Initial movement is random; until the second observation is made the gradient cannot be detected. See ***spatial*** and ***pseudospatial*** sensing mechanisms. Bacterial chemotaxis (so-called) is based on this mechanism.

tenascin (myotendinous antigen; cytotactin) Protein of the extracellular matrix (240 kD subunit: usually as a hexabrachion, a 6-armed hexamer of more than 1000 kD) selectively present in mesenchyme surrounding foetal (but not adult) rat mammary glands, hair follicles and teeth. Found in the matrix surrounding mammary tumours of rat. Tenascin contaminates cell-surface ***fibronectin*** and accounts for most of the haemagglutinating activity of extracellular matrix protein. Contains 14 repeats of the ***EGF-like domain***.

tensegrity The hypothesis that cells can behave like structures in which shape results from balancing tensile and hydrostatic forces.

tensin Actin-binding component of ***focal adhesions*** and submembranous cytoskeleton. Has SH2 domain and can undergo ***tyrosine phosphorylation***; it is speculated that it may link signalling systems with the cytoskeleton.

tenuin Subplasmalemmal protein (400 kD) from ***adherens junctions***, associated with membrane insertions of ***microfilament*** bundles, and membrane adjacent to circumferential microfilament bundles of epithelial cells.

TEP1 (TGF-α regulated and epithelial cell-enriched phosphatase) Also termed ***PTEN*** or MMAC1 (mutated in multiple advanced cancers 1).

teratocarcinoma Malignant tumour (teratoma) thought to originate from primordial germ cells or misplaced blastomeres that contains tissues derived from all three embryonic layers, eg. bone, muscle, cartilage, nerve, tooth-buds and various glands. Accompanied by undifferentiated, pluripotent epithelial cells known as embryonal carcinoma cells.

teratogen Agent capable of causing malformations in embryos. Notorious example is ***thalidomide***.

teratoma See ***teratocarcinoma***.

terminal bar Obsolete name for ***zonula occludens*** (tight junction).

terminal cisternae Regions of the ***sarcoplasmic reticulum*** adjacent to ***T-tubules***, and from which calcium is released when striated muscle is activated.

terminal web The cytoplasmic region at the base of microvilli in intestinal epithelial cells, a region rich in microfilaments from the microvillar core and from ***adherens junctions***, in myosin, and in other proteins characteristic of an actomyosin motor system.

termination codon The three codons, UAA known as *ochre*, UAG as *amber* and UGA as ***opal***, that do not code for an amino acid but act as signals for the termination of protein synthesis. They are not represented by any tRNA and termination is catalysed by protein release factors. There are two release factors in *E. coli*: RF1 recognizes UAA and UAG; RF2 recognizes UAA and UGA. Eukaryotes have a single GTP-requiring factor, eRF.

terminator DNA sequence at the end of a ***transcription unit*** that causes ***RNA polymerase*** to stop transcription.

terpene Lipid species, very abundant in plants. In principle terpenes are polymers of isoprene units. Function in plants is not clear. In animals ***dolichol***, an important carrier species in the formation of glycoproteins, is a terpenoid. Similarly squalene, an intermediate in the synthesis of cholesterol, is a terpene.

tertiary structure The third level of structural organization in a macromolecule. For example, the primary structure of a protein is the amino acid sequence, the secondary structure is the folding of the peptide chain (α-helical or β-pleated), and the tertiary structure is the way in which the helices or sheets are folded or arranged to give the three-dimensional structure of the protein. Quaternary structure refers to the arrangement of protomers in a multimeric protein.

testa Outer covering of a seed, also called the seed-coat; derived from the integument of the ovary.

testicular feminization If genetic males lack receptors for testosterone they develop as females and are unresponsive to male hormones.

testosterone Male sex hormone (androgen) secreted by the interstitial cells of the testis of mammals and responsible for triggering the development of sperm and of many secondary sexual characteristics.

tetanolysin Thiol-activated haemolysin released by the bacterium *Clostridium tetani*.

tetanospasmin See ***tetanus toxin***.

tetanus (lockjaw) Disease caused by the bacterium *Clostridium tetani*, spores of which persist in soil but can proliferate anaerobically in an infected wound. Disease entirely due to the ***tetanus toxin***, released by bacterial autolysis.

tetanus toxin (tetanospasmin) Neurotoxin released by *Clostridium tetani*; becomes active when peptide cleaved proteolytically to heavy (100 kD) and light (50 kD) chains held together by disulphide bond. Heavy chain binds to disialogangliosides (GD2 and GD1b), and part of the peptide (the amino-terminal B-fragment) forms a pore: light chain is a zinc endopeptidase that specifically attacks ***synaptobrevin***, to block neurotransmitters. See also ***botulinum toxin***.

tetracaine (amethocaine) Potent local anaesthetic.

tetracycline Broad-spectrum antibiotic that blocks binding of aminoacyl-tRNA to the ribosomes of both ***Gram-positive*** and ***Gram-negative*** organisms (and those of organelles). Produced by *Streptomyces aureofasciens*.

tetrad Four homologous chromatids paired together during first meiotic prophase. More generally, any group of four objects.

tetraethylammonium ion (TEA) A monovalent cation widely used in neurophysiology as a specific blocker of potassium channels. It is similar in size to the hydrated potassium ion, and gets stuck (reversibly) in the channels.

tetrahydrocannabinol (THC) A ***cannabinoid*** and one of the more psychoactive components of cannabis.

tetrahydrofolate See ***folate***.

Tetrahymena Genus of ciliate protozoa frequently used in studies on ciliary axonemes, self-splicing RNA and telomere replication.

tetramethythionine chloride See ***Methylene Blue***.

tetraploid Nucleus, cell or organism that has four copies of the normal ***haploid*** chromosome set.

tetrodotoxin (TTX) A potent ***neurotoxin*** (319 D) from the Japanese puffer fish. It binds to the sodium channel, blocking the passage of action potentials. Its activity closely resembles that of ***saxitoxin***.

tetrose General term for a monosaccharide with 4 carbon atoms.

textilotoxin Protein neurotoxin (70 kD) from venom of *Pseudonaja textilis textilis* that blocks neuromuscular transmission. All five subunits of the toxin have some phospholipase A2 activity.

TFIID (TBP) Transcription factor that binds to ***TATA box***.

TFIIX Any one of a number of accessory proteins involved in the binding of RNA polymerase II to DNA in association with ***TBP***.

TFIIIA Transcription factor, one of the first to be cloned and characterized. Has a crucial role in transcription of 5S ribosomal RNA. Multiple cysteine/histidine ***zinc finger*** motifs. Interacts with ***TFIIIB***.

TFIIIB Transcription factor consisting of ***TBP*** and two other proteins. An initiation factor required by RNA polymerase III; TFIIIA and TFIIIC assist its binding to the appropriate DNA sequence. Effectively acts as a positioning factor for polymerase.

TFIIIC Transcription factor; large (500 kD) complex containing at least five subunits.

TFP See ***trifluoperazine***.

TGF See ***transforming growth factor***.

TGGCA-binding proteins See ***CTF***.

TH domain (Tec homology domain) Proline-rich domain characteristic of ***Tec family*** protein kinases, probably ligand region for ***SH3*** domain.

thalassaemia Hereditary blood disease in which there is abnormality of the globin portion of haemoglobin. Widespread in Mediterranean countries.

thalidomide Sedative drug that when taken between the third and fifth week of pregnancy produces a range of malformations of the foetus, in severe cases complete absence of limbs (amelia), or much reduced limb development (phocomelia). A ***teratogen***.

thallus Simple plant body, not differentiated into stem, root, etc. Main form of the gametophyte generation of simpler plants such as liverworts.

thapsigargin Cell-permeable inhibitor of calcium ATPase of endoplasmic reticulum; leads to increase in cytoplasmic calcium ions. Acts independently of InsP3. A tumour promoter.

thaumatin Protein from the African plant *Thaumatococcus daniellii*. It tastes 10^5 times sweeter than sucrose.

theobromine (3,7-dimethyl xanthine) Principal alkaloid of cacao bean; has properties similar to theophylline and caffeine.

theophylline (1,3-dimethylxanthine) Inhibits cAMP ***phosphodiesterase***, and is often used in conjunction with exogenous dibutyryl cyclic-AMP to raise cellular cAMP levels. Other less potent methylxanthines are caffeine, theobromine and aminophylline.

thermal analysis Form of calorimetry in which the rate of heat flow (or some other property) to a solid is measured as a function of temperature.

thermal melting profile In general a record of the phase state of a system over a temperature range. Phase changes can be detected by exothermy or endothermy. Valuable in studying lipid and DNA structures.

thermodynamics The study of energy and energy flow in closed and open systems.

thermolysin Heat-stable metalloproteinase (EC 3.4.24.4.) produced by a strain of *Bacillus stearothermophilus*. Retains 50% of its activity after 1h at 80°C.

thermophile An organism that thrives at high temperature. The most extreme examples (hyperthermophiles) are ***cyanobacteria*** from hot springs that have optima of 50–55°C, and will tolerate temperatures of 90°C.

thermophilic See ***thermophile***.

thermotaxis A directed motile response to temperature. The grex of ***Dictyostelium*** *discoideum* shows a positive thermotaxis.

Thermus aquaticus Aerobic ***Gram-negative*** bacillus that lives in hot springs and was the source of ***Taq polymerase***.

thiamine pyrophosphatase (TPP; carboxylase) The coenzyme form of vitamin B1 (thiamine), deficiency of which causes beri-beri. Forms the prosthetic group of pyruvate dehydrogenase, α-ketoglutarate dehydrogenase and transketolase, in which it is involved in transfer of a two-carbon unit. Marker for the *trans* cisternae of the ***Golgi apparatus***.

thiamine pyrophosphate Co-carboxylase. A cofactor that has an unusually acidic carbon atom able to form carbon–carbon bonds. Found in pyruvate dehydrogenase and transketolase.

thick filaments Bipolar ***myosin***-II filaments (12–14nm diameter, 1.6μm long) found in striated muscle. Myosin filaments elsewhere are often referred to as 'thick filaments', although their length may be considerably less. The myosin heads project from the thick filament in a regular fashion. There is a central 'bare' zone without projecting heads, the core being formed from antiparallel arrays of ***LMM*** regions of the myosin heavy chains. Thick filaments will self-assemble ***in vitro*** under the right ionic conditions.

thigmotropism Tendency of an organism or part of an organism to turn towards or respond to a mechanical stimulus.

thin filaments Filaments 7–9nm in diameter attached to the ***Z discs*** of striated muscle; have opposite polarity in each half-sarcomere. Built of ***F-actin*** with associated ***tropomyosin*** and ***troponin***.

thin-layer chromatography (TLC) Chromatography using a thin layer of powdered medium on an inert sheet to support the stationary phase. Faster than paper chromatography, gives higher resolution, and requires smaller samples.

thioester Compounds of the type. R-CO-S-R′. See ***coenzyme A, palmitoylation***.

thioether The bond R-S-C, of which the best example is in methionine.

thiol endopeptidases Proteases that have an active thiol group. Includes papain and ficin.

thiol proteinase See ***thiol endopeptidases***.

thiol-activated haemolysins (oxygen-labile haemolysins) Cytolytic bacterial exotoxins that act by binding to cholesterol in cell membranes and forming ring-like complexes that act as pores. SH groups of these toxins must be in the reduced state for the toxin to function. Oxidation (to disulphide bridges) inactivates the toxin. Examples: ***tetanolysin, streptolysin O***, θ-toxin (***perfringolysin***), ***cereolysin***.

thionins Group of small, hydrophobic plant proteins of 45–50 residues that are toxic to animals.

thioredoxin Intercellular disulphide-reducing enzyme. Also secreted by a variety of cells, despite the lack of a ***signal sequence***, in a manner resembling the alternative secretory pathway for IL-1β, and a few other proteins.

Thomson's disease Glycogen storage disease in which the missing enzyme is phosphoglucomutase. See ***lysosomal diseases***.

thoracic duct The major efferent lymph duct into which lymph from most of the peripheral lymph nodes drains. Recirculating lymphocytes that have left the circulation in the lymph node return to the blood through the thoracic duct.

THP-1 Human monocytic cell line derived from peripheral blood of 1-year-old boy with acute monocytic leukaemia. Have Fc and C3b receptors and will differentiate into macrophage-like cells.

Thr See ***threonine***.

threonine (Thr; T) The hydroxylated polar amino acid (119 D). See Table A2.

threose A four-carbon sugar in which the two central hydroxyl groups are in *trans* orientation (*cis* in erythrose).

thrombasthenia Condition in which there is defective platelet aggregation, though adherence is normal. See ***Glanzmann's thrombasthenia***.

thrombin Protease (34 kD) generated in blood clotting that acts on ***fibrinogen*** to produce ***fibrin***. Consists of two chains, A and B, linked by a disulphide bond. The B chain has sequence homology with pancreatic serine proteases: cleaves at Arg-Gly. Thrombin is produced from prothrombin by the action either of the extrinsic system (tissue factor + phospholipid) or, more importantly, the intrinsic system (contact of blood with a foreign surface or connective tissue). Both extrinsic and intrinsic systems activate plasma factor X to form factor Xa which then, in conjunction with phospholipid (tissue derived or ***platelet factor 3***) and factor V, catalyses the conversion. See also Table F1.

thrombocyte Archaic name for a blood ***platelet***.

thrombocytopenia Gross deficiency in platelet number, consequently a tendency to bleeding.

thrombocytopenic purpura In severe ***thrombocytopenia***, bleeding into skin leads to small petechial haemorrhages. In primary thrombocytopenic purpura an autoimmune mechanism seems to cause platelet destruction; secondary thrombocytopenic purpura may be a result of drug-induced type II ***hypersensitivity*** in which platelets coated with antibody to the drug (which is acting as a ***hapten***) are destroyed in a complement-mediated reaction.

thromboglobulin (β–thromboglobulin) Protein derived from ***platelet basic protein***.

thrombomodulin Specific endothelial cell receptor (100 kD: luminal surface only) that forms a 1:1 complex with thrombin. This complex then converts ***protein C*** to Ca, which in turn acts on factors Va and VIIIa. Structurally similar to ***coated pit*** receptors.

thromboplastin Traditional name for substance in plasma that converts prothrombin to ***thrombin***. Now known not to be a single substance.

thrombopoietin (TPO) Growth factor (19D) that regulates the proliferation of ***megakaryocytes*** and production of platelets (thrombopoiesis). Receptor is c-mpl, a cytokine receptor that can cause phosphorylation of ***STAT***3 and STAT5 through Jak3.

thrombosis Formation of a solid mass (a ***thrombus***) in the lumen of a blood vessel or the heart.

thrombospondin Homotrimeric glycoprotein (450 kD) from α granules of ***platelets***, and synthesized by various cell types in culture. Also found in extracellular matrix of cultured endothelial, smooth muscle, and fibroblastic cells. May have autocrine growth-regulatory properties: involved in platelet aggregation.

thrombosthenin Obsolete name for platelet contractile protein: now known to be actomyosin (which makes up 15–20% of the total platelet protein).

thromboxanes Arachidonic acid metabolites produced by the action of thromboxane synthetase on prostaglandin cyclic endoperoxides. Thromboxane A2 (TxA2) is a potent inducer of platelet aggregation and release, and although unstable, the activation of platelets leads to the further production of TxA2. Also causes arteriolar constriction. Another endoperoxide product, ***prostacyclin***, has the opposite effects.

thrombus Solid mass that forms in a blood vessel, usually as a result of damage to the wall. The first aggregate is of platelets and fibrin, but the thrombus may propagate by clotting in the stagnant downstream blood.

thuringolysin O ***Cholesterol-binding toxin*** from *Bacillus thuringiensis*.

thy1 (CDw90; formerly theta-antigen) Differentiation antigen (19 KD glycoprotein) on surface of T-cells, neurons, endothelial cells and fibroblasts. GPI-anchored and a member of the ***immunoglobulin superfamily*** with only one V-type (variable) domain.

thylakoids Membranous cisternae of the chloroplast, found as part of the ***grana*** and also as single cisternae interconnecting the grana. Contain the photosynthetic pigments, reaction centres and electron-transport chain. Each thylakoid consists of a flattened sac of membrane enclosing a narrow intrathylakoid space.

thymectomy The excision of the thymus by operation, radiation or chemical means.

thymic aplasia (hypoplasia) A lack of T-lymphocytes, due to failure of the thymus to develop, resulting in very reduced cell-mediated immunity though serum immunoglobulin levels may be normal. See also ***DiGeorge syndrome***.

thymidine Term that is always used in practice for the nucleoside thymine deoxyriboside; not the riboside which naming of the other nucleosides might lead one to expect.

thymidine block A method for synchronizing cells in culture. In the absence of thymidine, DNA synthesis cannot occur, so cells are blocked before S phase; release of the block allows synchronous entry into cycle.

thymidine kinase (TK) Enzyme of pyrimidine salvage, catalysing phosphorylation of thymine deoxyriboside to form its 5′ phosphate, the nucleotide thymidylate. Animal cells lacking this enzyme can be selected by lethal synthesis, eg. by resistance to bromodeoxyuridine, and can be used as parentals in somatic hybridization, since they are unable to grow in ***HAT medium***.

thymine (2,6-di-hydroxy, 5-methylpyrimidine; 5-methyluracil) Pyrimidine base found in DNA (in place of uracil of RNA).

thymine dimer Dimer that can be formed in DNA by covalent linkage between two adjacent (*cis*) thymidine residues, in response to ultraviolet irradiation. Occurrence potentially mutagenic, although repair enzymes exist that can excise thymine dimers. See ***xeroderma pigmentosum***.

thymocyte Lymphocyte within the thymus; the term is usually applied to an immature lymphocyte.

thymoma A tumour of thymic origin.

thymopentin Biologically active pentapeptide corresponding to residues 32–36 of thymopoietin. Will induce prothymocytes and activate peripheral T-cells.

thymosin Peptide (28 amino acids) that restores aspects of immune function ***in vivo*** and ***in vitro***. Possibly a thymic hormone.

thymosin β-4 Small protein (5 kD: 43 residues) found in large amounts in many vertebrate cells (approximately 0.2 mM in neutrophils) and that binds ***G-actin*** thereby inhibiting polymerization.

thymus The lymphoid organ in which T-lymphocytes are educated, composed of stroma (thymic epithelium) and lymphocytes, almost entirely of the T-cell lineage. In mammals the thymus is just anterior to the heart within the ribcage; in other vertebrates in rather undefined regions of the neck or within the gill chamber in teleost fish. The thymus regresses as the animal matures.

thymus-derived lymphocyte See ***T-cell***.

thyroglobulin The 650 kD protein of the thyroid gland that binds thyroxine.

thyroid hormones Thyroxine and tri-iodothyronine are hormones secreted by the thyroid gland in vertebrates. These iodinated aromatic amino acid compounds influence growth and metabolism and, in amphibia, metamorphosis. The hormone ***calcitonin*** which has hypocalcaemic effects is also of thyroid origin but is not usually classed with thyroxine and tri-iodothyronine as a thyroid hormone. See also Tables H2 and H3.

thyroid-stimulating antibodies Long-acting thyroid stimulator is an autoantibody found in many cases of primary thyrotoxicosis which causes hyperplasia of the thyroid by undetermined mechanisms. Human thyroid-stimulating immunoglobulin is a different antibody found in all or nearly all cases of primary thyrotoxicosis and may act by binding to the thyrotropin (TSH) receptor site, causing increased synthesis of ***thyroglobulin***.

thyroid-stimulating hormone (TSH; thyrotropin) Polypeptide hormone (28 kD), secreted by the anterior pituitary gland, that activates ***cyclic AMP*** production in thyroid cells leading to production and release of the ***thyroid hormones***.

thyroiditis Disease of the thyroid, especially Hashimoto's disease, in which autoimmune destruction of the thyroid takes place.

thyroliberin See ***thyrotropin-releasing hormone***.

thyrotropin-releasing hormone (protirelin; TRH; thyroliberin; TRF) Tripeptide (pyroGlu-His-Pro-NH_2) that releases ***thyrotropin*** from the anterior pituitary by stimulating ***adenylate cyclase***. May also have ***neurotransmitter*** and ***paracrine*** functions.

thyrotropin See ***thyroid-stimulating hormone***.

thyroxine (T4; tetra-iodothyronine) See ***thyroid hormones*** and Table H3.

Ti plasmid Plasmid of ***Agrobacterium tumefaciens***, transferred to higher plant cells in crown gall disease, carrying the T-DNA that is incorporated into the plant cell genome. Used as a vector to introduce foreign DNA into plant cells.

Tie (Tie1; Tie2/Tek) Endothelium-specific receptor tyrosine kinase required for normal embryonic vascular development and tumour angiogenesis. Associates with p85 of PI3kinase. Ligand for Tie2 is ***angiopoietin***. ***VEGF*** is a ligand for Tie1.

tight junction See ***zonula occludens***.

tim *See* ***timeless***.

time-lapse Technique applied to speed up the action in a film or videotape sequence. In filming by taking a frame every few seconds and projecting at conventional speed (16 or 24 frames per second), the movements of cells can be greatly speeded up, and then become conspicuous. With videotape, the recording is made at slow tape speed and replayed at full speed. The opposite of slow-motion.

time-resolved fluorescence Method to avoid interference by autofluorescence. Using an emitter fluorochrome that has slow decay characteristics coupled to the reagent of interest and temporally separating excitation and measurement, the signal can be arranged to derive almost entirely from the reporter fluorophore. (Autofluorescence decays very rapidly.) See ***HTRF***.

timeless *(tim) Drosophila* gene essential for the production of circadian rhythms. The protein product, TIM, may be necessary for the accumulation of the PER protein, the product of the ***period*** gene. TIM and PER associate with one another and the regulated interaction seems to determine the entry of PER into the nucleus: both TIM and PER are produced in a circadian cycle.

TIMP See ***tissue inhibitors of metalloproteinases***.

TIP See ***tonoplast intrinsic protein***.

tissue Group of cells, often of mixed types and usually held together by extracellular matrix, that perform a particular function. Thus, tissues represent a level of organization between that of cells and of organs (which may be composed of several different tissues). Sometimes used in a more general sense, eg. epithelial tissue, where the common factor is the pattern of organization, or connective tissue, where the common feature is the function.

tissue culture Originally the maintenance and growth of pieces of explanted tissue (plant or animal) in culture away from the source organism. Now usually refers to the (much more frequently used) technique of cell culture, using cells dispersed from tissues, or distant descendants of such cells.

tissue culture plastic Polystyrene that has been rendered wettable by oxidation, a treatment that increases its adhesiveness for cells from animal tissues, and without which ***anchorage-dependent*** cells will not grow. Commercially achieved by treatment known as glow discharge.

tissue factor Integral membrane glycoprotein of around 250 residues that initiates blood clotting after binding factors VII or VIIa.

tissue inhibitors of metalloproteinases (TIMP) Family of proteins of around 200 residues that can inhibit metalloproteinases, eg. collagenase, by binding to them.

tissue plasminogen activator (TPA; tPA) Plasma serine protease, one of a closely related group of ***plasminogen activators***. Contains an ***EGF-like domain*** and multiple copies of the ***kringle*** domain.

tissue-typing The process of determining the allelic types of the antigens of the ***major histocompatibility complex*** (MHC) that determine whether a tissue graft will be accepted or rejected. At present carried out either by use of polyclonal or monoclonal antibodies against MHC antigens, or less usually by tests of MHC-restricted cell function or skin grafting (the latter not in humans).

titin (connectin) Family of enormous proteins (2000–3500 kD) found in the sarcomere of striated muscle. Form a scaffolding of elastic fibres that may be important for correct assembly of the sarcomere. Each titin molecule spans from ***M line*** to ***Z disc***.

TL antigens The mouse antigens coded for by the ***TLa complex***; in normal animals only found on intrathymic lymphocytes, but also seen on leukaemic cells (hence, thymus leukaemia antigen) in certain forms of the disease in mice. The molecules have structures similar in some ways to Class I MHC products but are disulphide bonded tetramers of two 45 kD chains and two 12 kD chains of ***beta-2-microglobulin*** type.

TLa complex Genes coding for and controlling ***TL antigens***; the complex is situated close to the H-2 complex on mouse chromosome 17 and resembles H-2 in several ways.

TLC (thin-layer chromatography) Chromatographic separation method in which a thin layer of the solid phase (often silica, aluminium oxide or cellulose) is fixed onto a glass or plastic sheet. Can be run one-dimensionally or in a second dimension with a different solvent system. Much used in separation of lipids. Visualization can be by staining or radioactive labelling and autoradiography.

TLCK (tosyl lysyl chloromethylketone) Protease inhibitor, particularly effective against trypsin and papain.

TMB-8 Inhibitor of the release of calcium from intracellular stores.

TMV See ***tobacco mosaic virus***.

TNF receptor (TNF-R; CD120) There are two receptors for ***TNFα***. Type I (CD120a; 55 kD) is present on most cell types and type II (CD120b; 75 kD) is mainly restricted to haematopoietic cells. Both types bind TNFα and lymphotoxin (TNFβ) and are members of the NGF receptor family. The two types have substantial sequence homology except in their cytoplasmic domains, and have different signalling capacities. TNFα RI contains a ***death domain*** and interacts with a number of cytoplasmic proteins (***TRADD, TRAF, FADD***).

TNFα See ***tumour necrosis factor***.

tobacco mosaic virus (TMV) Plant RNA virus, the first to be isolated. Consists of a single central strand of RNA (a helix of 6500 nucleotides) enclosed within a coat consisting of 2130 identical capsomeres that, in the absence of the RNA, will self-assemble into a cylinder similar to the normal virus but of indeterminate length. Causes mottling of the leaves of the tobacco plant.

Toc complex Transport system of the outer membrane of the chloroplast. Analogous to ***Tom complex*** though proteins are not the same. Toc75 seems to form the pore, Toc159 and Toc 34 are thought to be

GTP-regulated import receptors on the cytoplasmic side, Toc 34 function is unknown.

tocopherol (α-tocopherol; vitamin E) Protects unsaturated membrane lipids from oxidation and may prevent free-radical damage.

Togaviridae Class IV viruses with a single positive strand RNA genome. Bullet-shaped capsid, enveloped by a membrane formed from the host cell plasma-membrane; the budded membrane contains host lipids and viral ('spike') glycoproteins. The group can be divided into two main groups: alphaviruses, which include ***Semliki Forest virus*** and ***Sindbis virus***; and ***Flaviviridae***, which include yellow fever virus and rubella (German measles) virus. Many are transmitted by insects and were previously classified as ***arboviruses***.

tolbutamide A sulphonylurea that will bind to ***SUR*** and enhance insulin release from pancreatic ***B-cells***. Used in type II diabetes (as are a number of similar drugs).

tolerance The development of specific non-reactivity to an antigen. See ***immunological tolerance***.

toll *Drosophila* gene required for dorsoventral polarity determination. Protein (124 kD) is a transmembrane receptor with leucine-rich repeat. Interacts downstream with ***pelle*** and ***tube*** and defines dorsoventral polarity in the embryo. Toll, which is present over the entire surface of the embryo, is activated ventrally by interaction with a spatially restricted, extracellular ligand.

toluidine blue (CI Basic Blue 17) A thiazin dye related to Methylene Blue and Azure A in structure; often used for staining thick resin sections. Typically exhibits metachromasia.

Tom complex (translocase of outer membrane) Transport complex of the outer membrane of mitochondrion. The complex contains eight different proteins: Tom40 (40 kD) forms the 2.2nm hydrophilic pore and spans the outer membrane; Tom5, Tom6 and Tom7 are embedded within the membrane adjacent to Tom40; Tom20, Tom 22, Tom37, and Tom70 are on the cytosolic face with Tom22 on the inner face as well. The comparable system in chloroplasts is the ***Toc complex***.

tonic See ***adaptation***.

tonofilaments Cytoplasmic filaments (10nm diameter: ***intermediate filaments***) inserted into ***desmosomes***.

tonoplast Membrane that surrounds the vacuole in a plant cell.

tonoplast intrinsic protein (TIP) Plant protein, closely related to ***major intrinsic protein***. Found in plant storage vacuolar membranes.

tophus Mass of urate crystals surrounded by a chronic inflammatory reaction: characteristic of gout.

topographic map The spatially ordered projection of neurons onto their target; eg. in the retino-tectal projection, retinal ganglion cell axons project along the ***optic nerve*** to the contralateral tectum where they ramify to form terminal arbors. The target sites of the terminal arbors are ordered: neurons from a specific region of the retina consistently project to a specific region of the tectum, forming a map of the retina on the tectum.

topographical control Those phenomena of cell behaviour in which the shape of the local substrate of the cell affects its behaviour, see eg. ***contact guidance***.

topoinhibition Term used to describe the inhibition of cell proliferation as the cells become closely packed on a culture dish: generally superseded by the term ***density-dependent inhibition***.

topoisomerases Enzymes that change the degree of supercoiling in DNA by cutting one or both strands. Type I topoisomerases cut only one strand of DNA; type I topoisomerase of ***E. coli*** (omega protein) relaxes negatively supercoiled DNA and does not act on positively supercoiled DNA. Type II topoisomerases cut both strands of DNA; type II topoisomerase of *E. coli* (DNA gyrase) increases the degree of negative supercoiling in DNA and requires ATP. It is inhibited by several antibiotics, including nalidixic acid and ovobiocin.

TOR (target of rapamycin) Components of the ***ras/MAP kinase*** signalling pathway, originally characterized in yeast. Inhibited by ***rapamycin*** that is bound to ***FKBP*** 12.

Torres body Intranuclear inclusion body in liver cells infected with yellow fever virus (***Togaviridae***).

torso *(tor)* A ***receptor tyrosine kinase*** (EC. 2.7.1.112) activated at the poles of the *Drosophila* embryo. Activation of *torso* triggers expression of gap genes that operate in these areas by antagonizing Gro-mediated repression (see ***groucho***).

torus Structure found at the centre of a bordered ***pit***, especially in conifers, forming a thickened region of the pit membrane. When subjected to a pressure gradient, it seals the pit by pressing against the pit border.

totipotent Capable of giving rise to all types of differentiated cell found in that organism. A single totipotent cell could, by division, reproduce the whole organism.

toxic shock syndrome Endotoxic shock caused by bacterial contamination of tampons; the toxin responsible is produced by some strains of *Staphylococcus aureus*.

toxigenicity The ability of a pathogenic organism to produce injurious substances that damage the host.

toxin A naturally produced poisonous substance that will damage or kill other cells. Bacterial toxins are frequently the major cause of the pathogenicity of the organism in question. See ***endotoxins*** and ***exotoxins***.

toxoid Non-toxic derivative of a bacterial exotoxin produced by formaldehyde or other chemical treatment: useful as a vaccine because it retains most antigenic properties of the toxin.

Toxoplasma A genus of parasitic protozoa. *T. gondii* is an intracellular parasite whose intermediate hosts includes humans, the final host being felines of many species. Causes toxoplasmosis in humans in which the parasite finally locates in tissues such as brain, heart, the eye causing serious and sometimes fatal lesions.

TP-1 See ***trophoblast protein 1***.

TPA (1) See ***tissue plasminogen activator***. (2) A ***phorbol ester*** tumour promoter, 12-O-tetradecanoyl-phorbol-13-acetate also known as PMA, phorbol myristyl acetate.

TPCK (tosyl phenyl chloromethyl ketone) Nonspecific protease inhibitor, interacts with histidine residues and will inactivate many enzymes by interfering with the active site.

TphiCG loop (T ϕ CG) See ***T loop of RNA***.

TPR motif (tetratricopeptide motif) Degenerate consensus sequence of 34 residues found in various proteins that are involved in the regulation of RNA synthesis, protein import and *Drosophila* development.

trabecular bone See ***cancellous bone***.

tracheid Water-conducting cell forming part of the plant ***xylem***. Contains thick, lignified secondary cell walls, with no protoplast at maturity. Interconnects with neighbouring tracheids through pits; the end walls are not perforated (cf. ***vessel elements***).

TRADD (TNF receptor 1-associated death domain protein) Binds to TNF-R cytoplasmic domain and to FADD and RIP though does not itself seem to have any catalytic activity. Contains a ***death domain***.

TRAF (TNF receptor-associated protein) One of the various proteins that associates with the cytoplasmic domain of the TNF receptor.

TRAIL (TNF-related apoptosis-inducing ligand; Apo2L) An orphan member of the TNF ligand family that can be expressed either as a transmembrane protein (32 kD) or in soluble form; induces apoptosis in haematopoietic cell lines. Structurally similar to CD95 ligand.

tram (translocating chain-associating membrane protein) Transmembrane glycoprotein (probably crosses eight times) of endoplasmic reticulum (36 kD) apparently required for the translocation of nascent proteins into the cisternal space. A component of the ***translocon***. Abundant: potentially as many TRAM molecules as there are associated ribosomes.

TRAMP (DR3; wsl; Apo-3) Member of the ***death receptor*** family. Ligand not identified.

trans*-Golgi network** A complex of membranous tubules and vesicles, near the *trans* face of the ***Golgi apparatus, which is thought to be a major intersection for intracellular traffic of vesicles.

trans-splicing (of RNA) Splicing of two different pre-mRNA molecules together. Seems to rely on intron-like sequences. Contrasts with the normal *cis*-splicing of conventional RNA molecules.

transactivation Stimulation of transcription by a ***transcription factor*** binding to DNA and activating adjacent proteins.

transacylase An enzyme that transfers an acyl group, eg. transacetylase that transfers an acetyl group from acetyl-lipoamide to coenzyme A.

transaldolase Together with transketolase, links the pentose phosphate pathway with glycolysis by converting pentoses to hexoses.

transaminases Enzymes that convert amino acids to keto acids in a cyclic process using pyridoxal phosphate as cofactor, eg. aspartate aminotransferase catalyses the reaction: aspartate + α-ketoglutarate = oxaloacetate + glutamate. See ***amino transferases***.

transcriptase See ***reverse transcriptase***.

transcription Synthesis of RNA by RNA polymerases using a DNA template.

transcription factor Protein required for recognition by RNA polymerases of specific stimulatory sequences in eukaryotic genes. Several are known that activate transcription by RNA polymerase II when bound to ***upstream*** promoters. Transcription of the 5S RNA gene in *Xenopus* by RNA polymerase III is dependent on a 40 kD protein TFIIIA which binds to a regulatory site in the centre of the gene, and was the first protein found to exhibit the metal-binding domains known as ***zinc fingers***. See also Table T1.

transcription squelching Anomalous suppression of transcription of a gene by overexpression of a transcription factor that would be expected to raise transcription levels. Thought to be caused by sequestration of a limiting cofactor by the overexpressed transcription factor.

transcription unit A region of DNA that is transcribed to produce a single primary RNA transcript, ie. a newly synthesized RNA molecule that has not been processed. Transcription units can be mapped by kinetic studies of RNA synthesis, and in some instances directly visualized by electron microscopy.

transcriptional control Control of gene expression by controlling the number of RNA transcripts of a region of DNA. A major regulatory mechanism for differential control of protein synthesis in both pro- and eukaryotic cells.

transcriptional silencing Mechanism of transcriptional control where DNA is bundled into ***heterochromatin*** in order to make it permanently inaccessible for future transcription. Effectively, this allows for memory in the ***determination*** of cell fate in developing organisms. In *Drosophila*, ***homeotic genes*** are silenced by members of the **Polycomb** group of genes.

transcytosis Process of transport of material across an epithelium by uptake on one face into a coated vesicle, which may then be sorted in the endosomal compartment, and then delivery to the opposite face of the cell, still within a vesicle.

Table T1. Transcription factors

Superclass: basic domains
Class:leucine zipper factors (bZIP)

Family	Subfamily	Examples
AP-1(-like) components		
	Jun	XBP-1, v-Jun, c-Jun, JunB, JunD, dJRA
	Fos	v-Fos, c-Fos, FosB, Fra-1, Fra-2, dFRA, LRF-1
	Maf	v-Maf, c-Maf, MafB, MafK, MafF, MafG, NRL, kreisler
	NF-E2	NF-E2 p45, Nrf1, Nrf2, ECH, Cnc
	fungal AP-1-like factors	GCN4, CPC1, yAP-1, yAP-2, Pap1+
	CRE-BP/ATF	CREB-2, ATF-3, CRE-BP. CRE-BPa, ATF-a. yATF
	Others	Zta, CYS3
CREB		
	CREB	
	ATF-1	
	CREM	
	BBF-2	
	dCREB2	
	SKO1	
	HAC1	
	Pcr1	
C/EBP-like factors		
	C/EBPα	
	C/EBPβ	
	C/EBPγ	
	C/EBPδ	
	C/EBPε	
	CHOP-10	
	slbo	
	AcC/EBP	
bZIP/PAR		
	DBP	
	VBP	
	Hlf	
	TEF	
Plant G-box binding factors		
	CPRF-2 ('V')	CPRF-2
	EmBP-1 ('E')	HBP-1a(1), EmBP-1a, EmBP-1b, GBF1 (maize), GBF2, GBF3, CPRF-1, TAF-1
	HBP-1a ('Q')	HBP-1a, HBP-1a(c14), GBF9, GBF1 (A.t., G.m.), GBF4, GBF12, CPRF-3
	TGA1a ('L/M')	TGA1a, HBP-1b, HBP-1b(c1)
	TGA1b ('R')	TGA1b, O2
ZIP only		
	SWI6	
	SWI4	
	STE4	
	IREBF-1	
	GCF	
Other bZIP factors		
	Giant	
	OPI1	
Class: helix-loop-helix factors (bHLH)		
Ubiquitous (class A) factors		
	E2A	
	E2-2	
	m3	
	HEB / SCBP	
	Daughterless	

Table T1. (Continued)

Superclass: basic domains
Class: helix-loop-helix factors (bHLH) (continued)

Family	Subfamily	Examples
Myogenic transcription factors		
	MyoD	
	Myogenin	
	Myf-5	
	MRF4	
	SUM-1	
	CeMyoD	
	Nau	
	Esc1	
Achaete–Scute		
	Lethal of Scute	
	Scute	
	Achaete	
	Asense	
	MASH-1	
	MASH-2	
	ASH-3a	
	ASH-3b	
Tal/Twist/Atonal/Hen		
	Lymphoid factors	Tal-1, Tal-2, Lyl-1
	Mesodermal twist-like factors	Twist, M-Twist, X-Twist, Dermo-1, bHLH-EC2, Th1, SGC1
	HEN	HEN1, HEN2
	Atonal	NeuroD/BETA2, LIN-32, Ato, MATH-1, MATH-2
	Pancreatic factors	INSAF, BETA3
Hairy		
	Hairy	Hairy, Deadpan, HES-1, HES-2, HES-3, HES-5, Stra13
	Esp	E(spl)m5, E(spl)m7, E(spl)m8
	Fungal regulators	PHO4, NUC-1
Factors with PAS domain		
	AhR	
	Arnt	
	Single-minded	
INO		
	INO2	
	INO4	
HLH domain only		
	Emc	
	Id1	
	Id2	
	Id3	
	Id4	
	Olf-1	
Other bHLH factors		
	Delilah	
	Lc	
	CBF1	
Class: helix-loop-helix/leucine zipper factors (bHLH-ZIP)		
Ubiquitous bHLH-ZIP factors		
	TFE3	TFE3, TFEB, TFEC, Mi
	USF	SpF1, USF, USF2
	SREBP	SREBP-1, SREBP-2
	AP-4	AP-4

Table T1. (Continued)

Superclass: basic domains
Class: helix-loop-helix/leucine zipper factors (bHLH-ZIP) (continued)

Family	Subfamily	Examples
Cell-cycle controlling factors		
	Myc	c-Myc, N-Myc, L-Myc, L-Myc2 (X), B-Myc, v-Myc
	Mad/Max	Max, Mad1, Mxi1, Mad3, Mad4
	E2F	E2F-1, E2F-2, E2F-3, E2F-4, E2F-5, dE2F
	DRTF	DRTF1/DP-1, DP-2, dDP
Class: NF-1		
NF-1		
	NF-1A /mNF-1B	
	NF-1B	
	NF-1C	
	NF-1X	
Class: RF-X		
	RF-X	RF-X1, RF-X2, RF-X3, RF-X5
Superclass: Zinc-coordinating DNA-binding domains		
Class: Cys4 zinc finger of nuclear receptor type		
Steroid hormone receptors		
	Corticoid receptors	GR, MR
	Progesterone receptor	PR
	Androgen receptor	AR
	Estrogen receptor	ER
Thyroid hormone receptor-like factors		
	Retinoic acid receptors	RAR-α, RAR-β, RAR-γ, RAR-δ
	Retinoid X receptors	RXR-α, RXR-β, RXR-γ, USP
	Thyroid hormone receptors	T3R-α, T3R-β 851, 840, 852, 853, Rev-ErbAα
	Vitamin D receptor	VDR
	NGFI-B	NGFI-B
	FTZ-F1	SF-1, FTZ-F1
	PPAR	PPARα, PPARβ, PPARγ
	EcR	EcR
	ROR	HR3, RORα/RZRα, RZRβ, RORγ
	Tll/COUP	Tailless, Tlx, TR2,, TR4, COUP-TFI, ARP-1/COUP-TFII
	HNF-4	HNF-4α, HNF-4β, HNF-4γ
	CF1	CF1
	Knirps	Knirps
Class: diverse Cys4 zinc fingers		
GATA-Factors		
	vertebral GATA-factors	GATA-1, GATA-2, GATA-3, GATA-4
	fungal metabolic regulators	AREA/NIT-2, GLN3, UGA 43, NTL 1
Trithorax		
	Ttx	
	Hrx	
Other factors		
	BUF2	

Table T1. (Continued)

Superclass: Zinc-coordinating DNA-binding domains		
Family	Subfamily	Examples
Class: Cys2His2 zinc finger domain		
Ubiquitous factors		
	TFIIIA	
	Sp1	
	Sp3	
	Sp4	
	YY1	
Developmental/cell cycle regulators		
	Egr/Krox	SWI5, Egr-1, Egr-2, Egr-3
	Kruppel-like	Kruppel, Hunchback, Glass, Odd-skipped, Ovo, Snail, CF2, Evi-1, Ikaros, MZF-1, Sdc-1, NRSF, NRSF form 1, NRSF form 2, Gfi-1
	GLI-like	GLI, GLI3, WT1, Tra-1, RME1, BrlA
	Others	Teashirt, Tramtrack, RGM1
Metabolic regulators in fungi		
	ACE2	
	ADR1	
	MIG1	
	CreA	
	MSN2	
	MSN4	
Large factors with NF-6B-like binding properties		
	HIV-EP1	
	HIV-EP2	
	MBP-2	
	KBP-1	
	αA-CRYBP1	
	AGIE-BP1	
Viral regulators		
		T-Ag
Class: Cys6 cysteine-zinc cluster		
Metabolic regulators in fungi		
	ARG RII	
	CAT8	
	GAL4	
	LAC9	
	HAP1	
	MAL63	
	LEU3	
	UGA3	
	qa-1F	
	UME6	
	AmdR	
	PUT3	
Class: zinc fingers of alternating composition		
Cx7Hx8Cx4C zinc fingers		
	BAF1	
Cx2Hx4Hx4C zinc fingers		
	Byr3	
Superclass: Helix-turn-helix		
Class: homeo domain		
Homeo domain only		
	AbdB	Abd-B, Ceh-11, HOXA9, HOXB9, HOXC9, HOXD9, HOXA10, HOXC10, HOXD10, HOXA11, HOXC11, HOXD11, HOXC12, HOXD12, HOXA13, HOXC13, HOXD13

Table T1. (Continued)

Superclass: Helix-turn-helix
Class: homeo domain (continued)

Family	Subfamily	Examples
	Antp	abd-A, Antp, Ceh-15, Dfd, Flh, Ftz, HOXA2, HOXB2, HOXA3, HOXB3, (m), (m), HOXD3, HOXA4, HOXB4, HOXC4, HOXD4, HOXA5, HOXB5, HOXC5, HOXA6, HOXB6, HOXC6, HOXA7, (X.l.), (X.l.), HOXB7, HOXB8, HOXC8, HOXD8, IPF1, Mab-5, NK-1, Pb, Scr, Ubx, Zen-1, Zen-2
	Cad	Cad, Cdx-1, Cdx-2, Cdx-3
	Cut	CDP, Cut
	Dll	Dll, Dlx-1
	Ems	Ems, EMX1, EMX2
	En	En, En-1, En-2
	Eve	Eve, Evx-1
	Prd	al, Alx3, Gsc, K-2, S8, Otd, Otx1, Otx2, Phox-2, Unc-4
	HD-ZIP	HAT1, HAT2, HAT3, HAT4, HAT7, HAT9, HAT14, HAT22, Athb-1
	H2.0	HB24
		Hox11/Hlx
	HNF1	HNF-1, vHNF-1
	Lab	HOXA1, HOXB1, HOXD1, Lab
	Msh	Msx-1, Msx-2
	NK-2	NK-2, NK-3, NK-4, Nkx-2.2, Nkx-2.5, Nkx-6.1, Tinman, TTF-1
	Bcd	Bcd
	XANF	Hesx1/XANF-1
	PBC	Ceh-20, Exd, MATa1, Pbx1, Pbx2, Pbx3
	not assigned	Gtx, KN1, Knox3, MATα1, MATα2, Pc, PHO2, Prh, Ro, Zeste, Zmhox1a, HAT24, Unc-30, HB9, BarH1, BarH2, Aα Y1, Aα Y2, Aα Y3, α2-1, β2-1, d1-1
POU domain factors		
	I	Pit-1
	II	Oct-1, Oct-2, Oct-11, dOct-2, PDM-1, PDM-2, Nrl-16, Oct-, Oct-3, Oct-7, Oct-2
	III	POU-M1, N-Oct-3, Oct-6, Brn-4, Cf1a, Brn-1, Nrl-19, Nrl-20, XLPOU-1/Nrl-22
	IV	Brn-3a, Brn-3b, Brn-3c, I-POU, Unc-86
	V	Oct-3/4, Oct-3C ?, pou2, Oct-25, Oct-60, Oct-91
	VI	Brn-5, TCFbeta1, pou[c]
	other POU factors	CEH-18, Sprm-1, b1-1
Homeo domain with LIM region		
	Homeo domain with LIM region	Lin-11, Isl-1, Lmx-1, Lim-1, Lim-3, LH-2, Ap, Mec-3, Isl-2, Lim-2, Lmx2
	LIM-only transcription (co-)factors	MLP, DMLP1
Homeo domain plus zinc finger motifs		
	ATBF1	ATBF1-B, ATBF1-A
	Zfh1	
	Zfh2	
Class: paired box		
Paired plus homeo domain		
	Prd	
	Pax-3	
	Pax-6	
	Pax-7	
	Gsb	
	Gsbn	

Table T1. (Continued)

Superclass: Helix-turn-helix
Class: paired box (continued)

Family	Subfamily	Examples
Paired domain only		
	Pax-1	
	Pax-2	
	Pax-5	
	Pax-8	
	Poxn	
Class: forkhead/winged helix		
Developmental regulators		
		Fkh, Slp1, Slp2, lin-31, XFD-1, BF-1, v-Qin, c-Qin, Axial, Croc, Whn
Tissue-specific regulators		
		HNF-3α, HNF-3β, HNF-3γ, HFH-4, SGF-1
Other regulators		
		ILF, FKHR, HTLF, QRF-1, HCM1, FD1, FD2, FD3, FD4, FD5, HFH-1, HFH-2, HFH-3, HFH-4, HFH-5, HFH-6, HFH-7, HFH-B2, HFH-B3, Fkh-1, Fkh-2, Fkh-3, Fkh-4, Fkh-5, Fkh-6, BF-2
Class: heat-shock factors		
HSF		
	HSF1	
	HSF2	
	HSF3	
	dHSF	
	HSF24	
	HSF30	
	HSF8	
	fungal HSF	
Class: tryptophan clusters		
Myb		
	Myb-factors	c-Myb, A-Myb, B-Myb, v-Myb, GL1, MybSt1, P, C1, MYB.Ph2, MYB.Ph3, ATMYB1, ATMYB2
	Myb-like factors	BAS1, REB1, FlbD, Adf-1, RAP1
Ets-type		
		c-Ets-1, Ets-2, v-Ets, PEA3, Elk-1, SAP-1, SAP-2, Erg-1, Fli-1, PU.1, Spi-B, E4TF1-60/GABP-alpha, Elf-1, Tel, E74, E74A, E74B, yan, Pnt, P1, P2, Elg, D-Ets-3, D-Ets-4, D-Ets-6, N-Ets-3, ER71, ER81
Interferon-regulating factors		
		IRF-1, IRF-2, IRF-3, Pip, ICSBP, LSIRF-2, ISGF-3γ
Class: TEA domain		
TEA		
		TEF-1, Sd, TEC1, abaA
Superclass: β-scaffold factors with minor groove contacts		
Class: RHR (Rel homology region)		
Rel/ankyrin		
	NF-κB1, NF-κB2, RelA, RelB, c-Rel, v-Rel, dorsal	
ankyrin only		
	IκBα, IκBβ, IκBγ, IκBR, Bcl-3, cactus	

Table T1. (Continued)

Superclass: β-scaffold factors with minor groove contacts
Class: RHR (Rel homology region) (continued)

Family	Subfamily	Examples
NF-AT		
	NF-ATc	
	NF-ATp	
	NF-ATx	
	NF-ATc3	
Class: p53		
p53		
	p53	
Class: MADS box		
Regulators of differentiation		
	MEF-2	MEF-2A, MEF-2B, MEF-2C, MEF-2D, D-MEF2
	homeotic genes	AG, AP1, DEF A, GLO, NMH7, ZEM, PI, PMADS3, Fbp2, Fbp3, AGL1, AGL2, AGL3, AGL4, AGL5, AGL6, SQA, O-MADS, TAG1, TDR3, TDR4, TDR5, TDR6, NAG1, Tobmads1, MADS1
	yeast regulators	MCM1, YBR182C
Responders to external signals		
	SRF	
	RLM1	
Metabolic regulators		
	ARG RI	
Class: **β*-Barrel*** **α*-helix transcription factors***		
E2		
	E2, EBNA-1	
Class: TATA-binding proteins		
TBP		
	TBP	
Class: HMG		
SOX		
		SRY, Sox-2, Sox-4, Sox-5, Sox-9, Sox-18, SOX-LZ, Sox-8, Sox-10, Sox-11, Sox-12, Sox-13, Sox-14, Sox-15, Sox-16
TCF-1		
	TCF-1α, TCF-1, TCF-3, TCF-4	
HMG2-related		
	SSRP1, Dm-SSRP1, Ixr1, DSP1	
UBF		
	UBF, xUBF1	
MATA		
	mat-Mc	
Other HMG box factors		
	IRE-ABP, MNB1b, Rox1	
Class: heteromeric CCAAT factors		
Heteromeric CCAAT factors		
	CP1A/HAP3, CP1B/HAP2, CBF-C/HAP4	
Class: grainyhead		
Grainyhead		
		CP2, LBP-1a, Grainyhead

Table T1. (Continued)

Family	Subfamily	Examples
Superclass: β-scaffold factors with minor groove contacts		
Class: cold-shock domain factors		
csd		
		DbpA, DbpB, FRG Y1, FRG Y2
Class: runt		
Runt		
	PEBP2αA	PEBP2αA/AML3
	PEBP2αB	PEBP2αB/AML1, Ch-runtB2
	PEBP2αC1/AML2	
	Runt	Runt
	Lozenge	Lozenge
Superclass: other transcription factors		
Class: copper fist proteins		
Fungal regulator		
	ACE1/CUP2, AMT1	
Class: HMGI(Y)		
	HMGI(Y)	HMG I(Y), HMGI-C, cHMGI
Class: STAT		
STAT		
	STAT1	STAT2, STAT3, STAT4, STAT5A, STAT5B, STAT6
Class: pocket domain		
Rb		
	Rb, p107	
CBP		
	CBP	
Class: E1A-like factors		
E1A		
	E1A	

Classification is based on TRANSFAC, Release 3.4 [http://transfac.gbf-braunschweig.de/TRANSFAC/index.html]. T Heinemeyer, E Wingender, I Reuter, H Hermjakob, AE Kel, OV Kel, EV Ignatieva, EA Ananko, OA Podkolodnaya, FA Kolpakov, NL Podkolodny and NA Kolchanov on Transcriptional Regulation: TRANSFAC, TRRD, and COMPEL. *Nucleic Acids Res.* (1998); 26: 362–367.

transcytotic vesicle Membrane-bounded vesicle that shuttles fluid from one side of the endothelium to the other. There is some controversy as to whether or not transcytotic vesicles form pores.

transdetermination Change in determined state observed in experiments on *Drosophila* ***imaginal discs***. These can be cultured for many generations in the abdomen of an adult, where they proliferate but do not differentiate. If transplanted into a larva, they differentiate after pupation according to the disc from which they were derived; they maintain their determination. Occasionally the disc will differentiate into a structure appropriate to another disc. This is termed transdetermination. It is a rare event, involves a population of cells, and certain changes are more common than others, eg. leg to wing is more frequent than wing to leg.

transdifferentiation Change of a cell or tissue from one differentiated state to another. Rare, and has mainly been observed with cultured cells. In newts the pigmented cells of the iris transdifferentiate to form lens cells if the existing lens is removed.

transducin A ***GTP-binding protein*** found in the disc membrane of ***retinal rods*** and ***cones***: of the part of the cascade involved in transduction of light to a nervous impulse. A complex of three subunits; α (39 kD), β (36 kD) and γ (8 kD). Photoexcited rhodopsin interacts with transducin and promotes the exchange of GTP for GDP on the α subunit. The GTP-α subunit dissociates from the complex and activates a cGMP-phosphodiesterase by removing an inhibitory subunit. The α subunit of transducin can be ADP-ribosylated by cholera toxin and pertussis toxin.

transduction (1) The transfer of a gene from one bacterium to another by a ***bacteriophage***. In generalized transduction any gene may be transferred as a result of accidental incorporation during phage packaging. In specialized transduction only specific genes can be transferred, as a result of improper recombination out of the host chromosome of the ***prophage*** of a ***lysogenic*** phage. Transduction is an infrequent event but transducing phages have proved useful in the genetic analysis of bacteria. (2) The conversion of a signal from one form to another. For example, various types of sensory cells convert or transduce light, pressure, chemicals, etc. into nerve impulses and the binding of many hormones to receptors at the cell surface is transduced into an increase in cAMP within the cell.

transfection The introduction of DNA into a recipient eukaryote cell and its subsequent integration into the recipient cell's chromosomal DNA. Usually accomplished using DNA precipitated with calcium ions though a variety of other methods can be used (eg. ***electroporation***). Only about 1% of cultured cells are normally transfected. Transfection is analogous to bacterial transformation but in eukaryotes ***transformation*** is used to describe the changes in cultured cells caused by ***tumour viruses***. Though originally used to describe the situation in which the transfected DNA is integrated, it is now frequently used just to mean introduction of DNA into a target cell, hence the necessity to specify ***stable transfection***.

transfer cell Parenchyma cell specialized for transfer of water-soluble material to or from a neighbouring cell, usually a phloem sieve tube or a xylem tracheid. Elaborate wall ingrowths greatly increase the area of plasma membrane at the cell face across which transfer occurs.

transfer factor A dialysable factor obtained from sensitized T-cells by freezing and thawing, that may possibly immunopotentiate animals. The transfer of specific immunity from one animal to another has been claimed.

transfer RNA See ***tRNA***.

transferase A suffix to the name of an enzyme indicating that it transfers a specific grouping from one molecule to another, eg. acyl transferases transfer acyl groups.

transferrin The iron storage protein (80 kD) found in mammalian serum; a β-globulin. Binds ferric iron with a K_{ass} of around 21 at pH 7.4; 18.1 at pH 6.6. An important constituent of growth media. Transferrin receptors on the cell surface bind transferrin as part of the transport route of iron into cells.

transformasome Membranous extension responsible for binding and uptake of DNA; found on the surface of transformation-competent *Haemophilus influenzae* bacteria.

transformation Any alteration in the properties of a cell that is stably inherited by its progeny. Classical example was the transformation of *Diplococcus pneumoniae* to virulence by DNA, achieved in 1944 by Avery, MacLeod & McCarty. Currently usually refers to malignant transformation, but is used in other senses also, such as blast transformation of lymphocytes, which can be distinguished only by context. Malignant transformation is a change in animal cells in culture that usually greatly increases their ability to cause tumours when injected into animals. (It is assumed that parallel changes occur during carcinogenesis ***in vivo***). Transformation can be recognized by changes in growth characteristics, particularly in requirements for macromolecular growth factors, and often also by changes in morphology.

transformed cell See ***transformation***.

transforming genes Genes, originally of tumour viruses, responsible for their ability to transform cells. The term now serves as an operational definition of ***oncogenes***.

transforming growth factor (TGF) Proteins secreted by transformed cells that can stimulate growth of normal cells. Unfortunate misnomer, since they induce aspects of transformed phenotype, such as growth in semi-solid agar, but do not actually transform. TGF-α, a 50 amino acid polypeptide originally isolated from viral-transformed rodent cells, contains ***EGF-like domain*** and binds to EGF receptor. Stimulates growth of microvascular endothelial cells, ie. is angiogenic. TGF-β polypeptide, a homodimer of two 112 chains, is secreted by many different cell types, stimulates wound healing but *in vitro* is also a growth inhibitor for certain cell types. The TGF family includes many of the bone morphogenetic proteins (BMPs).

transforming virus Viruses capable of inducing malignant transformation of animal cells in culture. Among the ***Oncovirinae***, non-defective viruses that lack oncogenes can induce tumours such as leukaemias in animals, but cannot transform ***in vitro***. On acquisition of oncogenes they become (acute) transforming viruses.

transgelin Transformation and shape change-sensitive isoform of 21 kD actin binding protein. Highly conserved (as far back as yeast), binds ***F-actin*** (1: 6 transgelin: ***G-actin***) and causes gelation. Similar, but not identical, to ***calponin***.

transgene A ***gene*** or DNA fragment from one organism that has been stably incorporated into the ***genome*** of another organism (usually plant or animal).

transgenic Adjective describing an organism (usually plant or animal) that contains a ***transgene***.

transgenic organisms Organisms that have integrated foreign DNA into their germ line as a result of the *experimental* introduction of DNA.

transglutaminase An important extracellular enzyme that catalyses the formation of an amide bond between side chain glutamine and side chain lysine residues in proteins with the elimination of ammonia. The linkage is stable and plays an important role in many extracellular assembly processes.

transglycosylation Transfer of a glycosidically bound sugar to another hydroxyl group.

transhydrogenase Direct transfer of a hydrogen atom from NADH to NADPH as catalysed by enzymes in mitochondria from liver or heart.

transient expression (transient transfection) In ***transfection*** of animal cells, cells in which the transgene has not been physically incorporated into the genome (***stable transfection***), but is carried as an episome that can be lost. This means that expression levels will not be constant over time, and will eventually fall away.

transin Protease secreted by carcinoma cells: carboxy-terminal domain has distant homology to haemopexin (haem binding protein), and the N-terminal domain has the proteolytic activity. May be involved in digestion of extracellular matrix.

transition probability model A model to account for the apparently random variation in cell cycle time between individual animal tissue cells in culture, that postulates that transition from G1 to S phase is probabilistic. Contrasts with hypotheses that require the accumulation of critical levels of particular proteins.

transition proteins In ***spermatogenesis***, a group of proteins that displace ***histones*** from nuclear DNA, and that are in turn displaced by ***protamines*** to produce the transcriptionally inactive nuclear DNA characteristic of the sperm nucleus.

transition temperature The temperature at which there is a transition in the organization of, for example, the phospholipids of a membrane where the transition temperature marks the shift from fluid to more crystalline. Usually determined by using an ***Arrhenius plot*** of activity against the reciprocal of absolute temperature, the transition temperature being that temperature at which there is an abrupt change in the slope of the plot. In membranes such phase-transitions tend to be inhibited by the presence of cholesterol.

transitional elements Region at the boundary of the ***rough endoplasmic reticulum*** (RER) and the Golgi. ***Transport vesicles*** are responsible for the transfer of secretory proteins from this part of the RER to the Golgi system.

transitional endoplasmic reticulum See ***transitional elements***.

transitional epithelium An epithelial sheet made up of cells that change shape when the epithelium is stretched. Usually a ***stratified epithelium***: best-known example is in the bladder

transketolase See ***transaldolase***.

translation The process that occurs at the ribosome whereby the information in mRNA is used to specify the sequence of amino acids in a polypeptide chain.

translational control Control of protein synthesis by regulation of the translation step, eg. by selective usage of preformed mRNA or instability of the mRNA.

translocase (elongation factor G; EF-G) The enzyme that causes peptidyl-tRNA to move from the ***A site*** to the ***P site*** in the ***ribosome*** and the mRNA to move so that the next codon is in position for usage.

translocating chain-associating membrane protein (TRAM) Transmembrane glycoprotein (probably crosses eight times) of endoplasmic reticulum (36 kD) apparently required for the translocation of nascent proteins into the cisternal space. A component of the ***translocon***. Abundant: potentially as many TRAM molecules as there are associated ribosomes.

translocation Rearrangement of a chromosome in which a segment is moved from one location to another, either within the same chromosome or to another chromosome. This is sometimes reciprocal, when one fragment is exchanged for another.

translocon The complex of proteins associated with the translocation of nascent polypeptides into the cisternal space of the ***endoplasmic reticulum*** (ER). The translocon is a multifunctional complex involved in regulating the interaction of ribosomes with the ER as well as regulating translocation and the integration of membrane proteins in the correct orientation. ***Tram, signal peptidase*** and signal recognition protein are among the proteins associated with the translocon.

transmembrane protein A protein subunit in which the polypeptide chain is exposed on both sides of the membrane. The term does not apply when different subunits of protein complex are exposed at opposite surfaces. Most integral membrane proteins are also transmembrane proteins.

transmembrane transducer A system that transmits a chemical or electrical signal across a membrane. Usually involves a transmembrane receptor protein that is thought to undergo a conformation change that is expressed on the inner surface of the membrane. Many such transducing species are dimeric and the conformation change may involve interaction between the two components.

transmigration Migration of cells from one surface of a monolayer of cells to the other side; used particularly of the migration of leucocytes from the lumen of a blood vessel across vascular endothelium and into tissue, and by extension ***in vitro*** models of this process.

transmissible mink encephalopathy One of the transmissible spongiform encephalopathies, though originally thought to be an 'unconventional' type of ***slow virus*** infection. Similar to ***kuru, scrapie***, and ***Creutzfeldt–Jakob disease***. See ***prion***.

transmission electron microscopy (TEM) Those forms of electron microscopy in which electrons are transmitted through the object to be imaged, suffering energy loss by diffraction and to a small extent by absorption.

transpeptidase An enzyme that catalyses the formation of an amide linkage between a free amino group and a carbonyl group within an existing peptide linkage.

transpiration Loss of water vapour from land plants into the atmosphere, causing movement of water through the plant from the soil to the atmosphere via roots, shoot and leaves. Occurs mainly through the ***stomata***.

transplantation antigen Any antigen that is antigenically active in graft rejection. In practice the ***major histocompatibility complex*** and the H-Y antigens, and to a lesser extent minor histocompatibility antigens.

transplantation reaction The set of cellular phenomena observed after an allogeneic (mismatched) graft is made to an organism that leads to destruction, detachment or isolation of the graft. In mammals this includes the invasion and destruction of the graft by cytotoxic lymphocytes, inhibition of ***angiogenesis*** and other processes.

transport diseases Single-gene defect diseases in which there is an inability to transport particular small molecules across membranes. Examples are aminoacidurias such as cystinuria, ***iminoglycinuria, Hartnup disease, Fanconi syndrome***.

transport protein A class of transmembrane protein that allows substances to cross plasma membranes far faster than would be possible by diffusion alone. A major class of transport proteins expend energy to move substances (***active transport***); these are transport ATPases. See ***facilitated diffusion, symport, antiport***.

transport vesicle Vesicles that transfer material from the ***rough endoplasmic reticulum*** (RER) to the receiving face of the Golgi.

transportase See ***transport protein***.

transporter See ***transport protein***.

transportin Component (101 kD) of the nuclear pore complex. Exports RNA from the nucleus (whereas yeast counterpart, Kap104p, imports).

transposable element See ***transposon***.

transposase See ***transposon***.

transposition Movement form one location to another, particularly the movement of a DNA sequence (***transposon***) within the genome.

transposon (transposable element) Small, mobile DNA sequences that can replicate and insert copies at random sites within chromosomes. They have nearly identical sequences at each end, oppositely oriented (inverted) repeats, and code for the enzyme, transposase, that catalyses their insertion. Bacteria have two types of transposon: simple transposons that have only the genes needed for insertion: and complex transposons that contain genes in addition to those needed for insertion. Eukaryotes contain two classes of mobile genetic elements: the first are like bacterial transposons in that DNA sequences move directly; the second (retrotransposons) move by producing RNA that is transcribed by reverse transcriptase into DNA which is then inserted at a new site.

transthyretin Plasma protein (4.5mM in plasma) that transports ***thyroxine***. Tetrameric with four identical 127 residue subunits. Transthyretin forms a complex under physiological conditions with retinol binding protein (2mM in plasma) so that RBP is not lost by filtration in the kidney.

transudate Plasma-derived fluid that accumulates in tissue and causes ***oedema***. A result of increased venous and capillary pressure, rather than altered vascular permeability (which leads to cellular exudate formation).

transverse tubule See ***T-tubule***.

transversions ***Point mutation*** in which a purine is substituted by a pyrimidine or *vice versa*.

tre Human ***oncogene*** probably encoding a transcription factor.

TRE (thyroid hormone response element) DNA sequence recognized by the thyroid hormone receptor.

treadmilling Name given to the proposed process in microtubules in which there is continual addition of subunits at one end and disassembly at the other, so that the tubule stays of constant length but individual subunits move along. Could in principle be used as a transport mechanism, although this is not currently favoured as a possibility. Has also been suggested for microfilaments.

trefoil motif (P domain) Domain found in various secretory polypeptides that has highly conserved cysteine residues that are disulphide bonded in such a way as to generate a trefoil structure (bonded 1–5, 2–4, 3–6). There are also highly conserved A, G and W residues.

trehalose A disaccharide sugar (342 D) found widely in invertebrates, bacteria, algae, plants and fungi, formed by the dimerization of glucose. Yields glucose on acid hydrolysis.

trephones Substance supposedly released at a wound that stimulate mitosis: the opposite of ***chalones***.

Treponema Genus of bacteria of the spirochaete family (Spirochaetaceae). *T. pallidum* causes syphilis. Cells are corkscrew-like (6–15μm long, 0.1–0.2μm wide), motile, anaerobic, and with a ***peptidoglycan*** cell wall and a capsule of glycosaminoglycans similar to hyaluronic acid and chondroitin sulphate in composition. Membrane has ***cardiolipin***.

TRF Alternative name for ***thyrotropin-releasing hormone***.

TRF1 (telomerase repeat binding factor 1) Major protein component (33 kD) of ***telomeres***, colocalizes with telomeres in interphase and is located at chromosome ends during mitosis.

triacyl glycerols See ***triglycerides***.

triad (triad junction) The junction between the ***T-tubules*** and the ***sarcoplasmic reticulum*** in striated muscle.

tricarboxylic acid cycle (TCA cycle; citric acid cycle; Krebs cycle) The central feature of oxidative metabolism. Cyclic reactions whereby acetyl CoA is oxidized to carbon dioxide providing reducing equivalents (NADH or $FADH_2$) to power the electron transport chain. Also provides intermediates for biosynthetic processes.

trichocyst Small membrane-bounded vesicle lying below the pellicle of many ciliates. Fusion of the trichocyst with the plasma membrane occurs at a predictable site which can therefore be examined for membrane specialization.

trichohyalin (THH) Major structural protein of inner root sheath cells and medulla of hair follicle, present in small amounts in other specialized epithelia. Trichohyalin is a high molecular weight α-helix-rich protein that forms rigid structures as a result of postsynthetic modification by transglutaminases that crosslink the proteins and peptidyl-arginine deiminase that converts arginine to citrulline and modifies structure of the protein. Modified trichohyalin is thought to serve as a ***keratin*** intermediate filament-associated matrix protein, like ***filaggrin***.

Trichonympha Genus of flagellated protozoans symbiotic in the intestine of some cockroaches and termites where they are responsible for the digestion of cellulose.

trichothecenes (T-2 toxin; HT-2 toxin; diacetoxyscirpenol; deoxynivalenol) Mycotoxins produced by various species of fungi that contaminate various agricultural products. Are toxic for granulocytic and erythroblastic progenitor cells.

triflavin See ***disintegrin***.

trifluoperazine (TFP; trifluperazine; Stellazine) Antipsychotic drug that inhibits ***calmodulin*** at levels just below those at which it kills cells.

trigeminal system Neurons associated with the fifth or trigeminal nerve, the largest cranial nerve. The trigeminal system provides sensory innervation to the face and mucous membrane of the oral cavity, along with motor innervation to the muscles of mastication. It is called trigeminal because it has three major peripheral branches, the opthalmic, the maxillary and the mandibular nerves.

trigger protein See ***U-protein***.

triglycerides Storage fats of animal adipose tissue where they are largely glycerol esters of saturated fatty acids. In plants they tend to be esters of unsaturated fatty acids (vegetable oils). Present as a minor component of cell membrane. Important energy supply in heart muscle.

trigramin See ***disintegrin***.

trimethoprim A drug that inhibits the reduction of dihydrofolate to tetrahydrofolate (a later step than that inhibited by ***sulphonamides***). Selective for some bacterial DHF reductases and often used in conjunction with sulphonamides.

trinucleotide repeat Repetitive part of a genome that may form part of the coding sequence of a gene. The length of such repeats is frequently ***polymorphic***, and unstably amplified repeats appear to be the major cause of such genetic diseases as ***Huntington's chorea***, ***fragile X syndrome***, spinobulbar muscular atrophy and myotonic dystrophy.

trio Multidomain protein (2861 residues) that binds ***LAR*** transmembrane tyrosine phosphatase, has a serine/threonine protein kinase domain and separate rac- and rho-specific ***GEF***-domains.

triodobenzoic acid (TIBA) An inhibitor of basipetal ***auxin*** transport in plants.

triple response The vascular changes in the skin in response to mild mechanical injury, an outward-spreading zone of reddening (flare) followed rapidly by a weal (swelling) at the site of injury. Redness, heat, and swelling, three of the 'cardinal signs' of inflammation, are present.

triploid Having three times the haploid number of chromosomes.

triskelion A three-legged structure assumed by clathrin isolated from ***coated vesicles***. A trimer of clathrin (180 kD) with three light chains is probably the physiological subunit of clathrin coats in coated vesicles.

trisomy An additional copy of a chromosome so that there are three copies instead of two in a diploid organism. Best known example is trisomy 21 in ***Down's syndrome***.

tritium(^3H) Long-lived radioactive isotope of hydrogen (half-life 12.26 years). Weak β-emitter, very suitable for autoradiography, and relatively easy to incorporate into complex molecules.

Triton X-100 Non-ionic detergent used in isolating membrane proteins: the detergent replaces the phospholipids that normally surround such a protein. Other detergents of the Triton group are occasionally used so the full name should be quoted.

Triturus Genus of newts, much studied for their ***lampbrush chromosomes***.

trk ***Oncogene***, from human colon carcinoma, encoding a ***receptor tyrosine kinase***. See Table O1. The *trk* gene product is a receptor for ***NGF***, that of the *trkB* gene the receptor for ***neurotrophin*** 4 and ***BDNF***, and the *trkC* gene the receptor for NT-3.

tRNA (transfer RNA; sRNA; 4S RNA) The low molecular weight RNAs that specifically bind amino acids by aminoacylation to form ***aminoacyl tRNA***, and which possess a special nucleotide triplet, the anticodon, sometimes containing the base inosine. They recognize codons on mRNA. By this recognition the appropriate tRNAs are brought into alignment in turn in the ribosome during protein synthesis (translation), there being at least one species of tRNA for each amino acid. In practice most cells possess about 30 types of tRNA. The amino acids are bound at the 3′ terminus that is always 3′-ACC. The anticodon is around 34–38 nucleotides from the 5′ end and the total length of the various tRNAs is 70–80 bases.

trochophore Free-living ciliated larval form of several different invertebrate phyla.

trophectoderm The extra-embryonic part of the ectoderm of mammalian embryos at the blastocyst stage before the mesoderm becomes associated with the ectoderm.

trophic Concerning food or nutrition. Not to be confused with tropic (stimulatory).

trophoblast Extra-embryonic layer of epithelium that forms around the mammalian blastocyst, and attaches the embryo to the uterus wall. Forms the outer layer of the chorion, and together with maternal tissue will form the placenta.

trophoblast protein 1 (TP-1) Protein secreted by ***trophoblasts,*** which prolongs the lifetime of the corpus luteum, thus signalling pregnancy. Structurally related to ***interferons***.

trophozoite The feeding stage of a protozoan (as distinct from reproductive or encysted stages).

tropocollagen Subunit from which collagen fibrils self-assemble: generated from ***procollagen*** by proteolytic cleavage of the extension peptides.

tropomodulin Tropomyosin-binding protein that weakens tropomyosin–actin interaction. Found in erythrocyte membrane skeleton and in various non-erythroid cells.

tropomyosin Protein (66 kD) associated with actin filaments both in cytoplasm and (in association with ***troponin***) in the thin filament of striated muscle. Composed of 2 elongated α-helical chains (each about 33 kD), 40nm long, 2nm diameter. Each chain has six or seven similar domains and interacts with as many ***G-actin*** molecules as there are domains. Not only does the binding of tropomyosin stabilize the ***F-actin,*** but the association with troponin in striated muscle is important in control by calcium ions.

troponin Complex of three proteins, troponins C, I and T, associated with ***tropomyosin*** and actin on the thin filament of striated muscle, upon which it confers calcium sensitivity. There is one troponin complex per tropomyosin. Troponin C (18 kD) binds calcium ions reversibly, has a variable number of ***EF-hand*** motifs and is the least variable of the subunits. TnC binds TnI and TnT, but not actin. Troponin I (23 kD) binds to actin and at 1: 1 stoichiometry can inhibit the actin–myosin interaction on its own. Troponin T (37 kD) binds strongly to tropomyosin.

trp (1) ***Tryptophan***. (2) Operon encoding tryptophan metabolism genes in *E. coli.* (3) Transient receptor potential mutant in *Drosophila,* gene codes for a constitutively active calcium permeable channel activated by ***thapsigargin*** (though trp1 is insensitive to thapsigargin). (4) Human tyrosinase-related protein 1.

Trp See ***tryptophan***.

TRP proteins (transient receptor potential proteins) Family of proteins that may be the channels that are activated by depletion of intracellular calcium. Probably have six transmembrane regions (based on sequence analysis) and the limited sequence homology with voltage-gated calcium channel α1 subunits. Human Trp3 is the homologue of *Drosophila* TRP and TRPL, proteins involved in light sensitive conductance in the eye. Mammalian Trp6 encodes an ion channel that mediates calcium entry stimulated by a Gq-coupled receptor. ***Vanilloid receptor-1*** has some structural similarity.

TRT (telomerase reverse transcriptase; hEST2) The protein subunit (123 kD) that catalyses telomeric DNA extension has sequence and functional characteristics of a ***reverse transcriptase*** related to retrotransposon and retroviral reverse transcriptases. The *Saccharomyces cerevisiae* homologue has been found and subsequently identified as EST2 (ever shorter telomeres).

Trypan blue Biological stain used to determine cell viability. Trypan blue is unable to cross intact plasma membranes, and so only labels dead cells.

Trypanosoma (trypanosome) Genus of Protozoa that causes serious infections in humans and domestic animals. African trypanosomes of the *brucei* group are carried by Tsetse flies (*Glossina*) and, when they enter the bloodstream of the mammalian host, go through a complex series of stages. Perhaps the most interesting feature is that there are recurrent bouts of parasitaemia as the parasite alters its surface antigens to evade the immune response of the host (see ***antigenic variation***). The repertoire of antigenic variation is considerable. The South American trypanosomes (of which *T. cruzi* is the best known) are carried by reduviid bugs, and cause a chronic and incurable disease (Chagas disease). Other interesting features of trypanosomes are the kinetoplast DNA and glycosomes (organelles containing enzymes of the glycolytic chain).

trypanosomiasis Disease caused by ***Trypanosoma***.

trypsin Serine protease (EC 3.4.21.4; 23 kD) from the pancreas of vertebrates. Cleaves peptide bonds involving the amino groups of lysine or arginine.

tryptophan (Trp; W) One of the 20 amino acids found in proteins (204 D). Essential dietary component in humans. Precursor of nicotinamide. See Table A2.

TSG-14 (TNF-stimulated gene 14) TNF-inducible gene of fibroblasts encoding a protein of the ***pentraxin*** family.

TSH See ***thyroid-stimulating hormone***.

TSH-releasing factor A tripeptide produced by the hypothalamus that stimulates the anterior pituitary to release ***thyroid-stimulating hormone*** (***TSH***).

Tst Gene found as part of a 15.2 kb genetic element in some strains of *Staphylococcus aureus*; codes for toxic shock syndrome toxin 1.

TTX See ***tetrodotoxin***.

tube *Drosophila* mutant. Tube is a maternally encoded protein that, together with ***pelle*** transduces the signal from ***toll***. Toll, Cactus and Dorsal, along with Tube and Pelle, participate in a common signal transduction pathway to specify the embryonic dorsal-ventral axis.

tubercle Chronic inflammatory focus, a ***granuloma***, caused by *Mycobacterium tuberculosis*.

tuberculin skin test See ***Mantoux test***, ***Heaf test***.

tuberin Putative ***GTPase-activating protein*** (GAP) for ***Rap1*** and rab5. Encoded by tumour suppressor gene TSC2. See ***tuberous sclerosis***.

tuberous sclerosis Autosomal dominant disorder caused by mutation in tumour suppressor genes TSC1 or TSC2. Disease characterized by range of features including seizures, mental retardation, renal dysfunction and dermatological abnormalities. TSC1 encodes ***hamartin***, TSC2 encodes ***tuberin*** which has rap- and rab-***GAP*** activity.

tubocurarine An alkaloid that acts as a muscle relaxant by blocking acetylcholine (ACh) receptors.

tubulin Abundant cytoplasmic protein (55 kD), found mainly in two forms, α and β . A tubulin ***heterodimer*** (one α, one β), constitutes the ***protomer*** for microtubule assembly. Multiple copies of tubulin genes are present (and are expressed) in most eukaryotic cells studied so far. The different tubulin isoforms seem, however, to be functionally equivalent. γ-tubulin is localized in the ***centrosome*** and is involved in nucleation of microtubule assembly during the cell cycle. Highly conserved from yeast to mammals.

tularemia Disease of rodents and rabbits caused by *Pasteurella tularense*. Can infect humans (either transmitted by the deerfly or by direct contact with the bacterium).

tumorigenic Capable of causing tumours. Can refer either to a carcinogenic substance or agent such as radiation that affects cells, or to transformed cells themselves.

tumour (USA tumor) Strictly, any abnormal swelling, but usually applied to a mass of neoplastic cells.

tumour angiogenesis factor (TAF) Substance(s) released from a tumour that promotes vascularization of the mass of neoplastic cells. Once a tumour becomes vascularized, it will grow more rapidly, and is more likely to metastasize. TAF is almost certainly more than one substance. See ***angiogenin***.

tumour cell Cell derived from a tumour in an animal. Refers to a tumour-causing malignant cell, and not an adventitious normal cell. Loosely, a transformed cell able to give rise to tumours.

tumour initiation First stage of ***tumour*** development. See also ***tumour progression***.

tumour necrosis factor (TNF) TNFα or ***cachectin***, originally described as a tumour-inhibiting factor in the blood of animals exposed to bacterial ***lipopolysaccharide*** or ***Bacille Calmette–Guerin*** (BCG). Preferentially kills tumour cells ***in vivo*** and ***in vitro***, causes necrosis of certain transplanted tumours in mice and inhibits experimental metastases. Human TNFα is a protein of 157 amino acids and has a wide range of pro-inflammatory actions. Usually considered a cytokine. Soluble TNFα is released from the cell surface by the action of TACE (TNFα–converting enzyme), a metalloproteinase. TNFβ (***lymphotoxin***) has 35% structural and sequence homology with TNFα and binds to the same ***TNF receptors***. Unlike TNFα, TNFβ has a conventional signal sequence and is secreted from activated T- and B-cells.

tumour progression Second stage of ***tumour*** development. See also ***tumour initiation***.

tumour promoter Agent that in classical studies of carcinogenesis in rodent skin was able to increase the sensitivity of tumour formation by a previously applied primary carcinogen, but was unable to induce tumours when used alone. Important example was croton oil, active ingredients of which are now believed to be phorbol esters. These are believed to act as analogues of diacylglycerols, and may activate protein kinase C. Strictly speaking, not the same as a co-carcinogen, which is defined as being active when administered at the same time. Tumour promoters generally are carcinogens when tested more stringently.

tumour-specific antigen (tumour-specific transplantation antigen; TSTA) Antigen on tumour cells detected by cell-mediated immunity. For virus-transformed cells TSTA (unlike ***T-antigen***) is found to differ for different individual tumours induced by the same virus. May consist of fragments of T-antigens exposed at the cell surface.

tumour suppressor (anti-oncogene; cancer susceptibility gene) A gene that encodes a product that normally negatively regulates the cell cycle, and that must be mutated or otherwise inactivated before a cell can proceed to rapid division. Examples: ***p53***, RB (retinoblastoma), WT-1 (Wilms' tumour), DCC (deleted in colonic carcinoma), NF-1 (neurofibrosarcoma) and APC (adenomatous polyposis coli).

tumour virus Virus capable of inducing tumours.

tumourigenesis The creation of a ***tumour***.

TUNEL method (transferase-mediated dUTP nick-end labelling) Enables the visualization of cells undergoing apoptosis by labelling the ends of their fragmented DNA.

tunicamycin Nucleoside antibiotics from *Streptomyces lysosuperificus* that act in eukaryotic cells to inhibit N-glycosylation. Tunicamycin inhibits the first step in synthesis of the dolichol-linked oligosaccharide, by preventing the addition of N-acetyl glucosamine to dolichol phosphate.

turgor The pressure within cells, especially plant cells, derived from osmotic pressure differences between the inside and outside of the cell giving rise to mechanical rigidity of the cells. Turgor drives cell expansion and certain movements such as the closing or opening of stomata.

Turner's syndrome Genetic defect in humans in which there is only one X chromosome (affected individuals are therefore phenotypically female), probably as a result of meiotic non-disjunction.

turnover number Equivalent to Vmax, being the number of substrate molecules converted to product by one molecule of enzyme in unit time, when the substrate is saturating.

TW-240/260 kD protein Protein (240/260 kD) found in the terminal web of intestinal epithelial cells. Probably an isoform of ***spectrin*** and ***fodrin***.

twinfilin Yeast actin-depolymerizing factor containing two ***ADF-H domains***. Localizes to the cortical actin cytoskeleton. Will sequester ***G-actin*** by forming tight 1:1 complex but does not seeem to crosslink filaments. Human homologue has been identified in fibroblasts.

twitch muscle Striated muscle innervated by a single motoneuron and having an electrically excitable membrane that exhibits an all-or-none response (cf. tonic muscle): in mammals almost all skeletal muscles are twitch muscles. Physiologists often divide muscles into fast- and slow-twitch types, the fast-twitch muscles being associated with fast motor units.

twitchin Large protein (667 kD) associated with myosin and important in muscle assembly. Has multiple fibronectin III-homology repeats. Product of *unc-22* gene in *Caenorhabditis elegans*, mutations which produce animals that show twitching movements.

two-dimensional gel electrophoresis A high-resolution separation technique in which protein samples are separated by isoelectric focusing in one dimension and then laid on an SDS gel for size-determined separation in the second dimension. Can resolve hundreds of components on a single gel.

two-hybrid system Screening system to identify genes encoding proteins which interact specifically with other proteins. One gene is expressed in yeast as a ***fusion protein*** with the DNA-binding site of the ***GAL4*** transcription factor, and the other gene co-expressed as a fusion with the transcriptional activator domain of GAL4. Only if the two proteins interact directly are the two GAL4 domains held in close enough proximity to trigger expression of a ***reporter gene*** (usually ***Lac Z***) downstream of the UASG promoter recognized by GAL4.

TxA2 See ***thromboxanes***.

Ty element Transposable element of yeast, *Saccharomyces cerevisiae*. Each consists of a central region of around 5.6kb flanked by direct repeats of around 330bp. There are multiple Ty elements in each haploid genome.

TY-5 See ***macrophage inflammatory protein 1α***.

tyk-2 Protein tyrosine kinase (135 kD) that has some similarity in its kinase domain with ***JAK*** 1. Important in IFN α/β signalling

tylose A parenchyma cell outgrowth that wholly or partly blocks a ***xylem*** vessel. It grows out from an axial or ray parenchyma cell through a pit in the vessel wall.

type-2 astrocyte See ***astrocyte***.

Tyr See ***tyrosine***.

tyrosinase A copper-containing protein (a monoxygenase) that catalyses the oxidation of tyrosine, and sets in train spontaneous reactions that yield melanin, the black pigment of skin, hair and eyes. The first intermediate is 3,4-dihydroxyphenylalanine (DOPA). Lack of tyrosinase activity is responsible for albinism.

tyrosine (Tyr; Y) One of the 20 amino acids directly coded in proteins (181 D). Non-essential in humans since can be synthesized from phenylalanine. See Table A2.

tyrosine hydroxylase Enzyme required for the synthesis of the neurotransmitters ***noradrenaline*** and ***dopamine***.

tyrosine kinase Kinases that phosphorylate protein tyrosine residues. These kinases play major roles in mitogenic signalling, and can be divided into two subfamilies: receptor tyrosine kinases, which have an extracellular ligand-binding domain, a single transmembrane domain and an intracellular tyrosine kinase domain; and non-receptor tyrosine kinases, which are soluble, cytoplasmic kinases.

tyrosine kinase phosphorylation site Substrates of tyrosine protein kinases are generally characterized by a lysine or arginine 7 residues to the N-terminal side of the phosphorylated tyrosine, and an acidic residue (Asp or Glu) 3 or 4 residues to the N-terminal side of the tyrosine. There are, however, a number of exceptions to this rule, such as the tyrosine phosphorylation sites of enolase and lipocortin II.

tyrosine phosphorylation See ***tyrosine kinase***.

U

U937 Human myelomonocytic cell line frequently used as a model for myeloid cells – though is rather undifferentiated. Derived from a patient with histiocytic leukaemia.

U1 (sRNP) One of the classes of ***small nuclear RNAs***. There are six U-types that have a high uridylic acid content, U1–U5 are synthesized by RNA polymerase II, U6 by RNA polymerase III.

U-protein Hypothetical protein thought to regulate the transition of cells from G0 to G1 phase of the cell cycle, and thus inevitably into S phase. The idea would be that the concentration of this unstable (U) protein would have to exceed a threshold level for triggering progression through the cycle, and that this would only happen if the cell had adequate access to growth factors or to nutrients. Also known as trigger protein.

UAS$_G$ Promoter sequence recognized by the ***GAL4*** transcription factor. Used to control expression of a wide range gene products in ***transgenic*** plants and animals under GAL4 control.

UBC proteins Family of proteins involved in conjugating ***ubiquitin*** to proteins. UBC1, UBC4 and UBC5 have a role in targeting proteins for degradation, but others have more complex roles, including an involvement in cell cycle control (UBC3) and the secretory pathway (UBC6).

ubiquinone (coenzyme Q) Small molecule with a hydrocarbon chain (usually of several isoprene units) that serves as an important electron carrier in the respiratory chain. The acquisition of an electron and a proton by ubiquinone produces ubisemiquinone (a free radical); a second proton and electron convert this to dihydroubiquinone. Plastoquinone, which is almost identical to ubiquinone, is the plant form.

ubiquitin A protein (8.5 kD) found in all eukaryotic cells. Can be linked to the lysine side chains of proteins by formation of an amide bond to its C-terminal glycine in an ATP-requiring process. The protein–ubiquitin complex is subject to rapid proteolysis. Ubiquitin also has a role in the heat-shock response.

ubiquitin carboxy-terminal hydrolase L1 (UCH-L1) An abundant protein in the brain (up to 2% of total protein). A thiol protease that is thought to cleave polymeric ubiquitin to monomers and hydrolyze bonds between ubiquitin and small adducts. Found in ***Lewy bodies*** and mutations in UCH-L1 have been associated with familial forms of Parkinsonism.

ubiquitinoylation The covalent addition of ***ubiquitin*** residues to proteins. Single or multiple residues can be added and bound ubiquitin can also be a site for further addition of ubiquitin residues.

ubisemiquinone See ***ubiquinone***.

UDG (uracil-DNA glycosylase) Abundant ubiquitous enzyme responsible for repair of U–G mispairings.

UDP-galactose (uridine diphosphate-galactose) Sugar nucleotide, active form of galactose for galactosyl transfer reactions.

UDP-glucose (uridine diphosphate-glucose) Sugar nucleotide, active form of glucose for glucosyl transfer reactions.

ulcer Inflamed area where the epithelium and underlying tissue is eroded.

ulcerative colitis Inflammation of the colon and rectum: cause unclear, although there are often antibodies to colonic epithelium and *E. coli* strain 0119 B14.

Ultrabithorax *(Ubx) Drosophila* ***homeotic*** gene that is part of the ***bithorax complex***.

Mutations in *Ubx* affect parasegments 5–6, corresponding to the posterior thorax and anterior abdomen of the adult.

ultracentrifugation Centrifugation at very high *g*-forces: used to separate molecules, eg. mitochondrial from nuclear DNA on a caesium chloride gradient.

ultrafiltration Filtration under pressure. In the kidney, an ultrafiltrate is formed from plasma because the blood is at higher pressure than the lumen of the glomerulus. Also used experimentally to fractionate and concentrate solutions in the laboratory using selectively permeable artificial membranes.

ultrastructure General term to describe the level of organization that is below the level of resolution of the light microscope. In practice, a shorthand term for 'structure observed using the electron microscope', although other techniques could give information about structure in the submicrometre range.

ultraviolet (UV) Continuous spectrum beyond the violet end of the visible spectrum (wavelength less than 400nm), and above the X-ray wavelengths (greater than 5nm). Glass absorbs UV, so optical systems at these wavelengths have to be made of quartz. Nucleic acids absorb UV most strongly at around 260nm, and this is the wavelength most likely to cause mutational damage (by the formation of thymine dimers). It is the UV component of sunlight that causes actinic keratoses to form in skin, but which is also required for vitamin D synthesis.

umbelliferone (7-hydroxycoumarin) Common in many plants. Can be used as a fluorescent pH indicator.

unc-6 Gene identified in studies of neuronal development of the worm, ***Caenorhabditis elegans***. Homologous to ***netrin***.

uncoupling agent Agents that uncouple electron transport from oxidative phosphorylation. Ionophores can do this by discharging the ion gradient across the mitochondrial membrane that is generated by electron transport. In general the term applies to any agent capable of dissociating 2 linked processes.

underlapping Possible outcome of collision between two cells in culture, particularly head-side collision: one cell crawls underneath the other, retaining contact with the substratum, and obtaining traction from contact with the rigid substratum (unlike ***overlapping***, where traction must be gained on the dorsal surface of the other cell).

unequal crossing over ***Crossing over*** between homologous chromosomes that are not precisely paired, resulting in non-reciprocal exchange of material and chromosomes of unequal length. Favoured in regions containing tandemly repeated sequences.

unineme theory Theory that proposes that each chromosome (before S phase) consists of a single strand of DNA. Now generally accepted, and being non-controversial the term has fallen into disuse.

uniport A class of transmembrane ***transport proteins*** that conveys a single species across the plasma membrane.

unit membrane The 3-ply, approximately 7nm-wide membrane structure found in all cells, composed of a fluid lipid bilayer with intercalated proteins. The unit membrane theory carries with it the presumption that all biological membranes have basically the same structure.

unsaturated fatty acid Fatty acid with one or more double bonds. See Table L3.

untranslated region (UTR) The regions of a cDNA, typically that 5′ to the initiation (ATG) site and that 3′ to the stop site, which are not translated to make a peptide. Their functions are not well understood.

unwindase See ***DNA helicase***.

upstream Refers to nucleotide sequences that precede the codons specifying the mRNA or that precede (are on the 5′ side of) the protein-coding sequence. Also used of the early events in any process that involves sequential reactions.

uracil (2,6-dihydroxypyrimidine) The pyrimidine base from which uridine is derived.

uranyl acetate Uranium salt that is very electron-dense, and that is used as a stain in electron microscopy, usually for staining nucleic acid-containing structures in sections.

ure2p Yeast prion-like protein.

urea The final nitrogenous excretion product of many organisms.

uric acid The final product of nitrogenous excretion in animals that require to conserve water, such as terrestrial insects, or have limited storage space, such as birds and their eggs. Uric acid has very low water-solubility, and crystals may be deposited in, eg. butterflies' wings to impart iridescence. See also ***tophus***.

uridine The ribonucleoside formed by the combination of ribose and uracil.

urocanic acid Intermediate in degradation of L-histidine.

Urodela Amphibians of the order Caudata that have tails: newts and salamanders.

urogastrone A peptide isolated from human urine that inhibits gastric acid secretion. Now known to be identical to epithelial growth factor.

uroid Tail region of a moving amoeba.

urokinase (uPA) ***Serine protease*** from kidney that is a ***plasminogen activator***. Contains ***EGF-like domain***.

uromodulin A naturally occurring immunosuppressant (85 kD) originally found in the urine of pregnant women. Identical to Tamm–Horsfall glycoprotein but cell surface-linked by glycosylphosphatidylinositol.

uronic acids Carboxylic acids related to hexose sugars, etc. by oxidation of the primary alcohol group, eg. glucuronic, galacturonic acid.

uroporphyrinogen I synthetase An enzyme of haem biosynthesis that is defective in the inherited (autosomal dominant) disease, acute intermittent porphyria. UP I is isomerized to UP III by UP III synthetase, defective in the autosomal recessive disease, congenital erythropoietic porphyria.

uteroglobin *Progesterone*-binding protein found in lagomorphs (rabbits, hares, etc.), which is also a potent inhibitor of ***phospholipase*** A2. Forms an antiparallel dimer, linked by disulphide bonds at either end. A structurally related protein, CC10, is found in lining of pulmonary airways, and is also a PLA2 inhibitor.

UTR See ***untranslated region***.

utrophin (dystrophin associated protein) Autosomal homologue of ***dystrophin*** (395kD) localized near the neuromuscular junction in adult muscle, though in the absence of dystrophin (ie. in ***Duchenne muscular dystrophy***) utrophin is also located on the cytoplasmic face of the sarcolemma.

uvomorulin Glycoprotein (120 kD) originally defined as the antigen responsible for eliciting antibodies capable of blocking compaction in early mouse embryos (at the morula stage), and inhibiting calcium-dependent aggregation of mouse teratocarcinoma cells. May be the mouse equivalent of LCAM, the chick ***cell adhesion molecule***.

V

V8 protease Protease from *Staphylococcus aureus* strain V8. Cleaves peptide bonds on the carboxyl side of aspartic and glutamic acid residues. Used experimentally for selective cleavage of proteins for amino acid sequence determination or peptide mapping.

V See ***valine***.

V-gene See ***variable gene***.

v-onc General abbreviation for the viral form of an [oncogene], cf. c-*onc*, the normal, cellular [proto-oncogene].

V region Those regions in the amino acid sequence of both the heavy and the light chains of immunoglobulins where there is considerable sequence variability between one immunoglobulin and another of the same class, in contrast to constant sequence (C) regions. The V regions are associated with the antigen-binding areas. They contain hypervariable regions of particularly high sequence diversity.

V-type ATPase (vacuolar ATPase) One of three major classes of ion transport ATPase, characterized by a multi-subunit structure, and a lack of a phosphorylated intermediate. Pumps H^+. Found in intracellular acidic vacuoles and in some proton-pumping epithelia (eg. intercalated cells of kidney). Sensitive to ***bafilomycin***. Related to the ***F-type ATPase***. See also ***P-type ATPase***.

VAC See ***annexin***.

VacA (vacuolating cytotoxin) Protein (600–700 kD: hexamers or heptamers of identical 140 kD monomers) released into culture supernatant by type I ***Helicobacter pylori***. The 140 kD precursor is cleaved to form a 95 kD VacA monomer which is further cleaved to produce 37 and 58 kD fragments that behave as an AB toxin. Causes formation of large vacuoles in epithelial cells.

vaccination The process of inducing immunity to a pathogenic organism by injecting either an antigenically related but non-pathogenic strain (attenuated strain) of the organism or related non-pathogenic species, or killed or chemically modified organism of low pathogenicity. In all cases the aim is to expose the human or animal being vaccinated to an antigenic stimulus that leads to immune protection against disease, without inducing appreciable pathogenesis from the injection.

vaccine An antigen preparation that when injected will elicit the expansion of one or more clones of responding lymphocytes so that immune protection is provided against a disease.

vaccinia Virus of the Orthopoxvirus family used in vaccination against smallpox. Related to, but not identical to, cowpox virus. Also used as a vector for introducing DNA into animal cells.

vacuolar ATPase See ***V-type ATPase***.

vacuole Membrane-bounded vesicle of eukaryotic cells. Secretory, endocytotic, and phagocytotic vesicles can be termed vacuoles. Botanists tend to confine the term to the large vesicles found in plant cells that provide both storage and space-filling functions.

Val See ***valine***.

valine (Val; V) An essential amino acid (117 D). See Table A2.

valinomycin A potassium ***ionophore*** antibiotic, produced by *Streptomyces fulvissimus*. Composed of three molecules (L-valine, D-α-hydroxyisovaleric acid, L-lactic acid) linked alternately to form a 36-membered ring that folds to make a cage shaped like a tennis ball seam. This wraps specifically around potassium ions, presenting them with a hydrophilic interior and a lipid bilayer with a hydrophobic exterior. Potassium is thus free to diffuse through the lipid bilayer. Highly ion-specific, valinomycin is used in ***ion-selective electrodes***.

valvidones Tricyclic compounds with anti-inflammatory properties.

van der Waals' attraction Electrodynamic forces arise between atoms, molecules and assemblies of molecules due to their vibrations giving rise to electromagnetic interactions; these are attractive when the vibrational frequencies and absorptions are identical or similar, repulsive when non-identical. Other interactions originally proposed by van der Waals were included in this name, but these are usually separated into the Coulomb force, the Keesom force and the London force. Only the last is of electrodynamic nature. Probably important in holding lipid membranes into that structure and possibly in other interactions, eg. cell adhesion. Electrodynamic forces between large-scale assemblies can be of relatively long-range nature.

vanadate (VO_4^{3-}) Powerful inhibitor of many, but not all enzymes that cleave the terminal phosphate bond of ATP. The vanadate ion is believed to act as an analogue of the transition state of the cleavage reaction. ***Dynein*** is very sensitive to inhibition by vanadate, whereas ***kinesin*** is relatively insensitive. Similarly, ***tyrosine kinases*** are sensitive to vanadate, but threonine/serine ***protein kinases*** are insensitive.

vancomycin Complex glycopeptide antibiotic produced by actinomycetes. Inhibits ***peptidoglycan*** synthesis. Active against many ***Gram-positive bacteria***.

vanilloid receptor 1 (VR1) Receptor found selectively on sensory neurons, resembling (distantly) receptors of the TRP-type. Protein (95 kD) with six transmembrane domains having some similarity with ***SOCS*** (store-operated calcium channels), though VR-1 does not seem to be a selective calcium channel. Binding of ***capsaicin*** activates the receptor, which acts as a non-specific cation channel and induces death of the cell. Heat will also activate the receptor which has response characteristics similar to thermal nociceptors.

variable antigen Term usually applied to the surface antigens of those parasitic or pathogenic organisms that can alter their antigenic character to evade host immune responses. (See ***antigenic variation***).

variable gene See ***V region***.

variable region See ***V region***.

Varicella zoster Member of the Alphaherpesvirinae: human herpes simplex virus type 3, causative agent of chickenpox and ***shingles***.

variola virus Virus responsible for smallpox. Said to have been completely eradicated. Large DNA virus ('brick-like', 250–390nm × 20–260nm) with complex outer and inner membranes (not derived from plasma membrane of host cell).

vas See ***vasa***.

vasa *(vas) Drosophila* gene involved in oogenesis and embryonic positional specification; a ***DEAD-box helicase***.

vascular anticoagulant See ***annexin***.

vascular bundle Strand of vascular tissue in a plant, composed of xylem and phloem.

vascularization Growth of blood vessels into a tissue with the result that the oxygen and nutrient supply is improved. Vascularization of tumours is usually a prelude to more rapid growth and often to metastasis; excessive vascularization of the retina in diabetic retinopathy can lead indirectly to retinal detachment. Vascularization seems to be triggered by 'angiogenesis factors' that stimulate endothelial cell proliferation and migration. See ***angiogenin, tumour angiogenesis factor***.

vasculitis Inflammation of the blood vessel wall. May be caused by immune complex deposition in or on the vessel wall.

vasoactive intestinal contractor Mouse homologue of ***endothelin***-2.

vasoactive intestinal peptide (VIP) Peptide of 28 amino acids, originally isolated from porcine intestine but later found in the central nervous system, where it acts as a neuropeptide and is released by specific ***interneurons***. May also affect behaviour of cells of the immune system.

vasodilator stimulated phosphoprotein (VASP) A 46/50 kD protein that is a substrate for both cAMP- and cGMP-dependent protein kinases and which is associated with microfilament bundles in many tissue cells. Abundant in platelets; phosphorylation of VASP will inhibit platelet activation.

vasopressin (antidiuretic hormone; ADH) A peptide hormone released from the posterior pituitary lobe but synthesized in the hypothalamus. There are two forms, differing only in the amino acid at position 8: arginine vasopressin is widespread, while lysine vasopressin is found in pigs. Has antidiuretic and ***vasopressor*** actions. Used in the treatment of ***diabetes insipidus***.

vasopressor Any compound that causes constriction of blood vessels (vasoconstriction), thereby causing an increase in blood pressure.

vasotocin Cyclic nonapeptide, related to vasopressin but found in birds, reptiles and some amphibians.

vasp See ***vasodilator stimulated phosphoprotein***.

vault Large cytoplasmic ribonucleoprotein particle that has an eight-fold symmetry with a central pore and petal-like structures giving the appearence of an octagonal dome. May be related to the central plug of the nuclear pore complex.

vav An ***oncogene***, identified in humans, encoding a serine/threonine ***protein kinase***. See Table O1.

VCAM (vascular cell adhesion molecule; CD106) Cell adhesion molecule (90–110 kD) of the immunoglobulin superfamily expressed on endothelial cells, macrophages, dendritic cells, fibroblasts and myoblasts. Expression can be upregulated by inflammatory mediators (IL-1β, IL-4, TNFα, IFN-γ) and it is the ligand for the integrin VLA4.

VDAC Voltage-dependent anion channel.

vector (1) Mathematical term to describe something that has both direction and magnitude. (2) Common term for a plasmid that can be used to transfer DNA sequences from one organism to another. See ***transfection***. Different vectors may have properties particularly appropriate to give protein expression in the recipient, or for cloning, or may have different selectable markers.

vectorette method A method for ***PCR*** cloning an unknown sequence of DNA attached to a known sequence. To the end of the unknown sequence is attached a vectorette, a double-stranded sequence that includes a region of mismatch; primer for the known sequence drives the formation of a second primer site from the mismatch region and the unknown sequence is thus flanked with known sequences and can be specifically PCR amplified.

vectorial synthesis Term usually applied to the mode of synthesis of proteins destined for export from the cell. As the protein is made it moves (vectorially) through the membrane of the ***rough endoplasmic reticulum***, to which the ribosome is attached, and into the cisternal space.

vectorial transport Transport of an ion or molecule across an epithelium in a certain direction (eg. absorption of glucose by the gut). Vectorial transport implies a non-uniform distribution of ***transport proteins*** on the plasma membranes of two faces of the epithelium.

vegetal pole The surface of the egg opposite to the animal pole. Usually the cytoplasm in this region is incorporated into future endoderm cells.

VEGF (vascular endothelial growth factor; vascular permeability factor; VPF) Growth factor of the PDGF family that stimulates mitosis in vascular endothelium, angiogenesis, and also increases permeability of endothelial monolayers. Tissue-specific splice variants (VEGF121, VEGF165, VEGF-C) are found, VEGF165 having heparin-binding activity which VEGF121 lacks. VEGF121 only binds to flk-1. Functional form is a dimer (or heterodimer of splice variants) that binds to flt-1 (VEGF-R1), flt-4 (VEGF-R3, binds only VEGF-C) or flk-1 (VEGF-R2/KDR) receptor tyrosine kinases. Can form heterodimer with placental growth factor (PLGF). See ***Tie*** and ***flt***.

veiled cell A cell type found in afferent lymph and defined (rather unsatisfactorily) on the basis of its morphology. Probably an accessory cell (a ***dendritic cell***) migrating from the periphery (where it is referred to as a ***Langerhans cell*** if in the skin) to the draining lymph node. In the lymph node known as an interdigitating cell and found in the T-dependent areas of spleen or lymph nodes, involved in antigen presentation (Class II MHC positive). Has high levels of surface Ia antigens.

vein (1) Blood vessel that returns blood from the microvasculature to the heart; walls thinner and less elastic than those of artery. (2) In leaves, thickened portion of leaf containing ***vascular bundle***; the pattern, venation, is characteristic for each species.

veliger Free-living larval form of some invertebrates, develops from the trochophore larva.

venom A toxic secretion in animals that is actively delivered to the target organism, either to paralyse or incapacitate or else to cause pain as a defence mechanism. Commonly include protein and peptide toxins.

ventral nervous system defective (*vnd*) A *Drosophila* gene encoding an integral membrane glycoprotein related to ***amyloidogenic glycoprotein***.

verapamil A calcium-channel blocking drug (454 D), used as a coronary vasodilator and anti-arrhythmic.

Vero cells Cell line derived from kidney of African green monkey. Susceptible to a range of viruses.

versene Trivial name for ***EDTA***.

versican (large fibroblast glycoprotein; chondroitin sulphate core protein) Protein (264 kD) involved in cell signalling. N-terminal region similar to glial hyaluronic acid binding protein, centre has glycosaminoglycan attachment sites, C-terminal region has EGF-like repeats.

vesicle A closed membrane shell, derived from membranes either by a physiological process (budding) or mechanically by sonication. Vesicles of dimensions in excess of 50nm are believed to be important in intracellular transport processes. See also ***coated vesicles***.

vesicular stomatitis virus (VSV) ***Rhabdovirus*** causing the disease 'soremouth' in cattle. Widely used as a laboratory tool especially in studies on the spike glycoprotein as a model for the synthesis, post-translational modification and export of membrane proteins.

vesiculin Highly acidic protein (10 kD) found in ***synaptic vesicles***.

vessel Water-conducting system in the ***xylem***, consisting of a column of cells (vessel elements) whose end walls have been perforated or totally degraded, resulting in an uninterrupted tube.

vessel element Part of a ***xylem*** vessel in a higher plant, arising from a single cell. The end walls are perforated and may completely disappear, giving rise to a continuous tube. The remaining walls are thickened and lignified, and there is no protoplast.

VH and VL genes/domains VH and VL genes define in part the sequences of the variable heavy and light regions of immunoglobulin molecules. VH and VL domains are the regions of amino acid sequence so defined. J genes and, in the case of the heavy chain, a D gene (D = diversity) also define these regions. Gene rearrangement also plays a role in determining the sequences in which the genes are joined as the DNA of the immunoglobulin producing cell matures.

VHDL (very high density lipoprotein) Plasma lipoprotein with density greater than 1.21 g/ml. Protein content about 57%, 21% phospholipid, 17% cholesterol and 5% triacylglycerols. Molecular weights between 1.5–2.8 × 10^5kD.

viability test Test to determine the proportion of living individuals, cells or organisms, in a sample. Viability tests are most commonly performed on cultured cells and usually depend on the ability of living cells to exclude a dye (an exclusion test) or to specifically take it up (inclusion test).

Vibrio cholerae Bacterium that causes cholera, the life-threatening aspects of which are caused by the exotoxin (see ***cholera toxin***). Short, slightly curved rods, highly motile (single polar flagellum), ***Gram-negative***. Adhere to intestinal epithelium (adhesion mechanism unknown), and produce enzymes (neuraminidase, proteases) that facilitate access of the bacterium to the epithelial surface.

VIC See ***vasoactive intestinal contractor***.

Vicia faba Broad bean. Often used in plant genetics because cells have only six large chromosomes.

vicilin Seed storage protein of legumes. Protein from *Pisum sativum* is a trimer of 50 kD subunits. High proportion of ***beta-pleated sheet*** (40–50%) and only about 10% α-helix.

vidarabine Adenine arabinoside (Ara-A): nucleoside analogue with antiviral properties that has been used to treat severe herpes virus infections.

villin Microfilament-severing and -capping protein (95 kD) from microvillar core of intestinal epithelial cells. Severs at high calcium concentrations, caps at low.

vimentin ***Intermediate filament*** protein (58 kD) found in mesodermally derived cells (including muscle).

vinblastine Alkaloid (818D) isolated from *Vinca* (periwinkle): binds to ***tubulin*** heterodimer and induces formation of paracrystals rather than tubules. Net result is that microtubules disappear as they disassemble and are not replaced. Used in tumour chemotherapy.

vinca alkaloids See ***vinblastine*** and ***vincristine***.

vincristine Cytotoxic alkaloid that binds to ***tubulin*** and interferes with microtubule assembly. See ***vinblastine***, a related compound.

vinculin Protein (130 kD) isolated from muscle (cardiac and smooth), fibroblasts and epithelial cells. Associated with the cytoplasmic face of ***focal adhesions***: may connect microfilaments to plasma membrane integral proteins through ***talin***.

VIP See ***vasoactive intestinal peptide***.

Vipera ammodytes Western sand viper. See ***ammodytoxins***.

viral antigens Those antigens specified by the viral genome (often coat proteins) that can be detected by a specific immunological response. Often of diagnostic importance.

viral transformation Malignant transformation of an animal cell in culture, induced by a virus.

virgin lymphocyte A lymphocyte that has not, and whose precursors have not, encountered the antigenic determinant for which it possesses receptors.

virion A single virus particle, complete with coat.

viroid Extremely small viruses of plants. Their genome is a 240–350 nucleotide circular RNA strand, extensively base-paired with itself, so they resist RNAase attack. At one time the term was also used casually of self-replicative particles such as the ***kappa particle*** in *Paramecium*.

viropexia The non-specific phagocytosis of virus particles bound to surface receptors.

virus Viruses are obligate intracellular parasites of living but non-cellular nature, consisting of DNA or RNA and a protein coat. They range in diameter from 20–300nm. Class I viruses (Baltimore classification) have double-stranded DNA as their genome; class II have a single-stranded DNA genome; class III have a double-stranded RNA genome; class IV have a positive single-stranded RNA genome, the genome itself acting as mRNA; class V have a negative single-stranded RNA genome used as a template for mRNA synthesis; and class VI have a positive single-stranded RNA genome but with a DNA intermediate not only in replication but also in mRNA synthesis. The majority of viruses are recognized by the diseases they cause in plants, animals and prokaryotes. Viruses of prokaryotes are known as bacteriophages.

viscoelastic Of substances or structures showing non-Newtonian viscous behaviour, ie. elastic and viscous properties are demonstrable in response to mechanical shear.

viscous-mechanical coupling Method by which adjacent ***cilia*** are synchronized in a field. Coupling is through the transmission of mechanical forces, rather than of a synchronizing signal.

visinin See ***recoverin***.

Visna-maedi virus A ***retrovirus*** of sheep and goats. A member of the ***Lentivirinae***, related to ***HIV***. First identified in Iceland when it was introduced by sheep imported from Germany, and causes two diseases: maedi, the most common, is a pulmonary infection (maedi is Icelandic for shortness of breath); visna is due to infection of the nervous system, causing a paralysis similar to ***multiple sclerosis*** (visna is Icelandic for wasting).

visual purple See ***rhodopsin***.

vital stain (vital dye) A stain that is taken up by live cells and that can be used to stain, eg. a group of cells in a developing embryo in order to try to determine a ***fate map***.

vitamin A See ***retinoic acid***.

vitamin C See ***ascorbic acid***.

vitamin D See ***calciferol***.

vitamin E See ***tocopherol***.

vitamin H See ***biotin***.

vitamins Low molecular weight organic compounds of which small amounts are essential components of the food supply for a particular animal or plant. For humans Vitamin A, the B series, C, D_1 and D_2, E, and K are required. Deficiencies of one or more vitamins in the nutrient supply result in deficiency diseases. See Table V1.

vitellin Most abundant protein in egg-yolk.

vitelline membrane The membrane, usually of protein fibres, immediately outside the plasmalemma of the ovum and the earlier stages of the developing embryo. Its structure and composition vary in differing animal groups.

vitellogenic Giving rise to yolk of an egg.

vitellogenin A protein, precursor of several yolk proteins, especially phosvitin and lipovitellin in the eggs of various vertebrates, synthesized in the liver cells after oestrogen stimulation. Also found in large amounts in the haemolymph of female insects, synthesized and released from the fat-body during egg formation.

vitronectin Serum protein (70 kD) also known as serum spreading factor from its activity in promoting adhesion and spreading of tissue cells in culture. Contains the cell-binding sequence Arg-Gly-Asp (RGD) first found in ***fibronectin***.

vitrosin Old term for collagen isolated from embryonic chick vitreous. Synthesized by neural retina at early developmental stages, and by cells of the vitreous body later.

VLA proteins (very late antigens) VLA-1 and VLA-2 were originally defined as antigens appearing on the surfaces of T-lymphocytes 2–4 weeks after ***in vitro*** activation; they are now know to be part of the β ***integrin*** family. Additional

Table V1. Vitamins

Vitamin	Full name	Occurrence	Action	Deficiency disease
Fat-soluble				
A	Retinol (11-cis retinal)	Vegetables	phototransduction, morphogen	Night blindness, xerophthalamia
D	1,25-dihydroxy-cholecalciferol	Action of sunlight on 7-dehydrocholesterol in skin	Ca^{++} regulation, Phosphate regulation	Rickets
E	α-tocopherol	Plants, esp. seeds, wheatgerm	Antioxidant	Failure to grow to maturity, infertility
K_1		Higher green plants		
K_2	Range of molecules	Intestinal bacteria		
Water-soluble				
B_1	Thiamine	Degradation of α-keto acids	Beriberi	
	Folic acid (tetrahydofolic acid)	Plants	Purine biosynthesis	Anaemia
	Nicotinic acid (niacin)	Can be made		Pellagra
	Pantothenic acid (CoA)	Plants, microorganisms		
B_2	Riboflavine	Plants, microorganisms	Constituent of flavoproteins	
B_6	Pyridoxine			
	Pyridoxal			
	Pyridoxamine		Transamination	Acrodynia in rats, convulsions
B_{12}	Cobalamine	Intestinal microorganisms	Hydrogen transfer reactions	Pernicious anaemia
C	Ascorbic acid	Plants, esp. citrus fruits	Cofactor	Scurvy
H	Biotin	Intestinal bacteria	Protects against avidin toxicity, intermediate CO_2 carrier	

members of the subset are now known (VLA-3, VLA-4, VLA-5 and VLA-6), the β subunits all being identical. Some of the VLA proteins are receptors for collagen, laminin or fibronectin, and many are now known to be expressed on cells other than leucocytes.

VLDL (very low density lipoprotein) Plasma lipoproteins with density of 0.94–1.006 gm cm^{-3}; made by the liver. Transport triacylglycerols to adipose tissue. Apoproteins B, C and E are found in VLDL. Protein content, about 10%, much lower than in ***VHDL***.

Vmax The maximum initial velocity of an enzyme-catalysed reaction, ie. at saturating substrate levels.

vnd See ***ventral nervous system defective***.

voltage clamp A technique in electrophysiology, in which a ***microelectrode*** is inserted into a cell, and current injected through the electrode so as to hold the cell's membrane potential at some predefined level. The technique can be used with separate electrodes for voltage sensing and current passing; for small cells, the same electrode can be used for both. Voltage clamp is a powerful technique for the study of ***ion channels***. See ***patch clamp***.

voltage gradient Literally, the electric field in a region, defined as the potential difference between two points divided by the distance between them. Used more loosely, the potential difference across a plasma membrane.

voltage-gated ion channel A transmembrane ***ion channel*** whose permeability to ions is extremely sensitive to the transmembrane potential difference. These channels are essential for neuronal signal transmission, and for intracellular signal transduction. See ***sodium channel***.

voltage-sensitive calcium channels (VSCC) A variety of voltage-sensitive calcium channels are known and on the basis of electrophysiological and pharmacological criteria are grouped into six classes. The general function is to allow calcium influx into the cell as a result of membrane depolarization. The majority (**L-type**, **N-type**, **P-type** and **Q-type channels**) require substantial depolarization and are sometimes collectively known as high voltage-activated types. The ***R-type channels*** activate after moderate depolarization and the ***T-type channel*** opens at relatively negative potentials.

volutin granule Metachromatic granules containing polyphosphate, a linear phosphate polymer found in bacteria, fungi, algae and some higher eukaryotes that may serve as a stock of phosphate.

Volvox A genus of colonial flagellates. The colony is a hollow sphere about 0.5mm in diameter comprising about 50 000 cells embedded in a gelatinous wall and the cells are sometimes connected by cytoplasmic bridges. Each cell has a chloroplast and two flagella.

von Willebrand factor (vWF) Plasma factor involved in platelet adhesion through an interaction with factor VIII. See ***von Willebrand's disease***.

von Willebrand's disease Autosomal dominant platelet disorder in which adhesion to collagen, but not aggregation, is reduced. Both bleeding time and coagulation are increased. Factor VIII levels are secondarily reduced.

Vorticella Genus of ciliate ***protozoa***. It has a bell-shaped body with a belt of cilia round the mouth of the bell, to sweep food particles towards the 'mouth' and a long stalk, connecting it to the substratum, which contains the contractile ***spasmoneme***.

VSCC See ***voltage-sensitive calcium channel***.

VSG Variant surface glycoprotein of trypanosomes. See ***antigenic variation***.

VSV-G tag ***Epitope tag*** (YTDIEMNRLGK) derived from the vesicular stomatitis virus G-protein.

W

w See ***white***.

W See ***tryptophan***.

W locus Mouse coat colour locus, equivalent to the *kit* protooncogene, that encodes a ***receptor tyrosine kinase*** essential for development of haematopoietic and germ cells.

Waardenburg's syndrome Autosomal dominant disorder with deafness and pigmentary disturbances probably as a result of defects in function of ***neural crest***. Various forms of the syndrome are recognized. Waardenburg syndrome 1 (WS1) and WS3 (also known as Klein–Waardenburg syndrome) are caused by mutation in *Pax3* – a homologous defect to the mouse mutant *Splotch* that also has defective *Pax*-3. Waardenburg-Shah syndrome (WS4), in which Waardenburg's syndrome is associated with Hirschsprung's disease, is due to mutation in *Sox10* and there is a homologous mutation in *Dom* mice (dominant megacolon), piebald-lethal and lethal spotting. WS2 is heterogeneous with mutation in the microphthalmia (*MITF*) gene.

Waf-1 (p21; cip1) Inhibitor (21 kD) of ***cdk*** activity; found in a complex with cyclin D, cdk4 and ***PCNA***. Can bind to and inhibit all members of the cdk family though affinity varies. Expression regulated by p53 tumour suppressor. Also found in active cyclin/cdk complexes: multiple copies of Waf1 may be necessary to produce inhibition.

Walker motif Highly conserved ATP-binding region found in ***ABC proteins*** and other ATP-binding proteins.

warfarin Synthetic inhibitor of prothrombin activation and therefore an inhibitor of blood clotting. Also used as a rat poison.

warm antibodies Most IgG antibodies react better at 37°C than at lower temperatures, especially against red cell antigens. These are the warm antibodies as contrasted with ***cold agglutinins***, especially IgM, that agglutinate below 28°C.

wart Benign tumour of ***basal cell*** of skin, the result of the infection of a single cell with wart virus (papillomavirus). Virus is undetectable in basal layer, but proliferates in keratinizing cells of outer layers.

water potential The chemical potential (ie. free energy per mole) of water in plants. Water moves within plants from regions of high water potential to regions of lower water potential, ie. downgradient.

WBC (white blood cell) Term includes neutrophils, eosinophils, basophils, monocytes and lymphocytes.

WD-repeat proteins (WD40) The WD motif is a conserved sequence of approximately 40 amino acids usually ending with tryptophan and aspartic acid (WD). The motif is implicated in protein–protein interactions. Crystal structure of one WD-repeat protein (GTP-binding protein β subunit) reveals that the seven repeat units form a circular propeller-like structure with seven blades.

weal-and-flare See ***triple response***.

weaver In the murine mutation *weaver* there is early apoptotic death during development of cells in testes, cerebellum and midbrain. The defect is caused by a base pair substitution in the G-protein coupled inwardly rectifying potassium channel 2 gene. Up to 70% of the mesostriatal dopaminergic neurons are lost and major alterations of the dopaminergic dendrites of the substantia nigra have been described. The defect does not seem to be due to reduced neurotrophin levels.

wee Cell cycle checkpoint genes found in *Schizosaccharomyces pombe*. Mutants in *wee-1* and *wee-2* have normal growth rate but divide earlier so cells are smaller.

Wegener's granulomatosis A granulomatous vasculitis characterized by upper and lower respiratory tract granulomas and necrotising focal glomerulonephritis. Usually associated with autoantibodies to neutrophil cytoplasmic protease 3 (c-*ANCA*).

Wehi 3b cells Mouse myelomonocytic cells derived from Balb/c mouse. Cells produce IL-3.

Weibel–Palade body Cytoplasmic organelle found in the vascular endothelial cells of some animals, though not in the endothelium of all vessels. Although markers for endothelium, their absence does not necessarily mean the cells are not of endothelial origin.

Weil's disease See ***leptospirosis***.

Weismann's germ plasm theory The theory that organisms maintain genetic continuity from organism to offspring through the germ line cells (germ plasm) and that the other (somatic) cells play no part in the transmission of heritable factors.

Werner's syndrome Rare human genetic disorder characterized by genomic instability and predisposition to cancer. Defect is in protein that has a central domain with homology to helicase and yeast Sgs1 protein but defect does not seem to be due to altered helicase activity. A similar protein is defective in ***Bloom's syndrome***.

Western blot An electroblotting method in which proteins are transferred from a gel to a thin, rigid support (nitrocellulose) and detected by binding of labelled antibody. See ***blots***.

wg See ***wingless***.

Wharton's jelly Viscous hyaluronic acid-rich jelly found in the umbilical cord.

wheatgerm The embryonic plant at the tip of the seed of wheat. Wheatgerm has been used as the starting material for a cell-free translation system and is also the source of ***wheatgerm agglutinin***.

wheatgerm agglutinin (WGA) Lectin from wheat germ that binds to N-acetylglucosaminyl and sialic acid residues. See ***lectins***.

white (*w*) Eye colour gene of *Drosophila*, wild type product essential for red eyes.

white blood cells See ***leucocytes*** and specific classes (***basophils, coelomocytes, eosinophils, haemocytes, lymphocytes, neutrophils, monocytes***).

whole cell patch A variant of ***patch clamp*** technique, in which the patch electrode seals against the cell, with direct communication between the interior of the electrode and the cytoplasm.

Williams–Beuren syndrome Neurodevelopmental disorder with multisystem manifestations caused by heterozygosity for partial deletion of chromosome 7 band 7q11.23 near the ***elastin*** locus

Wilms' tumour Childhood kidney tumour. Like ***retinoblastoma***, both sporadic and inherited forms occur. Believed to be caused by development of homozygosity for a ***deletion mutation*** of the tip of the short arm of chromosome 11, which is presumed to contain a ***tumour suppressor*** gene.

Wilson's syndrome Rare autosomal recessive disease with degenerative changes in the brain and cirrhosis of the liver. Due to a ***ceruloplasmin*** deficiency that leads to excessive deposits of copper in liver, brain and kidney.

winged helix transcription factors Family of transcription factors characterized by a conserved DNA-binding domain found in *Drosophila* homeotic gene ***forkhead*** and rat hepatocyte nuclear factor 3 (HNF-3β). At least 80 genes with this motif are known, many with developmentally-specific patterns of expression. ***FAST*-1** is a member of this family.

wingless (*wg*) *Drosophila* homologue of ***int-1***, functions in pattern formation.

Wiskott–Aldrich syndrome ***Thrombocytopenia*** with severe immunodeficiency (both cell-mediated and IgM production).

Associated with increased incidence of leukaemia.

wnt Multigene family encoding various secreted signalling molecules important in morphogenesis. First member was *Drosophila wingless*, but many vertebrate homologues are now known. *Wnt*-1 induces accumulation of ***β-catenin*** and ***plakoglobin*** and affects the association of APC ***tumour suppressor*** protein with catenin.

wobble hypothesis Explains why the base inosine is included in position 1 in the anticodons of various t-RNAs, why many m-RNA codon words translate to a single amino acid, why there are appreciably fewer t-RNAs than m-RNA codon types, and why the redundant nature of the genetic code translates into a precise set of 20 amino acids. Inosine in position 1 in the anticodon can base pair with A, U or C in position 3 in the m-RNA codon, so that for example UCU, UCC, UCA all code for serine using an inosine anticodon.

Wollaston prism Prism composed of two wedge-shaped prisms with optical axes at right angles. Light emerging has two beams of opposite polarization. Used in differential interference contrast microscopes.

wortmannin Inhibitor of the PI(3) kinases, and possibly other points in the Ras/MAP kinase signalling pathway.

X chromosome A sex chromosome. In mammals paired in females, in amphibia paired in males.

X chromosome inactivation centre (*Xic*) Site on X chromosome responsible for the inactivation of that X chromosome in female mammals (see ***Lyon hypothesis***). Gene responsible, *Xist*, maps to this region and seems to code for a nuclear RNA that colocalizes with the inactivated X chromosome. *Xist* introduced on an autosome is capable of inactivation in *cis* and the *Xist* RNA becomes localized at that autosome.

X-inactivation The inactivation of one or other of each pair of X chromosomes to form the Barr body in female mammalian somatic cells. Thus tissues whose original zygote carried heterozygous X-borne genes should have individual cells expressing one or other but not both of the X-borne gene products. The inactivation is thought to occur early in development and leads to mosaicism of expression of such genes in the body. See also ***Lyon hypothesis***.

X-linked diseases Any inherited disease whose controlling gene or at least part of the relevant genome is carried on an X chromosome, eg. haemophilia. Most known conditions are recessive and thus since males have only one X chromosome they will express any such recessive character. Few dominants are known and the homozygous states are very rare so that female expression of such diseases is uncommon.

X-MAP (*Xenopus* microtubule assembly protein) Protein (215 kD) that promotes microtubule assembly, found in oocytes and early embryos.

X-ray diffraction Basis of powerful technique for determining the three-dimensional structure of molecules, including complex biological macromolecules such as proteins and nucleic acids, that form crystals or regular fibres. Low-angle X-ray diffraction is also used to investigate higher levels of ordered structure, as found in muscle fibres.

X-ray microanalysis See ***electron microprobe***.

xanthine (2,6-dihydroxypurine) A purine, the starting point for purine degradation. Its methylated derivatives (theophylline, theobromine, caffeine) are potent cAMP phosphodiesterase inhibitors.

xanthine oxidase Dehydrogenases involved in conversion of hypoxanthine to xanthine, and xanthine to uric acid, as the final catabolism of purines. Deficient in the human disease xanthinuria.

xanthoma Localized lesion of subcutaneous tissues in which there is an accumulation of cholesterol-filled macrophages. Characteristic of primary biliary cirrhosis.

xanthophylls ***Carotenoid*** pigments involved in photosynthesis. Consist of oxygenated carotenes, eg. lutein, violaxanthin and neoxanthine.

xanthopterin Yellow pterin pigment found in some insect wings.

xenobiotic Any substance that does not occur naturally but that will affect living systems.

xenobiotic response element (XRE) DNA regulatory sequence that binds the transcription factors that regulate genes coding for enzymes involved in detoxification.

xenogeneic Literally, of foreign genetic stock; usually applied to tissue or cells from another species, as in xenogeneic transplantation.

xenograft A graft between individuals of unlike species, genus or family.

Xenopus The genus of African clawed toads, *X. laevis* is widely used in developmental biology and was formerly used in pregnancy diagnosis. Ovulates easily under the influence of luteinizing hormone.

xenosome (1) A bacterial endosymbiont of certain marine protozoans. (2) Inorganic particles in various testate amoebae.

xenotropic Literally, 'growing in a foreign environment'. Especially of endogenous retroviruses transmitted genetically in the host of their origin but that can only replicate in the cells of a different species.

xenotropic virus A virus that can be grown on cells of a species foreign to the normal host species.

xeroderma pigmentosum Inherited (autosomal recessive) disease in humans associated with increased sensitivity to ultraviolet-induced mutagenesis, and thus to skin cancer. Sensitivity can be demonstrated in cultured cells, and appears to be due to deficiency in DNA repair, specifically in excision of ultraviolet-induced ***thymine dimers***.

Xgal (5-bromo-4-chloro-3-indolyl-β-D-galactoside) Chromogenic substrate for the enzyme ***beta-galactosidase***, yielding an intense blue colour. Very widely used in ***blue-white colour selection*** of bacteria containing recombinant plasmids, and in histochemistry of tissues/cells expressing β-galactosidase as a reporter gene.

Xic See ***X chromosome inactivation centre***.

XLP (X-linked lymphoproliferative disease.) The X-linked lymphoproliferative disease (XLP), one of six described X-linked immunodeficiencies, stems from a mutation at Xq25 which renders males impotent to mount an effective immune response to the ubiquitous EBV. Proliferation of EBV-infected B-cells is apparently unregulated and invariably results in fatal mononucleosis, agammaglobulinemia or malignant lymphoma. See ***SAP***.

xylan Plant cell wall polysaccharide containing a backbone of β(1-4)-linked xylose residues. Side chains of 4-*O*-methylglucuronic acid and arabinose are present in varying amounts (see ***glucuronoxylan*** and ***arabinoxylan***), together with acetyl groups. Found in the ***hemicellulose*** fraction of the wall matrix.

xylem Plant tissue responsible for the movement of water and inorganic solutes from the roots to the shoot and leaves. Contains ***tracheids***, ***vessels***, ***fibre cells*** and ***parenchyma***. Also provides structural support for the plant, especially in wood.

xyloglucan Plant cell wall polysaccharide containing a backbone of β(1-4)-linked glucose residues to most of which single xylose residues are attached as side chains. Galactose, fucose and arabinose may also be present in smaller amounts. It is the major hemicellulose of dicotyledonous primary walls, and acts as a food reserve in some seeds.

xylose Monosaccharide pentose that is found in xylans, very abundant components of hemicelluloses.

xylulose A 5-carbon ketose sugar, whose 5-phosphate is an intermediate in the ***pentose phosphate pathway*** and the ***Calvin–Benson cycle***.

Y See ***tyrosine***.

Y chromosome Chromosome found only in the heterogametic sex. Thus in mammals the male has one Y chromosome and one X chromosome. One region of the Y chromosome, the pseudoautosomal region, is homologous to and pairs with the X chromosome. The primary determinant of male sexual development is found on the unpaired, differentiated segment of the Y chromosome.

yabavirus A poxvirus of African monkeys that causes benign tumours.

yeast Yeast is the colloquial name for members of the fungal families, ascomycetes, basidiomycetes and imperfect fungi, that tend to be unicellular for the greater part of their life cycle. Commercially important yeasts include *Saccharomyces cerevisiae*; pathogenic yeasts include the genus *Candida*. See also ***Schizosaccharomyces pombe***.

yeast artificial chromosome (YAC) A vector system that allows extremely large segments of DNA to be cloned. Useful in chromosome mapping; contiguous YACs covering the whole *Drosophila* genome and certain human chromosomes are available.

yeast two-hybrid screening Strategy for screening for proteins that interact with a particular protein. A cDNA library is constructed such that candidate proteins are expressed as translational fusions with part of (typically) the GAL4 gene. Yeast cells are then cotransfected with a 'bait' construct consisting of the cDNA of interest fused in-frame to the other part of the GAL4 gene. Only if both expressed proteins physically interact will the two parts of the GAL4 protein come cloes enough to produce detectable β-galactosidase activity. Similar systems have now been developed that tag bait and targets with heterodimeric proteins other than GAL4.

yellow fever virus A togavirus (class IV) with an RNA genome responsible for the disease of the name whose symptoms include fever and haemorrhage. Transmitted by the mosquitoes *Aedes aegypti* and *Haemagogus*. Only one antigenic type of the virus known.

Yersinia Genus of ***Gram-negative bacteria*** of the ***Enterobacteriaceae***; all are parasites or pathogens. *Y. pestis* (formerly *Pasteurella pestis*) was the cause of plague (black death).

yes An ***oncogene***, identified in avian ***sarcoma***, encoding a tyrosine ***protein kinase*** (p375). See Table O1.

yolk cells In those eggs in which the yolk is not distributed evenly (telolecithal eggs) the cells formed when cleavage reaches the yolk region can be termed yolk cells.

yolk sac One of the set of extra-embryonic membranes, growing out from the gut over the yolk surface, in birds formed from the splanchnopleure, an outer layer of splanchnic mesoderm and an inner layer of endoderm.

Z

Z disc Region of the ***sarcomere*** into which ***thin filaments*** are inserted. Location of ***alpha-actinin*** in the sarcomere.

Z-DNA Form of DNA adopted by sequences of alternating ***purines*** and ***pyrimidines***. It is a left-handed helix with the phosphate groups of the backbone zig-zagged (hence Z) and a single deep groove. It is still not clear whether Z-DNA occurs in genomic DNA.

Z scheme of photosynthesis A schematic representation of the ***light-dependent reactions*** of ***photosynthesis***, in which the photosynthetic reaction centres and electron carriers are arranged according to their electrode potential (free energy) in one dimension and their reaction sequence in the second dimension. This gives a Z-shape, the two reaction centres (of ***photosystems*** I and II) being linked by the photosynthetic electron transport chain.

Zantac (ranitidine) Proprietary name for the anti-ulcer drug (H2 blocker) that was, until recently, the mainstay of Glaxo's success as a pharmaceutical company.

zap70 (zeta chain-associated protein; zap-70) Protein tyrosine kinase (70 kD) that associates with the zeta chain of the T-cell receptor following ligand binding. Mutation in zap70 can cause a form of ***SCID***.

Zea mays Maize or Indian corn. In the USA, 'corn' is taken to mean maize; in Europe this is not the case and corn usually means wheat or barley.

zeatin A naturally occurring ***cytokinin***, originally isolated from maize seeds. Its riboside is also a cytokinin.

zebrafish *(Brachydanio rerio)* Freshwater fish, easily reared in an aquarium. Its transparent embryo makes it possible to follow progeny of single cells until quite late stages of development. This, together with the availability of mutant lines, makes it an important preparation for the study of vertebrate ***cell lineage***.

zeiosis Blebbing of the plasma membrane; sometimes referred to as 'cell boiling'.

Zenker's fluid Fixative that is good for preserving cytoplasmic structure but needs to be made up freshly. Contains mercuric chloride.

zeta potential The electrostatic potential of a molecule or particle such as a cell, measured at the plane of hydrodynamic slippage outside the surface of the molecule or cell. Usually measured by electrophoretic mobility. Related to the surface potential and a measure of the electrostatic forces of repulsion the particle or molecule is likely to meet when encountering another of the same sign of charge. See ***cell electrophoresis***.

zf9 ***Zinc finger*** transcription factor from rat stellate cells (see ***Ito cells***) activated *in vitro*. A member of the ***Kruppel*** family.

zidovudine (AZT; azido-deoxythymidine; retrovir) Thymidine analogue; phosphorylated form acts as an inhibitor of viral reverse transcriptase. Used in HIV treatment.

Zigmond chamber See ***orientation chamber***.

zinc An essential 'trace' element being an essential component of the active site of a variety of enzymes. Zn^{2+} has high affinity for the side chains of cysteine and histidine. Zinc is present in tissues at levels of about 0.1mM, but intracellular levels must be much lower.

zinc finger A motif associated with ***DNA-binding proteins***. A loop of 12 amino acids contains either 2 cysteine and 2 histidine groups (a 'cysteine–histidine' zinc finger), or 4 cysteines (a 'cysteine-cysteine' zinc finger), that directly coordinate a zinc atom. The loops (usually present in multiples) intercalate directly into the DNA helix. Originally identified in the RNA polymerase III transcription factor TFIIIA.

zipper See ***leucine zipper***.

zippering Process suggested to occur in phagocytosis in which the membrane of the phagocyte covers the particle by a progressive adhesive interaction. The evidence for such a mechanism comes from experiments in which capped B-lymphocytes are only partially internalized, whereas those with a uniform opsonizing coat of anti-IgG are fully engulfed.

ZO-1 High molecular weight protein (225 kD in mouse, 210 kD in MDCK cells) associated with ***zonula occludens*** (tight junction) in many vertebrate epithelia. ***Cingulin***, which is distinct, is found in the same region.

zona pellucida A translucent non-cellular layer surrounding the ovum of many mammals.

zone of polarizing activity (ZPA) The small group of mesenchyme cells in avian limb buds that is located at the posterior margin of the developing bud and that produces a substance, possibly retinoic acid, that provides positional information to the developing limb bud.

zonula adherens Specialized intercellular junction in which the membranes are separated by 15–25nm, and into which are inserted microfilaments. Similar in structure to two apposed ***focal adhesions***, though this may be misleading. Microfilaments inserted into the zonula adherens may interact (via myosin) with other microfilaments to generate contraction. Constitute mechanical coupling between cells.

zonula occludens (tight junction) Specialized intercellular junction in which the two plasma membranes are separated by only 1–2nm. Found near the apical surface of cells in simple epithelia; forms a sealing 'gasket' around the cell. Prevents fluid moving through the intercellular gap and the lateral diffusion of intrinsic membrane proteins between apical and basolateral domains of the plasma membrane.

zonula occludens toxin Toxin (45 kD) released by *Vibrio cholerae*. Binds to a receptor on intestinal epithelial cells and triggers a cascade of events that result in alterations in the tight junction permeability barrier.

zoochlorellae Intracellular symbiotic algae usually of the genus *Chlorella* found in some lamellibranch molluscs, protozoans, flatworms, sponges and corals.

zoonosis An infectious disease of humans whose natural reservoir is a non-human animal, eg. psittacosis, a viral disease of birds, occasionally infecting humans.

zootype Postulated pattern of gene expression shared by all animal phyla. The hypothesis implies that six *hox*-type ***homeobox***-containing genes should be present in all metazoa.

zooxanthellae Intracellular photosynthetic symbiotic dinoflagellates found in a variety of marine invertebrates. Systematics uncertain.

zovirax (acyclovir) Nucleoside analogue (hydroxyethoxymethyl-guanine) with antiviral properties; active against both type 1 and 2 herpes virus. Inactive until phosphorylated by specific viral enzyme, ***thymidine kinase***, and then blocks replication.

ZPA See ***zone of polarizing activity***.

zwitterions A molecule carrying a positive charge at one end and a negative charge at the other. Also known as ampholyte or dipolar ions.

zygonema See ***zygotene***.

zygospore Fungal spore produced by the fusion of two similar ***gametes*** or ***hyphae***.

zygote ***Diploid*** cell resulting from the fusion of male and female gametes at fertilization.

zygotene Classic term for the second stage of the ***prophase*** of ***meiosis*** I, during which the homologous chromosomes start to pair.

zygotic-effect gene A gene whose ***phenotype*** is dependent on the genotype of the ***zygote***, rather than on the genotype of the mother. See ***maternal-effect gene***.

zymogen Inactive precursor of an enzyme, particularly a proteolytic enzyme. Synthesized in the cell and secreted in this safe form, then converted to the active form by limited proteolytic cleavage.

zymogen granule ***Secretory vesicle*** containing an inactive precursor (zymogen). The contents are often very condensed.

zymogenic cell Cells of the basal part of the gastric glands of the stomach. They contain extensive rough ***endoplasmic reticulum*** and ***zymogen granules*** and secrete pepsinogen, the inactive precursor of ***pepsin***, and rennin.

zymogram Electrophoretic gel (or other separation) in which the position of an enzyme is revealed by a reaction that depends upon its enzymic activity.

zymosan Particulate yeast cell-wall polysaccharide (mannan-rich) that will activate ***complement*** in serum through the alternate pathway. Becomes coated with C3b/C3bi and is therefore a convenient opsonized particle; also leads to C5a production in the serum.

zyxin Protein (82 kD) found at the ***adherens junction***. Interacts with ***alpha-actinin***. Found in fibroblasts, smooth muscle and pigmented retinal epithelium.

Useful sources

Although the books listed have been useful, the most valuable source, and the one where many usages were checked is PubMed, the NCBI online version of Medline at *http://www.ncbi.nlm.nih.gov/htbin-post/Entrez/query*. Usage was taken as definitive – even if strictly speaking the meaning ought to be wider or narrower.

The Leucocyte Antigen Facts Book, 2nd Edition. Barclay, AN, Brown, MH, Law SKA, McKnight AJ, Tomlinson MG, van der Merwe PA (1997) Academic Press.

Cytokine FactsBook. Callard RE and Gearing AJH (1994) Academic Press.

Dictionary of Immunology, 4th Edition. Herbert, WJ, Wilkinson PC and Stott DI (1995) Academic Press.

Dictionary of Science and Technology. Edited by Morris, C (1992) Academic Press.

Dictionary of Virology, 2nd Edition. Mahy BWJ (1997) Academic Press.

Encylopedia of Molecular Biology. Edited by Kendrew J (1994) Blackwell Science.

Guidebook to Protein Toxins and their use in Cell Biology. Edited by Rappuoli R and Montecucco C (1997) Oxford University Press.

A Dictionary of Genetics, 5th Edition. King RC and Stansfield WD (1997) Oxford University Press.

Oxford Dictionary of Biochemistry and Molecular Biology (1997) Oxford University Press.

Appendix

Prefixes for SI Units

Factor	*Prefix*	*Symbol*	*Factor*	*Prefix*	*Symbol*
10^{24}	yotta	Y	10^{-1}	deci	d
10^{21}	zetta	Z	10^{-2}	centi	c
10^{18}	exa	E	10^{-3}	milli	m
10^{15}	peta	P	10^{-6}	micro	μ
10^{12}	tera	T	10^{-9}	nano	n
10^{9}	giga	G	10^{-12}	pico	p
10^{6}	mega	M	10^{-15}	femto	f
10^{3}	kilo	k	10^{-18}	atto	a
10^{2}	hecto	h	10^{-21}	zepto	z
10^{1}	deca	da	10^{-24}	yocto	y

Greek Alphabet

Α	α	alpha
Β	β	beta
Γ	γ	gamma
Δ	δ	delta
Ε	ε	epsilon
Ζ	ζ	zeta
Η	η	eta
Θ	θ	theta
Ι	ι	iota
Κ	κ	kappa
Λ	λ	lambda
Μ	μ	mu
Ν	ν	nu
Ξ	ξ	xi
Ο	ο	omicron
Π	π	pi
Ρ	ρ	rho
Σ	σ	sigma
Τ	τ	tau
Υ	υ	upsilon
Φ	φ	phi
Χ	χ	chi
Ψ	ψ	psi
Ω	ω	omega

Useful constants

Avogadro's number (N)	6.022×10^{23} mol^{-1}
Boltzmann's constant (k)	1.318×10^{-23} J deg^{-1}
	3.298×10^{-24} cal deg^{-1}
Faraday constant (F)	9.649×10^{4} Coulomb mol^{-1}
Curie (Ci)	3.7×10^{10} disintegrations s^{-1}
Gas constant(R)	8.314 J mol^{-1} deg^{-1}
π	3.14159
e	2.71828

$\log_e x = 2.303\log_{10} x$

Single-letter codes for amino acids

A	Ala	Alanine
R	Arg	Arginine
N	Asn	Asparagine
D	Asp	Aspartic acid
B		Asparagine or aspartic acid
C	Cys	Cysteine
Q	Glu	Glutamine
E	Glu	Glutamic acid
Z		Glutamine or glutamic acid
G	Gly	Glycine
H	His	Histidine
I	Ileu	Isoleucine
L	Leu	Leucine
K	Lys	Lysine
M	Met	Methionine
F	Phe	Phenylalanine
P	Pro	Proline
S	Ser	Serine
T	Thr	Threonine
W	Trp	Tryptophan
Y	Tyr	Tyrosine
V	Val	Valine